IET MATERIALS, CIRCUITS AND DEVICES SERIES 48

Gyrators, Simulated Inductors and Related Immittances

Other volumes in this series:

Volume 2	**Analogue IC Design: The current-mode approach** C. Toumazou, F.J. Lidgey and D.G. Haigh (Editors)
Volume 3	**Analogue-Digital ASICs: Circuit techniques, design tools and applications** R.S. Soin, F. Maloberti and J. France (Editors)
Volume 4	**Algorithmic and Knowledge-Based CAD for VLSI** G.E. Taylor and G. Russell (Editors)
Volume 5	**Switched Currents: An analogue technique for digital technology** C. Toumazou, J.B.C. Hughes and N.C. Battersby (Editors)
Volume 6	**High-Frequency Circuit Engineering** F. Nibler *et al.*
Volume 8	**Low-Power High-Frequency Microelectronics: A unified approach** G. Machado (Editor)
Volume 9	**VLSI Testing: Digital and mixed analogue/digital techniques** S.L. Hurst
Volume 10	**Distributed Feedback Semiconductor Lasers** J.E. Carroll, J.E.A. Whiteaway and R.G.S. Plumb
Volume 11	**Selected Topics in Advanced Solid State and Fibre Optic Sensors** S.M. Vaezi-Nejad (Editor)
Volume 12	**Strained Silicon Heterostructures: Materials and devices** C.K. Maiti, N.B. Chakrabarti and S.K. Ray
Volume 13	**RFIC and MMIC Design and Technology** I.D. Robertson and S. Lucyzyn (Editors)
Volume 14	**Design of High Frequency Integrated Analogue Filters** Y. Sun (Editor)
Volume 15	**Foundations of Digital Signal Processing: Theory, algorithms and hardware design** P. Gaydecki
Volume 16	**Wireless Communications Circuits and Systems** Y. Sun (Editor)
Volume 17	**The Switching Function: Analysis of power electronic circuits** C. Marouchos
Volume 18	**System on Chip: Next generation electronics** B. Al-Hashimi (Editor)
Volume 19	**Test and Diagnosis of Analogue, Mixed-Signal and RF Integrated Circuits: The system on chip approach** Y. Sun (Editor)
Volume 20	**Low Power and Low Voltage Circuit Design With the FGMOS Transistor** E. Rodriguez-Villegas
Volume 21	**Technology Computer Aided Design for Si, SiGe and GaAs Integrated Circuits** C.K. Maiti and G.A. Armstrong
Volume 22	**Nanotechnologies** M. Wautelet *et al.*
Volume 23	**Understandable Electric Circuits** M. Wang
Volume 24	**Fundamentals of Electromagnetic Levitation: Engineering sustainability through efficiency** A.J. Sangster
Volume 25	**Optical MEMS for Chemical Analysis and Biomedicine** H. Jiang (Editor)
Volume 26	**High Speed Data Converters** A.M.A. Ali
Volume 27	**Nano-Scaled Semiconductor Devices** E.A. Gutiérrez-D (Editor)
Volume 28	**Security and Privacy for Big Data, Cloud Computing and Applications** L. Wang, W. Ren, K.R. Choo and F. Xhafa (Editors)
Volume 29	**Nano-CMOS and Post-CMOS Electronics: Devices and modelling** S.P. Mohanty and A. Srivastava
Volume 30	**Nano-CMOS and Post-CMOS Electronics: Circuits and design** S.P. Mohanty and A. Srivastava
Volume 32	**Oscillator Circuits: Frontiers in design, analysis and applications** Y. Nishio (Editor)
Volume 33	**High Frequency MOSFET Gate Drivers** Z. Zhang and Y. Liu
Volume 34	**RF and Microwave Module Level Design and Integration** M. Almalkawi
Volume 35	**Design of Terahertz CMOS Integrated Circuits for High-Speed Wireless Communication** M. Fujishima and S. Amakawa
Volume 38	**System Design with Memristor Technologies** L. Guckert and E.E. Swartzlander Jr.
Volume 39	**Functionality-Enhanced Devices: An alternative to Moore's law** P.-E. Gaillardon (Editor)
Volume 40	**Digitally Enhanced Mixed Signal Systems** C. Jabbour, P. Desgreys and D. Dallett (Editors)

Volume 43	**Negative Group Delay Devices: From concepts to applications** B. Ravelo (Editor)
Volume 45	**Characterisation and Control of Defects in Semiconductors** F. Tuomisto (Editor)
Volume 47	**Understandable Electric Circuits: Key concepts, 2nd Edition** M. Wang
Volume 49	**Advanced Technologies for Next Generation Integrated Circuits** A. Srivastava and S. Mohanty (Editors)
Volume 51	**Modelling Methodologies in Analogue Integrated Circuit Design** G. Dundar and M. B. Yelten (Editors)
Volume 53	**VLSI Architectures for Future Video Coding** M. Martina (Editor)
Volume 54	**Advances in High-Power Fiber and Diode Laser Engineering** Ivan Divliansky (Editor)
Volume 55	**Hardware Architectures for Deep Learning** M. Daneshtalab and M. Modarressi
Volume 58	**Magnetorheological Materials and Their Applications** S. Choi and W. Li (Editors)
Volume 60	**IP Core Protection and Hardware-Assisted Security for Consumer Electronics** A. Sengupta and S. Mohanty
Volume 64	**Phase-Locked Frequency Generation and Clocking: Architectures and circuits for modem wireless and wireline systems** W. Rhee (Editor)
Volume 65	**MEMS Resonator Filters** Rajendra M Patrikar (Editor)
Volume 67	**Frontiers in Securing IP Cores; Forensic detective control and obfuscation techniques** A Sengupta
Volume 68	**High Quality Liquid Crystal Displays and Smart Devices: Vol 1 and Vol 2** S. Ishihara, S. Kobayashi and Y. Ukai (Editors)
Volume 69	**Fibre Bragg Gratings in Harsh and Space Environments: Principles and applications** B. Aïssa, E.I. Haddad, R.V. Kruzelecky, W.R. Jamroz
Volume 70	**Self-Healing Materials: From fundamental concepts to advanced space and electronics applications, 2nd Edition** B. Aïssa, E.I. Haddad, R.V. Kruzelecky, W.R. Jamroz
Volume 71	**Radio Frequency and Microwave Power Amplifiers: Vol 1 and Vol 2** A. Grebennikov (Editor)
Volume 73	**VLSI and Post-CMOS Electronics Volume 1: VLSI and post-CMOS electronics and Volume 2: Materials, devices and interconnects** R. Dhiman and R. Chandel (Editors)

Gyrators, Simulated Inductors and Related Immittances

Realizations and applications

Raj Senani, Data Ram Bhaskar, Vinod Kumar Singh
and Abdhesh Kumar Singh

The Institution of Engineering and Technology

Published by The Institution of Engineering and Technology, London, United Kingdom

The Institution of Engineering and Technology is registered as a Charity in England & Wales (no. 211014) and Scotland (no. SC038698).

First published 2020

The Institution of Engineering and Technology
Michael Faraday House
Six Hills Way, Stevenage
Herts, SG1 2AY, United Kingdom

www.theiet.org

British Library Cataloguing in Publication Data
A catalogue record for this product is available from the British Library

ISBN 978-1-78561-670-9 (Hardback)
ISBN 978-1-78561-671-6 (PDF)

Typeset in India by MPS Limited

Dedication

This monograph is dedicated to:
The Legendary Indian Music Composer-Duo

The Melody Twins: Shankar–Jaikishan whose highly creative, melodious and ageless musical compositions provided immense inspiration in the research work of the authors over the past four decades in general and during the writing of this monograph in particular.

Contents

About the authors xvii
Preface xxi
Acknowledgement xxiii

1 Gyrators, integrated inductors and simulated inductors 1
Abstract 1
1.1 Prologue 1
1.2 Basic one-port circuit elements 2
1.3 Basic two-port circuit elements 3
1.3.1 The transformer 3
1.3.2 The gyrator 4
1.3.3 The two-port impedance converters and inverters 6
1.4 The pathological elements 8
1.5 Multi-terminal gyrator 10
1.6 Multiport inverters/converters 12
1.7 Commercially available inductors and Coilcraft 12
1.8 Basic difficulties in micro-miniaturization of inductors 13
1.9 Integrated inductors and transformers on the chip 15
1.10 Power gyrators 16
1.11 Use of ANSYS and COMSOL for the analysis of inductor designs 17
1.12 The need for simulated inductors 17
1.13 Concluding remarks 18
References 19

2 Gyrators and simulated inductors using op-amps 27
Abstract 27
2.1 Introduction 27
2.2 The gyrator 28
2.3 Op-amp gyrators and related circuits 31
2.3.1 NIC-based gyrator 31
2.3.2 VCCS-based gyrator 32
2.3.3 Generalized impedance converters (GIC)/gyrators 33
2.3.4 Two-op-amp resistively variable capacitance simulators 36
2.3.5 Two-op-amp lossless inductance simulator 40
2.3.6 Tripathi–Patranabis lossless grounded inductor 41

2.3.7 Lossless GI using summer/subtractor circuits 42
2.3.8 Two modified forms of the GIC and their applications 42
2.4 Single-op-amp lossless inductance simulators 44
2.4.1 Orchard–Willson gyrator 44
2.4.2 Schmidt–Lee circuit 45
2.4.3 Ramsey gyrator 46
2.4.4 Horn–Moschytz circuit 47
2.5 Economic inductance simulators and resonators 48
2.5.1 Ford–Girling circuit 48
2.5.2 Prescott circuit 49
2.5.3 Berndt–Dutta Roy circuit 49
2.5.4 Caggiano circuit 51
2.5.5 Patranabis circuit 51
2.5.6 The parallel/series RL inductors derived by Rao–Venkateswaran 52
2.5.7 Ahmed–Dutta Roy technique of deriving grounded-capacitor lossy GI 53
2.5.8 Senani–Tiwari circuit 55
2.5.9 Soliman–Awad tunable active inductor 56
2.5.10 Nagarajan–Dutta Roy–Choudhary circuit 57
2.5.11 Senani's single-resistance-tunable GIs 58
2.6 Lossless floating impedance simulators using four op-amps 59
2.6.1 Riordan's method of creating a lossless FI 59
2.6.2 GIC method of simulating FI 60
2.6.3 Tripathi–Patranabis FI 61
2.6.4 Mutator-simulated floating inductors 62
2.7 The multi-port immittance converters/inverters and multi-port gyrators 62
2.8 Three-op-amp-based floating immittance simulators 62
2.8.1 Three-op-amp-single-capacitor FIs based on GIC-type networks 63
2.8.2 FI realizations using three op-amps along with a grounded capacitor 65
2.8.3 Senani's single-resistance-controllable lossless FI 66
2.8.4 Patranabis–Paul capacitance floatation circuit 67
2.9 Lossless FIs using only two op-amps 68
2.9.1 The–Yanagisawa circuit 68
2.9.2 Sudo–Teramoto circuit 69
2.10 Economic op-amp-based lossless/lossy FIs 70
2.10.1 The cascade back-to-back approach to FI realization 71
2.10.2 Parallel back-to-back approach to FI realization 72
2.10.3 Senani–Tiwari FI 73
2.11 The active-*R* simulation of grounded/floating impedances and resonators 74
2.12 Switched-capacitor simulated inductors 76

2.13 FI realization using four-terminal floating nullors (FTFNs) 80
2.14 Non-ideal behaviour of simulated impedances due to finite GBP of the op-amps used 84
2.15 Concluding remarks 84
References 86

3 The operational transconductance amplifier based gyrators and impedance simulators 99
Abstract 99
3.1 The OTAs and their advantages in analog circuit design 99
3.2 Integrated OTAs 100
3.3 OTA–C gyrators, inductors and related impedances 102
3.4 OTA–*RC* impedance converters/inverters 110
3.5 Other OTA-based lossless/lossy inductors and FDNRs 112
3.6 Synthetic impedances using OTAs and op-amps 115
3.7 OTA-based capacitance multipliers 123
3.8 Inductor and FDNC simulators using OTAs and unity gain adders/subtractors 126
3.9 Active-only simulators using op-amps and OTAs 128
3.10 Electronically controllable resistors using OTAs 133
3.11 Simulation of mutually coupled circuits and transformers 138
3.12 Multi-port gyrators using OTAs: retrospection 138
3.13 Concluding remarks 140
References 141

4 Synthetic impedances using current conveyors and their variants 147
Abstract 147
4.1 Introduction 147
4.2 Realization of grounded and floating negative impedances 149
4.3 Realization of synthetic grounded inductors, FDNRs and related elements 151
4.3.1 Grounded inductance simulation using CCs 151
4.3.2 Single-CCII-based active gyrators 152
4.3.3 Single CCII-based grounded impedance simulators 153
4.4 Synthetic floating impedances without component-matching requirements 156
4.4.1 The first ever CC-based FI simulators without requiring any component matching 156
4.4.2 Two other single-CC-based FIs without component matching 165
4.4.3 GPIC/GPII/three-port gyrator configurations using CCs 166
4.4.4 Additional three-CC-based floating inductor/FDNR simulators 170
4.4.5 Floating impedance realization using two DOCCs 172
4.4.6 Floating impedances using CCIIs and op-amps/OTAs 173

4.4.7 Economical floating impedance circuits synthesized using the 'CCII-nullor' equivalence 176
4.4.8 Realization of mutually coupled circuits 179
4.5 Impedance simulation using CCCII 181
4.5.1 Current-controlled positive/negative resistance realization 184
4.5.2 Electronically tunable grounded/floating impedances 187
4.5.3 Electronically tunable synthetic transformer 192
4.6 Immittance simulation using different variants of current conveyors 193
4.6.1 Lossless FI realization employing only two DOCCs and three passive components 194
4.6.2 DVCC-based floating inductance/FDNR realization 194
4.6.3 CCIII-simulated inductors 195
4.6.4 Single-DVCC-based grounded *RL* and *CD* immittance simulators 196
4.6.5 DXCCII-based electronically controllable gyrator/inductor 197
4.6.6 Grounded inductor realized with modified inverting CCII 198
4.6.7 Synthetic floating immittances realized with DOCCII 199
4.6.8 Another FI with improved low-frequency performance realized with only two DOCCIIs 200
4.6.9 Floating impedance simulator realized with a DOCCII and an OTA 201
4.6.10 Lossless FI realization employing a DOCCCII and grounded capacitor 202
4.6.11 AN FI employing only a single DODDCC and three passive elements 203
4.6.12 External resistorless FI realization using DXCCII 203
4.6.13 Electronically tunable MOSFET-C FDNR using a DXCCII 204
4.6.14 Grounded inductance simulation using a DXCCII 205
4.6.15 FI realization using only two DVCCs/DVCCCs 206
4.6.16 FDCCII-based lossless grounded inductor employing three grounded passive elements 209
4.6.17 DXCCII-based grounded inductance simulators 211
4.6.18 Grounded-capacitor-based floating capacitance multiplier using CCDDCCs 213
4.6.19 Single-DODDCC-based grounded lossy inductance simulators employing a grounded capacitor 215
4.6.20 DCCII-based grounded inductance simulator 216
4.6.21 Miscellaneous FI simulators using ICCII, DVCC and FDVCC elements 217
4.7 Higher order filter design using nonideal simulated impedances 219
4.8 Concluding remarks 222
References 223

5 Current feedback-op-amp-based synthetic impedances **249**
Abstract 249
5.1 Introduction 249
5.2 The IC CFOA AD844 251
5.3 Systematic synthesis of gyrators/grounded inductance simulators 253
5.4 Lossless FI simulators using CFOAs 255
5.5 Grounded/floating generalized positive impedance converters/inverters (GPIC/GPII) and generalized negative impedance converters/inverters (GNIC/GNII) 263
5.6 Economic simulation of lossy grounded inductors 269
5.7 Low-component-count lossy FI simulation 271
5.8 Single resistance-controllable single CFOA simulators 273
5.9 Inductors and resonators using CFOA poles 279
5.10 GI/FI simulators using modified CFOAs 282
5.11 Concluding remarks 290
References 290

6 Applications of FTFN/OFA and OMAs in impedance synthesis **295**
Abstract 295
6.1 Introduction 295
6.2 An overview of nullors, FTFN/OFA and OMAs 296
6.3 Generation of FTFN-based floating immittances 305
6.3.1 Realization of floating generalized impedance converters/inverters 306
6.3.2 Generation of lossless FI circuits using a single FTFN 308
6.3.3 Single-resistance-tunable lossy FI simulation using only a single FTFN 310
6.4 Operational mirrored amplifiers (OMA)–based simulators 312
6.4.1 OMA-based floating GIC 312
6.4.2 A floating GIC using OMA 314
6.4.3 Floating impedance realization using a dual OMA 315
6.4.4 Three OMA-based floating impedance simulators 317
6.4.5 OMA-based FI using op-amp pole 320
6.4.6 OMA-based FI with extended frequency range 322
6.5 Operational floating amplifiers and their use in floating gyrator and FI realization 323
6.5.1 The OFA-based floating gyrator 324
6.5.2 The DD-OFA and its use in FI synthesis 325
6.6 Concluding remarks 327
References 328

7 Realization of voltage-controlled impedances **335**
Abstract 335
7.1 Introduction 335
7.2 Grounded VCZ realization using op-amps 337

7.2.1 Nay–Budak voltage-controlled resistors with extended dynamic range 337
7.2.2 Senani–Bhaskar VCZ configurations 339
7.2.3 Leuciuc–Goras VCZ configurations based upon GIC 342
7.2.4 Three-op-amp-based VCZ structure by Senani–Bhaskar 345
7.2.5 Senani's universal VCZ structure with only two op-amps 346
7.2.6 Ndjountche configuration using MOS resistive circuit 348
7.2.7 Economical VCZ configurations 349
7.3 Grounded VCZ configurations using CFOAs 351
7.4 VCZ configurations using current conveyors 352
7.5 The floating VCR 352
7.6 The floating/grounded voltage-controlled GIC/GIIs using CFOAs 355
7.7 Floating VC-negative-impedance realization using OMAs 357
7.8 Floating/grounded VCZ structures using CFOAs and analogue multipliers 359
7.9 Concluding remarks 363
References 364

8 Impedance synthesis using modern active building blocks 367
Abstract 367
8.1 Introduction 367
8.2 An overview of modern electronic circuit building blocks 368
8.3 Grounded impedance and floating impedance synthesis using modern building blocks 368
8.3.1 Unity gain VF/CF-based circuits 368
8.3.2 OTRA-based circuits 374
8.3.3 DDA-based circuits 375
8.3.4 Translinear operational current amplifier 380
8.3.5 CDTA-based circuits 380
8.3.6 CDBA-based circuits 382
8.3.7 CFTA-based circuits 385
8.3.8 VDTA-based impedance simulators 388
8.3.9 CCTA-based circuits 392
8.3.10 VDBA-based circuits 396
8.3.11 VD-DIBA-based circuit 399
8.3.12 CFCC-based circuits 402
8.3.13 Inductance simulation using OTA–COA 406
8.3.14 FDNR realization using capacitive gyrators 406
8.3.15 Lossless grounded inductance simulator using VDCC 408
8.3.16 Inductance simulation using VDIBA 408
8.3.17 General floating immittance simulator using CBTA 410
8.3.18 Simulation of inductor using current differential amplifiers (CDA) 412
8.3.19 GI/FI using other miscellaneous active elements 412
8.4 Concluding remarks 414
References 415

9 Transistor-level realization of electronically controllable grounded and floating resistors **423**
Abstract 423
9.1 Introduction 423
9.2 BJT-based translinear current-controlled resistors 424
9.2.1 Current-controllable grounded/floating resistors 424
9.2.2 A translinear current-controlled floating resistor due to Barthelemy and Fabre 426
9.2.3 A translinear current-controllable floating negative resistor 428
9.2.4 Translinear floating current-controlled positive resistance due to Senani, Singh and Singh 430
9.2.5 A circuit to realize current-controllable floating positive/negative resistance 433
9.2.6 A low transistor count current-controlled grounded/floating, positive/negative resistor due to Arslanalp, Yuce and Tola 434
9.2.7 Current-controlled-resistor based upon a new eight-transistor mixed-translinear-cell (MTC) 437
9.2.8 Electronically tunable active resistance circuits based upon differential amplifiers 438
9.3 CMOS linear voltage/current-controlled grounded/floating resistors 439
9.3.1 A two-MOSFET-based linear voltage-controlled resistor devised by Han and Park 440
9.3.2 Some general techniques of realizing linear MOS-resistive circuits 441
9.3.3 The two MOSFET transresistor due to Wang 444
9.3.4 Banu–Tsividis linear voltage-controlled floating linear resistor 445
9.3.5 Linear transconductor due to Park and Schaumann 445
9.3.6 Linear floating VCR due to Nagaraj 447
9.3.7 Wilson and Chan grounded VCR 448
9.3.8 Wang's grounded linear VCR 449
9.3.9 Positive/negative linear grounded VCRs due to Wang 450
9.3.10 Positive/negative linear grounded VCRs and voltage-controlled gyrator using NICs 452
9.3.11 Floating linear resistor proposed by Elwan, Mahmoud and Soliman 454
9.4 Concluding remarks 463
References 464

10 Bipolar and CMOS active inductors and transformers **473**
Abstract 473
10.1 Introduction 473
10.2 BJT-based gyrators and inductance simulators 474
10.2.1 The early attempts of devising transistor-based gyrators/simulated inductors 474
10.2.2 A two-transistor semiconductor FI simulator due to Takahashi, Hamada, Watanabe and Miyata 475

10.2.3 A direct-coupled fully integratable gyrator due to Chua and Newcomb 476
10.2.4 The integrated gyrator due to Haykim, Kramer, Shewchun and Treleaven 476
10.2.5 Synthesis of three transistor gyrators 477
10.2.6 The translinear floating inductance simulator 478
10.3 CMOS active inductors 480
10.3.1 CMOS inductor proposed by Uyanik and Tarim 480
10.3.2 CMOS grounded inductor proposed by Reja, Filanovsky and Moez 481
10.3.3 Constant-Q active inductor proposed by Tang, Yuan and Law 483
10.3.4 CMOS active inductors due to Krishnamurthy, El-Sankary and El-Masry 484
10.3.5 CMOS high-Q active grounded inductor due to Li, Wang and Gong 486
10.3.6 Tunable CMOS inductor using MOSFETs 487
10.3.7 CMOS inductor proposed by Sato and Ito 489
10.4 CMOS active transformers 492
10.5 Concluding remarks 495
References 496

11 Recent developments and concluding remarks 505
Abstract 505
11.1 Introduction 505
11.2 Retrospection 505
11.3 Recent developments on inductance simulation and related impedances 512
11.3.1 Evolution of single-active-element-based floating impedance configurations 512
11.3.2 Floating impedance configurations having electronic tunability and temperature-insensitivity 519
11.3.3 Impact of the circuits and techniques of impedance simulation on the area of memristive circuits 521
11.3.4 Impact of the circuit techniques of impedance simulation on the area of fractional-order circuits 527
11.4 Concluding remarks 531
11.5 Epilogue 531
References 532

Further reading 537
Index 543

About the authors

Raj Senani received B.Sc. from Lucknow University, B.Sc. Engg. from Harcourt Butler Technological Institute, Kanpur, M.E. (Honours) from Motilal Nehru Regional Engineering College, Allahabad (now Motilal Nehru National Institute of Technology (MNNIT), Allahabad) and Ph.D. in Electrical Engg. from the University of Allahabad.

Dr. Senani held the positions of Lecturer (1975–1986) and Reader (1986–1988) at the EE Department of MNNIT, Allahabad. He joined the ECE Department of the then Delhi Institute of Technology (later renamed as Netaji Subhas Institute of Technology (NSIT)) in 1988 and became a Full Professor in 1990. Since then, he has served as the Head, ECE Department; Head, Applied Sciences; Head, Manufacturing Processes and Automation Engineering; Dean Research, Dean Academic, Dean Administration, Dean Post Graduate Studies and Director of NSIT, a number of times and, lastly, during October 2008 to December 2014. Since April 01, 2015, he has been re-employed as the Professor of ECE at Netaji Subhas University of Technology (formerly NSIT).

Professor Senani's teaching and research interests are in the areas of Bipolar/CMOS Analog Integrated Circuits, Current-mode Circuits, Analog Filter Design, Electronic Instrumentation and Chaotic Nonlinear Circuits. He has authored/co-authored over 150 research papers in various international journals, 5 book chapters and 3 monographs, namely *Current feedback operational amplifiers and their applications* (Springer Science+Business Media, New York, 2013), *Current conveyors—variants, applications and hardware implementations* (Springer International Publishing, Switzerland, 2015) and *Sinusoidal oscillators and waveform generators using modern electronic circuit building blocks* (Springer International Publishing,

Switzerland, 2015). He served as the Editor-in-Chief for *IETE Journal of Education* during 2012–1017 and has been serving as an Associate Editor for Circuits, Systems and Signal Processing, Birkhauser, Boston (United States) since 2003, besides being on the editorial boards of several other journals and acting as an Editorial Reviewer for over 30 international journals.

Professor Senani is a Senior Member of IEEE, a Fellow of the Institution of Engineers and a Life Fellow of the Institute of Electronics and Telecommunication Engineers (IETE), India. He served as the Editor-in-Chief of the *IETE Journal of Education* during 2012–2017 and co-edited (along with Professors Dalibor Biolek and D. R. Bhaskar) a special issue of the journal *Advances in Electrical and Electronics Engineering* published in December 2017. He was elected as a Fellow of the National Academy of Sciences, India, in 2008. He is the recipient of Second Laureate of the 25th Khwarizmi International Award for the year 2012. Since 1990 Professor Senani's biography has been included in several editions of *Marquis Who's Who* series (published from NJ, United States) and a number of other international biographical directories.

D. R. Bhaskar received B.Sc. degree from Agra University, B.Tech. degree from Indian Institute of Technology (IIT), Kanpur, M.Tech. from IIT, Delhi, and Ph.D. from the University of Delhi. Dr. Bhaskar held the positions of Assistant Engineer in DESU (1981–1984), Lecturer (1984–1990) and Senior Lecturer (1990–1995) at the EE Department of Delhi College of Engineering and Reader in ECE Department of Jamia Millia Islamia (1995–2002). He became a Full Professor in January 2002 and has served as the Head of the Department of ECE during 2002–2005. Since August 05, 2016, he has been re-employed as the Professor of ECE at Delhi Technological University.

Professor Bhaskar's teaching and research interests are in the areas of Analog Integrated Circuits and Signal Processing, Communication Systems and Electronic Instrumentation. He has authored/co-authored over 90 research papers in various International journals, 4 book chapters and 3 monographs, namely *Current feedback operational amplifiers and their applications* (Springer Science+Business Media, New York, 2013), *Current conveyors—variants, applications and hardware implementations* (Springer International Publishing, Switzerland, 2015) and *Sinusoidal oscillators and waveform generators using modern electronic circuit building blocks* (Springer International Publishing, Switzerland, 2015).

Professor Bhaskar is a Senior Member of IEEE, a Fellow of the Institution of Engineers and a Life Fellow of the Institute of Electronics and Telecommunication Engineers (IETE), India. He served as an Editor of the *IETE Journal of Education* during 2012–2017 and co-edited (along with Professor Dalibor Biolek and Professor Raj Senani) a special issue of the journal *Advances in Electrical and Electronics Engineering* published in December 2017. He has acted/has been acting as a Reviewer for several international journals. His biography is included in *Marquis Who's Who* and a number of other international biographical directories.

Vinod Kumar Singh obtained B.E. and M.E. degrees in Electrical Engineering from MNR Engineering College, Allahabad, and Ph.D. in Electronics and Communication Engineering from Uttar Pradesh Technical University, India. He worked as a Research Assistant (1979–1980) at EE Department of MNR Engineering College, Allahabad, as Teaching Assistant (1980–1981) and Assistant Professor at EE Department of GB Pant University, as a Lecturer (1986–1992) and Assistant Professor at Institute of Engineering and Technology (IET) Lucknow (1992–2004), where he became a Full Professor in 2004. He served as the Head of the ECE Department at IET Lucknow between 1986 and 1988 and 2007 and 2010, where he functioned as the Dean of Research and Development since 2007. He is now the Vice-Chancellor of the International Institute of Management and Technology University, Meerut.

His teaching and research interests are in the areas of Analog Integrated Circuits, Analog Signal Processing, Network Analysis and Synthesis, Digital Electronics, Logic Design and Instrumentation Engineering. Dr. Singh has authored/co-authored over 30 research papers in various international journals, 2 book chapters and 2 monographs, namely *Current feedback operational amplifiers and their applications* (Springer Science+Business Media, New York, 2013) and *Sinusoidal oscillators and waveform generators using modern electronic circuit building blocks* (Springer International Publishing, Switzerland, 2015). He is an Editorial Reviewer for a number of international journals.

Abdhesh Kumar Singh received M.Tech. in Electronics and Communication Engineering from IASED and Ph.D. in the area of Analog Integrated Circuits and Signal Processing, from Netaji Subhas Institute of Technology (NSIT), University of Delhi. Dr. Singh held the positions of Lecturer and Senior Lecturer (June 2000–August 2001) at the ECE Department, AKG Engineering College, Ghaziabad. He joined ECE Department of Inderprastha Engineering College, Ghaziabad, India, as a Senior Lecturer in August 2001, where he became Assistant Professor in April 2002 and Associate Professor in 2006. Dr. Singh became a Full Professor in June 2009 and worked as Professor and Director of Delhi Technical Campus, Guru Gobind Singh Indraprastha University, New Delhi, India, till August 7, 2019. He is now a Visiting Professor in the ECE Department of Netaji Subhas University of Technology, New Delhi.

His teaching and research interests are in the areas of Bipolar and CMOS Analog Integrated Circuits and Signal Processing. Dr. Singh has authored/co-authored over 45 research papers in various international journals, 3 book chapters and 2 monographs, namely *Current feedback operational amplifiers and their applications* (Springer Science+Business Media, New York, 2013) *and Current conveyors—variants, applications and hardware implementations* (Springer International Publishing, Switzerland, 2015). He is an Editorial Reviewer for a number of international journals.

Preface

B. D. H Tellegen[1] is credited for formally introducing the *fifth* basic passive element named the *gyrator* in 1948 into the domain of circuit theory wherein only three[2] one-port (two-terminal) network elements (the *resistor*, the *capacitor* and the *inductor*) and only one two-port (four-terminal) passive element (the *transformer*) were known earlier. Thus, it was widely believed that with the introduction of this fifth basic passive element, the set of all passive elements was, therefore, complete! It may, however, be noted that while the transformer is passive and reciprocal, by contrast, the gyrator is passive and yet *non-reciprocal*. It was proclaimed by Tellegen himself that:

> A fifth element is conceivable which violates the reciprocity relation. We have denoted it by the name the '*ideal gyrator*'. By its introduction, the system of network elements is completed and network synthesis is much simplified.

Indeed, the gyrator was found to be a useful network element for network synthesis. However, while the original four passive elements were practically available quite easily, the gyrator had to be realized using electronic circuits. Thus, with the advent of semiconductor electronic devices, namely the bipolar junction transistors (BJTs) and *field-effect transistors* (FETs), and later the metal-oxide-semiconductor FETs (MOSFETs), there was a surge of interest taken by electronic circuit designers and academic researchers in formulating methods of realizing gyrators using BJTs and FETs. Further impetus to this area came by the commercial availability of integrated circuit (IC) op-amps as off-the-shelf ICs, which led to the development of a number of useful and immediately usable op-amp-based gyrator circuits.

A particularly interesting property of the gyrator is its *impedance-inverting* property in contrast to the transformer which has an *impedance-converting* property. Thus, a capacitively loaded gyrator simulates the behaviour of a lossless inductor, and it was this application which received much wider attention in the literature than any other direct application of the gyrator as a circuit element. Consequently, the area of inductance simulation by gyrator-based methods or otherwise became a very prominent area of analog circuits research when Orchard[3] postulated that an active filter, designed from a doubly terminated passive LC

[1]Tellegen B. D. H. 'The gyrator, a new electric network element'. Philips Research Reports. 1948; **3**(2):81–101.
[2]Later on in 1971 Leon. O. Chua postulated the *fourth* basic missing circuit element, namely the *memristor* whose first practical realization was much later produced by Hewlett-Packard Labs. in 2008.
[3]Orchard H. J. Inductor-less filter design. Electronics Letters. 1966; **6**(6):224–225.

ladder by replacing all passive inductors by simulated ones, leads to a minimum-sensitive design if the simulated inductor circuit has been well designed.

Thus, a lot of research was conducted on searching/developing gyrators/inductance simulation circuits using various building blocks such as op-amps, operational transconductance amplifiers, current conveyors and current-feedback operational amplifiers. With the advent of numerous new composite active building blocks recently, the area of inductance simulation is still continued to remain a very fertile and popular area of analog circuits' research.

The aim of this book/monograph is to provide an exhaustive coverage of prominent gyrator circuits, simulated inductors and related synthetic impedances using a variety of active devices and building blocks. These include BJTs and MOS transistors, the ubiquitous IC op-amps, operational transconductance amplifiers and a host of other modern electronic circuit building blocks. In the last category come current conveyors, current feed-back op-amps and a variety of other newer building blocks which have been widely investigated in the recent technical literature around the world during the last 15 years.

An exhaustive literature survey revealed that no book which deals with the mentioned topic on the scale envisaged was available. Therefore, the scope of this book is to include a comprehensive and wide coverage of this area based upon about 1,500 research papers published in reputed international journals around the world so far. The monograph contains a critical discussion of over 500 circuits of gyrators, simulated inductors and other related synthetic impedances, along with their relevant design equations, performance features, merits–demerits and applications. Thus, the monograph covers the entire gamut of circuits evolved so far but places more emphasis on *state-of-the-art* circuits, techniques and applications.

It is believed that this book should be useful to all academicians, educators, practising engineers and analog designers who are interested in electronic circuits in general and in electronic gyrators, synthetic inductors and other related impedances, in particular. It is also hoped that the practising engineers and analog designers too may find the book to be useful due to a comprehensive repertoire of over 500 circuits to choose from according to an application at hand.

The book also highlights several open problems and further ideas for research which are provided in most of the chapters.

Lastly, we would highly appreciate if comments and suggestions on any aspect of the contents of this book may kindly be mailed to the first author (senani@ieee.org). It would be our pleasure to modify/amend the concerned portions of the book accordingly in the subsequent revised version of the book.

New Delhi, India — Raj Senani
Delhi, India — Data Ram Bhaskar
Lucknow, India — Vinod Kumar Singh
Noida, India — Abdhesh Kumar Singh

August 15, 2019

Acknowledgement

The first author (RS) of this monograph got an opportunity to conduct his independent study on *active-RC networks* (thanks to his supervisor (late) Professor R. N. Tiwari) way back in 1974 while he was working on his Master's thesis entitled *Design of a high-Q, sharp-cutoff bandpass active filter* at Motilal Nehru Regional Engineering College (MNREC), Allahabad (now known as Motilal Nehru National Institute of Technology, Allahabad). This stint was instrumental in cultivating his interest in the area of *Active Networks* to the extent that he decided to take up an academic career only to enable him to carry out his research in this area. Thus, in June 1975, he joined the Electrical Engineering Department of MNREC, Allahabad, as a faculty member (Lecturer), where he enjoyed, besides undergraduate and post-graduate teaching, complete freedom in carrying out his research on *inductance simulation.* This research resulted in the discovery of a number of interesting circuits/techniques for the realization of grounded/floating, lossless/ lossy inductors and three-port gyrators using IC op-amps and the new active elements called *current conveyors* which were fast emerging as versatile alternative active building blocks for analog signal processing.

While at MNREC, Allahabad, the first author (RS), in his research, derived immense inspiration from the works and the words of Professor(s) R. W. Newcomb (University of Maryland, US), S. C. Dutta Roy (IIT, Delhi, India), D. Patranabis (Jadavpur University, India), H. K. Kim (University of Manitoba, Canada), S. Pasupathy (University of Toronto, Canada), (Late) B. B. Bhattacharyya (Concordia University, Canada) and several other senior researchers. Professor R. W. Newcomb was particularly kind in arranging to provide a copy of the excellent *Bibliography on Gyrators* (by J. Miller and R. W. Newcomb) xeroxed from his last remaining copy, way back in 1978! The first author (RS) became a Reader in the EE Department of MNREC in 1986 and continued his research on *Inductance simulation. Active filters and oscillators* till July 1988 after which he moved to Delhi Institute of Technology (DIT). He joined DIT as an Assistant Professor in the Department of Electronics and Communication Engineering, where he became a Full Professor in May 1990. It was at DIT where the first Research Lab named 'Linear Integrated Circuits' (LIC) Lab came into being.

It was at DIT (later named as Netaji Subhas Institute of Technology (NSIT)) that the second author of this monograph (DRB), who was himself a well-established faculty member at adjoining Delhi College of Engineering, joined as the first Ph.D. research scholar at the LIC Lab of DIT in 1989. Subsequently, the third author (VKS), who was a former student from MNREC, Allahabad, and was equally well-established as a faculty member at Institute of Engineering and Technology (IET), Lucknow, became a sort of visiting Ph.D. research scholar at the Lab beginning 1989. Finally, the fourth author of this monograph (AKS) too joined as a Ph.D. scholar at the ECE Department of DIT in 1990. Thus, the LIC Lab at DIT became the place where our research group (comprised the four authors of this monograph along with three other colleagues) started working in the areas mentioned above. Our research group at LIC Lab of DIT enjoyed the patronage and immense support of the founder Director of the Institute Professor B. N. Mishra, who was generous enough to provide unconditional support to consolidate the LIC Lab with the very best equipment and software needed for our work. Subsequently, the second author (DRB) moved to Jamia Millia Islamia in 1995, a central university, where he continued his independent research and was elevated to the rank of Professor and then the Head of the Department in 2002. On the other hand, while the third author (VKS) continued his academic work at IET Lucknow, became a Full Professor in 2004 and was elevated to the position of Vice-Chancellor of International Institute of Technology and Management University, Meerut; the fourth author (AKS) served at a number of institutes, rose to become a Full Professor at RD Engineering College, Ghaziabad, in 2009 and subsequently became Professor and Director of Delhi Technical Campus, Greater Noida.

In 1998, when NSIT shifted to its new campus, the research wing of the then LIC Lab was renamed as Analog Signal Processing (ASP) Research Lab where the research work of our group continued. Now and then, all the co-authors maintain their links with the ASP Research Lab and we have been continuing to collaborate on several research problems/projects of mutual interest till date. All the authors are thankful to their respective former employers for facilitating their continued linkages with ASP Research Lab at NSIT.

The first author (RS) thankfully acknowledges the support of the former Directors of NSIT Professor Yogesh Singh and Professor J. P. Saini, the former and the current Heads of the ECE Department Professor Maneesha Gupta and Professor S. P. Singh and the current Faculty-in-Charge of the ASP Research Lab. Dr. Tarun Kumar Rawat for facilitating the continued research work of our group at the ASP Research Lab. Thanks are also due to the founder Vice-Chancellor of Netaji Subhas University of Technology (formerly, NSIT) Professor J. P. Saini for his strong support. The second author (DRB) wishes to thank Professor Yogesh Singh, Vice-Chancellor of Delhi Technological University (DTU), Professor Madhusudan Singh (Head EE Department), Professor Pragati Kumar, Professor S. Indu (Head, ECE Department) and his colleagues Professor Neeta Pandey and Professor Rajeshwari Pandey for their moral support and encouragement. The third author (VKS) wishes to thank Professor Subodh Wairya (Head ECE, IET Lucknow) for his administrative help. He also wishes to thank Dr. Manoj Kumar Jain for his technical support from time to time. The fourth author (AKS) wishes to thank Professor J. P. Saini, Vice-Chancellor of NSUT, for his continued support and good wishes for the last two decades.

All the authors of this monograph wish to thank Dr. Paul Deards, Ms. Sarah Lynch and Ms. Olivia from IET for their strong support and advice during the course of writing this monograph and for their understanding in appreciating our difficulties and un-anticipated hurdles. Thanks are also due to S. Rawat for extending help in the preparation of the manuscript. The moral support of the other members of our research group, namely Professor Pragati Kumar, Dr. S. S. Gupta and Professor R. K. Sharma, is also gratefully acknowledged.

Last but not the least, we express our profound gratitude to all our respective family members for their understanding and unflinching moral support, without which this work could not have been carried out.

New Delhi, India	Raj Senani
Delhi, India	Data Ram Bhaskar
Lucknow, India	Vinod Kumar Singh
Noida, India	Abdhesh Kumar Singh

August 15, 2019

Chapter 1
Gyrators, integrated inductors and simulated inductors

Abstract

This chapter first reviews the basic one- and two-port circuit elements and then highlights the basic characterization and impedance inverting property of the *gyrator*. To place the scope and the contents of this monograph in the right perspective, this chapter then discusses several other issues such as commercially available passive inductors and *Coilcraft*, difficulties in micro-miniaturization of inductors, integrated circuit spiral inductors on the chip and use of ANSYS and COMSOL software in the analysis and characterization of on-chip inductors. The chapter ends by highlighting the need for and the importance of the simulated inductors which constitute a dominating topic in this monograph.

1.1 Prologue

Gyrators, simulated impedances and their applications have been a very prominent and popular area of research for well over five decades, because of the need for simulated inductors in microelectronic realization of filters, oscillators and several other analogue signal processing/signal generating applications. The importance and popularity of the gyrators, simulated impedances and their applications can be well understood by noting that over 1500 research papers have so far been published on these topics, spread over almost all leading international journals of repute around the world, with a majority of them having appeared during the last three decades.

In view of the above, the authors were convinced that the time was now ripe to write a book/monograph on this topic to fill this void and to make it available as a comprehensive source of information in this area.

While writing this monograph, a number of special features have been incorporated which include the following:

- The book is aimed to be a single source of reference to almost everything connected with gyrators, simulated inductors and a variety of other related impedances and their applications. To meet this end, it has been our endeavour to provide a comprehensive list of references at the end of each chapter. This is expected to be particularly useful for research scholars.

- To make it useful to all academicians and educators interested in electronic circuits in general and electronic gyrators, synthetic inductors and other impedances, in particular, a tutorial-review format of writing has been adopted. We believe that in this way, the material should become suitable for study even for the undergraduate student community of Electrical, Electronics and Communication Engineering.
- The various chapters of this monograph though collectively cover the entire gamut of circuits evolved so far but place more emphasis on the state-of-the-art developments and important research results on the respective topics.
- Researchers working in analogue circuit design, in general, and synthetic impedances, in particular, should find all the significant references at one source. In addition, a number of open problems and further ideas for research have been provided in almost every chapter. A list of supplementary references has been provided in Further reading. All these are expected to be useful to the prospective researchers.
- Since a comprehensive repertoire of over 500 circuits has been provided in this book, it is expected to be useful to the practising engineers/analogue circuit designers too who may find the book a useful reference from where they can choose a specific circuit according to the requirement for a given application.

Since this book is about gyrators, simulated inductors and related impedances and their applications, in the introductory chapter, we will discuss about the basic circuit elements, both one- and two-port, with particular focus on those which have a bearing on the topics covered. Contrary to the popular belief that somehow inductors cannot be fabricated on an integrated circuit (IC) chip, some type of inductors and even transformers are indeed fabricated on the silicon chip and are widely used specially for applications in GHz range but with very small values of the inductance (typically in nH ranges) and very modest quality (Q) factors (typically a few tens). So, we do give a brief account of the spiral inductors and transformers realizable on the chip also. The popular software ANSYS and COMSOL, used for the analysis and characterization of inductors, are also discussed briefly.

For applications in very low frequency ranges (<10 Hz range), low frequency range (10 Hz–100 kHz range) and medium frequency range (100 kHz–10 MHz or so), simulated inductors are preferred. It is in this context that the gyrators are used as a direct means of simulating inductors while there are several other methods of simulating them. Hence, the need for and the advantages accrued from the active simulation of inductors have been highlighted.

1.2 Basic one-port circuit elements

Although a majority of the classical circuit theory books, for instance, see [1–3], mention and deal with only three two-terminal elements, namely the resistor, the capacitor and the inductor, we take a modern view and also include the recently added fourth basic circuit element the *memristor* [4–7] into this discussion.

Since charge (q), flux (ϕ), current (i) and voltage (v) are the four fundamental electrical variables, in principle, four constitutive relations between them can be formulated:

1. *Resistor*: $v(t) = f[i(t)]$ or $i(t) = \Psi[v(t)]$ both define a resistor which is a linear resistor if the functions f and Ψ are both linear, e.g. $v(t) = Ri(t)$. In the event of these functions being non-linear, both of these constitutive relations define a non-linear resistor which is a current-controlled nonlinear resistor in case of the former and a voltage-controlled non-linear resistor in case of the latter.
2. *Inductor*: $\phi(t) = f[i(t)]$ or $i(t) = \psi[\phi(t)]$ both define a linear inductor if the functions involved in these two forms of the constitutive relations between flux and current are linear, e.g. $\phi(t) = Li(t)$. On the other hand, if the functions involved are non-linear then these relations define a current-controlled non-linear inductor and flux-controlled nonlinear inductor, respectively.
3. *Capacitor*: $q(t) = f[v(t)]$ or $v(t) = \psi[q(t)]$ both define a linear capacitor if these constitutive relations are linear, e.g. $q(t) = Cv(t)$; however, if they are nonlinear then they define, respectively, a nonlinear voltage-controlled capacitance and a nonlinear charge-controlled capacitance.
4. *Memristor*: $\phi(t) = f[q(t)]$ or $q(t) = \psi[\phi(t)]$ when both the functions are non-linear they define a charge-controlled memristor and a flux-controlled memristor, respectively. However, if these constitutive relations are linear, then differentiating both the sides of these equations, with respect to time, leads to a linear relation between voltage and current, noting that $d\phi(t)/dt = v(t)$ and $dq(t)/dt = i(t)$. It has been well established now that a signature characteristic of the memristor is that it must exhibit *a pinched* hysteresis 'If it's pinched; it's a memristor' [7]. Memristors and their applications are currently being widely investigated [4–7].

The inter-relationship between the four basic variables and their constitutive equations are normally depicted by the diagram of the type shown in Figure 1.1.

1.3 Basic two-port circuit elements

From the viewpoint of classical circuit theory, there are only two basic two-port circuit elements, namely the ideal transformer and the ideal gyrator. However, in this chapter, more general two-port elements known as impedance converters and inverters [1–3] have also been included because of being versatile active elements which are very closely related to the topics included in this book.

1.3.1 The transformer

Among two-port elements, the ideal transformer is characterized by the following transmission matrix:

$$\begin{bmatrix} v_1 \\ i_1 \end{bmatrix} = \begin{bmatrix} (N_1/N_2) & 0 \\ 0 & (N_2/N_1) \end{bmatrix} \begin{bmatrix} v_2 \\ -i_2 \end{bmatrix} \tag{1.1}$$

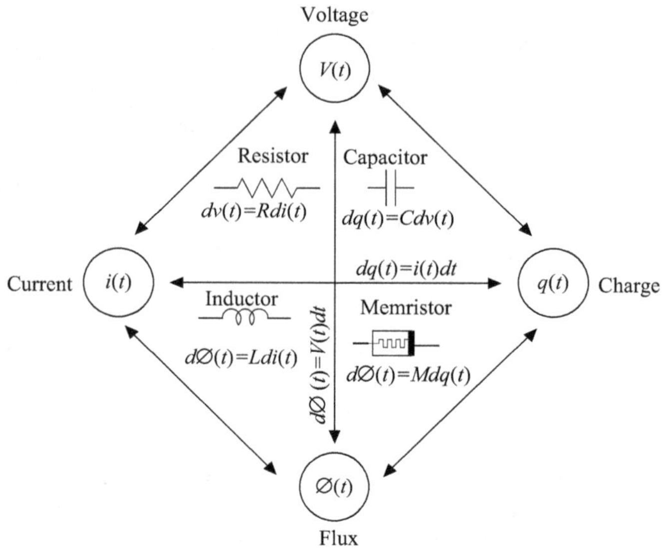

Figure 1.1 Pictorial representation of the four fundamental circuit elements

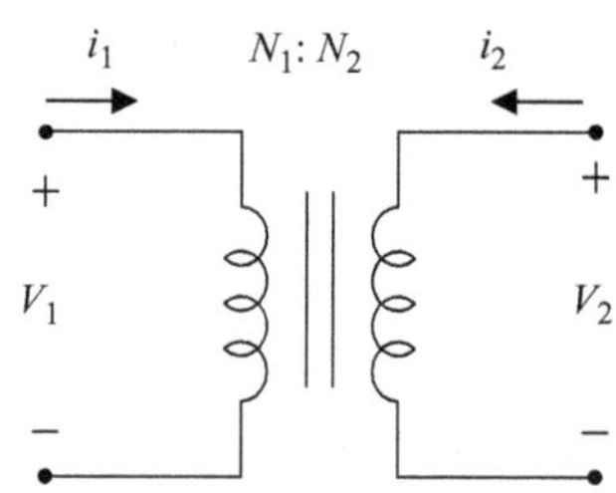

Figure 1.2 The ideal transformer

where (N_2/N_1) is the turns ratio of the transformer. When the secondary winding of the transformer (the output port) is terminated into a load impedance Z_L, the input impedance seen from the primary end (the input port) is found to be $Z_{in} = (N_1/N_2)^2 Z_L$, which implies the *impedance converting* property of the transformer (Figure 1.2).

1.3.2 The gyrator

Tellegan [8] is credited for formally introducing the *fifth* basic passive element named the *gyrator* in 1948 into the domain of circuit theory wherein only three one-port (two-terminal) network elements (namely, the *resistor*, the *capacitor* and the *inductor*) and only one two-port (four-terminal) passive element (namely, the *transformer*) were known earlier. Thus, it was widely believed that with the introduction of this fifth basic passive element, the set of all passive elements was, therefore, complete! It may, however, be noted that while the transformer is passive and reciprocal, by contrast, the gyrator is passive and yet *non-reciprocal*. It was proclaimed by Tellegan himself:

> A fifth element is conceivable which violates the reciprocity relation. We have denoted it by the name the *ideal gyrator*. By its introduction, the system of network elements is completed and network synthesis is much simplified.

Indeed, the gyrator was found to be a useful network element for network synthesis. However, while the original four passive elements (namely the resistor, capacitor, inductor and transformer) were practically available/constructed easily, the gyrator had to be realized using electronic circuits. Thus, with the advent of semiconductor electronic devices, namely the bipolar junction transistors (BJTs) and the *field effect-transistors* (FETs), and later the metal-oxide-semiconductor FETs (MOSFETs), there was a surge of interest taken by electronic circuit designers and academic researchers in formulating methods of realizing gyrators using BJTs and FETs. A good treatise on transistor-based gyrator circuits is [9].

Further impetus to this area came by the commercial availability of the IC op-amps as off-the-shelf ICs which lead to the development of a number of useful and immediately usable op-amp-based gyrator circuits [1–3]. A number of bibliographies on the work done on gyrators and simulated inductors have been compiled; see [10–12], which contain exhaustive list of references on the realization and application of gyrators till 1976.

A particularly interesting and useful property of the gyrator is its *impedance-inverting* property, in contrast to the transformer which has *impedance-converting* property, in that a capacitively loaded gyrator simulates the behaviour of a lossless inductor. It was this application which received a much wider attention in the literature in comparison to the other applications of the gyrator as a circuit element.

Consequently, the area of inductance simulation by gyrator-based methods or otherwise became a very prominent area of analogue circuits research, particularly more so when Orchard [13] postulated that an active filter designed from a doubly terminated passive LC ladder by replacing all the passive inductors by simulated ones would lead to a minimum-sensitive design if the simulated inductor circuit has been well designed.

An ideal gyrator is a two-port non-reciprocal passive network element which is characterized by the following short circuit admittance matrix ([Y]-matrix):

$$\begin{bmatrix} i_1 \\ i_2 \end{bmatrix} = \begin{bmatrix} 0 & \pm G \\ \mp G & 0 \end{bmatrix} \begin{bmatrix} v_1 \\ v_2 \end{bmatrix} \tag{1.2}$$

or an open circuit impedance matrix ([Z]-matrix) characterized by

$$\begin{bmatrix} v_1 \\ v_2 \end{bmatrix} = \begin{bmatrix} 0 & \pm R \\ \mp R & 0 \end{bmatrix} \begin{bmatrix} i_1 \\ i_2 \end{bmatrix} \tag{1.3}$$

where G represents the gyration conductance, and similarly, R represents gyration resistance of the gyrator.

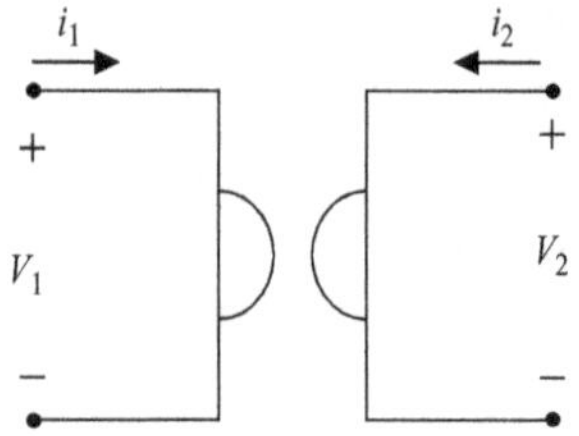

Figure 1.3 The ideal gyrator

From the above two characterizations, it can be easily deduced that the total power absorbed by both the ports of the gyrator is $P = v_1 i_1 + v_2 i_2 = v_1(\pm G v_2) + v_2(\mp G v_1) = 0$, and thus, an ideal gyrator is *passive* and yet non-reciprocal since $y_{12} \neq y_{21}$ and $z_{12} \neq z_{21}$ (Figure 1.3).

In the analysis of some physical gyrator circuits, it may be easier to find the transmission matrix (*ABCD*-matrix) which, for the ideal gyrator, can be represented as

$$\begin{bmatrix} v_1 \\ i_1 \end{bmatrix} = \begin{bmatrix} 0 & \pm b \\ \pm c & 0 \end{bmatrix} \begin{bmatrix} v_2 \\ -i_2 \end{bmatrix} \tag{1.4}$$

where parameter b has the dimension of impedance and c has the dimension of admittance.

It is now well understood that the gyrator is a special case of a more generalized two-port network element known as positive impedance inverter (PII) defined by the matrix equation (1.4).

If in equations (1.2) and (1.3), the gyration conductances or the gyration resistances are not equal then $P = v_1 i_1 + v_2 i_2 = v_1(\pm G_1 v_2) + v_2(\mp G_2 v_1) = \pm(G_1 - G_2) v_1 v_2$ could be either positive or negative thereby such a gyrator could also be *active* when $P < 0$.

If a gyrator is terminated into a capacitor C at port-2, then input impedance looking into port-1 would be

$$Z_{in} = s\frac{C}{G^2}; \quad \text{using the characterization of (1.2)} \tag{1.5}$$

$$Z_{in} = sCR^2; \quad \text{using the characterization of (1.3)} \tag{1.6}$$

$$Z_{in} = sC\frac{b}{c}; \quad \text{using the characterization of (1.4)} \tag{1.7}$$

1.3.3 The two-port impedance converters and inverters

Impedance converters and inverters are another class of important network elements which have wide applications in impedance simulation.

Let us now consider a general two-port active element as shown in Figure 1.4. Let this network be characterized by the transmission matrix:

$$\begin{bmatrix} v_1 \\ i_1 \end{bmatrix} = \begin{bmatrix} A & B \\ C & D \end{bmatrix} \begin{bmatrix} v_2 \\ -i_2 \end{bmatrix} \tag{1.8}$$

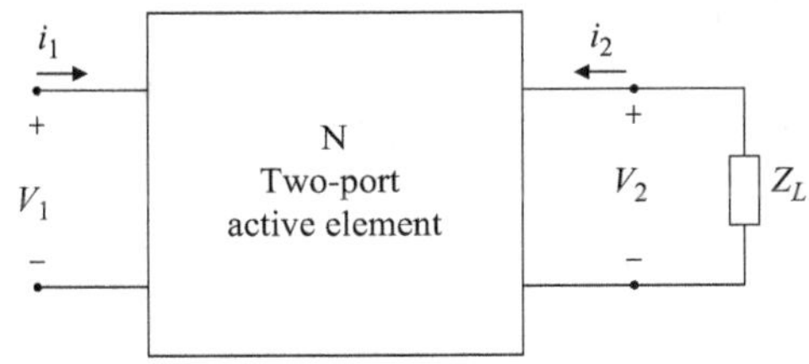

Figure 1.4 The general representation of two-port active elements

If this two-port is terminated into a load impedance Z_L, the impedance looking onto the input port is given by

$$Z_{in} = \frac{AZ_L + B}{CZ_L + D} \tag{1.9}$$

Consider now the following cases.

Case 1 If the two-port is designed to have the parameters $B = 0$ and $C = 0$ then Z_{in} becomes

$$Z_{in} = \left(\frac{A}{D}\right)Z_L \tag{1.10}$$

Thus, the circuit becomes an impedance converter. The circuit would become a positive impedance converter (PIC) if the parameters A and D carry the same sign, and it becomes a negative impedance converter (NIC) if these parameters are of opposite signs. Thus, the PIC is a two-port element characterized by either of the two types of transmission matrices:

$$\text{Type-1 PIC: } [T_1] = \begin{bmatrix} +A & 0 \\ 0 & +D \end{bmatrix} \text{ or Type-2 PIC: } [T_2] = \begin{bmatrix} -A & 0 \\ 0 & -D \end{bmatrix} \tag{1.11}$$

Similarly, two types of NICs can be characterized by the following transmission matrices:

$$\text{Type-1 NIC: } [T_1] = \begin{bmatrix} +A & 0 \\ 0 & -D \end{bmatrix} \text{ or Type-2 NIC: } [T_2] = \begin{bmatrix} -A & 0 \\ 0 & +D \end{bmatrix} \tag{1.12}$$

In the special case, when the magnitudes of both A and D are unity, the former is called a current inversion NIC (CNIC), while the latter is called a voltage inversion NIC (VNIC) [2].

Case 2 If the two-port is designed to have the parameters $A = 0$ and $D = 0$, then Z_{in} becomes $Z_{in} = \left(\frac{B}{C}\right)\left(\frac{1}{Z_L}\right)$.

Since the input impedance is now inversely proportional to the load impedance Z_L, or Z_L gets *inverted* when looking from port 1, such a two-port is called an impedance inverter which can be classified as a PII or negative impedance inverter (NII) depending upon whether the constant of proportionality is positive or negative. Thus, there could be two varieties of PII which can be characterized by the following transmission matrices:

$$\text{Type-1 PII: } [T_1] = \begin{bmatrix} 0 & +B \\ +C & 0 \end{bmatrix} \text{ or Type-2 PII: } [T_2] = \begin{bmatrix} 0 & -B \\ -C & 0 \end{bmatrix} \tag{1.13}$$

Similarly, the two types of NIIs can be characterized by the following transmission matrices:

$$\text{Type-1 NII: } [T_1] = \begin{bmatrix} 0 & +B \\ -C & 0 \end{bmatrix} \text{ or Type-2 NII: } [T_2] = \begin{bmatrix} 0 & -B \\ +C & 0 \end{bmatrix} \tag{1.14}$$

All the above-mentioned two-port elements have been well understood in the literature and their realizations in terms of electronic devices too have been known in the literature since long [1–3]. Both impedance inverters and converters are relevant to the theme of this monograph and would be dealt as and when needed in further chapters.

1.4 The pathological elements

The discussion of network elements will not be complete until we talk about the so-called pathological elements which are also called *degenerate* elements. Two well-known pathological elements, namely the nullator and the norator, were introduced in the domain of circuit theory by Carlin and Youla [14] and since then have been found to be particularly useful in (i) modelling of the controlled sources and various analogue circuit building blocks, (ii) realization of the various circuits using such models, (iii) unifying the realizations of the same function in terms of different active elements, (iv) generating equivalents of a circuit realized with devices which can be modelled with nullators and norators and (v) developing systematic synthesis methods in terms of devices which can be modelled by these pathological elements. A comprehensive compilation of work done on nullators and norators and their applications in circuit synthesis and design can be found in [15].

The two basic pathological elements are defined as follows:

Figure 1.5 Nullator and norator representation: (a) nullator and (b) norator

Nullator: The nullator is defined as a one-port (two-terminal) element which is characterized by the following equations: $v = 0; i = 0$ (Figure 1.5(a)).

Norator: The norator is defined as a one-port element which is characterized by $v = arbitrary; i = arbitrary$ *(Figure 1.5b).*

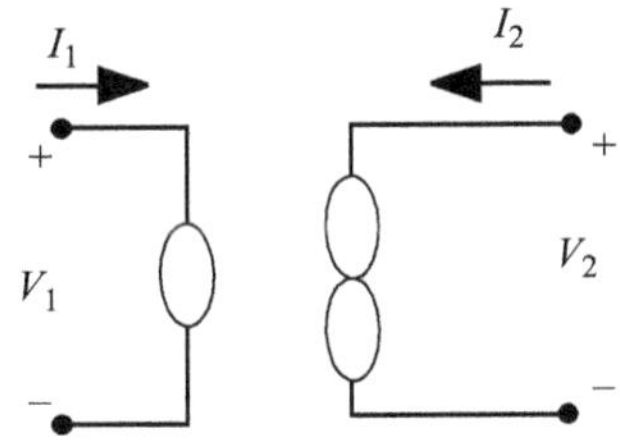

Figure 1.6 The nullor

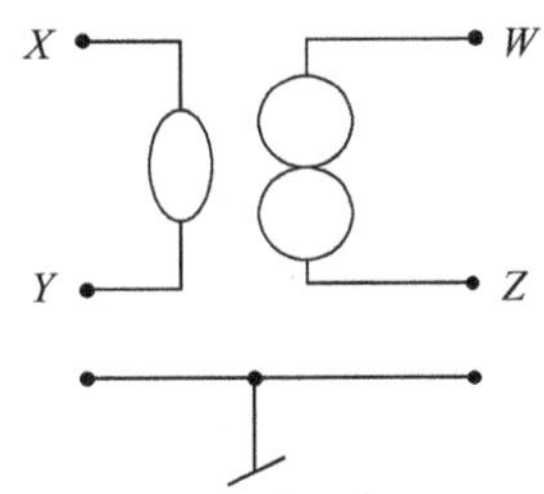

Figure 1.7 The four-terminal floating nullor

It is apparent that none of the two elements are physically realizable on their own as stand-alone elements. However, if one makes a two-port with the nullator constituting its input port and the norator constituting its output port, it would be characterized by the following transmission matrix (Figure 1.6):

$$\begin{bmatrix} V_1 \\ I_1 \end{bmatrix} = \begin{bmatrix} 0 & 0 \\ 0 & 0 \end{bmatrix} \begin{bmatrix} V_2 \\ -I_2 \end{bmatrix} \tag{1.15}$$

It is well known that in contrast to the nullator and a norator, a nullor is physically realizable by a number of ideal active elements in a variety of ways. Thus, it is well documented in the literature that an ideal op-amp is equivalent to a nullor [2,3] with one terminal of the norator connected to ground, while both an ideal BJT and MOSFET are equivalent to a three-terminal nullor. On the other hand, a *Second generation Current Conveyor* (CCII) [16] is also equivalent to a three-terminal nullor with nullator and the norator sharing a common terminal, see [17] and the references cited therein.

Lastly, the so-called four-terminal floating nullor [18,19] (Figure 1.7) which can be regarded to be the most general form of the nullor is realizable with an interconnection of either two CCII+ or two CCII– or two current feedback operational amplifiers (of AD844 type) [19].

Two other pathological elements, namely the voltage mirror (VM) and the current mirror (CM), were introduced by Awad and Soliman in [20].

The CM is a lossless two-port element which is defined by the following equations:

$$v_1 = arbitrary;\; v_2 = arbitrary;\; i_1 = i_2 = arbitrary \tag{1.16}$$

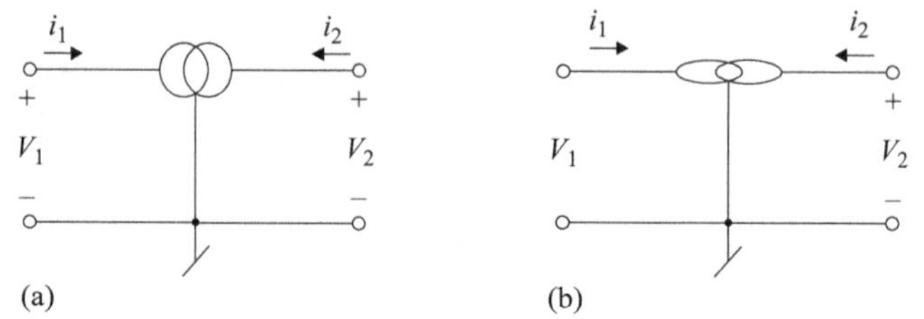

Figure 1.8 Two other pathological elements introduced by Awad and Soliman [20]: (a) current and (b) voltage mirror

On the other hand, a VM is a lossless two-port element characterized by

$$v_1 = -v_2; \; i_1 = i_2 = 0 \tag{1.17}$$

It may be noted that (i) the symbol of CM is same as the symbol of a transistorized CM; however, the present one is a bidirectional element and it is not realizable by any physical means. (ii) both CM and VM have a reference node which is shown by the ground connection; however, in various combinations of these elements, they are used as two-terminal elements with reference node generally not being shown/ used (Figure 1.8).

Like nullator and norator, the VM and CM also constitute a pair of universal active elements as demonstrated by Soliman and Saad [21] who showed that a CM–VM pair can not only represent a nullor but can also represent a CCII+ which could not be represented by nullors only; additional resistors are needed in this case.

1.5 Multi-terminal gyrator

As an extension to the idea of the two-port gyrator, the multi-terminal gyrator was introduced by Holt and Linggard in [22,23]. The multi-terminal gyrator is defined as having n-input terminals and m-output terminals with a common ground as shown in Figure 1.9.

The admittance matrix of multi-terminal gyrator having n-inputs and m-outputs and thereby $(n + m + 1)$ terminals is given by

$$\begin{bmatrix} I_1 \\ I_2 \end{bmatrix} = \begin{bmatrix} 0 & [G]^T \\ -[G]^T & 0 \end{bmatrix} \begin{bmatrix} V_1 \\ V_2 \end{bmatrix} \tag{1.18}$$

where the submatrix $[G]$ is called multi gyration matrix and may be a real $n \times m$ matrix.

The impedance inverting property of the two-port gyrator is available in the multi-terminal gyrator also, on multi-terminal basis. Thus, if the m-output terminals of the multi-gyrator are loaded into a m-terminal network having admittance matrix $[Y]$, the n-terminal network looking into the input side would have an admittance matrix $[Y]^*$ given by

$$[Y]^* = [G][Y]^{-1}[G]^T \tag{1.19}$$

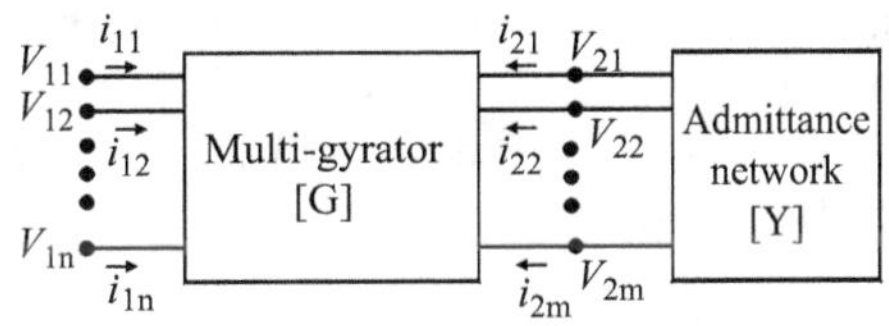

Figure 1.9 A multi-terminal gyrator terminated into a n-terminal network characterized by a [Y]-matrix

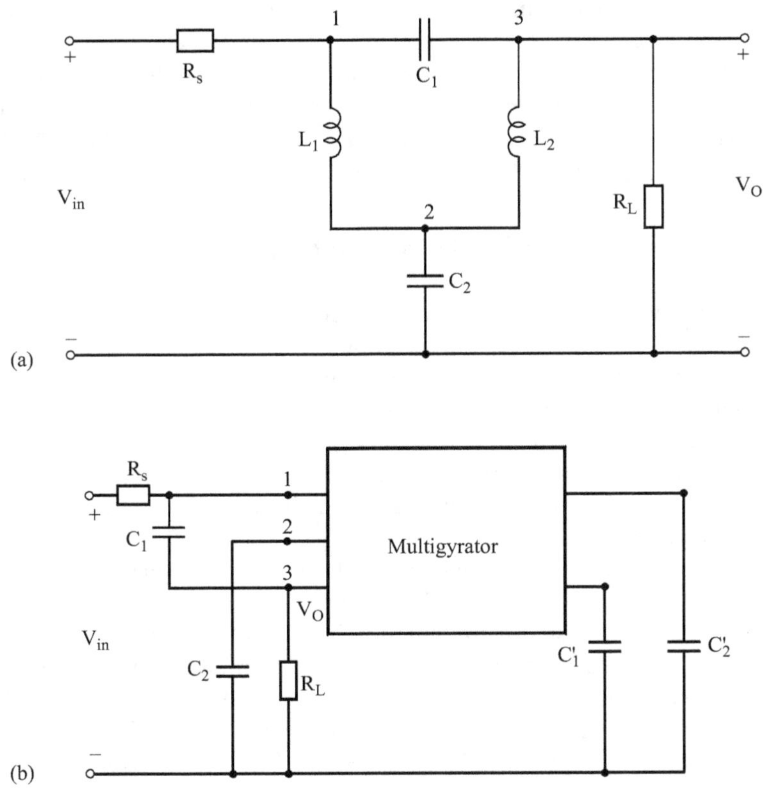

Figure 1.10 Active RC filter design using multi-terminal gyrator: (a) the passive RLC (L_1 and L_2 are mutually coupled) and (b) a multi-gyrator implementation of the given filter

From the above, it follows that if the loading network is purely capacitive, then at the *n*-input terminals of the multi-terminal gyrator characterized by $[Y]^*$ would be purely inductive.

As an example, consider a second-order LC filter shown in Figure 1.10.

The two mutually coupled ungrounded inductors of this circuit can be simulated by a (3 + 2) terminal gyrator as shown in Figure 1.10(b). Thus, a multigyrator has the capability to simulate complex mutually coupled inductive networks using a minimum number of grounded capacitors. In [22,23], it was demonstrated that a multi-gyrator can be systematically synthesized from unity gain voltage followers, unity gain current amplifiers and resistors.

1.6 Multiport inverters/converters

Silva in [24] has extended the idea of two-port impedance invertors/converters[1] to multi-ports. Although no practical realizations were proposed in [24], it was demonstrated that although in the case of two-port impedance converters/inverters are reciprocal of admittance converters/invertors, this is not, in general, true for multiport converters/invertors. Also, it was shown that multi-port gyrators and most of the circuits that can be used to simulate floating inductors can be shown to be special cases of multiport converters/invertors. Some of these ideas would be explained in this monograph in the subsequent chapters.

1.7 Commercially available inductors and Coilcraft

One of the best, well-known and popular sources of procuring inductors as off-the-shelf components is *Coilcraft*[2] who design and manufacture inductors in multiple sizes and constructions to meet a variety of applications such as filtering, tuning and impedance matching. Coilcraft provides RF chip inductors in compact size – which are useful for high frequency, high Q radio frequency (RF) chokes; air core inductors – which are characterized by high frequency, very high Q, narrowband and bias chokes – and power inductors – which have high inductance and are suitable for wideband filtering.

Coilcraft also provides assistance through its Filter Design Software which is freely available on their website and provides a tool for designing a variety of passive RLC filters choosing the required inductors from their catalogue and also displaying the expected frequency response. As an example, a third-order low pass elliptic filter was designed using Coilcraft, as per the following specifications: passband return loss = 10 dB, stop-band attenuation = 60 dB, impedance level = 50 Ω, Passband frequency = 1 MHz. The designed filter using *Coilcraft* 0402 inductor family, obtained from their Filter Design Software, looks as shown in Figure 1.11(a). It may be mentioned that the 0402 size chip inductors offer inductors in the range 2.8 nH to 120 nH, 1.0 nH to 220 nH and 1.0 nH to 120 nH and offer highest Q in these ranges.

The frequency response of the designed filter using the low pass filter designer looks as shown in Figure 1.11(b) which is seen to meet the frequency domain design specifications.

[1]For an interesting exposition on minimal realization of impedance converters/inverters, see [112].
[2]https://www.coilcraft.com

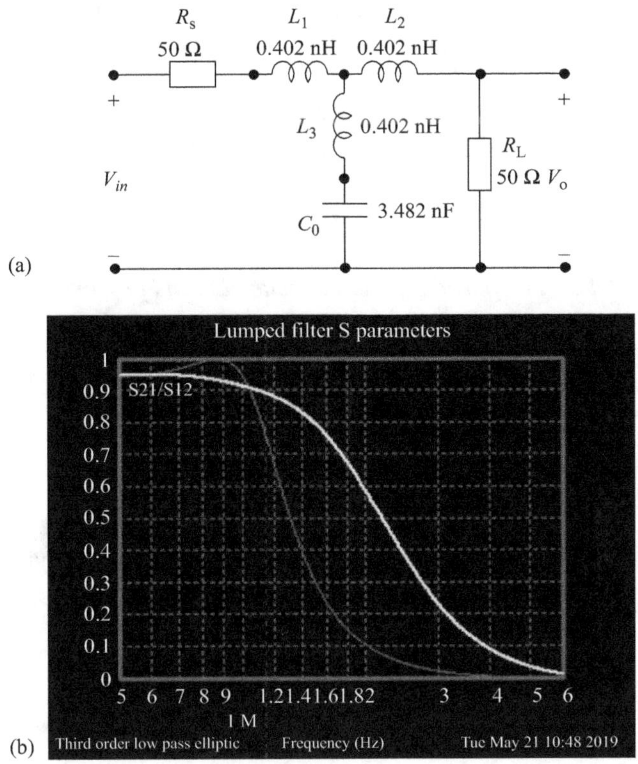

Figure 1.11 Third-order low pass elliptic filter designed using the Coilcraft Low Pass Filter Designer: (a) the designed circuit and (b) the frequency response

1.8 Basic difficulties in micro-miniaturization of inductors

In this section, we show why there are fundamental difficulties in micro-miniaturization of inductors as compared to that of the capacitors. Consider first the parallel plate capacitor shown in Figure 1.12.

We know that the value of the capacitance of the parallel plate capacitor is given by

$$C = \varepsilon \frac{A}{d} \tag{1.20}$$

where ε is the permittivity of the dielectric material. The associated losses are represented by a conductance G in parallel with the capacitance C which is given by

$$G = \sigma \frac{A}{d} \tag{1.21}$$

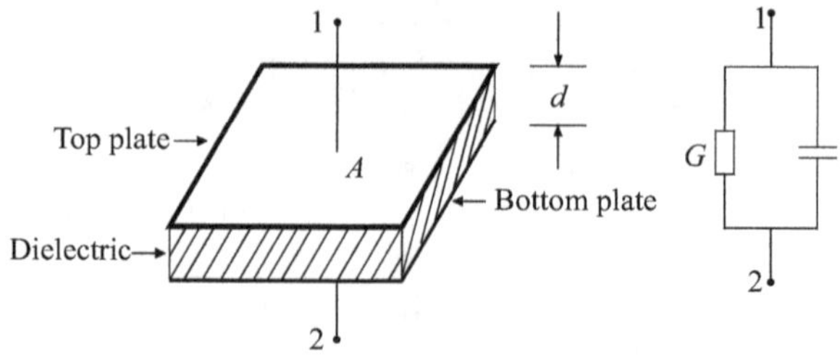

Figure 1.12 The parallel plate capacitor and its passive equivalent

where σ is the conductivity of the dielectric material. The quality factor Q of the capacitor is given by

$$Q = \omega \frac{\varepsilon}{\sigma} \tag{1.22}$$

Now if all the physical dimensions of the capacitor are scaled down by a factor x (where $x \ll 1$) for micro-miniaturization, the new value of the capacitance will be given by

$$C_x = \varepsilon \frac{(x^2 A)}{xd} = x\left(\varepsilon \frac{A}{d}\right) = xC \tag{1.23}$$

On the other hand, the modified expression for the quality factor, since it is independent of the dimensions, continues to remain the same, i.e.

$$Q_x = \omega \frac{\varepsilon}{\sigma} \tag{1.24}$$

From the above, it may be seen that when miniaturized, the capacitance value realizable would be reduced to very low values (say in pF); however, this micro-miniaturized capacitance can be still realized with high-Q, since Q remains unaffected by scaling down the size!

Let us now consider the issue of micro-miniaturization of an inductor.

Consider the following expression for the inductance of a solenoidal inductor (Figure 1.13):

$$L = \mu n^2 F\left(\frac{d}{l}\right) d \tag{1.25}$$

where μ is the permeability of the core material, n represents the number of turns, d is the diameter of the coil and $F(d/l)$ represents the form factor.

The value of the associated series resistance R, representing the losses, is given by

$$R = \rho \frac{l_w}{A} = \rho \frac{\pi d n}{A} \tag{1.26}$$

where ρ is the resistivity of the material of the coil, d is the axial diameter of the coil and A is the cross-sectional area of the coil.

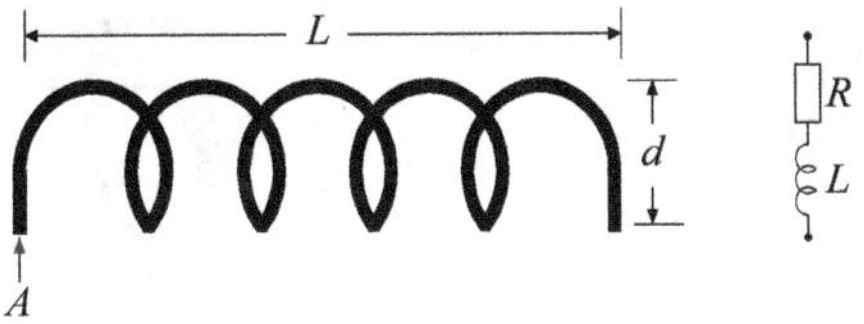

Figure 1.13 Solenoidal inductor and its passive equivalent

From the above, the quality factor of the inductor is given by

$$Q = \omega \frac{L}{R} \tag{1.27}$$

Now if all the linear dimensions are reduced by the factor x, the modified values L_x, R_x, and Q_x are given by

$$L_x = \mu n^2 F\left(\frac{xd}{xl}\right)xd = xL; \quad R_x = \rho\pi\left(\frac{xdn}{x^2A}\right) = \frac{R}{x}; \quad Q_x = \omega\left(\frac{xL}{R/x}\right) = x^2 Q \tag{1.28}$$

It is, thus, seen that for $x \ll 1$, the micro-miniaturization not only reduces the value of the realizable L but also degrades the quality factor of the realizable inductance drastically!

Hence, for applications requiring medium to high Q inductors, simulated inductors are important and desirable.

1.9 Integrated inductors and transformers on the chip

In spite of the fundamental limitations of micro-miniaturization of inductors, integrated on-chip passive inductors (together with on-chip transformers and capacitors) have been widely used in several RF ICs and millimetre wave ICs in many applications such as in the implementation of impedance-matching networks, noise-matching networks, differential-to-single-ended conversion, inter-stage filtering, bandwidth enhancement and a variety of other functions.

For such applications in the ICs, inductors and transformers are constructed by placing one or more metal windings on top of the bulk silicon substrate with a dielectric insulator (SiO_2) in between. Some of the commonly employed layout geometries for such inductors are square, hexagonal, octagonal, circular etc. as shown in Figure 1.14. The layout of the coil is defined by a few geometrical parameters which are the shape, the coil trace metal width (w), the spacing between the adjacent turns (s), number of turns (n), the inner diameter (d_{in}), the outer diameter (d_{out}) etc.

Square inductors were in vogue commonly in the late 1990s, but subsequently, it was realized that hexagonal, octagonal and circular shapes offer lower DC resistance than the former, at a given inductance value. An equivalent circuit of a

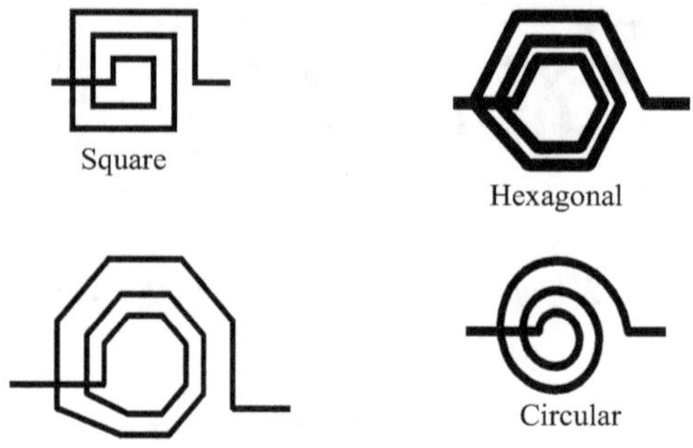

Figure 1.14 Some common planar inductor structures

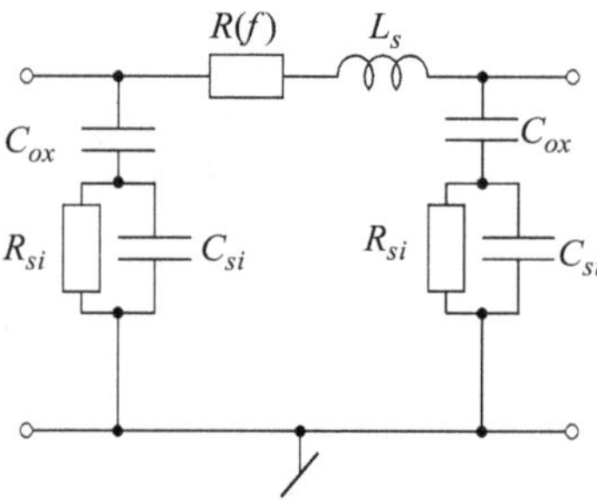

Figure 1.15 Equivalent π-model of a spiral inductor

spiral planar inductor is as shown in Figure 1.15. Here, $L(s)$ represents the main inductance and $R(f)$ represents the series losses occurring in the metal; and the two shunt branches at each end of the inductor represent the models of the parasitic elements of the substrate, while the capacitor C_{ox} models the capacitive parasitic between metal layers and substrate, and the parallel combination of R_{si} and C_{si} represent the substrate itself.

On-chip planar inductors usually can realize inductance values only in the nano-Henry ranges with a modest Q, typically less than a few tens, but nevertheless, planar IC inductors and transformers have been widely studied as is reflected by the publication of a large number of research papers and books summing up the state-of-the-art of on-chip spiral inductors and transformers; see [25–100].

1.10 Power gyrators

Apart from their realizations and numerous applications in low voltage ($\leq \pm 15$ V) and low power (less than or equal to a few mW) circuits and signal processing, gyrators also found inroads into electrical power systems where they could be realized using power converters and used for the purpose of energy transfer.

The introduction of the 'gyrator' concept to power processing circuits is due to the pioneering work of Singer [101–104]. Singer related the power gyrator to a general class of networks termed as *power output–power input* (POPI) networks. It was shown that power gyrators are useful to model inverse dual converters and that double bridge converters can be shown to behave as gyrators. Over the years, considerable work has been done on the classification, synthesis and applications of the power gyrators. Since the treatment of power gyrators is considered to be beyond the scope of this monograph, interested readers are referred to [101–111] and the references cited therein, for further information on this topic.

1.11 Use of ANSYS and COMSOL for the analysis of inductor designs

In order to analyse the behaviour of the on-chip spiral inductors, one needs to carry out a variety of analyses and characterizations for which, out of a variety of available software tools, ANSYS and COMSOL are the two relatively more popular ones. In COMSOL, the AC/DC module provides a number of tools which can be used for modelling and design of inductors and in calculating the device characteristics needed for a particular application.

It should be noted that real inductors are always non-ideal; in that, not only they have a resistive effect (e.g. small non-zero resistance of the metallic wire constituting the coil) which affect low frequency performance; they also have capacitive effect which plays a dominant role in high frequency behaviour of the inductors. Thus, in the simplest form, the non-ideal equivalent circuit of an inductor is a parallel combination of a series RL branch and a capacitor. Thus, practical inductor behaviour can be understood well from its RLC equivalent model or some extension thereof. Due to the RLC elements featuring in the equivalent model, the *self-resonance* is an important phenomenon in all practical inductors.

1.12 The need for simulated inductors

It has been explained earlier that spiral IC inductors on the chip are lossy, can realize only very small values of the order of nano-Henries and have exceptionally poor quality factors (a very modest value of about 10 at a frequency of the order of 1 GHz), hence are not preferred in fully integratable systems for applications in the medium frequency range (10 kHz–10 MHz), low frequency range (10 Hz–10 kHz) or very low frequency range (less than 10 Hz; as in several bio-medical and geophysical applications). In fact, the use of either the integrated inductors or discrete passive inductors for applications in the above-quoted frequency ranges is usually not preferred due to a number of disadvantages, which are as follows:

1. Passive inductors of reasonable values are bulky, occupy more space and are costly.
2. Unlike resistors and capacitors which are commercially available for all possible assorted values, the inductors need to be specifically designed.

3. Unlike resistors and capacitors, the variable-valued inductors are not very common; moreover, even for the variable inductors, their inductance value may not be possible to be varied over a very wide range.
4. Passive inductors are lossy, and hence, their quality factor is not theoretically infinite.
5. Nonlinear effects of the passive inductors can result in the production of harmonics leading to harmonic distortion in waveforms.
6. If there are a multiple number of inductors in a circuit, the unwanted and unaccounted for mutual inductive effect between them can deteriorate the performance of the circuit which may differ significantly from the intended performance.
7. Passive inductors are generally not suitable/possible to be integrated except spiral inductors/transformers for operations in GHz range where inductors of very small values with very modest Q values of the order of 10 or less are acceptable.

For applications in the frequency ranges mentioned above, therefore, the simulated inductors are preferred which, if properly designed, offer following several advantages:

1. Simulated indictors use active elements and devices and, hence, are small in size, are compact and are much cheaper than passive ones.
2. Simulated inductance, because its value is a function of RC elements or device parameter/gains, can be easily designed to have any required value; even very large inductance values (of the order of several Henries) can be easily simulated.
3. If the simulated inductance circuit is properly devised, the inductance value can be easily varied through a variable resistance or in several cases through an external control voltage/current.
4. It is relatively easy to attain high-quality factors of the order of 10^5 or more, in some cases, active loss cancellation schemes come handy for achieving very high Q.
5. Active synthetic inductors do not replicate magnetic effect rather they simulate only the v–i relationship of an inductor, and hence, in the event of there being a number of simulated inductors in a given circuit, no such thing as the mutual inductive effect exists.
6. When devised properly, a simulated inductance circuit may be highly suitable for IC implementation in both bipolar and CMOS technologies.

1.13 Concluding remarks

This monograph is mainly concerned with the gyrators, simulated inductors and related immittances, their realizations and applications. Hence, among the various kind of inductors outlined above, we will be mainly concerned with the electronic realizations of these elements using bipolar and MOS transistors or other analogue circuit building blocks like operational amplifiers, operational transconductance amplifiers and others, which are either commercially available as off-the-shelf ICs or can be realized using them.

We would also be discussing gyrators, simulated inductors and other related synthetic impedance circuits employing new electronic circuit building blocks of more recent origin which are although not yet available commercially but whose integratable versions implementable in both bipolar and CMOS technologies have already been proposed in the literature.

References

[1] Newcomb R. W. 'Active Integrated Circuit Synthesis'. Englewood Cliffs, NJ: Prentice Hall Inc.; 1968, Ch. 5, 6.

[2] Mitra S. K. 'Analysis and Synthesis of Linear Active Networks'. John Wiley and Sons Inc.; 1968, Ch. 2, 10.

[3] Bruton L. T. 'RC-Active Circuits: Theory and Design'. Englewood Cliffs, NJ: Prentice Hall Inc.; 1980, Ch. 2, 3.

[4] Chua, L. O. 'Memristor—the missing circuit element'. IEEE Transactions on Circuit Theory. 1971; **CT-18**(5):507–19.

[5] Strukov D. B., Snider G. S., Stewart D. R., Williams, R. S. 'The missing memristor found'. Nature. 2008; **453**:80–3.

[6] Chua L. O. 'The fourth element'. Proceedings of the IEEE. 2012; **100**(6): 1920–7.

[7] Chua L. O. 'If it's pinched it's a memristor'. Semiconductor Science and Technology. 2014; **29**(10):1–42.

[8] Tellegan B. D. H. 'The gyrator, a new electric network element'. Philips Research Reports. 1948; **3**(2):81–101.

[9] Adams K. M., Deprettere E. F. A., Voorman, J. O. 'The gyrator in electronic systems'. Advances in Electronics and Electron Physics. 1975; **37**: 79–179.

[10] Beddoes M. P., Morin, K. R. 'Bibliography on inductance simulation by gyrator methods'. IEEE Transactions on Circuit Theory. 1967; **14**(1): 107–11.

[11] Miller J. A., Newcomb, R. W. 'An annotated bibliography of gyrators in network theory: Circuits and uses'. Technical Report No. R-72-01, Electrical Engineering Department, University of Maryland, College Park, MD, 1972.

[12] Dutta Roy S. C. 'Bibliography of inductance simulation by active RC methods'. Microelectronics Reliability. 1976; **15**(6):637–9.

[13] Orchard H. J. 'Inductor-less filters'. Electronics Letters. 1966; **2**(6):224–5.

[14] Carlin H. J., Youla, D. C. 'Network synthesis with negative resistors'. Proceedings of the IRE. 1961; **49**(5):907–20.

[15] Kumar P., Senani, R. 'Bibliography on nullors and their applications in circuit analysis, synthesis and design'. Analog Integrated Circuits and Signal Processing. 2002; **33**(1):65–76.

[16] Sedra A. S., Smith K. C. 'A second generation current conveyor and its applications'. IEEE Transactions on Circuit Theory. 1970; **17**(1):132–5.

[17] Senani R. 'Floating impedance realisations: the nullor approach'. Electronics Letters. 1988; **24**(7):403–5.

[18] Senani R. 'Generation of new two-amplifier synthetic floating inductors'. Electronics Letters. 1987; **23**(22):1201–3.
[19] Senani R. 'A novel application of four-terminal floating nullor'. Proceedings of the IEEE. 1987; **75**(11):1544–6.
[20] Awad I. A., Soliman A. M. 'On the voltage-mirrors and the current-mirrors'. Analog Integrated Circuits and Signal Processing. 2002; **32**(1):79–81.
[21] Soliman A. M., Saad R. A. 'The voltage mirror-current mirror pair as a universal element'. International Journal of Circuit Theory and Applications. 2010; **38**(8):787–95.
[22] Holt A. G. J., Linggard R. 'The multi terminal gyrator'. Proceedings of the IEEE. 1968; **57**(8):1354–5.
[23] Holt A. G. J., Linggard R. 'Multi terminal gyrator'. Proceedings of the IEE. 1970; **117**(8):1591–8.
[24] Silva M. M. 'Multiport converters and inverters'. International Journal of Circuit Theory and Applications. 1978; **6**(3):243–52.
[25] Scuderi A., Ragonese E., Biondi T., Palmisano G., 'Integrated Inductors and Transformers: Characterisation, Design and Modelling for RF and mm-Wave Applications'. Finland: CRC Press; 2011.
[26] Normak U. 'Integrated Transformer'. Kista: KTH Kista; 1998.
[27] Niknejad A. M., Meyer R. G. 'Design, Simulation and Applications of Inductors and Transformers for Si RF ICs'. Boston, MA: Kluwer Academic Publisher; 2000.
[28] Fei Y. 'CMOS Active Inductors and Transformers: Principle, Implementation, and Applications'. Boston, MA: Springer; 2008. p. 230.
[29] Niknejad A. M., Meyer R. G. 'Analysis, design and optimization of spiral inductors and transformers for Si RF ICs'. IEEE Journal Solid-State Circuits. 1998; **33**(10):1470–81.
[30] Haobijam G., Palathinkal R. P. 'Design and Analysis of Spiral Inductors'. New Delhi, India: Springer; 2014. p. 107.
[31] Greenhouse H. M. 'Design of planar rectangular microelectronics inductors'. IEEE Transactions on Parts, Hybrids, and Packaging. 1974; **10**(2):101–9.
[32] Nguyen N. M., Meyer R. G. 'Si IC-compatible inductors and LC passive filters'. IEEE Journal of Solid-State Circuits. 1990; **25**(4):1028–31.
[33] Chang J. Y. C., Abidi A. A., Gaitan M. 'Large suspended inductors on silicon and their use in a 2 mm CMOS RF amplifier'. IEEE Electron Devices. 1993; **14**(5):246–8.
[34] Kuhn W. B., Elshabini-Riad A., Stephenson F. W. 'Centre-tapped spiral inductors for monolithic bandpass filters'. Electronics Letters. 1995; **31**(8): 625–6.
[35] Soyuer M., Burghartz J. N., Jenkins K. A., Ponnapalli S., Ewen J. F., Pence W. E. 'Multilevel monolithic inductors in silicon technology'. Electronics Letters. 1995; **31**(5):359–60.
[36] Ashby K. B., Koullias I. A., Finley W. C., Bastek J. J., Moinian S. 'High-Q inductors for wireless applications in a complementary silicon bipolar process'. IEEE Journal of Solid-State Circuits. 1996; **31**(1):4–9.

[37] Johnson R. A., Chang C. E., Asbeck P. M., Wood M. E., Garcia G. A., Lagnado I. 'Comparison of microwave inductors fabricated on silicon-on-sapphire and bulk silicon'. IEEE Microwave and Guided Wave Letters. 1996; **6**(9):323–5.

[38] Burghartz J. N., Soyuer J. N., Jenkins K. A. 'Microwave inductors and capacitors in standard multilevel interconnect silicon technology'. IEEE Transactions on Microwave Theory and Techniques. 1996; **44**(1):100–4.

[39] Burghartz J. N., Jenkins K. A., Soyuer M. 'Multilevel-spiral inductors using VLSI interconnect technology'. IEEE Electronic Device Letters. 1996; **17**(9): 428–30.

[40] Jang Z., Excell P. S., Hejazi Z. M. 'Calculation of distributed capacitances of spiral resonators'. IEEE Transactions on Microwave Theory and Techniques. 1997; **45**(1):139–42.

[41] Choong-Mo N., Young-Se K. 'High-performance planar inductor on thick oxidized porous silicon (OPS) substrate'. IEEE Microwave and Guided Wave Letters. 1997; **7**(8):236–8.

[42] Craninckx J., Steyaert M. 'A 1.8-GHz low-phase-noise CMOS VCO using optimized hollow spiral inductors'. IEEE Journal of Solid-State Circuits. 1997; **32**(5):736–44.

[43] Kim K., Kenneth K. O. 'Characteristics of an integrated spiral inductor with an underlying n-well'. IEEE Transactions on Electron Devices. 1997; **44**(9):1565–7.

[44] Long J. R., Copeland M. A. 'The modelling, characterisation, and design of monolithic inductors for silicon RF ICs'. IEEE Journal of Solid-State Circuits. 1997; **32**(3):357–69.

[45] Yim S. M., Chen T., Kenneth, K. O. 'The effects of a ground shield on the characteristics and performance of spiral inductors'. IEEE Journal of Solid-State Circuits. 2002; **37**(2):237–44.

[46] Burghartz J. N., Edelstein D. C., Jenkinns K. A., Kwark Y. H. 'Spiral inductors and transmission lines in silicon technology using copper-damascene interconnects and low-loss substrates'. IEEE Transactions on Microwave Theory and Techniques. 1997; **45**(10):1961–8.

[47] Yue C. P., Wong S. S. 'On-chip spiral inductors with patterned ground shields for Si-based RF ICs'. IEEE Journal of Solid-State Circuits. 1998; **33**(5): 743–52.

[48] Okabe H., Yamada H., Yamasaki M., Kagaya O., Sekine K., Yamashita K. 'Characterization of a planar spiral inductor on a composite-resin low-impedance substrate and its application to microwave circuits'. IEEE Transactions on Components, Packaging, and Manufacturing Technology. 1998; **21**(3):269–73.

[49] Pun A. L. L., Yeung T., Lau J., Clement J. R., Su D. K. 'Substrate noise coupling through planar spiral inductor'. IEEE Journal of Solid-State Circuits. 1998; **33**(6):877–84.

[50] Niknejad A. M., Meyer R. G. 'Design and optimization of spiral inductors and transformers for Si RF ICs'. IEEE Journal of Solid-State Circuits. 1998; **33**(10):1470–81.

[51] Park J. Y., Allen M. G. 'High Q spiral-type micro inductors on silicon substrates'. IEEE Transactions on Magnetics. 1999; **35**(5):3544–6.

[52] Mohan S. S., Hershenson D. D. M., Boyd S. P., Lee T. H. 'Simple accurate expressions for planar spiral inductances'. IEEE Journal of Solid-State Circuits. 1999; **34**(10):1419–24.

[53] Lopez-Villegas J. M., Samitier J., Cane C., Losantos P., Bausells J. 'Improvement of the quality factor of RF integrated inductors by layout optimization'. IEEE Transactions on Microwave Theory and Techniques. 2000; **48**(1):76–83.

[54] Post J. E. 'Optimizing the design of spiral inductors on silicon'. IEEE Transactions on Circuits and Systems II. 2000; **47**(1):15–7.

[55] Yue C. P., Wong S. S. 'Physical modelling of spiral inductors on silicon'. IEEE Transactions on Electron Devices. 2000; **47**(3):560–8.

[56] Koutsoyannopoulos Y. K., Papananos Y. 'Systematic analysis and modelling of integrated inductors and transformer in RF IC design'. IEEE Transactions on Circuits and Systems II. 2000; **47**(8):699–713.

[57] Ribas R. P., Lescot J., Leclercq J. L., Karam J. M., Ndagijimana F. 'Micromachined microwave planar spiral inductors and transformers'. IEEE Transactions on Microwave Theory and Techniques. 2000; **48**(8):1326–35.

[58] Hongrui J., Wang Y., Yeh J., Lien N. C. 'On-chip spiral inductors suspended over deep copper-lined cavities'. IEEE Transactions on Microwave Theory and Techniques. 2000; **48**(12):2415–23.

[59] Niknejad A. M., Meyer R. G. 'Analysis of eddy-current losses over conductive substrates with applications to monolithic inductors and transformers'. IEEE Journal of Solid-State Circuits. 2001; **49**(1):166–76.

[60] Zolfaghari A., Chan A., Razavi B. 'Stacked inductors and transformers in CMOS technology'. IEEE Journal of Solid-State Circuits. 2001; **36**(4): 620–8.

[61] Kuhn W. B., Ibrahim N. M. 'Analysis of current crowding effects in multi turn spiral inductors'. IEEE Transactions on Microwave Theory and Techniques. 2001; **49**(1):31–8.

[62] Bahl I. J. 'High-performance inductors'. IEEE Transactions on Microwave Theory and Techniques. 2001; **49**(4):654–64.

[63] Farina M., Rozzi T. 'A 3-D integral equation-based approach to the analysis of real-life MMICs—application to micro-electromechanical systems'. IEEE Transactions on Microwave Theory and Techniques. 2001; **49**(12):2235–40.

[64] Danesh M., Long J. R. 'Differentially driven symmetric micro-strip inductors'. IEEE Transactions on Microwave Theory and Techniques. 2002; **50**(1): 332–41.

[65] Andersen R. B., Jorgensen T., Laursen S., Kolding T. E. 'EM-simulation of planar inductor performance for epitaxial silicon processes'. Analog Integrated Circuits and Signal Processing. 2002; **30**(1):51–8.

[66] Jenei S., Nauwelaers B. K. J. C., Decoutere S. 'Physics-based closed-form inductance expression for compact modeling of integrated spiral inductors'. IEEE Journal of Solid-State Circuits. 2002; **37**(1):77–80.

[67] Yim S. M., Chen T., Kenneth K. O. 'The effects of a ground shield on the characteristics and performance of spiral inductors'. IEEE Journal of Solid-State Circuits. 2002; **37**(2):237–44.

[68] Melendy D., Francis P., Pichler C., Hwang K., Srinivasan G., Weisshaar A. 'A new wide band compact model for spiral inductors in RF ICs'. IEEE Electron Devices Letters. 2002; **23**(5):273–5.

[69] Lakdawala H., Zhu X., Luo H., Santhanam S., Carley L. R., Fedder G. 'Micro-machined high-Q inductors in a 0.18 μm copper interconnect low-k dielectric CMOS process'. IEEE Journal of Solid-State Circuits. 2002; **37**(3): 394–403.

[70] Tang C. C., Wu C. H., Liu S. I. 'Miniature 3-D inductors in standard CMOS process'. IEEE Journal of Solid-State Circuits. 2002; **37**(4):471–80.

[71] Burghartz J. N., Rejaei B. 'On the design of RF spiral inductors on silicon'. IEEE transactions on Electron Devices. 2003; **50**(3):718–29.

[72] Aguilera J., Berenguer R. 'Design and Test of Integrated Inductors for RF Applications'. Boston, MA: Kluwer Academic Publishers; 2003.

[73] Choi Y. S., Yoon J. B. 'Experimental analysis of the effect of metal thickness on the quality factor in integrated spiral inductors for RF ICs'. IEEE Electron Devices Letters. 2004; **25**(2):76–9.

[74] Chen J., Liou J. J. 'On-chip spiral inductors for RF applications: an overview'. Journal of Semiconductor Technology and Science. 2004; **4**(3):149–67.

[75] Scuderi A., Biondi T., Ragonnese E., Palmisano G. 'A lumped scalable model for silicon integrated spiral inductors'. IEEE Transactions on Circuits and Systems I. 2004; **51**(6):1203–9.

[76] Hsieh M. C., Fang Y. K., Chen C. H., Chen S. M., Yeh W. K. 'Design and fabrication of deep submicron CMOS technology compatible suspended high-Q spiral inductors'. IEEE Transactions on Electronic Devices. 2004; **51**(3): 324–31.

[77] Teo T. H., Choi Y. B., Liao H., Xiong Y. Z., Fu J. S. 'Characterization of symmetrical spiral inductor in 0.35 μm CMOS technology for RF application'. Solid State Electronics. 2004; **48**(9):1643–50.

[78] Biondi T., Scuderi A., Ragonese E., Palmisano G. 'Sub-nH inductor modelling for RF IC design'. IEEE Microwave and Wireless Components Letters. 2005; **15**(12):922–4.

[79] Dickson T. O., LaCroix M. A., Boret S., Gloria D., Beerkens R., Voinigescu S. P. '30-100-GHz inductors and transformers for millimetre-wave (Bi) CMOS integrated circuits'. IEEE Transactions on Microwave Theory and Techniques. 2005; **53**(1):123–33.

[80] Rotella F., Cismaru C., Tkachenko Y., *et al.* 'A broad-band lumped element analytic model incorporating skin effect and substrate loss for inductors and inductor like components for silicon technology performance assessment and RFIC design'. IEEE Transactions on Electron Devices. 2005; **52**(7):1429–41.

[81] Murphy O. H., McCarthy K. G., Delabie C. J. P., Murphy A. C., Murphy P. J. 'Design of multiple-metal stacked inductors incorporating an extended physical model'. IEEE Journal of Solid-State Circuits. 2005; **53**(6):2063–72.

[82] Sia C. B., Ong B. H., Chan K. W., Yeo K. S., Ma J. G., Do M. A. 'Physical layout design optimization of integrated spiral inductors for silicon-based RFIC applications'. IEEE Transactions on Electron Devices. 2005; **52**(12):2559–67.
[83] Mukherjee S., Mutnury B., Dalmia S., Swaminathan M. 'Layout level synthesis of RF inductors and filters in LCP substrates for Wi-Fi applications'. IEEE Transactions on Microwave Theory and Techniques. 2005; **53**(6):2196–210.
[84] Guo L. H., Zhang Q. X., Lo G. Q., Balasubramanian N., Kwong D. L. 'High-performance inductors on plastic substrate'. IEEE Electron Device Letters. 2005; **26**(9):619–21.
[85] Wu C. H., Kuo C. Y., Liu S. I. 'Selective metal parallel shunting inductor and its VCO application'. IEEE Transactions on circuits and Systems I. 2005; **52**(5):1811–8.
[86] Tsui H. Y., Lau J. 'An on-chip vertical solenoid inductor structure design for multigigahertz CMOS RFIC'. IEEE Transactions on Microwave Theory and Techniques. 2005; **53**(6):1883–90.
[87] Cheung T. S. D., Long J. R. 'Shielded passive devices for silicon-based monolithic microwave and millimeter-wave integrated circuits'. IEEE Journal of Solid-State Circuits. 2006; **41**(5):1183–200.
[88] Wang Y. Y., Li Z. F. 'Group-cross symmetrical inductor (GCSI): a new inductor structure with higher self-resonance frequency and Q factor'. IEEE Transactions on Magnetics. 2006; **42**(6):1681–6.
[89] Nieuwoudt A., Massoud Y. 'Variability-aware multilevel integrated spiral inductor synthesis'. IEEE Transactions on Computer-Aided Design of Integrated Circuits and Systems. 2006; **25**(12):2613–25.
[90] Biondi T., Scuderi A., Ragonese E., Palmisano G. 'Characterization and modeling of silicon integrated spiral inductors for high-frequency applications'. Analog Integrated Circuits and Signal Processing. 2007; **51**(2):89–100.
[91] Weon D. H., Kim J-II., Mohammadi S. 'Design of high-Q 3-D integrated inductors for high frequency applications'. Analog Integrated Circuits and Signal Processing. 2007; **50**(2):89–93.
[92] Nieuwoudt A., Massoud Y. 'Analytical wide-band modeling of high frequency resistance in integrated spiral inductors'. Analog Integrated Circuits and Signal Processing. 2007; **50**(2):133–6.
[93] Haobijam G., Paily R 'Efficient optimization of integrated spiral inductor with bounding of layout design parameters'. Analog Integrated Circuits and Signal Processing. 2007; **51**(3):131–40.
[94] Mandal S. K., Sural S., Patra A. 'ANN- and PSO-based synthesis of on-chip spiral inductors'. IEEE Transactions on Computer-Aided Design and Integrated Circuits and Systems. 2008; **27**(1):188–92.
[95] Haobijam G., Paily R. 'Performance study of fixed value inductors and their optimization using electromagnetic simulator'. Microwave and Optical Technology Letters. 2008; **50**(5):1205–10.
[96] Yin W. Y., Xie J. Y., Kang K., Shi J., Mao J. F., Sun X. W. 'Vertical topologies of miniature multi-spiral stacked inductors'. IEEE Transactions on Microwave Theory and Techniques. 2008; **56**(2):475–86.

[97] Ragonese E., Scuderi A., Biondi T., Palmisano G. 'Scalable lumped modelling of single-ended and differential inductors for RF IC design'. International Journal of RF and Microwave Computer-Aided Engineering. 2009; **19**(1): 110–9.

[98] Haobijam G., Paily R. 'Design of multilevel pyramidically wound symmetric inductor for CMOS RFICs'. Analog Integrated Circuits and Signal Processing. 2010; **63**(1):9–21.

[99] Bakkali M. E., Po F. C. W., Foucauld E. D., Viala B., Michel J. P. 'Design of a RF matching network based on a new tunable inductor concept'. Microelectronics Journal. 2011; **42**(1):233–8.

[100] Kang B. G., Park C. S., Chung S. K. 'Integrated transformer using magnetic sheet for LLC resonant converter'. Electronics Letters. 2014; **50**(10):770–1.

[101] Singer S. 'Power conversion and control with zero ac current harmonics by means of a time variable transformers'. Proceedings of the IEEE. 1984; **131**(4):147–50.

[102] Singer S. 'Canonical approach to energy processing network synthesis'. IEEE Transactions on Circuits and Systems. 1986; **CAS-33**(8):767–74.

[103] Singer S. 'Gyrator application in power processing circuits'. IEEE Transactions on Industrial Electronics. 1987; **IE-34**(3):313–8.

[104] Singer S. 'Loss-free gyrator realization'. IEEE Transactions on Circuits and Systems. 1988; **35**(1):26–34.

[105] Anderson B. D., Spaulding D. A., Newcomb R. W. 'The time-variable transformer'. Proceedings of the IEEE. 1965; **53**(6):634.

[106] Singer S. 'The application of loss-free resistors in power processing circuits'. IEEE Transactions on Power Electronics. 1991; **6**(4):595–600.

[107] Singer S., Ozeri S., Shmilovitz D. 'A pure realization of loss-free resistor'. IEEE Transactions on Circuits and Systems I. 2004; **51**(8):1639–47.

[108] Cid-Pastor A., Martinez-Salamero L., Alonso C., Estibals B., Alzieu J., Schweitz G., Shmilovitz D. 'Analysis and design of power gyrators in sliding-mode operation'. IEE Proceedings-Electric Power Applications. 2005; **152**(4):821–6.

[109] Cid-Pastor A., Martinez-Salamero L., Alonso C., Schweitz G., Calvente J., Singer S. 'Classification and synthesis of power gyrators'. IEE Proceedings-Electric Power Applications. 2006; **153**(6):802.

[110] Barazarte R. Y., Gonzalez G. G., Ehsani M. (Mark). 'Generalized gyrator theory'. IEEE Transactions on Power Electronics. 2010; **25**(7):1832–7.

[111] Kiss D., Hisakado T., Matsushima T., Wada O. 'Peer-to-peer energy transfer by power gyrators based on time-variable-transformer concept'. IEEE Transactions on Power Electronics. 2019; **34**(8):8230–40.

[112] Silva M. M. 'On the realization of immittance inverters with a minimum number of active components'. IEEE Transactions on Circuits and Systems. 1979; **26**(11):931–5.

Chapter 2

Gyrators and simulated inductors using op-amps

Abstract

This chapter deals with the realization of gyrators and other impedance simulation circuits using integrated circuit (IC) operational amplifiers (op-amps) as the main building blocks. The traditional voltage-mode op-amp is well recognized as the workhorse of analog circuit design, and therefore, it is no surprise that IC op-amps were considered to be the most appropriate building blocks for realizing gyrators and various other kinds of synthetic impedances. In this chapter, we present a wide variety of op-amp circuits for realizing grounded and floating synthetic impedances of various kinds. A number of general methods of synthesizing grounded and floating synthetic impedances have also been elaborated.

2.1 Introduction

With the invention of the bipolar junction transistor, its use in designing amplifiers, oscillators and various other electronic circuits, all of which were earlier realized by vacuum tube electronic devices, was undertaken by academicians, researchers and electronic circuit designers around the world. However, although a number of transistorized gyrators were developed by various researchers, with a few exceptions, the difficulty of incorporating the essential dc biasing circuitry was a dominant factor against the use of some of such transistorized gyrators in practical applications. With the advent of integrated circuit (IC) operational amplifiers (op-amps) as the off-the-shelf components, there was a surge of interest in realizing various electronic circuit functions using them. Of various IC op-amps developed, the μA741 type designed by David Fullagar of Fairchild Corporation found favour of circuit designers as a general-purpose IC op-amp because of not requiring any external compensation components to be added over its predecessors like μA702 and μA709 and its competitors like LM101 from National Semiconductors. Thus, the internally compensated μA741 op-amp soon became an extremely popular component in analogue circuit design because of its widespread use, low cost and reliable and hassle-free operation.

In view of the above, there was an immediate attention on trying to realize gyrators and simulated inductances using the IC op-amps. This chapter gives a detailed account of the various developments on gyrator realization and inductor simulation as reflected from the publications [1–207] with a major focus on only the significant works done in this area.

Before proceeding further, in the following we give a brief overview of the work reported in [1–207]. In [1,2,4,6,9,11,12,14,16–18,20–22,24,25,27,29,30,33–35,38,40–42,44,45,47,49,50,55–57,60,61,63,64,66–68,73,79,83,84,90,92–96,98–101,104,105,109,110,116–118,120,121,125,127–129,134,136,145,146,148–151,155–160,165,166,168,169,174,175], the authors have dealt with specific circuits for gyrator realization, inductance simulation and/or impedance inverters/converters and the capacitance multipliers; the methods of creating floating inductances and other floating impedances have been dealt in [3,7,11,15,19,46,48,54,59,70,76,78,80–82,84,92,102,103,106,108,112,114,115,119,122,132,135,138,139,141–144,153,177,178]; the analysis of performance of gyrators and impedance converters/inverters has been carried in [8,23,26,39,52,69,72,74,75,85,87,97,146,147,157]; the applications of gyrators and simulated inductors have been considered in [5,31,32,37,45,51,58,73,77,83,88–90,93,113,116,124,126,133,137,150,158,159,161,164,167,170–173,176,181]; bibliographies on gyrators and inductance simulation have been compiled in [12,71]; the frequency-dependent negative resistance (FDNR) was first introduced in [28]; the extension of two-port gyrators and impedance converters/inverters to their multi-port counterparts, namely the multi-gyrator and multi-port converters/inverters, was carried out in [36,91,107]; op-amp-based switched-capacitor inductors and several miscellaneous other ideas related to the realization of inductance simulation and FDNR circuits have been dealt in [43,53,68,123,130,131,140,152,154,179,180,198–206].

The organization of the remaining chapter is as follows: in Section 2.2 the basic concepts related to the gyrator have been outlined; Section 2.3 deals with some prominent op-amp-based gyrators and related circuits; Section 2.4 deals with single-op-amp-based lossless inductance simulators; Section 2.5 deals with economic inductance simulators and resonators; Section 2.6 is concerned with lossless floating impedance simulators; Section 2.7 discusses multi-gyrators and multi-port converters/inverters; Section 2.8 deals with three-op-amp-based floating impedance simulators, lossless FIs using only two op-amps are dealt in Section 2.9; economic op-amp-based lossless/lossy FIs are elaborated in Section 2.10; Section 2.11 discusses some prominent grounded/floating impedance simulators and resonators using the so-called active-R approach; Section 2.12 outlines the most general methods of FI simulation using the concept of four-terminal floating nullors (FTFNs) and lastly concluding remarks have been made in Section 2.13.

2.2 The gyrator

An ideal gyrator is a two-port non-reciprocal network element which is characterized by the following short circuit admittance matrix ([Y]-matrix)

$$\begin{bmatrix} i_1 \\ i_2 \end{bmatrix} = \begin{bmatrix} 0 & \pm G \\ \mp G & 0 \end{bmatrix} \begin{bmatrix} v_1 \\ v_2 \end{bmatrix} \tag{2.1}$$

or an open circuit impedance matrix ([Z]-matrix) characterized by

$$\begin{bmatrix} v_1 \\ v_2 \end{bmatrix} = \begin{bmatrix} 0 & \pm R \\ \mp R & 0 \end{bmatrix} \begin{bmatrix} i_1 \\ i_2 \end{bmatrix} \tag{2.2}$$

where G represents the gyration conductance, and similarly, R represents gyration resistance of the gyrator.

From the above two characterizations, it can be easily deduced that the total power absorbed by both the ports of the gyrator is $P = v_1 i_1 + v_2 i_2 = v_1(\pm G\, v_2) + v_2(\mp G\, v_1) = 0$, and thus, an ideal gyrator is *passive* and yet non-reciprocal since $y_{12} \neq y_{21}$ and $z_{12} \neq z_{21}$.

In the analysis of some physical gyrator circuits, it may be easier to find out the transmission matrix (*ABCD*-matrix) which for the ideal gyrator can be represented as

$$\begin{bmatrix} v_1 \\ i_1 \end{bmatrix} = \begin{bmatrix} 0 & \pm b \\ \pm c & 0 \end{bmatrix} \begin{bmatrix} v_2 \\ -i_2 \end{bmatrix} \tag{2.3}$$

where parameter b has the dimension of impedance and c has the dimension of admittance.

It is well known that the gyrator is a special case of a more generalized two-port network element known as positive impedance inverter (PII) defined by the matrix equation (2.3).

If in (2.1) and (2.2), the gyration conductances or the gyration resistances are not equal then $P = v_1 i_1 + v_2 i_2 = v_1(\pm G_1 v_2) + v_2(\mp G_2 v_1) = \pm(G_1 - G_2) v_1 v_2$ could be either positive or negative, thereby such a gyrator could be called *active* when $P < 0$.

If a gyrator is terminated into a capacitor C at port-2, then input impedance looking into port-1 would be

$$Z_{in} = s\frac{C}{G^2}; \text{ using the characterization of (2.1)} \tag{2.4}$$

$$Z_{in} = sCR^2; \text{ using the characterization of (2.2)} \tag{2.5}$$

$$Z_{in} = sC\frac{b}{c}; \text{ using the characterization of (2.3)} \tag{2.6}$$

Furthermore, any of the y-parameters out of y_{11} and y_{22} may be non-zero thereby leading to two possible forms of non-ideal gyrators which can be characterized as follows:

Type-I non-ideal gyrator:

$$\begin{bmatrix} i_1 \\ i_2 \end{bmatrix} = \begin{bmatrix} G_2 & \pm G_1 \\ \mp G_1 & 0 \end{bmatrix} \begin{bmatrix} v_1 \\ v_2 \end{bmatrix} \tag{2.7}$$

Type-II non-ideal gyrator:

$$\begin{bmatrix} i_1 \\ i_2 \end{bmatrix} = \begin{bmatrix} 0 & \pm G_1 \\ \mp G_1 & G_3 \end{bmatrix} \begin{bmatrix} v_1 \\ v_2 \end{bmatrix}, \tag{2.8}$$

It can be easily derived that in the first case if such a non-ideal gyrator is terminated into a capacitor at port-2, the input admittance Y_{in} looking into port-1 is given by

$$Y_{in} = G_2 + \frac{G_1 G_2}{sC} \tag{2.9}$$

thereby representing a parallel *RL* admittance.

Similarly, if the non-ideal gyrator has a characterization of (2.8), then the input impedance looking into the port-1, when port-2 is terminated into a capacitor *C*, would be given by

$$Z_{in} = \frac{1}{G_1} + \frac{sC}{G_1 G_3} \tag{2.10}$$

which represents series *RL* impedance.

This is a simple matter to verify that type-II non-ideal gyrator can be obtained from type-I non-ideal gyrator (and vice versa) by the simple artifice of swapping the input and output ports.[1]

Lastly, there could be yet another type[2] of non-ideal gyrator in which case both y_{11} and y_{22} may be non-zero.

Type-III non-ideal gyrator:

$$\begin{bmatrix} i_1 \\ i_2 \end{bmatrix} = \begin{bmatrix} G_2 & \pm G_1 \\ \mp G_1 & G_3 \end{bmatrix} \begin{bmatrix} v_1 \\ v_2 \end{bmatrix} \tag{2.11}$$

In this case, with output port terminated into a capacitor *C*, the input impedance and admittance looking into port-1 are given by

$$Z_{in} = \frac{sC + G_3}{sCG_2 + G_2 G_3 + G_1^2} \tag{2.12}$$

$$Y_{in} = G_2 + \frac{G_1^2}{sC + G_3} \tag{2.13}$$

In general, the above input impedance/admittance represents a *bilinear* inductor which has two possible canonic representations: (i) a series combination of a resistor and a parallel-*RL* impedance and (ii) a parallel combination of a resistor and series-*RL* impedance. In either case, when $G_2 \to 0$ and $G_3 \to 0$, a lossless grounded inductor is simulated with $L_{eq} = C/G_1^2$.

[1]Curiously, this does not appear to be very-widely understood or recognized as would be evident from numerous instances pointed out in this monograph at several places.

[2]At least four such single op-amp-gyrators are known in the literature earlier which have been discussed in subsequent sections of this chapter.

2.3 Op-amp gyrators and related circuits

The very early gyrators designed for possible realization with IC op-amps needed as many as four op-amps and thus, were not very economical. However, subsequently, a number of researchers demonstrated that two op-amps are enough to realize a lossless gyrator and many schemes for doing this were proposed. In this section, we highlight the most significant two-op-amp gyrators and lossless grounded inductor simulators.

2.3.1 NIC-based gyrator

Although now it is well known that even though the negative impedance converters (NICs) may appear to be very attractive active elements realizations using them often are problematic in respect of their stability properties, it was not so well appreciated earlier. Consequently, a number of researchers came up with NIC-based gyrators. One such NIC-based gyrator is shown in Figure 2.1.

It may be noticed that in this circuit both the op-amps along with two equal resistors are configured as current inversion NICs. The first NIC terminated into a resistor realizes a grounded negative resistance '–*R*'; on the other hand, the second NIC ensures that the two terminal voltages at its input and output are equal as well as the currents entering into the input and output port of this NIC are also equal (each equal to *i*). With this identification, a straightforward analysis of this circuit gives the following equations:

$$i_1 = \frac{v_1}{-R} + i; \quad \text{where } i = \frac{v_1 - v_2}{R} \tag{2.14}$$

$$i_2 = \frac{v_2}{R} + i \tag{2.15}$$

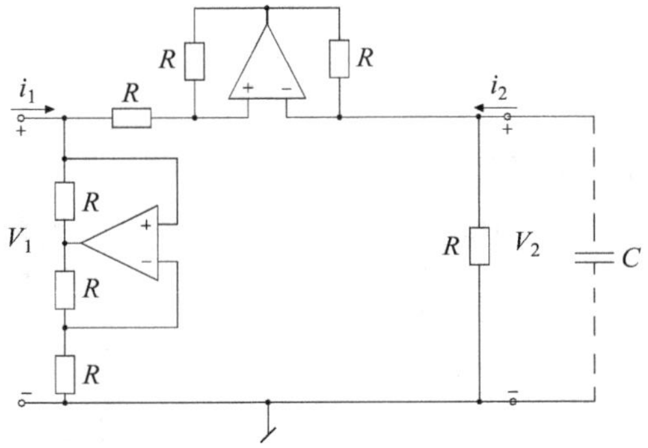

Figure 2.1 Lossless grounded inductor based upon ideal gyrator realized with two NICs [27]

The above equations lead to the following [*Y*]-matrix

$$[Y] = \begin{bmatrix} 0 & -\dfrac{1}{R} \\ \dfrac{1}{R} & 0 \end{bmatrix} \tag{2.16}$$

With port-2 terminated into a capacitor, the input impedance looking into port-1 is given by $Z_{in}(s) = sCR^2$, thereby representing a grounded inductance of value $L = CR^2$.

This realization suffers from the drawback of requiring as many as seven perfectly matched resistors and the associated sensitivity problems resulting due to any possible mismatch between them as well as the stability problems because of the two NICs. Consequently, this circuit is not very attractive from a practical viewpoint. Nevertheless, the key idea employed in this realization was subsequently found to be useful in the synthesis of a number of lossless floating inductance circuits and is even useful in the context of CMOS-compatible MOSFET-C gyrators and MOSFET-C inductors as will be demonstrated in Chapter 7 of this monograph.

2.3.2 VCCS-based gyrator

Another two-op-amp-based gyrator was proposed by Inigo [42] which consists of two single-op-amp-based voltage-controlled current source (VCCS) circuits in a parallel-back-to-back connection. This circuit is shown in Figure 2.2.

A straightforward analysis of this circuit yields the following [*Y*]-matrix:

$$[Y] = \begin{bmatrix} \left(\dfrac{1}{R_4} + \dfrac{1}{R_d} - \dfrac{R_b}{R_a R_c}\right) & \left(-\dfrac{1}{R_4} + \dfrac{R_b}{R_a R_c}\right) \\ \left(-\dfrac{1}{R_4} - \dfrac{1}{R_a}\right) & \left(\dfrac{1}{R_4} - \dfrac{R_2}{R_1 R_3} + \dfrac{1}{R_a}\right) \end{bmatrix} \tag{2.17}$$

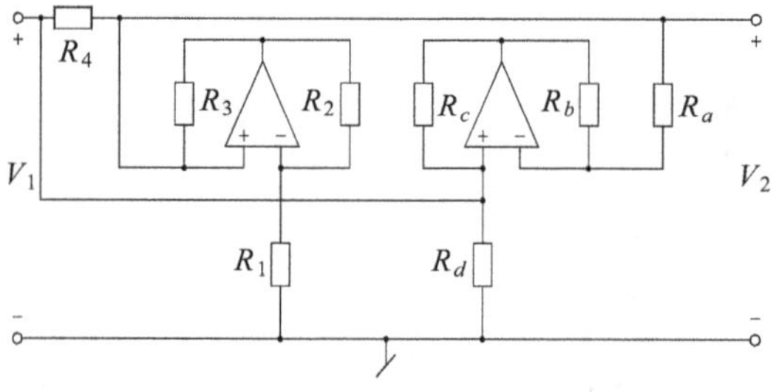

Figure 2.2 Gyrator realization using two VCCS [42]

A lossless gyrator is, therefore, realized by choosing the resistance values as $R_1 = R_2 = R_3 = R_c = R_d = R$; $R_4 = R_a = 2R$ and $R_b = 3R$. The [Y]-matrix, therefore, reduces to

$$[Y] = \begin{bmatrix} 0 & \frac{1}{R} \\ -\frac{1}{R} & 0 \end{bmatrix} \tag{2.18}$$

It is obvious that this gyrator also suffers from the drawback of requiring close matching of several resistor values.

2.3.3 Generalized impedance converters (GIC)/gyrators

A number of researchers including Riordan [10], Gorski-Popiel [16], Antoniou [17], Bruton [28], Mitra [27] and others came up with new types of two-op-amp-based circuits, at about the same time, which were initially termed as gyrators and had their input port grounded but had their output port floating. Subsequently, it was recognized that a more appropriate description of these circuits should be as *positive impedance converter* (PIC) (as used by Gorski-Popiel [16]) with both input and output ports having the ground reference. In later popular usage, however, all the circuits belonging to this class were popularly referred to as *generalized impedance converter* (GIC).

The first such circuit was proposed by Riordan [10] which is shown in Figure 2.3.

Although in Riordan's paper [10] the circuit was basically used to simulate a grounded inductance, he himself showed by analysis that with general values of the five impedances, the input impedance of the circuit is given by

$$\frac{V_{in}}{I_{in}} = Z_{in} = \frac{Z_1 Z_3 Z_5}{Z_2 Z_4} \tag{2.19}$$

The above expression is readily associated more with a GIC than a gyrator. Thus, in the original Riordan gyrator [10] a lossless inductor is simulated by choosing Z_2 as

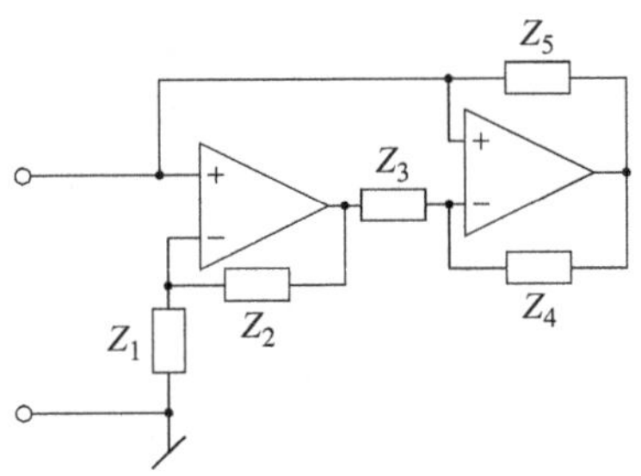

Figure 2.3 Riordan's Gyrator [10] (shown with general impedances rather than RC elements)

a capacitor with all the remaining impedances chosen as resistors. The input impedance is, thus, given by

$$Z_{in} = \frac{sC_2R_1R_3R_5}{R_4} \tag{2.20}$$

which represents a lossless grounded inductance of value

$$L_{eq} = \frac{C_2R_1R_3R_5}{R_4} \tag{2.21}$$

The other two-op-amp GICs proposed/derived by several researchers at about the same time, namely Gorski-Popiel [16], Antoniou [17] and Mitra [27], are shown in Figure 2.4.

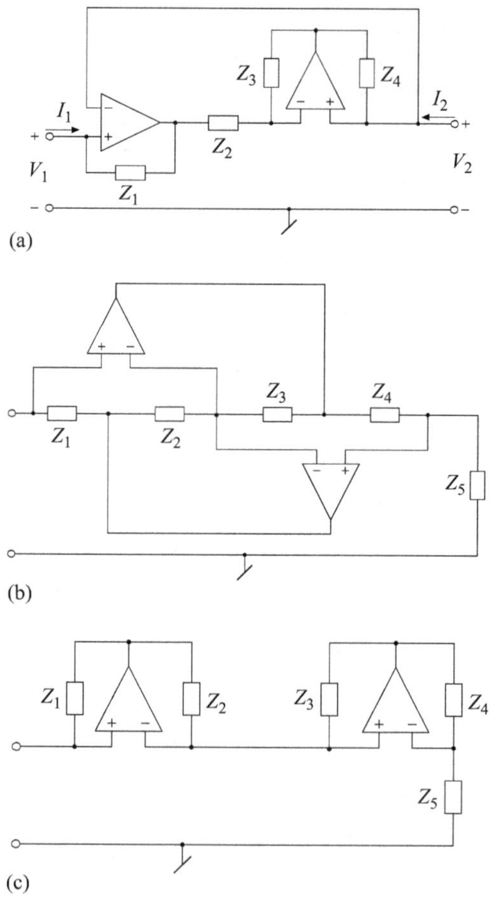

Figure 2.4 Other two op-amp-based GICs: (a) Gorski-Popiel [16], (b) Antoniou [17,29] and (c) Mitra [27]

The various two op-amp-based GICs shown in Figures 2.3 and 2.4 have been extensively studied in the literature considering the finite gain-bandwidth product (GBP) of the two op-amps employed, and it has been conclusively established that the cross-coupled GIC of Figure 2.4(b), popularly known as Antoniou's GIC, is guaranteed to offer stability unconditionally and has, therefore, been used extensively as the preferred GIC configuration.

Since Antoniou's GIC circuit has been exhaustively used in the realization of an inductor or FDNR, the various possibilities have also been extensively studied in the literature. In particular, the results described next are worth mentioning.

In inductance simulation mode, the Antoniou's GIC can be used in two different ways which are shown as type-I and type-II inductance simulation circuit in Figure 2.5.

The detailed non-ideal analysis of these circuits has been carried out [26,39] by taking the gains of the op-amps to be characterized by an integrator model, i.e. $A_i(s) \cong \omega_{ti}/s$ where $\omega_{ti} = A_{0i}\omega_{pi}; i = 1, 2$. Here, A_{0i} represents the DC gain and ω_{pi} represents the pole frequency of the ith op-amp. The above approximation is valid for frequency range constrained by $\omega \gg \omega_{pi}$. A detailed analysis of these circuits [39] reveals that the Q enhancement condition for type-I inductance

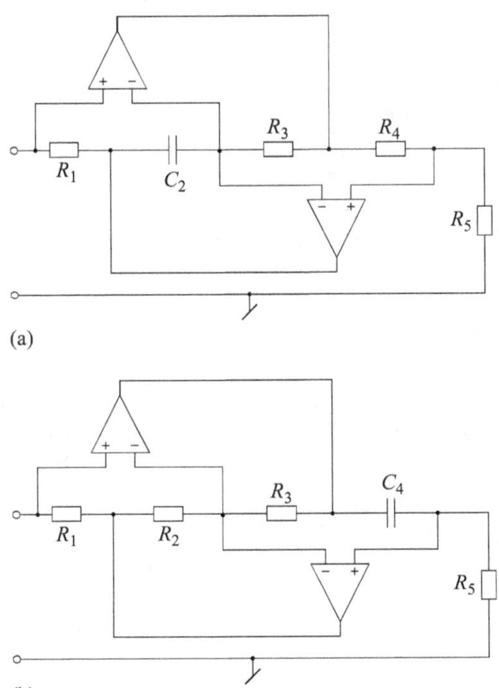

Figure 2.5 Two alternatives for inductance simulators based on Antoniou's GIC [29,171]: (a) type-I inductance simulator and (b) type-II inductance simulator

simulator is a function of ω_{t1} and ω_{t2} also which calls for either the use of identical op-amps or a priori measurement of the GBP of the two op-amps employed. By contrast, in type-II inductance simulator, the enhancement condition is independent of ω_{t1} and ω_{t2} and can be satisfied by only choosing two resistor values to be identical. Thus, type-II inductance simulator is superior of the two.

Another element of great interest in the context of op-amp *RC* filters is the so-called FDNR [28] which is a one-port or two-terminal element having $Z(s)$ of type $Z(s) = 1/Ds^2$. Its other counterpart is called a frequency-dependent negative conductance (FDNC) characterized by $Z(s)$ of the type $Z(s) = Ms^2$. Of the two, FDNR has been widely employed in a number of active filter realization methods. For realizing a grounded FDNR and grounded FDNC too, the Antoniou's GIC is the preferred choice.

For FDNR simulation, a GIC offers three possible choices, thereby leading to type-I FDNR, type-II FDNR and type-III FDNR which are shown in Figure 2.6.

A non-ideal analysis of the three types of the FDNRs reveals [26,39] that although in the first two cases the Q-enhancement condition is dependent on ω_{t1} and ω_{t2} only, in case of type-III FDNR, the enhancement condition is independent of GBPs of the op-amps and can be fulfilled, in practice, by matching of two resistor values only. Thus, type-III FDNR is the preferred choice for FDNR realization.

2.3.4 *Two-op-amp resistively variable capacitance simulators*

Resistively variable capacitors (RVCs) find a number of applications in measurement and control areas. Based upon the four two-op-amp GICs elaborated in the previous section, six possible two-op-amp-based capacitance simulators were exhaustively studied by Carlosena *et al.* [149] which are shown in Figure 2.7.

Ideally, all the six circuits are characterized by the same expression for their input impedances; however, their non-ideal expressions, taking into account the finite GBP of the op-amps, are all different. These have been extensively studied to enumerate all possible cases which lead to a capacitive input impedance. A rigorous theoretical and experimental study, considering a number of design aspects such as biasing, stability, frequency response and dynamic range, yielded three optimum RVC circuits which are shown in Figure 2.8. It may be pointed out that a detailed stability analysis of these circuits was also carried out subsequently by Abuelma'atti and Khan [157].

Application: An interesting application of the RVC of Figure 2.8(b) in a passive *RC* attenuator was demonstrated in [149] (see Figure 2.8(d)) wherein a passive *RC* 1:10 attenuator was devised to provide exact compensation condition (equal time constants, i.e. $R_1C_1 = R_2(C_2 + C_{eq})$). Figure 2.8(d) demonstrates the waveforms for the cases of overcompensation and compensation, which proves the utility of the RVC of Figure 2.8(b).

2.3.4.1 The capacitance multiplier

A low-component-count capacitance multiplier, which employs a buffered inverting amplifier with a capacitive feedback and which has been popularly employed in

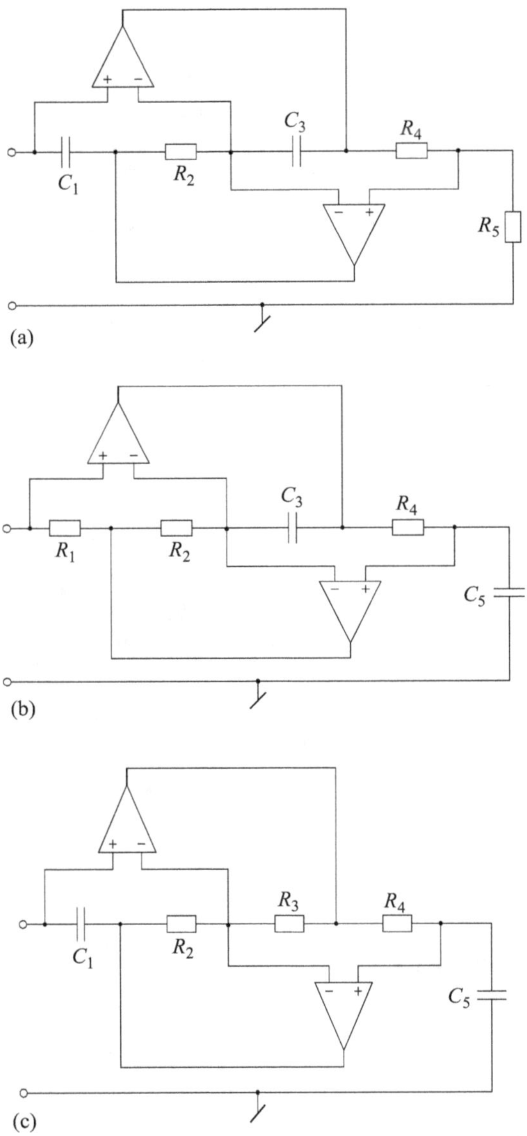

Figure 2.6 Three possible FDNR simulation circuits using Antoniou's GIC: (a) type-I FDNR, (b) type-II FDNR and (c) type-III FDNR

several situations, is shown in Figure 2.9. This circuit can be seen to be an embodiment of the capacitance-enhancement using Millers' theorem. As per Miller's theorem, if there is an amplifier with gain G (in this case, $G = -R_b/R_a$) with an impedance Z_f (in this case, the capacitor C_0) connected between its input and output, then the input impedance looking into the input port is given by $Z_{in} = Z_f/(1 - G)$.

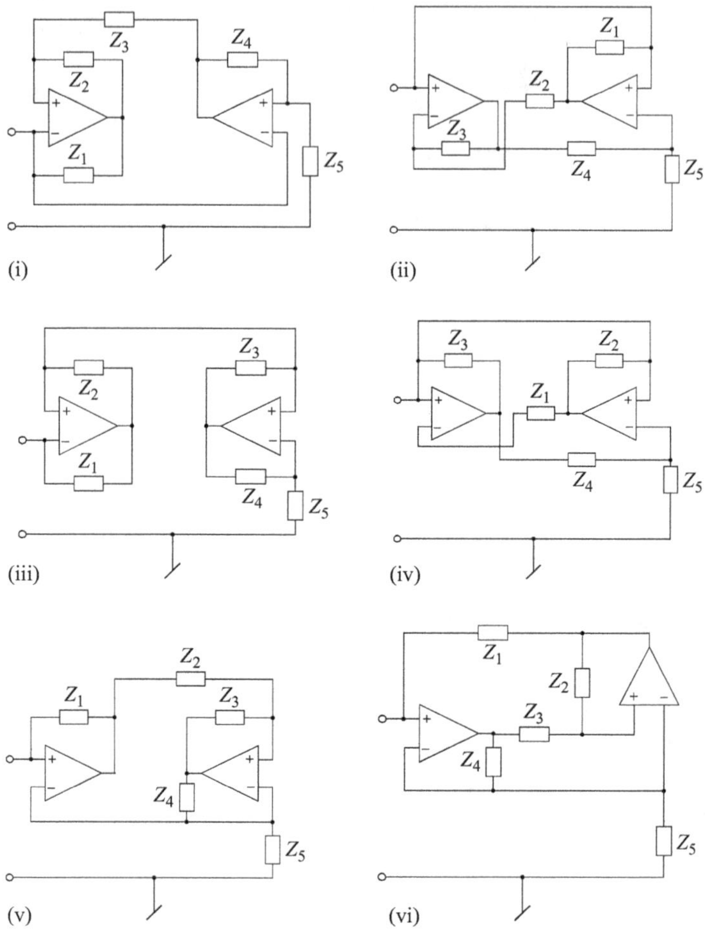

Figure 2.7 Six possible configurations (i)–(vi) for realizing resistively-variable capacitors [149]

Using this analysis, it is straight away found that the circuit realizes a grounded capacitance of value

$$C_{eq} = C_0\left(1 + \frac{R_b}{R_a}\right) \tag{2.22}$$

It is easy to visualize that using MOS amplifiers, this scheme may lead to a novel MOSFET-C version suitable for integration in CMOS technology.

2.3.4.2 Lossless capacitance simulator

Another lossless capacitance simulator realized with only two op-amps, extracted from a linear resistance-to frequency converter presented in [126], is shown in Figure 2.10.

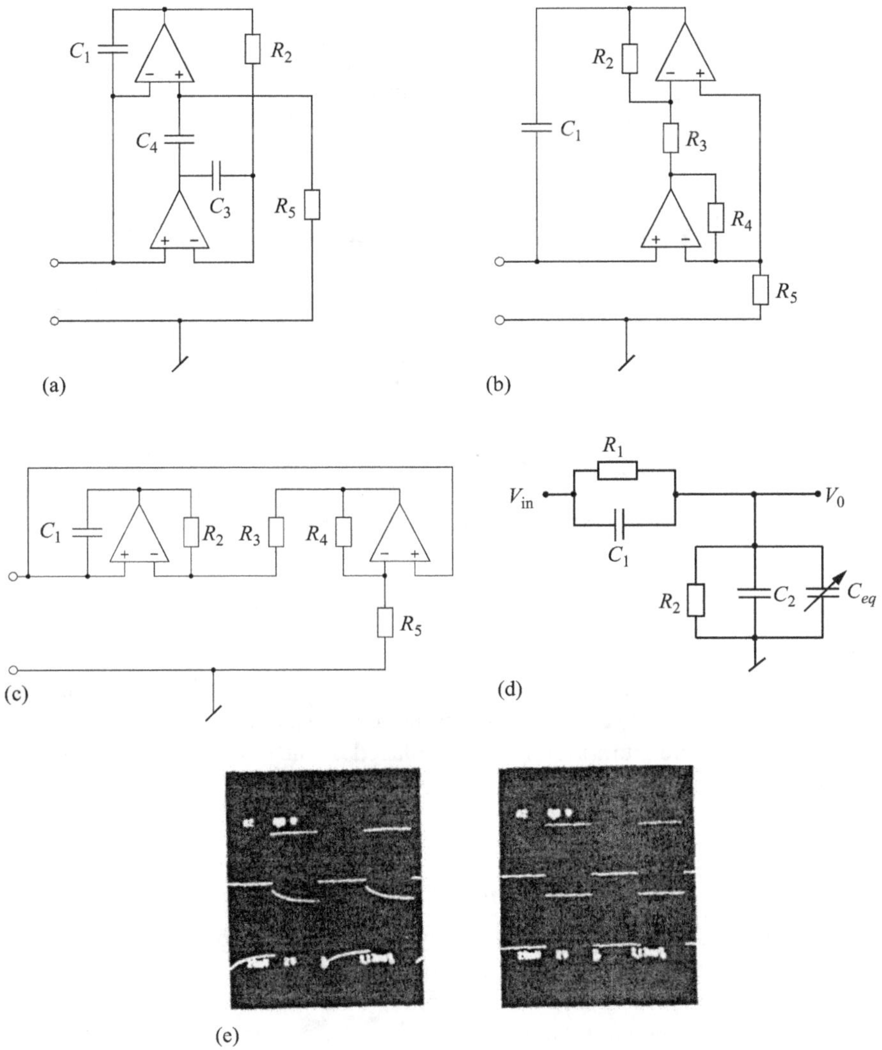

Figure 2.8 Three optimal RVC circuits [149] and an application: (a) $C_{eq} = C_1C_3R_2/C_4R_5$, (b) $C_{eq} = C_1R_2R_4/R_3R_5$, (c) $C_{eq} = C_1R_2R_4/R_3R_5$, (d) the attenuator and (e) measured square-wave response of overcompensated and compensated attenuator. Reprinted, with permission of [149] © IEE

Analysis of this circuit reveals that, for $C_1 = C_2 = C_0$ and $R_1 = R_2 = R_0$, equivalent capacitance realized by this circuit is given by

$$C_{eq} = C_0\left(\frac{R_0}{R_3}\right) \tag{2.23}$$

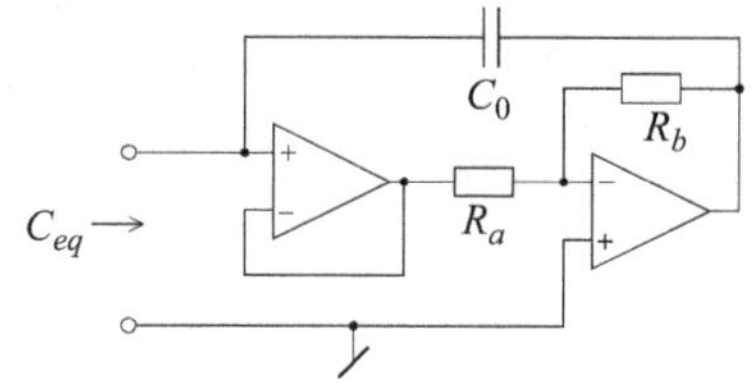

Figure 2.9 A well-known op-amp-based capacitance multiplier

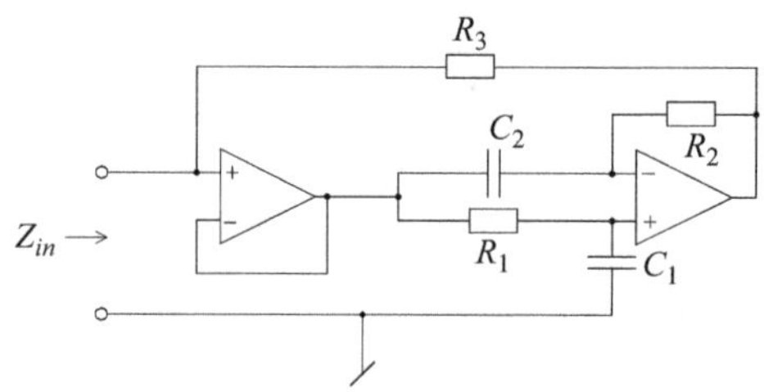

Figure 2.10 Lossless capacitance simulator [126]

Application: This circuit was shown to lead to an interesting linear resistance-to-frequency converter in [126] when the grounded capacitor (GC) used in a single-op-amp-based astable multi-vibrator was replaced by the circuit of Figure 2.10. The modified circuit generated a square wave whose frequency is linearly controllable by R_3.

2.3.5 Two-op-amp lossless inductance simulator

Rao and Venkateswaran [30] presented a systematic method of synthesizing inductive impedances using VCVS. This procedure resulted in a number of inductance simulators which included a two-op-amp circuit for lossless grounded inductance simulation which is shown in Figure 2.11.

This circuit is interesting even though it uses two capacitors because in the published literature, this circuit has been proposed and re-proposed[3] at least two more times not realizing that it existed in the literature as earlier as 1971. Analysis of this circuit reveals that its input impedance is given by

$$Z(s) = sC_1R_1R_3\frac{(1+sC_2R_2)}{(1+sC_1R_1)} \tag{2.24}$$

[3]In retrospection, it is interesting to find that the same circuit was derived in 1979 through another systematic derivation procedure by Ganguly in [101], and almost four decades later, in 2011, the same circuit was presented again by Maundy and Gift in [169] believing it to be new! Curiously, the authors of neither [101] nor [169] appear to have been aware of Rao–Venkateswaran's earlier original work of 1970 [30].

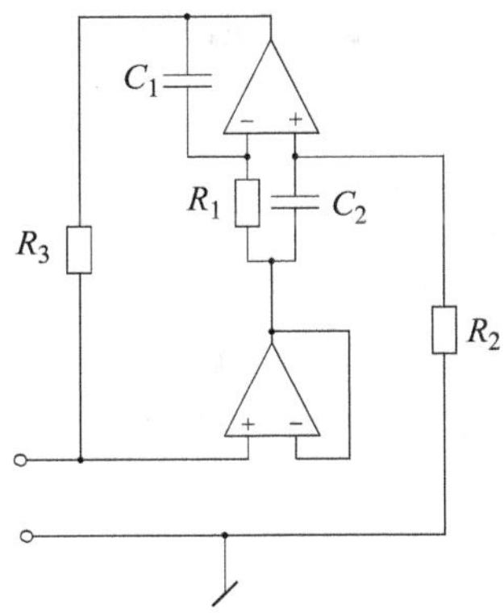

Figure 2.11 Circuit for realizing ideal inductor [30]

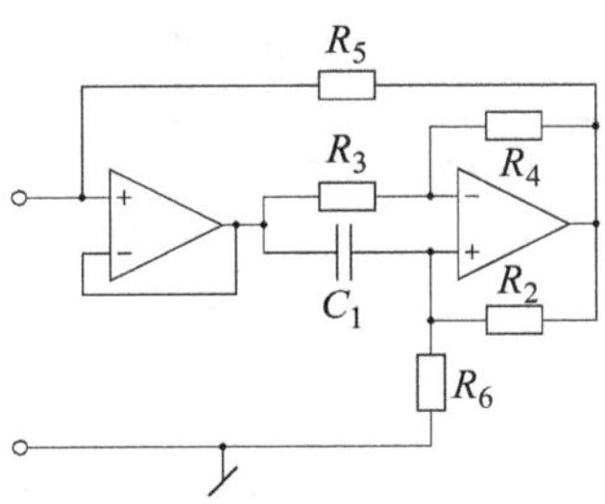

Figure 2.12 Circuit of the grounded inductor proposed by Tripathi and Patranabis [84]

which by choosing $C_1R_1 = C_2R_2$ reduces to

$$Z(s) = sC_1R_1R_3 \tag{2.25}$$

Thus, the circuit realizes a lossless inductance of value $L_{eq} = CR_1R_3$ which is single-resistance-tunable through a variable resistance R_3.

2.3.6 Tripathi–Patranabis lossless grounded inductor

Whereas most of the inductor simulation schemes can be looked upon as being obtained through a direct or buffered resistive feedback around some kind of integrator circuit, Tripathi and Patranabis [84], by contrast, presented a circuit which is based upon an isolated resistive feedback around a differentiator. This circuit is shown in Figure 2.12.

By straightforward analysis of the circuit, the value of realized grounded inductor is found to be

$$L_{eq} = \frac{C_1R_3R_5R_6}{R_3 + R_4} \tag{2.26}$$

provided $R_4R_6 = R_2R_3$. In fact, the exact equivalent of this circuit is a parallel RL admittance such that the resistive part can be adjusted with R_2 independently and can be made positive, negative or zero, while on the other hand, the inductive part is also independently variable through the resistor R_6.

2.3.7 Lossless GI using summer/subtractor circuits

A grounded lossless GI based upon a two-port gyrator derivable from a three-port gyrator [113] employs op-amp-based unity gain summers (UGSs) and subtractor and is shown in Figure 2.13.

This gyrator is characterized by the $[Y]$-matrix

$$[Y] = \begin{bmatrix} 0 & -\dfrac{1}{R_1} \\ +\dfrac{1}{R_2} & 0 \end{bmatrix} \tag{2.27}$$

Thus, with port-2 terminated into a capacitance the circuit simulates a GI of value $L_{eq} = CR_1R_2$. Although for realization in terms of conventional IC op-amps, both the subtractor and the summer require one op-amp along with four high-valued and matched resistors, using alternative building blocks such as fully integratable differential difference amplifiers and only MOSFET-based summers and subtractors, more efficient implementations can be obtained. This appears to be a worthwhile problem for further investigations!

Application: It was demonstrated in [113] that by replacing the capacitor of an op-amp-based astable multi-vibrator, by this gyrator with an unknown inductance L connected to its input port, an interesting linear inductance-to-time-period converter is obtainable which generates a square wave whose time period is linearly proportional to an unknown inductance L.

2.3.8 Two modified forms of the GIC and their applications

An interesting modification of the Antoniou's GIC circuit was proposed by Von Wangenheim [158] which is shown in Figure 2.14(a).

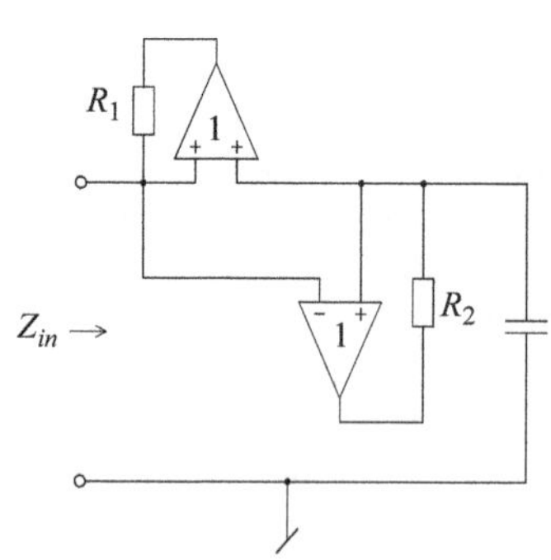

Figure 2.13 Gyrator realization extracted from a linear L/T converter [113]

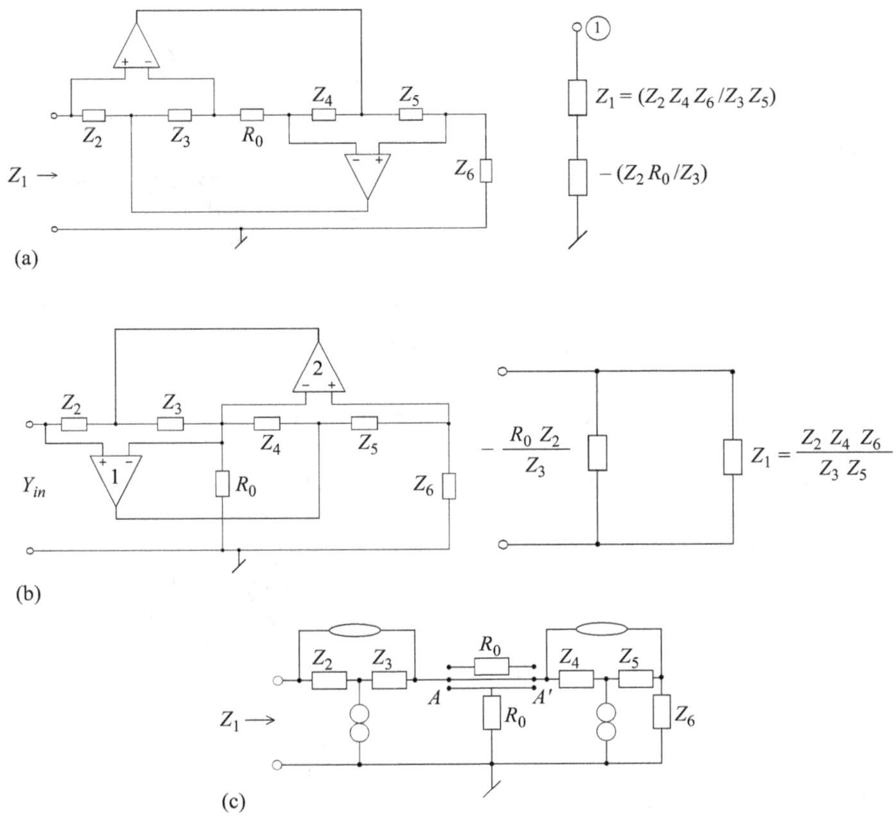

Figure 2.14 Two modifications of the GIC and their explanations: (a) Von Wangenheim's modification [158], (b) Senani's modification [160] and (c) explanation of the mechanisms of modified GICs through a nullor model. Reprinted, with permission of [160] IEE

The normal GIC has five impedances. If a sixth impedance, a resistor R_0 is inserted in between as shown in Figure 2.14(a), the input impedance gets modified to

$$Z_{in} = \frac{Z_2 Z_4 Z_6}{Z_3 Z_5} - R_0 \frac{Z_2}{Z_3} \tag{2.28}$$

It was argued that this modified circuit could be especially suitable for some applications such as oscillator realization in which the additional negative term would be particularly useful to start and maintain oscillations without requiring any additional circuitry.

Subsequently, Senani [160] demonstrated that an alternative modification of the GIC structure is also possible which is shown in Figure 2.14(b) for which the input admittance is given by

$$Y_{in} = \frac{Z_3 Z_5}{Z_2 Z_4 Z_6} - \frac{Z_3}{Z_2 R_0} \tag{2.29}$$

This alternative circuit is also useful in the context of sinusoidal oscillator realization based on parallel $-R$, L, C or parallel R, $-C$, D models [160]. It was demonstrated in [160] that the mechanism of the two modified GICs can be explained from the nullor model of Figure 2.14(c) where one nullator and a grounded norator represent an op-amp. If now one inserts a resistor R_0 between A and A' as in Figure 2.14(a), this makes the load impedance to the generalized NIC (GNIC1) as $[R_0 - (Z_4Z_6/Z_5)]$. This impedance gets multiplied by the impedance conversion factor $(-Z_2/Z_3)$ of GNIC2 which gives the input impedance of the circuit Figure 2.14(a) as given by (2.28). On the other hand, if a resistor R_0 is connected from A–A' to ground, by a similar process, this results in the input admittance of the circuit of Figure 2.14(b) given by (2.29).

2.4 Single-op-amp lossless inductance simulators

Till about 1974, a number of gyrators and other inductance simulation circuits had been known which needed at least two op-amps along with a number of passive components, some of the prominent ones from which have already been described in the earlier sections of this chapter. However, not even a single circuit was discovered which could realize a lossless two-port gyrator, with *both ports grounded* (having a ground reference) using a single op-amp. The complete absence of any single-op-amp lossless gyrator prompted Wilson and Orchard [62] to think that possibly such a gyrator may not exist. Motivated by this deduction, they searched for a mathematical proof and eventually found one. Consequently, Wilson and Orchard in [62] concluded, through a rigorous mathematical proof, that it is impossible to realize a lossless gyrator with both ports grounded using only a single differential-input–single-output type op-amp, given any number of resistors. However, this proof also leads to a surprising revelation that the possible existence of a lossless gyrator with one port grounded but with the other port *floating*, however, is not ruled out! This motivated them to find a single-op-amp active gyrator of this kind which they eventually found and published in [61]. Subsequently, however, several other researchers [64,90,100] too came up with a number of single-op-amp-based lossless grounded inductor circuits using a varying number of resistors and realization conditions but all using only a single capacitor (obviously, floating) in all the cases.

In this section, we present the various single op-amp-based active gyrators/ lossless grounded inductor simulators known so far.

2.4.1 Orchard–Willson gyrator

The first-ever single-op-amp-based active gyrator was proposed by Orchard and Wilson [61] and is shown in Figure 2.15.

By straightforward analysis, it is found that subject to the fulfilment of the following conditions:

$$R_2 = R_3 \text{ and } \left(1 + \frac{R_1}{R_2}\right)\left(1 + \frac{R_3}{R_4}\right) = \frac{R_1R_5}{R_4R_6} \tag{2.30}$$

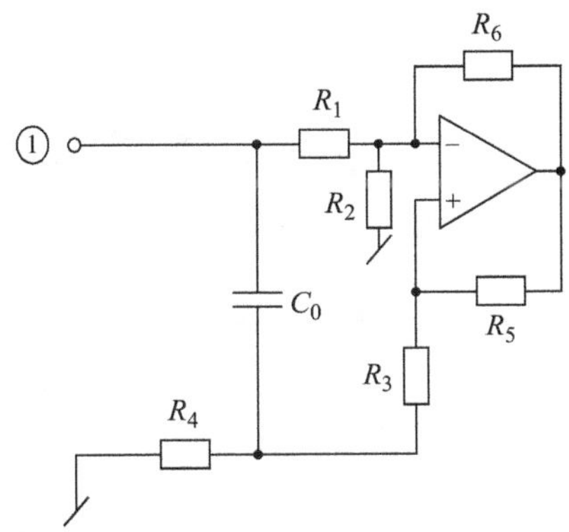

Figure 2.15 Orchard–Willson gyrator [61]

the input impedance for this circuit is given by

$$Z_{in}(s) = sC_0R_1R_4\frac{(1 + (R_1/R_2))}{(1 + (R_3/R_4))} \tag{2.31}$$

It is obvious that the circuit suffers from the drawback of requiring two conditions of realization which necessitates the employment of closely matched resistances. A detailed analysis of the mismatches carried out by the authors [61] shows that the highest Q-factor achieved could be about 100 which may be considered to be satisfactory in some applications, whereas the usable frequency range of the simulated inductance is determined primarily by the finite GBP of the op-amp employed.

2.4.2 Schmidt–Lee circuit

Subsequent to the discovery of the first single-op-amp active gyrator by Orchard and Willson [61], a multipurpose simulation network, realized with a single op-amp, was presented by Schmidt and Lee [64] which is shown in Figure 2.16. This circuit could realize a variety of impedances as special cases which is explained in the following equation. The expression for the input admittance is given by

$$Y_{in} = \frac{(Y_1 + Y_2)(Y_3Y_6 - Y_4Y_5) + Y_7\{Y_6(Y_1 + Y_3) - Y_5(Y_2 + Y_4)\}}{Y_3(Y_2 + Y_6) - Y_4(Y_1 + Y_5)} + Y_8 \tag{2.32}$$

where $Y_i = 1/R_i$, $i = 1$–8.

Schmidt and Lee [64] demonstrated that by appropriate choice(s) of different admittances, this circuit can realize a parallel R–C–D admittance, parallel C–D admittance, a series R–D resonator and also a lossless inductor, the details of which can be found in [64]. For the case of lossless inductance realization, the following choice of component values is required:

$$\frac{1}{R_1} = \frac{1}{R_2} = \frac{1}{2R_3} = \frac{1}{R_4} = \frac{1}{5R_5} = \frac{1}{3R_6} = \frac{1}{2R_7} = \frac{1}{R} \text{ and } R_8 = \infty \tag{2.33}$$

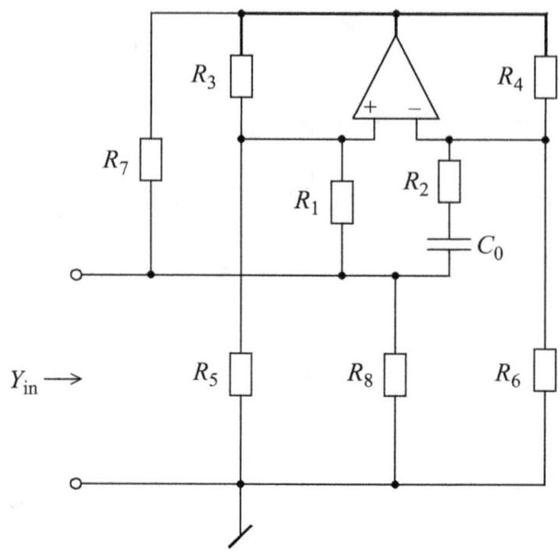

Figure 2.16 Schmidt–Lee circuit [64]

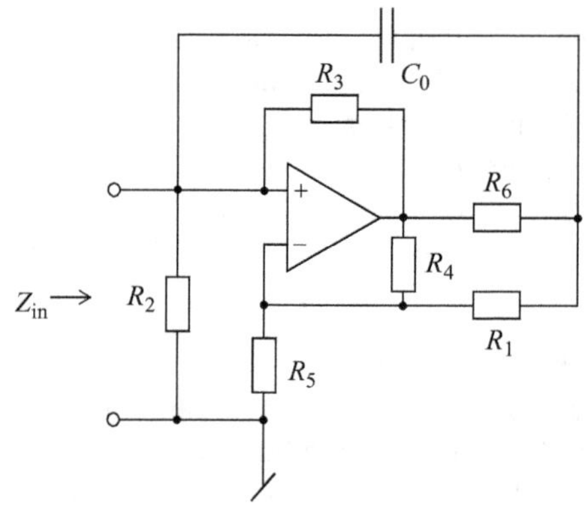

Figure 2.17 Ramsey gyrator [90]

The input impedance then becomes

$$Z_{in}(s) = \frac{1}{Y_{in}(s)} = sC_2\left(\frac{2}{9}\right)R^2 \tag{2.34}$$

Hence, a lossless grounded inductor of value $L_{eq} = (2/9)C_2R^2$ is realized.

2.4.3 Ramsey gyrator

Another single amplifier single capacitor circuit capable of simulating a lossless grounded inductor for use in active *RC* highpass and bandpass filters was presented by Ramsey [90] and is shown in Figure 2.17.

Analysis of this circuit reveals that the circuit realizes an inductance value given by

$$L_{eq} = \frac{C_0 R_3 R_5}{((R_3 R_5/R_1 R_2) - (R_4/R_1)) + (R_3 R_5/R_2 R_6)(1 + (R_4/R_1)) - (R_4/R_6)} \quad (2.35)$$

subject to the fulfilment of two conditions as follows:

$$R_5 = R_2 R_4 \left(\frac{1}{R_3} + \frac{1}{R_6}\right) \text{ and } R_1 R_3 = R_5 R_6 \quad (2.36)$$

It may be noted that like the earlier two circuits, this circuit also requires two conditions to be fulfilled and, thus, requires close matching of resistors values to realize a lossless inductance. As a consequence, the realization is quite sensitive to resistor value mismatches.

2.4.4 Horn–Moschytz circuit

Having observed that several papers were published describing single-op-amp-based realizations of both lossy and lossless inductors as well as other kinds of elements such as FDNRs, CD and RCD impedances and as some second-order immittance functions, Horn and Moschytz [100] presented a method for the design of general active *RC* driving point impedances of arbitrary order using a network containing only a single op-amp.

This generalized topology based upon a single op-amp had as many as ten admittances. It was shown that as special cases, this circuit could yield a variety of impedances such as lossless *L*, parallel *RLC* resonator and series *RLC* resonator, an ideal FDNR, parallel *RCD*/series *RCD* resonators and a number of other cases. From our viewpoint, the special case of realizing a lossless inductor is of particular interest and the same is shown in Figure 2.18. Analysis of this circuit

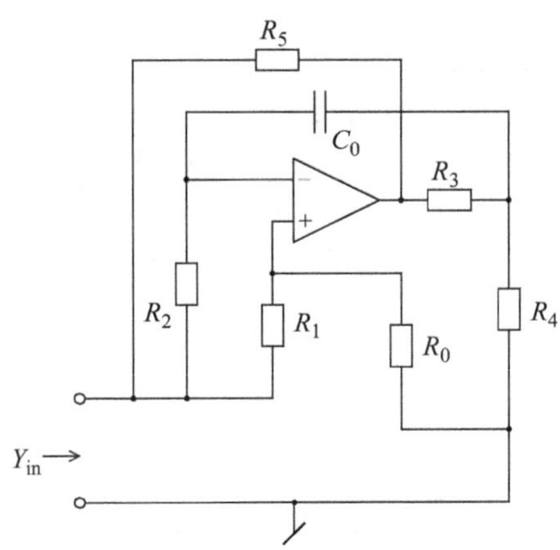

Figure 2.18 Horn–Moschytz lossless grounded inductor circuit [100]

reveals that a lossless inductance is realized, subject to fulfilment of the following condition:

$$\frac{R_0}{R_4} = \frac{R_1 + R_5}{R_3} + \frac{R_1}{R_2}\left(1 + \frac{R_5}{R_3}\right) \tag{2.37}$$

The realized inductance value is given by

$$L_{eq} = \frac{C_0 R_2 R_5 (1 + (R_0/R_1))}{1 + (R_3/R_4)} \tag{2.38}$$

It may be noted that in contrast to the previous three circuits, the circuit of Figure 2.18 offers the interesting feature of requiring only a single realization condition.

2.5 Economic inductance simulators and resonators

From the previous discourse, it is clear that if one tries to simulate a lossless inductor, ideal FDNR or FDNC elements using a single op-amp, the price to be paid is not only the use of six to eight resistors but one needs to fulfil the required conditions for the realization also. With the exception of Horn and Moschytz [100] circuit, which has only one condition to be fulfilled, all the three remaining circuits require at least two conditions to be fulfilled. Thus, all the single-op-amp-based gyrators suffer from the requirement of close matching of resistor values and, thus, are highly sensitive to component mismatches.

Several researchers demonstrated that, by contrast to the ideal inductor, a lossy one (either a series *RL* or parallel *RL*) can be realized not only more economically (i.e. using a fewer number of resistors) but may also not need any realization constraints or critical component-matching conditions. In this section, we highlight a number of significant contributions made in this direction.

2.5.1 Ford–Girling circuit

This is one of the two first-ever circuits for inductance simulation realizable with a single op-amp and only three passive components [4]. The circuit can be seen to be obtainable by a resistive feedback (through R_2) around the single-op-amp inverting Miller integrator having transfer function $(V_0(s)/V_{in}(s)) = -(1/sC_1R_1)$ such that the input admittance of the overall circuit is given by

$$Y_{in} = \left(\frac{1}{R_1} + \frac{1}{R_2}\right) + \frac{1}{sC_0R_1R_2} \tag{2.39}$$

thereby simulating a lossy inductor with $R_{eq} = R_1 \parallel R_2$ and $L_{eq} = C_0R_1R_2$. A lossless version of this circuit was also presented by Ford–Girling [4] which required another op-amp configured as an NIC to cancel the resistive loss of the original circuit and is shown in Figure 2.19.

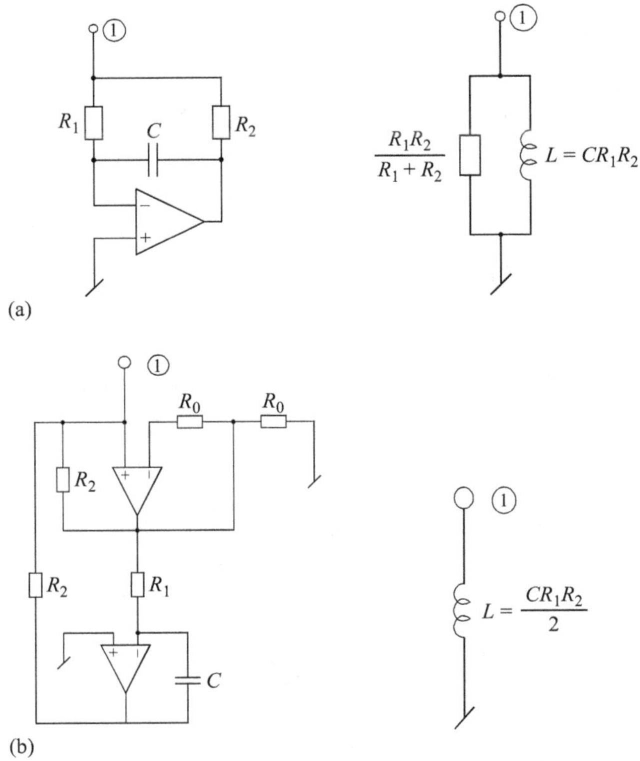

Figure 2.19 Ford–Girling circuits [4]: (a) the lossy inductor and (b) the loss-compensated ideal inductor

2.5.2 Prescott circuit

The circuit of Figure 2.20 introduced by Prescott [6] is another low-component count circuit presented about the same time as the Ford–Girling circuit [4]. This circuit can be looked upon as being made of a non-ideal differentiator with a resistive feedback around it. By a straightforward analysis, the input impedance of this circuit is found to be

$$Z_{in} = (R_1 + R_2) + sC_0R_1R_2 \tag{2.40}$$

thereby representing a series *RL* circuit with $R_{eq} = (R_1 + R_2)$ and $L_{eq} = C_0R_1R_2$. The attractive feature of this circuit is the employment of the op-amp as a unity gain amplifier which offers the widest possible closed-loop bandwidth.

2.5.3 Berndt–Dutta Roy circuit

Yet another circuit which realizes a grounded inductor using only a single unity gain amplifier (UGA) was presented by Berndt–Dutta Roy [25] which is shown in Figure 2.21. This circuit, however, realizes a frequency-dependent resistance and inductance.

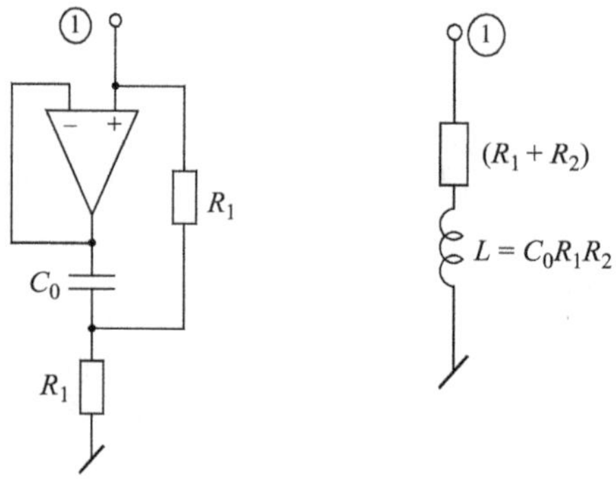

Figure 2.20 Prescott circuit for realizing lossy grounded inductor [6]

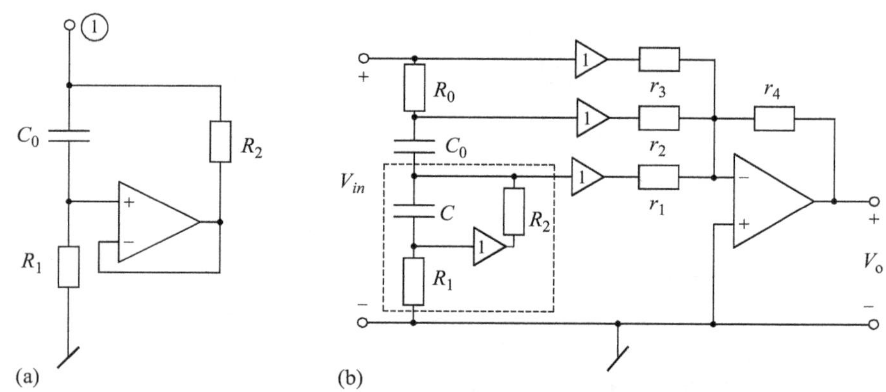

Figure 2.21 Berndt–Dutta Roy circuit [25] and its application: (a) the lossy grounded inductor and (b) the general configuration of Leach and Chan [173] employing Berndt–Dutta Roy [25] inductor

Analysis of this circuit [25] reveals that the input impedance of this circuit is given by

$$Z(s) = \frac{R_2(1 + sCR_1)}{(1 + sCR_2)} \tag{2.41}$$

from which the equivalent inductance and resistance are given by

$$L_{eq} = \frac{C_0R_2(R_1 - R_2)}{1 + \omega^2 C_0^2 R_2^2}; \quad R_1 > R_2; \quad R_{eq} = \frac{R_2\left(1 + \omega^2 C_0^2 R_1 R_2\right)}{1 + \omega^2 C_0^2 R_2^2} \tag{2.42}$$

and the quality factor is given by

$$Q_0 = \frac{\omega C_0(R_1 - R_2)}{1 + \omega^2 C_0^2 R_1 R_2}; \quad R_1 > R_2 \tag{2.43}$$

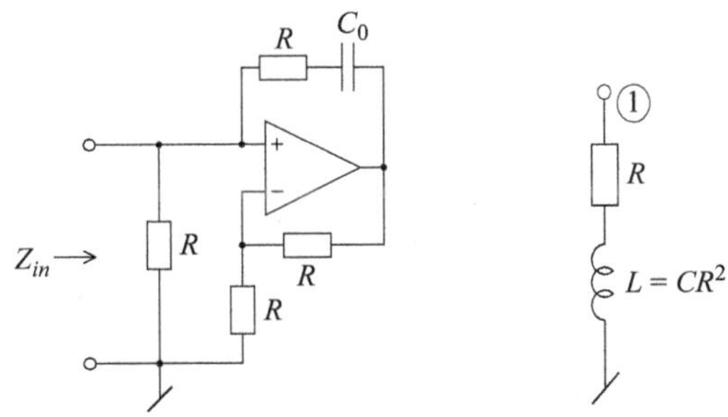

Figure 2.22 Caggiano inductance simulation circuit [174]

The circuit attains a maximum $Q_0 = 1/2(R_1/R_2)^{1/2}$ at a frequency $\omega_0 = 1/C_0(R_1R_2)^{1/2}$.

Application: Dutta Roy and Ahmed [58] demonstrated that the use of this circuit in Leach and Chan's [173] method of synthesizing a general biquadratic filter results in a minimum sensitive realization of all the five basic filter functions (namely low pass, high pass, bandpass, band stop and all pass) from the same configuration.

2.5.4 Caggiano circuit

The basic concept behind Caggiano circuit [174] appears to be the idea that a parallel combination of a resistor R and negative *series RC* impedance would turn out to be equivalent impedance $Z = R + sCR^2$ as follows:

$$
\begin{aligned}
Z &= R \,||\, -\left(R + \frac{1}{sC}\right) = R \,||\, -\left(\frac{1 + sCR}{sC}\right) \\
&= \frac{-R(1 + sCR)}{sCR - (1 + sCR)} = R + sCR^2
\end{aligned}
\tag{2.44}
$$

Thus, it is seen that a series RC placed in the feedback path of a noninverting amplifier of gain 2 would give equivalent impedance as $-(R + (1/sC))$. Hence, the circuit of Figure 2.22 realizes grounded series RL impedance as worked out above.

2.5.5 Patranabis circuit

Patranabis [34] approach was theoretically more formal than that of Caggiano [174]. Following the Patranabis formulation [34], a NIC terminated into a parallel RC admittance, with another resistor R put in series can be used to realize an inductance as follows:

$$
\begin{aligned}
Z &= R - \left\{R \,||\, \frac{1}{sC}\right\} = R - \frac{R/sC}{(R + (1/sC))} \\
&= R - \left(\frac{R}{1 + sCR}\right) = \frac{sCR^2}{(1 + sCR)}
\end{aligned}
\tag{2.45}
$$

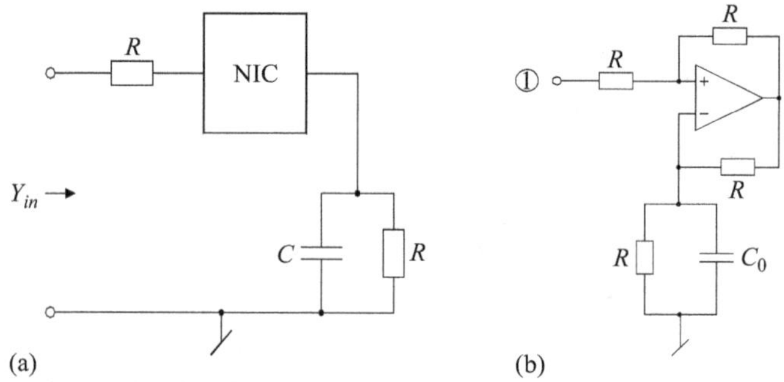

Figure 2.23 A simplified NIC-based circuit derived from Patranabis scheme [34] of inductance simulation using an NIC [34]: (a) the basic scheme and (b) an op-amp implementation (although not explicitly claimed in [34])

thereby yielding

$$Y_{in} = \frac{1}{R} + \frac{1}{sCR^2} \tag{2.46}$$

which represents a parallel *RL* admittance. It must, however, be pointed out that in the publication [34], no op-amp circuit implementing this idea was explicitly mentioned. However, if one uses an op-amp-based NIC, in the general schematic of Patranabis [34], the resulting circuit would take the form of the circuit shown in Figure 2.23.

2.5.6 The parallel/series RL inductors derived by Rao–Venkateswaran

Rao and Venkateswaran in [30] presented a systematic method of generating circuits realizing different types of driving point impedances using op-amps. The method involved a buffered resistive feedback around a two-port active network realizing some form of ideal/non-ideal integrator-type circuit. It may be pointed out that, about the same time, Patranabis [66] also presented a method of synthesizing inductors via resistive feedback around a first-order filter. Rao and Venkateswaran [30] presented a number of grounded inductance simulators via their proposed method including a lossless inductor requiring as many as three op-amps. One of the two particularly appealing circuits therein was the one shown in Figure 2.24. This circuit can be considered to be a modified version of Ford–Girling [4] circuit with a buffered resistive feedback around an inverting integrator. Consequently, the equivalent input admittance of this circuit is given by

$$Y_{in} = \frac{1}{R_2} + \frac{1}{sC_0R_1R_2} \tag{2.47}$$

Thus, as compared to the Ford–Girling circuit [4], this circuit provides a single-resistance-tunable inductance value (controllable by R_1).

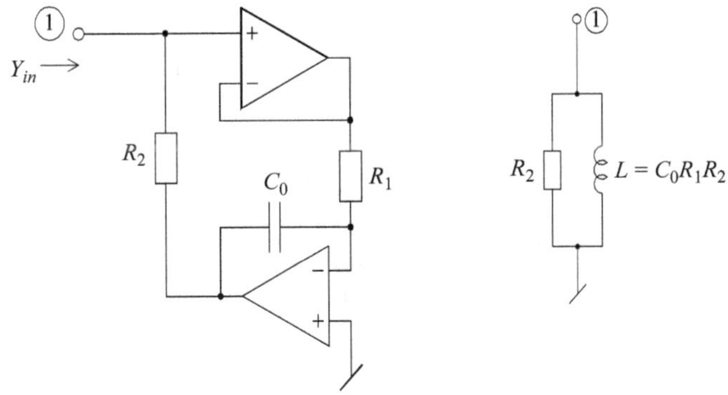

Figure 2.24 The parallel RL inductor proposed by Rao–Venkateswaran [30]

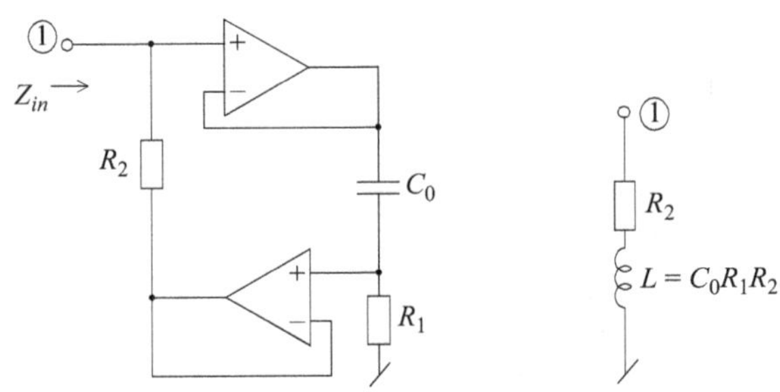

Figure 2.25 Grounded series RL impedance derived by Rao–Venkateswaran [30]

The other appealing circuit derived by Rao and Venkateswaran [30], using the same general method, was the two-UGA-based series *RL* simulator shown in Figure 2.25. This circuit can be seen to be a modified form of Prescott circuit [6] such that the resistive feedback around the first-order *RC* high pass filter now becomes a buffered resistive feedback (through R_2) which, however, changes the nature of impedance realized; it now becomes a single-resistance-controllable grounded series *RL* impedance given by

$$Z(s) = R_2 + sCR_1R_2 \tag{2.48}$$

Note that in this circuit also, the inductive part is single resistance controllable through R_1.

2.5.7 *Ahmed–Dutta Roy technique of deriving grounded-capacitor lossy GI*

In [67] Ahmed and Dutta Roy presented a method of grounding the capacitor [172,181] in a class of those op-amp-based GI circuits which originally did not employ a GC. Their methodology was based upon their proposed equivalence shown in Figure 2.26(a) as per which, a first-order *RC* circuit containing a floating

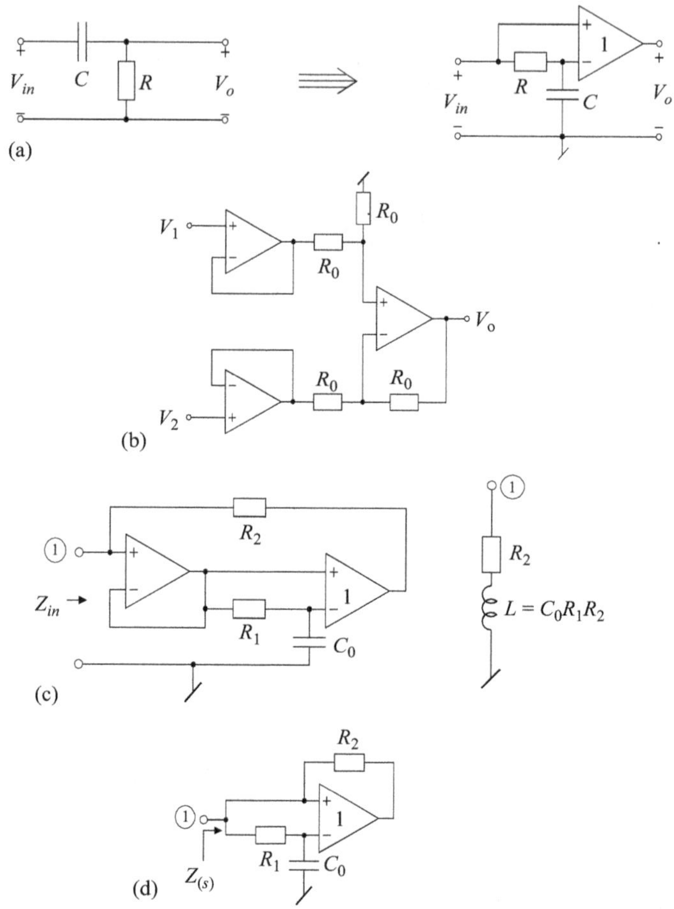

Figure 2.26 Ahmed–Dutta Roy's technique of grounding the capacitors in low-component-count lossy inductor simulation circuits: (a) a RC circuit and its GC version, (b) the realization of the UGDA, (c) GC version of Rao–Venkateswaran circuit d GC version of Berndt–Dutta Roy circuit

capacitor and the transformed circuit using a GC when appropriately embedded into a network containing a unity gain differential amplifier (UGDA) as shown, both realize exactly the same transfer function.

Thus, using this approach, the grounded-capacitor version of the Rao–Venkateswaran [30] circuit turned out to be as shown in Figure 2.26(c) which realizes the input impedance $Z_{in}(s) = R_2 + sC_0R_1R_2$. However, the penalty to be paid to obtain a GC-version is that the required UGDA itself requires more than one op-amp, four matched resistors and also two UGAs to ensure ideally infinite input impedance at both of its input terminals (as shown in Figure 2.26(b)). Alternatively, the two UGAs can be dispensed with by having the four resistors named R_0 to be of large values. Similarly, the circuit of Figure 2.26(d) is the GC-version of the Berndt–Dutta Roy [25] circuit.

2.5.8 Senani–Tiwari circuit

Till 1977 or so, although a number of single-op-amp lossy inductance simulation circuits were known they all employed a floating capacitor. It was, therefore, being attempted if such impedance can be realized using a single op-amp with a GC. Although Ahmed–Dutta Roy [67] technique of grounding the capacitor did yield such a circuit which has already been described in the previous section, the circuit in fact required a UGDA which itself needed four equal and high-valued matched resistors due to which the UGDA input terminals did not have the required ideally infinite input impedances, and furthermore, a matching constraints and use of overall six resistors nullified the intention of keeping low-component-count. Senani and Tiwari [92] proposed a new single-op-amp circuit which needed a smaller number (only four) resistors and a GC to realize a series *RL* type lossy inductor using a single op-amp configured as an NIC. This circuit is shown in Figure 2.27. With all the four resistors taken equal, i.e. $R_1 = R_2 = R_3 = R_4 = R$, the circuit simulates $Z_{in} = R + sC_0R^2$. The circuit obviously needs a cancellation constraint and stability considerations require the resistors to be kept slightly mismatched to ensure the pole of the non-ideal impedance in the left half of the s-plane to ensure stability. Furthermore, the circuit has large Q-sensitivities which can be kept low only for moderate Q values.

Subsequently, Stoyanov [127] pointed out that the same circuit was published by him in a Bulgarian patent disclosure. However, Senani apprised in his

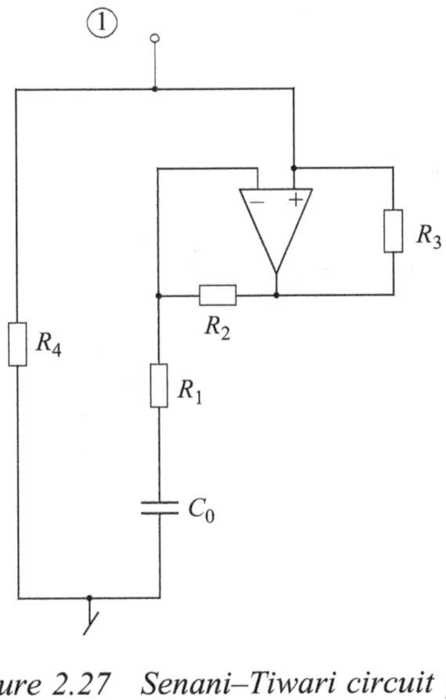

Figure 2.27 Senani–Tiwari circuit [92]

reply in [127] that this patent publication was mentioned neither in Engineering Index nor in Electrical and Electronics Abstracts, both of which were considered to be the most comprehensive and exhaustive resources of worldwide research done in the field of engineering those days. Due to this reason, Senani–Tiwari [92] believed this circuit to be new and forwarded the same to Proc IEEE for publication in November 1977 which was subsequently published in the July 1978 issue of Proc IEEE.

2.5.9 Soliman–Awad tunable active inductor

While single-capacitor lossless inductors or active gyrators using only one capacitor were proposed by Orchard–Wilson [61], Ramsey [90], Schmidt–Lee [64] and Horn–Moschytz [100]. By contrast, Soliman–Awad [93] presented a circuit which although employed two capacitors, the type of admittance realized was somewhat less complex than the earlier quoted circuits, in which it represented a parallel *RL* circuit which could yield a lossless grounded inductor upon fulfilment of only a single condition. This circuit is shown in Figure 2.28. Analysis of this circuit reveals its input admittance as

$$Y_{in} = \left(\frac{1+(R_1/R_3)}{R_1+R_2}\right) + \frac{(1-(R_2/(R_1+R_2))(1+(C_1/C_2)))}{R_4} + \frac{1}{sC_2R_3R_4(1+(R_2+R_1))} \tag{2.49}$$

Thus, without imposing any conditions on the resistance values the circuit realizes a parallel *RL* admittance and can realize a lossless inductor subject to the fulfilment of the following condition:

$$\frac{C_1}{C_2} = \frac{R_4}{R_2} + \frac{R_1}{R_2}\left(1+\frac{R_4}{R_3}\right) \tag{2.50}$$

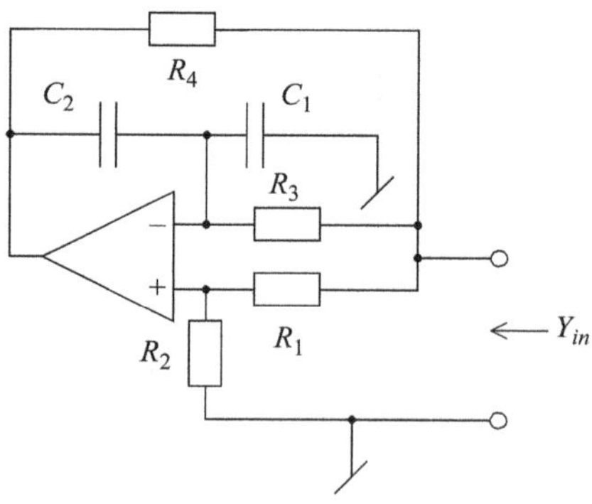

Figure 2.28 Soliman–Awad [93] tunable active inductor

The value of the realized inductance is given by

$$L_{eq} = C_2 R_3 R_4 \left(1 + \frac{R_2}{R_1}\right) \tag{2.51}$$

Application: This circuit can be readily converted into a single-capacitor-controlled sinusoidal oscillator by connecting an external variable capacitor C_0 across the input port. This application was subsequently devised by Soliman and Awad themselves in a subsequent publication [189].

2.5.10 Nagarajan–Dutta Roy–Choudhary circuit

During the same period, a number of authors proposed single-op-amp circuits which, however, realized second-order resonators such as those of Molo [63] and Soliman–Awad [83] and, hence, are not included here. However, a particularly interesting circuit was the one proposed by Nagarajan *et al.* [41] which employed only three resistors and two capacitors along with a single UGA which is shown in Figure 2.29. With $R_1 = R_2 = R_3 = R$ and $C_1 = C_2 = C$, they demonstrated that the input impedance of the circuit is given by

$$Z_{in}(j\omega) = R\left(3 - \omega^2 C^2 R^2\right) + j\omega\left(4CR^2\right) \tag{2.52}$$

Thus, it was shown that a high-Q (infinite) is achievable at a specific frequency at which the inductance value realized becomes $L_{eq} = 4CR^2$.

Application: In retrospection, it is interesting to note that for general values of the circuit elements as shown, the input impedance actually represents a series *RLM* network which can be easily converted into an interesting sinusoidal oscillator as demonstrated by Senani in [176]. Thus, with a capacitor C_0 connected across the input port, the condition of oscillation of the resulting oscillator would be

$$C_0 = \frac{1}{(R_1 + R_2 + R_3)}\left[\frac{1}{R_1 C_1} + \frac{1}{R_2 C_2} + \frac{1}{R_2 C_1} + \frac{1}{R_3 C_2}\right] \tag{2.53}$$

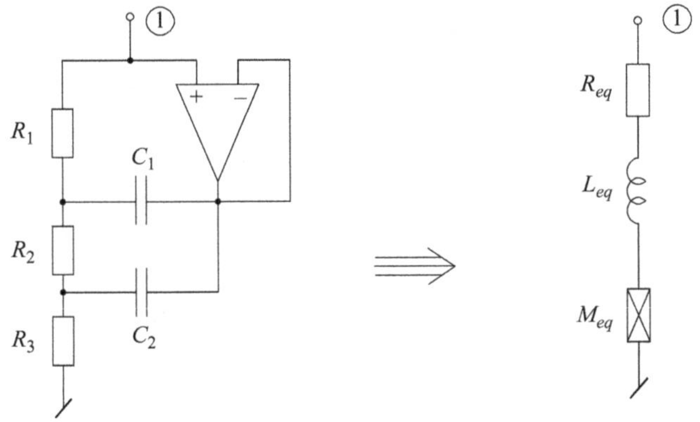

Figure 2.29 Nagarajan–Dutta Roy–Choudhary circuit [41]

whereas the frequency of oscillation is given by

$$f = \frac{1}{2\pi}\left[\frac{1}{C_1C_2}\left(\frac{1}{R_1R_2} + \frac{1}{R_2R_3} + \frac{1}{R_1R_3}\right)\right]^{\frac{1}{2}} \tag{2.54}$$

This is one of the two possible sinusoidal oscillators, using a single-op-amp-based UGA, derivable from a parallel LC tank circuit using the network transformations proposed by Senani in [176].

2.5.11 Senani's single-resistance-tunable GIs

From the circuits described earlier, it is seen that although it is possible to realize a lossless series *RL* type and parallel *RL* type grounded inductors using only a single op-amp, in none of the circuits described so far the inductance value can be varied through a single variable resistance. Two circuits to achieve this feat were proposed for the first time by Senani in [175]. The circuits from [175] are presented in Figure 2.30 which were systematically derived through an analytical procedure in [175].[4]

The first circuit realizes series *RL* impedance given by

$$Z_{in}(s) = \left(R_1 + R_2 + \frac{R_1R_2}{R_3}\right) + sC_0R_1R_2\left(1 + \frac{R_4}{R_3}\right) \tag{2.55}$$

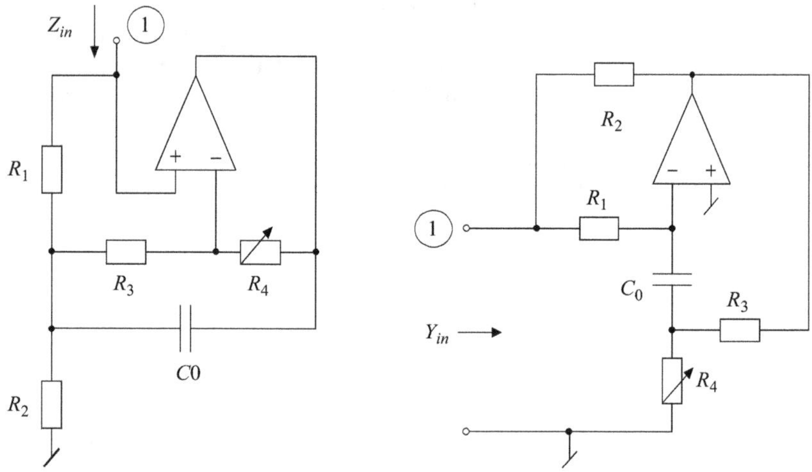

Figure 2.30 Senani's single-resistance-tunable GIs [175]: (a) grounded series RL circuit and (b) grounded parallel RL circuit

[4]Senani R. 'Realisation of a Class of Active Networks' (Ph.D. thesis). 1987; University of Allahabad [190]; later published as: Senani, R. 'Novel Analog Circuits using Op-amps, OTAs and Current Conveyors', Lambert Academic Publishing, Germany, 2014

from where it is seen that the inductance value is single-resistance-controllable through a single variable resistance R_4. On the other hand, the circuit of Figure 2.30(b) realizes a parallel RL admittance given by

$$Y_{in}(s) = \left(\frac{1}{R_1} + \frac{1}{R_2} + \frac{R_3}{R_1 R_2}\right) + \frac{(1 + (R_3/R_4))}{sC_0 R_1 R_2} \tag{2.56}$$

In this case also, the inductance value is single-resistance-controllable through a single variable resistance R_4.

2.6 Lossless floating impedance simulators using four op-amps

The problem of synthesizing a lossless floating inductance has been considered to be relatively difficult than that of its grounded counterpart. During 1967–90, a number of op-amp-based FI circuits were developed which either require a large number of amplifiers or else if the number of amplifiers could be reduced, then the number of resistors employed was large and the circuit design involved either requirement of identical resistors or certain design constraints to be fulfilled in terms of the values of these resistors or both. In this section, we will give an account of some significant contributions made in this area.

2.6.1 Riordan's method of creating a lossless FI

It is interesting to observe that the earliest solution to the problem of realizing lossless FI was presented by none other than Riordan himself [10] which is shown in Figure 2.31. This circuit can be seen to be consisting of two identical active RC blocks connected back-to-back, each derived from the two-op-amp Riordan

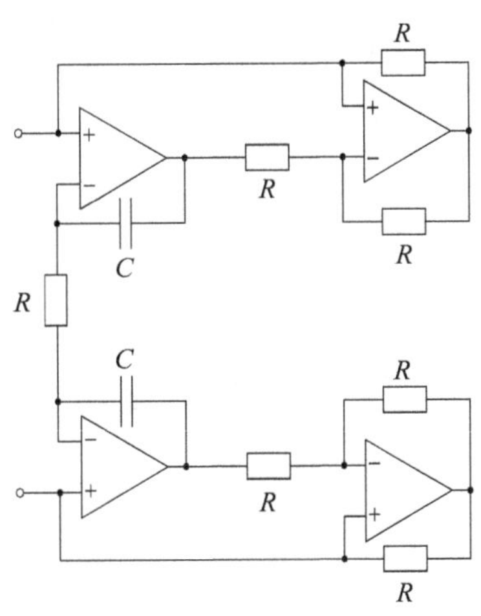

Figure 2.31 The floating inductance realizable from two Riordan gyrators [10]

gyrator. By straightforward analysis, it may be found that this two-port network simulates a floating impedance of value $Z_{eq}(s) = sC_0R^2$.

2.6.2 GIC method of simulating FI

It was subsequently realized that two GICs connected back-to-back with an impedance embedded in between them can be used to realize a more general floating impedance from which a variety of floating impedances could be realized as special cases by appropriate (resistive/capacitive) choice of various impedances as per the general scheme shown in Figure 2.32. In this scheme, although any of the known forms of GICs can be employed, because of the established superiority of Antoniou's GIC [29,171] over the other alternatives, the same is shown in Figure 2.32.

The methodology employed in formulating this configuration can be seen to be exactly the same as the one employed in the floating version of Riordan gyrator. By a straightforward analysis, the transmission matrix of this circuit can be easily written by considering the circuit to be a cascade of three blocks, namely a GIC, impedance Z_5 and another GIC (with its ports reversed), as follows:

$$[T] = \begin{bmatrix} 1 & 0 \\ 0 & \dfrac{Z_2Z_4}{Z_1Z_3} \end{bmatrix} \begin{bmatrix} 1 & Z_5 \\ 0 & 1 \end{bmatrix} \begin{bmatrix} 1 & 0 \\ 0 & \dfrac{Z_1Z_3}{Z_2Z_4} \end{bmatrix} = \begin{bmatrix} 1 & \dfrac{Z_1Z_2Z_3}{Z_2Z_4} \\ 0 & 1 \end{bmatrix} \tag{2.57}$$

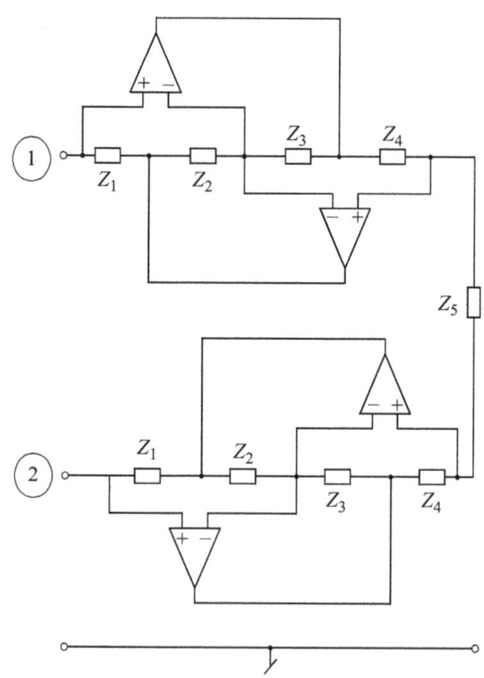

Figure 2.32 Simulating floating impedance using two identical GICs

The above matrix implies that the circuit simulates floating impedance $Z_{eq} = Z_1Z_3Z_5/Z_2Z_4$ between ports 1 and 2. It is interesting to point out that in the event of the two identical set of passive components are mismatched, the above transmission matrix would get modified as

$$[T] = \begin{bmatrix} 1 & \dfrac{Z_1Z_2Z_3}{Z_2Z_4} \\ 0 & \left(\dfrac{Z_2Z_4}{Z_1Z_3}\right)\left(\dfrac{Z_1'Z_3'}{Z_2'Z_4'}\right) \end{bmatrix} \tag{2.58}$$

It, therefore, follows that the condition required for floatation is that $(Z_2Z_4/Z_1Z_3)(Z_1'Z_3'/Z_2'Z_4') = 1$. This condition can be fulfilled in practice by setting all the components equal and keeping one resistance as a variable which should be adjusted such that the voltages at the two ports are indistinguishable.

From (2.57) it also follows that, like its grounded counterpart, from the structure of Figure 2.32, a variety of floating impedances such as floating lossless FI, floating FDNR and floating FDNC can be realized as special cases by appropriate (resistive/capacitive) choice of the various circuit elements.

2.6.3 Tripathi–Patranabis FI

The circuit of Figure 2.33 is obtained from the Tripathi–Patranabis [84] circuit for grounded inductance simulation circuit by arranging two identical active-RC blocks back-to-back with a resistor of value $2R_6$ embedded in between. The overall transmission matrix of the circuit is found to be

$$[T] = \begin{bmatrix} 1 & \dfrac{2R_5\{sC_1R_2R_3R_6 + R_2R_3 - R_4R_6\}}{R_2(R_3 + R_4)} \\ 0 & 1 \end{bmatrix} \tag{2.59}$$

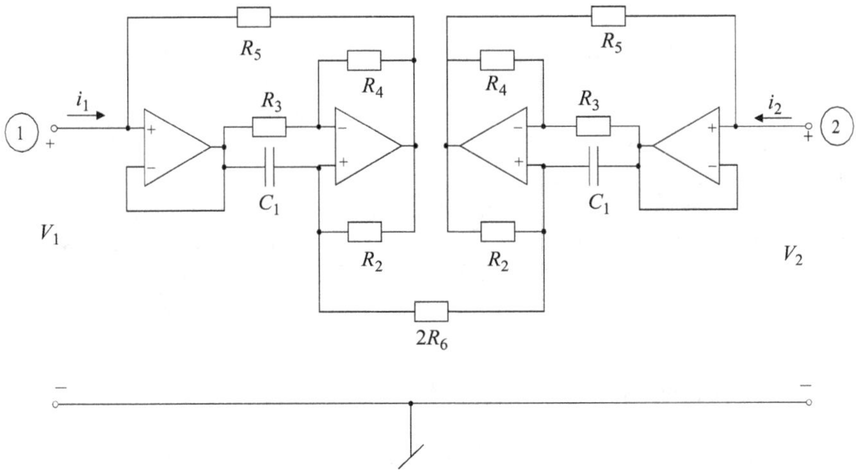

Figure 2.33 FI circuit proposed by Tripathi and Patranabis [84]

From (2.59), it is found that subject to fulfilment of the condition $R_2R_3 = R_4R_6$, the simulated inductance value is given by

$$L_{eq} = \frac{2C_1R_3R_5R_6}{(R_3 + R_4)} \tag{2.60}$$

It may be noted that as compared to the GIC-based FI, which employs 9 passive components, this one requires 11. Moreover, in contrast to the former which requires only a matching between two identical blocks, it requires a matching condition as well as another cancellation condition to realize the intended floating impedance.

2.6.4 Mutator-simulated floating inductors

In 1975–76, Murata and Rikoski [177], and Rikoski and Choudhry [178] presented methods of simulating inductors using mutators. It may be mentioned that mutators are linear active two-port elements for transforming one type of nonlinear element into another which were introduced by Chua [179]. Out of the various types of mutators, the so-called type-I *L–R* and *C–R* mutators both can be used to simulate floating inductors. Murata and Rikoski [177] did present a four-op-amp circuit comprising two-op-amp-based *C–R* mutators which are actually same as two-op-amp-based GIC. On the other hand, Rikoski and Choudhry in [178] presented an *L–R* mutator based circuit which unfortunately requires as many as four op-amps; the resulting FIs will need as many as eight op-amps and, hence, are not of much practical interest. In a subsequent chapter, it will be shown that as compared to the op-amps, the mutators are more economically realizable using other active building blocks like CCs and CFOAs. The authors of this monograph believe that the subject area of mutators has not been explored sufficiently and deserves further investigations.

2.7 The multi-port immittance converters/inverters and multi-port gyrators

In [91] Silva extended the concepts of impedance conversion and impedance inversion to the multi-port case. Thus, multi-port impedance and admittance converters and inverters were introduced as special cases of hybrid converters, and it was shown that multi-winding transformers, multi-port gyrators and most of the circuits which have been employed for simulating FIs can be shown to be special cases of multi-port converters and inverters. However, except the notion of multi-gyrators, any practical realizations of multi-port inverters and converters do not appear to have been presented explicitly ever. This appears to be a left-out area which has enough scope of new innovations.

2.8 Three-op-amp-based floating immittance simulators

Since both the GIC-based FIs as well as the so-called mutator-based FIs require at least four op-amps, with there being two identical op-amp *RC* sub-networks connected back-to-back with a resistor embedded between them, a drawback of these approaches is that both result in circuits which employ double the number of passive components as compared to their grounded counterparts and which are non-canonical in respect of capacitors (since in both the cases, two capacitors are employed to create a first-order inductive impedance).

Consequently, any method/circuit which can create a lossless FI using a single capacitor along with a reduced number of op-amps was considered to be a desirable alternative, and many researchers were in lookout for such a circuit. In the following, we describe a number of circuits which employ a two-op-amp GIC type circuit and which are modified in such a manner that, subject to the fulfilment of some conditions, the modified circuit can simulate a lossless FI.

2.8.1 Three-op-amp-single-capacitor FIs based on GIC-type networks

In the literature, there are three distinctly different approaches presented by three different groups of researchers from three different institutions, namely Reddy [70], Rathore and Singhi [114] and Grunigen, Ramseir and Moschytz (GRM) [135] who proposed three-op-amp-based FIs employing a single capacitor which are shown in Figure 2.34. All the three teams of authors quoted above devised quite different looking methods to arrive at these circuits; however, Senani in [144] demonstrated, through nullor models, that the GRM circuit is a nullor equivalent of one of Reddy's circuit, whereas the Rathore and Singhi circuit as well as Reddy's second circuit can be shown to be derivable from an alternative form of this nullor model. In fact, Senani [144] demonstrated that in all 18 such three-op-amp FIs are possible, out of which only 4 were known earlier, and consequently, the 14 remaining circuits identified in [144] were all new. Evaluation of all these 18 circuits to find out the best realization (s) has, however, never been attempted in the literature so far and, thus, appears to be a worthwhile problem for further investigation. Analysis reveals that the [*Y*]-matrix of all these 18 circuits turns out to be of the following form:

$$[Y] = \begin{bmatrix} \dfrac{Z_2Z_4}{Z_1Z_3Z_5} & -\dfrac{Z_2Z_4}{Z_1Z_3Z_5} \\ \left\{\left(\dfrac{Z_0}{Z_6Z_7} - \dfrac{1}{Z_5}\right) - \dfrac{Z_0Z_2Z_4}{Z_3Z_5Z_6Z_7}\right\} & \left\{\left(-\dfrac{Z_0}{Z_6Z_7} + \dfrac{1}{Z_5}\right) + \dfrac{Z_0Z_2Z_4}{Z_3Z_5Z_6Z_7}\right\} \end{bmatrix} \tag{2.61}$$

so that for floating impedance realization, one needs to fulfil the following two conditions:

$$\left(\frac{Z_0}{Z_6Z_7} - \frac{1}{Z_5}\right) = 0 \quad \text{and} \quad \frac{Z_0}{Z_6Z_7} = \frac{1}{Z_1} \tag{2.62}$$

Thus, with $Z_5 = Z_6 = Z_1$ and $Z_7 = Z_0$, the circuit realizes a floating impedance

$$Z_f = \frac{Z_1^2 Z_3}{Z_2 Z_4} \tag{2.63}$$

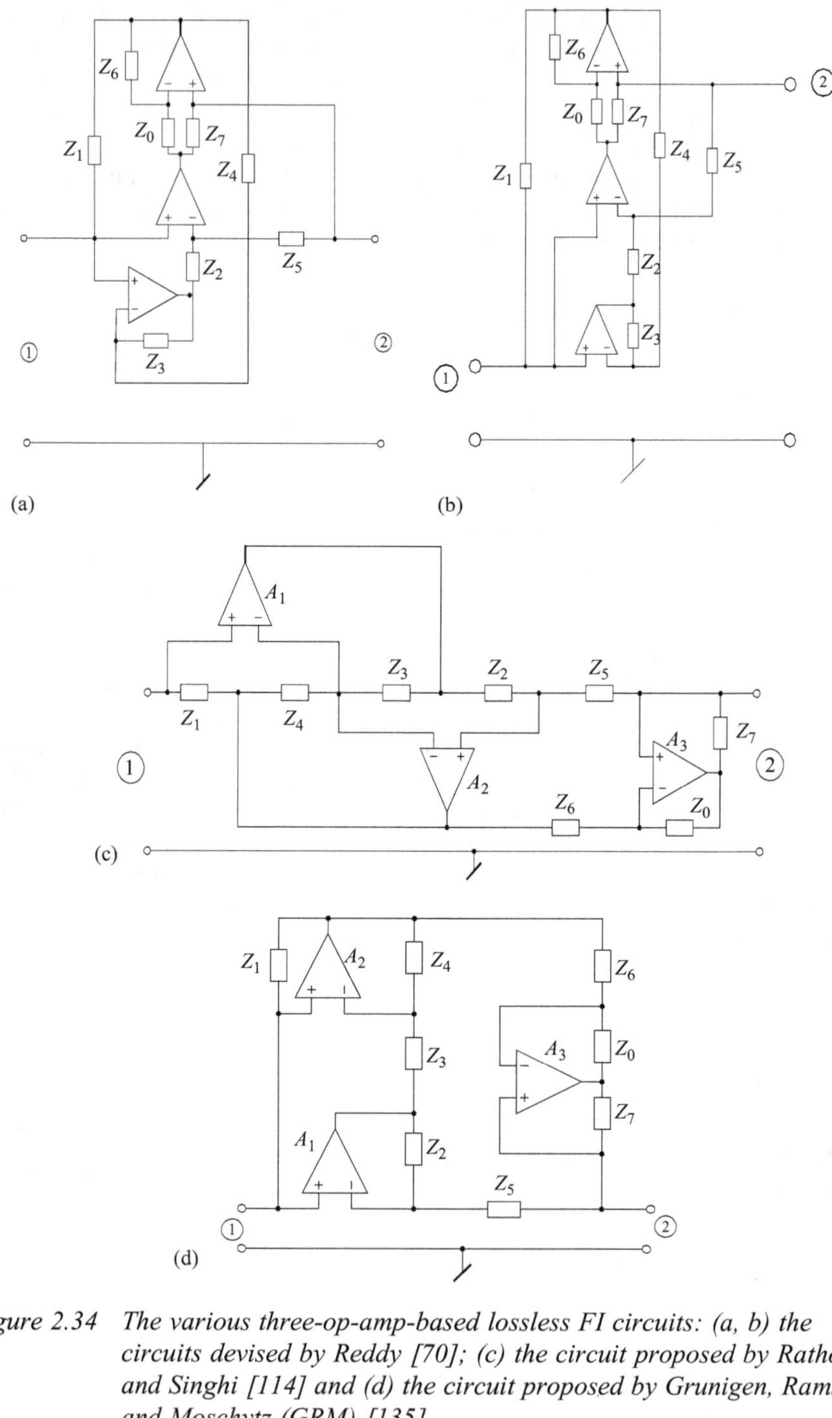

Figure 2.34 The various three-op-amp-based lossless FI circuits: (a, b) the circuits devised by Reddy [70]; (c) the circuit proposed by Rathore and Singhi [114] and (d) the circuit proposed by Grunigen, Ramseir and Moschytz (GRM) [135]

The practical viability of at least 3 of these 18 circuits has been demonstrated in [70,135].

2.8.2 FI realizations using three op-amps along with a grounded capacitor

The three-op-amp FIs discussed in the previous section although employ a single capacitor, it is floating. Since a GC is usually considered to be more desirable from the point of view of IC implementation, it was thought that a circuit which could realize a lossless FI using three op-amps and a GC would, therefore, be more suitable from the point of view of IC implementation. The earliest attempt in this direction was made by Deboo [9]. The Deboo's circuit, however, suffers from the drawback of having non-unity resistance spread. This difficulty was overcome by the circuit proposed by Glover [76] which is shown in Figure 2.35. This circuit employs three op-amps – all configured as NICs having parameters $A = 1, D = -1$ with $B = 0 = C$. A close look at this circuit reveals that it can be considered to be a cascade of seven sections. Consequently, its transmission matrix can be found by the multiplication of the transmission matrices of the seven sub-sections as follows:

$$[T] = \begin{bmatrix} 1 & 0 \\ \dfrac{1}{R_1} & 1 \end{bmatrix} \begin{bmatrix} 1 & 0 \\ 0 & -1 \end{bmatrix} \begin{bmatrix} 1 & R_2 \\ 0 & 1 \end{bmatrix} \begin{bmatrix} 1 & 0 \\ sC & 1 \end{bmatrix} \begin{bmatrix} 1 & 0 \\ 0 & -1 \end{bmatrix} \begin{bmatrix} 1 & R_3 \\ 0 & 1 \end{bmatrix} \begin{bmatrix} 1 & 0 \\ -\dfrac{1}{R_4} & 1 \end{bmatrix} \tag{2.64}$$

In the above equation if $R_1 = R_2 = R_3 = R_4 = R$, the transmission matrix reduces to

$$[T] = \begin{bmatrix} 1 & sCR^2 \\ 0 & 1 \end{bmatrix} \tag{2.65}$$

which represents an FI of value $L_f = CR^2$.

However, in spite of the advantage of having unity resistance spread, this circuit too is not devoid of problems; due to the employment of three NICs, there are stringent requirements of stability. It has been shown in [76] that these

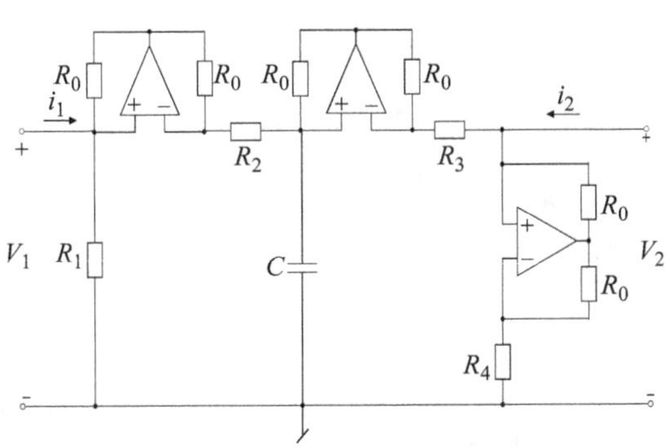

Figure 2.35 Three op-amps GC FI proposed by Glover [76]

requirements can be met only by a very judicious choice of polarities of the input terminals of the three op-amps along with very careful considerations of correct balance of impedances on the input and output ports of all the three NICs.

2.8.3 Senani's single-resistance-controllable lossless FI

Deboo's circuit [9] as well as Glover's circuit [76] which employ three op-amps and a single GC but have a common limitation that the realized floating inductance value can be varied only by varying the capacitance value which is not a very convenient option. To circumvent this difficulty, a new three-op-amp FI was introduced by Senani in [97]. Senani's three op-amp-GC FI [97] shown in Figure 2.36 overcame this difficulty and made it possible to have the realized FI value single-resistance-controllable. In this circuit, three op-amps, all configured as UGSs, along with the four associated resistors actually implement three VCCS[5]

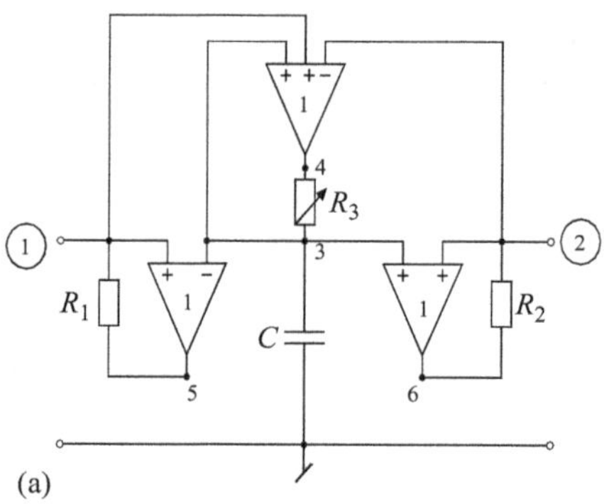

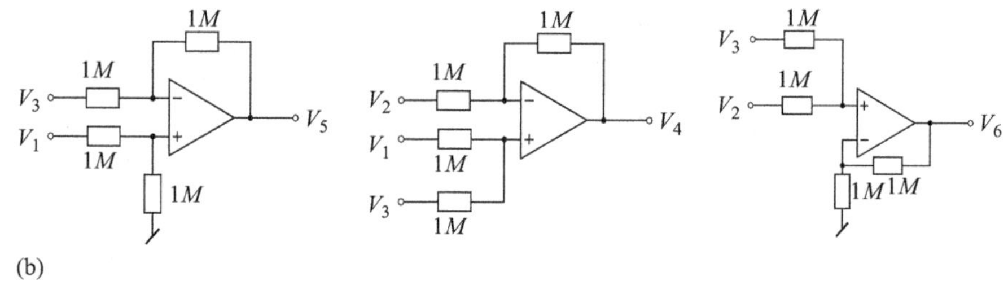

Figure 2.36 Single-resistance-controlled lossless FI employing a GC [97]: (a) the FI circuit, (b) circuits for realizing various kinds of unity gain summers: $V_5 = V_1 - V_3$; $V_4 = V_1 - V_2 + V_3$; $V_6 = V_2 + V_3$

[5]This circuit was actually derived from a three-OTA-based three-port gyrator circuit by replacing each OTA by an op-amp based unity gain summer and a resistor, performing the same functions (R. Senani, 'Floating inductance simulation using operational-transconductance-amplifiers'. Proceedings of the IEEE. November 22, 1979, unpublished).

circuits, and the circuit is, in fact, a three-port gyrator with ports 1 and 2 making the two ports of the FI and port-3 being the terminals where the GC is connected. The detailed circuit implementations of the three UGS are shown in Figure 2.36.

A straightforward analysis reveals that with the capacitor C deleted, the resulting three-port (assuming that all the UGSs are realized using high-valued resistors so that current drawn by their various input terminals is negligibly small) has the following characterization:

$$I_1 = \frac{\{V_1 - (V_1 - V_3)\}}{R_1}, \quad I_2 = \frac{\{V_2 - (V_2 + V_3)\}}{R_2}, \quad I_3 = \frac{(V_3 - V_4)}{R_3};$$
$$\text{where } V_4 = V_1 - V_2 + V_3 \tag{2.66}$$

Solving these equations, the $[Y]$-matrix of the circuit is found to be

$$[Y] = \begin{bmatrix} 0 & 0 & \frac{1}{R_1} \\ 0 & 0 & -\frac{1}{R_2} \\ -\frac{1}{R_3} & \frac{1}{R_3} & 0 \end{bmatrix} \tag{2.67}$$

Therefore, the circuit represents a three-port gyrator if $R_1 = R_2 = R$. Furthermore, with port-3 terminated into a capacitor C, the resulting two-port simulates a lossless FI of value $L_f = CRR_3$ which is single-resistance-tunable through R_3.

2.8.4 Patranabis–Paul capacitance floatation circuit

Using a somewhat similar approach, a capacitance floatation circuit was proposed by Patranabis and Paul [103], which is shown in Figure 2.37.

By an analysis similar to the one carried out in the previous section, it can be shown that this circuit realizes a lossless floating capacitance of value $C_{eq} = C_0R_1/R_2$ which is single-resistance-controllable through R_1.

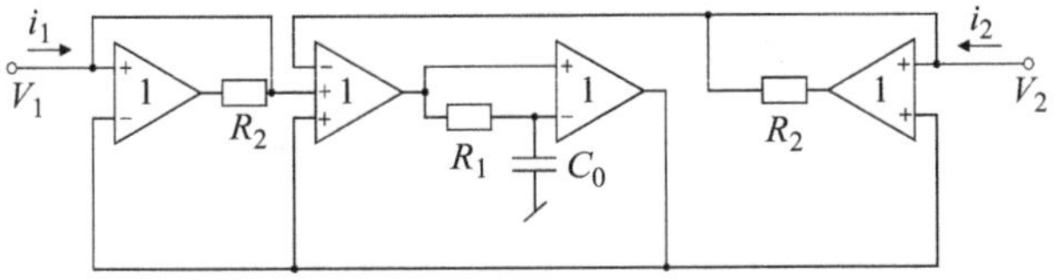

Figure 2.37 Capacitance floatation circuit proposed by Patranabis and Paul [103]

2.9 Lossless FIs using only two op-amps

As described in the previous few sub-sections, it was amply demonstrated by several researchers that three op-amps are needed to create a lossless FI, and that in some cases, it becomes possible to have the capacitor grounded too. Thus, the problem to realize a lossless FI using only two op-amps but no more than a single capacitor was open and waiting to be resolved. In this section, we present two such circuits which accomplish the desired objective using only two op-amps.

2.9.1 The–Yanagisawa circuit

The and Yanagisawa [81] were the first to demonstrate that it is, indeed, possible, and they presented two circuits each employing two op-amps along with a single capacitor to simulate a lossless FI which are shown in Figure 2.38(a) and (b). The analysis of both the circuits shows that they can realize lossless FIs subject to the fulfilment of the required relations between the various resistance values.

For the circuit of Figure 2.38(a), the following conditions are to be met to realize a lossless FI:

$$h = k = \frac{1}{2}; \quad R_1 = R_4 = R; \quad R_2 = \frac{R}{2}; \\ R_3 = \frac{R}{4}; \quad R_5 = \frac{6R}{11}; \quad R_6 = \frac{R}{3} \text{ and } R_7 = \frac{3R}{2} \tag{2.68}$$

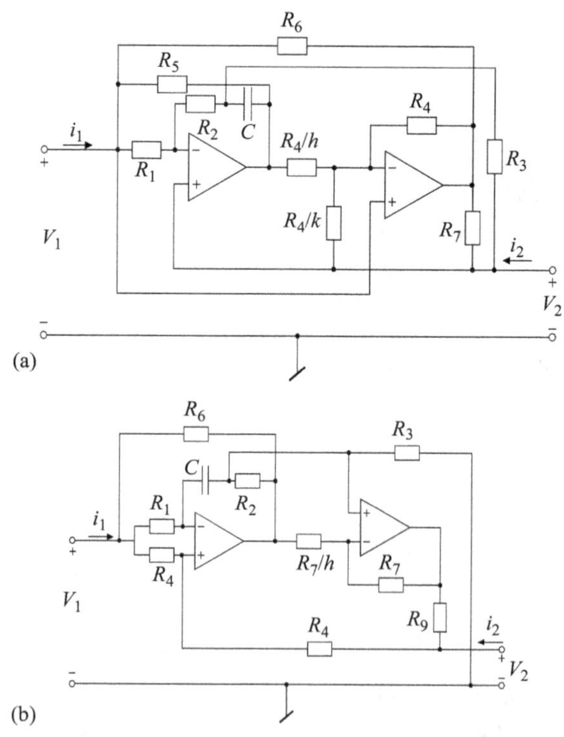

Figure 2.38 Two-op-amp FIs proposed by The and Yanagisawa [81] (a) the first proposition (b) the second proposition)

The simulated FI value is then found to be

$$L_{eq} = CR^2 \tag{2.69}$$

On the other hand, in the circuit of Figure 2.38(b), the required relation between various resistor values is

$$\begin{aligned} &R_1 = R_2 = R_4 = 2R; \quad R_3 = \frac{2R}{3}; \\ &R_6 = R_7 = R; \quad R_9 = \frac{R}{5}; \end{aligned} \tag{2.70}$$

$$m = 1 \text{ and } h = \frac{7}{10} \tag{2.71}$$

while the simulated inductance has a value $L = CR_2$

It may be mentioned that both the circuits have the theoretical novelty of realizing a lossless FI using two op-amps and a single capacitor, but the price to be paid is the fulfilment of several conditions/constraints in terms of the resistors employed in these circuits.

2.9.2 Sudo–Teramoto circuit

At about the same time, Sudo and Teramoto [78] also proposed a two-op-amp circuit which does not have the novelty of employing a single capacitor but employs a much smaller number of resistors (only three) as compared to The and Yanagisawa's [81] circuit (which needs as many as nine resistors). Moreover, in Sudo and Teramoto's [78] circuit (shown in Figure 2.39), it is possible to have all

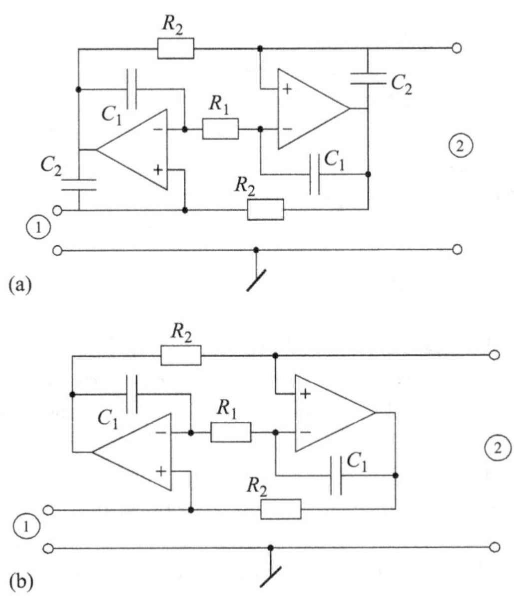

Figure 2.39 FI circuits proposed by Sudo and Teramoto [78]: (a) lossless FI and (b) single resistance controllable parallel FI

the resistors equal as well as all the capacitors equal. The [*Y*]-matrix of this circuit is found to be

$$[Y] = \begin{bmatrix} \left\{\dfrac{1}{sC_1R_1R_2} + \dfrac{1}{R_2} - \dfrac{C_2}{C_1R_1}\right\} & \left\{-\dfrac{1}{sC_1R_1R_2} - \dfrac{1}{R_2} + \dfrac{C_2}{C_1R_1}\right\} \\ \left\{-\dfrac{1}{sC_1R_1R_2} - \dfrac{1}{R_2} + \dfrac{C_2}{C_1R_1}\right\} & \left\{\dfrac{1}{sC_1R_1R_2} + \dfrac{1}{R_2} - \dfrac{C_2}{C_1R_1}\right\} \end{bmatrix} \tag{2.72}$$

Thus, with $C_1 = C_2 = C$ and $R_1 = R_2 = R$, it is found that the [*Y*]-matrix reduces to

$$[Y] = \begin{bmatrix} \dfrac{1}{sCR^2} & -\dfrac{1}{sCR^2} \\ -\dfrac{1}{sCR^2} & \dfrac{1}{sCR^2} \end{bmatrix} \tag{2.73}$$

thereby representing a lossless FI of value CR^2.

Another remarkable feature of this circuit is a specific simplified form of this circuit which is obtained by removing the two capacitors C_2 from the circuit, shown in Figure 2.39(b).

From (2.72), it may be verified that with the two capacitors named C_2 are removed, the [*Y*]-matrix of the resulting circuit reduces to

$$[Y] = \begin{bmatrix} \left\{\dfrac{1}{sC_1R_1R_2} + \dfrac{1}{R_2}\right\} & \left\{-\dfrac{1}{sC_1R_1R_2} - \dfrac{1}{R_2}\right\} \\ \left\{-\dfrac{1}{sC_1R_1R_2} - \dfrac{1}{R_2}\right\} & \left\{\dfrac{1}{sC_1R_1R_2} + \dfrac{1}{R_2}\right\} \end{bmatrix} \tag{2.74}$$

Thus, in this case, the circuit realizes a floating parallel *RL* admittance wherein the inductance value is single-resistance controllable through a variable resistance R_1.

It is interesting to point out that any two-op-amp-based circuit which can simulate floating series *RL* impedance with inductance value being similarly single-resistance controllable has not been discovered so far and, thus, constitutes an interesting problem which is open to investigation.

2.10 Economic op-amp-based lossless/lossy FIs

From the study of the literature on FI realization, it is observed that to create a lossless FI one is not only required to employ two to four op-amps along with a large number of resistors and has to compulsorily satisfy either a matching between several resistors or needs to fulfil some cancellation constraints or both. Consequently, several researchers toiled with the idea if the number of components can be reduced, and the design constraints may be simplified if instead of simulating a lossless FI one focuses on simulating a lossy FI instead. It was found that, indeed, lossy FIs could be simulated employing only two op-amps and a relatively smaller number of *RC* elements while not requiring very stringent conditions for realization. In the following section, we describe some prominent approaches to lossy FI simulations using op-amps.

2.10.1 The cascade back-to-back approach to FI realization

In 1974 Wise in [54] and Dutta Roy in [59] *independently* proposed two different looking methods for floating impedance simulation which yielded a number of new lossy FI circuits in [54] and one specific lossy FI circuit in [59]. From [54,59], undoubtedly the best circuit for series *RL* type lossy FI simulation is the one shown in Figure 2.40. It is apparent that the circuit in question is based upon the series *RL* type lossy GI circuit of Prescott [6]. Wise [54] demonstrated that by removing the grounded resistor of Prescott circuit, the resulting two-port can be considered to be a non-ideal PIC, and an FI can be then obtained by embedding a resistor between two such non-ideal PICs connected back-to-back. A straightforward analysis then yields the transmission matrix of the circuit to be

$$[T] = \begin{bmatrix} 1 & \dfrac{R_1}{sC_1R_1+1} \\ 0 & \dfrac{1}{sC_1R_1+1} \end{bmatrix} \begin{bmatrix} 1 & R_2 \\ 0 & 1 \end{bmatrix} \begin{bmatrix} 1 & R_1' \\ 0 & sC_1'R_1'+1 \end{bmatrix}$$

$$= \begin{bmatrix} 1 & \left(R_1 + R_1' + R_2 + sC_1'R_1'R_2\right) \\ 1 & \dfrac{\left(sC_1'R_1'+1\right)}{\left(sC_1R_1+1\right)} \end{bmatrix} \tag{2.75}$$

Thus, with $C_1R_1 = C_1'R_1'$ the transmission matrix reduces to

$$[T] = \begin{bmatrix} 1 & \left(R_1 + R_1' + R_2 + sC_1'R_1'R_2\right) \\ 1 & 1 \end{bmatrix} \tag{2.76}$$

which represents a series *RL* type lossy grounded inductor with $R_{eq} = (R_1 + R_1' + R_2)$ and $L_{eq} = sC_1R_1R_2$. It may be observed that in this case also, like GIC-based FIs, only a single condition is needed for achieving *floatation* which is good for post-design adjustment of the circuit.

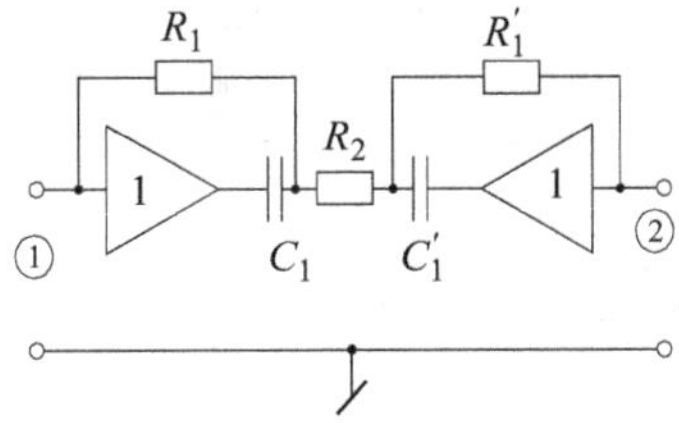

Figure 2.40 Simplified representation of a lossy FI (series RL) proposed by Wise [54] and Dutta Roy [59]

An alternative version of the circuit of Figure 2.40 is also possible (yielding exactly the same expressions) if the resistor R_1 is connected to the junction of $R_2 - C_1'$ and R_1' is connected to the junction of R_2–C_1. For further details of this *cross-coupled* version of the circuit of Figure 2.40, as well as the variants of many of the circuits using VFs which are also realizable with CFs including a lossless/lossy inductor and GICs realizable from the employment of CFs and VFs both, the reader is referred to [206] and Chapter 8 of this monograph.

Application: The classical methods of active filter design based upon impedance simulation approach require lossless inductance simulation and ideal FDNR simulation circuits and as such, lossy grounded/floating inductors/FDNRs do not find a place in these methods. In [176] however, Senani proposed four network transformations which make it possible to incorporate even the non-ideal immittance simulation networks directly into the design of active filters and oscillators. For the detailed applications of these transformations, the reader is referred to [176].

In one such transformation referred to as $T - 4$ in [176] all impedances of the *RLC* proto-type filter are multiplied by a frequency-dependent scaling factor $F(s) = (1 + s)/s$ which transforms a resistor into a series *RC*, an inductor into a series *RL* and a capacitor into a series CD branch. When the floating series *RL* branches are realized by the circuit of Figure 2.40 (or its cross-coupled version) and series CD branches (coming in shunt arms) are realized by *RC*:CR transformed version of the Prescott circuit [6] (see Figure 2.20), the resulting op-amp *RC* simulated ladder network is realizable with all unity gain amplifiers. Furthermore, the number of these unity gain amplifiers is reducible by invoking the notion of *amplifier sharing* [206]. Since unity gain connected op-amp offers the maximum closed-loop bandwidth, the resulting filters are capable of operating at relatively higher frequency ranges as compared to other op-amp *RC* filters.

2.10.2 Parallel back-to-back approach to FI realization

In retrospection, the Riordan-gyrator-based FI, the GIC-based FI simulation methods, the mutator-simulated FIs, the method proposed by Wise [54] and also that of Dutta Roy [59] for lossy FI simulation all can be regarded to belong to a common approach which can be called 'cascade back-to-back' approach. In an analogous manner, a 'parallel back-to-back approach'[6] to FI simulation is also feasible (although the class of circuits on which it can be applied may not be as large). One specific single-op-amp circuit on which it is applicable is the Ford and Girling [4] circuit described earlier in this chapter. It can be readily worked out that if the Ford–Girling [4] circuit is converted into a two-port network by un-grounding

[6]This approach has been formally published in 1991 [146]. However, the claim made therein that "this method is new and to the best of authors knowledge is not available in the literature" is not correct. The first author of this monograph would like to bring on record that this approach was investigated earlier in 1980 (i.e. more than a decade before the publication of [146]) in M. K. Jain "A study of active-RC floating inductance simulators employing operational amplifiers". M. Tech. thesis, July 1980 (M.N.R. Engineering College, Allahabad, India) supervised by the first author (RS) of this monograph.

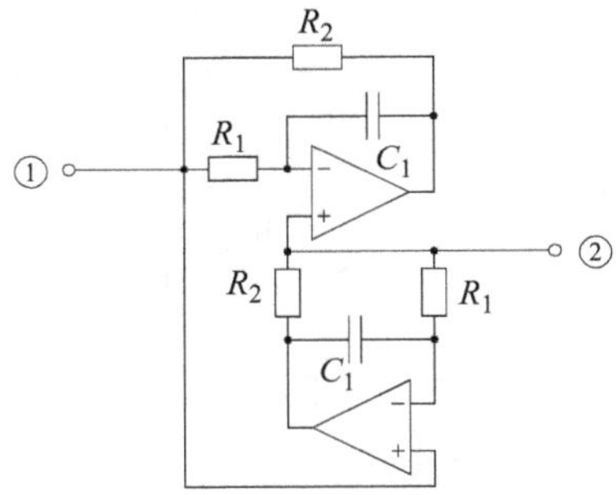

Figure 2.41 Floating parallel RL realization

the non-inverting input terminal of the op-amp used therein, the resulting circuit will have the following [Y]-matrix:

$$[Y_a] = \begin{bmatrix} \left(\frac{1}{R_1}+\frac{1}{R_2}+\frac{1}{sC_1R_1R_2}\right) & -\left(\frac{1}{R_1}+\frac{1}{R_2}+\frac{1}{sC_1R_1R_2}\right) \\ 0 & 0 \end{bmatrix} \tag{2.77}$$

On the other hand, if the ports of the same network are reversed, the modified circuit will have [Y]-matrix:

$$[Y_b] = \begin{bmatrix} 0 & 0 \\ -\left(\frac{1}{R_1}+\frac{1}{R_2}+\frac{1}{sC_1R_1R_2}\right) & \left(\frac{1}{R_1}+\frac{1}{R_2}+\frac{1}{sC_1R_1R_2}\right) \end{bmatrix} \tag{2.78}$$

Now, if the original two-port network and the modified one, as described above, are connected in parallel, the final circuit would take the form shown in Figure 2.41.

Since the circuit of Figure 2.41 has been obtained by a parallel connection of two identical back-to-back arranged sub-networks, the [Y]-matrix of the circuit would be $[Y] = [Y_a] + [Y_b]$, i.e.

$$[Y] = \left(\frac{1}{R_1}+\frac{1}{R_2}+\frac{1}{sC_1R_1R_2}\right)\begin{bmatrix} 1 & -1 \\ -1 & 1 \end{bmatrix} \tag{2.79}$$

which represents a parallel-*RL* type FI.

Another circuit on which this approach can be applied is the modified version of Ford–Girling [4] circuit derived by Rao–Venkateswaran [30] (described earlier); however, the resulting circuit would have two drawbacks: (i) it would employ four op-amps and (ii) the single-resistance-tunability prevalent in the grounded version would be lost in the floating version.

2.10.3 Senani–Tiwari FI

The floating series *RL*-type inductor shown in Figure 2.42 was derived [92] on the premise that if a grounded series *RL* can be realized by a parallel connection of a resistor *R* and a negative series *RC* branch realized from a grounded NIC, a floating

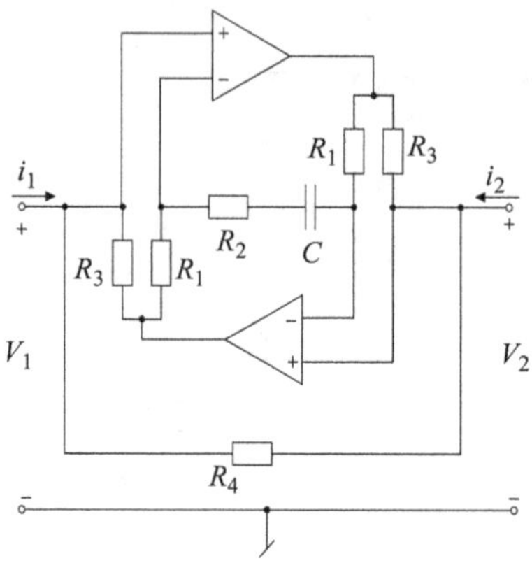

Figure 2.42 The canonic lossy FI proposed by Senani and Tiwari [92]

version should be obtainable by having a parallel connection of a resistor R and a *floating* series RC negative impedance (realized from a floating NIC) both connected between ports 1 and 2. The resulting circuit is shown in Figure 2.42.

By a straightforward analysis it can be easily found that with $R_1 = R_2 = R_3 = R_4 = R$, the circuit simulates an equivalent floating impedance $Z_{eq} = R + sCR^2$.

It is interesting to point out that Bryson and Wierzba [130] presented the notion of semi-indefinite networks and also illustrated its application in FI synthesis. The included examples, of some FIs in [130], however, employ a large number of active and passive components. The interested readers are referred to [130] for details.

On a retrospection, it is interesting to note that although a single-resistance-tunable floating lossy FI in the form of parallel-RL is known (derivable as a special case from the Sudo–Teramoto [78] lossless FI as in Figure 2.39(b)), any similar single-resistance-controllable series-RL type lossy FI using only two-op-amps is not known so far and, therefore, constitutes an interesting problem. If discovered, such a circuit would have useful applications in active simulation of passive RLC filters containing lossy FIs.

2.11 The active-*R* simulation of grounded/floating impedances and resonators

The so-called active-R approach[7] [89,94–96,99,110,116,122,129,131,134,145–148,150,151,156,162,166] to op-amp-based analog circuit design employs the

[7]In spite of the caution by Barrie Gilbert on the unreliability of the dominant pole of the internally compensated type op-amps which may restrict the usefulness of this approach [98,191–193,197], considerable amount of work was still done on this approach. It may also be pointed out that besides other things, the necessity of measuring op-amp poles, a priori, is another requirement in this approach, for any meaningful design of such circuits.

non-ideal open-loop gain of the op-amp as an integrator to devise various functional circuits using only external resistors without requiring any external capacitors. During 1975–90, this approach was widely investigated to derive a number of active-R biquads, active-R sinusoidal oscillators and a variety of impedance/resonator realizations. To complete the class of op-amp-based immittance simulators, in the following, we discuss some prominent active-R networks which realize different kinds of impedances.

We first discuss two simplest circuits for grounded impedance simulation which are shown in Figure 2.43. For the analysis of this circuit, we assume the op-amp gain to be represented by $A(s) = \omega_t/(s + \omega_p)$ where ω_p is the dominant pole of the internally compensated type op-amp and $\omega_t = A_0\omega_p$ is the gain-bandwidth-product where A_0 is the DC gain and ω_p is the pole frequency. Taking an approximation $A(s) \cong \omega_t/s$ (i.e. assuming operating frequency $\omega \ll \omega_p$) the first circuit (Figure 2.43(a)) is found to have its input admittance as $Y(s) = (1/R_0) + (\omega_t/sR_0)$, thereby representing a parallel-RL impedance. On the other hand, the second circuit (Figure 2.43(b)) has input impedance given by $Z(s) = R_0 + (R_0\omega_t/s)$ which represents grounded series-RC impedance. These simple circuits are quite useful in experimentally determining the GBP of the given op-amp, by converting them into RLC resonant circuits (by connecting an external capacitor across the first circuit and by connecting an external inductor in series with the second circuit [134]).

We now present some second-order active-R resonators realizable from two op-amps and a single resistor. Assuming $A_1(s) \cong \omega_{t1}/s$ and $A_2(s) \cong \omega_{t2}/s$ by

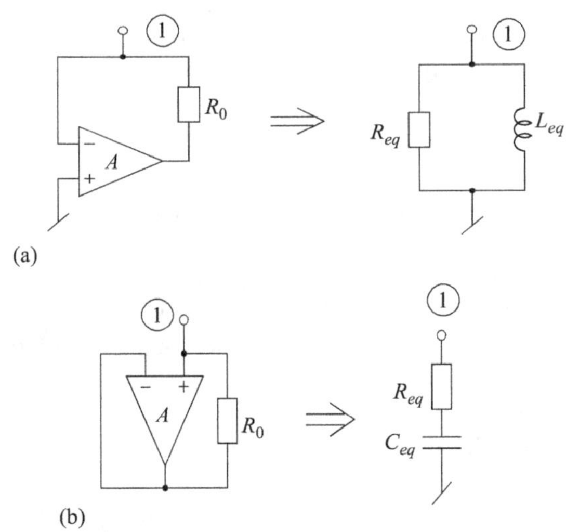

Figure 2.43 Two simple single-op-amp impedance simulators: (a) grounded parallel inductor due to Hamaoui [196] and (b) grounded series RC impedance [116]

straightforward analysis, it can be easily verified that the four circuits presented in Figure 2.44 simulate the type of impedances shown in the diagrams.

$$Y_{in}(s) = \frac{1}{R_0} + \frac{1}{s(R_0/\omega_{t2})} + \frac{1}{s^2(R_0/\omega_{t1}\omega_{t2})}$$

$$Z_{in}(s) = R_0 + \frac{1}{s(1/R_0\omega_t)} + \frac{1}{s^2\left(1/R_0\omega_t^2\right)} \qquad \text{for } \omega_{t1} = \omega_{t2} = \omega_t$$

$$Z_{in}(s) = R_0 + \frac{1}{s^2(1/R_0\omega_{t1}\omega_{t2})}$$

$$Y_{in}(s) = \frac{1}{R_0} + \left(\frac{1}{R_0\omega_{t1}\omega_{t2}}\right)s^2$$

Application: A number of authors have demonstrated the use of these second-order resonators in the realizations of filters and oscillators, for instance, the sinusoidal oscillator using op-amp compensation poles [150] can be related to the *RLM* resonator of Figure 2.44(a), on the other hand, the multifunction biquad of [161] can also be related to the same resonator with resistor function replacement by OTAs along with some other necessary modifications.

In Figure 2.44(e) we show the *RLCM* oscillator and its physical realization obtained by putting a capacitor C_0 at the front end of this resonator (Figure 2.44(a)). The characteristic equation of the oscillator (with $R_0C_0 = T$) is found to be $Ts^3 + s^2 + \omega_{t2}s + \omega_{t1}\omega_{t2} = 0$. Thus, the condition and frequency of oscillation are given by $(\omega_{t1} - (1/T)) \leq 0$ and $f_0 = \sqrt{f_{t1}f_{t2}}$. Employing 741 type op-amps with measured values of GBPs as $f_{t1} = 0.988$ MHz and $f_{t2} = 0.935$ MHz with bias voltages taken as $V_1 = \pm 15$ V and $\pm V_2$ varied from ± 2.5 V to ± 16 V have shown the frequency range to be variable from 502 kHz to 941 kHz (see Figure 2.44(f)). A typical waveform obtained from this circuit with a frequency of 781.25 kHz is shown in Figure 2.44(e).

2.12 Switched-capacitor simulated inductors

Although this monograph deals with only continuous-time analog circuits for realizing gyrators, synthetic inductors and other related impedances, it does not appear to be out of place to briefly discuss switched-capacitor simulated inductors too, even though they are basically discrete-time circuits.

It may be recalled that switched-capacitor circuits were developed because of the need for realizing fully integratable active filters and other analog signal processing circuits in CMOS technology so that analog and digital both kinds of circuits could be integrated on the same chip. Consequently, there was tremendous research activity on the evolution of a variety of switched-capacitor circuits and techniques during the period 1977–90 or so. Considerable attention was also given on switched-capacitor gyrators and simulated inductors, for which interested readers can refer to [198–205] and those cited therein. However, in this section,

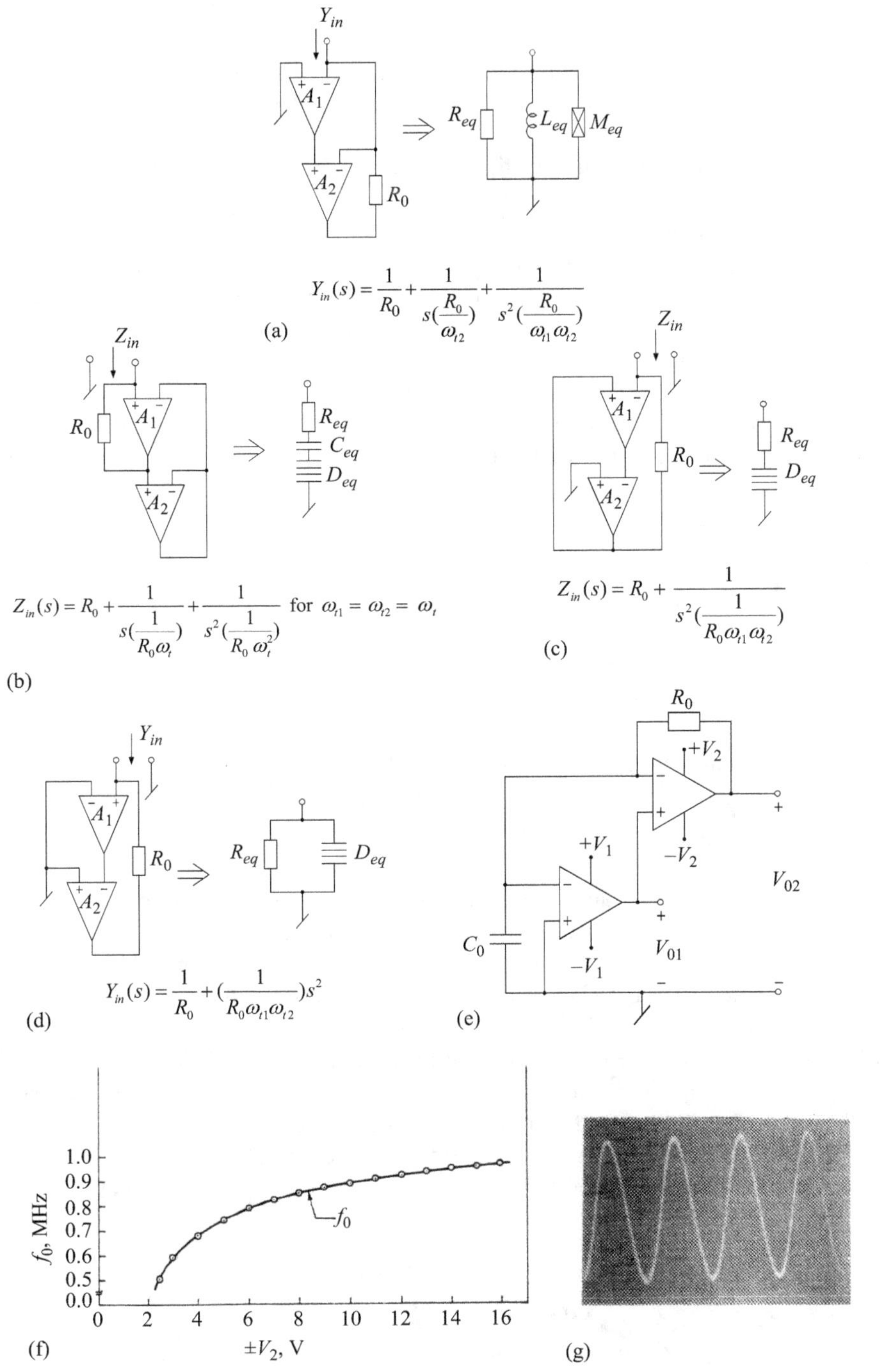

Figure 2.44 Some two-op-amp-based active-R resonators [162]: (a–d) various resonators, (e) Senani's sinusoidal oscillator using op-amp compensation poles [150], (f) variation of oscillation frequency with the bias voltage $\pm V_2$ and (g) a typical waveform of the oscillator (frequency = 781.25 kHz; amplitude = 0.116 V peak-to-peak). Reprinted, with permission from [162] © 1993 IEE

we highlight only some representative switched-capacitor inductors which are closely related to some op-amp *RC* inductance simulators which we have already elaborated in this chapter.

First, we consider an interesting single-op-amp-based inductance simulation circuit which is shown in Figure 2.45.

Identifying that a periodically switched-capacitor *C* functions as a simulated resistor of value $R = \tau/C$ where τ is the time period of the two-phase non-overlapping clock which is applied to the gates of the MOSFETs acting like switches, it is easy to see that this circuit is analogous to the Ford–Girling grounded parallel *RL* simulator [4] described earlier. The circuit, in fact, can be said to be a switched-capacitor analog of the op-amp *RC* Ford–Girling circuit.

Analysis of this circuit shows that, as expected, this circuit simulates a lossy grounded inductor (parallel *RL*) with

$$L_{eq} = \frac{\tau^2 C_2}{C_1 C_3} \tag{2.80}$$

and

$$R_{eq} = \frac{\tau C_2}{C_1 + C_3 - (C_1 C_3 / C_2)} \tag{2.81}$$

However, it is interesting to note that contrary to its active-*RC* counterpart, the switched-capacitor version can have even infinite Q provided one chooses

$$C_2 = \frac{C_1 C_3}{C_1 + C_3} \tag{2.82}$$

In this case, the circuit can be looked upon to be acting as a gyrator with the port where capacitor C_2 is connected to be port-2 of such a gyrator.

Consider now the circuit of Figure 2.46, which can be seen to be a switched-capacitor analogue of the Hamaoui inductor [196] described earlier in this chapter.

Assuming the gain of the op-amp to be represented as $A(s) = GB/(s + \omega_p)$ where $GB = A_0\omega_p$ is the gain-band width-product, ω_p is the 3-dB frequency of the open-loop gain and A_0 is the DC gain of the op-amp, from a routine analysis, it follows that Z_{in} of the circuit would be a parallel combination of a resistor $R = \tau/C$ and a series *RL* branch where equivalent resistance would be $R_s = R/A_0$ and the equivalent value of simulated inductance as $L_s = \tau/CA_0\omega_p$. Alternatively, it can be

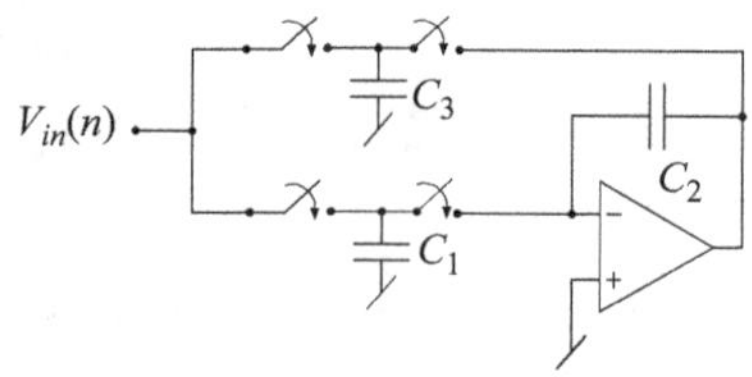

Figure 2.45 Switched-capacitor grounded inductor due to [198]. © 1978 IEE

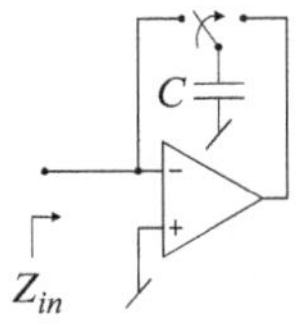

Figure 2.46 Switched-capacitor inductance simulator [199]. © 1979 IEE

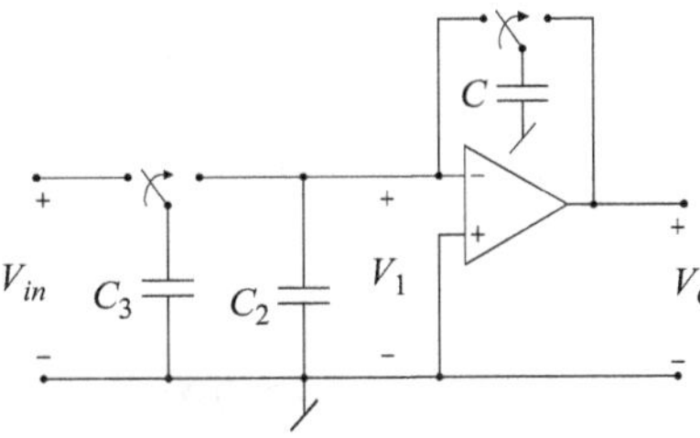

Figure 2.47 A bandpass and low pass filter [199]. © 1979 IEE

derived that Z_{in} of this circuit represents a lossy inductor with equivalent inductance given by

$$L_{eq} = \frac{\tau^2}{C_1 A_0 (C/C_0)} \tag{2.83}$$

taking $\omega_p = 1/R_1 C_0$ where R_1 and C_0 are the internal components of the op-amp. The frequency-dependent Q_{eq} is given by [199]

$$Q_{eq} = \frac{A_0 C_1}{(1 - Z^{-2}) C_0} \tag{2.84}$$

Application: Sanchez-Sinencio and Gomez-Osorio [199] demonstrated an interesting application of this inductor in realizing a low-component-count low pass/bandpass filter which was obtained by augmenting this circuit by a capacitor C_2 and another switched-capacitor C_3 as shown in Figure 2.47. This circuit realizes a BP filter at V_1 and a LP filter at V_0.

In [200], Lee presented an interesting switched-capacitor floating inductor which is shown in Figure 2.48. In this circuit, the floating voltage V_L is sensed by the capacitors C_b and C_b' which is then integrated and stored on the capacitor C_0. Capacitors C_b and C_b' are then charged to the output voltage V_0 of the op-amp. Subsequently, the two capacitors C_b and C_b' are discharged across the two voltage followers. This creates an average current at the input terminals which is proportional to the integral of the input voltage, thereby simulating an equivalent floating inductance whose value is given by

$$L_{eq} = \frac{\tau^2 C_0}{C_a C_b} \tag{2.85}$$

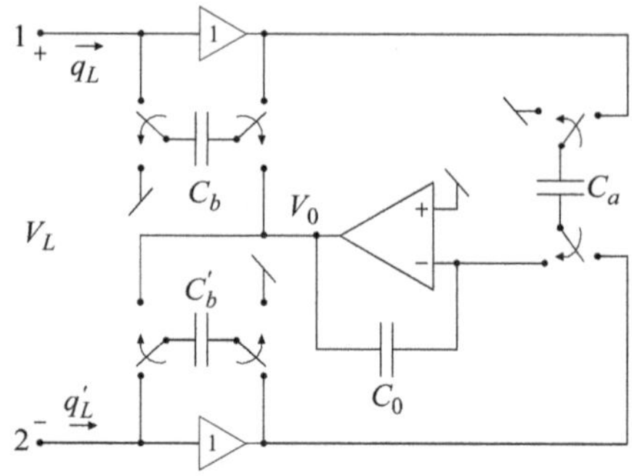

Figure 2.48 Switched-capacitor floating inductance simulation due to Lee [200]. © *1979 IEE*

2.13 FI realization using four-terminal floating nullors (FTFNs)

In [139] Senani showed that a FTFN provides a novel solution to the problem of *RC*-active synthesis of floating immittances. An FTFN is symbolically shown on Figure 2.49(a) and is the most generalized abstraction of the concept of a nullor since in an FTFN none of the four of its terminals has a ground connection.

The technique presented in [139] converts a given one port *RC*-active circuit simulating an arbitrary grounded admittance $Y(s)$ into a two-port *RC* active network which would be simulating the same impedance function $Y(s)$ in floating form. The methodology proposed in [139] involves the following steps:

1. From the given *RC*-active grounded impedance simulation network obtain its nullor model.
2. Formulate a two-port network from the above-quoted nullor model by disconnecting all the elements and terminals from ground. Since no ground connection is left, this two-port will have $i_2 = -i_1$ and would realize a floating impedance.
3. A physical circuit can be obtained from the two-port of the preceding step by replacing the FTFN by a physical circuit for which it was proposed to realize the same by an op-amp-OTA-based implementation of Figure 2.49(b).

As an example, if the given active-*RC* inductor be the circuit of Figure 2.50(a) [175] which simulates a single-resistance-controllable grounded series-*RL* impedance whose impedance is

$$Z(s) = \left(R_1 + R_2 + \frac{R_1R_2}{R_3}\right) + sCR_1R_2\left(1 + \frac{R_4}{R_3}\right) \tag{2.86}$$

then replacing the op-amp by its nullor model, the nullor model of the entire circuit will be as shown in Figure 2.50(b). Now the two-port network obtained from this nullor model would be as shown in Figure 2.50(c). Finally, the physical realization of the intended FI circuit would be as shown in Figure 2.50(d).

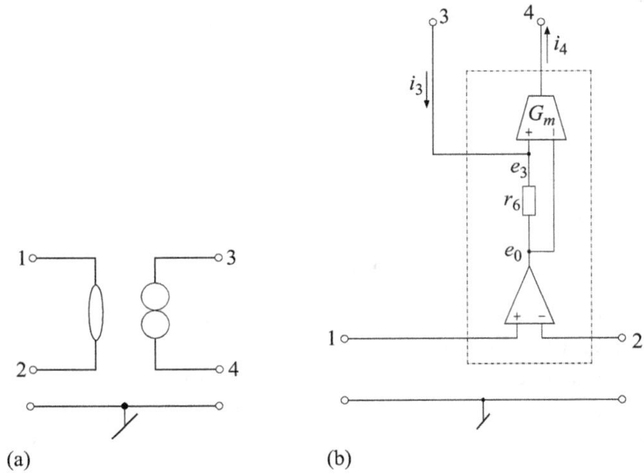

Figure 2.49 The FTFN: (a) symbolic representation of an FTFN and (b) FTFN implementations using an op-amp and OTA

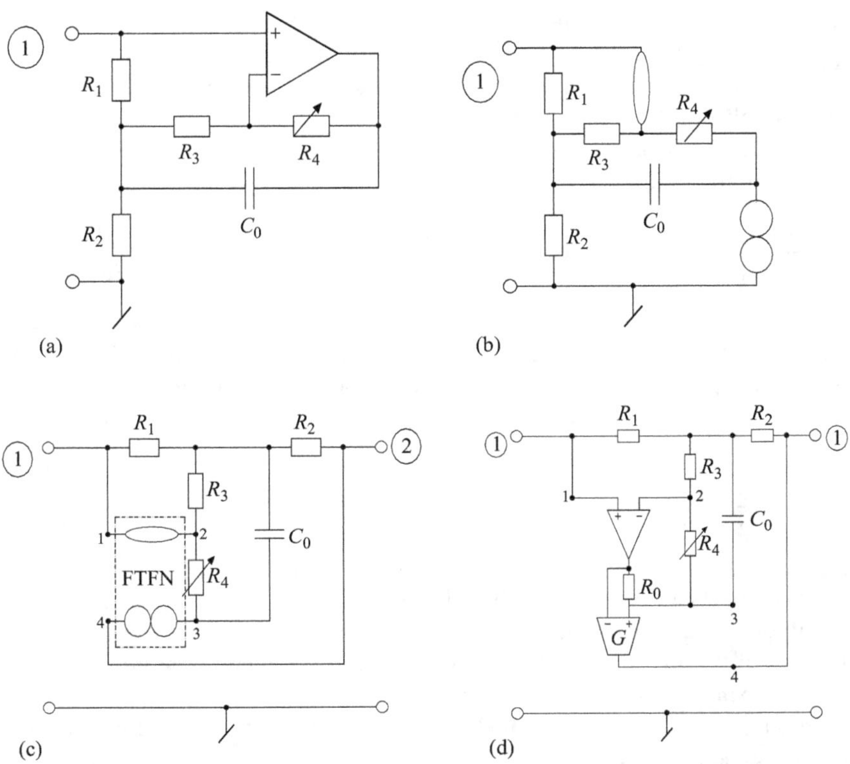

Figure 2.50 Procedure of converting grounding impedance into a floating impedance using FTFN [175]: (a) an exemplary grounded inductance network, (b) nullor model of the chosen network, (c) a two-port network obtained from the nullor model and (d) the FTFN-based FI circuit

The circuit of Figure 2.50(d) realizes a $[Y]$-matrix given by

$$[Y(s)] = \frac{1}{(R_1 + R_2 + (R_1R_2/R_3)) + sCR_1R_2(1 + R_4/R_3)} \begin{bmatrix} 1 & -1 \\ -1 & 1 \end{bmatrix} \quad (2.87)$$

provided $GR_0 = 1$

The novel features of this methodology are as follows:

1. It is the most general approach to FI realization applicable to any given grounded impedance simulation network.
2. It retains the basic characteristic features of the given circuit, i.e. if the original circuit is canonic (in terms of the number of capacitors), the resulting FI too retains the same feature; if the given grounded impedance circuit is single resistance tunable, the resulting FI too preserves the same feature. It may be noted that neither the cascaded back-to-back approach nor the parallel back-to-back approach can offer these advantages.

We present here another example of the usefulness of this approach as follows. Recall that among the various kinds of FIs presented so far, The and Yanagisawa [81] circuits only realize a lossless FI using two op-amps and a single capacitor. It was shown in [138] that a number of two-op-amp-based lossless FIs can be generated as follows. We first choose the single-op-amp-based lossless grounded inductance circuit which is shown in Figure 2.51(a) [138]. This circuit simulates a lossless grounded inductance given by

$$L = \frac{CR_3R_6(1 + R_2/R_1)}{(1 + (R_5/R_4))} \quad \text{provided that} \quad \frac{R_2}{R_4} = \left(\frac{R_1 + R_6}{R_5}\right) + \frac{R_1}{R_3}\left(1 + \frac{R_6}{R_5}\right) \quad (2.88)$$

We derive a corresponding lossless FI circuit using the technique described above which is as shown in Figure 2.51(b) [138] from which the final circuit can be obtained by replacing the FTFN by the two-op-amp-based implementation shown in Figure 2.51(c) [138]. This two-op-amp circuit implements an FTFN subject to the fulfilment of the following condition

$$\frac{r_6}{r_1} = \frac{r_3}{r_2} = \frac{r_5}{r_4} \quad (2.89)$$

In [138], it has been shown that from the two op-amps-based FI circuit described above as many as 32 two-op-amp-based lossless FI circuits are derivable; however, any analytical and experimental study of all these 32 circuits, to determine the best realization of this class, has not been carried out in the open literature as yet and is, therefore, open to investigation.

An observation of the published literature reveals that after the publication of this most generalized approach of FI simulation using the notion of FTFN as described here, there has been no other publication disclosing any new[8] general

[8] In retrospect, the seemingly different methods of realizing floating impedances published in [141–143] may be seen to be quite similar to the FTFN-based approach described in Section 2.13.

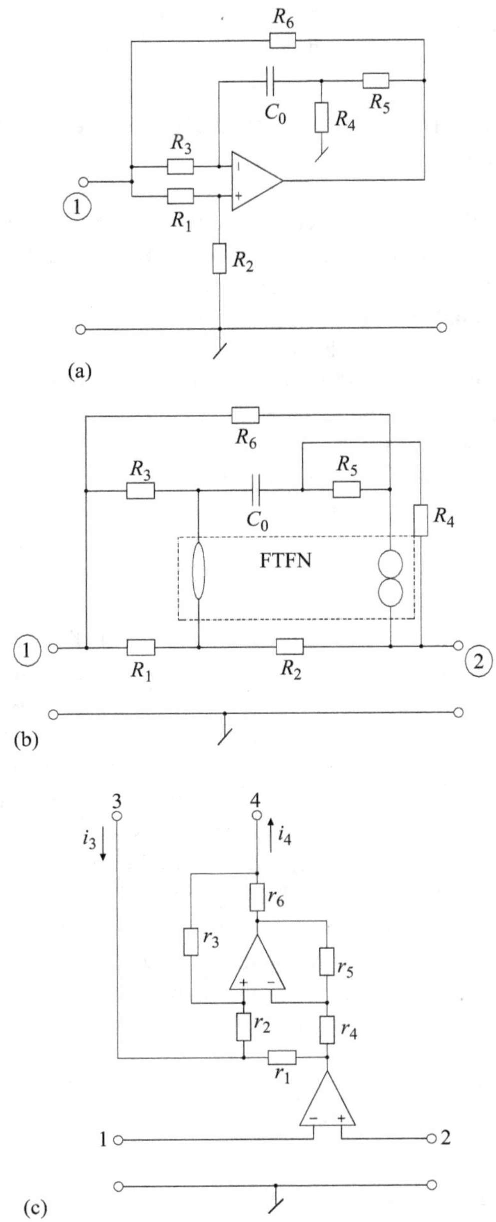

Figure 2.51 *Derivation of two-op-amp single capacitor lossless FI [138]: (a) a lossless grounded inductor, (b) an FTFN based lossless FI realizing the same function and (c) an entirely op-amp-based implementation*

method of FI simulation using op-amps! Thus, the general methods of [138,139] appear to have been the last contributions on FI simulation using IC op-amps.

2.14 Non-ideal behaviour of simulated impedances due to finite GBP of the op-amps used

A dominant non-ideal effect to be considered in op-amp *RC* circuits is the finite GBP of the op-amps. To account for this, the op-amp is usually modelled either by a one pole transfer function or by an integrator model, as explained in Section 2.11. A detailed non-ideal analysis of the GIC simulated lossless inductors and FDNRs is available even in standard textbooks, now the crux of which has been mentioned in Section 2.3 of this chapter. On the other hand, a critical study of a class of low-component count grounded inductance simulators as well as three specific FI simulators based upon cascade back-to-back approach has been carried out by Dutta Roy [69] and by Ahmed and Dutta Roy in [67,74,87].

Some general conclusions of such studies are: (i) due to the effect of the finite GBP, the non-ideal immittance function yields a frequency-dependent inductive behaviour such that the simulated inductance remains intact only up to a certain maximum frequency, depending upon in which mode the op-amp is being employed (i.e., as infinite gain op-amp, unity gain amplifier or finite gain inverting/non-inverting amplifier); (ii) on the other hand, the quality factor of the realized inductor, instead of varying linearly with frequency, generally exhibits a resonance type characteristics with Q attaining a maximum value at some specific frequency; and (iii) many inductance simulators are found to possess very high active Q sensitivities (often proportional to Q^2), thereby making them unsuitable for high Q applications.

Unfortunately, the detailed non-ideal analysis is not available for all the op-amp *RC* grounded and floating impedance simulators discussed in this chapter, and therefore, a comparative study of all the circuits cannot be made. However, as a thumb rule, most of the GI and FI circuits using op-amps are useful up to a frequency which is a small fraction[9] (typically 1%) of the GBP of the op-amp used beyond which large deviations in the performance are expected due to the finite frequency-dependent gain of the op-amps.

On the other hand, the active-*R* impedance simulators are capable of operating over much higher frequency ranges typical several hundreds of kHz even if they are realized using the same type of op-amps as considered above.

2.15 Concluding remarks

In this chapter, we discussed methods and circuits to realize gyrators, simulated inductors and related impedances in both grounded and floating forms using IC op-amps. It must be mentioned that arrival of many other active analog building blocks

[9]As an example, if the GBP of the op-amps used is 1 MHz, the usable frequency range of the simulated inductor realized with such op-amps would be around 10 kHz.

such as operational transconductance amplifiers (OTA), current conveyors (CC), current feedback op-amps (CFOA), etc. notwithstanding, the role and utility of the IC op-amp (particularly the internally compensated types) has not diminished by any means. The reason is that although the newer building blocks have caught the attention and imagination of the academia, the practising engineers are still more comfortable with the traditional and time-tested IC op-amps whose advantages and limitations are well understood and well documented.

With this background in mind, we have focused mainly on those circuits which employ the least possible number of IC op-amps to realize any given function. Thus, we have discussed a number of two-op-amps gyrator circuits. It is found that a lossless gyrator with both ports grounded although can be realized by two-op-amps, but without exception, all such circuits require matching constraints or cancellation conditions in terms of the resistors used in the realization. On the other hand, a number of impedance converter type circuits, which can also be looked upon as gyrators with appropriate selection of ports, have been evolved by several researchers which typically require only two op-amps and five circuit impedances. A remarkable advantage of these circuits is that they do not require any component-matching condition for realizing various kinds of impedances as special cases. However, these circuits can be looked upon as gyrators/impedance inverters too, but the second port of such a gyrator is floating, and therefore, these circuits cannot be used as grounded two-port gyrators. Of the six basic circuits of this kind which can be related to each other via nullor models, the Antoniou's two-op-amp GIC has been established as the best circuit of this class. Consequently, the best circuits to simulate an inductance, an FDNR and RVCs are now well recognized.

Although at least four distinctly different attempts have been made to realize a single-op-amp gyrator, none of them has both the ports grounded. In fact, it is impossible to realize a single-op-amp gyrator with both the ports grounded; hence, these four circuits lead to four different lossless grounded inductor realizations employing a floating capacitor. The penalty to be paid is the requirement of several component matching/cancellation conditions in terms of the resistors used. On the other hand, it has been well established in the literature that compared to the simulation of ideal inductor, non-ideal inductors can be simulated more economically, quite often employing only a bare minimum of only three passive components and without requiring any component matching condition. A variety of such low-component-count circuits have been discussed in this chapter.

As compared to the grounded impedances simulation, the simulation of floating impedances using op-amps has been a comparatively difficult problem. A wide variety of circuits have been presented in the literature from which the most significant ones have been discussed in this chapter. Lossless FIs using GIC-type circuits need at least four op-amps and double the number of passive components than their GI counterparts. As a result, they require close matching between the two identical subnetworks and are non-canonic in the number of capacitors employed, though, by careful design, the realization constraints can be brought down to only one which is also helpful in adjusting the circuit for floatation. On the other hand, although canonic lossless FI circuits (i.e. those circuits which use no more than a single capacitor) can be realized with only two op-amps, they require more number of realization constraints, apart from using a larger number of resistors. If the capacitor employed is to be grounded also, then at least three op-amps are needed with the realization conditions still remaining quite critical. A number of circuits belonging to the various categories mentioned above have been discussed. It was subsequently shown that several researchers have demonstrated that, as compared to the lossless FIs, the lossy ones in the form of a series *RL* impedance or parallel *RL* admittance can be usually realized with no more than two op-amps and a fewer number of resistors than needed in their ideal counterparts, but such circuits still need two capacitors. Lastly, it was shown how the four-terminal-floating-nullors (FTFN) provided a novel solution to the problem of lossless FI realization. The FTFN-based method was shown to be a very general and versatile method of FI simulation.

Besides op-amp *RC* impedance simulators, two special classes of simulators, namely the active-*R* impedance simulators and switched-capacitor impedance simulation circuits and gyrators were also considered; however, only some representative circuits had been included in this chapter. The readers interested in studying more on these topics can find a number of relevant references for further study in the list of references given at the end of this chapter, as well as those given in Appendix.

Throughout the chapter and wherever applicable, a number of unsolved problems have been pointed out which are open to investigation.

References

[1] Bogert B. P. ‘Some gyrator and impedance inverter circuits’. Proceedings of the IRE. 1955; **43**(7):793–6.

[2] Morse A. S., Huelsman L. P. ‘A gyrator realization uses operational amplifiers’. IEEE Transactions on Circuit Theory. 1964; **11**(2):277–8.

[3] Holt A. G. J., Taylor J. ‘Method of replacing ungrounded inductors by grounded gyrators’. Electronics Letters. 1965; **1**(4):105.

[4] Ford R. L., Girling F. E. J. ‘Active filters and oscillators using simulated inductors’. Electronics Letters. 1966; **2**(2):52.

[5] Orchard H. J. ‘Inductorless filters’. Electronics Letters. 1966; **2**(6):224–5.

[6] Prescott A. J. ‘Loss-compensated active gyrator using differential-input operational amplifiers’. Electronics Letters. 1966; **2**(7):283–4.

[7] Sheahan D. F. ‘Gyrator-floatation circuit’. Electronics Letters. 1967; **3**(1): 39–40.

[8] Schwarz A. F. ‘On the stability properties of a negative-immittance converter’. IEEE Transactions on Circuit Theory. 1967; **14**(1):77.

[9] Deboo G. J. ‘Application of a gyrator-type circuit to realize ungrounded inductors’. IEEE Transactions on Circuit Theory. 1967; **14**(1):101–2.

[10] Riordan R. H. S. ‘Simulated inductors using differential amplifiers’. Electronics Letters. 1967; **3**(2):50–1.

[11] Keen A. W., Peters J. L. 'Operational differentiation using simulated inductance with single-ended amplifiers'. Electronics Letters. 1967; **3**(2):61–2.

[12] Beddoes M. P., Morin K. R. 'Bibliography on inductance simulation using gyrator methods'. IEEE Transactions on Circuit Theory. 1967; **14**(1): 107–11.

[13] Keen A. W., Peters J. L. 'Inductance simulation with a single differential-input operational amplifier'. Electronics Letters. 1967; **3**(4):136–7.

[14] Antoniou A. 'Gyrators using operational amplifiers'. Electronics Letters. 1967; **3**(8):350–2.

[15] Keen A. W., Peters J. L. 'Nonreciprocal representation of the floating inductor with grounded-amplifier realisations'. Electronics Letters. 1967; **3**(8): 369–71.

[16] Gorski-Popiel J. 'RC-active synthesis using positive-immittance converters'. Electronics Letters. 1967; **3**(8):38–82.

[17] Antoniou A. 'New gyrator circuits obtained by using nullors'. Electronics Letters. 1968; **4**(5):87–8.

[18] Bialko M. 'Realizations of inductive and capacitive gyrators'. IEEE Transactions on Circuit Theory. 1968; **15**(20):158–60.

[19] Keen A. W., Glover J. L. 'Synthesis of the ungrounded inductor with grounded amplifier realizations'. The Radio and Electronic Engineer. 1968; **36**(1):53–9.

[20] Keen A. W., Glover J. L. 'Ideal-transformer and converter realisations using operational amplifiers'. Proceedings of the IEE. 1968; **115**(8):1109–14.

[21] Kalinowski J. J. 'An inductance realization using two operational amplifiers'. Proceedings of the IEEE. 1968; **56**(9):1636–7.

[22] Holt A. J., Linggard R. 'The multi-terminal gyrator'. Proceedings of the IEEE. 1968; **56**(8):1354–5.

[23] Antoniou A. 'Stability properties of some gyrator circuits'. Electronics Letters. 1968; **4**(23):510–2.

[24] McGregor I. M. 'A fully integrable gyrator'. IEEE Journal of Solid-State Circuits. 1969; **4**(1):53–4.

[25] Berndt D. F., Dutta Roy S. C. 'Inductor simulation using a single unity gain amplifier'. IEEE Journal of Solid-State Circuits. 1969; **4**(3):161–2.

[26] Bruton L. T. 'Nonideal performance of a class of positive immittance converters'. IEEE Transactions on Circuit Theory. 1969; **16**(4):572–3.

[27] Mitra S. K. 'Analysis and Synthesis of Linear Active Networks'. London: John Wiley & Sons, Inc.; 1969. p. 492–5.

[28] Bruton L. T. 'Network transfer functions using the concept of frequency-dependent negative resistance'. IEEE Transactions on Circuit Theory. 1969; **16**(3):406–8.

[29] Antoniou A. 'Realisation of gyrators using operational amplifiers, and their use in RC-active-network synthesis'. Proceedings of the IEE. 1969; **116**(11): 1838–50.

[30] Rao K. R., Venkateswaran S. 'Synthesis of inductors and gyrators with voltage-controlled voltage sources'. Electronics Letters. 1970; **6**(2):29–30.

[31] Yanagisawa T., Kanemitu I., Shyoji Y. 'Constant-resistance network using a three-port gyrator'. Proceedings of the IEEE. 1970; **58**(5):836–7.

[32] Dutta Roy S. C. 'Driving-point-function synthesis using a unity-gain amplifier'. IEEE Transactions on Circuit Theory. 1970; **17**(2):266–8.

[33] Dutta Roy S. C., Nagarajan V. 'On inductor simulation using a unity-gain amplifier'. IEEE Journal of Solid-State Circuits. 1970; **5**(3):95–8.

[34] Patranabis D. 'Inductor realisation with a negative-impedance converter'. Electronics Letters. 1970; **6**(13):415–7.

[35] Patranabis D. 'Grounded inductor simulation with a differential amplifier'. International Journal of Electronics. 1970; **28**(5):481–3.

[36] Holt A. G. J., Linggard R. L. 'Multi-terminal gyrator'. Proceedings of the IEE. 1970; **117**(8):1591–8.

[37] Moschytz G. S. 'Inductorless filters: a survey'. IEEE Spectrum. 1970; **7**(9): 63–75.

[38] Rao K. S., Murti V. G. K. 'Active RC realization of bilinear RL impedance'. Proceedings of the IEEE. 1970; **58**(11):1860–1.

[39] Bruton L. 'Nonideal performance of two-amplifier positive-impedance converters'. IEEE Transactions on Circuit Theory. 1970; **CT-17**(4):541–9.

[40] Patranabis D. 'Inductor realization with RC elements'. Proceedings of the IEEE. 1971; **59**(1):103–5.

[41] Nagarajan V., Dutta Roy S. C., Choudhury N. 'A circuit for realizing high-quality solid state inductance'. Solid-State Electronics. 1971; **14**(4):345–7.

[42] Inigo R. M. 'Gyrator realization using two operational amplifiers'. IEEE Journal of Solid-State Circuits. 1971; **6**(2):88–9.

[43] Keen A. W., Glover J. L. 'Active RC equivalents of RCL networks by similarity transformation'. Electronics Letters. 1971; **7**(11):288–90.

[44] Dutta Roy S. C. 'Inductor realization with RC elements'. Proceedings of the IEEE. 1971; **59**(9):1380–1.

[45] Patranabis D., Sen P. C. 'A simulated inductor and an RC oscillator'. International Journal of Electronics. 1971; **31**(5):441–51.

[46] Antoniou A. 'Floating negative-impedance converters'. IEEE Transactions on Circuit Theory. 1972; **19**(2):209–12.

[47] Allen P. E., Means J. A. 'Inductor simulation derived from an amplifier rolloff characteristic'. IEEE Transactions on Circuit Theory. 1972; **19**(4): 395–7.

[48] Antoniou A. 'Realisation of polynomials as driving-point admittances'. Electronics Letters. 1972; **8**(23):557–8.

[49] Patranabis D. 'A single-transistor realization of bilinear RL impedance'. Proceedings of the IEEE. 1972; **60**(8):1012–4.

[50] Patranabis D., Roy S. B. 'Active RC realization of bilinear and linear RL impedances'. International Journal of Electronics. 1972; **33**(6):681–7.

[51] Rollett J. M. 'Economical RC active lossy ladder filters'. Electronics Letters. 1973; **9**(3):70–2.

[52] Antoniou A, Naidu K. S. 'Modeling of a gyrator circuit'. IEEE Transactions on Circuit Theory. 1973; **CT-20**(5):533–40.

[53] Hilberman D. 'Input and ground as complements in active filters'. IEEE Transactions on Circuit Theory. 1973; **CT-20**(5):540–7.
[54] Wise D. R. 'Active simulation of floating lossy inductances'. Proceedings of the IEE. 1974; **121**(2):85–7.
[55] Patranabis D. 'Inductor realization with phase-inverting amplifier'. International Journal of Electronics. 1974; **36**(4):487–93.
[56] Patranabis D., Gupta D. K. 'Bilinear inductance with RC feedback in a single grounded emitter transistor'. IETE Journal of Research. 1974; **20**(5): 194–7.
[57] Patranabis D. 'A note on inductor realization with RC elements'. International Journal of Electronics. 1974; **37**(6):761–4.
[58] Dutta Roy S. C., Ahmed M. T. 'Synthesis of RC active networks using nonideal simulated inductance'. IEEE Transactions on Circuits and Systems. 1974; **21**(2):250–1.
[59] Dutta Roy S. C. 'A circuit for floating inductance simulation'. Proceedings of the IEEE. 1974; **62**(4):521–3.
[60] Bialko M., Sienko W., Newcomb R. W. 'Active synthesis using the DVCCS/DVCVS'. International Journal of Circuit Theory and Applications. 1974; **2**(1):23–38.
[61] Orchard H. J., Willson A. N. 'New active-gyrator circuit'. Electronics Letters. 1974; **10**(13):261–2.
[62] Willson A. N., Orchard H. J. 'Realization of ideal gyrators'. IEEE Transactions on Circuits and Systems. 1974; **21**(6):729–32.
[63] Molo F. 'Parallel resonator with a resistance and a frequency-dependent negative resistance realized with a single operational amplifier'. IEEE Transactions on Circuits and Systems. 1974; **21**(6):783–8.
[64] Schmidt C. E., Lee M. S. 'Multipurpose simulation network with a single amplifier'. Electronics Letters. 1975; **11**(1):9–10.
[65] Willson A. N., Temes G. C., Gadenz R. N. 'On the active realization of lossless two-ports'. IEEE Transactions on Circuits and Systems. 1975; **22**(1): 63–5.
[66] Patranabis D. 'Inductor realization with resistive feedback in RC filters'. International Journal of Electronics. 1975; **38**(3):367–74.
[67] Ahmed M. T., Dutta Roy S. C. 'Active inductance simulation with grounded condensers'. International Journal of Circuit Theory and Applications. 1975; **3**(4):371–9.
[68] Patranabis D. 'On the realization of RL impedance'. IETE Journal of Research. 1975; **21**(8):399–402.
[69] Dutta Roy S. C. 'On operational amplifier simulation of a grounded inductor'. Archiv fur Elektronik und Uebertragungstechnik. 1975; **29**:107–15.
[70] Reddy M. A. 'Some new operational-amplifier circuits for the realization of the lossless floating inductance'. IEEE Transactions on Circuits and Systems. 1976; **23**(3):171–3.
[71] Dutta Roy S. C. 'Bibliography of inductance simulation by active RC methods'. Microelectronics and Reliability. 1976; **15**:637–9.

[72] Antoniou A. 'Influence of ambient-temperature variations in two gyrator circuits'. Proceedings of the IEE. 1976; **123**(6):490–4.
[73] Dutta Roy S. C. 'Some recent contributions to inductor simulation and applications to low sensitivity biquad design'. IETE Journal of Research. 1976; **22**(7):419–25.
[74] Ahmed M. T., Dutta Roy S. C. 'Evaluating the frequency limitations of operational amplifier RC networks: the dominant-pole technique and its application to inductance simulation'. IETE Journal of Research. 1976; **22**(11):703–8.
[75] Soundararajan K. 'Nonideal performance of single amplifier simulated frequency dependent negative resistances'. IEEE Transactions on Circuits and Systems. 1976; **23**(9):573–7.
[76] Glover J. 'Lossless ungrounded inductor realisation'. Electronics Letters. 1976; **12**(25):679–81.
[77] Bhattacharyya B. B., Das S. K., Swamy M. N. S. 'A simple inductance simulation scheme and its application in data communications'. Proceedings of the IEEE. 1977; **65**(2):268–9.
[78] Sudo S., Teramoto M. 'Constitution of floating inductance using operational amplifiers'. The Transactions of the IECE of Japan. 1977; **E60**(4):185–6.
[79] Kudo M., Noguchi S., Hayashibe R. 'Solid state inductor circuit, utilizing unity gain amplifier'. 1977; **E60**(6):321–327.
[80] Sudo S., Teramoto M. 'A constituting method of floating inductance and its application in all-pass-network'. The Transactions of the IECE of Japan. 1977; **E60**(8):403–4.
[81] The L. Q., Yanagisawa T. 'Some new lossless floating inductance circuits'. Proceedings of the IEEE. 1977; **65**(7):1071–2.
[82] Yukio I., Toshio S., Ginsaku Y. 'New floating inductance simulation using two operational amplifiers and its stability'. 1977; **1977**(2):63–70.
[83] Soliman A. M., Awad S. S. 'Canonical high-selectivity parallel resonator, using a single operational amplifier, and its applications in filters'. 1977; **1**(4): 145–8.
[84] Tripathi M. P., Patranabis D. 'Grounded and floating ideal inductors with RC elements'. IETE Journal of Research. 1977; **23**(6):323–6.
[85] Soundararajan K. 'Non-ideal performance of single-amplifier simulated inductances'. International Journal of Electronics. 1977; **42**(6):601–11.
[86] Schwartz E., Schwartz H.-D. 'On active RC synthesis of immittances'. Archiv fur Elektronik und Uebertragungstechnik. 1977; **31**:53–9.
[87] Ahmed M. T., Dutta Roy S. C. 'A critical study of some non-ideal floating inductance simulators'. Archiv fur Elektronik und Uebertragungstechnik. 1977; **31**:182–8.
[88] Nandi R. 'Realization of allpass or notch filters: a simulated inductor approach'. IEE-IERE Proceedings-India. 1977; **15**(6):254–8.
[89] Soliman A. M., Fawzy M. 'Active-R resonator realization'. Proceedings of the IEEE. 1978; **66**(2):254–5.

[90] Ramsey W. T. 'Highpass and bandpass filters using a new single-amplifier simulated inductor'. Electronic Circuits and Systems. 1978; **2**(3):79–85.

[91] Silva M. M. 'Multiport converters and inverters'. International Journal of Circuit Theory and Applications. 1978; **6**(3):243–52.

[92] Senani R., Tiwari R. N. 'New canonic active RC realizations of grounded and floating inductors'. Proceedings of the IEEE. 1978; **66**(7):803–4.

[93] Soliman A. M., Awad S. S. 'A tunable active inductance using a single operational amplifier and its applications in filters'. Archiv fur Elektronik und Uebertragungstechnik. 1978; **32**:44–8.

[94] Soliman A. M. 'A novel inductor simulation using the pole of the operational amplifier'. Frequenz. 1978; **32**(8):239–40.

[95] Allen J., Gumussoy M., Holt A. G. J. 'Inductance simulation and filter design using a single-pole amplifier approximation'. Electronics Letters. 1978; **14**(19):629–31.

[96] Siddiqi Z. A., Ahmed M. T. 'Realisation of grounded capacitor with operational amplifiers and resistance'. Electronics Letters. 1978; **14**(19): 633–4.

[97] Senani R. 'Realisation of single-resistance-controlled lossless floating inductance'. Electronics Letters. 1978; **14**(25):828–9.

[98] Gilbert B. 'Simulation of inductors and capacitors using operational-amplifier compensation pole: a caution'. Electronics Letters. 1978; **14**(25):832.

[99] Nandi R. 'Active-R realization of bilinear RL impedances and their applications in a high-Q parallel resonator and external capacitorless oscillator'. Proceedings of the IEEE. 1978; **66**(12):1666–78.

[100] Horn P., Moschytz G. S. 'Active RC single-opamp design of driving-point impedances'. IEEE Transactions on Circuits and Systems. 1979; **26**(1): 22–30.

[101] Ganguly U. S. 'On inductor simulation: reactive two-pole synthesis with all-pass and related transfers'. Proceedings of the IEEE. 1979; **67**(2):319–20.

[102] Patranabis D., Paul A. N. 'Floating ideal inductor with one DVCCS'. Electronics Letters. 1979; **15**(18):545–6.

[103] Patranabis D., Paul A. N. 'Novel capacitor floatation scheme'. Electronics Letters. 1979; **15**(21):688–9.

[104] Ahmed M. T., Siddiqi M. A. 'Low-sensitivity high-frequency inductance realisation without capacitor'. Electronics Letters. 1979; **15**(9):260–1.

[105] Patranabis D., Kanjilal K., Tripathi M. P., Paul A. N. 'Generalization of ideal inductor realization schemes with resistive feedback in active RC networks'. Journal of IETE. 1979; **25**(6):233–6.

[106] Patranabis D., Tripathi M. P., Roy S. B. 'A new approach for lossless floating inductor realization'. IEEE Transactions on Circuits and Systems. 1979; **26**(10):892–3.

[107] Silva M. De. 'On the realization of immittance inverters with a minimum number of active components'. IEEE Transactions on Circuits and Systems. 1979; **26**(11):931–5.

[108] Sanchez-Sinencio E., Alonso A. A., Baez D., Gomez-Osorio J. L. 'Switched-capacitor simulation of floating inductors using the operational amplifier pole'. Proceedings of the IEEE. 1979; **67**(10):1448–9.

[109] Rathore T. S., Singhi B. M. 'Simulation of RL impedances suitable for micro-miniaturization'. Microelectronics Reliability. 1979; **19**(3):233–5.

[110] Sanchez-Sinencio E., Baez-Lopez D. 'A general circuit for inductance simulation'. 22nd Midwest Symposium on Circuits and Systems; Philadelphia, PA, June 17–19, 1979; **22**:289–93.

[111] Shiro S., Mituo T. 'Floating inductance simulation using operational amplifiers and its applications'. Proceedings of the Faculty of Engineering of Tokai University. 1980; **1980**(1):1–13.

[112] Senani R. 'Comments on: 'Floating ideal inductor with one D.V.C.C.S.' and 'Novel capacitor floatation scheme''. Electronics Letters. 1980; **16**(4):117.

[113] Senani R. 'A new technique for inductance-to-time-period conversion using integrated circuit operational amplifiers'. IEEE Transactions on Industrial Electronics and Control Instrumentation. 1980; **27**(1):36–7.

[114] Rathore T. S., Singhi B. M. 'Active RC synthesis of floating immittances'. International Journal of Circuit Theory and Applications. 1980; **8**(2):184–8.

[115] Patranabis D., Paul A. N. 'Reply to: Comments on 'Floating ideal inductor with one D.V.C.C.S.' and 'Novel capacitor floatation scheme''. Electronics Letters. 1980; **16**(4):117.

[116] Ananda Mohan P. V. 'Novel active filters using amplifier pole'. Electronics Letters. 1980; **16**(10):378–80.

[117] Rathore T. S., Singhi B. M. 'Active RC simulation of floating RL-impedance'. Journal of IETE. 1980; **26**(7):323–4.

[118] Rathore T. S., Singhi B. M. 'A family of inductance simulations'. Journal of Institution of Engineers (I). 1980; **61**(2):58–9.

[119] Rathore T. S., Singhi B. M. 'Active cascade synthesis of floating immittances'. Journal of IETE. 1980; **26**(6):277–8.

[120] Nandi R. 'Novel insensitive lossless inductor simulation through inverse function generation'. Electronics Letters. 1980; **16**(12):481–2; Singh V. K. 'Comment: Novel insensitive lossless inductor simulation through inverse function generation'. Electronics Letters. 1980; **16**(23):901.

[121] Kudo M., Hiwada K., Noguchi S. 'High-*Q* inductance circuit using an integrator and its application'. IEEE Transactions on Circuits and Systems. 1980; **CAS-27**(12):1164–8.

[122] Ananda Mohan P. V. 'Active-*R* floating lossless inductance realization'. Archiv fur Elektronik und Uebertragungstechnik. 1981; **35**:455–6.

[123] Goras L. 'Linear and nonlinear mutators derived from GIC-type configurations'. IEEE Transactions on Circuits and Systems. 1981; **28**(2):165–8.

[124] Moni R. S., Rao K. R., Reddy M. A. 'Analysis of high-quality gyrator filter'. IEEE Transactions on Circuits and Systems. 1981; **28**(4):342–5.

[125] Abdallah Y. A. 'Comments on "Active RC single-opamp design of driving-point impedances"'. IEEE Transactions on Circuits and Systems. 1981; **28**(9):939.

[126] Senani R. 'Linear resistance-to-frequency conversion employing integrated circuit operational amplifiers'. International Journal of Electronics. 1981; **50**(6):485–91.

[127] Stoyanov, G 'Comments on "New canonic active RC realizations of grounded and floating inductors"'. Proceedings of the IEEE. 1982; **70**(1): 101–3.

[128] Patranabis D., Paul A. N., Dey A. N. 'Ideal grounded with one grounded capacitor and two operational amplifiers'. Journal of the Institution of Engineers (India). 1982; **62**:63–6.

[129] Siddiqi M. A., Ahmed M. T. 'Active-R simulation of ideal grounded capacitor'. IETE Journal of Research. 1982; **28**(8):424–5.

[130] Bryson P. R., Wierzba G. M. 'Theory and application of semi-indefinite networks'. IEE Proceedings G-Electronic Circuits and Systems. 1982; **129**(6):285–9.

[131] Ahmed M. T., Siddiqi M. A., Javed M. T. 'A general synthesis technique for active-R networks'. International Journal of Electronics. 1983; **54**(3): 417–25.

[132] Patranabis D., Paul A. N., Ghosh D. K. 'A floating NIC and floating ideal inductors'. IETE Journal of Research. 1983; **29**(1):1–3.

[133] Ghausi M. 'Analog active filters'. IEEE Transactions on Circuits and Systems. 1984; **CAS-31**(1):13–31.

[134] Ishida M., Fukui Y., Ebisutani K. 'Novel active-R synthesis of a driving-point impedance'. International Journal of Electronics. 1984; **56**(1):151–8.

[135] Grunigen D. C. V., Ramseir D., Moschytz G. S. 'Simulation of floating impedances for low-frequency active filter design'. Proceedings of the IEEE. 1986; **74**(2):366–8.

[136] Osowski S. 'Novel realisation of converter-type gyrator'. Electronics Letters. 1986; **22**(8):424–5.

[137] Senani R. 'Network transformation for active-RC realisation of RLM-immittances'. Frequenz. 1986; **40**(3):67–98.

[138] Senani R. 'Generation of new two-amplifier synthetic floating inductors'. Electronics Letters. 1987; **23**(22):1202–3.

[139] Senani R. 'A novel application of four-terminal floating nullor'. Proceedings of the IEEE. 1987; **75**(11):1544–6.

[140] Singh V. On Hilberman's theorem and the characterization of ideal Op Amp semi-indefinite networks'. IEEE Transactions on Circuits and Systems. 1988; **35**(2):251.

[141] Singh V. 'New floating point impedance simulation using operational amplifiers'. IEE Proceedings G-Electronic Circuits and Systems. 1988; **135**(3):116–8.

[142] Singh V. 'Correspondence: New floating point impedance simulation using operational amplifiers'. IEE Proceedings G-Electronic Circuits and Systems. 1989; **136**(3):155.

[143] Singh V. 'On floating impedance simulation'. IEEE Transactions on Circuits and Systems. 1989; **36**(1):161–2.

[144] Senani R. 'Three op amp floating immittance simulators: a retrospection'. IEEE Transactions on Circuits and Systems. 1989; **36**(11):1463–5.

[145] Senani R. 'Realisation of linear circuits using IC op-amps: some appraisals'. IETE Journal of Education. 1990; **31**(2):61–70.

[146] Siddiqi M. A., Ahmed M. T. 'Realisation and performance assessment of active-R inductors'. International Journal of Electronics. 1991; **70**(4): 713–28.

[147] Siddiqi M. A., Ahmed M. T. 'Direct form active-R synthesis techniques and their critical assessment'. International Journal of Electronics. 1991; **71**(4):621–35.

[148] Qiu D. 'Circuit design of an integrable simulated inductor and its applications'. IEEE Transactions on Instrumentation and Measurement. 1991; **40**(6):902–7.

[149] Carlosena A., Muller D., Moschytz G. S. 'Resistively variable capacitors using general impedance converters'. IEE Proceedings G-Circuits, Devices and Systems. 1992; **139**(4):507–16.

[150] Senani R. 'Simple sinusoidal oscillator using opamp compensation poles'. Electronics Letters. 1993; **29**(5):452–3.

[151] Serrano I., Carlosena A. 'GB-R impedances: new approach to impedance simulation'. Electronics Letters. 1995; **31**(9):689–90.

[152] Su W. J., Zhu Q. S., Lidgey F. J. 'Novel applications of row-semi-indefinite networks'. 1993 IEEE International Symposium on Circuits and Systems; Chicago, IL, USA, May 3–6, 1993; **4**:2275–8.

[153] Senani R. 'Floating GNIC/GNII configuration realised with only a single OMA'. Electronics Letters. 1995; **31**(6):423–5.

[154] Leuciuc A. 'Realisation of immittance functions with complex singularities by means of modified Antoniou GIC'. Electronics Letters. 1995; **31**(10): 770–1.

[155] Senani R. 'Universal linear voltage-controlled-impedance configuration'. IEE Proceedings G-Circuits, Devices and Systems. 1995; **142**(3):208.

[156] Van de Gevel M. 'Comment on: GB-R impedances: new approach to impedance simulation'. Electronics Letters. 1995; **31**(17):1398.

[157] Abuelma'atti M. T., Khan M. H. 'On the stability of resistively variable capacitors using general impedance converters'. Active and Passive Electronic Components. 1995; **18**:129–35.

[158] Von Wangenheim L. 'Modification of the classical GIC structure and its application to RC-oscillators'. Electronics Letters. 1996; **32**(1):6–8.

[159] Imai Y. 'Simulated inductance circuit with parallel negative conductance and its application for a sinusoidal oscillator'. IEICE Transactions on Fundamentals. 1996; **E79**(5):634–8.

[160] Senani R. 'Alternative modification of the classical GIC structure'. Electronics Letters. 1996; **32**(15):1329.

[161] Singh A. K., Senani R. 'Low-component-count active-only immittances and their application in realising simple multifunction biquads'. Electronics Letters. 1998; **34**(8):718–9.

[162] Singh A. K., Senani R., Tripathi M. P. 'Low-component-count, high-frequency resonators and their applications, using op-amp compensation poles'. Frequenz. 1999; **53**(7–8):1–9.

[163] Ghausi M. S., Laker K. R., 'Modern filter design: active RC and switched capacitor'. Materials, Circuits and Devices). North Carolina, USA: SciTech Publishing Inc.; 2003.

[164] Munoz D. R., Berga S. C., Escriva C. R., Freitas P. J. P. 'Generalized impedance converter as a new sensor signal conditioning circuit'. IMTC 2005-Instrumentation and Measurement Technology Conference; Ottawa, Canada, May 17–19, 2005:587–91.

[165] De Marcellis A., Ferri G., Stornelli V. 'NIC-based capacitance multipliers for low-frequency integrated active filter applications'. 2007 Ph.D. Research in Microelectronics and Electronics Conference; Bordeaux, France, July 2–5, 2007:225–8.

[166] Ishida M., Ebisutani K., Fukui Y., Kasai T. 'Grounded immittance simulation using single operational amplifier with a finite gain-bandwidth product'. Electronics and Communications in Japan (Part I: Communications). 2007; **71**(1):39–47.

[167] Munoz D. R., Moreno J. S., Escriva C. R., Berga S. C., Anton A. E. N. 'Constant current drive for resistive sensors based on generalized impedance converter'. IEEE Transactions on Instrumentation and Measurement. 2008; **57**(10):2290–6.

[168] Dutta Roy S. C. 'Comments on "Realization of an active inductance for a low power high bandwidth DC power line communication network transceiver"'. International Journal of Electronics and Communications (AEU). 2011; **65**(11):983.

[169] Maundy B., Gift J. G. 'Active grounded inductor circuit'. International Journal of Electronics. 2011; **98**(5):555–67.

[170] Yu C. H., Tseng C. Y. 'Sensorless control based on virtual neutral voltage for BLDC motor drive'. Applied Mechanics and Materials. 2015; **789–790**:962–6.

[171] Bruton L. T., 'RC-Active Circuits: Theory and Design'. Englewood Cliffs, NJ: Prentice-Hall, Inc.; 1980. p. 157–80.

[172] Newcomb R. W. 'Active Integrated Circuit Synthesis'. Englewood Cliffs, NJ: Prentice-Hall, Inc.; 1968. p. 12.

[173] Leach D. P., Chan S. P. 'A generalised method of active RC network synthesis'. IEEE Transactions on Circuit Theory. 1971; **CT-18**(6):643–50.

[174] Caggiano A. C. 'Operational amplifier simulates inductance'. Electronics. 1968; **41**(19):99.

[175] Senani R. 'Reply by R. Senani to the Comments on "New canonic active realizations of grounded and floating inductors"'. Proceedings of the IEEE. 1982; **70**(1):101–3.

[176] Senani R. 'Network transformations for incorporating nonideal simulated immittances in the design of active filters and oscillators'. IEE Proceedings, Pt. G. 1987; **134**(4):158–66.

[177] Murata T., Rikoski R. A. 'Mutator simulated floating inductors'. International Journal of Electronics. 1975; **39**(2):229–32.
[178] Rikoski R. A., Choudhry A. 'Mutator realized inductance'. Proceedings of the IEEE. 1976; **64**(6):1012–3.
[179] Chua L. O. 'Synthesis of new nonlinear network elements'. Proceedings of the IEEE. 1968; **56**(8):1012–3.
[180] Tanno Y., Watanabe M., Isozaki S. 'Realizations of simulation inductors using mutators with differentiators'. Electronics and Communications in Japan Part 2. 1986; **69**(8):1109–15.
[181] Bhushan M., Newcomb R. 'Grounding of capacitors in integrated circuits'. Electronics Letters. 1967; **3**(4):148–9.
[182] Singh V. 'Proof of a recent conjecture concerning impedance floatation'. IEEE Transactions on Circuits and Systems. 1989; **36**(8):1110–3.
[183] Daryanani G. 'Principles of Active Network Synthesis and Design'. Columbus, OH: John Wiley & Sons Inc.; 1976.
[184] Van Valkenberg M. E. 'Network Analysis'. New Delhi, India: Prentice-Hall of India Pvt. Limited; 1974.
[185] Temes G. C., Mitra S. K. 'Modern Filter Theory and Design'. New Delhi, India: John Wiley & Sons Inc.; 1973.
[186] Bowron P., Stephenson F. W. 'Active Filters for Communications and Instrumentation'. New York, USA: McGraw-Hill Inc.; 1979.
[187] Temes G. C., Lapatra J. W. 'Introduction to Circuit Synthesis and Design'. New York, USA: McGraw-Hill Inc.; 1980.
[188] Schaumann R., Ghausi M. S., Laker K. R. 'Design of Analog Filters: Passive, Active RC and Switched Capacitor'. Series in Electrical and Computer Engineering. Englewood Cliffs, NJ: Prentice-Hall; 1990.
[189] Soliman A. M., Awad S. S. 'A novel sine-wave generator using a single operational amplifier'. Proceedings of the IEEE. 1978; **66**(2):253–54.
[190] Senani R. 'Realization of Some Classes of Active Networks' (Ph.D. thesis). 1988; India: University of Allahabad.
[191] Srinivasan S., Radhakrishna K. R. 'Comment: Simulation of inductors and capacitors using operational amplifier compensation pole: a caution'. Electronics Letters. 1979; **15**(15):472.
[192] Gilbert B. 'Reply: Simulation of inductors and capacitors using operational amplifier compensation pole: a caution'. Electronics Letters. 1979; **15**(15): 472–3.
[193] Sutcliffe H., Wong C. Y. 'Operational amplifier with compensation pole as basis for active networks'. Electronics Letters. 1981; **17**(15):529–30.
[194] Genin R. 'A sine wave generator using a frequency-dependent negative conductance'. Proceedings of the IEEE. 1975; **63**(11):1611–2.
[195] Senani R., Tiwari R. N. 'New canonic active RC realizations of grounded and floating inductors'. Proceedings of the IEEE. 1978; **66**(7):803–4.
[196] Hamaoui H. 'Operational amplifiers as inductors'. Fairchild Semiconductor Application Note. 1973; app-321.

[197] Sanchez-Sinencio E., Baez-Lopez D. 'Comment: Inductance simulation and filter design using a single-pole amplifier approximation'. Electronics Letters. 1979; **15**(15):465–6.

[198] Hosticka B. J., Moschytz G. S. 'Switched-capacitor simulation of grounded inductors and gyrators'. Electronics Letters. 1978; **14**(24):788–90.

[199] Sanchez-Sinencio E., Gomez-Osorio J. L. 'Switched-capacitor simulation of grounded inductors using operational-amplifier pole'. Electronics Letters. 1979; **15**(6):169–70.

[200] Lee M. S. 'Switched-capacitor filters using floating-inductance simulation circuits'. Electronics Letters. 1979; **15**(20):644–5.

[201] Tems G. C., Jahanbegloo M. 'Switched-capacitor circuits bilinearly equivalent to floating inductor or F.D.N.R.'. Electronics Letters. 1979; **15**(3): 87–8.

[202] Brugger U. B., Hosticka B. J., Moschytz G. S. 'Switched-capacitor simulation of floating inductors using gyrators'. Electronics Letters. 1979; **15**(16): 494–6.

[203] Nossek J. A. 'Improved circuit for switched-capacitor simulation of an inductor'. Electronics Letters. 1980; **16**(4):141–2.

[204] Viswanathan T. R., Faruque S. M., Singhal K. 'Switched-capacitor transconductance and related building blocks'. IEEE Transactions on Circuits and Systems. 1980; **CAS-27**(6):502–8.

[205] Hou C. L., Chen C. C., Hwang C. M., Wu J. S. 'Switched-OTA technique for floating immittance function simulators'. Microelectronics Journal. 1995; **26**(7):691–5.

[206] Senani R., Bhaskar D. R., Singh A. K., Sharma R. K. 'Synthesis of analog circuits using only voltage and current followers as active elements'. In M. Fakhfakh, E. Tlelo-Cuautle and R. Castro-Lopez (Eds.). Analog/RF and Mixed Signal Circuit Systematic Design. Berlin: Springer-Verlag; 2013. p. 289–315.

[207] Senani R. 'Some observations concerning the methods of filter/oscillator realization using the concept of FDNR'. Proceedings of the IEEE. 1979; **67**(12):1665–6.

Chapter 3

The operational transconductance amplifier based gyrators and impedance simulators

Abstract

The operational transconductance amplifier–capacitor (OTA–C) circuits were evolved to provide analog signal processing/signal generation circuits which could be fully integrated in to bipolar or CMOS technology. Thus, apart from their applications in the realization of oscillators, multipliers, nonlinear waveform generators and analog continuous-time filtering, OTA–C circuits and sometimes a combination of OTAs and op-amps have been extensively employed in the realization of gyrators and synthetic impedances of various types. This chapter presents an account of the most prominent works done in this area and highlights certain common synthesis methodologies of deriving such circuits.

3.1 The OTAs and their advantages in analog circuit design

As against the traditional voltage-mode op-amp (VOA), which is a differential voltage-controlled voltage source (DVCVS), the operational transconductance amplifier (OTA) is a differential voltage-controlled current source (DVCCS). An OTA is characterized by ideally infinite input impedance and ideally infinite output impedance. Several IC OTAs are commercially available as off-the-shelf ICs in which the transconductance g_m is linearly controllable through an external dc bias current I_B over a large range (typically, as large as four decades, such as for LM3080). There are some commercially available versions like LM13600/ LM13700 which have two OTAs and two unity gain voltage-controlled voltage sources (VCVS) on the same chip. Such building blocks have also been referred to in the literature as integrated DVCCS/VCVS elements, for instance, see [1–5].

Although the OTA was introduced in 1969 [6], and Bialko and Newcomb [1] in their pioneering work had already well elaborated all finite linear circuits realizable with the integrated DVCCS in 1971, a major impetus in the research on the internal architecture and applications of OTAs in analog circuit design came after 1980 or so when attempts were being made to design electronically tunable continuous-time filters and oscillators suitable for implementation in IC technology as alternative to the then-prevalent discrete time switched-capacitor circuits. In this context, it was realized that the internal circuit of a bipolar or CMOS OTA does not contain any passive resistances and most of the analog functions could be accomplished with only OTAs and capacitors, often without requiring any external passive resistors. Furthermore, such OTA–C circuits offer wide range of electronic tunability and much higher operational frequency range as compared to VOA-based circuits. Because of these advantageous features, OTA–C circuits are equally suited for IC implementation in both bipolar and CMOS technologies. Consequently, the 1980s and 1990s witnessed vigorous research activities on OTAs and their applications which resulted in the development of a wide variety of electronically controllable filters, oscillators, impedance simulators, relaxation oscillators and multipliers, etc. suitable for IC implementation in bipolar and/or CMOS technology.

During the past several decades, OTAs or DVCCS/DVCVS elements have been extensively used in the realization of grounded impedances [7–22], floating impedances [9,11–13,23–42], gyrators [43,44], resonators [45], multiport gyrators/impedance converters/inverters [46–48], synthetic transformer [49], unilateral inductances [2,3,50–52], generalized impedance converters [53,54] and *active-only* impedances [55,56]. Several researchers have also considered OTA and op-amp combinations to devise different circuits [57–60], whereas the issues like temperature compensation have been dealt in [61,62], while other general aspects concerning OTA–C impedance simulation circuits have been dealt in [63–70].

In this chapter, we discuss the major contributions made during the last four decades, in the areas of gyrator realization and grounded/floating synthetic impedances of the various kinds.

3.2 Integrated OTAs

One of the earlier OTA IC was the CA3080 manufactured by RCA which has subsequently been made available by several other IC manufacturers such as National Semiconductors (LM3080) and now several others. A simplified schematic of the 3080-type OTA is shown in Figure 3.1.

It can be seen that the circuit consists of the differential amplifier made from *n–p–n* transistor pair Q_1–Q_2 which constitutes the core of the circuit for which an external dc bias current I_B is supplied from outside through the current mirror made from the *n–p–n* transistors Q_3–Q_4. The collector current I_{C1} is carried over to a *p–n–p* Wilson current mirror (WCM) made from Q_{11}–Q_{12}–Q_{13} whose output is fed to an *n–p–n* WCM made from Q_8–Q_9–Q_{10}. From the other side, the collector current I_{C2} is carried over to another *p–n–p* WCM made from Q_5–Q_6–Q_7. Thus, it is easy to decipher that

$$I_{out} = I_{C1} - I_{C2} \tag{3.1}$$

Furthermore, neglecting the base currents of Q_1 and Q_2

$$I_{C1} + I_{C2} \cong I_B \tag{3.2}$$

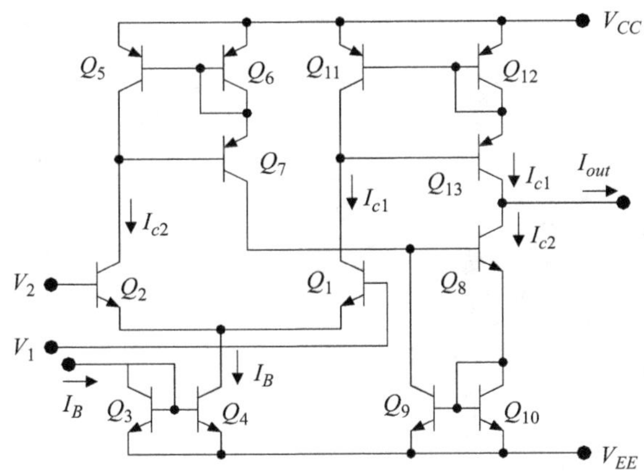

Figure 3.1 Simplified schematic of OTA LM3080 [67]

Also,

$$V_2 = V_{BE2} - V_{BE1} + V_1 \tag{3.3}$$

from where we can write

$$V_{id} = V_1 - V_2 = V_{BE1} - V_{BE2} \tag{3.4}$$

Also,

$$V_{BE} = V_T \ln\left(\frac{I_C}{I_{C0}}\right) \tag{3.5}$$

From the above equations it can be deduced that

$$I_{out} = I_B \tanh\left(\frac{V_{id}}{2V_T}\right) \tag{3.6}$$

If V_{id} is kept sufficiently small, such that $V_{id} \ll 2V_T$, then I_{out} can be approximated as

$$I_{out} \cong \left(\frac{I_B}{2V_T}\right) V_{id} \tag{3.7}$$

Thus,

$$g_m = \left(\frac{I_B}{2V_T}\right) \tag{3.8}$$

Figure 3.2 shows the simplified circuit schematic of another popular IC OTA, namely the LM13600/13700 which has two OTAs (nearly, having the same architecture as in Figure 3.1) along with two on-chip unity gain buffers. In realizing various functional circuits, quite often a unity gain voltage follower (VF) is needed

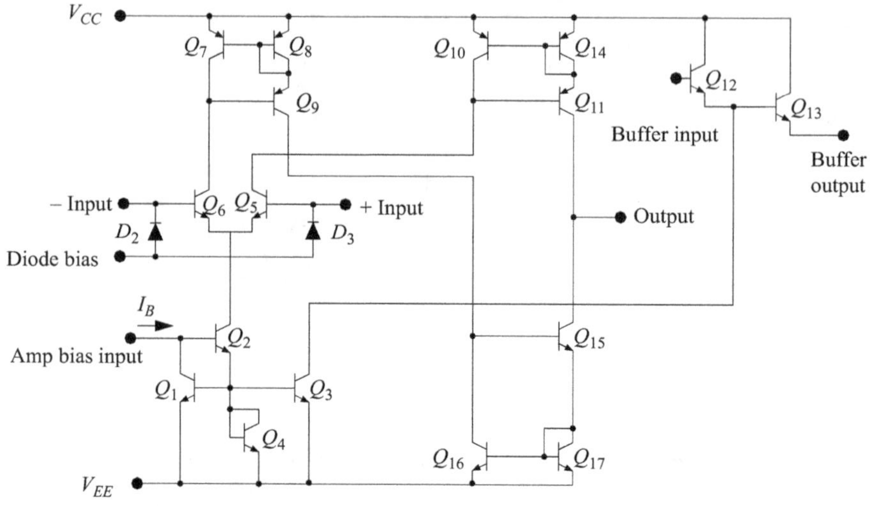

Figure 3.2 Simplified schematic of LM13600/LM13700 [66]

to isolate the load impedance and the OTA-based signal-processing circuits. It is in such applications that LM13600/13700 are particularly useful. The amplifier input bias current supplying mechanism is slightly different than that of LM3080. The externally injected dc bias current I_B is applied to Q_1. It is very easy to see that since V_{BE1}, V_{BE3} and V_{BE4} are equal, if there is perfect matching between Q_1, Q_4 (a diode-connected BJT) and Q_3 then $I_{C3} \approx I_B \approx I_{C12}$. At the output end, Q_{12} and Q_{13} both are configured as emitter followers which constitute the circuit of the VF.

3.3 OTA–C gyrators, inductors and related impedances

As explained in Chapter 2, a gyrator can be realized by connecting two voltage-controlled current sources (VCCS) of opposite polarities in a back-to-back connection. Thus, a gyrator can be simply made from two OTAs as shown in Figure 3.3. When the output port is terminated into an external capacitor C_0, the input impedance can be found as follows:

$$i_1 = g_{m2}v_2 \tag{3.9}$$

$$i_2 = -g_{m1}v_1 \tag{3.10}$$

Furthermore,

$$V_2(s) = -\frac{1}{sC_0}I_2(s) \tag{3.11}$$

Solving the above equations, one obtains

$$Z_{in}(s) = \frac{V_1(s)}{I_1(s)} = \frac{sC_0}{g_{m1}g_{m2}} \tag{3.12}$$

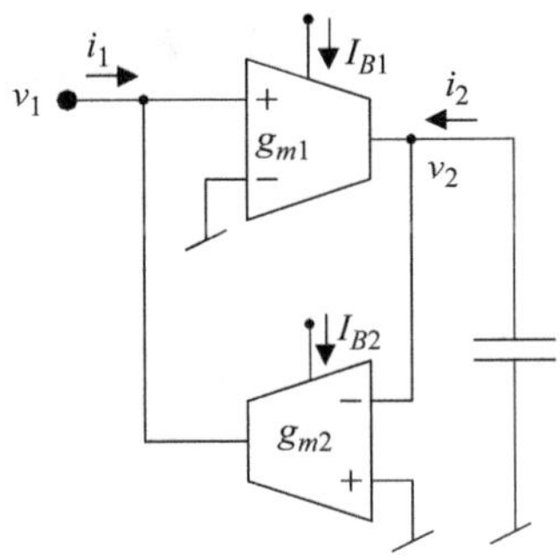

Figure 3.3 OTA–C realization of grounded inductor [1]

The circuit, thus, simulates a lossless grounded inductor of value

$$L_{eq} = \frac{C_0}{g_{m1}g_{m2}} \tag{3.13}$$

Since $g_{m1} = I_{B1}/2V_T$ and $g_{m2} = I_{B1}/2V_T$; L_{eq} is finally given by

$$L_{eq} = \frac{4V_T^2C_0}{I_{B1}I_{B2}} \tag{3.14}$$

There is another interesting way to systematically synthesize the lossless inductor of Figure 3.3 from first principles. From the results of numerous inductance simulation circuits of Chapter 2, it is readily seen that the input impedance of a lossless inductor circuit is given by an expression of the type

$$Z_{in}(s) = \frac{V_1(s)}{I_1(s)} = sC_0R_1R_2 \tag{3.15}$$

Replacing R_1 and R_2 by $1/g_{m1}$ and $1/g_{m2}$, respectively, we can write

$$I_1(s) = \frac{g_{m1}g_{m2}}{sC_0}V_1(s) \tag{3.16}$$

Let

$$V_2(s) = \frac{g_{m1}}{sC_0}V_1(s) \tag{3.17}$$

Equation (3.16) can then be written as

$$I_1(s) = g_{m2}V_2(s) \tag{3.18}$$

Equation (3.17) represents the transfer function of an integrator, hence, can be realized by an OTA having transconductance g_{m1} with its output terminal terminated into a capacitor C_0. Finally, to complete the circuit, another OTA with transconductance g_{m2} is needed to be connected from the output V_2 to the input terminal to create a current I_1 as implied by (3.18). This completes the synthesis of the circuit.

If we want to synthesize, a circuit having

$$Z_{in}(s) = \frac{s^2 C_1 C_2}{g_{m1} g_{m2} g_{m3}} \tag{3.19}$$

thereby representing a frequency-dependent negative conductance (FDNC) element defined by $Z(s) = s^2 M$, following the same procedure as explained above, the expression considered in (3.19) can be decomposed into the following two synthesis equations:

$$V_2(s) = \frac{g_{m1} g_{m2}}{s^2 C_1 C_2} V_1(s) \tag{3.20}$$

$$I_1(s) = g_{m3} V_2(s) \tag{3.21}$$

Thus, we need to have a cascade of two integrators to realize (3.20) each of which is realizable with one OTA and one grounded capacitor. Finally, from the output V_2 of this double integrator, the third OTA with transconductance g_{m3} needs to be connected to the input terminal to ensure satisfaction of (3.21). The resulting circuit is shown in Figure 3.4.

Finally, if we want to realize a frequency-dependent negative resistance (FDNR) element defined by $Z(s) = 1/s^2 D$, we must have a circuit with input impedance of the type

$$Z_{in}(s) = \frac{g_{m1} g_{m2} g_{m3}}{s^2 C_1 C_2 g_{m4} g_{m5}} \tag{3.22}$$

This can be realized by using the gyrator of Figure 3.3 and connecting, instead of a capacitor C_0, a circuit having $Z_{in}(s) = s^2 C_1 C_2 / g_{m1} g_{m2} g_{m3}$, i.e. the circuit of Figure 3.4. The resulting realization of the FDNR element is shown in Figure 3.5. It can be verified that this circuit, indeed, realizes the impedance given in (3.22).

Let us now turn our attention to the realization of floating versions of these elements. It may be noted that now we are essentially looking for a two-port which should have a $[Y]$-matrix characterization

$$\begin{bmatrix} i_1 \\ i_2 \end{bmatrix} = \frac{1}{Z(s)} \begin{bmatrix} 1 & -1 \\ -1 & 1 \end{bmatrix} \begin{bmatrix} v_1 \\ v_2 \end{bmatrix} \tag{3.23}$$

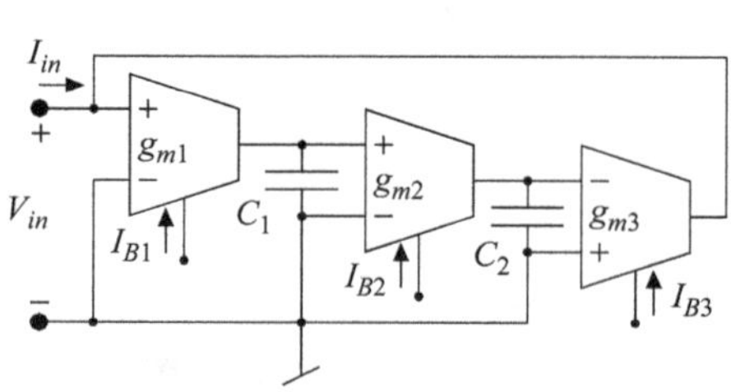

Figure 3.4 OTA–C realization of FDNC [21,22]

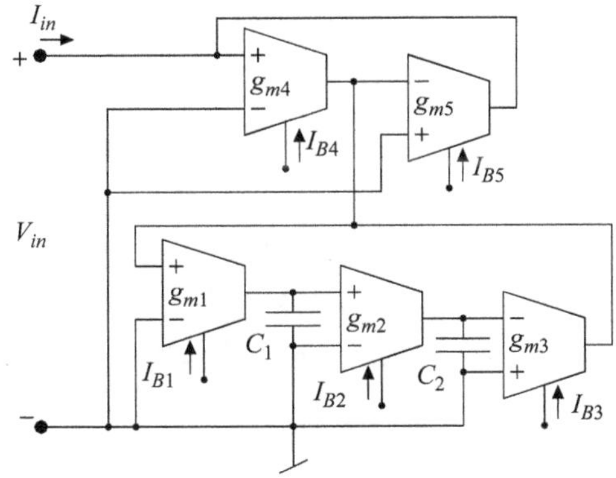

Figure 3.5 OTA–C realization of the FDNR [21,22]

or a transmission matrix characterization

$$\begin{bmatrix} v_1 \\ i_1 \end{bmatrix} = \begin{bmatrix} 1 & Z(s) \\ 0 & 1 \end{bmatrix} \begin{bmatrix} v_2 \\ -i_2 \end{bmatrix} \tag{3.24}$$

where $Z(s) = (sC_0/g_{m1}g_{m2})$ for a floating inductor (FI); $Z(s) = (s^2C_1C_2/g_{m1}g_{m2}g_{m3})$ for a floating FDNC (FFDNC) and $Z(s) = g_{m1}g_{m2}g_{m3}/s^2C_1C_2g_{m4}g_{m5}$ for a floating FDNR.

We now present a number of solutions for the OTA–C realization of lossless FIs (LLFIs) which have been evolved since 1971.

In the first instance, we may recall that a gyrator-based method for realizing a FI using two grounded gyrators (see Chapter 2) has been very well known and has also been documented in the classic paper by Bialko and Newcomb [1]. Since lossless gyrator is realizable using parallel back-to-back connection of two OTAs of opposite polarities, the four-OTA structure of Figure 3.6 which is a cascade of two identical gyrators with a grounded capacitor embedded in between them is straightforward. This circuit realizes a lossless floating inductance of value $L_{eq} = (C_0/g_{m1}g_{m2})$. For the analytical-minded reader, this can be proved quite easily as follows:

The first gyrator is characterized by the transmission matrix

$$[T_1] = \begin{bmatrix} 0 & \dfrac{1}{g_{m1}} \\ g_{m2} & 0 \end{bmatrix} \tag{3.25}$$

The grounded capacitor C_0 is characterized by

$$[T_{C_0}] = \begin{bmatrix} 1 & sC_0 \\ 0 & 1 \end{bmatrix} \tag{3.26}$$

while the second gyrator has a transmission matrix

$$[T_2] = \begin{bmatrix} 0 & \dfrac{1}{g_{m2}} \\ g_{m1} & 0 \end{bmatrix} \tag{3.27}$$

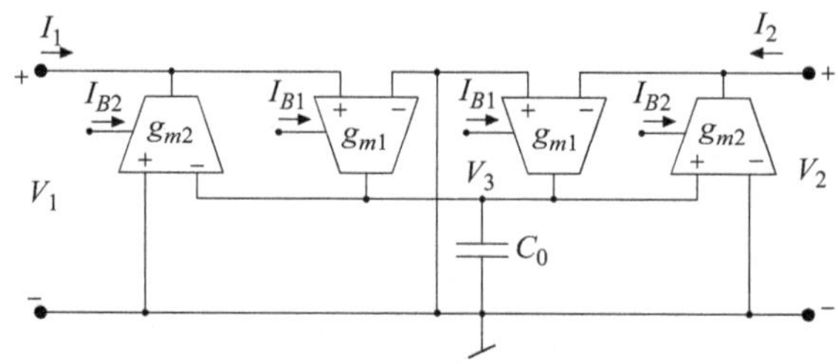

Figure 3.6 Simulation of lossless floating inductance [1]

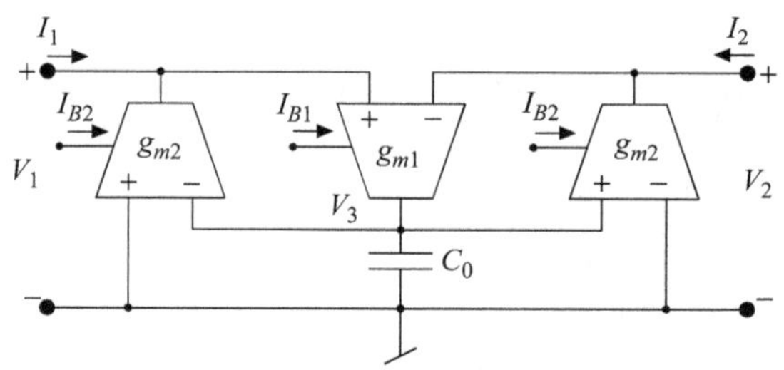

Figure 3.7 Simulation of lossless floating inductance [23]

Hence, the transmission matrix of the complete circuit of Figure 3.6 can be found as

$$[T] = \begin{bmatrix} 0 & \dfrac{1}{g_{m1}} \\ g_{m2} & 0 \end{bmatrix} \begin{bmatrix} 1 & sC_0 \\ 0 & 1 \end{bmatrix} \begin{bmatrix} 0 & \dfrac{1}{g_{m2}} \\ g_{m1} & 0 \end{bmatrix} = \begin{bmatrix} 1 & \dfrac{sC_0}{g_{m1}g_{m2}} \\ 0 & 1 \end{bmatrix} \tag{3.28}$$

which represents a floating inductance of value $L_{eq} = (C_0/g_{m1}g_{m2})$.

An interesting question now arises whether a lossless inductor could be realized with only three OTAs? The answer is in the affirmative and the first such circuit was published in [23] and is shown here in Figure 3.7.

This circuit could be systematically synthesized[1] from the following characterizing equations by following the procedure outlined earlier. An LLFI can be characterized by

$$I_1(s) = \frac{g_{m1}g_{m2}}{sC_0}\{V_1(s) - V_2(s)\} \tag{3.29}$$

and

$$I_2(s) = -I_1(s) \tag{3.30}$$

[1]The first author of this monograph (RS) would like to bring this on record that he had derived the same circuit a year before the publication of [23] and had, in fact, also communicated a manuscript disclosing this circuit as well as one more active-R circuit using two OTAs in: Senani R. 'Floating inductance simulation using operational-transconductance-amplifiers'. Proceedings of the IEEE. 1979, unpublished.

In (3.29) assume

$$V_3(s) = \frac{g_{m1}}{sC_0}\{V_1(s) - V_2(s)\} \tag{3.31}$$

and then (3.30) can be written as

$$I_1(s) = g_{m2}V_3(s) \text{ and } I_2(s) = -g_{m2}V_3(s) \tag{3.32}$$

Equation (3.31) can be seen to be representing a differential integrator which is realizable from OTA_1 along with capacitor C_0 in the circuit of Figure 3.7. The remaining two OTAs with their input terminals grounded in each case are connected such that they convert the voltage into currents $I_1(s)$ and $I_2(s)$ as per (3.32). This completes the synthesis of the circuit of Figure 3.7.

There is another very insightful way of devising this circuit from a different characterization of the two gyrator-based circuits of Figure 3.6. With the grounded capacitor C_0 deleted, the resulting three-port has the following short-circuit admittance matrix characterization:

$$\begin{aligned} &I_1(s) = g_{m2}V_3(s), \; I_2(s) = -g_{m2}V_3(s) \quad \text{and} \\ &I_3(s) = -g_{m1}V_1(s) + g_{m1}V_2(s) \end{aligned} \tag{3.33}$$

In matrix form, this can be written as

$$\begin{bmatrix} I_1(s) \\ I_2(s) \\ I_3(s) \end{bmatrix} = \begin{bmatrix} 0 & 0 & g_{m2} \\ 0 & 0 & -g_{m2} \\ -g_{m1} & g_{m1} & 0 \end{bmatrix} \begin{bmatrix} V_1(s) \\ V_2(s) \\ V_3(s) \end{bmatrix} \tag{3.34}$$

Equation (3.34) clearly characterizes a three-port admittance inverter [48,64]. Thus, it can be easily confirmed that with port 3 terminated into a grounded capacitor C_0, the resulting circuit would realize an LLFI of value $L_{eq} = (C_0/g_{m1}g_{m2})$. Interestingly, the three-port admittance inverter can now be synthesized with only three OTAs as implied by (3.33) each of which represents a VCCS. A circuit synthesized in this way leads to the circuit[2] of Figure 3.7.

In [37], Petchmaneelumka has presented a five-OTA-based circuit (Figure 3.8) which has been claimed to simulate an LLFI. The same author in [38] has presented

[2]An op-amp RC single-resistance-controllable lossless FI (which was actually derived from his three OTA-GC FI mentioned in footnote 1earlier) employing a grounded capacitor was actually synthesized in this manner by Senani [63] wherein the function of the VCCS required was carried out by three-op-amp-based unity gain summers (UGSs) and a resistor. Later, Patranabis and Paul [56,57] came up with nearly the same result with the differential integrator made from a DVCCS (same as an OTA) and a capacitor while the circuit still used two UGS and a resistor to realize each of the two VCCS required in the circuit. Subsequently, Nandi [23] came up with another circuit for the same function in which these two remaining UGSs (along with their associated resistors) were also replaced by DVCCS/OTA elements, thereby finally leading to the three OTA-C circuit of Figure 3.7. This sequential evolution of these three circuits was clearly brought out by Singh in his comment [24].

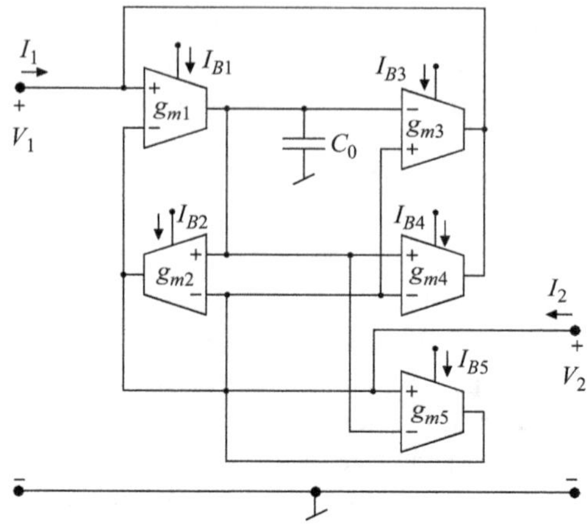

Figure 3.8 Lossless FI proposed by Petchmaneelumka [37]

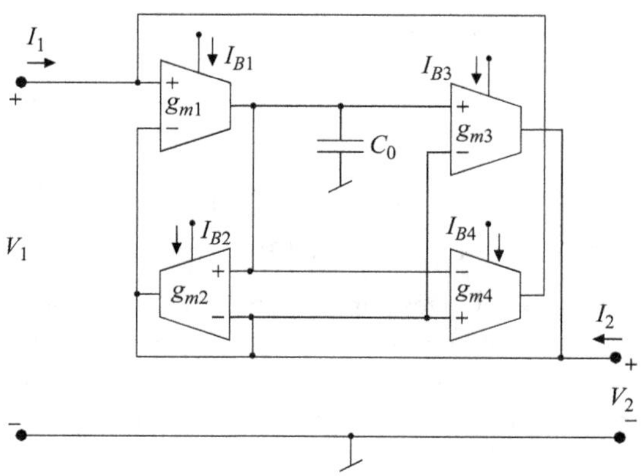

Figure 3.9 Lossless positive FI proposed by Petchmaneelumka [38]

four-OTA-based lossless positive FI circuit (Figure 3.9) and lossless negative FI (Figure 3.10) circuit.

At first one might wonder as to what could be the motivation for going for a five OTA structure for LLFI simulation when the function could be easily achieved by a four OTA circuit as in Figure 3.6 or even with a three-OTA circuit of Figure 3.7? However, from the analysis given in [37], it turns out that, subject to the fulfilment of the conditions $g_{m2} = g_{m3} = g_{m0}$ and $g_{m4} = g_{m5} = kg_{m0}$, the circuit simulates (approximately) a floating inductance given by

$$L_{eq} \cong \frac{C_0}{g_{m1} g_{m0} (1 - k)} \tag{3.35}$$

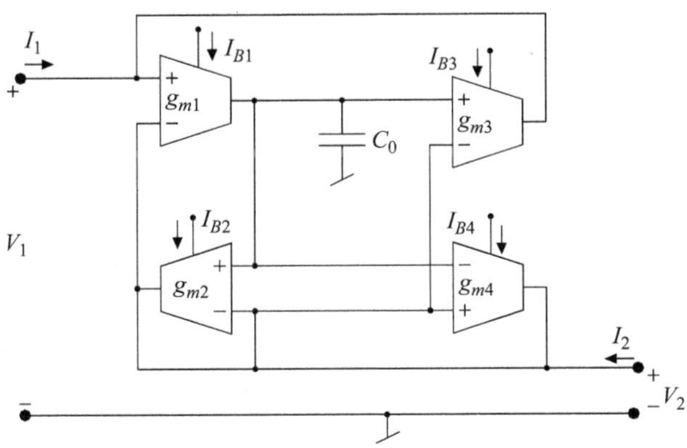

Figure 3.10 Lossless negative FI proposed by Petchmaneelumka [38]

which can be simplified to

$$L_{eq} = \frac{4C_0V_T^2}{I_{B1}I_{B0}\,(1-k)} \tag{3.36}$$

Thus, the main advantage of the circuit is its capability of simulating a positive FI (for $K < 1$) as well as a negative FI (for $K > 1$) from the same circuit without changing the topology.

By straightforward analysis, it can be deduced that the circuit of Figure 3.9 simulates a lossless positive FI with (assuming $I_{B4} = I_{B2} + I_{B3}$):

$$L_{eqp} = \frac{4C_0V_T^2}{(1+G)I_{B1}I_{B2}} \tag{3.37}$$

whereas the circuit of Figure 3.10 simulates a lossless negative FI with

$$L_{eqn} = -\frac{4C_0V_T^2}{GI_{B1}I_{B2}} \tag{3.38}$$

where $G = I_{B3}/I_{B2}$.

Jaikla and Siripruchyanun [71] have presented two LLFI circuits using one single output and one dual-output OTA. The circuit for realizing the positive LLFI is shown in Figure 3.11 and realizes an inductance of value $L_{eq} = (C_0/g_{m1}g_{m2})$. On the other hand, the circuit of Figure 3.12 realizes a negative LLFI of the same value.

Apart from LLFI realizations which may find applications in the design of higher order low-pass filters (LPFs), the FFDNC elements are also useful in the design of filters using inverse Bruton's transformation. A FFDNC circuit proposed by Senani [26] is shown in Figure 3.13 which realizes equivalent impedance given by $Z(s) = (s^2C_1C_2/g_{m1}g_{m2}g_{m3})$.

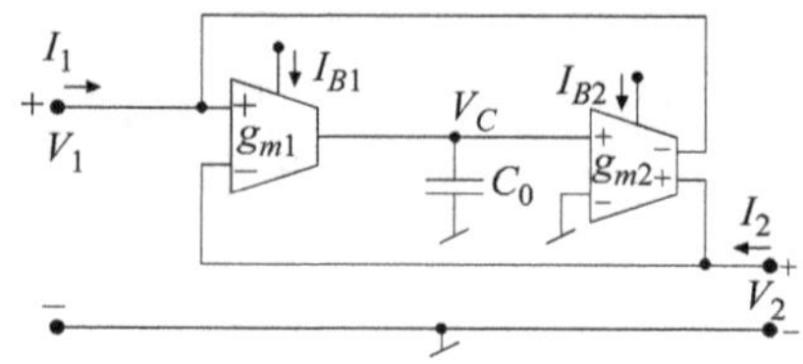

Figure 3.11 Positive lossless FI [71]

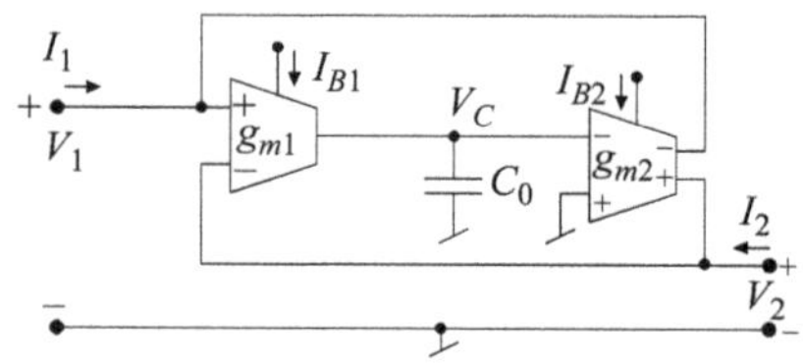

Figure 3.12 Negative lossless FI [71]

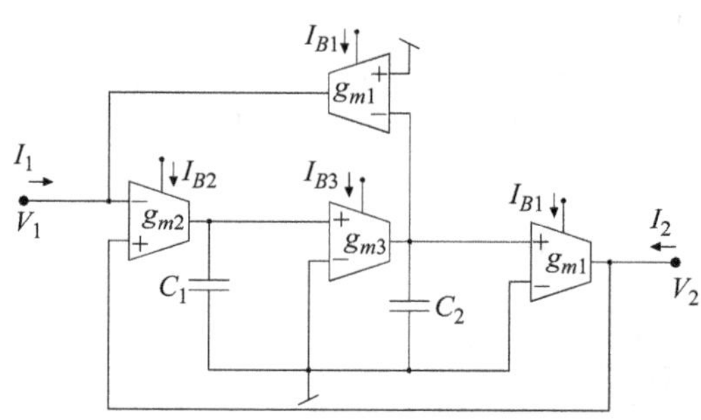

Figure 3.13 Floating FDNC circuit proposed by Senani [26]

3.4 OTA–*RC* impedance converters/inverters

Trimmel and Heinlein [53] presented a two-OTA-based topology for FI simulation with fully floating OTAs as shown in Figure 3.14. Analysis of this circuit reveals the following expression:

$$L_{eq} = \left(\frac{C_4R_1R_3R_5}{R_2}\right)\left[\frac{1+(1/g_{m1})((1/R_1)+(1/R_2)+(R_2/R_1R_3))+(1/g_{m2})((1/R_3)+(1/R_5))}{1-(1/g_{m1}R_2)}\right] \tag{3.39}$$

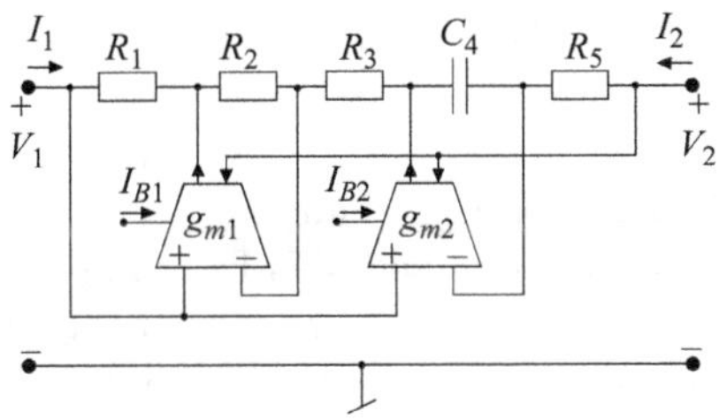

Figure 3.14 FI using Riordon-type gyrator with fully floating OTAs [53]

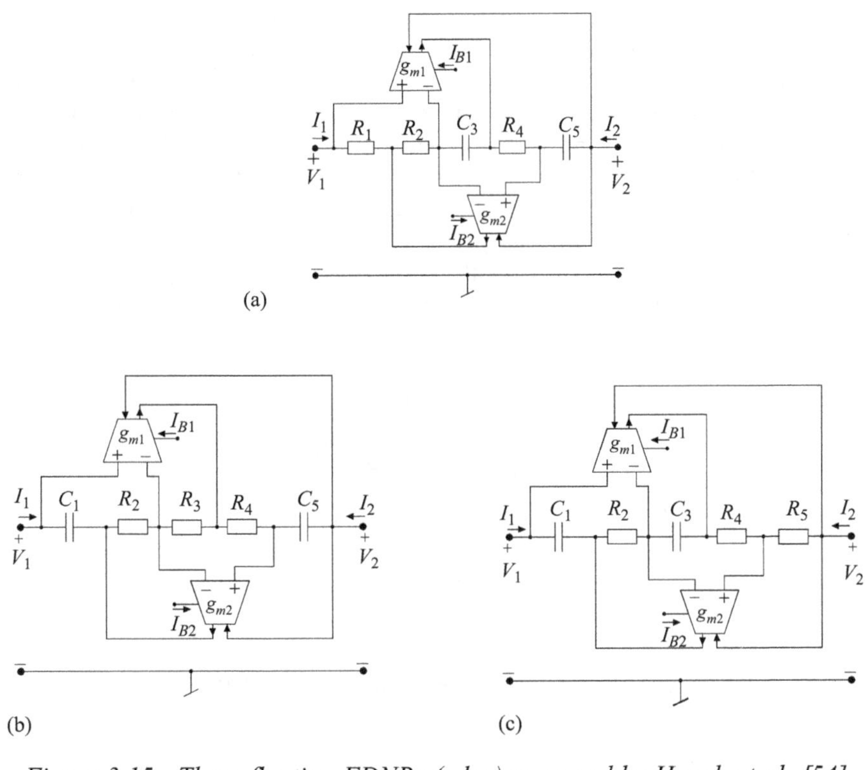

Figure 3.15 Three floating FDNRs (a,b,c) proposed by Huynh et al. [54]

Now if g_{m1} and g_{m2} are sufficiently large, (3.39) reduces to yield

$$L_{eq} \cong \left(\frac{C_4 R_1 R_3 R_5}{R_2}\right). \tag{3.40}$$

An extension of these ideas to the realization of FDNRs has been carried out by Huynh *et al.* [54]. These FDNR circuits are shown in Figure 3.15.

3.5 Other OTA-based lossless/lossy inductors and FDNRs

Since LM13600/LM13700 type OTAs contain an on-chip VF, in this section we present how a combination of OTAs and VFs has been exploited by a number of researchers in devising economical circuits (i.e. those circuits having reduced number of active and/or passive components). A circuit as shown in Figure 3.16 for lossless grounded inductance (GI) realization using a single OTA but along with two VFs was presented by Khan *et al.* [9] with equivalent inductance given by $L_{eq} = (2V_T R_0 C_0 / I_B)$. It has been experimentally demonstrated that changing the value of external bias current (I_B) from 0.1 μA to 1 mA, the value of L_{eq} could be changed from 0.22 mH to 2.2 H, thereby exhibiting a very wide tuning range.

An alternative lossless GI circuit from [11] is shown in Figure 3.17. It employs a single capacitor, a unity gain VF and two OTAs to realize an inductance $L_{eq} = (C_0 / g_{m1} g_{m2})$. The authors [11] have investigated the effect of the nonideal non-unity gain of the VF expressed as $(1-\delta)$ and found that it slightly modifies the input admittance as $Y(s) = \left((1/R_p) + (1/sL_p)\right)$ where $R_p = 1/\delta g_{m1}$ and $L_p = (C_0 / g_{m1} g_{m2})$. Thus, the nonideal quality factor of the simulated GI is found to be $Q_p = (g_{m2} / \delta\omega C_0)$. For $\delta \ll 1$ (typical values of δ being of the order of $\delta <$ 0.005), this circuit can be conveniently designed to achieve high-Q values. This inference has been substantiated by the authors by demonstrating an application of

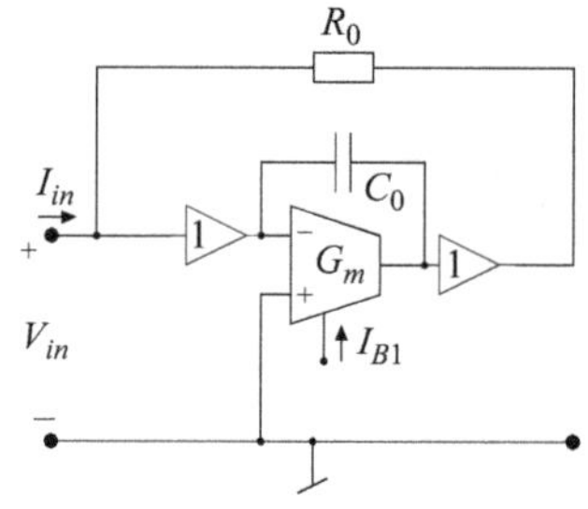

Figure 3.16 Simulation of lossless GI [9]

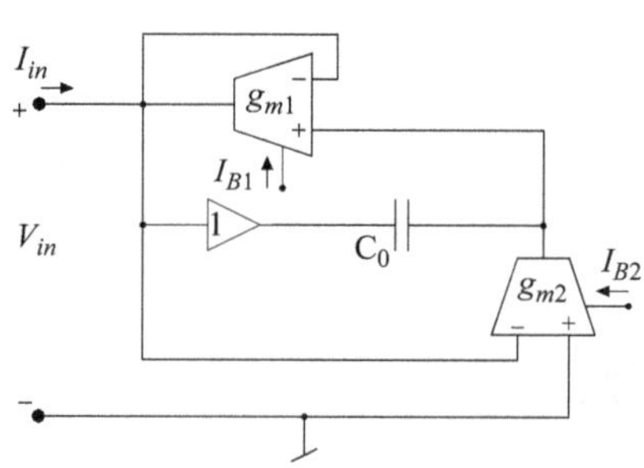

Figure 3.17 Simulation of lossless GI [11]

this circuit in a second-order band-pass filter (BPF) realization. Moreover, electronic tunability of the centre frequency of the BPF has also been confirmed by implementations based on LM13600E OTAs by varying I_B from 32 μA to 97.8 μA as a consequence of which ω_0 was found to vary from about 10 kHz to 30 kHz.

In the context of op-amp circuits, it had been amply demonstrated by numerous authors that as compared to lossless GI, the lossy ones (series *RL*, parallel *RL* and bilinear *RL*) could be realized more economically, typically with only one (in some cases, two) op-amp along with a minimum number of passive components, namely only two resistors and a single capacitor which, however, cannot be grounded (unless one employs additional passive and/or active components).

In view of the above observation and experience, it was logical for the researchers to try the simulation of lossy inductors using OTAs with the expectation of requiring a smaller number of active and passive components. This expectation fortunately turned out to be correct and one such circuit was first proposed by Soliman [3] which is shown in Figure 3.18. By straightforward analysis, its input admittance is found to be $Y_{in}(s) = ((1/R) + (g_m/sCR))$. Thus, the circuit simulates a parallel *RL* type GI with electronic tunability of the inductive part by g_m (hence, by the external dc bias current I_B). Also, the capacitor used is grounded as preferred for IC implementation.

It was demonstrated in [3] that by loading the input port into a capacitor C_1 the circuit can be used to realize a parallel *RLC* resonator and a sinusoidal oscillator. A nonideal analysis of the circuit representing the transconductance g_m as $g_m = g_{m0}e^{-j\omega\tau}$ reveals that R_{eq} and L_{eq} are modified as

$$\left.\begin{aligned} \frac{1}{R_{eq}} &= \frac{1}{R}\left(1 - \frac{g_{m0}\tau}{C}\right) \\ L_{eq} &= \frac{CR}{g_{m0}} \end{aligned}\right\} \text{for } \omega \leq \frac{1}{3\tau} \tag{3.41}$$

Thus, to a first-order approximation, the magnitude of the realized inductance is not affected by the parasitic phase shift of the transconductance. This can be used to realize an oscillator[3] (Figure 3.19) by connecting a capacitor C_1 at the input port resulting in the expression for the oscillation frequency as

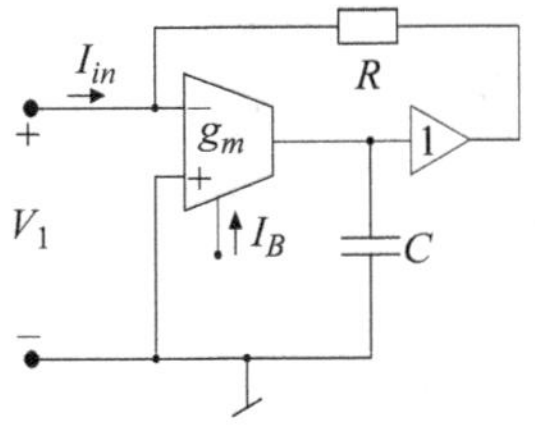

Figure 3.18 Simulation of lossless inductance [3]

[3]It is interesting to point out that the same circuit has been re-introduced by Khan *et al.* [12] also, not realizing that it was already presented much earlier by Soliman [3].

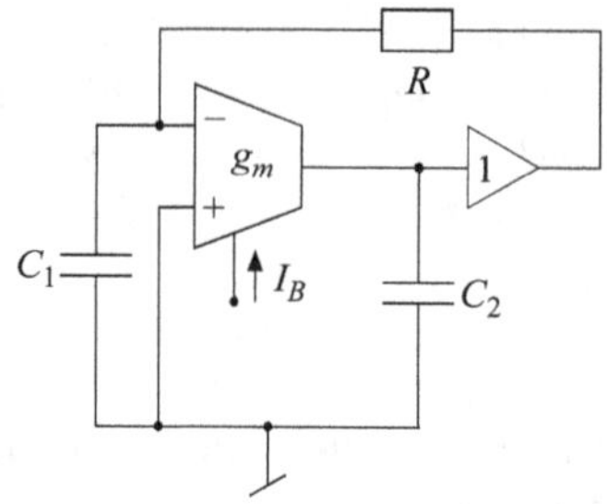

Figure 3.19 An oscillator circuit [3]

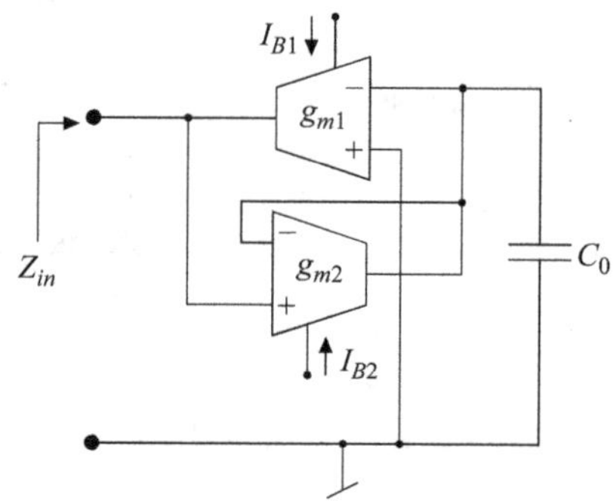

Figure 3.20 Simulation of series RL impedance

$$f_0 = \frac{1}{2\Pi\sqrt{C_1 R_1 \tau}}; \quad \text{for } \omega \leq \frac{1}{3\tau}. \tag{3.42}$$

The preferred output terminal of this oscillator is the output terminal of the VF whereas the condition of oscillation is set by C_2 and the oscillation frequency by C_1.

On another note, in retrospection, it must be emphasized that if the intention is to realize a series *RL*, parallel *RL* type GI using OTAs, it can be done rather systematically in an alternative manner as follows. If we consider a nonideal gyrator characterized by the following [*Y*]-matrix:

$$[Y_1] = \begin{bmatrix} 0 & g_{m1} \\ -g_{m2} & g_{m2} \end{bmatrix} \tag{3.43}$$

and terminate its output into a capacitor C_0, it can be easily deduced that the realized input impedance is given by

$$Z_{in}(s) = \frac{1}{g_{m1}} + \frac{sC_0}{g_{m1}g_{m2}}. \tag{3.44}$$

The matrix $[Y_1]$ can be easily synthesized from two OTAs as shown in Figure 3.20.

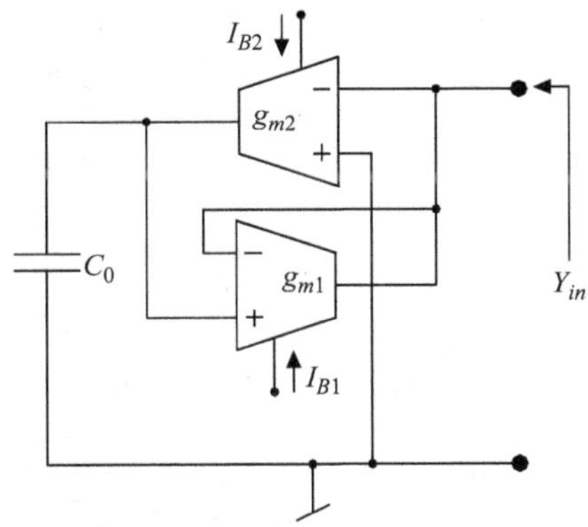

Figure 3.21 Simulation of parallel RL impedance

On the other hand, if we choose the nonideal gyrator to have the $[Y_2]$ matrix as

$$[Y_2] = \begin{bmatrix} g_{m1} & -g_{m1} \\ g_{m2} & 0 \end{bmatrix} \tag{3.45}$$

the resulting circuit would realize

$$Y_{in}(s) = \left(g_{m1} + \frac{g_{m1}g_{m2}}{sC_0} \right) \tag{3.46}$$

and would be shown in Figure 3.21. Note that in both the cases, the realized inductance value is electronically tunable through g_{m2} (and hence, by external dc bias current I_{B2}).

3.6 Synthetic impedances using OTAs and op-amps

In Chapter 2 of this monograph, it was shown that FI circuits using op-amps and *RC* elements usually require two to four op-amps and suffer from one or more of the following drawbacks: (i) requirement of critical component-matching conditions for the realization, resulting in pronounced sensitivity to component tolerances; (ii) use of more than one capacitor and (iii) use of a large number of resistors.

Long back, it was proved by Voorman[4] in his doctoral dissertation (this was brought to the attention of the first author by a reviewer of [59]) that it is impossible to simulate an FI with op-amp *RC* circuits without requiring any component matching. Thus, the configurations which require *no more than one matching-constraint* would be the most suitable ones if they are acceptable from other considerations.

In this section, we demonstrate that impedance simulation circuits realized with OTA-op-amp combinations are free from the drawbacks mentioned above and

[4]Voorman J. O. 'The gyrator as a Monolithic Component in Electronic Systems' (Ph.D. dissertation). June 1977; Katholieke Universtiet Nijmegen, The Netherlands.

which indeed require no more than a single component-matching constraint for the desired realization. In particular, we show how a one-port GI simulation network can be converted into a two-port network simulating the same kind of immittance in floating form, such that the resulting circuit requires no more than a single design constraint.[5]

The technique described in this section can be applied to all one-port networks, which are obtainable by a buffered resistive feedback (through a resistance R_1) around an active-*RC* network *N* such that the feedback resistance appears between the input terminal say, and the output terminal of one of the op-amps inside *N*. An active *RC* one-port of this type (Figure 3.22(a)) having driving point admittance $Y_{11}(s) = Y(s)$ is then required to be converted into a two-port network N^* having a short circuit-admittance matrix given by

$$\begin{bmatrix} I_1(s) \\ I_2(s) \end{bmatrix} = Y(s)\begin{bmatrix} 1 & -1 \\ -1 & 1 \end{bmatrix}\begin{bmatrix} V_1(s) \\ V_2(s) \end{bmatrix} \tag{3.47}$$

In this method, we first formulate a two-port network N_1 by a cascade connection of a unity gain amplifier and a two-port obtained from that of the circuit of Figure 3.22(a) by disconnecting from ground all grounded components and terminals 1, 2, 3, 4, …, *m* tying them together and creating a new node. This newly formed node with respect to the ground terminal is treated as port 2. The *Y*-parameters of N_1 are then given by

$$[Y_2(s)] = Y(s)\begin{bmatrix} 1 & -1 \\ 0 & 0 \end{bmatrix} \tag{3.48}$$

The desired two-port *N* is then obtained by connecting the input terminals of an OTA having transconductance G_m across the feedback resistor R_1 and connecting the output terminal at port 2 of N_1. The *Y*-matrix of the resulting network N^* is given by

$$[Y(s)]_{N^*} = Y(s)\begin{bmatrix} 1 & -1 \\ -G_mR_1 & G_mR_1 \end{bmatrix} \tag{3.49}$$

This realizes the desired FI with *Y*(*s*) characterized by (3.47) subject to the fulfilment of the condition $G_mR_1 = 1$. Applying this methodology, an LLFI circuit based upon the classical GIC based GI circuit is shown in Figure 3.22(d) [60] which realizes

$$\left.\begin{aligned} Y(s) &= \frac{R_2}{sC_4R_1R_3R_5} \\ L &= \frac{C_4R_1R_3R_5}{R_2} \end{aligned}\right\}; \text{ subject to the fulfilment of the condition } G_mR_1 = 1 \tag{3.50}$$

[5]Senani R. 'Circuit techniques for converting simulated grounded immittances into floating immittances'. IEE Proceedings-G Electronic Circuits and Systems. ECS674/NJ1/2, 15 September 1983 (unpublished).

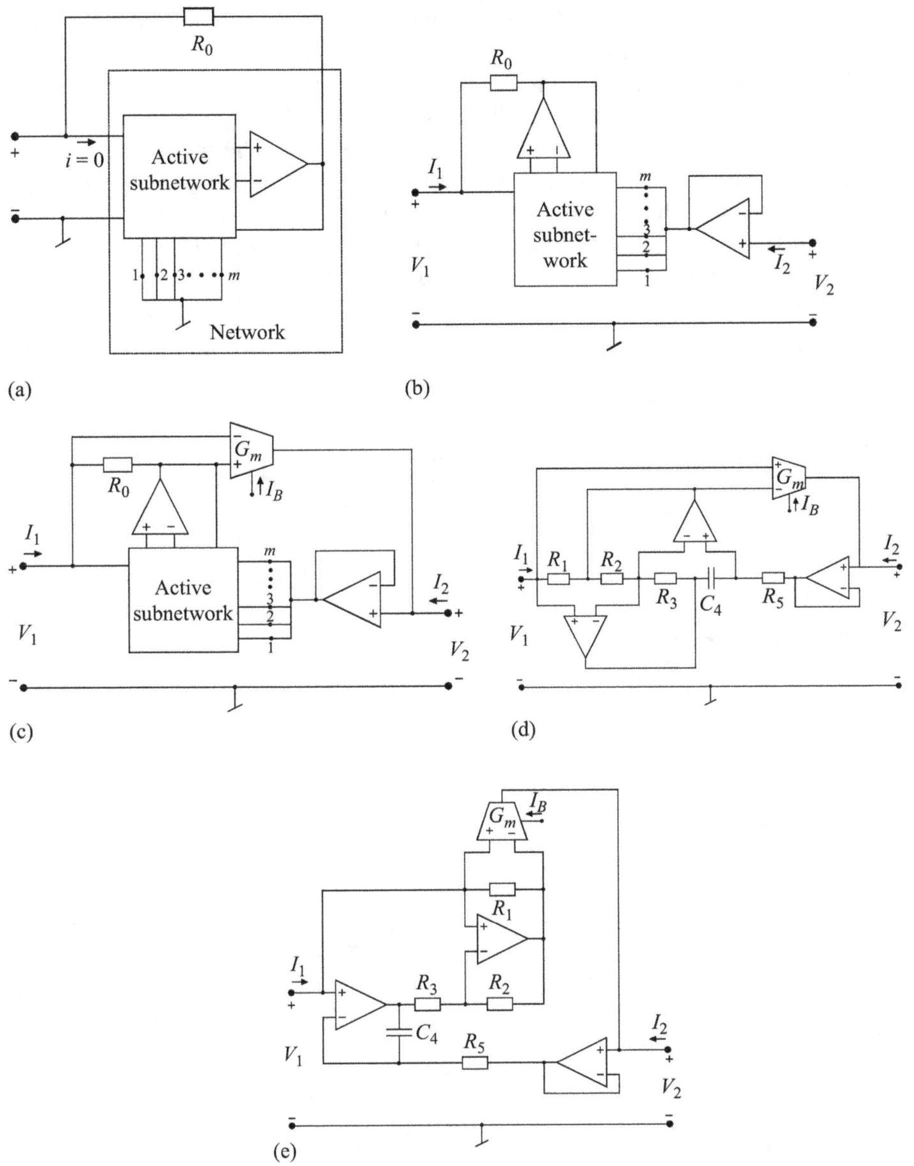

Figure 3.22 Creation of an FI from a given GI: (a) the general schematic representation of a given GI circuit, (b) creation of a two-port from the circuit of (a), (c) conversion of the circuit of (b) into an FI, (d) an exemplary FI circuit obtained from Antoniou's GIC [60], (e) a Riordon gyrator based FI [59]

An alternative LLFI corresponding to the classic Riordon gyrator obtained by this method is shown here in Figure 3.22(e).

We now present some FI circuits which provide single-resistance controlled inductance value with only one condition to be fulfilled for floatation of the type $G_m R_1 = 1$. These circuits are shown in Figures 3.23–3.25. The first circuit is based upon a single-resistance controlled single op-amp GI circuit discussed in Chapter 2 and follows the method of FI creation described earlier in Figure 3.22. The inductance value is tunable by a single resistor R_4 in this case. The second circuit (Figure 3.24) is based upon a modified (buffered) Ford–Girling circuit[6] and has

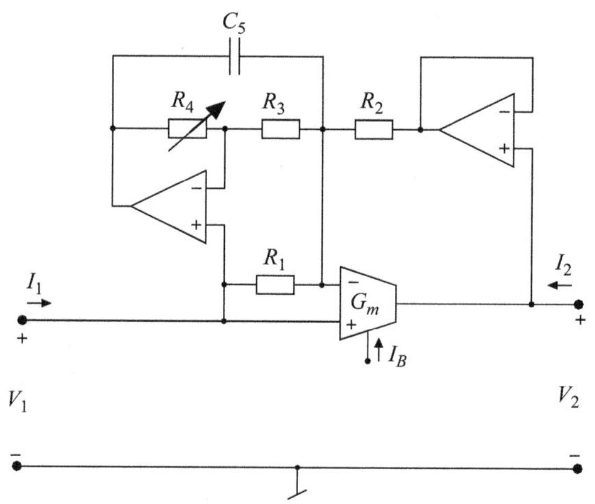

Figure 3.23 Single-resistance series-RL type tunable FI due to Senani [59]

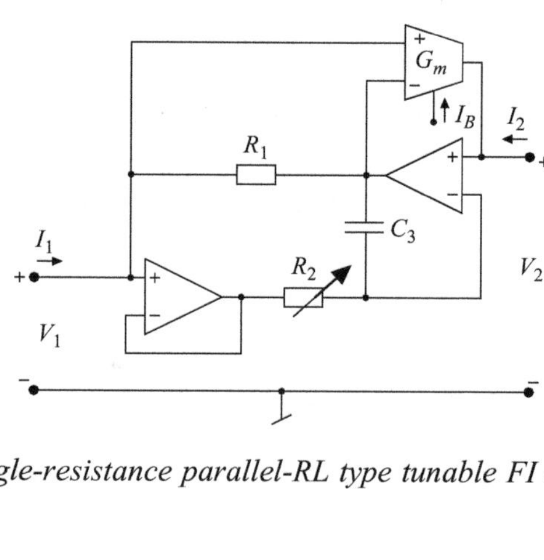

Figure 3.24 Single-resistance parallel-RL type tunable FI due to Senani [59]

[6]Ford R. L., Girling F. E. J., 'Active filters and oscillators using simulated inductance'. Electronics Letters. 1996; 2(2):52.

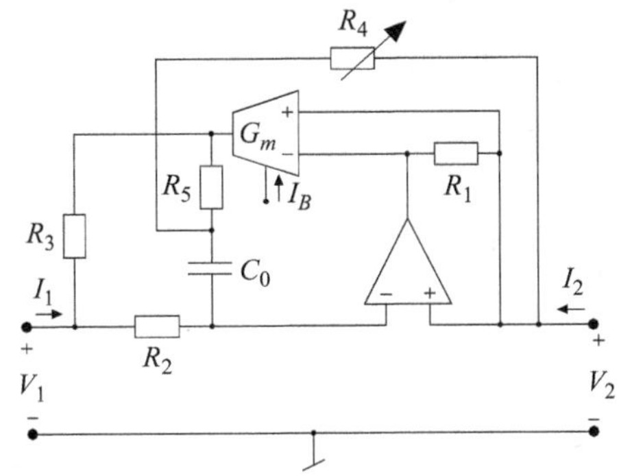

Figure 3.25 A reduced component count single-resistance tunable FI due to Senani [59]

tunability of L_{eq} through R_2. Lastly, the circuit of Figure 3.25 is based upon a single-resistance tunable version of Ford and Girling circuit realized with a single op-amp. The resulting FI, therefore, provides inductance control through variable resistor R_4 but this circuit has the advantage of employing only two active elements as compared to the earlier two, both of which require at least three active elements namely, two op-amps and an OTA.

We have presented so far a variety of circuits for various kinds of floating impedances based upon op-amp *RC* GI networks and have shown their conversion into FIs using an additional OTA (and sometimes a buffer too) according to the presented methodology.

We present a few more examples of the applications of this general technique of FI realization, which have been adopted by other researchers.

Consider first the conversion of the lossless GI of the circuit of Figure 3.17 which was presented in [11]. An FI corresponding to this is obtained by ungrounding the grounded noninverting input terminal of OTA_2 and thus, creating port 2 and then using an additional OTA_3 across input terminals of OTA_1 (which actually carries out the function of an unilateral floating resistor of value $1/G_{m1}$) and putting its output terminal at port 2, the resulting circuit is shown in Figure 3.26. Note that with polarities of the input terminals of this additional OTA (with transconductance G_{m0}) chosen appropriately, we can ensure $I_2(s) = -I_1(s)$ provided $G_{m0} = G_{m1}$. Thus, this circuit can be seen to employ the same method of floatation as has been demonstrated at the beginning of this section. The $[Y]$-matrix of the circuit of Figure 3.26, therefore, turns out to be

$$[Y] = \left(\frac{G_{m1}G_{m2}}{sC_0}\right)\begin{bmatrix} 1 & -1 \\ -1 & 1 \end{bmatrix}; \quad \text{subject to } G_{m0} = G_{m1} \tag{3.51}$$

and the circuit simulates an LLFI of value

$$L_{eq} = \left(\frac{C_0}{G_{m1}G_{m2}}\right) \tag{3.52}$$

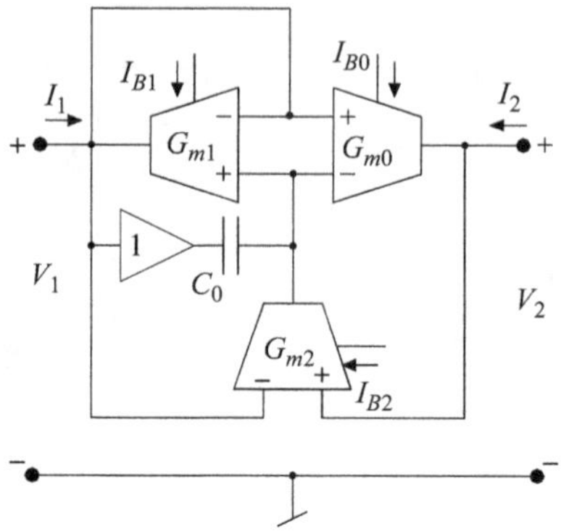

Figure 3.26 Simulation of lossless FI [11]

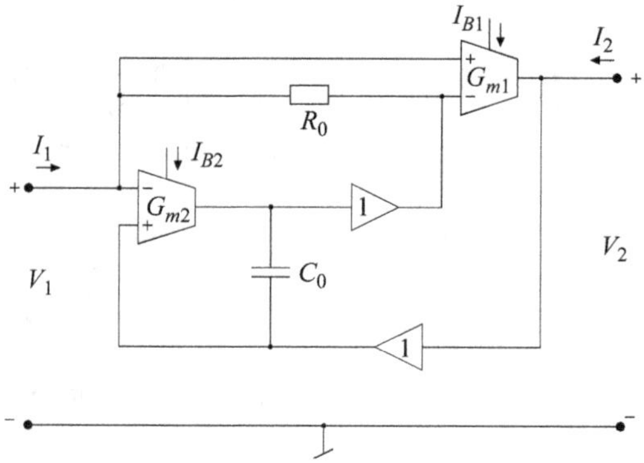

Figure 3.27 Simulation of nonideal FI [12]

The next example is the circuit of the parallel *RL* type GI presented by Soliman [3] which was also subsequently represented by Khan *et al.* [12] who also devised a floating version of this circuit as in Figure 3.27. It can be seen that what has been done is exactly the same as described earlier, i.e. first a two-port is created from the GI circuit and then a voltage buffer is added at port 2 and finally, an additional OTA is used to sense the input current $I_1(s)$ by putting an OTA across resistor R_0 and connecting its output current to port 2 such that by taking $G_{m1}R_0 = 1$, it becomes possible to $I_2(s) = -I_1(s)$ and thus, the circuit realizes a floating admittance:

$$Y_{eq} = \left(\frac{1}{R_0} + \frac{G_{m1}}{sC_0R_0}\right); \quad \text{if } G_{m1}R_0 = 1 \tag{3.53}$$

We now show three low-component count (LCC) FI circuits [72] which require only one op-amp, one OTA along with a single capacitor and two/three resistors.

These LCC circuits are shown in Figures 3.28–3.30. An analysis of the circuit of Figure 3.28 reveals its [Y] matrix as

$$[Y_1] = \left(\frac{1}{R_1} + \frac{1}{R_2} + \frac{1}{sC_0R_1R_2}\right)\begin{bmatrix} 1 & -1 \\ -\dfrac{1}{G_mR_0} & \dfrac{1}{G_mR_0} \end{bmatrix} \tag{3.54}$$

which represents floating parallel RL impedance with $L_{eq1} = CR_1R_2$; $R_{eq1} = R_1R_2/(R_1 + R_2)$ provided that $G_mR_0 = 1$.

On the other hand, the [Y] matrix of the circuit of Figure 3.29 is found to be

$$[Y_2] = \left(\frac{1}{R_1 + R_2 + sC_0R_1R_2}\right)\begin{bmatrix} 1 & -\left\{1 + sC_0R_2\left(1 - \dfrac{1}{G_mR_0}\right)\right\} \\ -1 & \left\{1 + sC_0R_2\left(1 - \dfrac{1}{G_mR_0}\right)\right\} \end{bmatrix} \tag{3.55}$$

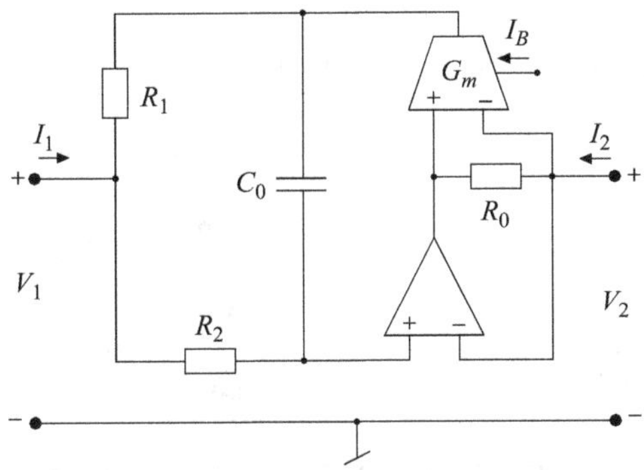

Figure 3.28 Simulation of floating parallel RL impedance [72]

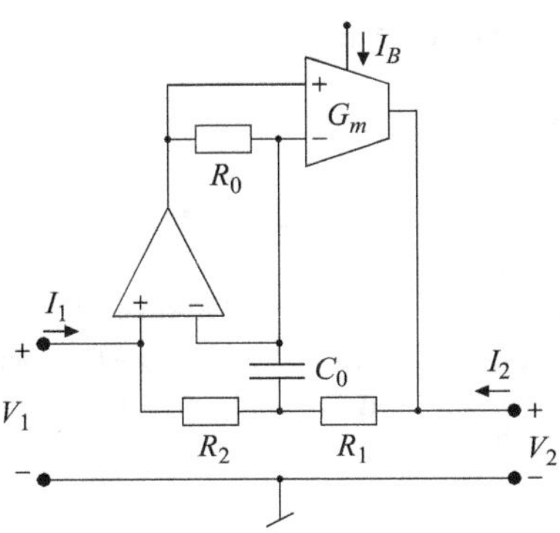

Figure 3.29 Simulation of floating series RL impedance [72]

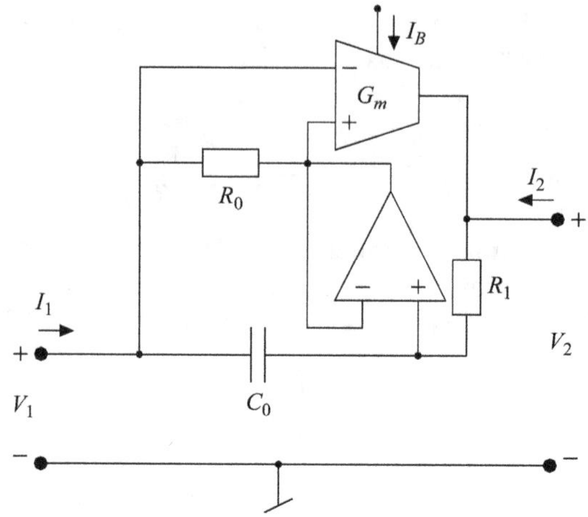

Figure 3.30 Simulation of bilinear floating parallel RL impedance [72]

This circuit realizes a series-*RL* type FI with equivalent values given by

$$L_{eq2} = C_0R_1R_2; \; R_{eq2} = (R_1 + R_2); \quad \text{provided } G_mR_0 = 1 \tag{3.56}$$

Lastly, the [*Y*] matrix of the circuit of Figure 3.30 is given by

$$[Y_3] = \left(\frac{1}{R_0(1 + sC_0R_1)}\right)\begin{bmatrix} (1 + sC_0R_0) & -(1 + sC_0R_0) \\ -(G_mR_0 + sC_0R_0) & (G_mR_0 + sC_0R_0) \end{bmatrix} \tag{3.57}$$

This circuit realizes bilinear *RL* impedance. The equivalent floating impedance consists of parallel *RL* impedance ($R_p \| 1/sL_0$) in series with a resistor (R_s) such that

$$R_s = R_0; \; R_p = (R_1 - R_0); \; L_0 = C_0R_0(R_1 - R_0); \\ \text{provided } G_mR_0 = 1 \text{ and } R_1 > R_0 \tag{3.58}$$

It may be noted that for all the three circuits, the condition of floatation is $G_mR_0 = 1$ which can be easily adjusted in practice by trimming the resistor R_0 in the circuits such that voltages at ports 1 and 2 are indistinguishable which implies, as required, $I_2(s) = -I_1(s)$.

It is worth mentioning that with the OTA and the resistor R_0 deleted, the circuit of Figures 3.28 and 3.29 could be seen to be two-port networks obtainable from the lossy GI circuits of Ford–Girling and Prescott, respectively.[7] On the other hand, the

[7]See [4,6] of Chapter 2

circuit of Figure 3.30 with only the OTA deleted can be seen to be the two-port obtained from Berndt–Dutta Roy (see Chapter 2). Furthermore, the circuit of Figure 3.30 is minimal FI due to the employment of only two resistors and a capacitor along with only two active elements.

3.7 OTA-based capacitance multipliers

Capacitance multipliers are active-*RC* circuits which can create large-valued grounded or floating capacitive input impedance. In Chapter 2, it was shown how a variety of resistively variable capacitive impedance circuits can be realized from the GIC-type two op-amp circuits.

In this section, we consider the realization of capacitance multipliers using OTAs. The basic motivation of doing this is to obtain capacitive input impedance whose value can be electronically controlled by varying the G_{m0} of an OTA through an external dc bias current. One such circuit was presented by Ahmed–Khan–Parveen [8] which is shown here in Figure 3.31. This circuit employs an OTA along with two voltage buffers and realizes input impedance given by

$$Z_{eq}(s) = \frac{V_1(s)}{I_1(s)} = \frac{1}{sC_0R_0G_{m0}} = \frac{2V_T}{s(I_{B0}R_0C_0)}; \quad \text{with } C_{eq} = \left(\frac{I_{B0}R_0}{2V_T}\right)C_0 \tag{3.59}$$

whose value can be varied by varying I_B.

Another circuit which employs only one OTA and one buffer [7] is shown in Figure 3.32. Simple analysis yields, input impedance as

$$Z_{in}(s) = \frac{V_1(s)}{I_1(s)} = \frac{1/sC_0}{(1 + G_{m0}R_0)}; \quad C_{eq} = \left(1 + \frac{I_{B0}}{2V_T}R_0\right)C_0 \tag{3.60}$$

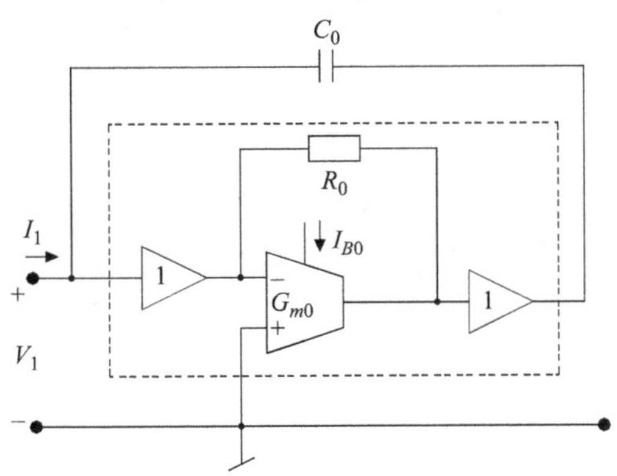

Figure 3.31 Capacitance multiplier [8]

The passive resistor R_0 in this circuit (Figure 3.32) can be replaced by another OTA configured as a grounded resistor which leads to the circuit of Figure 3.33. This circuit makes the equivalent capacitance as

$$C_{eq} = C_0\left(1 + \frac{I_{B1}}{I_{B2}}\right) \tag{3.61}$$

This circuit has the advantage of offering temperature-insensitivity of the realized capacitance multiplication factor which can be changed by varying I_{B1} or I_{B2}.

Figure 3.34 shows another capacitance multiplier [13] whose function is also temperature insensitive. A straightforward analysis of this circuit yields

$$Z_{in}(s) = \frac{G_{m2}}{sCG_{m1}}; \quad \text{which gives } C_{eq} = \left(\frac{G_{m1}}{G_{m2}}\right)C = C\left(\frac{I_{B1}}{I_{B2}}\right) \tag{3.62}$$

Thus, it is seen that the V_T has been cancelled out due to the two transconductances appearing as the ratio in the expression for C_{eq}.

A floating version of this circuit is obtained by using one more OTA and a voltage buffer, following the general method described in an earlier section, as shown in Figure 3.35 [13]. The condition required for floatation is $G_{m3} = G_{m1}$

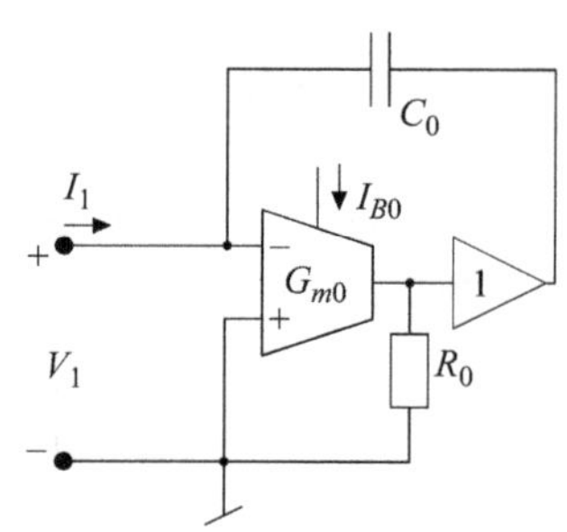

Figure 3.32 Capacitance multiplier [7]

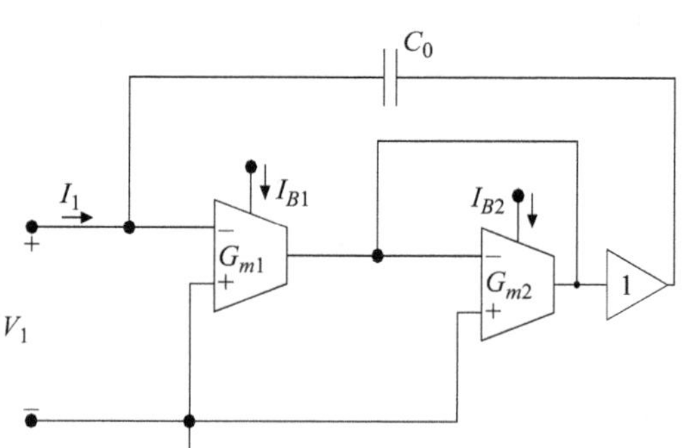

Figure 3.33 Temperature-independent capacitance multiplier [7]

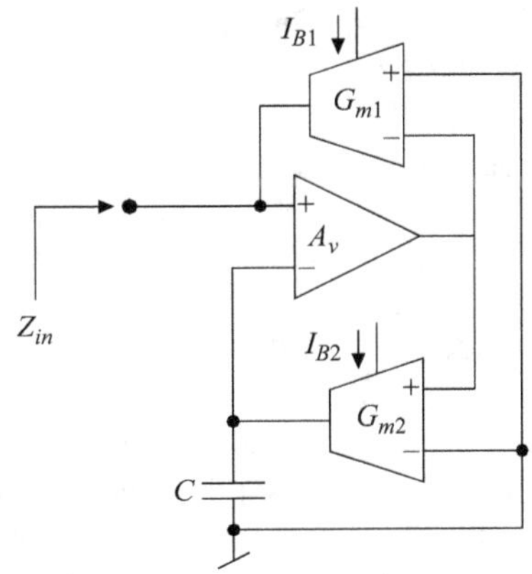

Figure 3.34 Grounded capacitance multiplier [13]

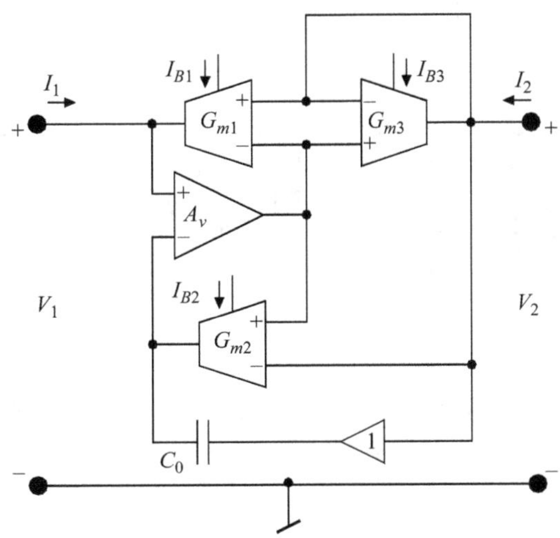

Figure 3.35 Floating capacitance multiplier [13]

which can be met easily in practice by joining the external dc bias current terminals of the two concerned OTAs and supplying a current $2I_B$ from an external current source.

It may be pointed out that in this section we have limited our discussion to a single floating capacitance multiplier circuit only which requires as many as four active elements and a capacitor. Of course, there are other circuits published in the literature which employ even a more number of active elements [42]; however, any floating capacitance multiplier circuit using a grounded capacitor along with a reasonable number of OTAs has not been evolved so far although a four/five OTA grounded capacitor circuit is easily conceivable using a three-port gyrator with a two-port gyrator terminated into a grounded capacitor – connected at port 3 of the three-port gyrator.

3.8 Inductor and FDNC simulators using OTAs and unity gain adders/subtractors

In Chapter 2 of this monograph, it was explained that ideal or lossy inductance simulation circuits can be shown to be obtainable by a direct or buffered resistive feedback around an ideal or nonideal integrator/differentiator circuits [70].

A number of inductors/FDNR circuits have been evolved in the literature based upon the above-mentioned idea (but without necessarily saying so) using OTAs also. In this section, we present some circuits which employ OTAs along with unity gain differential amplifiers (UGDAs) to realize inductor/FDNC simulators.

Consider first the circuit of Figure 3.36 [52]. The OTA along with the grounded capacitor clearly realizes an integrator so that we have voltage across this capacitor as $V_{in}(s)G_m/sC_0$. Now the UGDA creates an output which is $V_{in}(s)(1-(G_m/sC_0))$ which represents a nonideal integrator. When a resistive feedback through the resistor R_0 is applied, the input impedance is given by (using Miller's theorem):

$$Z_{in}(s) = \frac{R_0}{[1-(1-(G_m/sC_0))]} = \frac{sC_0R_0}{G_m} \tag{3.63}$$

thereby realizing a lossless inductor of value $L_{eq} = C_0R_0/G_m$ whose value is controllable by G_m and hence, by the external dc bias current I_B.

Application: In Figure 3.37, the application of this circuit as an oscillator is shown [52]. The oscillation frequency of this oscillator is determined from its characteristic equation which is given by

$$s^2 + \frac{G_m}{C_0C_1R_1} = 0 \tag{3.64}$$

Thus, although the oscillation frequency is $\omega_0 = \sqrt{G_m}/\sqrt{C_0C_1R_1}$ and controllable by I_B, a drawback of this circuit appears to be not having any control over the pole location [52].

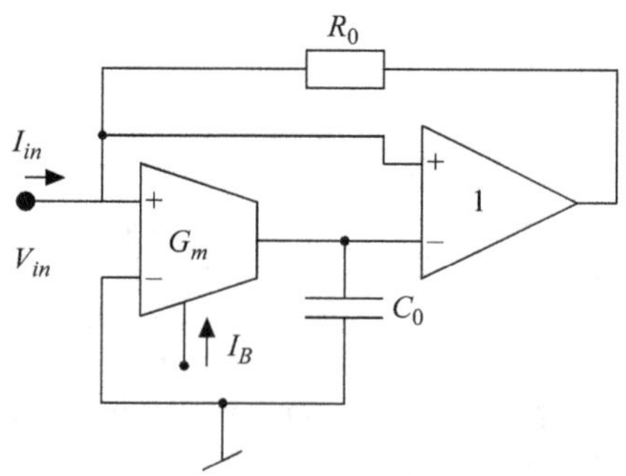

Figure 3.36 Simulation of an ideal grounded inductor [52]

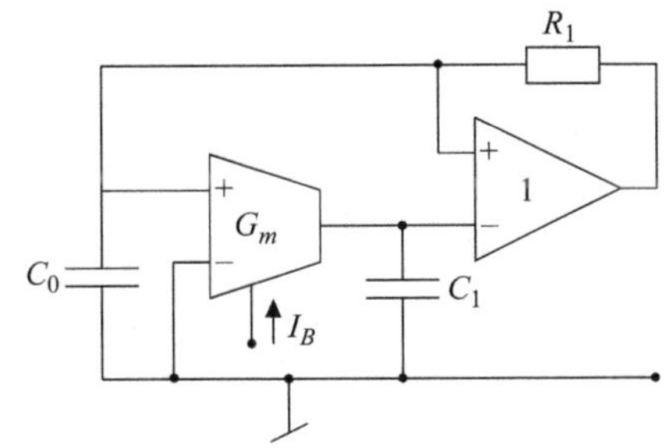

Figure 3.37 Sinusoidal oscillator using ideal inductor of Figure 3.36 [52]

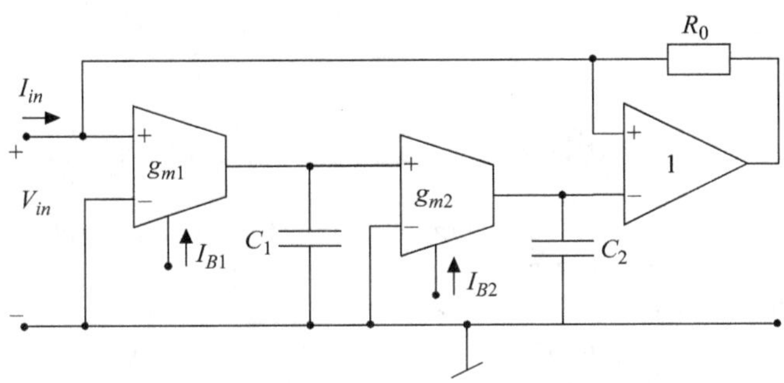

Figure 3.38 Simulation of an ideal FDNC [52]

Consider now the circuit of Figure 3.38. Here, instead of an integrator block, using two OTAs and two grounded capacitors a double integrator is made first, i.e. the output of the UGDA is made to be $(1 - (g_{m1}g_{m2}/s^2C_1C_2))V_{in}(s)$ so that a resistive feedback yields $Z_{in}(s) == s^2C_1C_2R_0/g_{m1}g_{m2}$, thereby simulating an FDNC.

As the last example of this approach we present in Figure 3.39, an FI simulator is obtained in this manner [25]. Analysis of this circuit gives the $[Y]$ matrix as

$$[Y(s)] = \left(\frac{G_mR_3}{sC_0R_1R_2}\right)\begin{bmatrix} 1 & -1 \\ -1 & 1 \end{bmatrix} \tag{3.65}$$

thereby realizing an LLFI of value $L_{eq} = C_0R_1R_2/G_mR_3$ which can be electronically tuned by I_B.

It may be noticed that the UGDAs as well as the UGSs as needed in this circuit would need an op-amp along with four matched resistors which appears to be a drawback of this, as well as the three other circuits described earlier in this section.

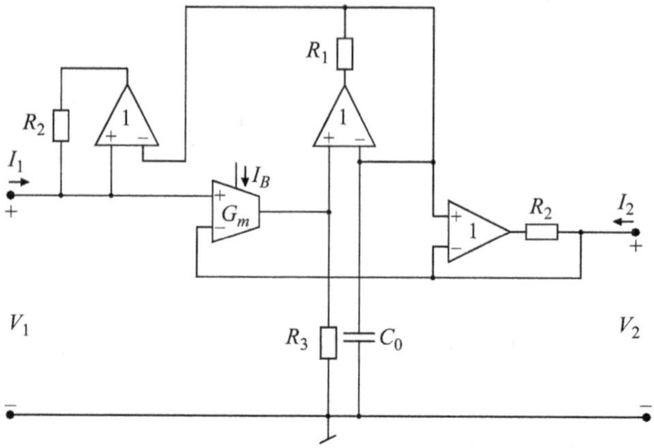

Figure 3.39 Simulation of a lossless FI [25]

3.9 Active-only simulators using op-amps and OTAs

In Chapter 2, we had presented a number of inductance simulation circuits which employed the otherwise parasitic, op-amp pole into the creation of inductive impedances by resistive feedback thus, resulting in the so-called active-R inductors and resonators. This approach was subsequently extended by a number of researchers by replacing the feedback resistor by an OTA. In this approach, the op-amp is represented by an integrator model $A(s) \cong \omega_t/s$; for $\omega \gg \omega_p$ where $\omega_t = A_0\omega_p$ is the gain bandwidth product and ω_p is the pole of the open-loop gain of the internally compensated type op-amp, having a dominant pole at $\omega = \omega_p$. The advantages accrued from such an approach are (i) complete elimination of external capacitors and (ii) electronic control of the realized impedance through the transconductance of the OTA which can be varied through an external dc bias current.

Such *active-only* impedance simulation circuits are claimed to be attractive for IC implementation as the only elements employed are internally compensated type op-amps (such as μA741 and LF356) and IC OTAs such as LM3080 and LM13600/LM13700. Moreover, from the viewpoint of CMOS implementation, a number of researchers have experimented with such circuits using internally compensated CMOS op-amps and CMOS OTAs (in which case G_m can be even controlled by an external dc *bias voltage*).

Singh and Senani [45] systematically derived a class of impedance simulation circuits, each employing two op-amps and an OTA, eight of which are shown in Figure 3.40. These circuits simulate series *RL*, series *LC*, series *RLC*, an FDNC or *M* element, series *LM*, series *RM*, parallel *LM*, series *RM* and series *RLM* impedances, the relevant parameters of which are given in Table 3.1.

In [45] it has been demonstrated that these circuits possess low sensitivity properties and can be designed for stable operation quite easily. By the addition of

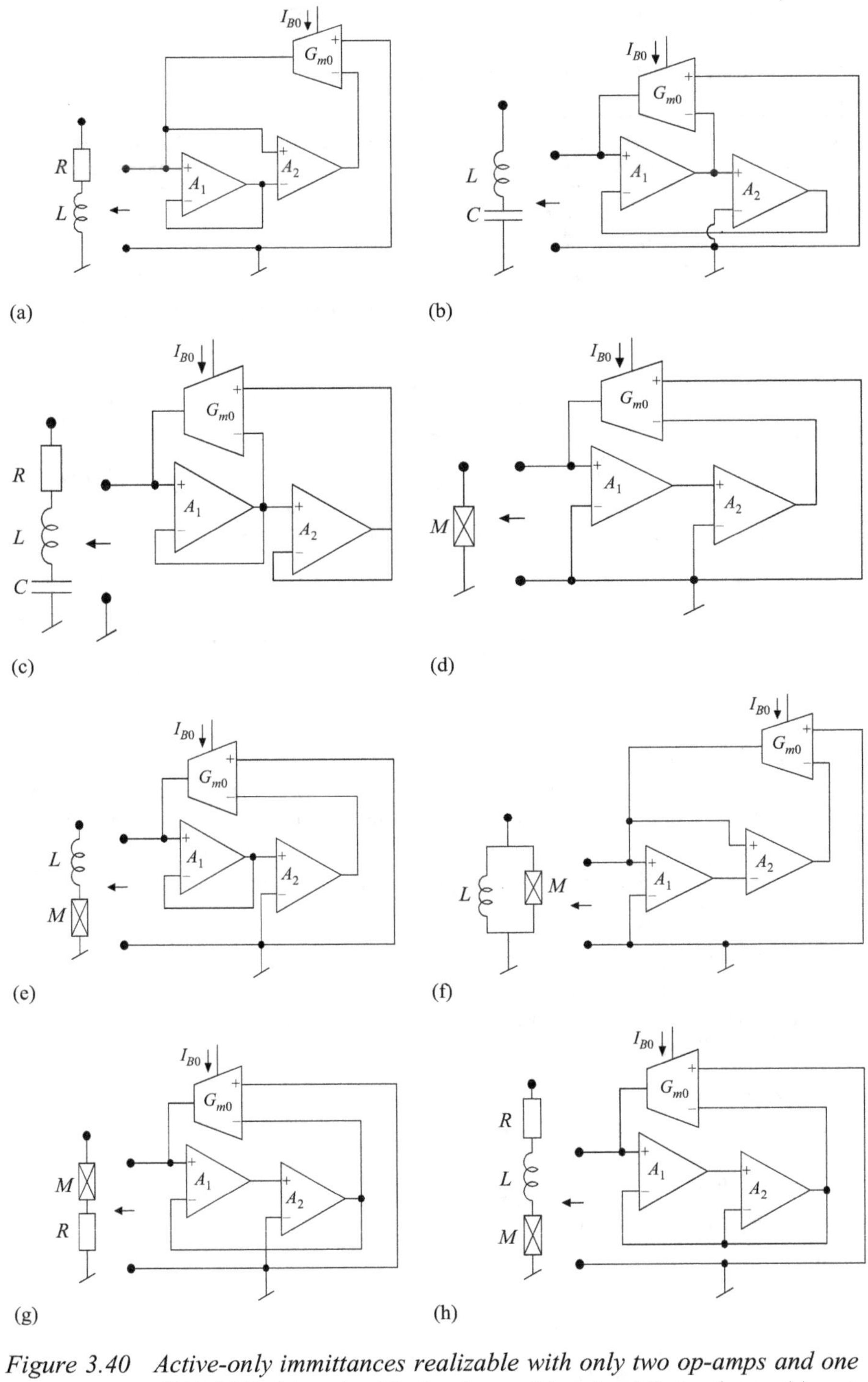

Figure 3.40 Active-only immittances realizable with only two op-amps and one OTA [45]: (a) series RL simulator, (b) series LC simulator, (c) series RLC simulator, (d) an FDNC simulator, (e) series LM simulator, (f) parallel LM simulator, (g) series RM simulator, (h) series RLM simulator

Table 3.1 Relevant parameters of impedance simulators

Circuit reference	Equivalent parameters	Circuit reference	Equivalent parameters
Figure 3.40(a)	$R = \omega_{t1}/\omega_{t2}G_{m0}$; $L = 1/\omega_{t2}G_{m0}$	Figure 3.40(b)	$L = 1/\omega_{t1}G_{m0}$; $C = G_{m0}/\omega_{t2}$
Figure 3.40(c)	$R = (1 + (\omega_{t2}/\omega_{t1}))/G_{m0}$; $L = 1/\omega_{t1}G_{m0}$; $C = G_{m0}/\omega_{t2}$	Figure 3.40(d)	$M = 1/\omega_{t1}\omega_{t2}G_{m0}$
Figure 3.40(e)	$L = 1/\omega_{t2}G_{m0}$; $M = 1/\omega_{t1}\omega_{t2}G_{m0}$	Figure 3.40(f)	$L = 1/\omega_{t2}G_{m0}$; $M = 1/\omega_{t1}\omega_{t2}G_{m0}$
Figure 3.40(g)	$R = 1/G_{m0}$; $M = 1/\omega_{t1}\omega_{t2}G_{m0}$	Figure 3.40(h)	$R = 1/G_{m0}$; $L = 1/\omega_{t1}G_{m0}$; $M = 1/\omega_{t1}\omega_{t2}G_{m0}$

an OTA or a capacitor, many of these resonators/immittances can be used to implement a variety of filter functions.

Application: A simple multifunction biquad realized from the LM immittance of Figure 3.40(f) which is obtained by ungrounding the grounded terminals and adding one more OTA, is shown in Figure 3.41(a). Out of the various possible ways of realizing different kinds of filter responses from Figure 3.41(a), the following have been obtained to be the best:

1. LPF: with $V_{in1} = V_{in3} = 0$ and $V_{in2} = V_{in}$
2. High-pass filter: with $V_{in2} = V_{in3} = 0$ and $V_{in1} = V_{in}$
3. BPF: with $V_{in1} = V_{in2} = 0$ and $V_{in2} = V_{in}$
4. Bandreject filter: with $V_{in1} = V_{in2} = V_{in}$ and $V_{in3} = 0$

With $A_i(s) = \omega_{ti}/s$; $i = 1{,}2$, (ω_{ti} is the gain bandwidth product of the ith op-amp), the resulting functions are found to be:

$$T_{LPF} = \frac{V_{03}}{V_{in}} = \frac{G_{m0}/G_{m1}\omega_{t1}\omega_{t2}}{D(s)} \tag{3.66}$$

$$T_{HPF} = \frac{V_{03}}{V_{in}} = \frac{s^2}{D(s)} \tag{3.67}$$

$$T_{BPF} = \frac{V_{01}}{V_{in}} = \frac{(G_{m0}/G_{m1}\omega_{t1})s}{D(s)} \tag{3.68}$$

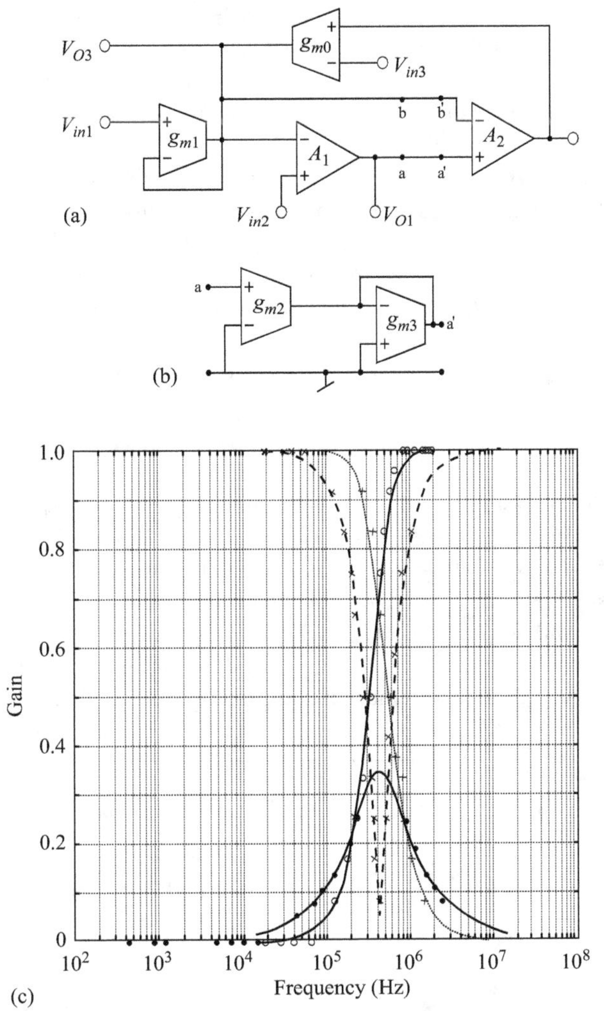

Figure 3.41 Multifunction voltage-mode biquad derived from the circuit of Figure 3.40(f) [45]: (a) circuit of multifunction biquad, (b) subcircuit for independent tunability of filter parameters, (c) frequency responses. Reprinted, with permission, from [45]

$$T_{BRF} = \frac{V_{03}}{V_{in}} = \frac{(s^2 + (G_{m0}/G_{m1}\omega_{t1}\omega_{t2}))}{D(s)} \tag{3.69}$$

where

$$D(s) = s^2 + \left(\frac{G_{m0}}{G_{m1}\omega_{t2}}\right)s + \frac{G_{m0}}{G_{m1}\omega_{t1}\omega_{t2}} \tag{3.70}$$

All the realized filters were experimentally verified using μA741 type op-amps having $\omega_{t1} = 2.050$ Mrad/s and $\omega_{t2} = 5.906$ Mrad/s and CA3080 type OTAs with $G_{m0} = 4.65$ mA/V and $G_{m1} = 6.64$ mA/V, designed for $Q = 0.707$ and $f_0 = 463$ kHz. The various frequency responses are shown in Figure 3.41(c) which are in good agreement with the theoretical results.

We now present two FI circuits using this approach. Consider first the circuit of Figure 3.42 [28] which uses an op-amp as an integrator giving an output voltage $V_0(s) = A(s)\{V_2(s) - V_1(s)\} \cong (\omega_t/s)\{V_2(s) - V_1(s)\}$. Now through two single-ended OTAs this voltage is converted into two equal and opposite currents $+G_{m1}V_0$ and $-G_{m2}V_0$ so that the port currents of the two-port circuit become

$$I_1(s) = -G_{m1}V_0(s) \cong \frac{G_{m1}\omega_t}{s}\{V_1(s) - V_2(s)\} \quad \text{and}$$

$$I_2(s) = G_{m2}V_0(s) \cong \frac{G_{m2}\omega_t}{s}\{V_2(s) - V_1(s)\} \tag{3.71}$$

The circuit, thus, simulates an LLFI of value $L_{eq} = (1/\omega_t G_{m0}) = (2V_T/I_{B0}\omega_t)$ which is electronically tunable by I_{B0} for $G_{m1} = G_{m2} = G_{m0}$. Since a better approximation of $A(s)$ is $A(s) = (\omega_t/(s + \omega_p))$, this modifies the realized admittance value as $Y_{eq}(s) = G_{m0}\omega_t/(s + \omega_p)$ which represents a series RL circuit with $L_{eq} = 1/\omega_t G_{m0}$ and $R_{eq} = \omega_p/\omega_t G_{m0}$ where from the quality factor is given by $Q_{eq} = (\omega L_{eq}/R_{eq}) = (\omega/\omega_p) = (f/f_p)$. Thus, at a nominal frequency of 100 kHz, Q_{eq} achievable is 20,000! Using the same approach, if a double integrator is used in

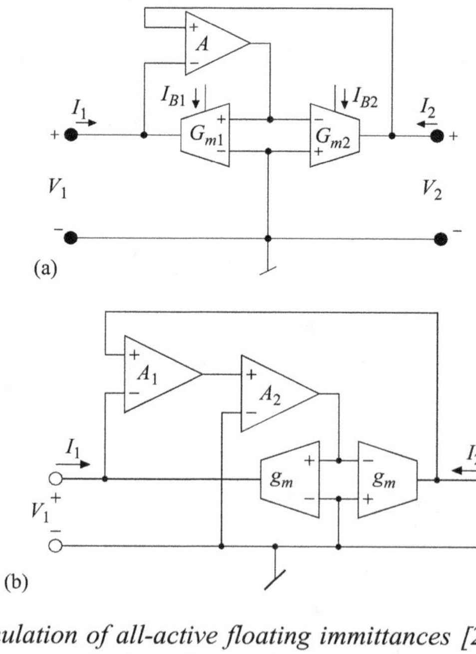

Figure 3.42 Simulation of all-active floating immittances [28]: (a) simulation of all-active ideal FI, (b) simulation of all-active ideal FDNC

the same scheme, the resulting circuit would simulate a FFDNC characterized by $Z_{eq}(s) = s^2/G_{m0}\omega_{t1}\omega_{t2}$ which is electronically controllable through I_{B0}. Considering a more accurate op-amp model, the realized floating impedance takes the following form:

$$Z_{eq}(s) = \frac{(s+\omega_{p1})(s+\omega_{p2})}{G_{m0}\omega_{t1}\omega_{t2}} = \frac{s^2}{G_{m0}\omega_{t1}\omega_{t2}} + \frac{(\omega_{p1}+\omega_{p2})s}{G_{m0}\omega_{t1}\omega_{t2}} + \frac{\omega_{p1}\omega_{p2}}{G_{m0}\omega_{t1}\omega_{t2}} \tag{3.72}$$

This indicates that this circuit realizes series *RLM* floating impedance with

$$R = \left(\frac{f_{p1}}{f_{t1}}\right)\left(\frac{f_{p2}}{f_{t2}}\right)\left(\frac{1}{G_{m0}}\right); \quad L = \frac{(f_{p1}+f_{p2})}{G_{m0}f_{t1}f_{t2}} \text{ and } M = \frac{1}{\omega_{t1}\omega_{t2}G_{m0}} \tag{3.73}$$

3.10 Electronically controllable resistors using OTAs

Realization of an electronically variable grounded resistor using OTAs is rather straightforward. The input resistance of the arrangement of Figure 3.43(a) is found to be $R_{in} = (1/G_m) = (2V_T/I_B)$, which indicates that by varying I_B from 0.1 μA to 1 mA (four decades), and assuming $V_T = 25$ mV, R_{in} can be varied over a wide-range of four decades (i.e. 50 Ω to 500 kΩ).

It is interesting to point out that a grounded negative resistance is obtainable from the same circuit by interchanging the polarities of the input terminals.

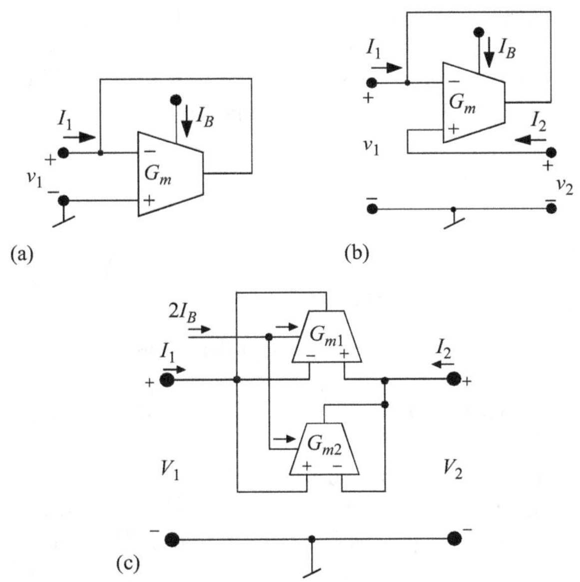

Figure 3.43 OTA-based resistors: (a) grounded resistor [1], (b) semi-floating resistor, (c) fully floating resistor [27]

If the noninverting terminal of the OTA is ungrounded and a two-port is created as shown in Figure 3.43(b), it is characterized by

$$[Y] = \begin{bmatrix} G_m & -G_m \\ 0 & 0 \end{bmatrix} \tag{3.74}$$

The above is semi-indefinite $[Y]$ matrices which represent a 'semi-floating' resistance. If two such circuits are connected in a parallel back-to-back connection as shown in Figure 3.43(c), this circuit has a $[Y]$ matrix

$$[Y] = \begin{bmatrix} G_m & -G_m \\ -G_m & G_m \end{bmatrix} \tag{3.75}$$

for $G_{m1} = G_{m2} = G_m$ the circuit realizes a fully floating positive resistance: $R_{eq} = (1/G_m) = (2V_T/I_B)$. From this circuit, a floating negative resistance can be obtained by the simple artifice of changing the polarities of both the OTAs which will make its $[Y]$ matrix as

$$[Y] = \begin{bmatrix} G_m & -G_m \\ -G_m & G_m \end{bmatrix} = -\left(\frac{I_B}{2V_T}\right)\begin{bmatrix} 1 & -1 \\ -1 & 1 \end{bmatrix} \tag{3.76}$$

which represents a floating negative resistance of value $-(2V_T/I_B)$. Here, it may be mentioned that in OTA-based analog circuit design, there are many situations where it may not be really necessary to employ a fully floating simulated resistance; in many cases even a semi-floating resistor may suffice.

As an example, consider the simple *RC*-LPF of Figure 3.44(a). It would appear that because the resistor is 'floating', it needs to be replaced by the two OTAs circuit of Figure 3.43(c). However, we show that in this case even the single-OTA semi-floating resistor can do the same job, if we write the node equation by replacing R by $1/G_m$, as follows: The node equation is

$$\frac{V_1(s) - V_2(s)}{R} = V_2(s)sC, \tag{3.77}$$

Now replacing R by $1/G_m$, we get

$$G_m\{V_1(s) - V_2(s)\} = V_2(s)sC \tag{3.78}$$

If this equation is synthesized by an OTA and a grounded capacitor circuit, we obtain the circuit of Figure 3.44(b).

Thus, it is seen that the OTA in this circuit has been used as a semi-floating resistance $R = 1/G_m$.

As another example, we show the utility of the semi-floating resistance in the generation of a number of new[8] grounded/floating inductance simulation circuits. If we were to obtain a resistorless version of the lossless GI circuit of Figure 3.16, it

[8]The circuits of Figure 3.44(c), (d), (f) and (g) have not been published in the open literature earlier and are being formally presented, for the first time, in this book only.

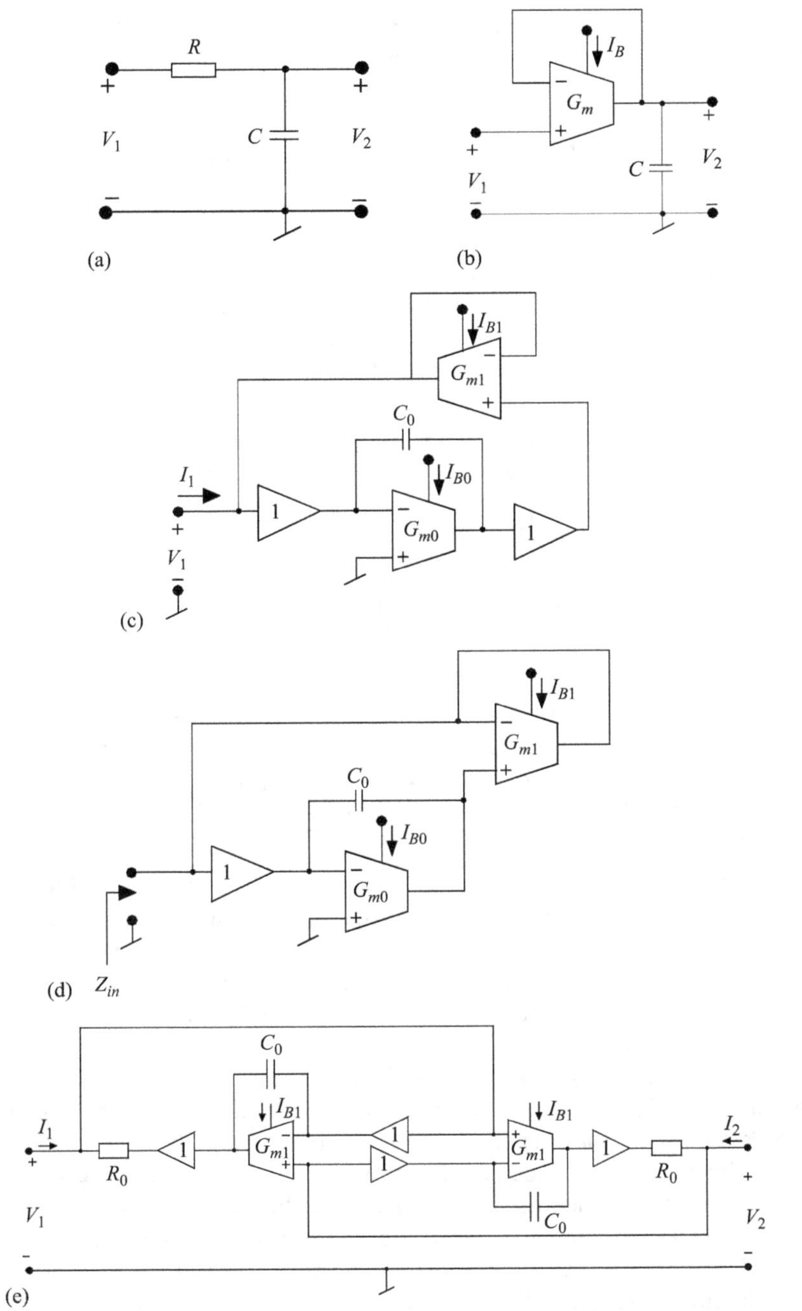

Figure 3.44 *Application examples: (a) simple RC–low pass filter, (b) OTA–C filter, (c) resistorless version of a GI circuit of Figure 3.16, (d) modified version of resistorless version of a GI circuit of Figure 3.16, (e) FI obtained by cascade back-to-back approach, (f) an alternative FI version of GI of Figure 3.16, (g) another version of FI without any resistor*

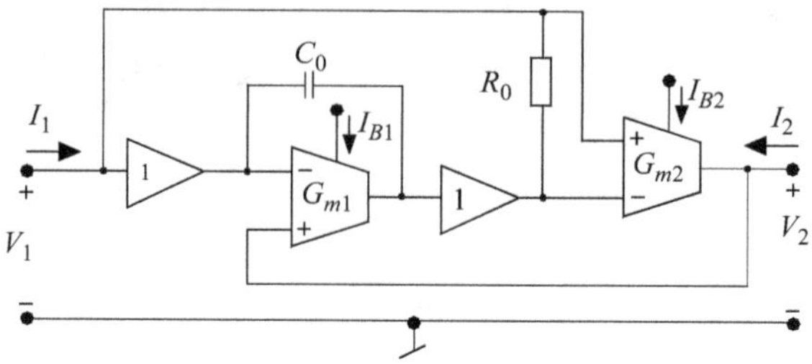

(f)

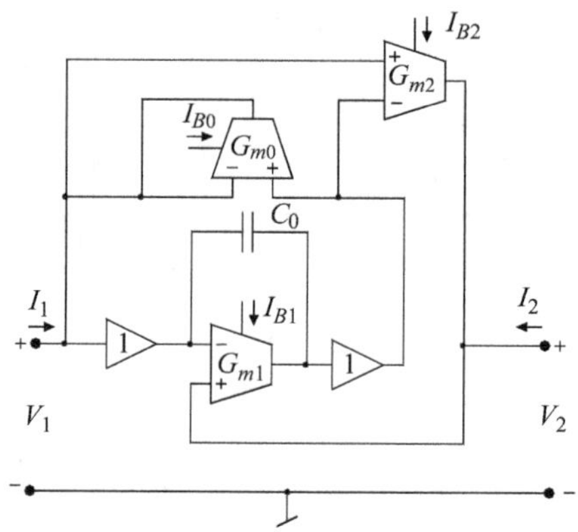

(g)

Figure 3.44 (Continued)

could be obtained by replacing the feedback resistor R_0 by a single OTA simulated semi-floating resistor of value $1/G_{m1}$ as shown in Figure 3.44(c). Note that the second voltage buffer is now rendered redundant and the circuit can be further simplified as shown in Figure 3.44(d). This circuit can be considered to be an alternative to the one shown in Figure 3.17 using exactly the same number of active and passive components.

It is interesting to observe that a floating version of the lossless GI of the circuit of Figure 3.16 was also presented by the authors of [9], which is based upon the cascaded back-to-back approach of FI simulation and the same is reproduced here in Figure 3.44(e). This circuit, however, suffers from the drawbacks of (i) employing two capacitors, (ii) using double the number of passive elements, (iii) requirement of component-matching between two identical subcircuits and (iv) employing a large number of active elements.

In the following, we show that an alternative and better circuit can be derived by following the methodology highlighted in this section. For this, we first show a circuit which simulates the same kind of FI which has been obtained by creating a two-port from the same GI circuit by un-grounding the noninverting terminal of OTA embedded between two VFs and then putting the input terminal of the second OTA across the resistor R_0 with its current output terminal connecting to port 2, thus, subject to the condition $G_{m2}R_0 = 1$, the circuit of Figure 3.44(f) simulates an

LLFI of value

$$L_{eq} = \frac{C_o R_0}{G_{m1}} = \frac{2V_T C_o R_0}{I_{B1}} \tag{3.79}$$

Lastly, if the resistor R_0 is also simulated by a semi-floating resistor using a third OTA as shown in Figure 3.44(g), the resulting circuit realizes an LLFI with

$$L_{eq} = \frac{4V_T^2 C_0}{I_{B1} I_{B0}} \tag{3.80}$$

subject to the fulfilment of the condition $I_{B2} = I_{B0}$. It may now be seen that when compared with the FI of Figure 3.44(e) the last two circuits use a single capacitor and employ a modest number of four/five active elements, respectively.

We now present a circuit which realizes a generalized resistance converter. The part of the circuit of Figure 3.45 shown in dotted box actually implements an electronically controlled current conveyor which is configured as an NIC by shorting its Y-terminal and Z-terminal. Therefore, when port x is terminated into a resistor R_2, the input resistance is given by

$$R_{in} = -R_2 \left(\frac{2V_T / R_1}{I_{B2}} \right) \quad \text{for} \;\; G_{m1} R_2 \gg 1 \tag{3.81}$$

A positive resistance of the same value can be obtained by interchanging the polarities of the input terminals of the OTA2.

Lastly, consider a somewhat more sophisticated electronically controlled resistance circuit employing three OTAs which is shown in Figure 3.46. By straightforward analysis, one can confirm that this two-port circuit realizes a floating resistance given by

$$R_{eq} = \frac{G_{m2}}{G_{m1} G_{m3}} \tag{3.82}$$

Clearly, a negative resistance is realizable by changing the polarities of the input terminals of the OTA3. The workability of this circuit has been confirmed by balanced CMOS OTAs for the details of which the reader is referred to [30].

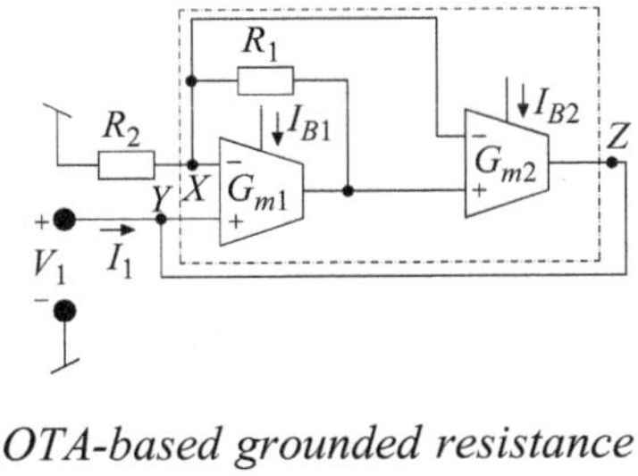

Figure 3.45 OTA-based grounded resistance converter [10]

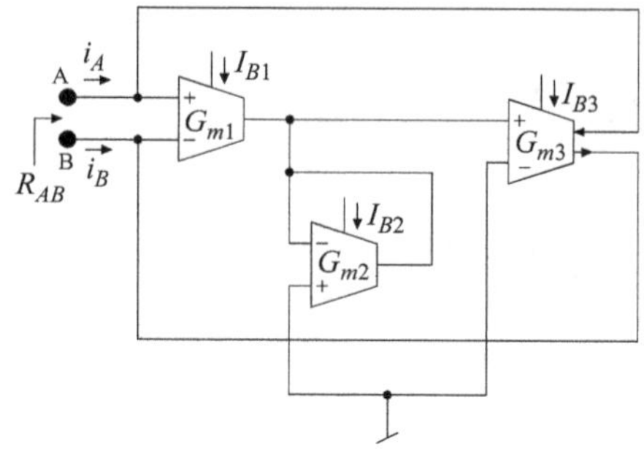

Figure 3.46 Electronically controllable floating resistor [30]

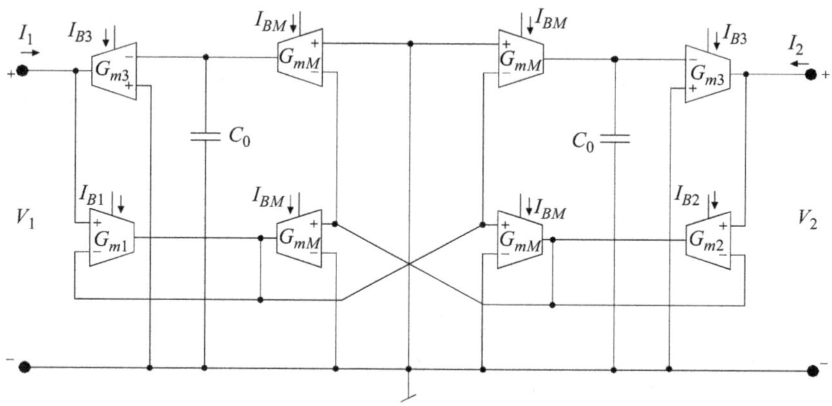

Figure 3.47 Electronically controllable OTA–C mutually coupled circuit [49]

3.11 Simulation of mutually coupled circuits and transformers

Figure 3.47 shows an OTA circuit which simulates a mutually coupled circuit. By straightforward analysis this OTA circuit is found to have the parameter values given by

$$L_1 = \frac{C_0}{G_{m1}G_{m3}}, \quad L_2 = \frac{C_0}{G_{m2}G_{m3}} \quad \text{and} \quad -M = \frac{-C_0}{G_{mM}G_{m3}} \tag{3.83}$$

3.12 Multi-port gyrators using OTAs: retrospection

In retrospect, it is interesting to point out that many of the developments/circuits included in this chapter can be traced back to the notion of multi-gyrators on one hand and to 'a simple method of producing floating inductors' published by Sewell

[73] as early as in 1969 which surprisingly got completely overlooked by most of the researchers working in the area of inductance simulation using OTAs.

We may recall that as an extension to the familiar two-port gyrator, multi-gyrators (three, four-port gyrator and so on) were already introduced in the circuit theory literature long back (1968, 1970) by Holt and Linggard [46,47].

Sewell [73] proposed two configurations: one for realizing a gyrator circuit using the so-called five-terminal amplifiers which are shown here in Figure 3.48. These five-terminal amplifiers are actually VCCS – which is the same thing as a differential input dual (complementary output) OTA. The circuit of Figure 3.48(a) is characterized by a definite admittance matrix given by

$$[Y] = \begin{bmatrix} 0 & 0 & g & -g \\ 0 & 0 & -g & g \\ -g & g & 0 & 0 \\ g & -g & 0 & 0 \end{bmatrix} \tag{3.84}$$

which can be partitioned as

$$[Y] = \begin{bmatrix} 0 & G \\ -G & 0 \end{bmatrix} \tag{3.85}$$

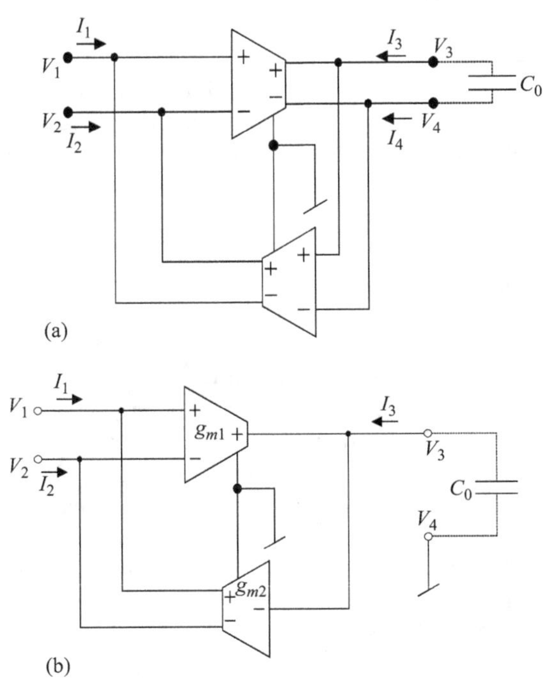

Figure 3.48 Gyrator circuits using five-terminal amplifiers [53]: (a) simulation of FI using four-port gyrator, (b) simulation of FI using three-port gyrator

Note that this latter $[Y]$ matrix is of the same form as that of a common two-port gyrator; only thing different is that both input port and output port are floating in this case. If output port is terminated into a capacitor, the circuit realizes a floating inductance $L = C_0/g^2$ (H).

On the other hand, for obtaining an FI from a grounded capacitor, one needs a three-port gyrator characterized by

$$[Y] = \begin{bmatrix} 0 & 0 & g \\ 0 & 0 & -g \\ -g & g & 0 \end{bmatrix} \tag{3.86}$$

Now if one connects a capacitor between port 3 and ground, the resulting circuit would simulate an LLFI between ports 1 and 2, the value of which would be still given by $L = C_0/g^2$ (H).

Looking back, now we can easily relate many earlier circuits to be different manifestations of either of the two schemes described above. This is particularly and more specifically true about the circuits of Figures 3.7, 3.11 and 3.12.

3.13 Concluding remarks

This chapter focussed on the realization of gyrators, inductance simulators and related impedances using IC OTAs as basic building blocks. First, gyrators and positive/negative grounded inductors were considered using only OTAs and capacitors. These circuits have the advantage of employing only OTAs and capacitors (in many cases no external resistors might be employed) while providing electronic control of the realized inductance value. These features make the OTA-based circuits attractive for IC implementation though they can be surely used effectively in several discrete applications too, using commercially available IC OTAs.

Besides entirely OTA–C circuits, a class of impedance simulation circuits using mixed source arrangements such as using op-amps and OTAs, OTAs and unity gain adder/subtractor circuits and the so-called active only circuits were also presented. Besides inductor simulators, circuits realizing FDNC elements, capacitance-multipliers and electronically variable positive/negative, grounded/floating resistance realizations were also elaborated upon. For some other works related to this approach, not explicitly dealt herewith, the readers may see [74,75].

Lastly, it was pointed out how many of the OTA–C FI circuits can be seen to be the new *avatars* of two classic FI creation schemes first introduced by Sewell [73] as early as in 1969 using the so-called five-terminal amplifiers which were functionally acting as VCCS.

It is hoped that this chapter has revealed many interesting and useful immittance simulation circuits which can be advantageously employed in the design of electronically tunable active filters, oscillators and any other analog signal processing/generation applications.

References

[1] Bialko M., Newcomb R. W. 'Generation of all finite linear circuits using the integrated DVCCs'. IEEE Transactions on Circuit Theory. 1971; **18**(6):733–6.

[2] Bialko M. 'High Q inductance simulation using a single operational amplifier or DVCCS/DVCVS element'. Bulletin De L' Academic Polonaise Des Sciences. 1975; **23**(6):521–6.

[3] Soliman A. M. 'A grounded inductance simulation using the DVCCS/DVCVS'. Proceedings of the IEEE. 1978; **66**(9):1089–91.

[4] Odess L. R. 'Comments on 'A grounded inductance simulation using the DVCCS/DVCVS''. Proceedings of the IEEE. 1979; **67**(12):1662–3.

[5] Soliman A. M. 'Reply: A grounded inductance simulation using the DVCCS/DVCVS'. Proceedings of the IEEE. 1979; **67**(12):1962.

[6] Wheatley C. F., Wittlinger H. A. 'OTA obsoletes op-amp'. Proceedings of the National Electronics Conference, 1969:152–7.

[7] Khan I. A., Ahmed M. T. 'OTA-based integrable voltage/current-controlled ideal C-multiplier'. Electronics Letters. 1986; **22**(7):365–6.

[8] Ahmed M. T., Khan I. A., Parveen T. 'Wide range electronically tunable component multipliers'. International Journal of Electronics. 1988; **65**(5): 1007–11.

[9] Khan I. A., Ahmed M. T., Parveen T. 'Novel wide-range electronically tunable ideal grounded inductance'. IEE Proceedings Pt. G. 1988; **135**(3):104–5.

[10] Surakampontorn W., Jutaviriya S., Riewruja V. 'OTA-based electronically tunable voltage-controlled resistance converter'. International Journal of Electronics. 1989; **67**(1):81–5.

[11] Khan I. A., Ahmed M. T., Minhaj N. 'Novel OTA-C ideal inductance simulators'. Proceedings of the 14th National System Conference; Aligarh Muslim University, Aligarh, March 1991:11–4.

[12] Khan I. A., Ahmed M. T., Parveen T. 'Some non-ideal component simulators using OTAs'. Proceedings of the 14th National System Conference; Aligarh Muslim University, Aligarh, March 1991:33–6.

[13] Ahmed M. T., Khan I. A., Minhaj N. 'Novel electronically tunable C-multipliers'. Electronics Letters. 1995; **31**(1):9–11.

[14] Abuelma'atti M. T., Khan M. H. 'Current-controlled OTA-based single-capacitor simulations of grounded inductors'. International Journal of Electronics. 1995; **78**(5):881–5.

[15] Abuelma'atti M. T., Khan M. H. 'New temperature-insensitive electronically-tunable grounded capacitor simulator'. Active and Passive Electronic Components. 1996; **19**(1):55–8.

[16] Ndjountche T., Unbehauen R., Luo F.-L. 'Electronically tunable generalized impedance converter structures'. International Journal of Circuit Theory and Applications. 1999; **27**(5):517–22.

[17] Gupta S. 'Universal impedance generator handles biomedical applications'. Electronic Design, 2001.

[18] Park Ji-Mann., Shin Hee-Jong., Chung, Won-Sup. 'Active-RC circuit synthesis for the simulation of current-controllable inductors and FDNRs'. Journal of the Institute of Electronics Engineers of Korea SD. 2003; **40**(12):54–62.

[19] Padilla-Cantoya I. 'Capacitor multiplier with wide dynamic range and large multiplication factor for filter applications'. IEEE Transactions on Circuits and Systems-II. 2013; **60**(3):152–6.

[20] Theingjit S., Tangsrirat W. 'Grounded series RL-type lossy inductance simulator using Gm-C'. KKU Engineering Journal. 2016; **43**(S2):228–30.

[21] Theingjit S., Pukkalanun T., Tangsrirat W. 'Grounded FDNC and FDNR realizations based on Gm-C technique and their applications to ladder filter design'. Engineering Letters. 2016; **24**(3):1–5.

[22] Theingjit S., Pukkalanun T., Tangsrirat W. 'FDNC realization and its application to FDNR and filter realizations'. Proceedings of the International Multi Conference of Engineers and Computer Scientists; Hong Kong, March 16–18, 2016, International Association of Engineers.

[23] Nandi R. 'Lossless inductor simulation: novel configurations using D.V.C.C.S.'. Electronics Letters. 1980; **16**(17):666–7.

[24] Singh V. K. 'Comment on lossless inductor simulation: novel configurations using DVCCS'. Electronics Letters. 1981; **17**(15):549–50.

[25] Nandi R. 'New grounded-capacitor simulation of grounded and floating inductances using DVCCS/DVCVS'. Circuit Theory and Applications. 1981; **9**(1):115–7.

[26] Senani R. 'On the synthesis of a class of immittances and filters using grounded capacitors'. International Journal of Circuit Theory and Applications. 1983; **11**(4):410–5.

[27] Khan I. A., Ahmed M. T. 'Realisation of tunable floating resistors'. Electronics Letters. 1986; **22**(15):799–800.

[28] Khan I. A., Ahmed M. T., Minhaj N. 'Novel technique for immittance simulation-realisation of some all-active simulators'. International Journal of Electronics. 1992; **72**(3):431–41.

[29] Hou C. L., Wang W. Y. 'Circuit transformation method from OTA-C circuits into CFA-based RC circuits'. IEE Proceedings-Circuits Devices Systems. 1997; **144**(4):209–12.

[30] Kaewdang K., Kumwachara K., Surakampontorn W. 'Electronically tunable floating CMOS resistor using OTA'. IEEE International Symposium on Communications and Information Technology; Beijing, China, October 12–14, 2005, IEEE.

[31] Springl V., Jaikla W., Siripruchyanun M. 'Floating positive/negative resistance simulators employing single dual-output OTA'. International Symposium on Communications and Information Technologies; Bangkok, Thailand, October 18–20, 2006, IEEE.

[32] Banchuin R., Chaisrcharoen R., Chipipop B., Sirinaovakul B. 'In depth analysis of the CMOS OTA-based floating inductors'. International Symposium on Intelligent Signal Processing and Communication Systems; Yonago Convention Centre, Tottori, Japan, December 12–15, 2006.

[33] Banchuin R., Chippipop B., Sirinaovakul B. 'The complete passive equivalent circuit of the practical OTA-based floating inductor'. Proceedings of the International Conference: Mixed Design of Integrated Circuits and Systems; Gdynia, Poland, June 22–24, 2006, IEEE.

[34] Banchuin R., Chipipop B., Sirinaovakul B. 'Novel practically applicable passive equivalent circuit model of the alternatively structured higher performance practical OTA-based floating inductor'. Proceedings of 2007 International symposium on intelligent signal processing and communication systems; Xiamen, China, November 28–December 01, 2007, IEEE.

[35] Riewruja V., Petchmaneelumka W. 'Floating current-controlled resistance converters using OTAs'. International Journal of Electronics and Communications. 2008; **62**(10):725–31.

[36] Uzunov I. S. 'Theoretical model of ungrounded inductance realized with two gyrators'. IEEE Transactions on Circuits and Systems-II. 2008; **55**(10):981–5.

[37] Petchmaneelumka W. 'OTA-based positive/negative floating inductance simulator'. Proceedings of the International Multiconference of Engineers and Computer Scientists; Hong Kong, March 18–20, 2009, International Association of Engineers.

[38] Petchmaneelumka W. 'Simple floating inductance simulators using OTAs'. International Instrumentation and Technology Conference; Singapore, May 5–7, 2009, IEEE.

[39] Pierzchala M., Fakhfakh M. 'Generation of active inductor circuits'. Proceedings of 2010 IEEE International Symposium on Circuits and Systems; Paris, France, May 30th–June 2nd, 2010, IEEE.

[40] Fard R. A., Pooyan M. 'A low voltage and low power parallel electronically tunable resistor with linear and nonlinear characteristics'. Microelectronics Journal. 2012; **43**(7):492–500.

[41] Katageri M. V., Mutsaddi M. M., Mathad R. S. 'Comparative study of simulated floating inductances using OTA in low pass filter'. IOSR Journal of Engineering. 2014; **04**(06):22–31.

[42] Ananda Mohan P. V. 'Capacitor floatation scheme using only OTAs and grounded capacitors'. Journal of Circuits, Systems, and Computers. 1995; **5**(2):181–97.

[43] Prodanov V. I., Green M. M. 'A current-mode FDNR circuit element using capacitive gyrators'. IEEE International Symposium on Circuits and Systems; London, UK, May 30th–June 2nd, 1994, IEEE.

[44] Li, Yong-an. 'NAM expansion method for systematic synthesis of OTA-based floating gyrators'. International Journal of Electronics and Communications. 2013; **67**(4):289–94.

[45] Singh A. K., Senani R. 'Low-component-count active-only immittances and their application in realising simple multifunction biquads'. Electronics Letters. 1998; **34**(8):718–9.

[46] Holt A. G. J., Linggard R. 'The multi-terminal gyrator'. Proceedings of the IEEE. 1968; **56**(8):1354–5.

[47] Holt A. G. J., Linggard R. 'Multi-terminal gyrator'. Proceedings of the IEE. 1970; **117**(8):1591–8.

[48] Silva M. M. 'Multiport converters and inverters'. Circuit Theory and Applications. 1978; **6**(3):243–52.
[49] Higashimura M., Fukui Y. 'Electronically tunable OTA-C mutually coupled circuit'. Electronics Letters. 1991; **27**(14):1251–2.
[50] Bialko M., Tomaszewski Z. 'Simulation of floating inductance having unilateral properties'. Archiwum Elektrotechniki. 1973; **22**(1):33–47.
[51] Bialko M., Tomaszewski Z. 'The fundamental properties of the one-and two-ports containing the L,C reactance elements and an unilateral inductance L'. Archiwum Elektrotechniki. 1974; **23**(4):1037–48.
[52] Nandi R. 'New ideal active inductance and frequency-dependent negative resistance using D.V.C.C.S./D.V.C.V.S.: applications in sinusoidal oscillator realisation'. Electronics Letters. 1978; **14**(17):551–3.
[53] Trimmel H. R., Heinlein W. E. 'Fully floating chain-type gyrator circuit using operational transconductance amplifiers'. IEEE Transactions on Circuit Theory. 1971; **18**(6):719–21.
[54] Huynh L. T., Knob A., Zufferey J. L. 'Realization of bandpass filter with floating frequency dependent-negative conductances using operational transconductance amplifiers'. Proceedings of the 1978 European Conference on Circuit Theory and Design. Lausanne, Switzerland; 1978. p. 155–9.
[55] Abuelma'atti M. T. 'Active–only immittance simulators'. Frequenz. 2003; **57**(1–2):8–11.
[56] Senani R. 'Comment: Floating ideal inductor with one DVCC'. Electronics Letters. 1980; **60**(4):117.
[57] Patranabis D., Paul A. N. 'Floating ideal inductor with one D.V.C.C.S.' Electronics Letters. 1979; **15**(18):545–6.
[58] Patranabis D., Paul A. N. 'Novel capacitor floatation scheme'. Electronics Letters. 1979; **15**(21):688–9.
[59] Senani R. 'Some new synthetic floating inductance configurations'. Archiv Fur Elektronik und Ubertragungstechnik. 1981; **35**(7/8):307–10.
[60] Senani R. 'New single-capacitor simulations of floating inductors'. Electrocomponent Science and Technology. 1982; **10**(1):7–12.
[61] Malvar H. S., Luettgen M. 'Temperature compensation of OTA-based filters and amplifiers'. Electronics Letters. 1987; **23**(17):890–1.
[62] Karybakas C. A., Kosmatopoulos C., Laopoulos Th. 'Improved temperature compensation of OTAs'. Electronics Letters. 1992; **28**(8):763–4.
[63] Senani R. 'Realisation of single-resistance-controlled lossless floating inductance'. Electronics Letters. 1978; **14**(25):828–9.
[64] Silva M. M. 'On the realization of immittance inverters with a minimum number of active components'. IEEE Transactions on Circuits and Systems. 1979; **CAS-24**(11):931–5.
[65] Geiger R. L., Sanchez-Sinencio E. 'Active filter design using operational transconductance amplifiers: a tutorial'. IEEE Circuits and Devices Magazine. 1985; **1**(2):20–32.
[66] LM13600. 'Dual operational transconductance amplifiers with linearizing diodes and buffers'. National Semiconductor. Data Sheet, 1995.

[67] LM3080. 'Operational transconductance amplifier'. National Semiconductor. Data Sheet, 1995.

[68] Hou C. L., Chen C. C., Hwang C. M., Wu J. S. 'Switched-OTA technique for floating immittance function simulators'. Microelectronics Journal. 1995; **26**(7):691–5.

[69] Ramus X. 'Demystifying the operational transconductance amplifier'. Application Report, SBOA117A, 2009.

[70] Patranabis D. 'Inductor realization with resistive feedback in RC filters'. International Journal of Electronics. 1975; **38**(3):367–374.

[71] Jaikla W., Siripruchyanun M. 'Floating positive and negative inductance simulators based on OTAs'. International Symposium on Communications and Information Technologies; Bangkok, Thailand, October 18–20, 2006, IEEE.

[72] Senani R. 'Canonic synthetic floating-inductance circuits employing only a single component-matching condition'. Journal of the Institution of Electronics and Telecommunications. 1981; **27**(6):201–4.

[73] Sewell J. I. 'A simple method for producing floating inductors'. Proceedings of the IEEE. 1969; **57**(12):2155–6.

[74] Sun Y. 'OTA-C filter design using inductor substitution and Bruton transformation methods'. Electronics Letters. 1998; **34**(22):2082–3.

[75] Kuntman H., Menekay S., Cicekoglu O., Kuntman A. 'Novel parallel lossy inductance simulation circuit employing DO-OTA'. First IEEE Balkan Conference on Signal Processing, Communications, Circuits and Systems; Maslak, Istanbul, Turkey, June 2–3, 2000.

Chapter 4

Synthetic impedances using current conveyors and their variants

Abstract

This chapter discusses prominent contributions made over the last five decades on the evolution of gyrators, synthetic inductances and other related immittances using current conveyors (CCs) and their numerous variants developed over the years. The various circuits using the most basic forms of CCs, namely the first-generation CCs (CCI) and the second-generation CCs (CCII) have been elaborated first. This is then followed by the circuits employing an electronically controllable version of CCII called controlled second-generation current conveyor. Finally, numerous recent circuits based on different variants of CCs such as dual-output CCIIs, differential voltage CCs, third-generation CCs, differential difference CCs, inverting CCII, dual-X CCII and fully differential CCII and numerous combinations thereof have been dealt with. A number of applications of the various synthetic impedances have been elaborated.

4.1 Introduction

In the open literature the credit for introducing the concept of current conveying and its embodiment as first-generation current conveyor (CCI) in 1968 and that of a more versatile the so-called second-generation current conveyor (CCII) in 1970 goes to Sedra and Smith.[1] However, some recent disclosures have indicated that basic current conveyor topology was invented independently earlier by Gilbert[2] who had also coined the term current conveyor for it.

The CCI is a three-port network element with the three ports called X, Y and Z with its terminal voltages and currents given by $I_Y = I_X, V_X = V_Y$ and $I_Z = \pm I_X$

[1]Smith K. C., Sedra A. 'The current conveyor – a new circuit building block'. Proceedings of the IEEE. 1968; **56**(8):1368–9; Sedra A., Smith K. C. 'A second generation current conveyor and its applications'. IEEE Transactions on Circuit Theory. 1970; **17**(1):132–4.

[2]It was brought to the attention of the first author (through a private communication) by Barrie Gilbert that he had independently invented a translinear *current conveyor topology* as a part of a monolithic IC op-amp which he had designed during 1967 but which was actually fabricated and mass produced by analog devices as AD844 much later. This fact has now been disclosed by Gilbert himself in a recent publication titled 'The Current-mode muddle' [1].

whereas CCII is a three-port network element characterized by $i_y = 0, v_x = v_y$ and $i_z = \pm i_x$. It was demonstrated that CCII was a more versatile element in the realization of various basic analog operations as well as in the realization of various controlled sources and other functional circuits. The versatility of the CCII stems from the fact that on one side, it has properties similar to those of an ideal op-amp, e.g. ideally infinite input impedance looking into terminal *Y* and equality of voltages on terminals *X* and *Y* but on the other side, it could accept a current input at terminal *X* (which has ideally zero input impedance) and *convey* the same current at port *Z* (which has ideally infinite output impedance). These superior terminal characteristics were subsequently exploited by several researchers to come up with functional circuits which could be realized using CCIIs with the least possible number of external elements without requiring any component-matching constraints which were unavoidable in the case of the realization of the same functions using the then prevalent IC op-amps. To take advantage of the superior terminal characteristics of the CCII elements, several applications of the CCIIs started appearing in the technical literature after 1970 which included several proposals on realizing inductors too. It may be mentioned, however, that in their second publication on CCII itself Sedra and Smith had shown how a *gyrator* could be made from two CCIIs of opposite polarities along with a bare minimum of only two external resistors!

As amply demonstrated in earlier chapters of this monograph, the active elements called Impedance converters and impedance inverters are not only useful building blocks in their own right but have been extensively employed to simulate inductors, FDNR (frequency-dependent negative resistance; an element having $Z(s) = 1/Ds^2$), FDNC (frequency-dependent negative conductance; an element having $Z(s) = Ms^2$) and several forms of negative elements such as negative resistances, negative capacitances and negative inductances. CCIIs are extremely versatile in realizing both impedance converters and inverters which are useful for the realization of synthetic impedances which have interesting applications in several areas such as in the realization of filters and oscillators (which include linear oscillators, relaxation oscillators as well as chaotic oscillators), in the techniques of loss-cancellation in gyrators/inductors/capacitors, in enhancing the frequency range of oscillators and several others.

Thus, during the last five decades, an abundance of inductance simulation circuits and those for realizing other kinds of related impedances (in both grounded and floating forms) have been developed by numerous researchers around the world; for instance, see [2–132].

Before the advent of the CCs, the analog circuit design has been very heavily dominated by the IC op-amps. However, of late it is being understood that CCs are not only as versatile as the op-amps but can outperform op-amps in many situations and applications. In particular, the CC-based circuits generally can be devised with a bare minimum number of external passive components, do not require any component matching, do not suffer from the effects of gain bandwidth conflict and do not have severe finite slew rate based effects (if realized with translinear circuits) exhibited by their op-amp counterparts. It is, therefore, widely believed now that CCs are the building blocks whose time has come. This belief has been further supported by the commercial availability of several bipolar CCs as off-the-shelf ICs coupled with excellent workability and versatility of several CC-based functional circuits which have been developed during the last five decades. On the other hand, numerous complementary metal oxide semiconductor (CMOS) architectures for the CCs have also been developed from time to time, for instance [133–149].

In this chapter, we take a stock of all works [1–413] done in this direction, however, we discuss *explicitly* only the prominent works done in this area along with their applications. Most of the circuits, elaborated here, can be made from the three-terminal CCs which are commercially available as ICs or can be built from other available ICs.

4.2 Realization of grounded and floating negative impedances

The basic two-port active element to create negative circuit impedances is the negative impedance converter (NIC). Although the use of negative impedances as employed in filters suffer from the drawbacks of associated notorious stability problems, still in several applications such as in oscillators, loss-compensation or frequency range extension, negative impedances can be used effectively by appropriately fulfilling stability constraints.

An NIC or a negative grounded element can be implemented by a CCI or CCII in a number of ways some of which are shown in Figure 4.1.

More elaborate designs of NICs employing more than one CC have also been known which exhibit a reduced effect of the parasitic impedances of the CCs, for instance, see [130]. On the other hand, a number of authors have dealt with the realization of floating negative impedances such as [33,44,45,47], some of which are shown in Figure 4.2.

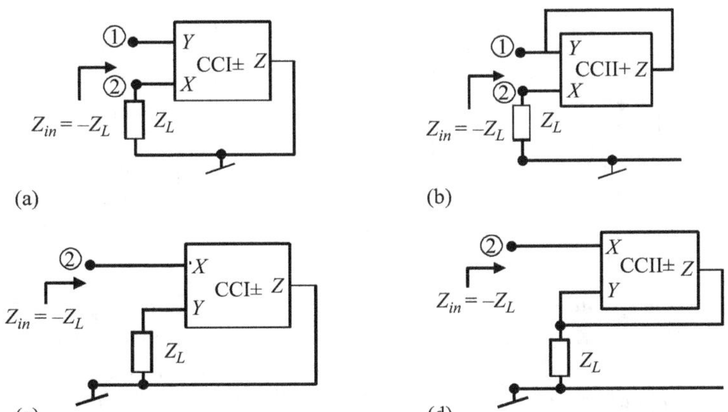

(a) (b) (c) (d)

Figure 4.1 Four different ways (a)-(d) of creating negative grounded impedance

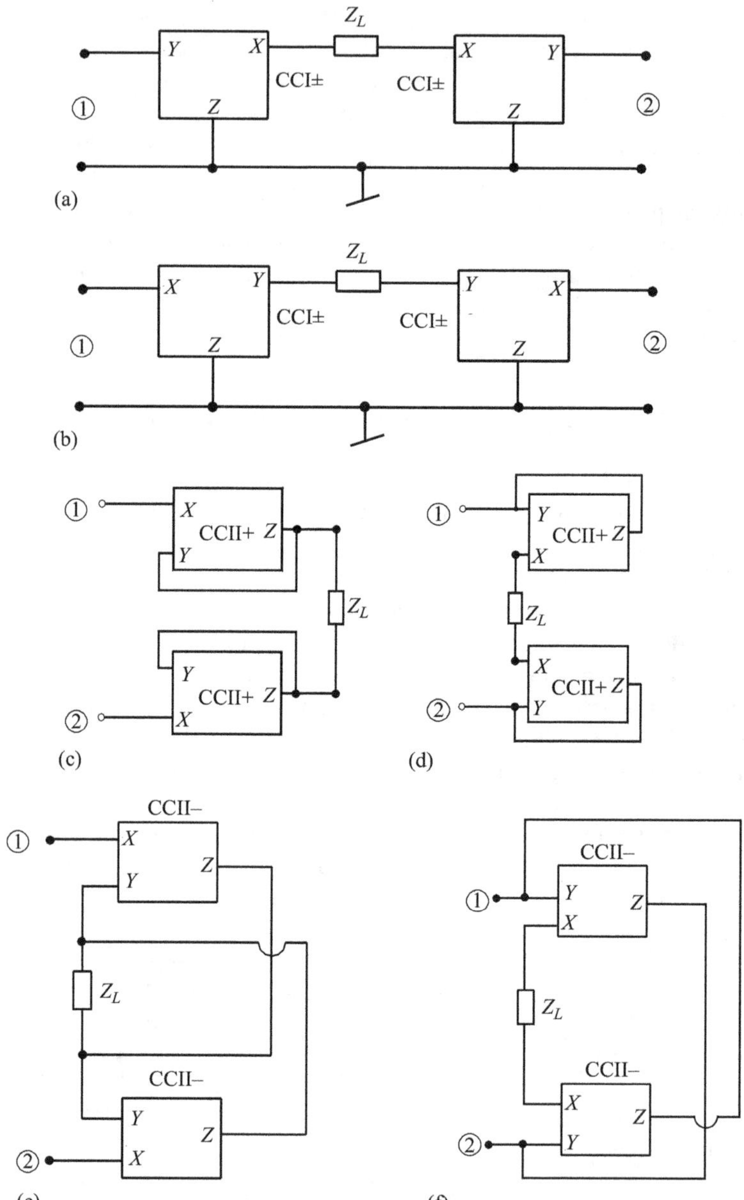

Figure 4.2 Some floating negative impedance circuits: (a) configuration for floating negative impedance using CCI± configured as NIC, (b) an alternative floating negative impedance, (c) floating negative impedance using CCII+ as NIC [44], (d) an alternative floating negative impedance using CCII+ as NIC, (e) a floating negative impedance using two CCII− [406,407], (f) an alternative floating negative impedance using CCII− [45]

If floating negative impedance is required, the same can be obtained by the well-known method of putting the load impedance in between two NICs; all three elements being connected in cascade. Thus, in principle, six such circuits, each realizable with only two CCs, are possible and are shown in Figure 4.2; out of which that of Figure 4.2(c) has been reported by Nandi [44], that of Figure 4.2(f) has been discussed by Nandi [45], whereas the one of Figure 4.2(e) has been presented by Patranabis *et al.* [406,407].

It is worth pointing out that some of these circuits evolve from CCII-nullor equivalence [47,50,172]. A related concept, the so-called four-terminal floating nullor (FTFN) [172,173] and its application in floating impedance realization would be explained subsequently in Chapter 6 of this monograph.

All the CCII−/CCII+ based structures of Figure 4.2 are characterized by the terminal equations: $i_1 = -i_2 = (-1/Z_L)(V_1 - V_2)$ and represent floating impedance $-Z_L$, between ports 1 and 2.

It is relevant to mention that four electronically controllable negative resistance converters have been presented by Surakampontorn and Thitimajshima [49]. While floating current-controlled resistance converter using electronically controlled conveyor (ECC) has been introduced by Riewruja and Petchmaneelumka [110].

4.3 Realization of synthetic grounded inductors, FDNRs and related elements

The use of CCs in simulating inductors, FDNRs and related elements has been one of the most prominent and significant applications of the CCs. This is attributed to the fact that the type of circuits for realizing these elements which can be made with CCs had not been possible to be made from other building blocks such as IC op-amps.

From the op-amp-based inductance simulation circuits presented in Chapter 2 of this monograph, it may be seen that using op-amps a lossless gyrator could not be made without requiring component-matching and cancellation constraints to realize the intended characterization, Furthermore, using op-amps, no circuit exists which can realize a *lossless* grounded inductor using a bare minimum number of passive components (namely only two resistors and a grounded capacitor).

In the following subsection, it will be shown that using CCs it is possible to devise synthetic grounded impedances which can overcome the above-mentioned difficulties and drawbacks.

4.3.1 Grounded inductance simulation using CCs

Sedra and Smith in their introductory paper on the second-generation CC itself had outlined several basic applications of CCII±. One of these was the interesting application of CCII in realizing PII/gyrator. It was shown that a gyrator can be made from a parallel back-to-back connection of two CCIIs configured as voltage-controlled current sources of the opposite polarity. With port 2 of such a two-port circuit terminated into impedance Z_3 (as in Figures 4.3 and 4.4) the input impedance of the resulting circuits is found to be $Z_{in} = Z_1Z_2/Z_3$.

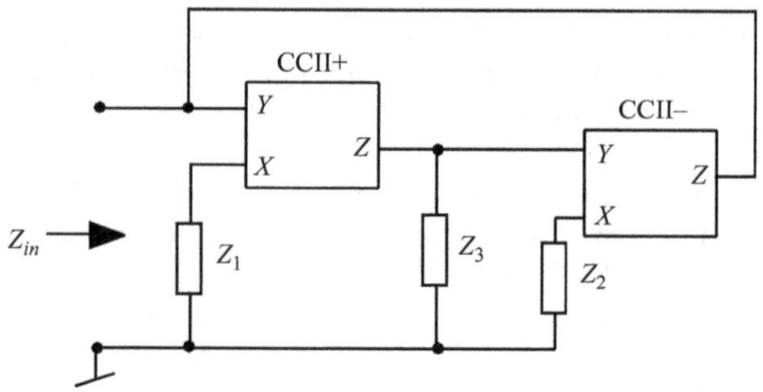

Figure 4.3 Sedra–Smith's lossless grounded impedance simulator

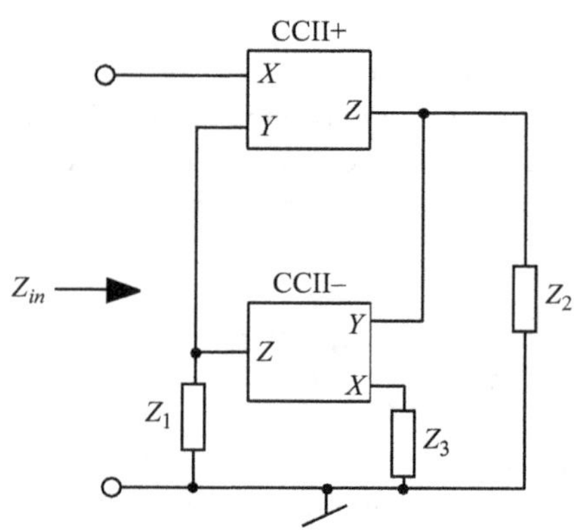

Figure 4.4 Lossless grounded impedance simulator proposed by Cicekoglu [76]

Both the circuits of Figures 4.3 and 4.4, with Z_3 as a capacitor C_0 and the remaining impedances being resistors R_1 and R_2, realize a pure inductance with $L_{eq} = C_0R_1R_2$. The circuit of Figure 4.4 was studied by Cicekoglu [76] considering the voltage and current tracking errors of the CCIIs, and it was found that this alternative version does offer better performance than the Sedra–Smith gyrator of Figure 4.3. Several other forms of lossless grounded inductors using CCs have been advanced from time to time such as [18,29,80,91,92]. However, most of these alternative circuits invariably require more than two CCs.

It may be mentioned that with identical polarities of CCIIs and appropriate selection of the circuit impedances (resistive/capacitive), the circuits of Figure 4.3 can realize a frequency-dependent positive resistance as in [13].

4.3.2 Single-CCII-based active gyrators

The literature on CCs contains a number of *active gyrators* such as those in [56,58,60,61,72,98,100,105,106,109,112,113]; however, the first single-CC active gyrator realizing a lossless inductance was proposed by Soliman [4] (and also *independently* by Paul and Patranabis [28]). The circuit contained in both [4] and [28] is shown in Figure 4.5.

By straightforward analysis, the input impedance of the circuit is found to be

$$Z_{in} = \frac{R_1R_2 + R_2(R_3 + R_4) - R_1R_4 - sC_1R_1R_2R_4}{sC\{(2R_2R_3) - R_1R_4\} - R_4} \tag{4.1}$$

Subject to the fulfilment of the following conditions:

$$2R_2R_3 = R_1R_4 \text{ and } R_2(R_1 + R_3 + R_4) = R_1R_4 \tag{4.2}$$

the input impedance becomes purely inductive and is given by

$$Z_{in} = sCR_1R_2 \tag{4.3}$$

It is, however, obvious from an inspection of expression (4.1) that by appropriate conditions imposed upon passive components, the same circuits can also be used to realize grounded series *RL* of parallel *RL* impedances.

Arslan *et al.* [91] have introduced two new lossless grounded inductance simulators using a single CCI. These circuits (shown in Figure 4.6) although employ a floating capacitor, like that in Soliman's circuit [4], would require one less resistor.

Application: A straightforward analysis shows that both the circuits of Figure 4.6 realize $Z_{in} = sCR^2/3$. The practical workability of these circuits has been successfully demonstrated in [91] by using them in the realization of a third-order high-pass Butterworth filter wherein a CMOS CCI was employed. The filter was realized for a cut-off frequency of 106 kHz.

4.3.3 Single CCII-based grounded impedance simulators

As amply demonstrated in Chapter 2, compared to lossless grounded inductance, a lossy inductor can often be simulated with a canonical number of components. This

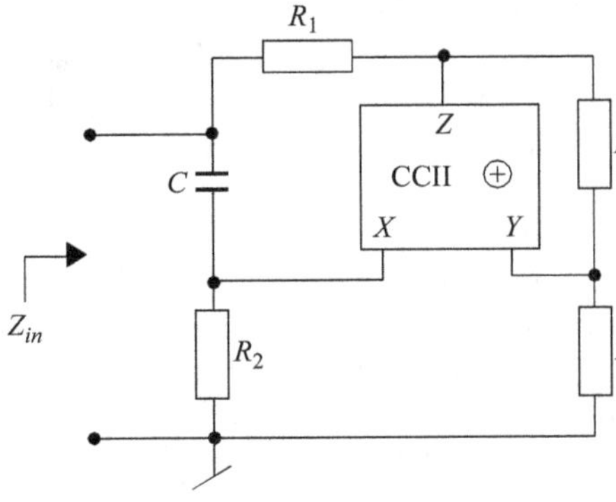

Figure 4.5 A single CCII+-based active gyrator proposed by Soliman [4] (also by Paul and Patranabis [28])

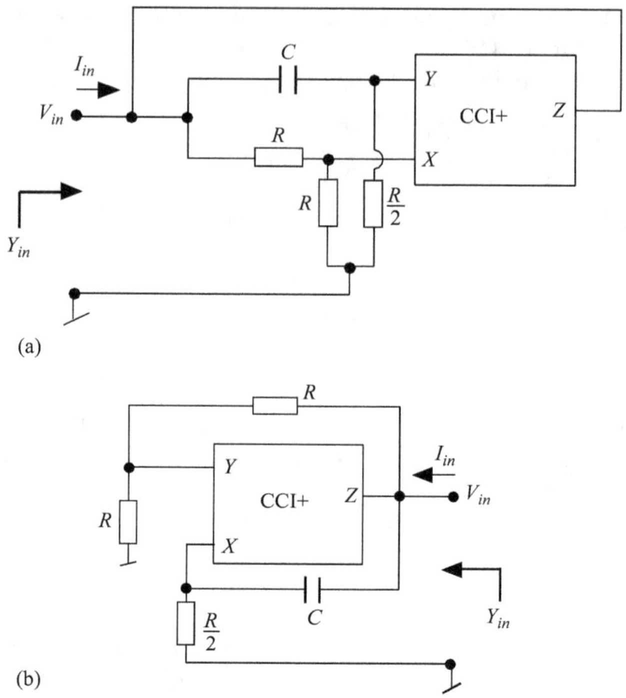

Figure 4.6 Two lossless grounded inductors (a), (b) using CCI

is indeed true in the case of CCII-based circuits too where a lossy grounded inductor can be simulated using only one CC, two resistors and a capacitor. A number of such circuits have been proposed in the technical literature by a number of researchers from time to time. However, as expected, several of these are analogous to the well-known economical op-amp *RC* simulators known earlier. A few of such circuits are shown in Figure 4.7, which were proposed by Nandi [3,6].

In the first circuit (Figure 4.7(a)), the CCII+ is configured as a unity gain voltage follower; thereby it is seen that the circuit is, in fact, a CC-version of the well-known Prescott circuit based on an op-amp [174]. The circuit simulates a series *RL*-type grounded inductor with $L_{eq} = CR_1R_2$ and equivalent resistance $R_{eq} = (R_1 + R_2)$.

On the other hand, the circuit Figure 4.7(b) although simulates a bilinear inductor with a series combination of a resistor R_a and a parallel-*RL* consisting of resistive part R_b and inductive part L_{eq} with their values given by $L_{eq} = CR_2 (R_1 - R_2)$, $R_a = R_2$ and $R_b = (R_1 - R_2)$, subject to $R_1 > R_2$. This circuit is a CC-analog of the Berndt–Dutta Roy circuit [176] but has the advantage of employing a grounded capacitor over the Berndt–Dutta Roy circuit [176] which employs a floating capacitance (FC).

Finally, in the circuit of Figure 4.7(c) exactly the same number of components have been employed as in the earlier two circuits while this circuit also simulates

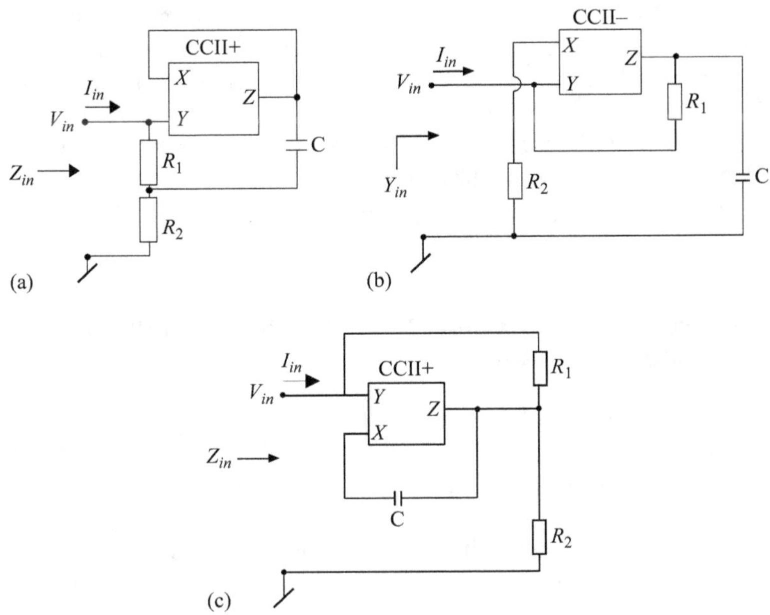

Figure 4.7 Low-component-count lossy inductance simulation circuits proposed by Nandi: (a) simulation of a lossy inductor (series RL) [3], (b) simulation of a bilinear inductor [3], (c) another circuit that realizes a series RL-type lossy inductor [6]

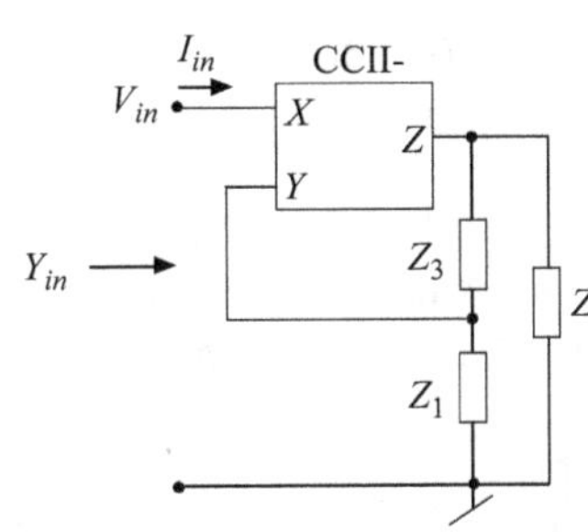

Figure 4.8 Generalized form of Soliman's CCII-based Ford-Girling equivalent circuit [5]

series RL inductor with $R_{eq} = (R_1 + R_2)$, but equivalent inductance is given by $L_{eq} = 2CR_1R_2$.

Lastly, Figure 4.8 shows the generalized form of a single-CC configuration which was proposed by Soliman [5] to realize the Ford-Girling [175] equivalent circuit using CCII. The realization of a lossy FDNR employing both grounded capacitors is possible from this circuit if the impedances Z_1 and Z_2 are chosen as capacitors and the impedance Z_3 is taken to be a resistor. This nonideal FDNR was shown [5] to be advantageous in realizing a low-pass (LP) filter section with transmission zero in the open left half of s-plane.

For some other CC-based lossy grounded inductance simulators or nonideal FDNR/FDNC simulators and related other circuits of interest, the readers are referred to [12,16,17,25,52,53,66,81–85,88,131,159–163,165–169,171,177–186, 190–192].

4.4 Synthetic floating impedances without component-matching requirements

A large number of contributions have been made on simulating floating lossless inductors using CCs such as [11,14,20,31,59,79,87,89,90,93,103,111,115,116,120, 121,126,127,129], as well as on realizing V–I scalars/FCs/FC multipliers such as [76,94,117,118].

In this subsection, we deal with some exemplary circuits from this repertoire, which are of specific significance, as will be demonstrated subsequently.

4.4.1 The first ever CC-based FI simulators without requiring any component matching

From the contents of Chapter 2 we understand that using op-amps, it had been conclusively proved, it is impossible to simulate a floating impedance of any kind (lossy or lossless) *without requiring any passive component matching.* Even with operational-transconductance amplifiers (OTAs), floating inductor (FI) realizations often call for equality of two transconductance values. Note that all available FI simulators using op-amps either require double the number of active and RC elements than their grounded counterparts or require a large number of resistors with critical component-matching conditions, for instance, see [151–158].

On the other hand, for simulating a lossless floating inductance either one can use CCII-based gyrators or CCII-based mutators (such as those in [2,57,86, 123,125]); however, in both the cases, the resulting circuits would require four or more CCs. Moreover, since these methods employ two identical gyrators/mutators to realize a lossless floating inductance; hence, the requirement of some prescribed component-matching conditions would still be needed for the intended realizations. As a consequence, any mismatch in the component values would, therefore, lead to inevitable degradation of the simulated inductance which leads to the appearance of undesirable parasitic components in its equivalent circuit. In 1979, it dawned upon the first author (RS) that the most desirable circuit for FI simulation will be the one which does not require any component matching or cancellation constraints. But it was difficult to comprehend at that time whether such a circuit was really possible using CCs. A very rigorous effort evidently did lead to the evolution of a class of such circuits which were published in a series of publications [11,22,23,30,35,41]. Here we present an account of the successive development of these circuits.

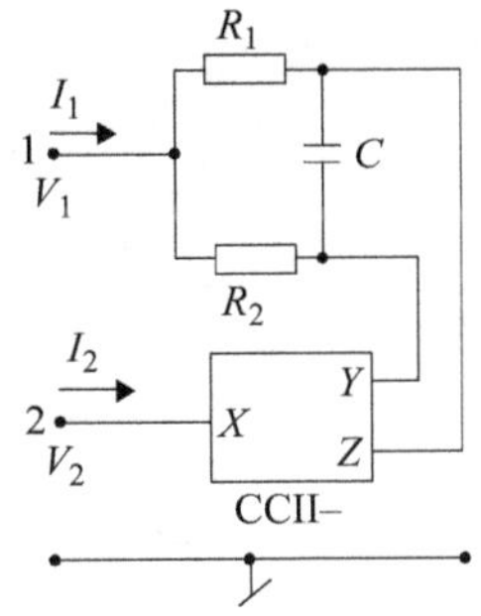

Figure 4.9 Senani's lossy floating inductance simulator without any component matching [11]

It was demonstrated *for the first time ever* by Senani in 1979 [11], *that indeed,* a lossy FI (parallel *RL* type) can be realized by using only one CCII– along with only three passive components (only two resistors and a capacitor) *without requiring any component-matching condition*. The circuit proposed in [11] is shown in Figure 4.9.

Assuming CCII– to be characterized by $i_y = 0$, $v_x = h_{21}\ v_y$ and $i_z = -h_{32}i_x$, where nominally $h_{21} = 1 = h_{32}$, the nonideal [*Y*]-matrix of the circuit is found to be

$$[Y'] = \left\{\left(\frac{1}{R_1} + \frac{1}{R_2}\right) + \frac{1}{sCR_1R_2}\right\}\begin{bmatrix} 1 & -\dfrac{1}{h_{21}} \\ -\dfrac{1}{h_{32}} & \dfrac{1}{h_{21}h_{32}} \end{bmatrix} \tag{4.4}$$

From the above, it can be seen that for $h_{21} = 1 = h_{32}$, the circuit simulates parallel *RL* floating impedance with R_{eq} and L_{eq} given by

$$R_{eq} = \frac{R_1R_2}{(R_1 + R_2)}, \quad L_{eq} = CR_1R_2 \tag{4.5}$$

The circuit enjoys very low active and passive sensitivities, all of which are given by

$$0 \le \|S_{xj}^F\| \le 1 \tag{4.6}$$

From the above circuit, a grounded inductor can be obtained in two different ways.

1. By shorting terminal 2 to ground in which case a grounded inductor is realized between the terminal 1 and ground, with equivalent inductance and resistance having the same values as shown above. However, the nonideal driving point admittance of the circuit in this case is given by

$$Y_{11}^{/}(s) = \left(\frac{1}{R_1} + \frac{1}{R_2}\right) + \frac{1}{sCR_1R_2} = Y_{11}(s)\text{ideal} \tag{4.7}$$

which implies that

$$S_{h_{21}}^{L_{eq}} = S_{h_{32}}^{L_{eq}} = S_{h_{21}}^{R_{eq}} = S_{h_{32}}^{R_{eq}} = 0 \tag{4.8}$$

Thus, an attractive feature of this circuit is that it is completely insensitive to active parameter variations!

2. It is interesting to observe that by shorting terminal 1 to ground, one obtains Soliman's Ford-Girling equivalent circuit [5] with equivalent inductance and resistance still given by the same values. It is worth mentioning that Soliman's circuit [5] is more advantageous in realizing a grounded FDNR in parallel with a capacitance, due to the use of both grounded capacitors which is an attractive feature from the point of view of integrated circuit implementation.

Soon after the publication of [11], it was demonstrated by Senani by presenting two new FI configurations in [23] how it is possible to devise single-resistance controllable lossy FIs too using CCIIs as active element, while ensuring that *no component-matching conditions are required.* These configurations are shown in Figure 4.10.

For the circuit of Figure 4.10(a), assuming ideal characterization of CCIIs, the parameters of the realized floating admittance are given by

$$\frac{1}{R_{eq}} = \frac{1}{R_1} + \frac{1}{R_2} + \frac{R_4}{R_1 R_2}; \quad L_{eq} = \frac{C R_1 R_2}{(1 + (R_4/R_3))} \tag{4.9}$$

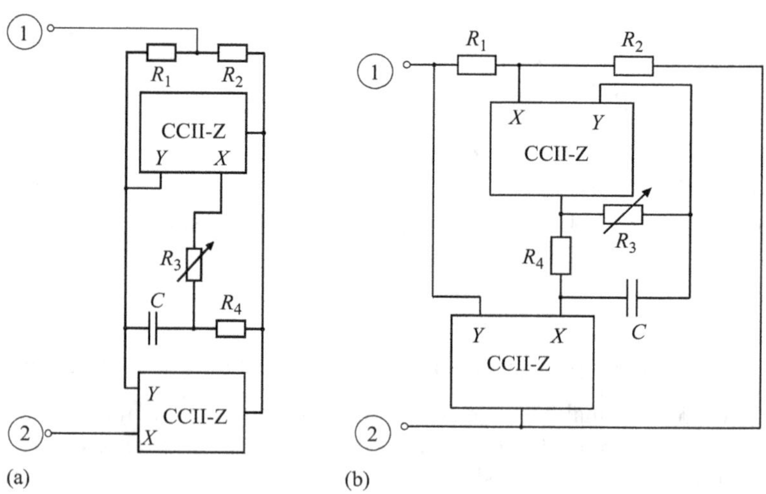

Figure 4.10 Single-resistance-controlled lossy FI simulations: (a) circuit for realizing tunable parallel RL floating admittance, (b) circuit for realizing tunable series RL floating admittance. Reprinted , with permission from [23] © 1980 IEE

and for the circuit of Figure 4.10(b), the parameters of the realized floating impedance are given by

$$R_{eq} = R_1 + R_2 + \frac{R_1 R_2}{R_4}; \quad L_{eq} = CR_1 R_2 \left(1 + \frac{R_3}{R_4}\right) \tag{4.10}$$

To further establish the important capability of the CCs in making it possible to realize FIs of various types without requiring any component matching, Senani in [22] presented three new FI configurations, all of which realize single-resistance-controllable FIs. One of these circuits is shown in Figure 4.11, which realizes a lossless FI of value $L_{eq} = CR_1 R_2$ without any constraints and component matching.

The remaining two circuits from [22] realize lossy FIs without any component matching. These are shown in Figure 4.12.

The relevant expressions of the realized immittances for the circuits of Figure 4.12(a) and (b) are, respectively, given by

$$Z(s) = R_1 + sCR_1 R_2; \quad Y(s) = \frac{1}{R_1} + \frac{1}{sCR_1 R_2} \tag{4.11}$$

Since a grounded capacitor is preferred for IC implementation, Pal [127] and, subsequently, Singh [29] presented four-CC-based circuits for realizing lossless floating inductance employing a grounded capacitance. Both these circuits, however, employed more than the minimum required number of resistors (four in Pal's circuit and three in Singh's circuit) and *did require* component-matching conditions (matching of two pairs of resistors in Pal's circuit [127] and matching of one pair of resistors in Singh's circuit [29]). From the above two developments, therefore, it was found that the idea that with CCs one can evolve FI circuits without requiring any component matching in a canonic manner as demonstrated by Senani in a series of publications [11,22,23]. Somehow was not comprehended by the other researchers yet!

To circumvent the above-mentioned deficiencies, therefore, Senani again demonstrated in [30], how even this task can be accomplished with CCs employing a grounded capacitor and not requiring any component-matching condition – a trait of employing CCs which he was trying to establish through his earlier publications

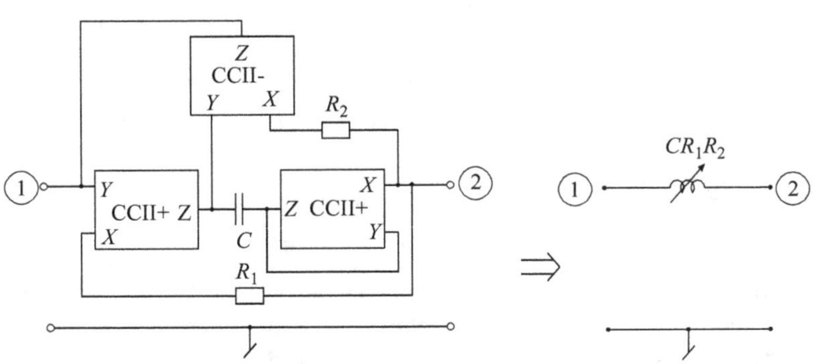

Figure 4.11 Lossless FI without any component matching. Reprinted, with permission from [22]. ©1980 IEE

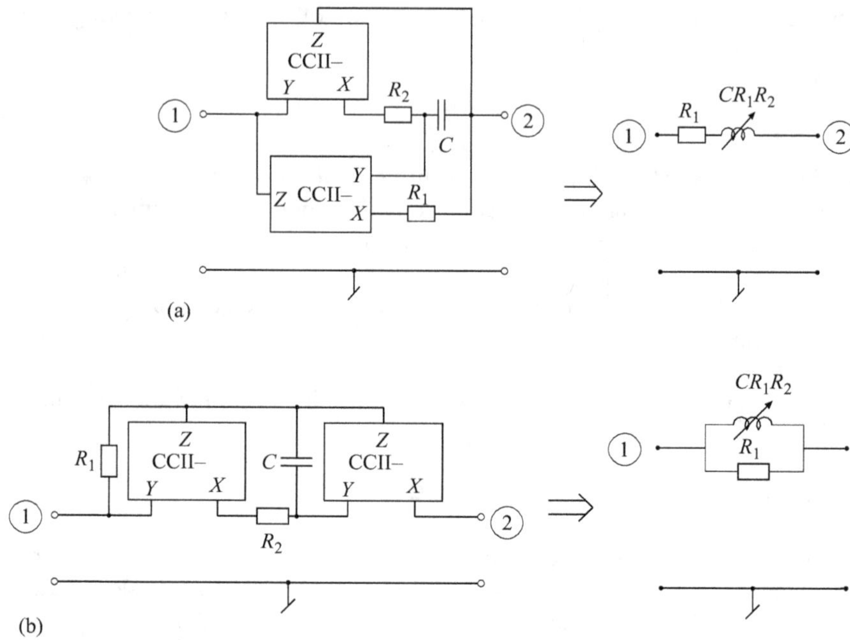

Figure 4.12 Lossy FIs without component-matching condition: (a) circuit for realizing a tunable floating series RL impedance, (b) circuit for realizing tunable floating parallel RL admittance. Reprinted , with permission from [22] © 1989 IEE

[11,22,23]. Thus, his circuit [30] shown here in Figure 4.13, like Pal and Singh circuits, uses four CCs and a grounded capacitor but needs no more than two resistors and provides a single-resistance control of the realized inductance but *does not require any component-matching* like his earlier circuits of [11,22,23].

It is useful to consider a nonideal analysis of this circuit taking the nonideal characterization of the CCs as

$$\left.\begin{array}{ll} i_y = 0, \quad v_x = h_{21}v_y, \quad \text{and} \quad i_z = \pm h_{32}i_x; & \text{for CCII}\pm \\ i_y = h_{12}i_x, \quad v_x = h_{21}v_y, \quad \text{and} \quad i_z = h_{32}i_x; & \text{for CCI} \end{array}\right\} \tag{4.12}$$

The nonideal Y-matrix of the circuit is given by

$$[Y] = \frac{h_{32C}h_{12D}h_{21D}h_{21C}h_{32B}}{sCR_1R_2}\begin{bmatrix} +h_{21B} & -h_{21A} \\ \dfrac{-h_{21B}h_{32D}}{h_{32C}h_{12D}} & \dfrac{+h_{21A}h_{32D}}{h_{32C}h_{12D}} \end{bmatrix} \tag{4.13}$$

From the above matrix, it can easily be seen that even nonideally; all the y-parameters of the circuit still represent lossless inductance. Moreover, from the nonideal matrix above, it can be easily confirmed that the circuit also enjoys very low active and passive sensitivities all of which are given by

$$0 \leq \|S_{xj}^{F}\| \leq 1 \tag{4.14}$$

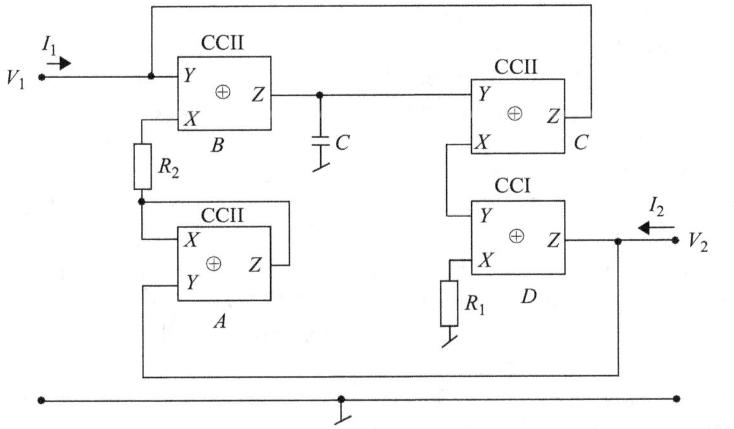

Figure 4.13 Senani's Single-resistance-controlled lossless FI employing a grounded capacitor; Reprinted, with permission from [30]. © 1982 IEE

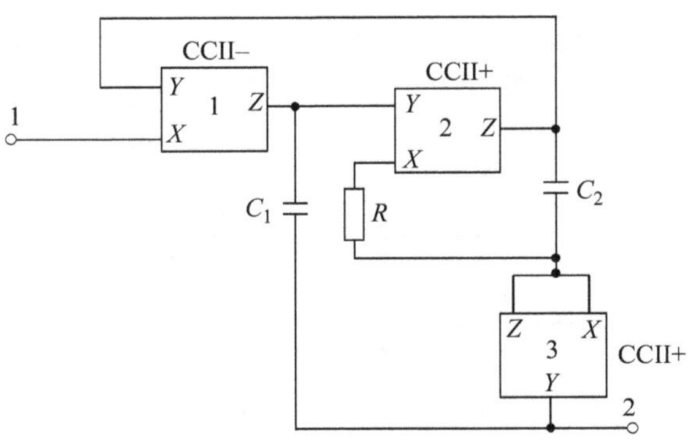

Figure 4.14 Floating FDNR circuit realization proposed by Nandi et al. [32]

It is important to note that if the resistor R_1 and the capacitor C are interchanged, the circuit simulates a floating capacitor with $C_{eq} = C(R_1/R_2)$ and therefore, in this mode, the circuit can be used for *realizing FC multipliers*. On the other hand, if R_1 and R_2 are replaced by capacitors C_1 and C_2 and the capacitor C is replaced by a resistor R, the circuit would be realizing an ideal floating FDNR ($Z(s) = 1/D_{eq}s^2$) with D_{eq} given by $D_{eq} = C_1C_2R$.

In retrospect it is found that soon after 1982, other researcher too picked up the most the most significant property of using CCs that if properly devised FI circuits using them should not require any component matching and that by judicious constructions, one can always manage to use only a bare minimum number of passive components.

The first evidence of this recognition came in 1983, where a floating ideal FDNR was reported by Nandi *et al.* [32] which is shown here in Figure 4.14 and realizes a floating impedance of value $Z(s) = 1/s^2C_1C_2R$, thereby representing an FDNR of value $D_{eq} = C_1C_2R$.

Another circuit (Figure 4.15) employing exactly the same number of CCs and *RC* elements to realize an ideal floating FDNR was subsequently published in 1984 by Nandi *et al.* [36] for simulating ideal tunable FDNR having value $D_{eq} = C_1C_2R$ where D_{eq} can be tuned by the only resistor *R* used in the circuit.

In contrast to Pal [127] and Singh [29] circuits, these two FDNR circuits did possess the characteristic feature of not requiring any component-matching condition while using canonical number of passive components. However, it appears to have skipped the attention of the authors of both [32] and [36] that a floating FDNR circuit obtainable from the *RC*:*CR* transformation of the lossless FI was already highlighted earlier in [22] itself.

In response to these circuits, Senani in [35] came up with a new floating FDNR configuration which could be realized with only two CCII− as shown in Figure 4.16 [35] without requiring any component-matching conditions.

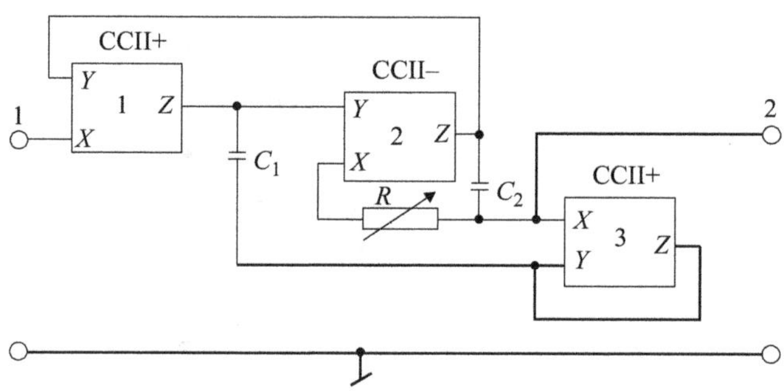

Figure 4.15 Lossless floating FDNR proposed by Nandi et al. [36]

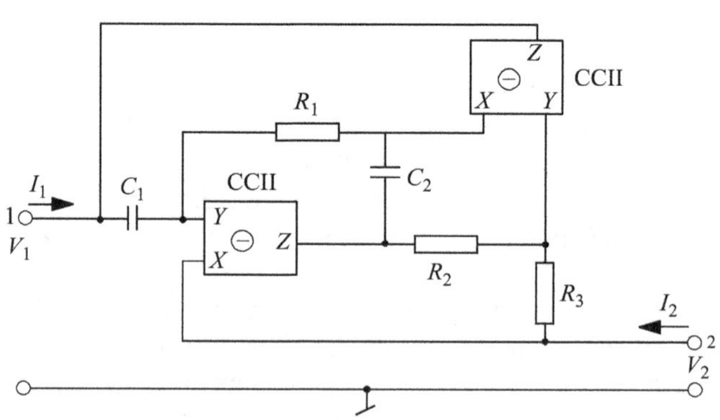

Figure 4.16 The first ideal floating FDNR realization using only two CCIIs proposed by Senani [35]

The circuit of Figure 4.16 simulates an ideal FDNR having value $D_{eq} = C_1C_2R_1R_2/R_3$. The practical validity of this floating FDNR was subsequently demonstrated by Wilson [39] who used this configuration to verify his CCII– implementation based on operational mirrored amplifiers.

Furthermore, Senani [41] subsequently demonstrated that the more versatile floating generalized positive immittance converter/inverter (FGPIC/FGPII) circuits too could be realized with only two CCII– without requiring any component-matching conditions. The circuits from [41][3] are shown here in Figure 4.17(a)–(c).

Analysis of these circuits, assuming ideal CCII–s, shows that the floating impedance realized between ports 1 and 2 is given by $(Z_1Z_4Z_L/Z_2Z_3)$, from which it is obvious that all the three circuits can be used to realize a variety of floating impedances using various choices (resistive/capacitive) of the circuit impedances.

A detailed nonideal analysis of the circuits of Figure 4.17(a)–(c) taking into account the nonideal h-parameters of the CCII–s reveals that the nonideal Y-parameters of the circuits of Figure 4.17(a) and (b) are more complex and unsymmetrical than those of circuit of Figure 4.17(c). Considering only the non-ideal current gains between ports X and Z of the two CCII–s, namely h_{32A} and h_{32B} of the two CCII–s, the nonideal Y-parameters of this circuit are found to be

$$Y_{11} = \frac{Y_1Y_4Y_L}{Y_2Y_3}h_{32A} + Y_1(1 - h_{32A}) = -Y_{12} \tag{4.15}$$

$$-Y_{21} = \frac{Y_1Y_4Y_L}{Y_2Y_3}\left(\frac{1}{h_{32B}}\right) + Y_1\left(\frac{Y_4}{Y_2}\right)\left(\frac{1}{h_{32B}} - 1\right) = Y_{22} \tag{4.16}$$

From the above expressions, it is seen that, other than the desirable first part, the equivalent circuit of the realized impedance will contain unwanted parasitic elements. However, these circuits can be successfully used in the design of filters as shown in [55] even though the CCII– was realized therein by the entirely op-amp-based CC-implementation of Huertas [95].

From the above-mentioned developments, it was well established that CCs provide novel solution to the problem of FI simulation, in contrast to FI circuits made from op-amps, OTAs or combinations thereof, in that by judicious arrangements with CCs, any kind of floating impedances could always be synthesized such that no component-matching conditions were needed and that one can invariably synthesize circuits which required only a bare minimum number of passive components.

These novel capabilities achievable with the use of CCs paved the way for a large number of new and innovative floating impedance circuits on which we shall dwell upon in the remaining sections of this chapter as well in several subsequent chapters of this monograph. Needless to say the above-quoted properties could easily be ensured in numerous FI circuits evolved using the CCIIs, the controlled CCIIs (CCCIIs) as well as in FI circuits realized with numerous other variants of the CCs. This will be demonstrated in the subsequent sections of this chapter.

[3]Obviously, these circuits are more economical end eminently more versatile than circuits evolved for similar purposes in [397].

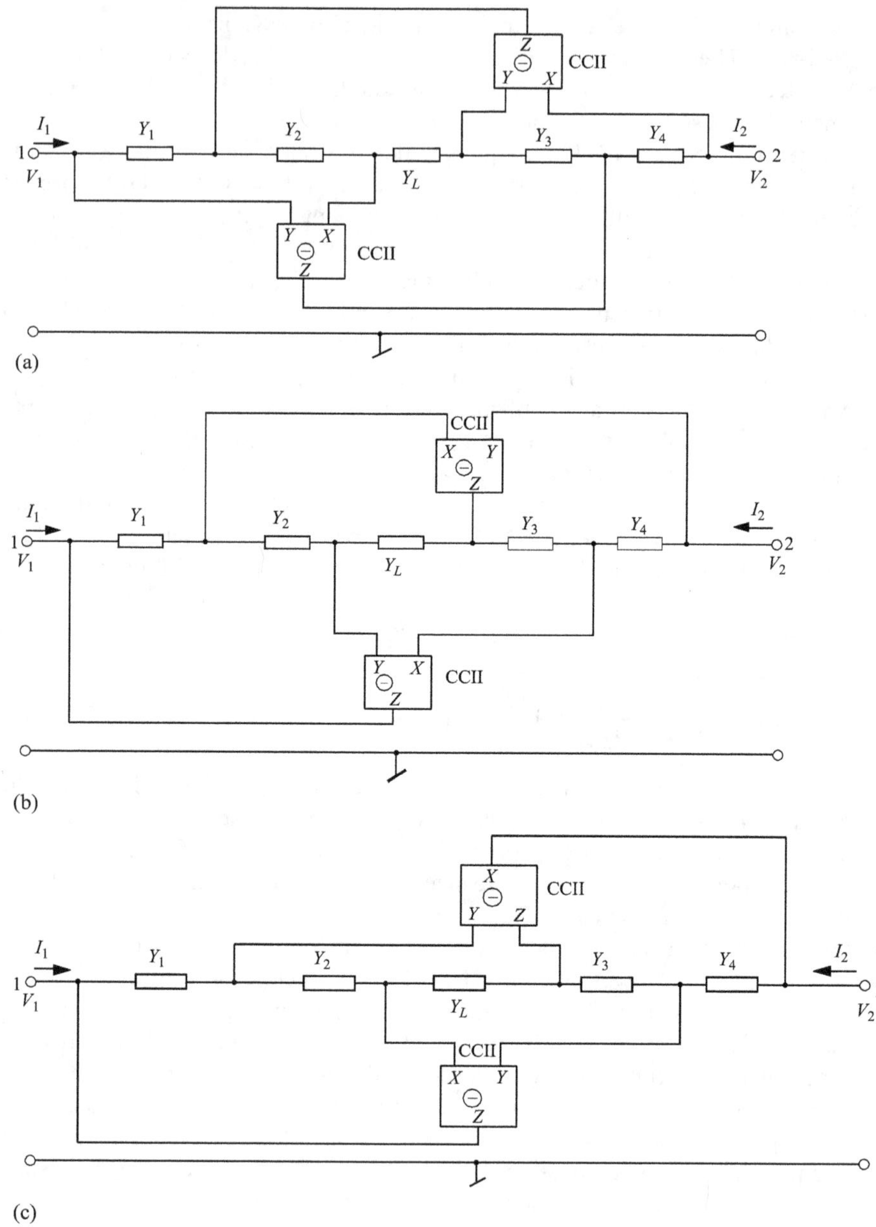

Figure 4.17 Three different floating generalized positive immittance converter/inverter (FGPIC/FGPII) circuits (a)-(c) proposed by Senani [41]

4.4.2 Two other single-CC-based FIs without component matching

Subsequent to the publication of the circuit of Figure 4.9, yet another single-CCII-based FI was reported by Singh [14] which employs exactly the same number of active and passive components but simulates a floating series *RL* impedance without requiring any component-matching condition (see Figure 4.18). The equivalent resistance (R_{eq}) and inductance (L_{eq}) for this circuit are given by $R_{eq} = (R_1 + R_2)$, $L_{eq} = CR_1R_2$.

From a nonideal analysis of this circuit, it is found that the nonideal parameters of this circuit are quite complex in contrast to the circuit of Figure 4.9, due to which all the sensitivity coefficients of this circuit *are not low*.

It is curious to observe in retrospect that although in 1983 Nandi and Nandi [31] reported another single-CCII-based FI circuit (see Figure 4.19), but the authors did not acknowledge as a reference either the paper [13] (which was published more than 4 years earlier) or any of the earlier references [11,23,30] (all of which were also published earlier to their paper [31]) wherein this significant capability of CCs (i.e. not requiring any component-matching condition for FI realization) was brought to the notice for *the first time ever*!

A limitation of their circuit is, however, that it can realize only a bilinear FI consisting of a resistance $r_a = R_1$ in parallel with a series *RL* consisting of a

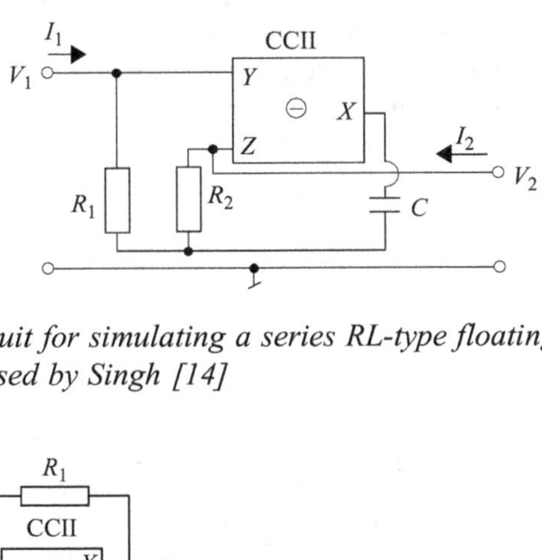

Figure 4.18 A circuit for simulating a series RL-type floating inductance proposed by Singh [14]

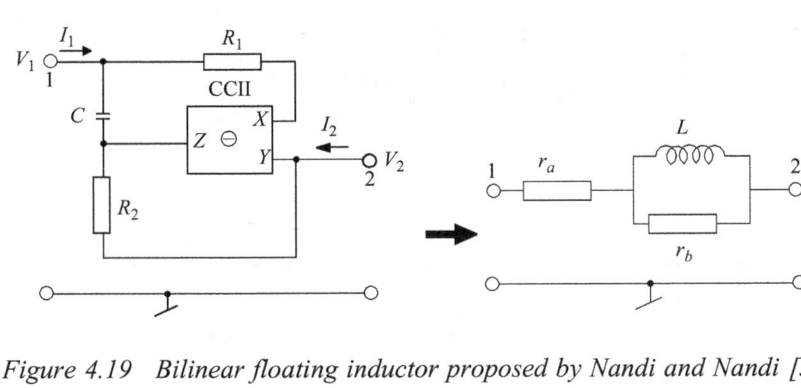

Figure 4.19 Bilinear floating inductor proposed by Nandi and Nandi [31]

resistance $r_b = (R_2 - R_1)$ and an inductor $L = CR_1R_2$, as per the equivalence shown in Figure 4.19 and therefore, has limited applications.

4.4.3 GPIC/GPII/three-port gyrator configurations using CCs

Floating GPIC/GPII constitutes a more general class of active elements which are useful for creating a variety of floating impedances as special cases. Since the introduction of CC, a number of CC-based configurations to realize them have so far been evolved; for instance, see [30,38,46,74,90,94,97].

Higashimura and Fukui [46] presented a number of circuits for general floating immittance realization using four CCs and four admittances out of which two admittances were required to be identical. Since these circuits required matching of two admittances for the desired realizations, these circuits are not being shown here explicitly; the interested readers may refer to [46] for details.

On the other hand, Yuce *et al.* [97] proposed an interesting circuit, which although employs four CCs but has the advantage of employing all CCII+ only. The circuit from [97] is shown in Figure 4.20 which is characterized by

$$[Y] = \left(\frac{Y_1Y_2}{Y_3} + \frac{Y_1Y_2}{Y_4}\right)\begin{bmatrix} 1 & -1 \\ -1 & 1 \end{bmatrix} = Y_{eq}\begin{bmatrix} 1 & -1 \\ -1 & 1 \end{bmatrix} \quad (4.17)$$

Although in this circuit matching of components is not needed but two capacitors would be required to simulate a lossless FI although both the capacitors are grounded. Similarly, in the case of floating FDNR simulation, two capacitors would be employed but both of these will be floating.

A practical advantage of this circuit is that all the CCII+s are readily implementable using commercially available off-the-shelf Integrated Circuit (IC) AD844.

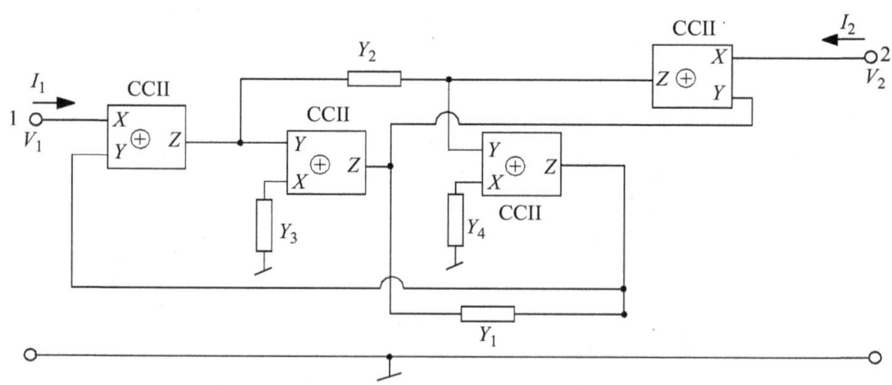

Figure 4.20 A general floating impedance simulation circuit proposed by Yuce et al. [97]

Another four-CC-based generalized floating impedance converter was introduced by Toumazou and Lidgey [38] which is based on an appropriate connection of two differential voltage-controlled current sources, as shown in Figure 4.21.

The floating impedance realized by this circuit is given by

$$I_1 = -I_2 = \frac{V_1 - V_2}{Z_{eq}}, \quad Z_{eq} = \frac{Z_1 Z_4}{(Z_2 + Z_3)} \tag{4.18}$$

Kiranon and Pawarangkoon [74] demonstrated that the same kind of lossless floating inductance can also be realized by a circuit comprising all CCII+ only along with only two resistors and a grounded capacitor as shown in Figure 4.22.

This circuit between ports 1 and 2 simulates a lossless floating inductance $L_{eq} = CR_1R_2$. For a systematic derivation of number of other current-mode lossless floating inductance circuits using CCs, the reader is referred to [71,73,75,101].

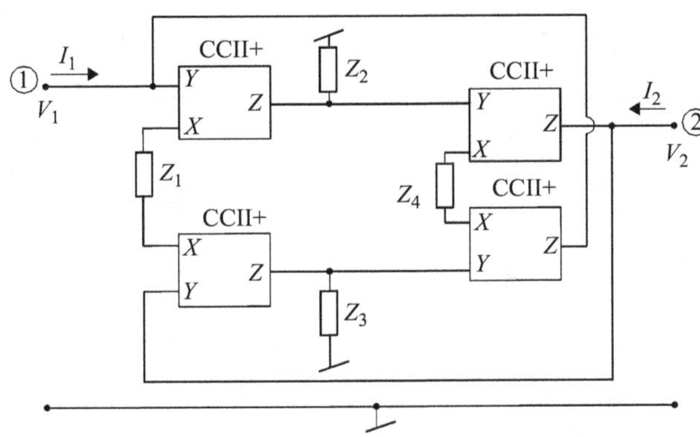

Figure 4.21 A generalized floating impedance realization proposed by Toumazou and Lidgey [38]

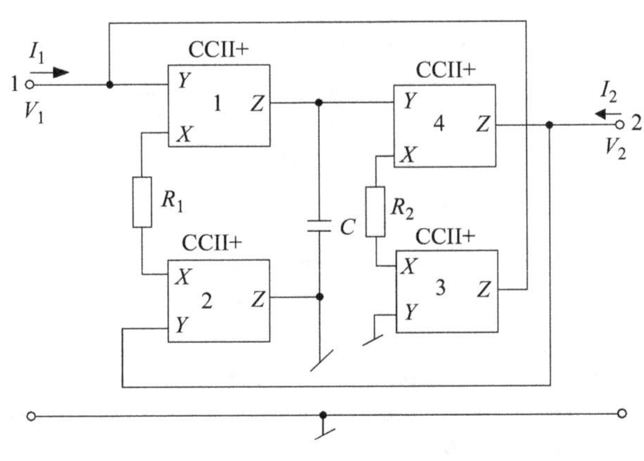

Figure 4.22 A lossless FI circuit proposed by Kiranon and Pawarangkoon [74]

Layos and Haritantis [71] presented a network synthetic derivation of a number of four-terminal (three-port) gyrators using CCs, OTAs and combinations thereof. Five specific circuits from the compilation of [71] are shown in Figure 4.23 which have different constraints/conditions of realizations. For instance, with a grounded capacitor connected at port 3, the circuit of Figure 4.23(a) realizes a lossless FI of value $L = CR_1R_3$ provided $R_1 = R_2$ and $R_3 = R_4$ [127]. On the other hand, with $R_1 = R_2$ the circuit of Figure 4.23(b) realizes $L = CR_1R_3$ with $R_1 = R_2$ [385]; the circuit of Figure 4.23(c) gives $L = (CR_1R_3)/2$ for $R_1 = R_2$ [29]; the circuit of Figure 4.23(d) simulates $L = CR_1R_2$ without any component matching [30] and so does the circuit of Figure 4.23(e) with $L = CR_1R_3$ without requiring any component matching [46].

A more versatile floating generalized impedance converter, although it employs five CCII+s along with five general impedances, was proposed by Pal [94]. This circuit simulates a floating impedance given by $Z_{1\text{-}2} = Z_1Z_3Z_5/Z_2Z_4$ and

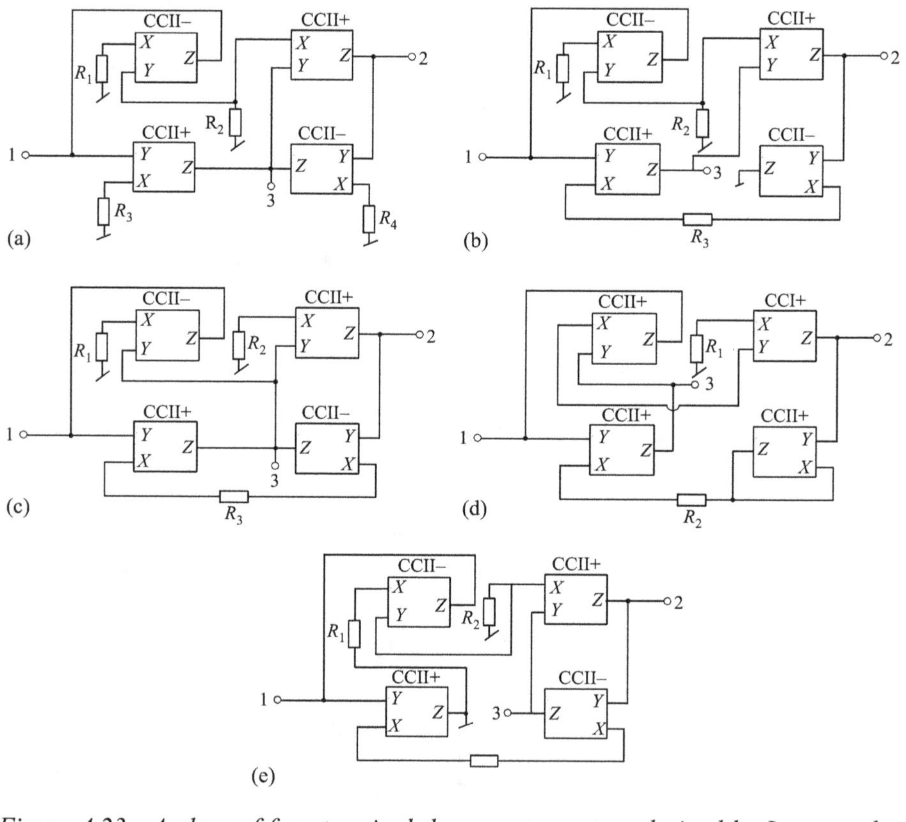

Figure 4.23 A class of four-terminal three-port gyrators derived by Layos and Haritantis [71] using only CCs: (a) based on the circuit by Pal [127], (b) based on the circuit by Pal [385], (c) based on the circuit by Singh [29], (d) based on the circuit by Senani [30], (e) based on the circuit by Higashimura and Fukui [46]

is shown here in Figure 4.24. It is interesting to note that this expression is identical to the one encountered in the classical Antoniou's GIC [150] (which, however, realizes a grounded impedance of the same kind). It is obvious that by judicious choice (resistive/capacitive) of the five impedances, the circuit can be used to realize a variety of floating impedances as follows:

(i) Floating inductor is obtained by choosing Z_4 as a capacitor and all the remaining impedances as resistors thereby leading to $Z_{1\text{–}2} = sC_4R_1R_3R_5/R_2$. (ii) Floating capacitor: choosing Z_3 as capacitor, the circuit can be seen to be a grounded capacitor to FC converter with the value of the FC given by $C_{1\text{–}2} = C_3R_2R_4/R_1R_5$. (iii) Floating FDNC: choosing Z_2 and Z_4 as capacitors, the input impedance becomes $Z_{1\text{–}2} = s^2C_2C_4R_1R_3R_5$ and the circuit, therefore, simulates a floating FDNC.

A compensated version of a simulated FI employing four CCs was presented by Ferri *et al.* [90] which involves an addition of two more CCs along with two resistors to the structure as shown in Figure 4.25.

By including the *Z*-port parasitics given by $R_z//(1/sC_z)$ and *Y*-port input capacitance as C_y, an analysis shows that the equivalent floating impedance realized by this circuit is given by

$$Z_{eq} = \left(\frac{R_1R_2}{R_Z} - 2R_x\right) + 2C_TR_1R_2 \tag{4.19}$$

where C_T represents the effective total capacitance at the node connecting *Z*-port of CCII+ 3 and *Y*-port CCII+ 6, i.e. $C_T = (C + C_{Z3} + C_{Y6})$. The condition for theoretical zeroing of the series resistance R_s is found to be

$$\frac{R_1R_2}{R_Z} = 2R_x \tag{4.20}$$

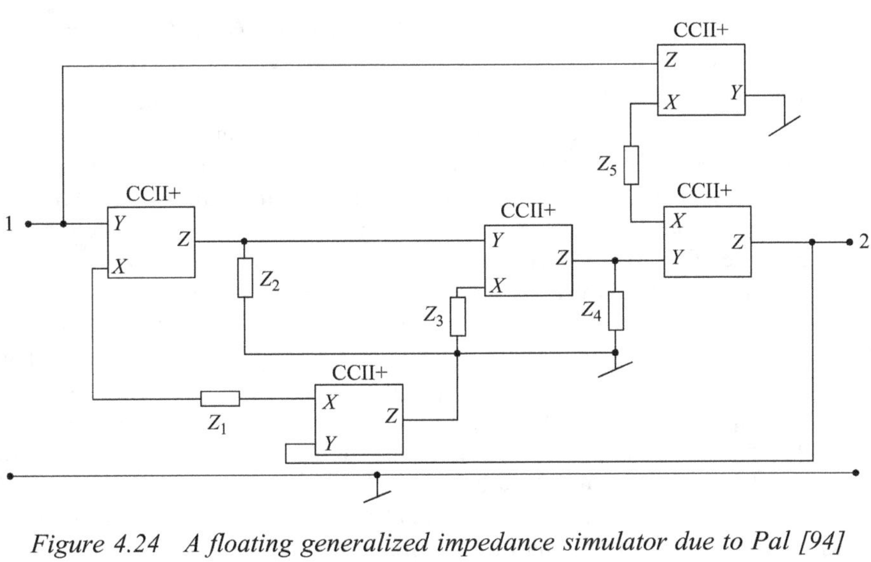

Figure 4.24 A floating generalized impedance simulator due to Pal [94]

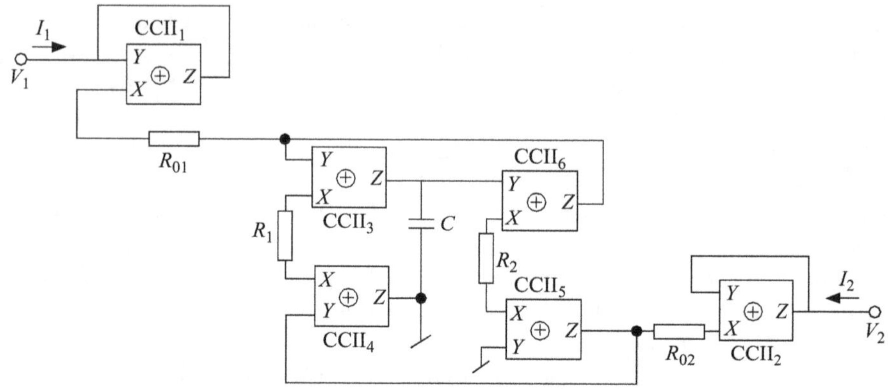

Figure 4.25 A compensated simulated floating inductor proposed by Ferri et al. Reprinted, with permission from [90]. © 2003 IEE

The effectiveness of the proposed compensation method has been demonstrated by realizing CCII+ with AD844 and choosing $R_1 = 100$ kΩ, $R_2 = 3.162$ kΩ and $C =$ 307 nF taking due cognisance of the current feedback operational amplifier (CFOA) parasitic impedances $R_x = 50$ Ω and $R_Z = 3$ MΩ. It has been shown that the series resistance reduces from 80 Ω to about 11 μΩ, which confirms the validity of the compensation and exhibits a considerable extension of the operating frequency range from two or three decades to about six decades.

Lastly, it may be mentioned that a circuit of Yuce [104] published in 2007 realized a lossless FI of value $L_{eq} = C_1R_1R_2$ but suffered from the drawback of requiring to fulfil a matching condition $C_2 = C_1(R_1 + R_2)/R_2$. For this and a number of other two-CC-based lossy FI simulators, the reader is referred to [71,101].

4.4.4 Additional three-CC-based floating inductor/FDNR simulators

It has been shown in [187,188] that if only two nullors are permitted then the realization of a gyrator requires at least five resistors and if only three resistors are allowed then at least three nullors would be needed to realize an ideal gyrator.

On the other hand, it is also known [189] that an impedance converter, with respect to some port, can also be looked upon as an impedance converter with respect to some other port and vice versa.

In view of the above, the two-CCII-based FGPIC/FGPII circuits of Section 4.4.1 can be looked upon as two nullor five impedance circuits with each CCII− being equivalent to a three-terminal nullor. It, therefore, follows that there should be a class of three-nullor three impedances, converters and inverters too. In other words, this means that a floating GPIC/GPII network should also be realizable by three CCs but no more than three passive elements (instead of five components). We now present a variety of such three-CC-based FGPIC/FGPII configurations or special cases thereof, which have been introduced in the literature as lossless FIs or lossless floating FDNR simulators.

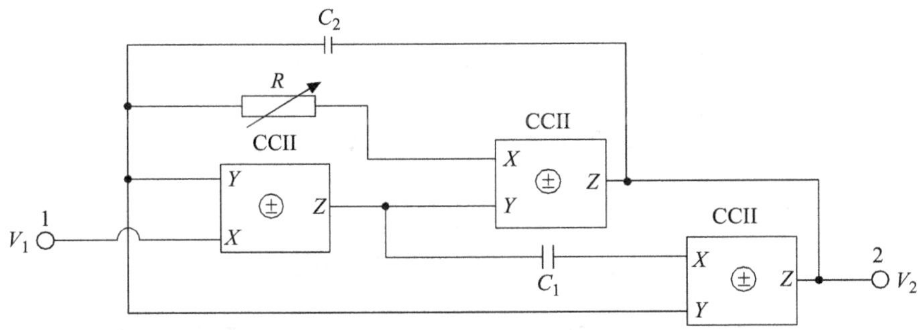

Figure 4.26 Floating ideal tunable FDNR simulation scheme proposed by Abdalla [37]; $Z_{1-2} = 1/s^2C_1C_2R$

In 1985 Abdalla [37] in his comment presented a modified form of the circuit by Nandi *et al.* [36], which is shown in Figure 4.26. It was claimed that in this circuit, the same result is obtained whether one takes all the three CCII+ or CCII− a result which, in a recheck, was not found to be correct.[4]

In 1987, Higashimura and Fukui [47] derived two new three-CC-based generalized floating immittance simulators based on the nullor model shown in Figure 4.27(a) using the equivalence of a CCII− as a three terminal nullor. This nullor model leads to two possible realizations of the general floating immittances as shown in Figure 4.27(b) and (c), respectively. Both the circuits simulate a floating admittance between ports 1 and 2 having value $Y_{1-2} = Y_1Y_3/Y_2$.

Higashimura and Fukui [48] presented four new lossless tunable floating FDNR simulation circuits, each using two CCIIs and a current inversion type NIC as active elements. These circuits also have the advantage of not requiring any component-matching conditions and providing tunability of the FDNR through a single-variable resistor. These circuits are shown in Figure 4.28(a)–(d).

All the circuits of Figure 4.28(a)–(d) realize the admittance matrix:

$$[Y] = s^2D_0\begin{bmatrix} 1 & -1 \\ -1 & 1 \end{bmatrix}, \quad D_0 = C_1C_2R \tag{4.21}$$

Higashimura and Fukui [42] presented two more ideal floating FDNR circuits, each employing two CCIIs, one inverting/non-inverting buffer, two capacitors, and a resistor. The circuit based on inverting buffer, therein however, have the drawback of requiring two more equal-valued resistors for realizing the inverting buffer using a CCII or an op-amp.

[4]Kumar P., Senani R. 'Novel tunable floating FDNR simulation using current conveyors: some appraisals', 2005, *unpublished.*

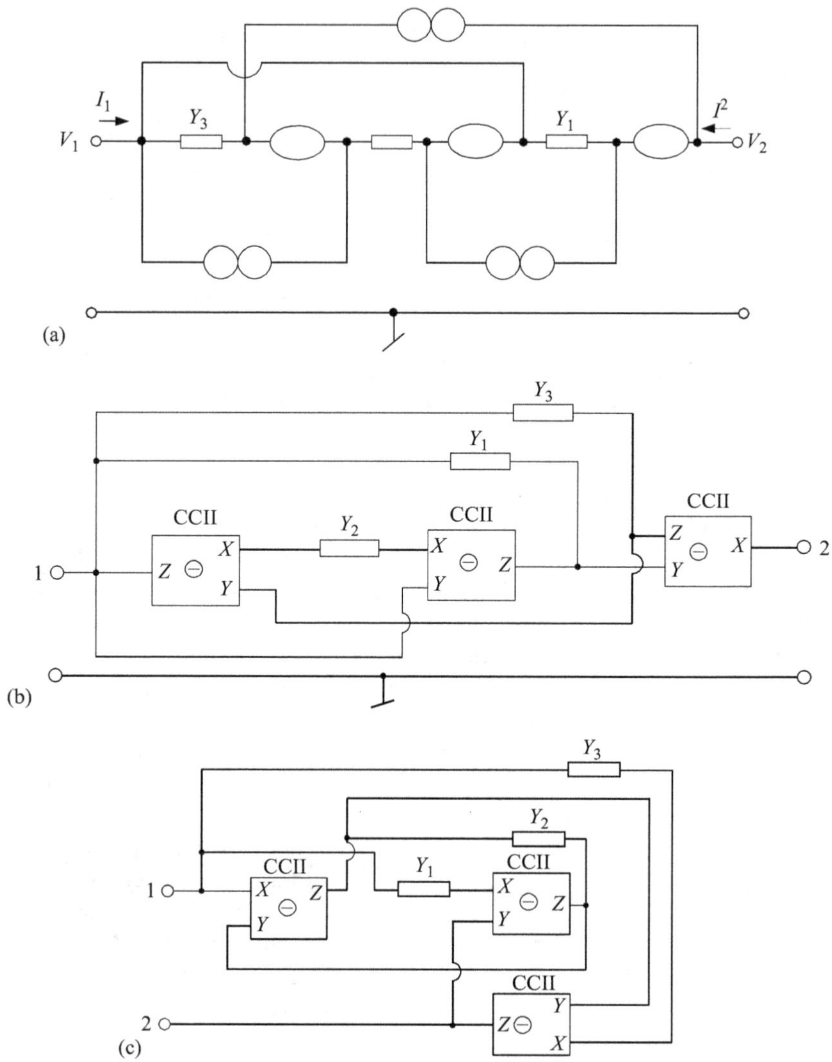

Figure 4.27 CC-based floating impedance realization derived by Higashimura and Fukui [47] from a nullor model: (a) nullor model of the FI, (b) and (c) CCII– based FIs derived from the nullor model

4.4.5 Floating impedance realization using two DOCCs

In the previous section, we have presented a variety of lossless floating inductance, lossless floating FDNR or floating generalized impedance converter/inverter circuits which require three CCs and three impedances or two CCs but five passive elements. In this section, we present a number of interesting circuits to demonstrate that if dual-output second-generation CCs (DOCCII±) are employed then a lossless

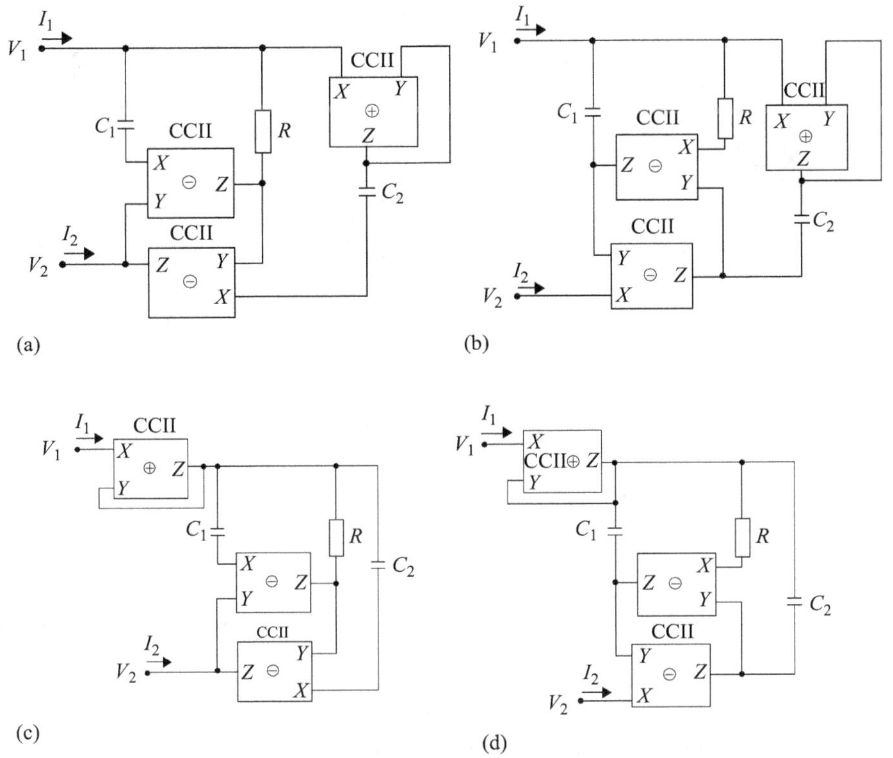

Figure 4.28 Four circuit configurations (a)-(d) for lossless tunable floating FDNR simulations equivalent to those proposed by Higashimura and Fukui [46]

FI could be realized with only two DOCCIIs and three passive elements with the additional advantage that the capacitor employed could even be grounded, which is attractive from the viewpoint of IC implementation.

We present two such circuits in Figure 4.29 out of which Figure 4.29(a) was proposed by Pal [363] in 1989 as an application example of the first modified CCII which he proposed in the same publication [363]. This circuit realizes a floating lossless inductor of value $L_{eq} = CR_1R_2$ assuming the modified CCII$\pm$ to be characterized by $i_y = 0,\ \ v_x = v_y,\ \ i_{z^+} = +i_x$ and $i_{z^-} = -i_x$.

The second circuit of Figure 4.29(b) was introduced by Ananda Mohan in [328] which also realizes a lossless floating inductance of the same value. Later on, in another publication [336] in 1998 Ananda Mohan showed that with C and R_2 interchanged the modified circuit realizes a FC multiplier using a grounded capacitor.

4.4.6 *Floating impedances using CCIIs and op-amps/OTAs*

A number of researchers have also proposed floating inductance configurations using CCs along with other building blocks such as op-amps and/or OTAs with the

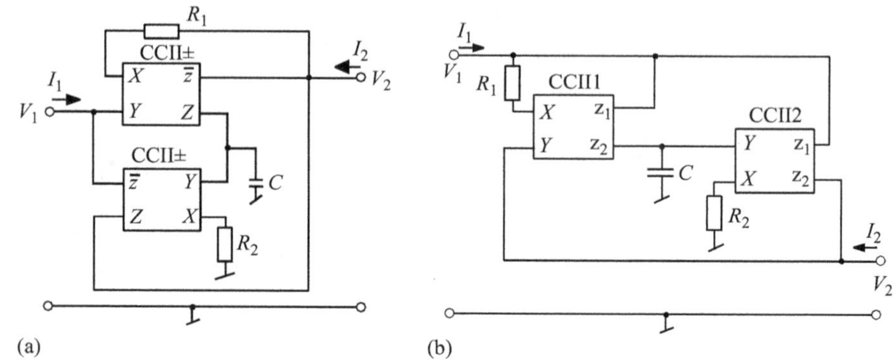

Figure 4.29 Lossless FI simulations employing a grounded capacitor: (a) FI proposed by Pal [363], (b) FI proposed by Ananda Mohan [328]

intention of achieving some interesting properties which may not be feasible with circuits realized with only CCs. Some such examples are outlined in this section.

Maundy *et al.* [106] presented a mixed-source arrangement for realizing a floating inductance using two CCs and two op-amps which is shown here Figure 4.30(a). In this circuit, the two CCs along with a resistor have been employed to create a differential voltage-controlled current source. By straightforward analysis, it is found that between ports 1 and 2, the circuit realizes a floating impedance of value $Z_{1-2} = sC_1R_1R_2$.

In the second circuit proposed by Higashimura and Fukui [55], which is shown in Figure 4.30(b), an OTA is used in conjunction with two CCs to create an inductance which is electronically tunable through the external bias current of the OTA. It has been found that this circuit realizes inductance of value $L_{eq} = CR_1/g_m$ which is electronically controllable through g_m since for a bipolar OTA (such as LM3080) its transconductance g_m is given by $g_m = I_{bias}/2V_T$ and is, thus, linearly controllable through the external dc bias current I_{bias}.

Layos and Haritantis [71] proposed a number of four-terminal (three-port) gyrators using either CCs only or employing a combination of CCs and OTAs, three of which are shown in Figure 4.31. The circuit of Figure 4.31 realizes a three-port gyrator characterized by the [Y] matrix:

$$\begin{bmatrix} I_1 \\ I_2 \\ I_3 \end{bmatrix} = \begin{bmatrix} 0 & 0 & \dfrac{1}{R_1} \\ 0 & 0 & -\dfrac{1}{R_1} \\ -G_m & G_m & 0 \end{bmatrix} \begin{bmatrix} V_1 \\ V_2 \\ V_3 \end{bmatrix} \tag{4.22}$$

so that with port 3 terminated into a grounded capacitor C_0, the circuit realizes a floating inductance of value $L = C_0R_1/G_m$.

The circuit of Figure 4.31(b) realizes a three-port gyrator using a CCII+ and an OTA and permits the grounding of the resistor R_1. Its [Y] matrix is found to be the same as that of the circuit of Figure 4.31(a).

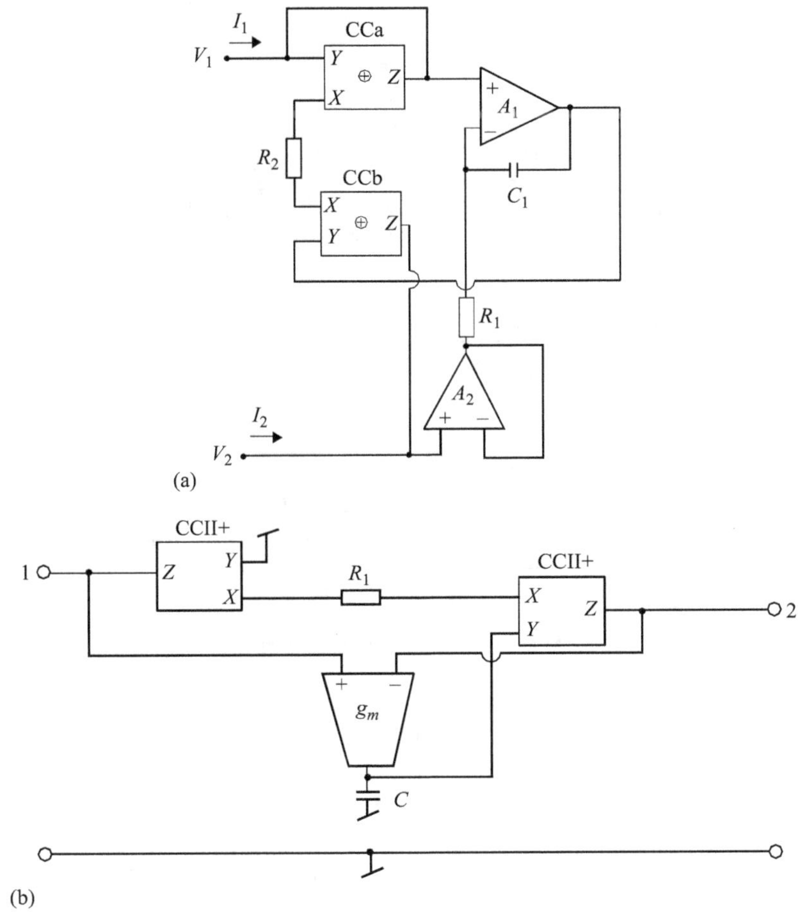

Figure 4.30 Some CCII–OTA-based FIs: (a) a floating bidirectional inductor proposed by Maundy et al. [106], (b) lossless FI proposed by Higashimura and Fukui [55]

Note that while with port 3 terminated into a capacitor C_0, the circuit would simulate a lossless FI with a grounded capacitor C_0, on the other hand, with position of C_0 and R_1 interchanged, the same circuit would realize lossless FC while still employing a grounded capacitor.

Lastly, the circuit of Figure 4.31(c) has $[Y]$ matrix given by

$$[Y] = \begin{bmatrix} 0 & 0 & +\dfrac{1}{R_1} \\ 0 & 0 & -\dfrac{1}{R_1} \\ -G_m & G_m & 0 \end{bmatrix} \tag{4.23}$$

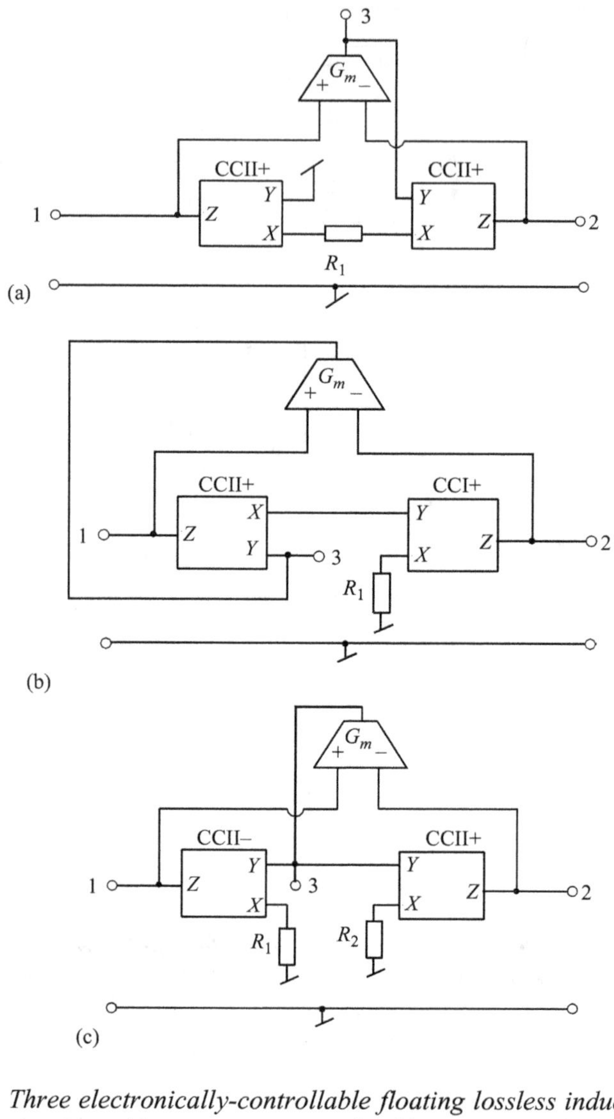

Figure 4.31 Three electronically-controllable floating lossless inductance simulators/three-port gyrators (a)-(c) using a combination of CCII+ and OTA proposed by Layos and Haritantis [71]

and thus would need the equality of the two resistors R_1 and R_2 to realize a three-port gyrator.

4.4.7 Economical floating impedance circuits synthesized using the 'CCII-nullor' equivalence

In view of the methods and circuits discussed earlier as well as from the previous expositions made in Chapter 2, it turns out that lossy/nonideal inductors and

FDNRs can be realized with a smaller number of CCs than their ideal counterparts. In fact, a number of such circuits using CCs have been proposed in the literature, for instance, see [7–9,15,18,23,24,26,34,40,43,99,357]. In the following, we describe a novel nullor[5]-based methodology to generate such circuits in a systematic manner.

The equivalence between a CCII− and three-terminal floating nullor was first employed in 1988 by Senani [50] to generate a class of new FI structures. These circuits provide the novel features of single-resistance tunability of the realized floating inductance value along with the employment of a bare minimum of only three passive components, namely two resistors and a capacitor. Later on, the nullor approach had also been used by other researchers for similar purposes as in [51,64,164].

Senani [50] considered the two nullor models shown in Figure 4.32 which simulate parallel and series lossy grounded inductors as shown. From these models, the nullor models of floating impedances were obtained by ungrounding all the passive elements, nullators and norators. Furthermore, from these FI models, two additional FI models were obtained by swapping nullators by norators and vice versa. Lastly, four additional nullor models were evolved by carrying out the

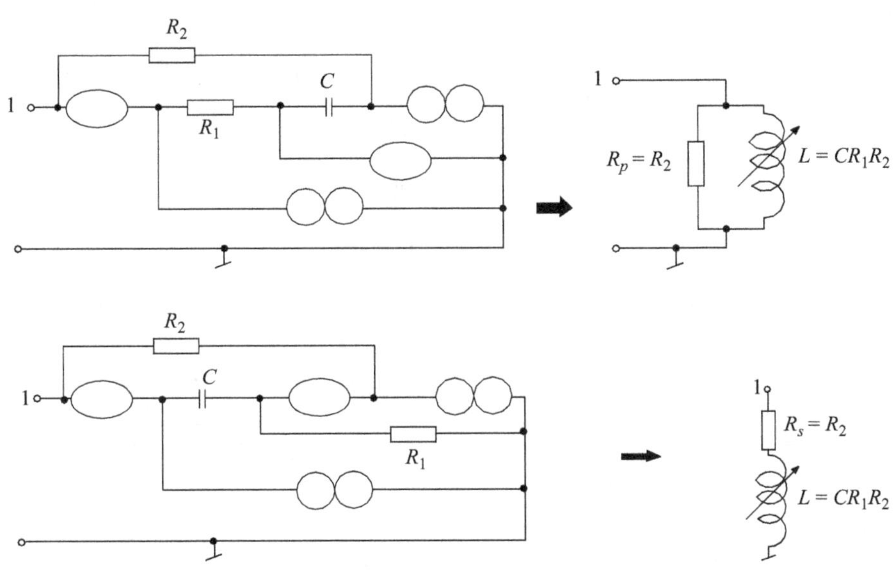

Figure 4.32 Nullors model employed for synthesizing CCII− based floating impedances [50]

[5]Nullator, norator and nullor were first introduced into the domain of circuit theory by Carlin [400]. For the use of nullors in FI synthesis and relation between CCs, op-amps and nullors readers may also see [50,401].

operation of *port-transposition* on the preceding four models. The operation of port-transposition, however, changes the nature of impedances realized from floating series *RL* to floating parallel *RL* and vice versa.

The various CCII− based inductance simulation circuits emanating from the above-described eight nullor models are shown in Figure 4.33. By straightforward analysis, it can be verified that the circuits of Figure 4.33(a), (c), (f) and (g) have the $[Y]$-matrix given by

$$[Y] = \left(\frac{1}{R_2} + \frac{1}{sCR_1R_2}\right)\begin{bmatrix} 1 & -1 \\ -1 & 1 \end{bmatrix} \tag{4.24}$$

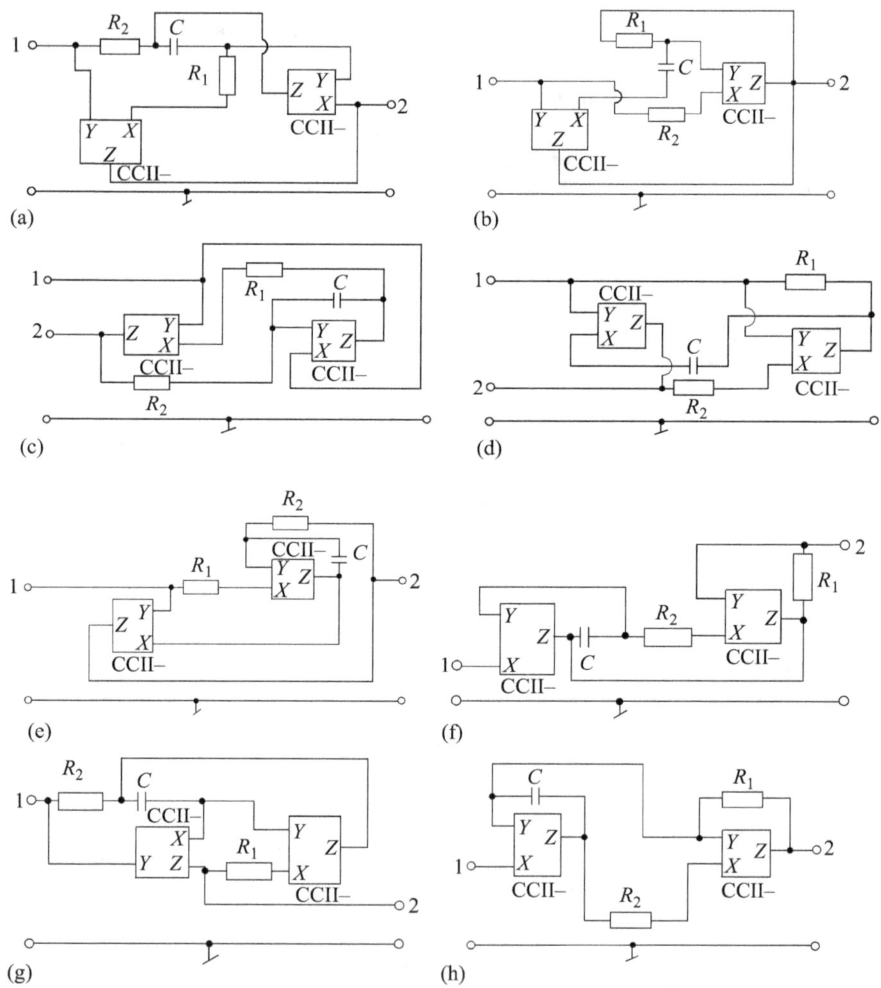

Figure 4.33 A class of eight low-component-count FI simulators (a)-(h) derived by Senani [50]

which represents floating parallel RL impedance with

$$L = CR_1R_2; \quad R_p = R_2 \tag{4.25}$$

On the other hand, the circuit configurations of Figure 4.33(b), (d), (e) and (h) are characterized by the $[Y]$-matrix:

$$[Y] = \frac{1}{(R_2 + sCR_1R_2)}\begin{bmatrix} 1 & -1 \\ -1 & 1 \end{bmatrix} \tag{4.26}$$

thereby representing a floating series RL impedance with

$$L = CR_1\bar{R}_2; \quad R_s = R_2 \tag{4.27}$$

From all the eight circuits of Figure 4.33(a)–(h), nonideal floating FDNRs can be realized by applying RC:CR transformation.

In this chapter, we have mainly been concerned with first-order impedances only; for CC-based circuits for simulating higher order immittances, in both grounded and floating forms, the readers are referred to [67,68,70,124,128].

4.4.8 Realization of mutually coupled circuits

Mutually coupled circuits find a number of applications in communication, instrumentation and control. A mutually coupled circuit essentially has a primary self-inductance, a secondary self-inductance and mutual inductance. Many researchers have proposed active circuits for the simulation of mutually coupled circuits using CCs, for instance, see [96,107,108].

A circuit for realizing equivalent of a mutually coupled circuit using only CCII+s was presented by Abuelma'atti *et al.* [96] which is shown here in Figure 4.34.

By a straightforward analysis, the voltages V_1 and V_2 of this circuit are given by

$$V_1 = sC_1R_1R_3I_1 + sC_1R_3R_mI_2 \tag{4.28}$$

$$V_2 = sC_1R_3R_mI_1 + sC_1R_2R_3I_2 \tag{4.29}$$

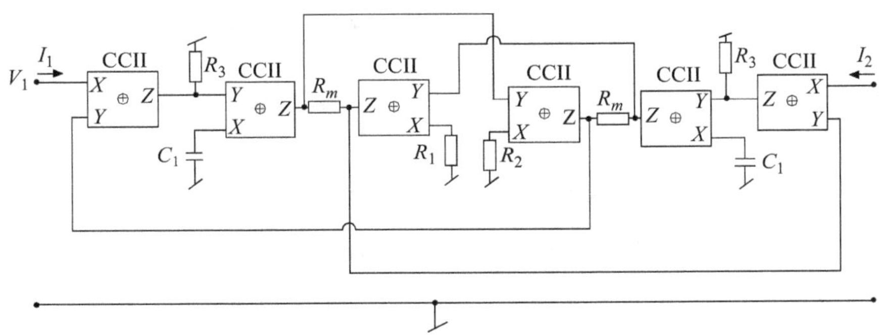

Figure 4.34 CCII-based simulated coupled circuit introduced by Abuelma'atti et al. [96]

From (4.28) and (4.29) it is seen that the two equivalent self-inductances and the equivalent mutual inductance realized by this circuit are given by

$$L_1 = C_1R_1R_3, \quad L_2 = C_1R_2R_3, \quad M = C_1R_mR_3 \tag{4.30}$$

While the circuit of Figure 4.34 requires as many as six CCIIs, a circuit for simulating a mutual-coupled circuit employing a reduced number of only four CCs was proposed by Yuce *et al.*[108] and is shown in Figure 4.35. This circuit is characterized by the following matrix equation:

$$\begin{bmatrix} V_1 \\ V_2 \end{bmatrix} = \begin{bmatrix} s(L_1 + M_{11}) & sM_{12} \\ sM_{21} & s(L_2 + M_{22}) \end{bmatrix} \begin{bmatrix} I_1 \\ I_2 \end{bmatrix} \tag{4.31}$$

where

$$L_1 = C_1R_1R_2, \quad L_2 = C_2R_3R_4 \tag{4.32}$$

$$M_{11} = M_{12} = M_1 = C_1R_1R_5; \quad M_{21} = M_{22} = M_2 = C_2R_3R_5 \tag{4.33}$$

If we take $C_1R_1 = C_2R_3$, $M_1 = M_2 = M$ is achievable.

Both the circuits, however, suffer from the drawback of requiring a large number of CCIIs (six in the circuit of Figure 4.34 and four in the circuit of Figure 4.35). Furthermore, unfortunately, both the circuits suffer from the drawback of having two capacitors at the X-terminal of the CC which is often associated with stability problems [100]. In the latter case, the two CCII− each would require two CCII+ for their implementation, thereby making the total CCII+ count as six. In view of this, there appears to be enough scope for evolving improved alternative realizations of the mutually coupled circuits, free from these difficulties and employing a reduced number of CCs.

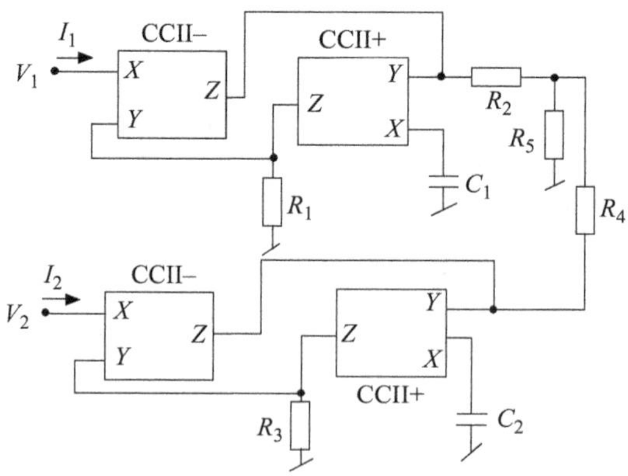

Figure 4.35 Mutually coupled circuit proposed by Yuce et al. [108]

4.5 Impedance simulation using CCCII

A comprehensive treatment of operational-transconductance-amplifier–capacitor (OTA–C) circuits for synthesizing electronically controllable impedances has been given in Chapter 3 of this monograph.

It is well known that the CCCII [133,134,193–326,328] provide electronic control of the X-port input resistance R_x through an external bias current. Because of this reason, CCCIIs have attracted lot of attention of researchers and circuit designers as alternative building blocks to synthesize a variety of electronically controllable waveform generators and other functional circuits.

The electronic-controllability in a CCCII is the outcome of a specific circuit known as a *mixed translinear cell* (MTC) which makes the front end of a typical CCII architecture. This MTC, apart from the electronic control, is also responsible for creating a higher slew rate and larger input signal handling capability. These features have been the primary reason for a lot of research on the use of CCCIIs in the realization of a variety of linear as well as nonlinear signal-processing circuits, for instance, see [133,134,193–326,328] and the references cited therein. This section focuses on the synthesis of various electronically controllable grounded and floating resistors inductors and other elements using CCCIIs. However, before discussing the various impedance simulation circuits, it is useful to know that a number of bipolar/CMOS/BiCMOS implementations of the CCCIIs have so far been evolved in the technical literature from time to time, for instance, [196–199,243,246,261,305,319,320].

The basic translinear CCCII circuit proposed by Fabre *et al.* [193,194] employs an MTC comprising transistors Q_1–Q_2–Q_3–Q_4 and is shown in Figure 4.36.

An analysis of the circuit of Figure 4.36 shows that it is characterized by

$$I_Y = 0 \tag{4.34}$$

$$\frac{V_Y - V_X}{V_T} = \sin\text{h}^{-1}\left(\frac{-I_X}{2I_B}\right) \cong -\frac{-I_X}{2I_B}, \quad \text{for } I_X \ll 2I_B \tag{4.35}$$

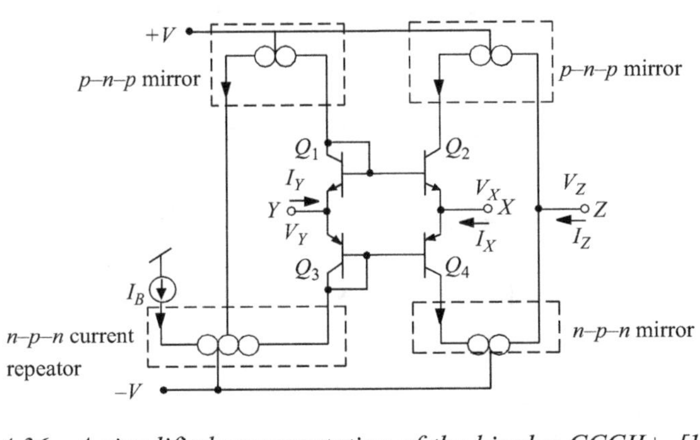

Figure 4.36 A simplified representation of the bipolar CCCII+ [193,194]

or

$$V_X \approx V_Y + I_X R_x \quad \text{where} \quad R_x = \frac{V_T}{2I_B} \quad \text{and} \quad I_z = I_x \tag{4.36}$$

From (4.36), it follows that R_x is *electronically tunable* through the external dc bias current I_B. With $\pm V = \pm 2.5$ V, this circuit exhibits a voltage gain of 0.9984, a current gain of 1.022 and a 3-dB bandwidth of about 615 MHz [194]. The symbolic notation and simplified equivalent circuit of a CCCII are shown in Figure 4.37.

From the circuit of Figure 4.36, a CCCII− can be obtained by breaking the link at the junction of Z-terminal and adding one more pair of current mirrors in a cross-coupled manner. Thus, it is easy to create a CCCII having complementary Z-output terminals or as many number of current output terminals having either positive or negative polarity as may be needed in a particular application.

A CMOS CCCII architecture analogous to that of Figure 4.36 as shown Figure 4.38 was introduced by Chaisricharoen *et al.* [196].

In this circuit the basic translinear cell consists of M_{n1}, M_{n2}, M_{p5} and M_{p6}. An analysis of this circuit shows that to get a linear relationship between V_X, V_Y and I_X, the following condition should be satisfied:

$$\frac{\mu_p W_p}{L_p} \cong \frac{\mu_n W_n}{L_n} \tag{4.37}$$

$$V_X \cong V_Y + \frac{I_x}{\sqrt{2I_B C_{ox}}\left(\sqrt{\mu_p W_p/L_p} + \sqrt{\mu_n W_n/L_n}\right)} \tag{4.38}$$

A detailed analysis [196] of the circuit of Figure 4.38 reveals that

$$R_x \cong \frac{1}{\sqrt{2I_B C_{ox}}\left(\sqrt{\mu_p W_p/L_p} + \sqrt{\mu_n W_n/L_n}\right)} \tag{4.39}$$

where R_x is the transresistance at port-X.

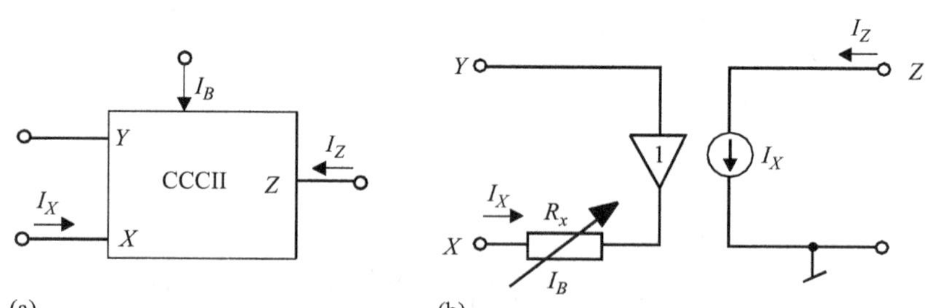

Figure 4.37 Representations of the CCCII: (a) symbolic notation, (b) simplified equivalent circuit representation

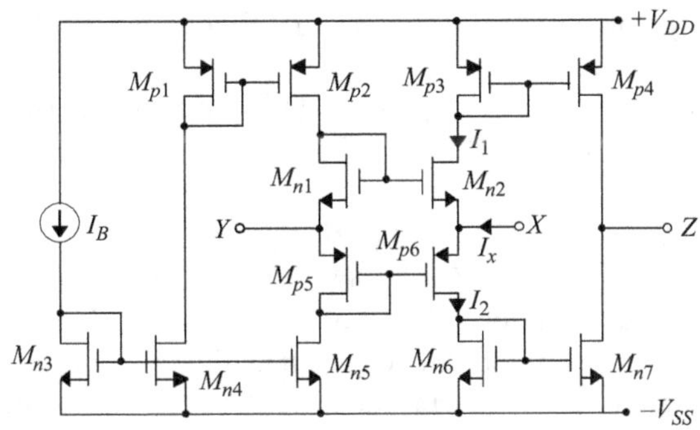

Figure 4.38 An exemplary CMOS realization of CCCII+ [196]

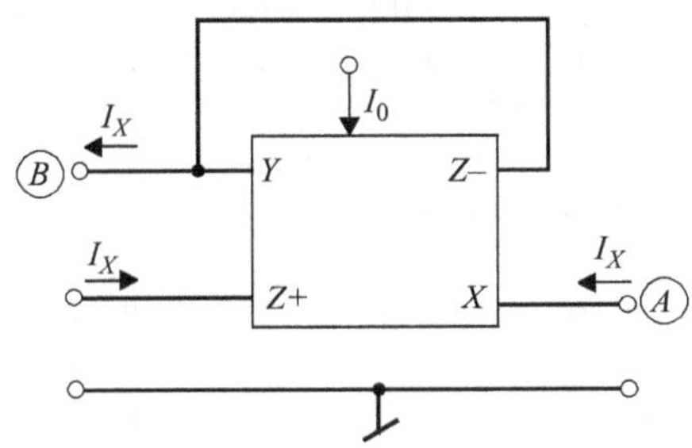

Figure 4.39 The circuit connections for realizing a floating negative controlled resistance [243] using a CCCII with negative intrinsic resistance R_x

Thus, in this case also, R_x is controllable by dc bias current I_B however, while in a bipolar CCCII $R_x \propto 1/I_B$, in a CMOS CCCII $R_x \propto 1/\sqrt{I_B}$. Several other CMOS-based CCCII configurations have also been developed, for the details of which the reader is referred to [133,196–198]. However, in the following, a brief discussion about these alternatives follows.

In this context, it may be mentioned that a 2.2 GHz BiCMOS CCCII in standard 0.8 μm BiCMOS technology was proposed by Seguin *et al.* [199].

Moreover, Barthelemy and Fabre [243] proposed a CCCII with negative intrinsic resistance whose value is given by $R_x = -V_T/2I_B$ which was shown to be particularly useful in realizing a floating negative controlled resistance (see Figure 4.39).

Minaei *et al.* [244] developed two new realizations of CCCII one in CMOS and the other in Bipolar technology, which provide good linearity, very high input impedance at port-*Y*, high output impedance at port-*Z*, with highly accurate current gain.

On the other hand, Chunhua *et al.* [245] studied a new CCCII architecture in which all Wilson current mirrors were employed in place of all simple mirrors.

Two alternative architectures based on differential pair with biasing circuitry and mechanism of duplicating the *X*-port current through CMOS transistors were proposed by Ercan and Alci for realizing a CCCII with positive R_x and CCCII with negative R_x in [246].

On the other hand, Chaisricharoen *et al.* [196] introduced balanced differential pair structure for realizing CCCII.

4.5.1 Current-controlled positive/negative resistance realization

The realization of a current-controlled positive resistor using CCCII is straightforward. If terminal-Y is grounded and terminal-Z is also grounded as shown in Figure 4.40, the circuit realizes a current-controlled resistance, since looking into terminal-X, the equivalent input impedance is found to be $R_{in} = V_T/2I_B$. If V_T is taken as 25 mV and I_B is changed from 0.1 μA to 1 mA, a variable resistance (R_{eq}) between 12.5 Ω and 125 kΩ is realizable from such a circuit.

A circuit for realizing a floating current controlled positive resistance (FCCPR) which simulates a floating resistance of value $R_{1\text{–}2} = V_T/2I_B$ between ports 1 and 2 was presented by Singh *et al.* in [200]. A circuit analogous to the circuit of [200] is shown in Figure 4.41(a) which realizes an FCCPR. A slightly modified (new) form is shown in Figure 4.41(b) which, however, realizes an FCCPR of value $V_T/4I_B$.

On the other hand, a grounded current-controlled negative resistance can be realized from the CCCII+ [194] by shorting terminals Y and Z, grounding terminal-X and then looking into the input node formed by joining Y–Z terminals as shown in Figure 4.42, thereby realizing a resistance given by $R_{in} = -V_T/2I_B$.

The above circuit can be converted into a two-port one in two different ways, thereby realizing a floating current controlled negative resistance (FCCNR) and FCCPR as shown in Figure 4.43. It may be easily verified that the circuit of Figure 4.43(a) realizes an FCCNR of value $R_{1\text{–}2} = -V_T/I_B$ while that of Figure 4.43(b) realizes the same value of FCCPR.

A new design for electronically tunable grounded/floating resistors was proposed by Pawarangkoon and Kiranon in [201] which was based on a multiple output CCCII characterized by the following equations: $I_y = 0$, $V_x = V_y + R_xI_x$, $I_{z1} = -I_x$, $I_{z2} = KI_x$ and $I_{z3} = -KI_x$ where K was equal to the ratio of two bias currents, i.e. $K = I_1/I_2$ (see Fig. 2a of [201]).

This modified CCCII can be configured either as a floating resistance or as a grounded resistor as shown in Figure 4.44(b) and (c), respectively.

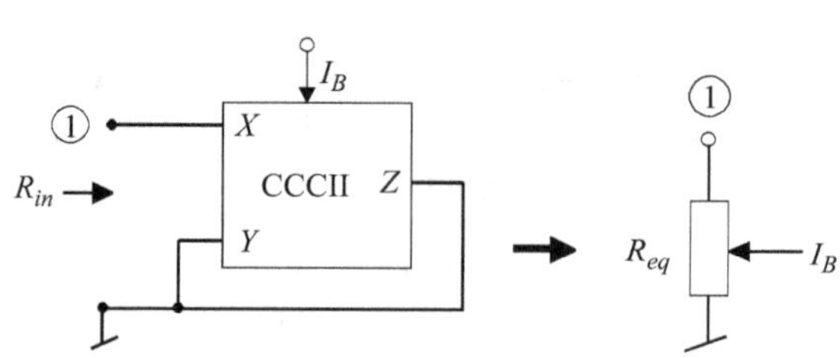

Figure 4.40 Current-controlled grounded resistance simulation using CCCII+

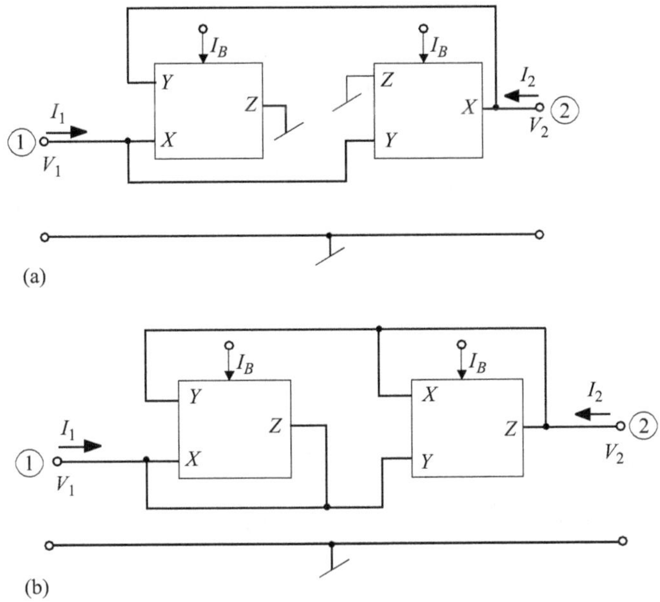

Figure 4.41 Two FCCPR configurations (a), (b) based on the circuit proposed by Senani et al. [200]

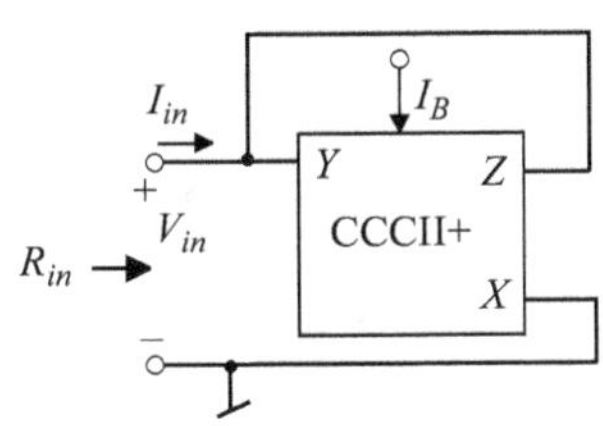

Figure 4.42 A negative current-controllable grounded resistance [194]

In both the cases, the equivalent resistance realized by the circuits is given by

$$R_{eq} = \frac{V_T}{2I_0}\left(\frac{1}{1-(I_1/I_2)}\right) \tag{4.40}$$

from where it is seen that by proper selection of the external dc bias currents I_1 and I_2, in the bipolar implementation of MO-CCCII given by Pawarangkoon and Kiranon [201], the realized R_{eq} can be made positive or negative and also the value of the realized resistance can be varied by varying the dc bias current I_0. The workability of this circuit has been demonstrated in [201] by realizing the circuit with bipolar transistors arrays NR100N and PR100N.

An external resistorless and electronically tunable CCCI based floating FDNR circuit was described by Metin and Minaei [357] which is shown here in

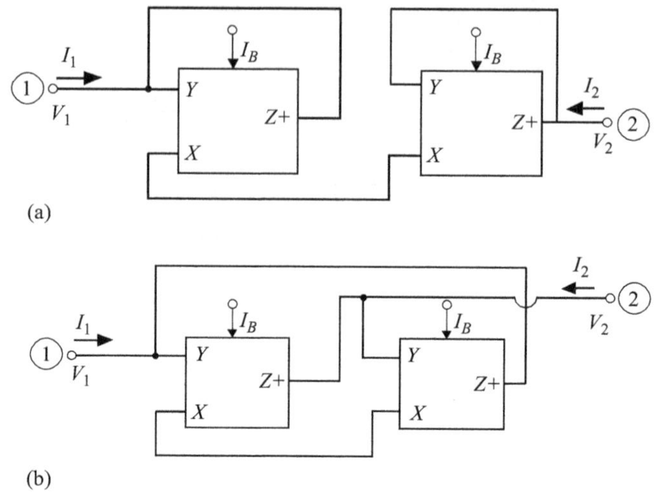

Figure 4.43 *(a) FCCNR and (b) FCCPR derived from the circuit of Figure 4.42 [200]*

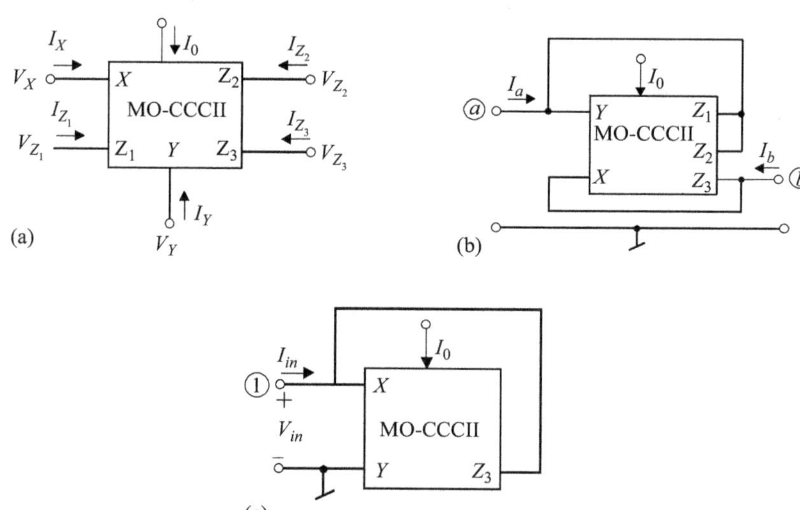

Figure 4.44 *The MO-CCCII proposed by Pawarangkoon and Kiranon [201]: (a) the symbolic notation, (b) electronically tunable floating resistor, (c) electronically tunable grounded resistor*

Figure 4.45. Assuming that the bias current for the first and third CCCI+s are large enough so that the X-terminal parasitic resistances R_{x1} and R_{x3} can be ignored, the analysis shows the floating FDNR to have a value $D_{eq} = C_1C_2R_{x2}$. On the other hand, if the ignored parasitic resistances are accounted for in the analysis and they are taken to be equal, i.e. $R_{x1} = R_{x3} = R_{x13}$, the modified analysis indicates that the

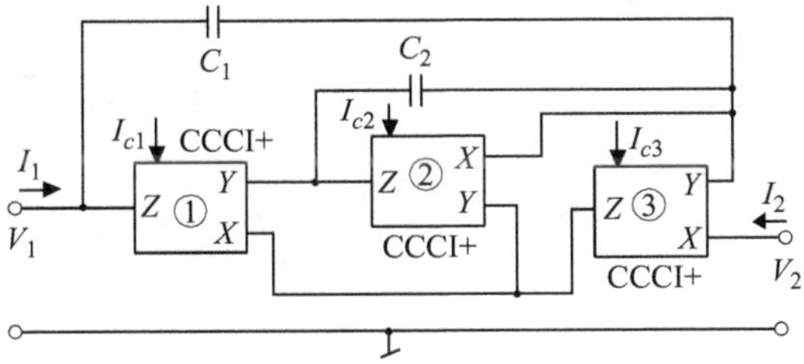

Figure 4.45 CCCI+ based floating FDNR simulator [357]

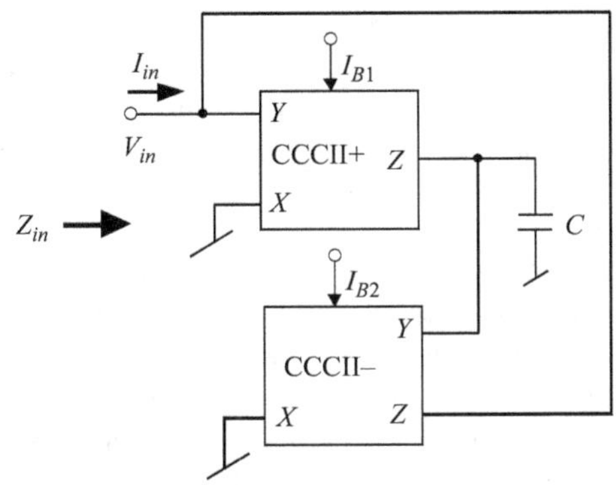

Figure 4.46 Electronically controllable grounded inductance based on Sedra–Smith gyrator

FDNR will have a series resistance and capacitance besides the FDNR which can however be compensated by adding a resistor R_C in series with the *Y*-terminal of the first and third CCCIs [357].

4.5.2 Electronically tunable grounded/floating impedances

In the previous subsections, it has been shown how electronically controlled resistances with both positive and negative values, in grounded as well as floating forms, are realizable by CCCIIs. In this subsection, we show some prominent circuits for realizing various other kinds of impedances with the feature of electronic-tunability.

At the outset it may be mentioned that if in the Sedra and Smith's two CCII−based gyrator, CCCII± are used instead of CCII±, along with a capacitor *C*, as shown in Figure 4.46, the resulting circuit would realize a current-controlled grounded inductor.

The circuit of Figure 4.46 has $Z_{in} = sCR_{x1}R_{x2}$ so that equivalent inductance realized by this circuit has a value $L_{eq} = CV_T^2/4I_{B1}I_{B2}$.

Kiranon and Pawarangkoon in [205] presented a four CCCII+ based floating inductance simulation circuit which employs two resistors and a grounded

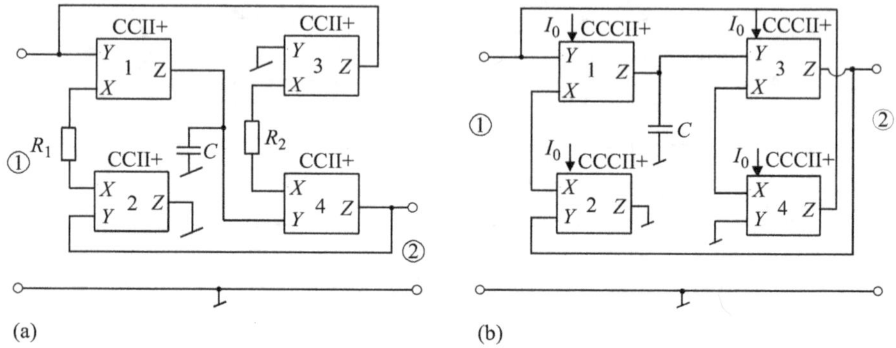

Figure 4.47 Electronically controllable lossless floating inductance circuit based on the circuit of Kiranon and Pawarangkoon [205]: (a) the basic FI circuit, (b) electronically controllable version with R_1 and R_2 both replaced by R_{x1} and R_{x2}

capacitor and simulates a lossless floating inductance of value $L_{eq} = CR_1R_2$ (Figure 4.47(a)). Since both the resistors are connected between X-terminals of the two CCCII+s, both can be replaced by the intrinsic input resistances R_x of the CCCIIs. Thus, by replacing one resistor at a time Kiranon and Pawarangkoon [205] presented two modified FI simulation circuits realizing a current controlled floating inductance (see Fig. 2 a and b of [205]). Of course, there is no restriction on replacing both the external resistors R_1 and R_2 by the corresponding intrinsic parasitic resistances of the CCCIIs. When this is done, this results in the circuit of Figure 4.47(b).

Since all CCIIs are biased with the same dc bias current I_0, the circuit of Figure 4.47(b) simulates a floating inductance between ports 1 and 2 having value

$$L_{1-2} = \left(\frac{V_T^2}{4I_0^2}\right)C \tag{4.41}$$

If dual outputs CCCIIs are permitted, Yuce *et al.* [207] demonstrated that not four but only three CCCIIs are required to simulate a lossless floating inductance employing a grounded capacitor whether one requires a positive inductor or negative inductor. These circuits are shown in Figure 4.48(a) and (b).

For the circuit Figure 4.48(a), an analysis shows that its Y-matrix is given by

$$\begin{bmatrix} I_1 \\ I_2 \end{bmatrix} = \frac{1}{sCR_{x1}(R_{x2}+R_{x3})}\begin{bmatrix} 1 & -1 \\ -1 & 1 \end{bmatrix}\begin{bmatrix} V_1 \\ V_2 \end{bmatrix} \tag{4.42}$$

from which it is apparent that the circuit simulates a positive floating inductance of value $L_{eq} = CR_{x1}(R_{x2} + R_{x3})$.

The Y-matrix for the circuit of Figure 4.48(b) too is of the same form as the above but with a negative sign which, therefore, represents a negative floating inductance of value $L_{eq} = -CR_{x1}(R_{x2} + R_{x3})$.

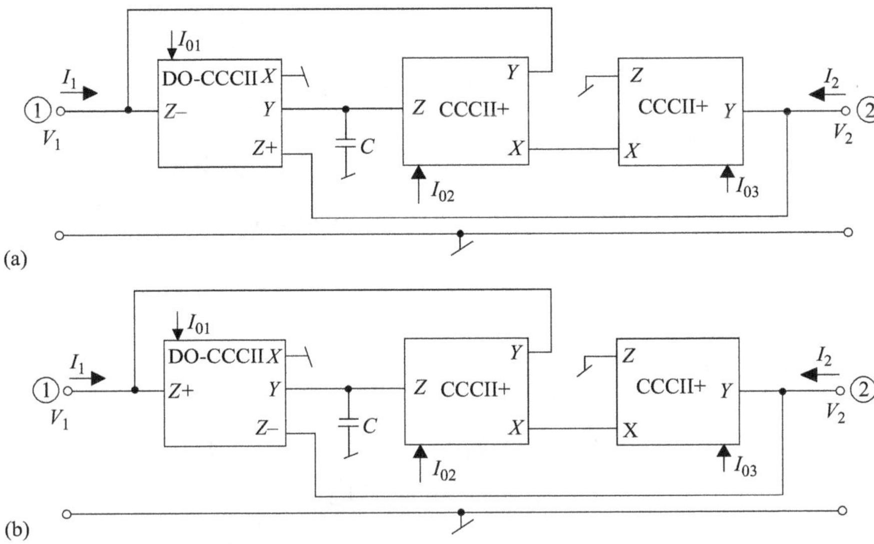

Figure 4.48 FI simulators proposed by Yuce et al. [207]: (a) positive FI, (b) negative FI

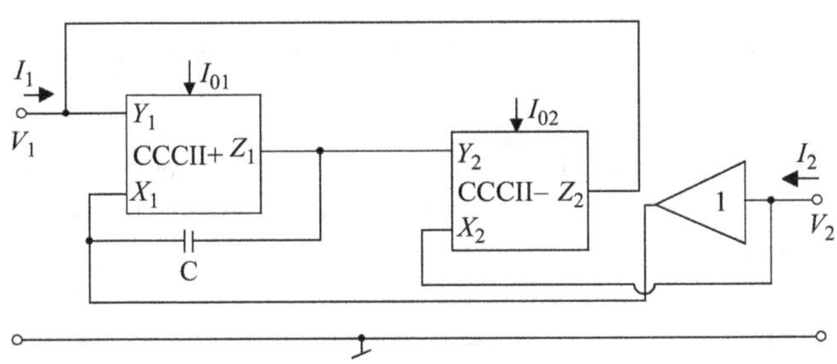

Figure 4.49 Lossless FI simulator proposed by Khan and Zaidi [333]

Khan and Zaidi [333] also presented a circuit employing only two CCCIIs and along with a voltage buffer as shown in Figure 4.49.

The *Y*-matrix of this circuit is given by

$$[Y] = \frac{1}{sCR_{x1}R_{x2}} \begin{bmatrix} 1 & -1 \\ -1 & 1 \end{bmatrix}; \quad \text{where } R_{xi} = \frac{V_T}{2I_{0i}}, \quad i = 1, 2 \tag{4.43}$$

In [264] Minas *et al.* demonstrated that using a dual CCCII in conjunction with an internally compensated type op-amp modelled for its open-loop gain as an integrator, i.e. $A \approx B/s$ (Figure 4.50), one can realize an external-capacitor-less floating inductance simulator with realized equivalent floating inductance value given by

$$L = \frac{V_T}{2I_0 B} \tag{4.44}$$

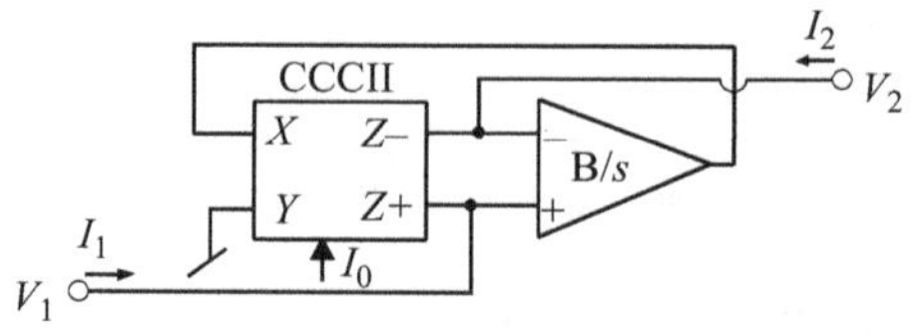

Figure 4.50 An FI proposed by Minas et al. [264]

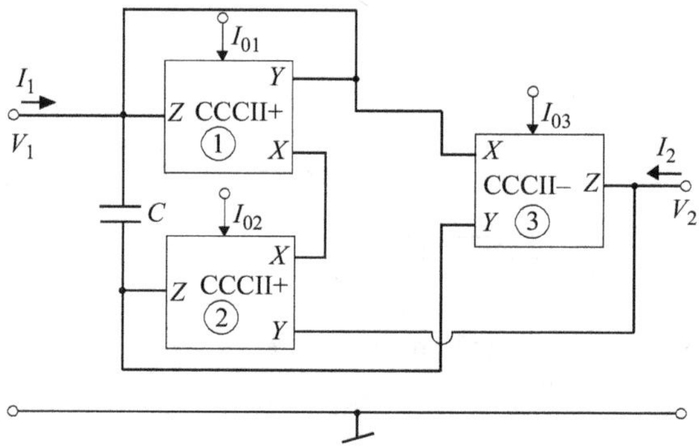

Figure 4.51 Electronically controllable FI reported by Minaei and Yuce [210]

and is, therefore, controllable through the external dc bias current I_0 of the CCCII.

Application: An application of this circuit was demonstrated in [264] by utilizing this in the design of a fourth-order elliptical filter employing two FIs. The circuit, however, exhibited a limited dynamical range of around 50 mV constrained by the op-amp, although the electronic control of realized floating inductance was found to be over a wide range of I_0 varied from 0.1 μA to over 100 μA or so [264].

Minaei and Yuce [210] presented a number of circuits for realizing tunable FI circuits out of which one circuit employed three CCCIIs along with only a single capacitor for lossless FI simulation. The quoted circuit from [210] is shown in Figure 4.51. For the circuit of Figure 4.51, the simulated inductance value is given by $L = CR_{x3}(R_{x1} + R_{x2})$ and is, thus, controllable through any/or all of I_{01}, I_{02} and I_{03}.

Application: This lossless FI [210] was used to construct a fifth-order Chebyshev LP filter encountering two FIs and three grounded capacitors besides two equal resistive terminations at source and load ends. The inductance value was observed to be tunable from about 0.5 mH to 100 mH with a good accuracy.

It was shown by Kiranon and Pawarangkoon in [205] that a lossless FI can be realized using a grounded capacitor along with four CCCII+. Some circuits employing CCCIIs were subsequently described in [209,210]. A circuit employing only two DOCCCIIs to realize lossless FI which was proposed by Sedef *et al.* [209] is shown in Figure 4.52.

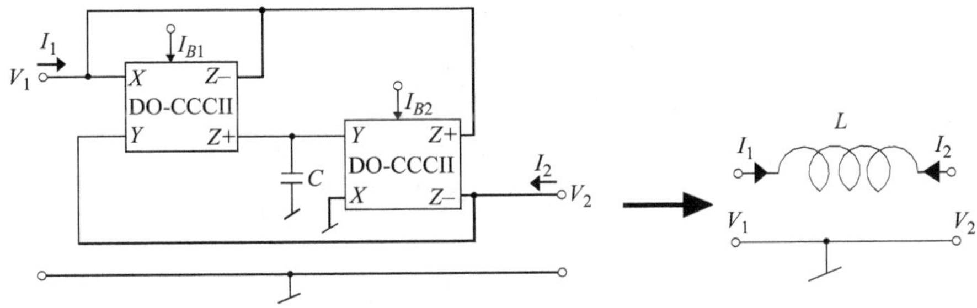

Figure 4.52 Current-controlled FI proposed by Sedef et al. [209]

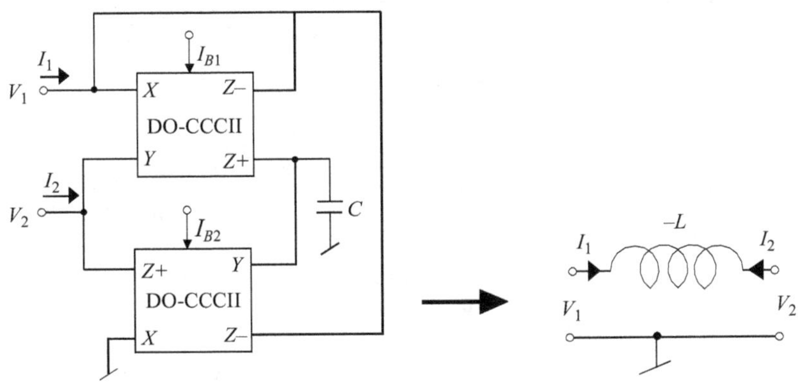

Figure 4.53 Floating negative inductance realization using two DOCCCIIs

This circuit is characterized by:

$$I_1 = -I_2 = \frac{V_1 - V_2}{sCR_{x1}R_{x2}} \tag{4.45}$$

thus, the inductance value realized between ports 1 and 2 is given by

$$L = CR_{x1}R_{x2} = \frac{CV_T^2}{4I_{B1}I_{B2}} \tag{4.46}$$

which shows that the equivalent L is controllable by any or both of I_{B1} and I_{B2}.

It is interesting to note that a negative FI can be realized from the same configuration (although not so indicated in [209]) by interchanging the $Z+$ and $Z-$ terminals of the second DOCCCII. Such a circuit is shown in Figure 4.53 for which the realized equivalent negative floating inductance is given by

$$L = -CR_{x1}R_{x2} \tag{4.47}$$

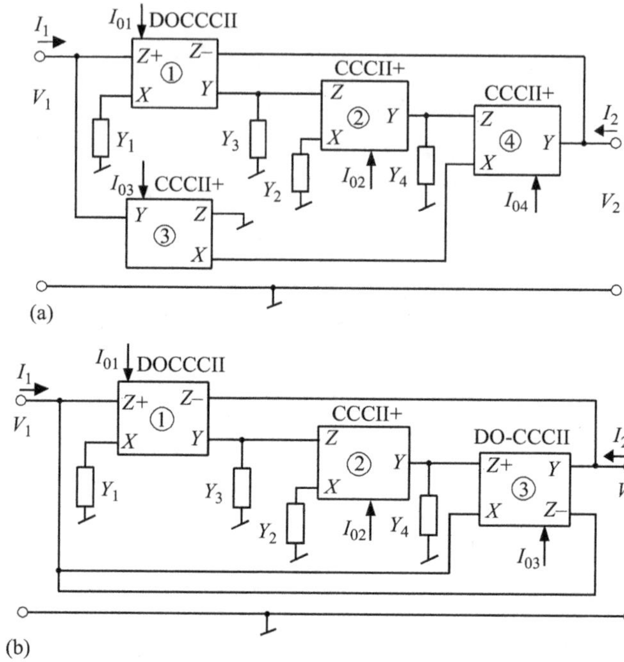

Figure 4.54 Generalized floating admittance simulators proposed by Yuce [339]

Two interesting generalized floating admittance simulators were introduced by Yuce [339] which are shown in Figure 4.54. Both the circuits realize the admittance matrix:

$$\begin{bmatrix} I_1 \\ I_2 \end{bmatrix} = \frac{Y_1 Y_2}{Y_3 Y_4 R_x} \begin{bmatrix} 1 & -1 \\ -1 & 1 \end{bmatrix} \begin{bmatrix} V_1 \\ V_2 \end{bmatrix} \tag{4.48}$$

by employing four CCCIIs and three CCCIIs, respectively. It is obvious that by appropriate selection (resistive/capacitive) of various circuit admittances, both the circuits can realize a variety of floating impedances such as lossless FI, lossless FC (while using a grounded capacitor for Y_2), floating FDNR (using both grounded capacitors for Y_1 and Y_2) and FDNC (still using both grounded capacitors in place of Y_3 and Y_4). Another notable feature of these realizations is the lack of any component-matching requirement which anyway is expected from CC-based circuits.

4.5.3 *Electronically tunable synthetic transformer*

A synthetic transformer is known to be a useful circuit element in the design of stagger-tuned filters. An active transformer when compared with a passive transformer provides a number of advantages such as suitability for IC implementation, ease of adjustment of the coupling coefficient and freedom of any magnetic interference. A synthetic transformer circuit based on CCCIIs, proposed by Yuce and Minaei [212], is presented here in Figure 4.55.

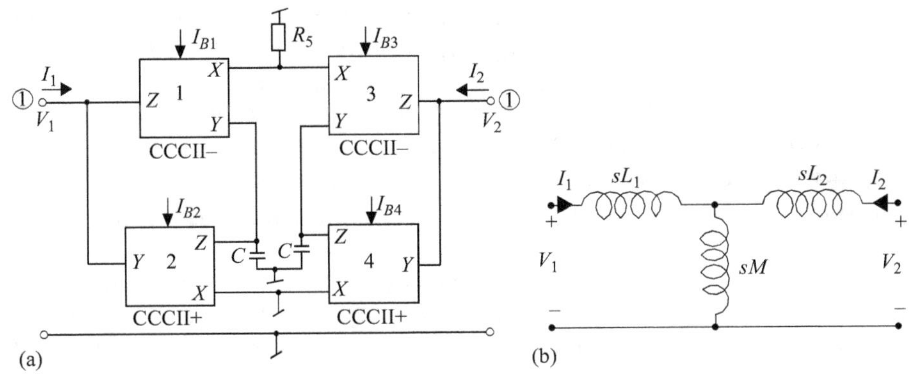

Figure 4.55 A synthetic transformer proposed by Yuce and Minaei [212]: (a) the CCCII-based active transformer circuit, (b) its passive equivalent

By straightforward analysis it can be shown that the circuit is equivalent to an inductive-T network with self-inductances L_1 and L_2 and mutual inductance M based on which its terminal equations can be written as the following matrix equation:

$$\begin{bmatrix} V_1 \\ V_2 \end{bmatrix} = \begin{bmatrix} sL_p & sM_{12} \\ sM_{21} & sL_s \end{bmatrix} \begin{bmatrix} I_1 \\ I_2 \end{bmatrix} \tag{4.49}$$

where $L_p = L_1 + M_{11}$ and $L_s = L_2 + M_{22}$. The equivalent parameters are thus, given by

$$L_1 = R_{x1}R_{x2}C_1; \quad L_2 = R_{x3}R_{x4}C_2; \quad M_{11} = M_{12} = R_{x2}R_5C_1 \quad \text{and} \\ M_{21} = M_{22} = R_{x4}R_5C_2 \tag{4.50}$$

To obtain a resistorless realization, the resistor R_5 can also be replaced by another CCCII+ configured as a current-controllable resistor.

4.6 Immittance simulation using different variants of current conveyors

During the past two decades, a large number of varieties of the basic CCs have been proposed by various researchers from time to time.

Thus, chosen from a vast amount of literature in the area of impedance simulation using different variants of CCs [1,135–148,329–395,402,406], a number of novel synthetic impedance circuits have been described is this subsection using the new variants of CCs (such as DOCCII, differential voltage CCs (DVCC), third-generation CC (CCIII), dual-X CCII (DXCCII), modified inverting second-generation CC (MICCII), differential difference CC (DDCC) and fully differential CCII (FDCCII), etc.) for realizing both grounded and floating forms of inductors and other related elements, which possess a number of interesting features. It may, however, be emphasized that these newer variants of the CCs have not only been employed in numerous applications other than impedance simulation, a number of their bipolar/CMOS/BiCMOS implementable architectures have also been advanced by various researchers, for instance, see [122,133–148,197].

4.6.1 Lossless FI realization employing only two DOCCs and three passive components

From the circuits described in the earlier section, it is apparent that if floating impedance is to be realized with a minimum of only three passive elements, then at least three CCs of the normal kind are needed.

On the other hand, if a realization using only two normal CCs is sought, then at least five passive elements are needed. Thus, using normal CCs any FI simulator using only two CCs and only three passive elements was not possible. However, if instead of normal CCs, DOCCIIs are permitted, then an FI using only two DOCCIIs and only three elements becomes possible. This was first demonstrated by Ananda Mohan in [328]. The circuit from [328] is shown in Figure 4.56.

By straightforward analysis it can be determined that the FI simulated by this circuit is given by $L_{eq} = C_0R_1R_2$. This circuit has the obvious advantage of employing a grounded capacitor when compared to the two-CC− based and three-CC− based FI simulator circuits described earlier, all of which employ a floating capacitor.

4.6.2 DVCC-based floating inductance/FDNR realization

Sedef and Acar [329] demonstrated that if one employs DVCCs[6] with dual complimentary outputs (characterized by $I_{y1} = 0$, $I_{y1} = 0$, $V_x = V_{y1} - V_{y2}$, $I_{z1} = +I_x$ and $I_{z2} = -I_x$), as per the circuit arrangement of Figure 4.57, a floating impedance can then be realized using only two DVCCs[7] along with only three impedances all of which can be grounded. Their circuit has the Y-matrix:

$$[Y] = \left(\frac{Z_3}{Z_1Z_2}\right)\begin{bmatrix} 1 & -1 \\ -1 & 1 \end{bmatrix} \tag{4.51}$$

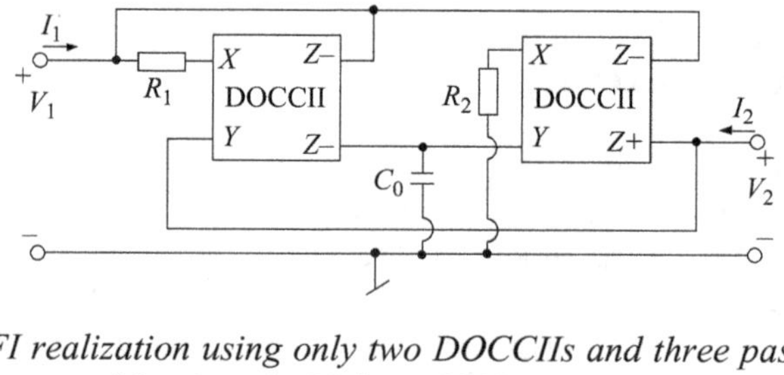

Figure 4.56 FI realization using only two DOCCIIs and three passive elements proposed by Ananda Mohan [328]

[6]It is interesting to point out that the DVCC was first introduced by Pal [363]. Although it was proposed again by Elwan and Soliman in [138], it was this latter publication where the first ever CMOS realization of the DVCC was presented.

[7]A number of floating inductors, using only three passive elements, using DVCC and other varieties of CCs have been dealt in [355].

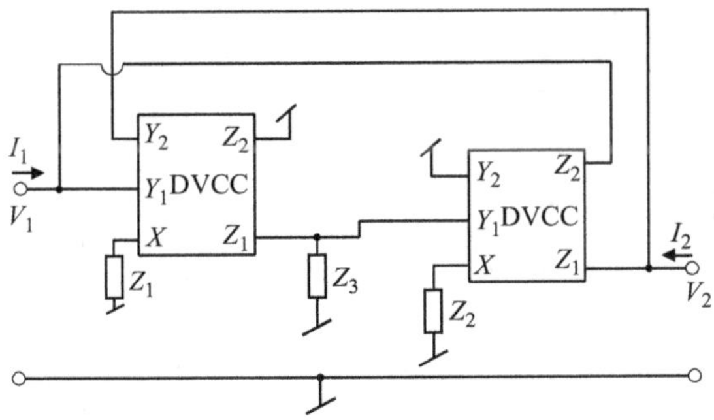

Figure 4.57 Floating impedance circuit employing two DVCCs as proposed by Sedef and Acar [329]

Looking into (4.51) it is obvious that from the same circuit, a lossless floating inductance can be realized by choosing Z_1 and Z_2 as resistors and Z_3 as a capacitor and an ideal FDNR is realizable by choosing Z_3 as a resistor and Z_1 and Z_2 as capacitors. In both the cases, the realized impedances are controllable through a single-variable resistance. In this context, if the resistor is replaced by MOSFET-based voltage-controlled grounded resistor as the one in [376], then electronically variable floating inductance or floating FDNR would be realizable.

Application: The operation of this circuit has been demonstrated by employing a CMOS DVCC (see Fig. 1b of [329]) based on 1.2 μm CMOS process parameters obtained through MOSIS. The FDNR was employed to design a fourth-order Butterworth band-pass (BP) filter with $f_0 = 100$ kHz and BW $= 20$ kHz. Good correspondence between theoretical and simulation results was claimed, thereby confirming the practical feasibility of the circuit of Figure 4.57.

4.6.3 CCIII-simulated inductors

A CCIII is a 3-port active element characterized by the terminal equations $I_y = -I_x$, $V_x = V_y$ and $I_z = \pm I_x$ [384].

Several researchers have demonstrated the utility of the CCIIIs in inductance simulation, for instance, see [327,330,343,348]. On the other hand, CMOS realizations of CCIII have also been presented in the literature, for example, see [135,137].

Four exemplary circuits derived by Wang and Lee [331] using CCIII are shown in Figure 4.58. By a straightforward analysis, Y_{in} and Z_{in} of these circuits are found to be

$$Y_{ina} = \frac{1}{R_1} + \frac{1}{R_2} + \frac{1}{sC_0R_1R_2} = Y_{inc} \tag{4.52}$$

and

$$Z_{inb} = R_1 + R_2 + sC_0R_1R_2 = Z_{ind} \tag{4.53}$$

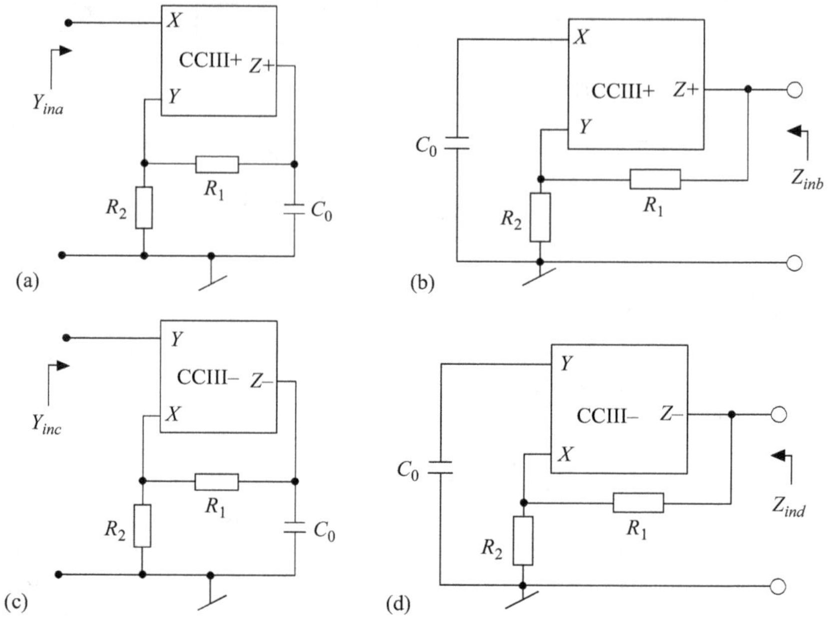

Figure 4.58 Four canonic RL immittance simulators (a)-(d) using a single CCIII as presented by Wang and Lee [331]

Application: For checking the operation of the circuits, a CMOS CCIII introduced by Piovaccari [137] with level 28 MOS model parameters provided by TSMC 0.6 μm CMOS technology was employed. The dc bias supply was taken as ±2.5 V and the circuits were used to realize filters whose SPICE simulations were shown to agree with the theory.

4.6.4 Single-DVCC-based grounded RL *and* CD *immittance simulators*

Incekaraoglu and Cam [334] presented two circuits for simulating lossy inductors (and also for lossy FDNR using *RC*:*CR* transformation) using a single DVCC (characterized by $I_Y = 0$, $V_X = V_{Y1} - V_{Y2}$, $I_Z = I_X$) and three passive components which are shown here in Figure 4.59. Analysis of these circuits yields

$$Z_{ina} = \frac{V_{in}}{I_{in}} = (R_1 + R_2) + sC_0R_1R_2 \tag{4.54}$$

and

$$Y_{ina} = \frac{I_{in}}{V_{in}} = \left(\frac{1}{R_1} + \frac{1}{R_2}\right) + \frac{1}{sC_0R_1R_2} \tag{4.55}$$

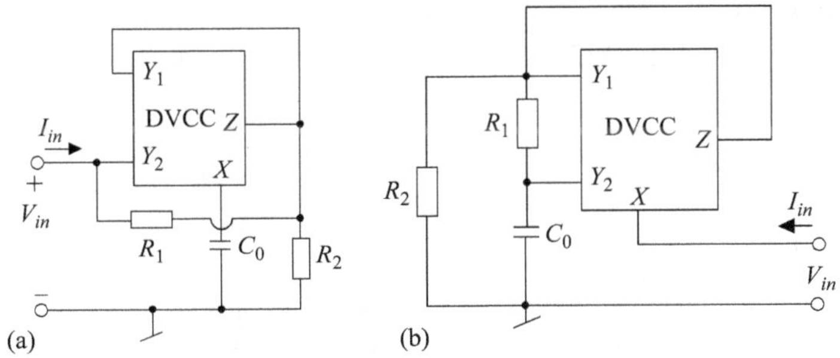

Figure 4.59 Lossy grounded inductors proposed by Incekaraoglu and Cam [334]: (a) simulation of series RL impedance, (b) simulation of parallel RL admittance

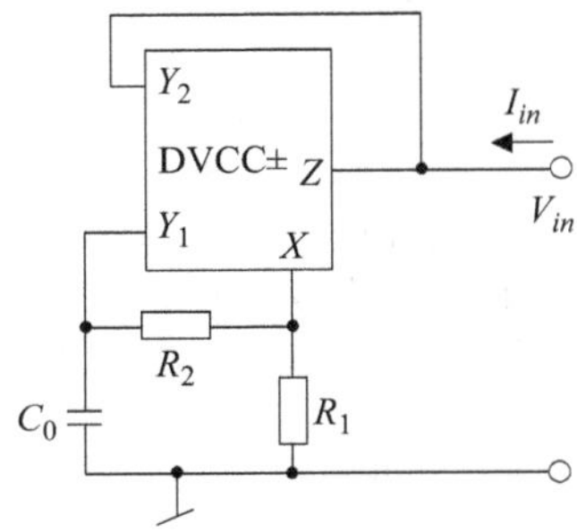

Figure 4.60 Yuce's circuit for positive/negative lossy inductors [338]

Yuce [338] also presented a similar circuit which is shown in Figure 4.60 and realizes

$$Y_{in} = \frac{I_{in}}{V_{in}} = \pm\left(\frac{1}{R_1} + \frac{1}{R_2}\right) \pm \frac{1}{sCR_1R_2} \tag{4.56}$$

Thus, this circuit can simulate a positive or a negative parallel *RL* admittance depending upon whether DVCC is characterized by $i_z = +\, i_x$ (DVCC+) or $i_z = -i_x$ (DVCC−).

All these circuits have an attractive feature of employing a grounded capacitor as preferred for IC implementation.

4.6.5 DXCCII-based electronically controllable gyrator/inductor

The DXCCII can be considered to be a building block which possesses the properties of both CCII and ICCII. DXCCII has found interesting applications in MOSFET-C filter design. The dual-*X* terminals make easy to cancel the square nonlinearity of the MOSFETs if the drain and source of the MOSFET are connected between two *X*-terminals. The basic idea has been found to be useful in the realization of MOSFET-based gyrators and hence, simulated inductors.

Zeki and Toker in [337] presented a MOSFET-C gyrator which is shown in Figure 4.61. By a routine analysis, the Y-matrix of this circuit is given by

$$[Y] = \begin{bmatrix} 0 & 2\mu C_{ox}\left(\frac{W}{L}\right)_2 (V_{G2} - V_{th}) \\ -2\mu C_{ox}\left(\frac{W}{L}\right)_1 (V_{G1} - V_{th}) & 0 \end{bmatrix} \tag{4.57}$$

This gyrator can be used to simulate an inductor by connecting a capacitor at either port 2 or port 1. With a capacitor C_1 connected at port 2, the input impedance of the resulting circuit is found to be

$$Z_{in} = \frac{sC_1}{4\mu^2 C_{ox}^2 (W/L)_1 (W/L)_2 (V_{G1} - V_{th})(V_{G2} - V_{th})} \tag{4.58}$$

Thus, the circuit realizes grounded inductor value (L_{eq}) is given by

$$L_{eq} = \frac{C_1}{4\mu^2 C_{ox}^2 (W/L)_1 (W/L)_2 (V_{G1} - V_{th})(V_{G2} - V_{th})} \tag{4.59}$$

It is, thus, seen that L_{eq} can be controlled by V_{G1} and/or V_{G2}.

Application: The application of the DXCCII-based gyrator of Figure 4.61 was demonstrated by realizing a current-mode biquad filter (see Fig. 2b of [337]) wherein the CMOS DXCCII from [377] was used to simulate the inductor. The variation of the cut-off frequency of the filter was found to be controllable as predicted by theory.

4.6.6 *Grounded inductor realized with modified inverting CCII*

A MICCII [340] was defined by the following equation:

$$\begin{bmatrix} V_x \\ I_y \\ I_z \end{bmatrix} = \begin{bmatrix} -\beta & 0 & 0 \\ 0 & 0 & 0 \\ 0 & \pm 2\alpha & 0 \end{bmatrix} \begin{bmatrix} V_y \\ I_x \\ V_z \end{bmatrix} \tag{4.60}$$

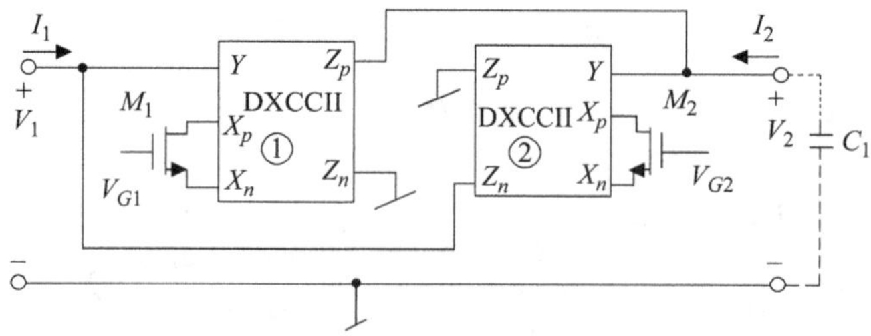

Figure 4.61 DXCCII-based electronically controllable gyrator proposed by Zeki and Toker [337]

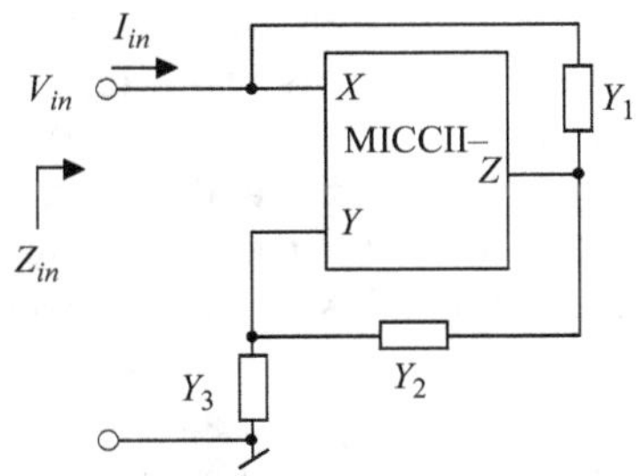

Figure 4.62 Realization of the grounded inductor employing MICCII– [340]

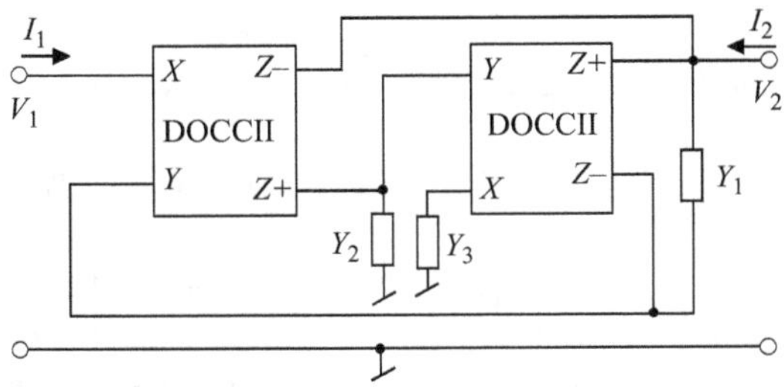

Figure 4.63 Generalized floating admittance inverter proposed by Minaei et al. [342]

where α is the nonideal voltage gain and β is the nonideal current gain of the MICCII (ideally equal to unity). The positive and negative signs of the parameter α denote the positive (MICCII+) and negative (MICCII–), respectively. The input admittance of the circuit of Figure 4.62 is found to be

$$Y_{in} = \frac{I_{in}}{V_{in}} = \frac{y_1 y_3}{2y_2} + y_1 - \frac{y_3}{2} \tag{4.61}$$

With $y_1 = 1/R$, $y_3 = 2/R$ and $y_2 = sC$, the circuit of Figure 4.62 simulates a lossless grounded inductance (L_{eq}) of value CR^2 employing only three of passive and one active elements.

4.6.7 Synthetic floating immittances realized with DOCCII

Minaei *et al.* [342] demonstrated that a general floating admittance could be realized using only two DOCCIIs and three passive components (see Figure 4.63). This configuration can be used to simulate an FI, FC, floating FDNR and floating admittance converter (FAC) by appropriate choice (resistive/capacitive) of various circuit admittances without requiring any passive components matching. With ideal DOCCIIs, a straightforward analysis of this circuit gives its short circuit admittance matrix as

$$[Y] = \frac{y_1 y_2}{y_3}\begin{bmatrix} 1 & -1 \\ -1 & 1 \end{bmatrix} \tag{4.62}$$

The realization of the following floating elements is of interest: Choosing y_1 and y_2 as resistors and y_3 as a capacitor, an FI is realized; if y_1 and y_3 are selected as resistors and y_2 is taken as a capacitor, an FC is realized; if y_1 and y_2 are capacitors with y_3 being a resistor, an FDNR is realized and finally, if y_1 and y_3 are selected as resistors, with $y_2 = y(s)$ as a general admittance, the circuit realizes a FAC. In the specific realizations of FC, FDNR and FAC, the only resistors are connected at the terminal-X. Since no capacitor (s) is (are) connected to terminal-X in any case (s), the circuit becomes highly suitable for high-frequency operation.

It may be mentioned that another general structure for converting a grounded admittance into a floating admittance was proposed by Yuce *et al.* [341] which, however, needs three CCIIs of various kinds.

It is also worth mentioning that two NICs using two CCs and a resistor and another modified compensated NIC employing three CCs were reported by Yuce [346] which may find interesting application in their own right.

4.6.8 Another FI with improved low-frequency performance realized with only two DOCCIIs

In several earlier sections, it has been amply demonstrated that a lossless FI can be realized by using two DVCCs or two DOCCs with only three passive elements (two resistors and a grounded capacitor), Yuce and Minaei [351] introduced another lossless FI configuration using exactly the same number of active and passive components. The circuit proposed by Yuce and Minaei [351] provides the following advantageous features: (i) feasibility of accommodating the effect of finite input resistance R_x of both the DDCCs and (ii) appearance of the Z-port parasitic directly across ports 1 and 2, respectively. The circuit of [351] is shown in Figure 4.64 and exhibits improved low-frequency performance. However, a drawback of this circuit is that it requires a component matching condition ($R_1 = R_2$) to realize a lossless FI.

In [351] as many as eight different topologies of FIs were presented, five of which require five CCs of various types. By contrast, only three configurations of [351] are implementable with only two CCs and the circuit of Figure 4.64 is one

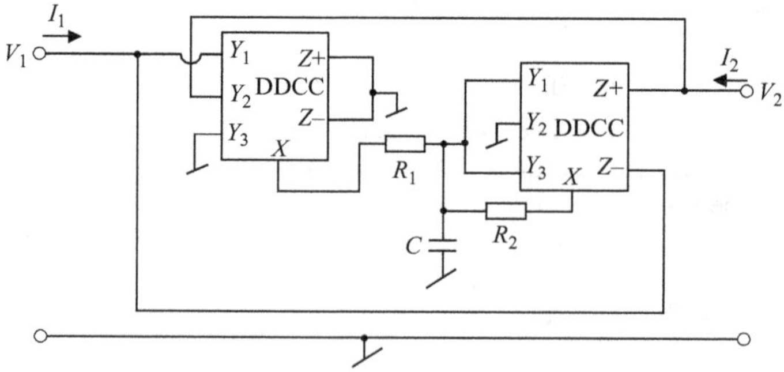

Figure 4.64 FI proposed by Yuce and Minaei [351]

of these. For details of all the FI circuits and their SPICE simulation results the readers may refer to [351].

4.6.9 Floating impedance simulator realized with a DOCCII and an OTA

Sagbas *et al.* introduced an FI simulator in [349] which employs one DOCCII, one OTA along with two grounded passive components. Their configuration can simulate an *electronically variable* FI which could be made either positive or negative. The same circuit can also realize either a floating capacitor or a floating resistor by suitable choice of the two admittances and has the advantage of not requiring any component-matching conditions.

Another advantage of the circuit is that the simulated floating elements are *electronically controllable* through the biasing current of the OTA and/or through the grounded variable resistor or a variable capacitor.

An analysis of the circuit of Figure 4.65 shows its short circuit admittance matrix to be

$$[Y] = \left(\frac{y_1}{y_2}\right) g_m \begin{bmatrix} 1 & -1 \\ -1 & 1 \end{bmatrix} \tag{4.63}$$

From the above, it follows that the circuit can simulate an FI, a floating capacitor or an electronically variable resistor by a suitable choice of the two circuit admittances y_1 and y_2.

Application: The application of the simulated FI realizable from the circuit of Figure 4.65 was demonstrated by employing this in the design of a fifth-order Chebyshev LP filter having a cut-off frequency of 1 MHz in which case, the FI was realized by taking component/parameter values as $R_1 = 1$ kΩ, $C_2 = 17$ pF and $g_m = 0.1$ ms. A CMOS DOCCII was employed therein (see Fig. 2 of [349]) which was based on the then-prevalent TSMC 0.35 μm CMOS technology parameters. The SPICE simulation results of the filter employing a passive inductor as well as the simulated FI realized from the circuit of Figure 4.65 were found to be in good agreement.

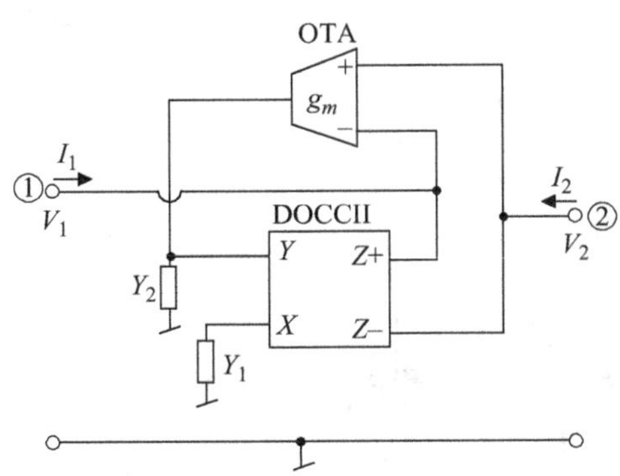

Figure 4.65 Floating impedance simulator proposed by Sagbas et al. [349]

4.6.10 Lossless FI realization employing a DOCCCII and grounded capacitor

CCCII has been used to realize a variety of grounded and floating impedances by many researchers, for instance, see [332,333,339,345,357,369]. Sagbas *et al.* [352] presented a circuit using one DOCCCII, one OTA and a grounded capacitor (Figure 4.66) which realizes a lossless FI employing a grounded capacitor with the advantage of not requiring any component-matching conditions. The realized FI is electronically controllable through the external dc bias current of the DOCCCII or through the variation of the bias current of the OTA. The analysis of the circuit of Figure 4.66 reveals the following [*Y*] matrix:

$$[Y] = \left(\frac{g_m}{sCR_x}\right)\begin{bmatrix} 1 & -1 \\ -1 & 1 \end{bmatrix} \tag{4.64}$$

where R_x is the parasitic input resistance of the *X*-terminal given by $R_x = V_T/2I_0$ and V_T is the thermal voltage with I_0 being the dc bias current of the DOCCCII. On the other hand, the OTA is assumed to have its transconductance as $g_m = I_B/2V_T$. It is

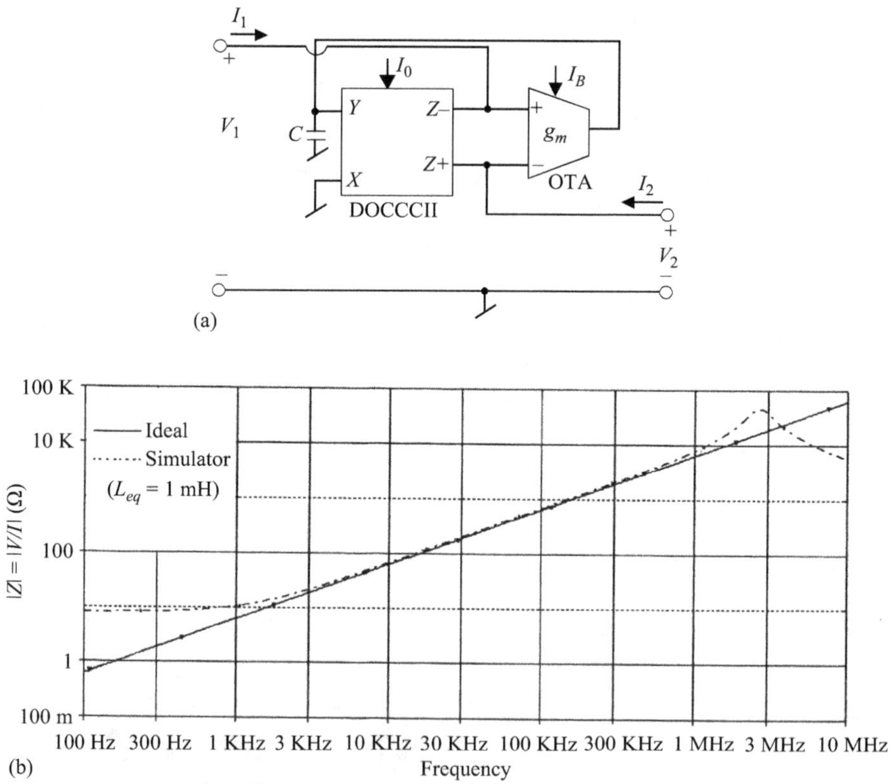

Figure 4.66 FI simulator with electronic controllability proposed by Sagbas et al. (a) the lossless FI circuit, (b) the magnitude response of the realized inductance. Reprinted, with permission from [352] © 2008 Elsevier GmbH

thus seen that the value of simulated FI is given by $L_{eq} = (CV_T^2/I_B I_0)$ and is controllable through the dc bias current I_B of the OTA.

Thus, it is seen that the inductance value can be controlled electronically through either I_B or I_0. By grounding either of the ports, the presented configuration can simulate an electronically adjustable grounded inductor.

Application: The application of this FI was demonstrated by employing the DOCCCII and OTA realizable by the bipolar circuit [332,383] based on PR100N and NR100N bipolar transistor arrays ALA400 from AT&T. The circuit was biased with ±2.5 V dc supplies. The FI simulator was employed to realize in a fourth-order BP filter whose workability was demonstrated by SPICE simulations. The simulated inductor showed good performance from 1 kHz to 1 MHz (see Figure 4.66(b)).

4.6.11 AN FI employing only a single DODDCC and three passive elements

A DODDCC is characterized by $I_{Y1} = I_{Y2} = I_{Y3} = 0$, $V_X = (V_{Y1} - V_{Y2} + V_{Y3})$, $I_{z+} = I_x$ and $I_{z-} = -I_x$. An interesting circuit of lossy FI using only one DODDCC along with a grounded capacitor was first proposed by Ibrahim *et al.* [368] and is shown in Figure 4.67.

This circuit realizes a parallel *RL* floating impedance with $R_{eq} = R_1//R_2$ and $L_{eq} = CR_1R_2$ with the attractive features of not requiring any component matching, using only three passive components and having the capacitor grounded.

4.6.12 External resistorless FI realization using DXCCII

A resistorless simulated FI using DXCCIIs and three NMOS transistors (operating in triode region) was proposed by Saad and Soliman [398] which is shown here in Figure 4.68 with drain and source terminals of all the three MOSFETs connected to complementary voltage, the square nonlinearities of the MOSFETs are eliminated in the expression for their drain currents. Due to this, each MOSFET, thus, realizes a linear voltage-controlled resistor with equivalent resistance given by

$$R_i = \frac{1}{2\mu_n C_{ox}(W/L)_i(V_{Gi} - V_{th})} \tag{4.65}$$

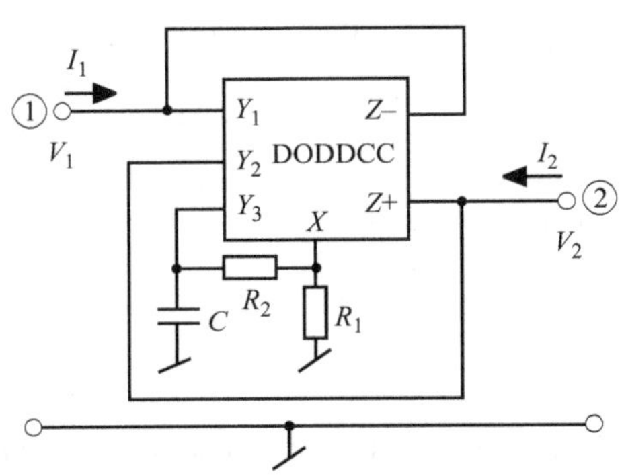

Figure 4.67 Realization of lossy FI employing single DODDCC proposed by Ibrahim et al. [368]

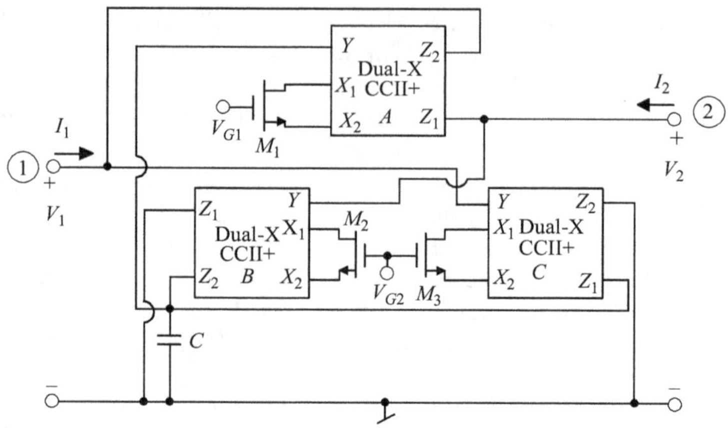

Figure 4.68 Resistorless simulation of FI using DXCCIIs proposed by Saad and Soliman [398]

Since M_2 and M_3 are given a common gate voltage V_G at their gates, it follows that $R_2 = R_3$. By straightforward analysis, the $[Y]$ matrix of this circuit is found to be

$$[Y] = \left(\frac{4}{sC}\right)\left\{\mu_n C_{ox}\left(\frac{W}{L}\right)\right\}^2 (V_{G1} - V_{th})(V_{G2} - V_{th})\begin{bmatrix} 1 & -1 \\ -1 & 1 \end{bmatrix} \quad (4.66)$$

where all MOSFETs are assumed to have the same (W/L) rations. In view of the above, the value of the simulated FI is given by

$$L_{eq} = \frac{C}{4(\mu_n C_{ox}(W/L))^2 (V_{G1} - V_{th})(V_{G2} - V_{th})} \quad (4.67)$$

which is electronically controllable through V_{G1} and/or V_{G2}.

It is interesting to observe that with capacitor C deleted, the circuit is an electronically controllable three-port gyrator.

4.6.13 *Electronically tunable MOSFET-C FDNR using a DXCCII*

Another electronically tunable MOSFET-C FDNR circuit based on a single DXCCII, two capacitors and a MOSFET (as a resistor) was presented by Kacar *et al.* [136] and is shown in Figure 4.69. The input impedance of this circuit is given by

$$Z_{in} = \frac{2}{s^2 C^2 R} \quad \text{where} \quad R = \frac{1}{2\mu_n C_{ox}(W/L)(V_c - V_{th})} \quad (4.68)$$

from where the value of the simulated FDNR (D) is found to be $D = C^2 \mu_n C_{ox}(W/L)(V_c - V_{th})$ which can be controlled through the control voltage V_c.

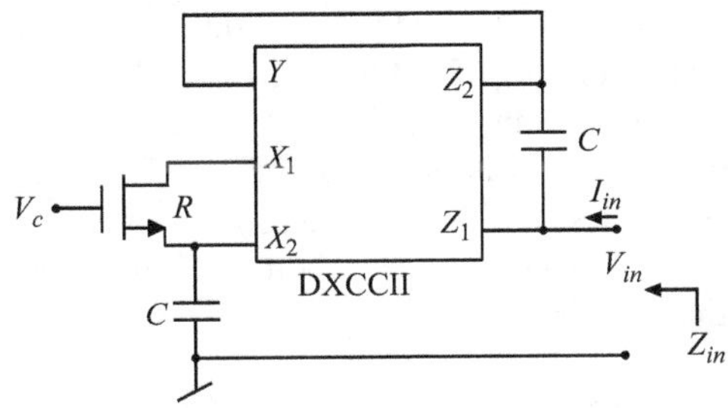

Figure 4.69 Electronically controllable MOSFET-C FDNR presented by Kacar et al. employing a single DXCCII [136]

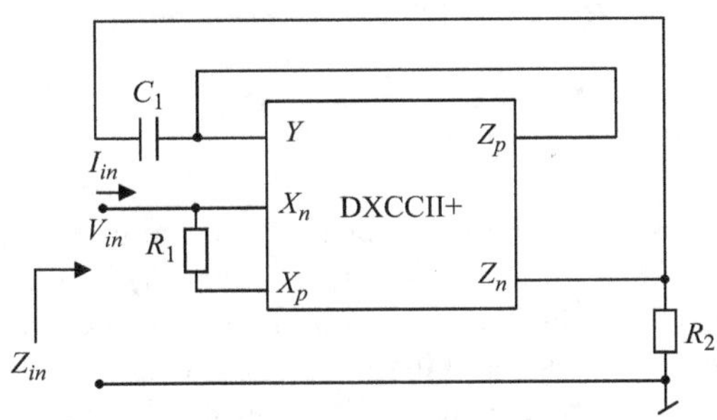

Figure 4.70 Simulation of grounded inductance using a single DXCCII proposed by Kacar and Yesil [354]

Application: The circuit of Figure 4.69 was used to realize a fifth-order elliptic filter employing two grounded FDNRs, both of which were realized from the circuit of Figure 4.69. Based on TSMC CMOS 0.35 μm process model parameters and the DXCCII circuit biased with ±1.5 V dc, the filter was realized for a cut-off frequency of 1.59 MHz and was demonstrated to work well.

4.6.14 Grounded inductance simulation using a DXCCII

Use of single DXCCII in realizing a grounded inductance simulator[8] was shown by Kacar and Yesil [354]. The circuit from [354] is shown in Figure 4.70 for which the input admittance is given by

$$Y_{in} = \frac{2}{sC_1R_1R_2} - \frac{(R_1 - 4R_2)}{R_1R_2} \tag{4.69}$$

By selecting $R_1 = 4R_2$, one gets $Z_{in} = (sCR_1R_2/2)$, which represents an inductance $L_{eq} = (C_1R_1R_2/2)$.

[8]For another work dealing with the use of DXCCII in realizing a positive/negative, lossy/lossless inductance simulation circuits, the readers are referred to [354].

Application: The workability of the inductance simulation circuit of Figure 4.70 was demonstrated in [359] by SPICE simulations based on the CMOS DXCCII from [141] by designing a third-order Butterworth high-pass filter and a BP filter. The CMOS DXCCII was biased with ±1.5 V dc supply and MOS transistor model was based on TSMC CMOS 0.35 process prevalent then and the filters were designed for $f_0 = 1.59$ MHz. The simulation results were found to be in good agreement with the theory.

4.6.15 FI realization using only two DVCCs/DVCCCs

An interesting FI configuration based on two DVCCs and three passive elements, including a grounded capacitor, was presented by Horng [359] which also did not require any component-matching conditions. The circuit from [359] is shown in Figure 4.71 with ideal DVCCs, the $[Y]$ matrix of the circuit is given by

$$[Y] = \left(\frac{1}{sC_1R_1R_2}\right)\begin{bmatrix} 1 & -1 \\ -1 & 1 \end{bmatrix} \tag{4.70}$$

which represents a lossless FI with its inductance value given by $L_{eq} = C_1R_1R_2$.

A novel feature of this circuit is that since R_1 and R_2 both are connected to the X-terminal of the DVCCs, these can be eliminated and the X-port parasitic resistances R_{x1} and R_{x2} can be considered in their places. This would, therefore, result in $L_{eq} = C_1R_{x1}R_{x2}$ which can be electronically controllable through the external bias currents I_{B1} and/or I_{B2} which control R_{x1} and R_{x2}, respectively.

Yuce [360] introduced another DVCC-based circuit configuration, shown in Figure 4.72. This circuit uses two active and three passive elements but employs all grounded components. This circuit is capable of simulating FI, FC and FDNR elements by appropriate selection of passive elements (resistor (s) or capacitor (s)). An analysis of this circuit gives the following $[Y]$ matrix:

$$[Y] = \left(\frac{y_1y_2}{y_3}\right)\begin{bmatrix} 1 & -1 \\ -1 & 1 \end{bmatrix} \tag{4.71}$$

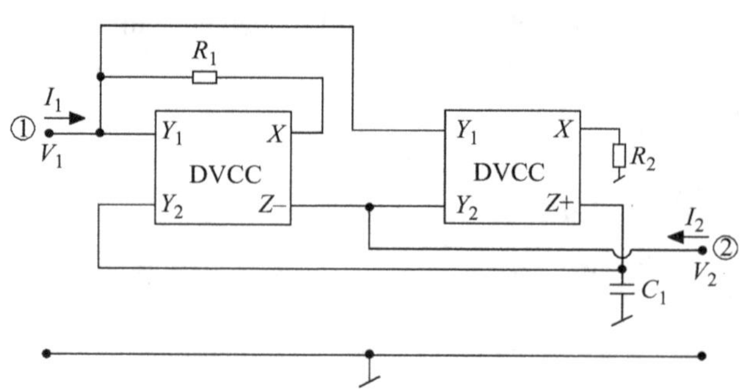

Figure 4.71 FI using DVCC proposed by Horng [359]

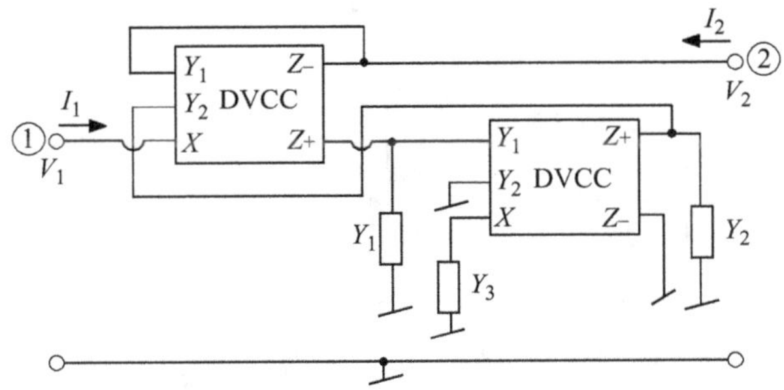

Figure 4.72 Floating impedance simulator due to Yuce [360]

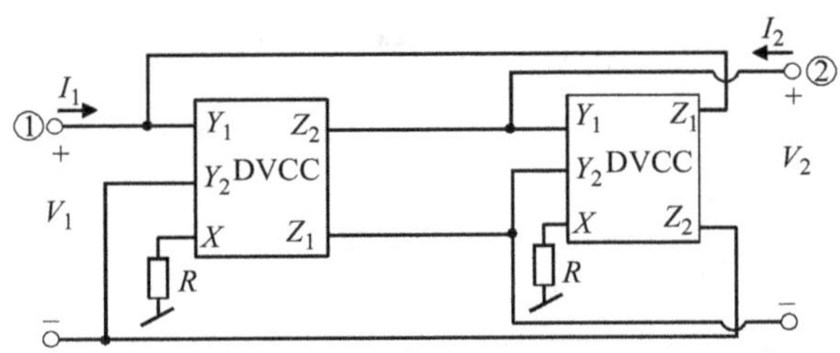

Figure 4.73 A fully floating gyrator proposed by Elwan and Soliman [138]

From the circuit of Figure 4.72 an FI is realized by selecting y_3 as a capacitor and the two remaining admittances as resistors; a floating capacitor is simulated by choosing y_1 as a capacitor with y_2, y_3 being resistors; and finally, an FDNR is obtainable by choosing y_1, y_2 as capacitors and y_3 as a resistor.

Elwan and Soliman [138] presented a number of basic applications of DVCC in analog signal processing out of which an interesting one was that of a fully floating gyrator which is shown here in Figure 4.73.

A somewhat similar circuit, which realizes a floating FDNR two grounded capacitors and a grounded resistor, was proposed by Sedef and Acar [329]. An interesting three-port gyrator is derivable from the quoted circuit (although not so conceived explicitly in [329]) which is shown here in Figure 4.74. By straightforward analysis, this three-port gyrator is found to be characterized by the following [Y] matrix:

$$\begin{bmatrix} I_1 \\ I_2 \\ I_3 \end{bmatrix} = \begin{bmatrix} 0 & 0 & \frac{1}{R} \\ 0 & 0 & -\frac{1}{R} \\ \frac{1}{R} & -\frac{1}{R} & 0 \end{bmatrix} \begin{bmatrix} V_1 \\ V_2 \\ V_3 \end{bmatrix} \tag{4.72}$$

Laoudias and Psychalinos [379] demonstrated that if the parasitic X-port intrinsic resistance R_x is accounted for in the formulation of the so-called differential voltage

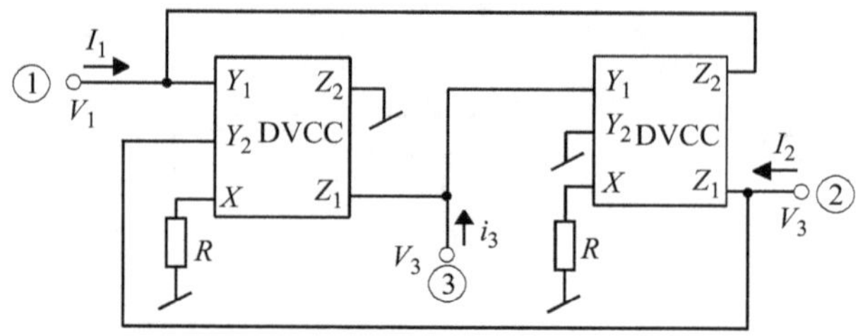

Figure 4.74 A three-port gyrator derivable from floating FDNR of Sedef and Acar [329]

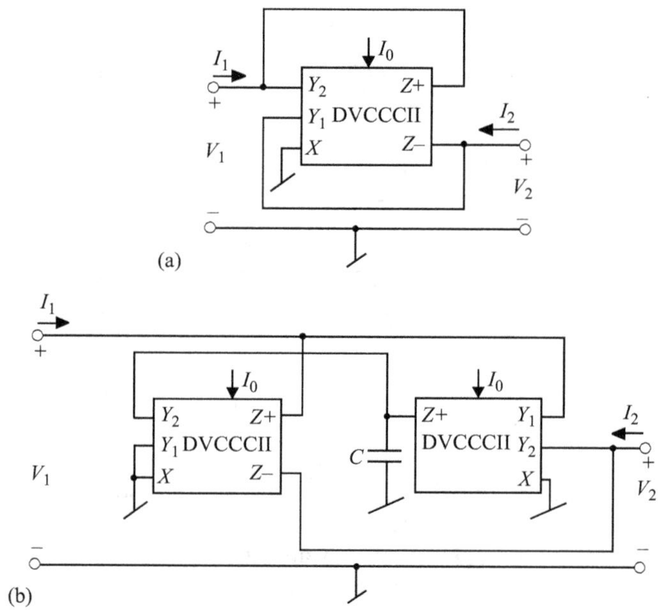

Figure 4.75 Electronically controllable floating impedance circuits proposed by Laoudias and Psychalinos [379]: (a) floating resistance, (b) floating inductance

controlled current conveyor II (DVCCCII) characterized by $V_x = (V_{Y1} - V_{Y2}) + R_x I_x$; $I_{Y1} = 0 = I_{Y2}$, $I_{z+} = I_x$ and $I_z = -I_x$ (where $R_x = V_T/2I_0$), then an interesting class of circuits can be derived. For example, an electronically controllable floating resistance can be easily formulated as shown in Figure 4.75(a). On the other hand, an electronically controllable lossless floating inductance can be realized as shown in Figure 4.75(b).

Application: The active simulators of Figure 4.75 have been shown [380] to be useful in realizing a third-order Leapfrog active filter based on a third-order LP RLC ladder resulting in a structure with only four DV-CCCII and all the three grounded capacitors. The filter had cut-off frequency equal to 2.5 MHz and performed well.

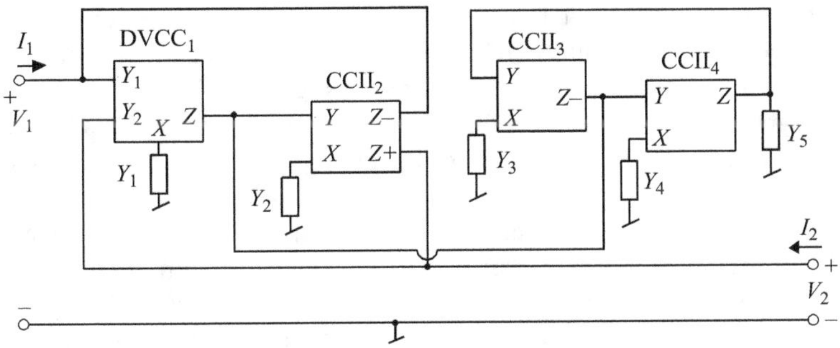

Figure 4.76 A generalized floating impedance proposed by Jaikla and Siripruchyanun [382]

Jaikla and Siripruchyanun [382] presented an interesting configuration (see Figure 4.76) as a generalized floating impedance simulator which although requires four active elements, namely one DVCC and three CCIIs, but this additional requirement is well compensated by the resulting versatility of the configuration. This configuration has not only all the five circuit admittances grounded, it permits apart from the realization of floating *L*, *C*, and FDNR elements, a floating FDNC element too. This feature has significant advantage since none of the previously described two/three active-elements-based FI simulators are able to achieve this functionality.

The various floating elements realizable from this configuration are as follows:

1. FI: With $Y_3 = sC_3$ and all other elements as resistors, yielding $L_{eq} = C_3R_1R_2R_5/R_4$ or using $Y_4 = sC_4$ yielding $L_{eq} = C_4R_3R_2R_5/R_3$
2. FC: Ether using $Y_1 = sC_1$, thereby giving $C_{eq} = C_1R_3R_4/R_2R_5$ or using $Y_2 = sC_2$, thereby obtaining $C_{eq} = C_2R_3R_4/R_1R_5$
3. F-FDNR: Using $Y_1 = sC_1$, $Y_2 = sC_2$, thereby resulting in $D = C_1C_2R_3R_4/R_5$
4. F-FDNC: By taking $Y_3 = sC_3$ and $Y_4 = sC_4$, thereby producing $M = C_3C_4R_1R_2R_5$

Application: Workability of the various modes of operation of this circuit was demonstrated in [382] through filter designs using appropriate bipolar DVCC and CCII implementation using PNP and NPN transistors based on parameters of PR200N and NR200N bipolar transistor arrays from AT&T with ± 1.5 V supply voltages and I_B set to 200 μA. Cut-off frequencies 1 kHz–9 MHz were realizable using the described FI simulators when they were used to design filters using floating elements.

4.6.16 FDCCII-based lossless grounded inductor employing three grounded passive elements

An FDCCII [143] is considered to be a more versatile form of CCII which has been found to be useful in many analog signal operations, for instance, see [140,143–148]. The application of the FDCCII when used in the realization of a grounded inductance leads to the circuit having a minimum number of grounded passive elements while offering the advantage of not requiring component-matching condition/constraint. A configuration using a single FDCCII and all grounded passive components was proposed by Kacar [361] which is shown here in Figure 4.77.

A straightforward analysis of this circuit (assuming ideal FDCCII) yields its input impedance as $Z_{in} = sCR_1R_2$, thereby giving an inductance of value $L_{eq} = CR_1R_2$.

Kacar and Yesil [362] presented three FDCCII-based grounded FDNR simulators all of which employ, other than a single FDCCII, a minimum number of only three passive elements, namely two capacitors and a resistor all of which are grounded. Theses circuits, however, do require design constraints/matching in terms of the two capacitors. Two of the three circuits presented in [362] having a simpler matching condition, namely $C_1 = C_2 = C$, which is also preferable from IC implementation viewpoint of minimizing total capacitance employed in the circuit.

A straightforward analysis of the circuits shown in Figure 4.78 reveals that in both the cases, the value of the realized D-element is given by $D = C^2R$.

Application: These FDNRs were employed to design a third-order passive RLC-based FDNR filter requiring a single FDNR which was realized using a CMOS FDCCII [140] in which active feedback cascade current minors were employed in the output stages of the CC due to their increased output resistance levels. Using the

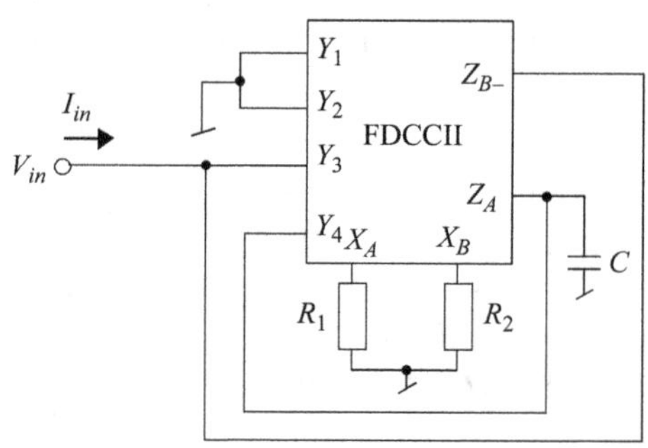

Figure 4.77 Simulated grounded inductor using a single FDCCII [361]

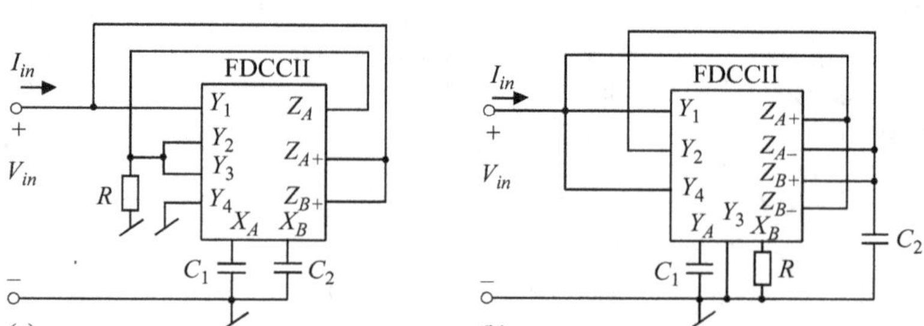

Figure 4.78 Two FDNR realizations (a), (b) using a single FDCCII [362]

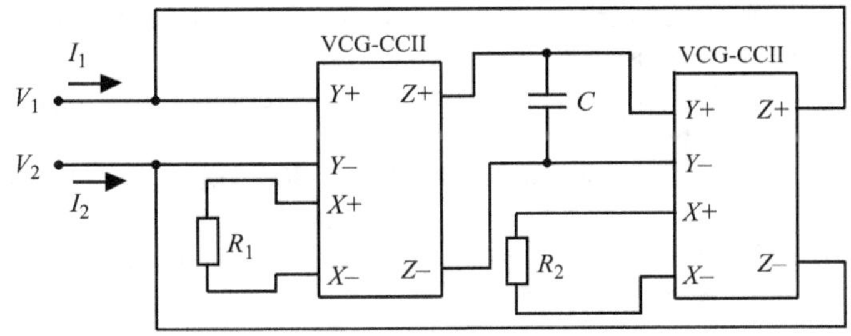

Figure 4.79 A floating inductance simulator based on VCE-CCII proposed by Marcellis et al. [381]

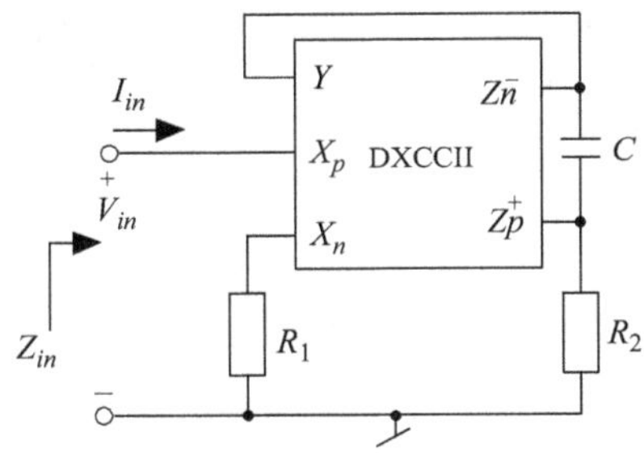

Figure 4.80 DXCCII-based grounded inductor by Myderrizi, Minaei and Yuce [366]

then-prevalent 0.35 μm CMOS technology parameters, the designed filter and the FDNR were found to be operating satisfactorily over a frequency range of 70 Hz–7 MHz [362].

Marcellis *et al.* [381] presented floating impedance simulators using a slightly different type of CCII which they chose to call voltage and current gained CCII (VCG-CCII). Their floating inductance simulator is shown in Figure 4.79 which simulates a lossless floating inductance of value $L_{eq} = CR_1R_2$.

4.6.17 DXCCII-based grounded inductance simulators

A DXCC [377] is yet another variant of CCII which is characterized by $I_Y = 0$, $V_{xp} = V_Y$, $V_{xn} = -V_Y$, $I_{zp} = +I_{xp}$ and $I_{zn} = +I_{xn}$. Metin [365] presented a number of circuits which employ a single DXCCII along with only three passive elements to realize series/parallel *RL*-type lossy grounded inductor. However, the grounded inductance simulator proposed by Myderrizi, Minaei and Yuce [366] realizes lossless inductors which were shown to be useful for applications in the 30 kHz–30 MHz frequency range. This circuit is shown in Figure 4.80. With ideal DXCCII characterization, an analysis of the circuit of Figure 4.80 yields

$$Z_{in}(s) = \frac{V_{in}}{I_{in}} = sCR^2 \quad \text{if} \quad R_1 = R_2 = R \tag{4.73}$$

Thus, the value of inductance can be controlled only by simultaneous variation of the grounded resistors R_1 and R_2.

Since $V_{xp} = V_y$ and $V_{xn} = -V_y$, these complementary voltages are very useful in cancelling the square nonlinearity of a FET if its drain and source terminals are connected at X_p and X_n terminals of the DXCCII [380]. This feature was very nicely exploited by Myderrizi and Zeki [380], who devised an interesting grounded-capacitance multiplier as shown in Figure 4.81 which simulates an equivalent capacitance C_{eq}, given by

$$C_{eq} = \left[1 + \frac{\beta_1(V_{c1} - V_{th})}{\beta_2(V_{c2} - V_{th})}\right] C \tag{4.74}$$

On the other hand, using similar nonlinearity-cancelled MOSFETs, Zeki and Toker [337] had devised an electronically tunable active gyrator shown in Figure 4.82 which, upon terminating its port 2 into a capacitor C_1, simulates a lossless grounded inductor looking into the input port.

This active gyrator is characterized by

$$\begin{bmatrix} I_1 \\ I_2 \end{bmatrix} = \begin{bmatrix} 0 & 2\mu C_{ox}\left(\frac{W}{L}\right)_2 (V_{G2} - V_{th}) \\ -2\mu C_{ox}\left(\frac{W}{L}\right)_1 (V_{G1} - V_{th}) & 0 \end{bmatrix} \begin{bmatrix} V_1 \\ V_2 \end{bmatrix} \tag{4.75}$$

whereas the simulated inductance value with its output port terminated into a capacitor C_1 is given by

$$L_{eq} = \frac{C_1}{4\mu^2 (W/L)_1 (W/L)_2 (V_{G1} - V_{th})(V_{G2} - V_{th})} \tag{4.76}$$

and is, hence, electronically tunable by V_{G1} and/or V_{G2}.

Application: Through SPICE simulations, it has been demonstrated [337] that the inductance simulation circuit of Figure 4.82 operated pretty well between 30 kHz and 10 MHz Furthermore, the authors have demonstrated an application of this gyrator in realizing an LP/BP current-mode filter which was obtained by loading the output port into a capacitor C_2 and connecting another grounded capacitor C_1 at the input port with an input current source I_{in} in parallel with it.

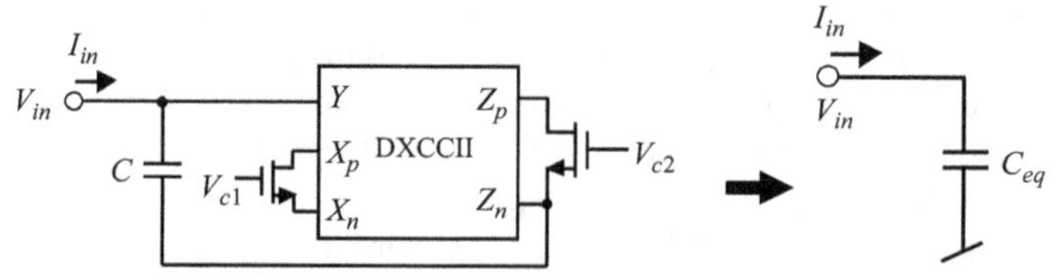

Figure 4.81 A MOSFET-C grounded-capacitance multiplier devised by Myderrizi and Zeki [380]

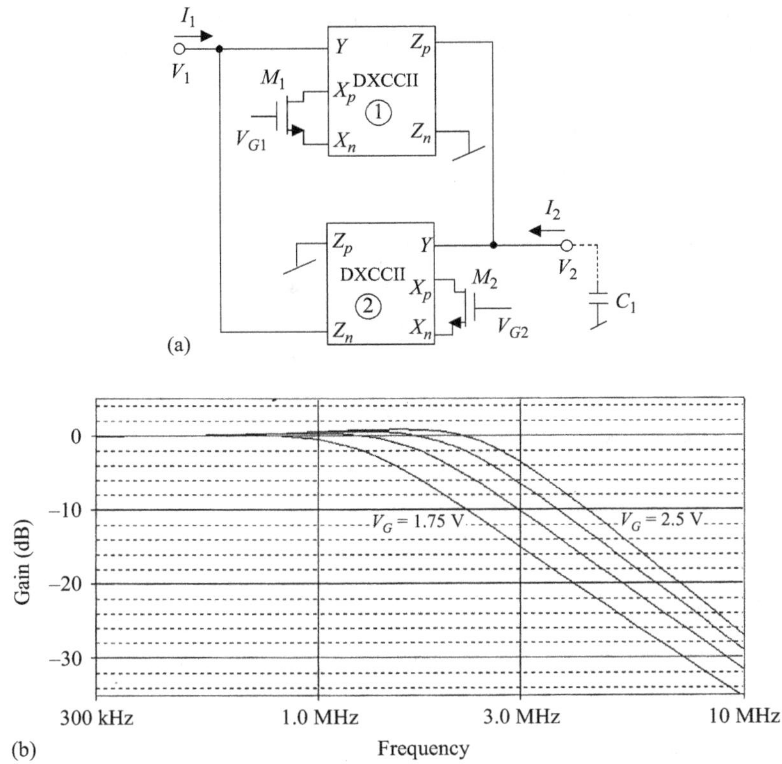

Figure 4.82 DXCCII-based active gyrator devised by Zeki and Toker: (a) the gyrator circuit, (b) application of the gyrator in realizing a current-mode filter LP/BP filter. Reprinted, with permission from [337] © 2004 Elsevier GmbH

The output currents coming out of the two Z_p output terminals realized current-mode LP and BP filters. It was confirmed that the tuning range of 1.4 MHz–2.9 MHz was achievable when the control voltages $V_{G1} = V_{G2} = V_G$ were swept from 1.75 V–2.5 V (see Figure 4.82(b)).

4.6.18 Grounded-capacitor-based floating capacitance multiplier using CCDDCCs

A CCDDCC can be characterized by the following matrix equation:

$$\begin{bmatrix} V_x \\ I_{y1} \\ I_{y2} \\ I_{y3} \\ I_z \end{bmatrix} = \begin{bmatrix} R_x & 1 & -1 & 1 & 0 \\ 0 & 0 & 0 & 0 & 0 \\ 0 & 0 & 0 & 0 & 0 \\ 0 & 0 & 0 & 0 & 0 \\ \pm 1 & 0 & 0 & 0 & 0 \end{bmatrix} \begin{bmatrix} I_x \\ V_{y1} \\ V_{y2} \\ V_{y3} \\ V_z \end{bmatrix} \tag{4.77}$$

Prommee and Somdunyakanok [367] presented an interesting lossless FI which is shown in Figure 4.83. An analysis of the circuit of Figure 4.83 gives the short circuit admittance matrix as

$$[Y] = sC\left(\frac{R_{x2}}{R_{x1}}\right)\begin{bmatrix} 1 & -1 \\ -1 & 1 \end{bmatrix} \tag{4.78}$$

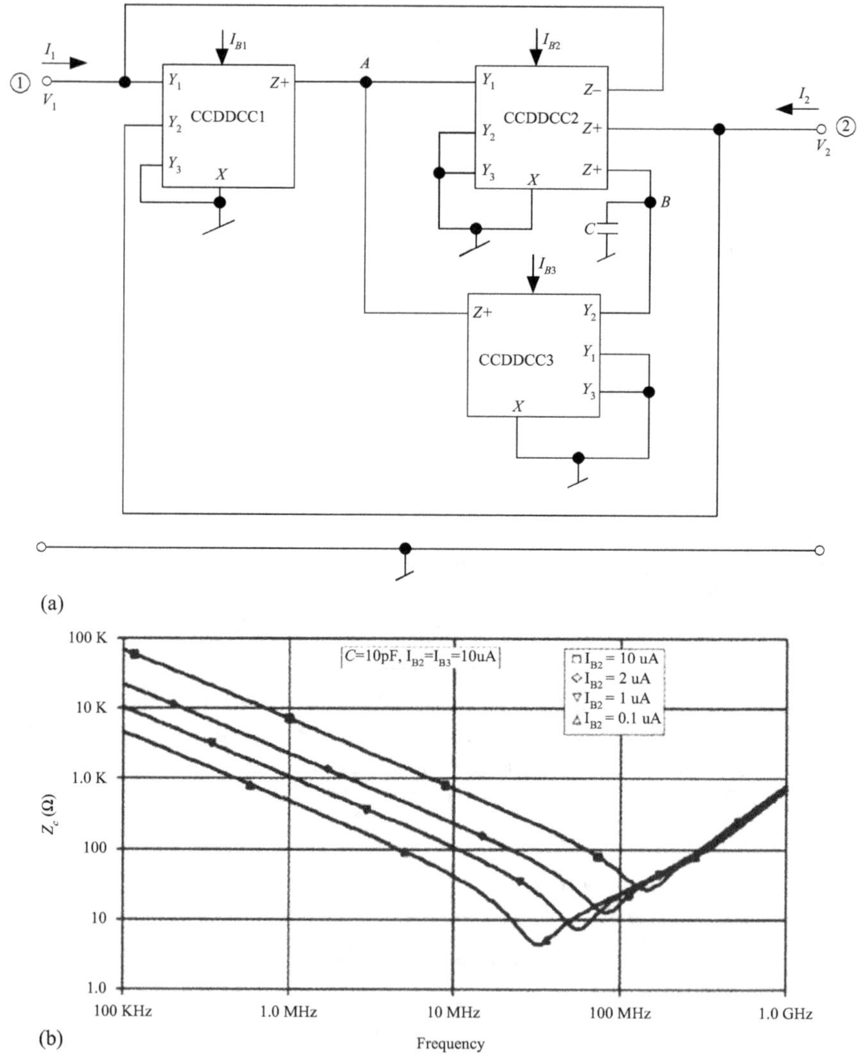

Figure 4.83 A floating capacitance multiplier, (a) the electronically variable floating capacitance multiplier, (b) the frequency response of the simulated capacitance for different values of the controlling currents. Reprinted, with permission from [367] © 2009 Elsevier GmbH

The circuit, thus, realizes a lossless FC with equivalent capacitance of value

$$C_{eq} = C\left(\frac{R_{x2}}{R_{x1}}\right) = C\sqrt{\frac{I_{B1}}{I_{B2}}} \tag{4.79}$$

which is controllable by the external dc bias currents of CCDDCC1 and CCDDCC2.

The validity of this capacitance multiplier circuit was demonstrated by realizing the circuit CMOS CCDDCCs [367] using the model parameters of the then-prevalent TSMC 0.25 μm CMOS technology with the circuit biased with ±1.25 V dc, of the variability of the realized capacitance value was observed to be variable through the external dc bias current I_{B2}.

4.6.19 Single-DODDCC-based grounded lossy inductance simulators employing a grounded capacitor

A grounded lossless inductance simulator was introduced by Ibrahim *et al.* [368] which employs a modified DODDCC (MDO-DDCC). The circuit from [368] is shown here in Figure 4.84 which simulates a lossless grounded inductance using the MDO-DDCC as an active element. A DODDCC is characterized by the terminal equations $I_{y1} = I_{y2} = I_{y3} = 0$, $V_x = V_{y1} - V_{y2} + V_{y3}$, $I_{z+} = I_x$ and $I_{z-} = -I_x$ whereas the modified DODDCC, on the other hand, has $I_{z+} = I_{x+}/2$ and $I_{z-} = -I_x$; the other equations being the same as those for DODDCC.

The input impedance of this circuit is given by

$$Z_{in} = V_{in}/I_{in} = 2sCR^2 \quad \text{for } R_1 = R_2 = R \tag{4.80}$$

from where it is observed that the simulated grounded lossless inductance is given by $L_{eq} = 2CR^2$.

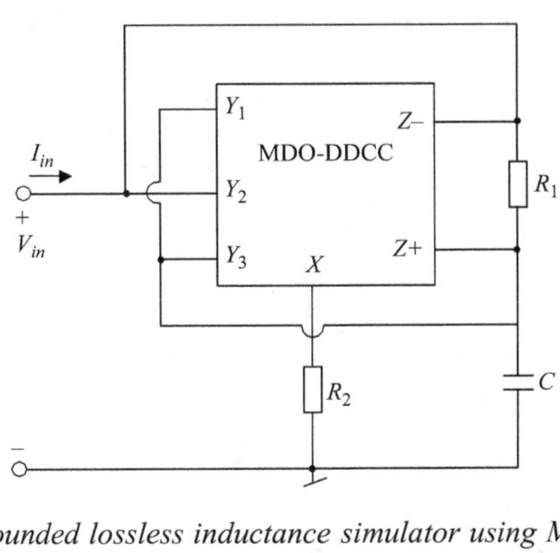

Figure 4.84 Grounded lossless inductance simulator using MDO-DDCC [368]

4.6.20 DCCII-based grounded inductance simulator

A CCII with current differencing capability is known as differential CC (DCCII). The DCCII is a five-port active analog building block which can ideally be characterized by the following matrix equation:

$$\begin{bmatrix} V_{XN} \\ V_{XP} \\ I_{Z+} \\ I_{Z-} \\ I_Y \end{bmatrix} = \begin{bmatrix} 0 & 0 & 1 \\ 0 & 0 & 1 \\ 1 & -1 & 0 \\ -1 & 1 & 0 \\ 0 & 0 & 0 \end{bmatrix} \begin{bmatrix} I_{XP} \\ I_{XN} \\ V_Y \end{bmatrix} \tag{4.81}$$

Two grounded inductance simulators each employing a single DCCII with three passive components were proposed by Metin [411]. A careful look into this circuit reveals that this circuit is actually a nonideal gyrator which becomes an ideal gyrator with $R_1 = R_2 = R$. With $R_1 \neq R_2$, its $[Y]$ matrix is given by

$$[Y] = \begin{bmatrix} 0 & -\dfrac{1}{R_1} \\ \dfrac{1}{R_2} & \left(\dfrac{1}{R_1} - \dfrac{1}{R_2}\right) \end{bmatrix} \tag{4.82}$$

so that with $R_1 = R_2 = R$ the matrix, as expected, reduces to

$$[Y] = \begin{bmatrix} 0 & -\dfrac{1}{R} \\ \dfrac{1}{R} & 0 \end{bmatrix} \tag{4.83}$$

which represents a gyrator. Now, with either port terminated into a grounded capacitor C, the input impedance looking into the other port turns out to be $Z(s) = sCR^2$. This explains why both the circuits presented in [411] (see Fig. 2 therein) realize a lossless grounded inductance of value $L_{eq} = CR^2$. When $R_1 \neq R_2$, then out of the two possible circuits, one realizes a series RL and other a parallel RL (Figure 4.85).

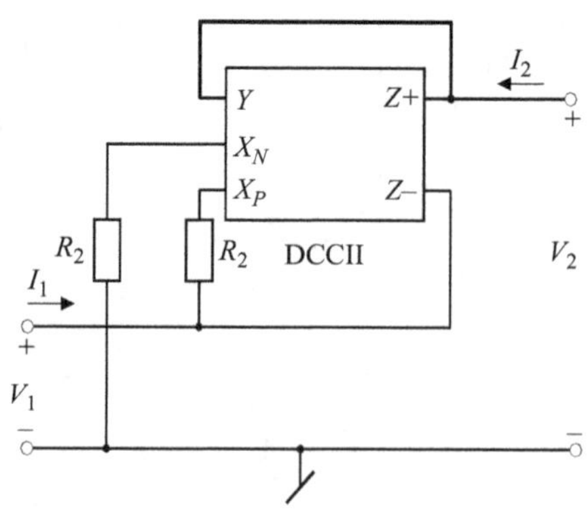

Figure 4.85 Active gyrator obtained from the grounded inductance simulators of Metin [411]

Some other DCCII-based circuits for inductance simulation have also been devised in [347], but the circuits of [347] therein do not have the advantage of employing a grounded capacitor.

4.6.21 Miscellaneous FI simulators using ICCII, DVCC and FDVCC elements

In view of the large number of references on impedance simulation as given at the end of this chapter, it is impossible to include discussion of all circuits in a limited space, so we have discussed only those circuits which, in our opinion, are the relatively more significant ones than those which are left out. Nevertheless, we would surely like to point out references [387,391–394,396,398,403,404] which, we believe, the readers may find interesting.

In the last, we present two interesting lossless FI circuits based on the employment of a combination of DVCC and the so-called FDVCC and another circuit based on the inverting second-generation current conveyors (ICCII) [399]. Use of DVCC in impedance simulation has already been considered in this chapter earlier. The FDVCC, on the other hand, is characterized by the terminal equations: $V_x = (V_{y1} - V_{y2}), I_{y1} = I_{y2} = 0,\ \ I_{z1+} = +I_x$ and $I_{z1-} = I_{z2-} = -I_x$. Lastly, it must be mentioned that the ICCIIs, along with their CMOS realizations, were formally introduced[9] by Sobhy and Soliman [386] and are characterized by the terminal equations: $i_y = 0, v_x = -v_y, i_z = \pm i_x$.

In [386], while discussing a CMOS realization of ICCII, Sobhy and Soliman [386] presented a four ICCII-based lossless FI which is shown here in Figure 4.86.

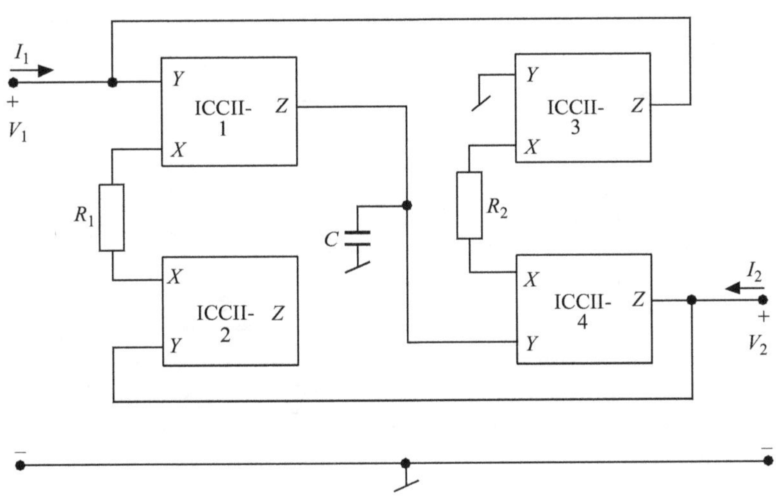

Figure 4.86 Lossless FI using four ICCII– by Sobhy and Soliman [386]

[9]A variant of CC characterized by the same terminal equations was, however, postulated by Chong and Smith as early as 1986 [405] wherein this was referred to as the type CCII+2 and CCII−2 (see Fig. 2 titled 'New Current Conveyor Classification' therein).

By a straightforward analysis, using the terminal characterization of ICCII, it can be easily derived that the equivalent FI realized by this circuit is given by $L_{eq} = CR_1R_2$. It may be mentioned that this circuit has striking resemblance with a four CCII+ based FI realization elaborated earlier in Figure 4.16, thereby indicating that what can be done with two CCIIs can as well be done by ICCIIs.

In [386] Soliman has presented a number of floating inductance circuits using a variety of CCs, several of which employ two different kinds of CCs, two particularly noteworthy circuits from this compilation which although employ two different kind of CCs but have the advantage of employing all the three passive elements connected to ground are shown in Figure 4.87. Both the circuits simulate a lossless FI of value $L_{eq} = CR_1R_2$.

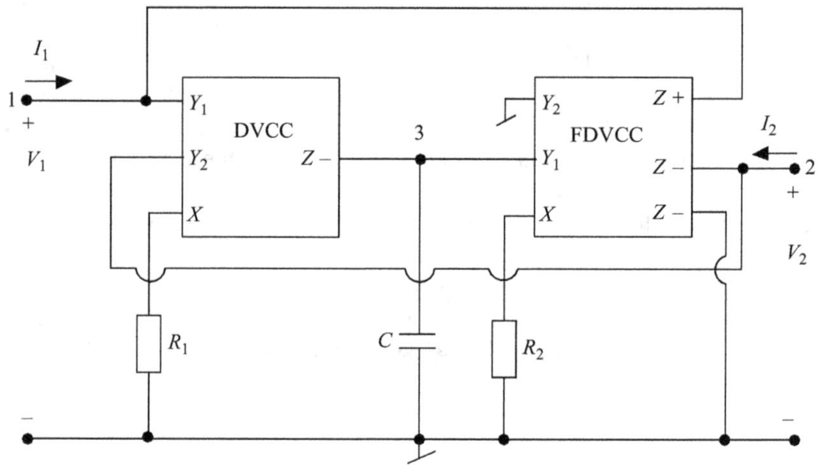

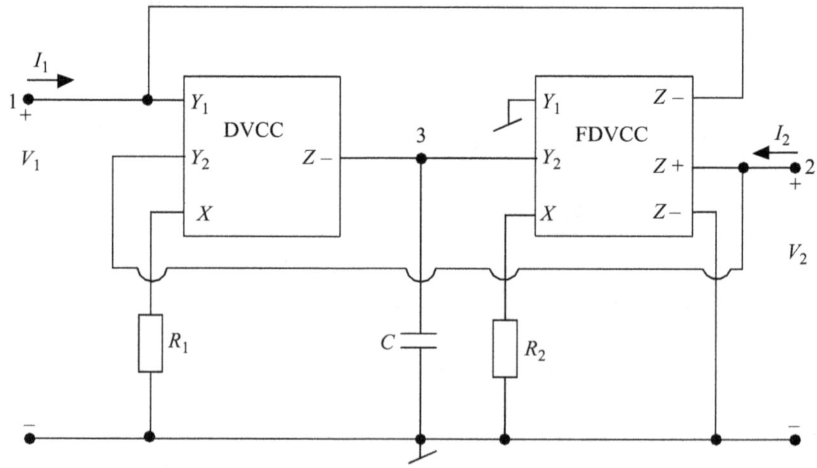

Figure 4.87 Lossless FI with all grounded passive elements using mixed types of CCs [386]

4.7 Higher order filter design using nonideal simulated impedances

This chapter has presented a large number of lossless simulated inductors, ideal FDNRs in both grounded and floating forms using the original type of CCs as well as using their numerous variants. Obviously, all these simulated impedance circuits can be used in the design of active filters and oscillators by simply replacing the passive inductors (both grounded and floating) and FDNRs (again both grounded and floating) by the simulated ones for which this chapter has provided an abundance of circuits to choose from depending upon the requirement. For example, there are methods and circuits which allow simulation of lossless inductors and FDNRs using all grounded passive elements not only in the case of grounded elements but also in the case of simulation of floating elements. Thus, choosing this kind of simulations one might design a filter circuit in which all reactive simulated elements have all passive elements grounded which may be attractive for IC implementation.

As another example, a number of simulated inductors and FDNRs have been descried which do not use any external resistors, employ grounded capacitors and provide electronic control of the simulated impedance. These circuits too could be used directly as replacements for the inductors/FDNRs in converting a given passive RLC filter into an electronically controllable active filter.

On the other hand, this chapter has also presented a number of simulated impedance circuits which are attractive due to the use of a smaller number of active and passive elements but realize lossy grounded/FIs and lossy grounded/floating FDNRs. It would appear that such circuits cannot be used to design higher order active filters based on the passive RLC proto-type by using the simulated inductor approach or FDNR approach. Fortunately, to deal with this difficulty, four network transformations were presented by Senani [43] which facilitate the direct incorporation of even such lossy inductance simulation circuits (both grounded and floating) and nonideal FDNRs (both grounded and floating) directly into the methods of higher order ladder filter design using the simulated inductance and FDNR approaches.

It was demonstrated in [11] by Senani that the single-CC-based lossy FI circuit of Figure 4.9 and *RC*:*CR* transformed version of Soliman's Ford-Girling equivalent [5] can be employed directly in higher order filter design as follows. Consider the passive RLC proto-type higher order LP filter of Figure 4.88(a). If we apply Senani's transformation T-2 from [11] which requires to multiply all the impedances by the frequency-dependent scaling factor $(1/1 + s)$, this transforms a resistor into parallel-*RC* impedance and an inductor into a parallel-*RL* impedance and a capacitor into a parallel-CD impedance. The resulting transformed filter now appears as shown in Figure 4.88(b). To convert this circuit into an active filter, all parallel-*RL* branches can be simulated by the single-CC-based circuit of Figure 4.9 whereas all parallel-CD branches can be simulated by the *RC*:*CR* transformed version of the circuit of Figure 4.9. These two circuits are shown in Figure 4.88(c), for ready reference. The final active filter would take the form as shown in Figure 4.88(d).

The following features of this methodology are worth pointing out: (i) since this method employs two specific impedance simulation circuits which are minimum

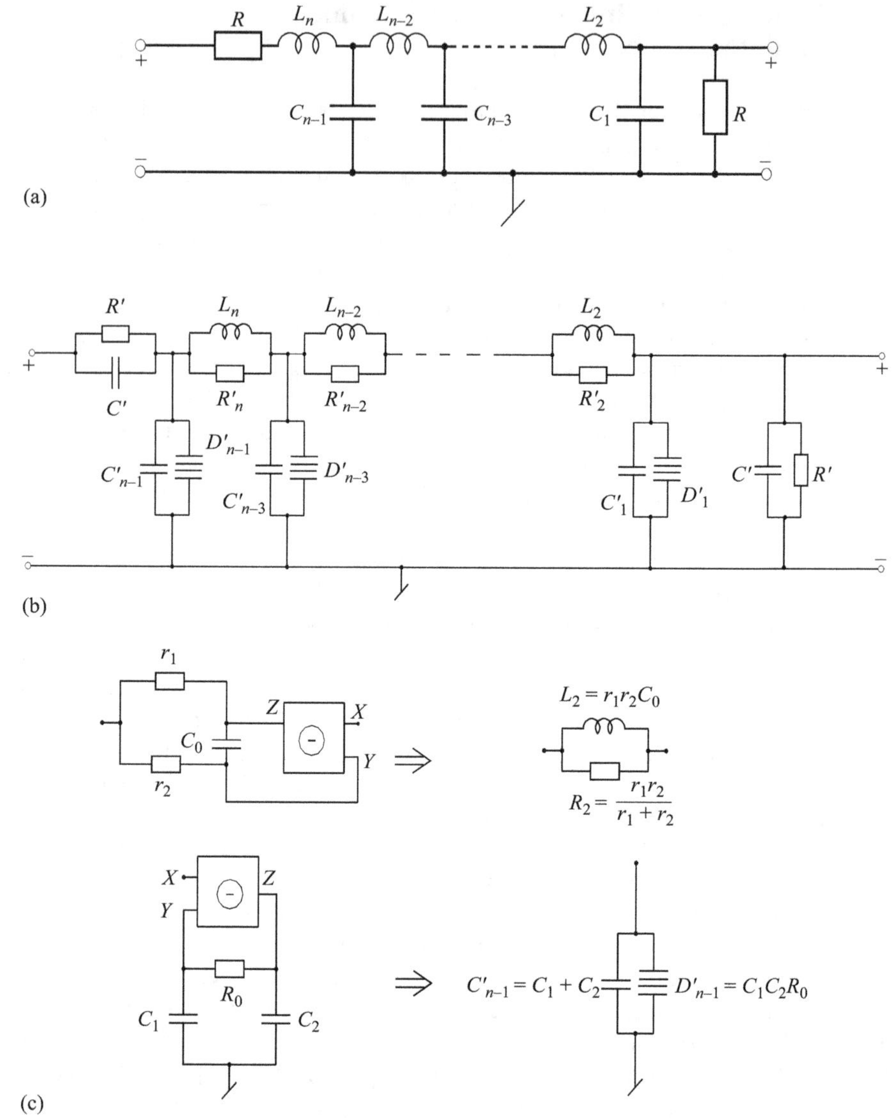

Figure 4.88 Application of CC-based nonideal simulated inductors and FDNRs in higher order filter design: (a) passive RLC proto-type low-pass filter, (b) circuit obtained by applying transformation T-4 from [43] on the circuit of (a), (c) subcircuits to realize a floating parallel impedance and grounded parallel CD impedance, (d) final nth-order active filter obtained by replacing various immittances by suggested CC simulations, (e) experimental results of fourth-order low-pass Butterworth filter designed to have a cut-off frequency of 1 kHz. Reprinted, with permission from [43] © 1985 IEE

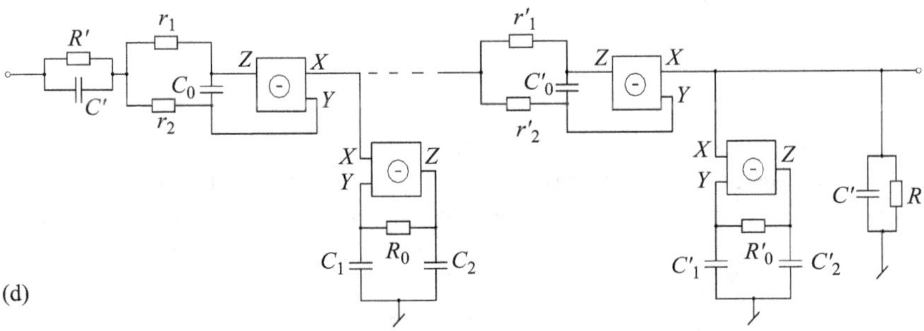

(d)

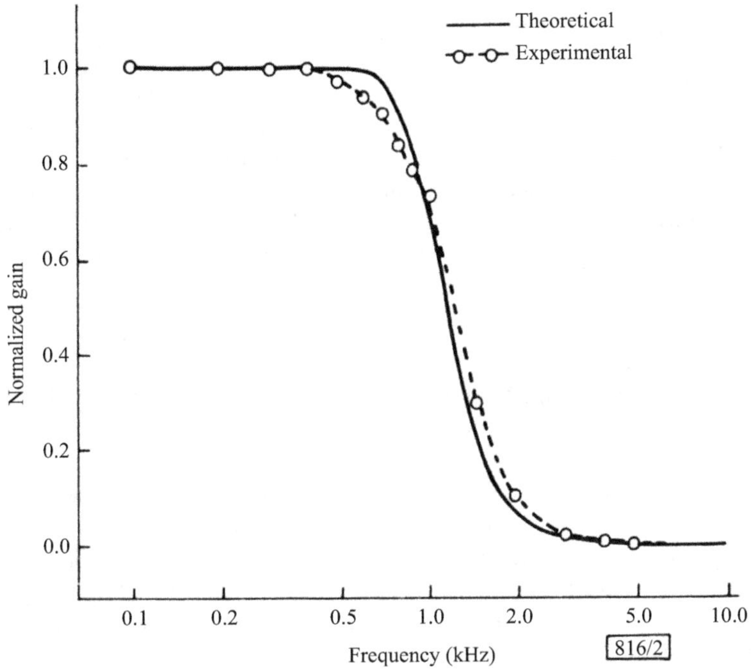

(e)

Figure 4.88 (Continued)

sensitive to component and parameter variations, the resulting active filter also inherits this minimum sensitivity property; (ii) in contrast to the usual FDNR approach based on Bruton's transformation, the LP filter of Figure 4.88(d) does not require any *low-frequency compensation*. This is because of the presence of the resistors R' shunting the capacitors C' at the source end as well as at the load end. This has been experimentally verified in [40,43]. Thus, this design method is inherently capable of providing a true LP filter response; (iii) this methodology is also applicable to the design of high pass, BP and band elimination filters obtained from a given LP proto-type by applying the well-known LP to high pass, LP to BP and LP to band elimination transformations; (iv) it is interesting to note that the number of CCs required in the illustrated example is exactly equal to the number of reactive elements in the considered passive RLC proto-type.

In Figure 4.88(e) we show the experimental results of a fourth-order LP Butterworth filter designed to have a cut-off frequency of 1 kHz which confirms the validity of this design approach. It may, however, be mentioned that in the absence of an integrated CCII− (which can now be realized by AD844s), the results obtained then were [40,43] based on an op-amp-OTA implementation of the CCII−.

Lastly, it may be mentioned that using Senani's transformation T-4 from [11], a single-CC-based floating series *RL* impedance simulation network (such as that of Figure 4.18 [14]), in conjunction with a single-CC-based grounded series CD impedance simulation circuit (obtained by *RC*:*CR* transformation of any of the grounded series *RL* simulation networks such as that of Figure 4.7(a) and (c) [3,6]) can be similarly used as direct elements in the higher order filter design to obtain equally economical alternative higher order filters using CCs. On the other hand, Metin [208] presented an interesting alternative method of using the series *RL* FI quoted above (the circuit of Figure 4.18 [14]) in the design of higher order filters. It may, however, be noted that the simulated inductance circuits used in the design example given above are characterized by very low passive as well as very low active sensitivities – a property which the alternative circuits pointed out above unfortunately does not possess.

4.8 Concluding remarks

This chapter has highlighted how CCs made it possible to devise a variety of synthetic floating impedance circuits *without requiring any component matching condition(s)* – a feat which was impossible to be achieved with the op-amp-based circuits and that the origination of this basic idea was first made in a publication by Senani in 1979 [11], followed by several other publications [22,23,30,35,41,50].

In this chapter, three categories of circuits to realize various kinds of grounded and floating impedances were included: (i) those which employ basic forms of CCIs and CCIIs, (ii) circuits employing CCCIIs and (iii) circuits based on the use of different variants of CCs.

The most significant contributions made by researchers working in this area can be summarized as follows:

1. It has been well established that grounded or floating lossy inductors (series *RL*, parallel *RL* or bilinear *RL*) all can be realized using only one CCII− and only three passive components. However, any single-resistance-controlled lossy FI (in the form of series[10] *RL* impedance) realizable with a single CCII± has not been published in the open literature yet and is open to investigation.

[10] A single-CCII-based circuit for single resistance controlled floating parallel RL admittance was given in the Ph.D. thesis of the first author of this monograph: Senani R. 'Realization of Some Classes of Active Networks'. University of Allahabad; 1987; Chapter 3: p. 143–6; also see [409].

2. A lossless FI using only two resistors and a capacitor or a floating generalized impedance converter/inverter using only three impedances needs at least three CCIIs. On the other hand, if two CCIIs are permitted then circuits to carry out both the quoted functions require at least five passive components, namely four resistors and a capacitor in the former case and five general impedances in the latter case.
3. Synthesizing an FI using a grounded capacitor needs at least four CCII of the normal kind however, if DOCCIIs are permitted, this task can be accomplished by only two DOCCIIs and a number of such circuits were elaborated.
4. For electronically variable grounded/floating impedance synthesis CCCII, DOCCCII or DXCCII were shown to be appropriate active elements and a number of prominent circuits employing them have been elaborated.
5. Apart from the basic CCs, numerous variants of them have been proposed by various researchers from time to time such as DVCCs, FDCCIIs, VCG-CCIIs, DDCCs, CCDDCCIIs, MDO-DDCCs and DCCIIs, MI-CCIIs, etc.[11] A number of prominent circuits of grounded and floating immittance simulation using these elements have been described in this chapter. However, unfortunately, no detailed comparative study has so far been made of all the available circuits so as to find out the best realizations of these different classes. This also constitutes an interesting problem worthy of further research.

References

[1] Gilbert B. 'The current-mode muddle'. AEU – International Journal of Electronics and Communications. 2018; **83**:398–406.

[2] Soliman A. M. 'Generalized immittance inverters and their realizations'. International Journal of Electronics. 1976; **41**(1):59–64.

[3] Nandi R. 'Inductor simulation using a current conveyor'. Proceedings of the IEEE. 1976; **65**(10):1511–12.

[4] Soliman A. M. 'New Active-Gyrator circuit using a single current conveyor'. Proceedings of the IEEE. 1978; **66**(11):1580–1.

[5] Soliman A. M. 'Ford-Girling equivalent circuit using CCII'. Electronics Letters. 1978; **14**(22):721–2.

[6] Nandi R. 'Active inductances using current conveyors and their application in a simple band pass filter realization'. Electronics Letters. 1978; **14**(12): 373–5.

[7] Senani R. 'Active simulation of inductors using current conveyor'. Electronics Letters. 1978; **14**(15):483–4.

[11]In retrospect, it may be observed that the general method of transforming grounded impedances into floating ones using the concept of FTFNs *explicitly* presented for the first time in 1988 [172,173] and a similar method using the operational floating amplifier (which is essentially, the same thing as an FTFN) and the other one using the floating CCII later presented in 2011 [392] (also see [393,394]) can also be related to some ideas *implicit* in references [403,408].

[8] Nedungadi A. 'Comment: Active simulation of inductors using current conveyor'. Electronics Letters. 1979; **15**(4):112–13.

[9] Senani R. 'Reply: Active simulation of inductors using current conveyor'. Electronics Letters. 1979; **15**(4):113–14.

[10] Nandi R. 'A new equal-valued grounded-capacitor resonator realization using current conveyor'. Proceedings of the IEEE. 1979; **67**(5):370–1.

[11] Senani R. 'Novel active RC circuit for floating-inductor simulations'. Electronics Letters. 1979; **15**(21):679–80.

[12] Nandi R. 'Insensitive grounded-capacitor simulation of grounded inductors using current conveyors'. Electronics Letters. 1979; **15**(21):693–4.

[13] Pookaiyaudom S., Srisarakham W. 'Realization of stable current controllable frequency-dependent positive resistance'. Proceedings of the IEEE. 1979; **67**(12):1660–2.

[14] Singh V. 'A new active-RC circuit realization of floating inductance'. Proceedings of the IEEE. 1979; **67**(12):1659–60.

[15] Singh V. 'Realization of floating inductance using current conveyor'. Proceedings of the IEEE. 1979; **67**(11):1573–4.

[16] Rathore T. S., Singhi B. M. 'Simulation of RL impedances suitable for micro-miniaturization'. Microelectronics Reliability. 1979; **19**(3):233–5.

[17] Rathore T. S. 'Active-RC synthesis of grounded and floating RL impedances employing current conveyors and minimum number of capacitors'. Indian Journal of Technology. 1982; **22**:244–5.

[18] Patranabis D., Paul A. N. 'Floating inductor with two current conveyors'. International Journal of Circuit Theory and Applications. 1980; **8**(4):457–68.

[19] Pal K. 'Realization of ideal grounded inductances and frequency dependent negative resistances using current conveyors and without matched components'. International Journal of Circuit Theory and Applications. 1981; **9**(2): 242–5.

[20] Singh V. 'Novel active RC realisations of tunable floating inductors'. Electronics Letters. 1980; **16**(19):758.

[21] Pal K. 'Novel F.D.N.C simulation using current conveyors'. Electronics Letters. 1980; **16**(16):639–41.

[22] Senani R. 'New tunable synthetic floating inductors'. Electronics Letters. 1980; **16**(10):382–3.

[23] Senani R. 'Novel active RC realisations of tunable floating inductors'. Electronics Letters. 1980; **16**(4):154–5.

[24] Rathore T. S., Singhi B. M. 'Active RC synthesis of floating immittances'. International Journal of Circuit Theory and Applications. 1980; **8**(2):184–8.

[25] Rathore T. S. 'Grounded capacitor synthesis of RL Impedances based on lagging current concept'. International Journal of Circuit Theory and Applications. 1980; **8**(4):461–5.

[26] Salawu R. I. 'Realization of frequency dependent negative resistance'. Microelectronics Reliability. 1980; **20**(6):853–7.

[27] Patranabis D., Paul A. N. 'Floating inductor with two current conveyors'. International Journal of Circuit Theory and Applications. 1980; **8**(4):457–61.

[28] Paul A. N., Patranabis D. 'Active simulation of grounded inductors using a single current conveyor. IEEE Transactions on Circuit and Systems. 1981; **28**(2):164–5.

[29] Singh V. 'Active RC single–resistance controlled lossless floating inductance simulation using single grounded capacitor'. Electronics Letters. 1981; **17**(24):920–1.

[30] Senani R. 'Novel lossless synthetic floating inductor employing a grounded capacitor'. Electronics Letters. 1982; **18**(10):413–14.

[31] Nandi R., Nandi S. 'Novel insensitive floating inductor simulation using single current conveyor'. AEU – International Journal of Electronics and Communication. 1983; **37**(11–12):393–6.

[32] Nandi S., Jana P. B., Nandi R. 'Floating ideal FDNR using current conveyors'. Electronics Letters. 1983; **19**(7):251.

[33] Paul A. N., Dey A. N., Patranabis D. 'A floating NIC without constraint and its application'. IEEE Transactions on Circuits and Systems. 1983; **30**(3):181–3.

[34] Senani R. 'Novel application of generalised current conveyor'. Electronics Letters. 1984; **20**(4):169–70; also see *ibid* (1984) 20(8):356.

[35] Senani R. 'Floating ideal FDNR using only two current conveyors'. Electronics Letters. 1984; **20**(5):205–6.

[36] Nandi S., Jana P. B., Nandi R. 'Novel floating ideal tunable FDNR simulation using current conveyors'. IEEE Transactions on Circuits and Systems. 1984; **31**(4):402–3.

[37] Abdalla Y. A. N. 'Comment on Novel floating ideal tunable FDNR simulation using current conveyors'. IEEE Transactions on Circuits and Systems. 1985; **32**(3):303.

[38] Toumazou C., Lidgey F. J. 'Floating-impedance converters using current conveyors'. Electronics Letters. 1985; **21**(15):640–2.

[39] Wilson B. 'Floating FDNR employing a new CCII-conveyor implementation'. Electronics Letters. 1985; **21**(21):996–7.

[40] Senani R. 'Novel higher-order active filter design uses current conveyors'. Electronics Letters. 1985; **21**(22):1055–7.

[41] Senani R. 'On the realization of floating active elements'. IEEE Transactions on Circuits and Systems. 1986; **33**(3):323–4.

[42] Higashimura M., Fukui Y. 'Novel lossless tunable floating FDNR simulation using two current conveyors and a buffer'. Electronics Letters. 1986; **22**(18): 938–9.

[43] Senani R. 'Network transformations for incorporating nonideal simulated immittances in the design of active filters and oscillators'. Proceedings of the IEE. 1987; **134**(4):158–66.

[44] Nandi R. 'Precise realisation of an insensitive floating negative admittance convertor'. Electronics Letters. 1987; **23**(15):775–7.

[45] Nandi R. 'Novel floating negative immittance convertor'. Proceedings of the IEE. 1987; **134**(5):236–8.

[46] Higashimura M., Fukui Y. 'Novel method for realising lossless floating immittance using current conveyors'. Electronics Letters. 1987; **23**(10):498–9.

[47] Higashimura M., Fukui Y. 'Realization of floating immittance using three current conveyors'. The Transactions of the IEICE. 1987; **J70**(8):1203–4.
[48] Higashimura M., Fukui Y. 'New lossless tunable floating FDNR simulation using two current conveyors and an INIC'. Electronics Letters. 1987; **23**(10): 529–31.
[49] Surakampontorn W., Thitimajshima P. 'Integrable electronically tunable current conveyors'. Proceedings of the IEE. 1988; **135**(2):71–7.
[50] Senani R. 'Floating immittance realisation: nullor approach'. Electronics Letters. 1988; **24**(7):403–5.
[51] Higashimura M., Fukui Y. 'Type 1 mutator using current conveyor and its application to immittance simulation'. International Journal of Electronics. 1988; **64**(3):377–83.
[52] Ishida M., Okazaki Y., Fukui Y., Ebisutani K. 'Realization of immittance function using current conveyors'. The Transactions of the IEICE. 1988; **J71-A**(5):1205–7.
[53] Higashimura M., Fukui Y. 'Realization of impedance function using current conveyors'. International Journal of Electronics. 1988; **65**(2):223–31.
[54] Senani R. 'Three op amp floating immittance simulators: a retrospection'. IEEE Transactions on Circuits and Systems. 1989; **36**(11):1463–5.
[55] Singh V. 'An implementation of CCII-current conveyor, with application'. IEEE Transactions on Circuits and Systems. 1989; **36**(9):1250–1.
[56] Higashimura M., Fukui Y. 'Simulation of lossless floating inductance using two current conveyors and an operational transconductance amplifier'. International Journal of Electronics. 1989; **66**(4):633–8.
[57] Himura A., Fukui Y., Ishida M., Higashimura M. 'Immittance function simulator using a single current conveyor'. The Transactions of the IEICE. 1989; **E72**(12):1279–84.
[58] Kumar U., Shukla S. K. 'The implementation and applications of current conveyors'. Microelectronics Journal. 1989; **20**(5):25–46.
[59] Wilson B. 'Recent developments in current conveyors and current-mode circuits'. IEE Proceedings G – Circuits, Devices and Systems. 1990; **137**(2): 63–77.
[60] Himura A., Fukui Y., Ishida M. 'Series impedance simulators using one CCII'. Electronics Letters. 1990; **26**(4):269–70.
[61] Fabre A., Alami M. 'Insensitive current-mode band pass implementation-based non-ideal gyrators'. IEEE Transactions on Circuits and Systems-I. 1992; **39**(2):152–5.
[62] Fabre A., Siarry P., Lameche M. 'Current controlled translinear impedance converter'. International Journal of Electronics. 1991; **70**(4):795–801.
[63] Tabei K., Czarnul Z., Takagi S. 'Realization of highly linear MOS circuits using negative impedance convertors'. Electronics Letters. 1991; **27**(16): 1416–7.
[64] Higashimura M., Fukui Y. 'Realization of immittance floatator using nullor'. IEICE Transactions on Fundamentals. 1992; **E-75**:644–9.

[65] Hou C. L., Chen R. D., Wu Y. P., Hu P. C. 'Realization of grounded and floating immittance function simulators using current conveyors'. International Journal of Electronics. 1993; **74**(6):917–23.

[66] Chang C-M., Wang H-Y., Chien, C-C. 'Realization of series impedance functions using one CCII+'. International Journal of Electronics. 1994; **76**(1): 83–5.

[67] Vrba K., Cajka J., Zeman V. 'Floating RC networks using current conveyors'. Radioengineering. 1996; **5**(2):8–11.

[68] Liu S-I., Yang, C-Y. 'Higher-order immittance function synthesis using CCIIIs'. Electronics Letters. 1996; **32**(25):2295–6.

[69] Saaid O., Fabre A. 'Class AB current-controlled resistor of high performance current-mode applications'. Electronics Letters. 1996; **32**(1):4–5.

[70] Vrba K., Cajka J. 'High-order one-port elements for low pass realization'. Journal of Electrical Engineering. 1997; **48**:31–4.

[71] Layos M. C., Haritantis I. 'On the derivation of current-mode floating inductors'. International Journal of Circuit Theory and Applications. 1997; **25**(1):29–36.

[72] Wang H. Y., Lee C. T. 'Immittance function simulator using a single current conveyor'. Electronics Letters. 1997; **33**(7):574–6.

[73] Al-Walaie Soliman A., Alturaigi M. A. 'Current mode simulation of lossless floating inductance'. International Journal of Electronics. 1997; **83**(6):825–9.

[74] Kiranon W., Pawarangkoon P. 'Floating inductance simulation based on current conveyors'. Electronics Letters. 1979; **33**(21):1748–9.

[75] Abuelma'atti M. T. 'Comment on Floating inductances simulation based on current conveyors'. Electronics Letters. 1998; **34**(11):1037.

[76] Cicekoglu O. 'New current conveyor based active-gyrator implementation'. Microelectronics Journal. 1998; **29**(8):525–8.

[77] Karybakas C. A., Horopanitis E. E. 'V-I scalar circuits based on CCIIs'. Electronics Letters. 1998; **34**(4):317–18.

[78] Ozoguz S., Acar C. 'On the realization of floating immittance function simulators using current conveyors'. International Journal of Electronics. 1998; **85**(4):463–75.

[79] Cabeza R., Carlosena A. 'Computational synthesis of arbitrary floating impedances'. International Journal of Circuit Theory and Applications. 1998; **26**(5):463–75.

[80] Cicekoglu M. O. 'Active simulation of grounded inductors with CCII+s and grounded passive elements'. International Journal of Electronics. 1998; **85**(4): 455–62.

[81] Abuelma'atti M. T., Tasadduq N. A. 'New negative immittance function simulators using current conveyors'. Microelectronics Journal. 1999; **30**(9): 911–15.

[82] Abuelma'atti M. T. 'Comment on Active simulation of grounded inductors with CCII+s and grounded passive elements'. International Journal of Electronics. 2000; **87**(2):177–81.

[83] Cicekoglu O. 'Reply to comment on 'Active simulation of grounded inductors with CCII+s and grounded passive elements''. International Journal of Electronics. 2000; **87**(2):183–4.
[84] Wang H. Y., Lee C. T. 'Systematic synthesis of R-L and C-D immittances using single CCIII'. International Journal of Electronics. 2000; **87**(3):293–301.
[85] Fabre A., Saaid O. 'Phase compensation of ideal inductances based second-generation current conveyors'. Analog Integrated Circuits and Signal Processing. 2000; **24**(2):153–62.
[86] Tomita Y., Agu M. 'A multi-port D-R mutator using CCIIs with current followers'. International Journal of Electronics. 2001; **88**(1):31–9.
[87] Ferri G., Guerrini N. 'High-valued passive elements simulation using low-voltage low-power current conveyors for fully integrated applications'. IEEE Transactions on Circuits and Systems-II. 2001; **48**(4):405–8.
[88] Cicekoglu O., Toker A., Kuntman H. 'Universal immittance function simulators using current conveyors'. Computers and Electrical Engineering. 2001; **27**(3):227–38.
[89] Minaei S., Cicekoglu O., Kuntman H., Turkoz S. 'Electronically tunable, active only floating inductance simulation'. International Journal of Electronics. 2003; **89**(12):905–12.
[90] Ferri G., Guerrini N. C., Diqual M. 'CCII-based floating inductance simulator with compensated series resistance'. Electronics Letters. 2003; **39**(22): 1560–2.
[91] Arslan E., Cam U., Cicekoglu O. 'Novel lossless grounded inductance simulators employing only a single first generation current conveyors'. Frequenz. 2003; **57**(9–10):204–6.
[92] Khan I. A., Zaidi M. H. 'A novel ideal floating inductor using translinear conveyors'. Active and Passive Electronic Components. 2003; **26**(2):87–9.
[93] Gift S. J. G. 'New simulated inductor using operational conveyors'. International Journal of Electronics. 2004; **91**(8):477–83.
[94] Pal K. 'Floating inductance and FDNR using positive polarity current conveyors'. Active and Passive Electronic Components. 2004; **27**(2):81–3.
[95] Huertas J. L. 'Circuit implementation of current conveyor'. Electronics Letters. 1980; **16**(6):225–6.
[96] Abuelma'atti M. T., Al-Shahrani S. M., Al-Absi M. K. 'Simulation of a mutually coupled circuit using plus-type CCIIs'. International Journal of Electronics. 2005; **92**(1):49–54.
[97] Yuce E., Cicekoglu O., Minaei S. 'Novel floating inductance and FDNR simulators employing CCII+s'. Journal of Circuits, Systems and Computers. 2006; **15**(1):75–81.
[98] Yuce E., Minaei S., Cicekoglu O. 'Limitations of the simulated inductors based on single current conveyors'. IEEE Transactions on Circuits and Systems-I. 2006; **53**(12):2860–7.
[99] Metin B., Cicekoglu O. 'A novel floating lossy inductance realization topology with NICs using current conveyors'. IEEE Transactions on Circuits and Systems-II. 2006; **53**(6):483–6.

[100] Yuce E., Cicekoglu O. 'The effects of non-idealities and current limitations on the simulated inductances employing current conveyors'. Analog Integrated Circuits and Signal Processing. 2006; **46**(2):103–10.

[101] Fakhfakh M., Loulou M., Tlelo-Cuautle E. 'Synthesis of CCIIs and design of simulated CCII-based floating inductance'. 14th IEEE International Conference on Electronics, Circuits and Systems; Marrakesh, Morocco: IEEE, December 11–14, 2007; 379–82.

[102] Yuce E., Cicekoglu O., Minaei S. 'CCII-based grounded to floating immittance converter and a floating inductance simulator'. Analog Integrated Circuits and Signal Processing. 2006; **46**(3):287–91.

[103] Yuce E. 'On the realization of the floating simulators using only grounded passive components'. Analog Integrated Circuits and Signal Processing. 2006; **49**(2):161–6.

[104] Yuce E. 'Inductor implementation using a canonical number of active and passive elements'. International Journal of Electronics. 2007; **94**(4): 317–26.

[105] Yuce E. 'Comment reply: The effects of non-idealities and current limitations on the simulated inductances employing current conveyors'. Analog Integrated Circuits and Signal Processing. 2007; **51**(1):55.

[106] Maundy B., Gift S., Aronhime P. 'A novel hybrid active inductor'. IEEE Transactions on Circuits and Systems-II. 2007; **54**(8):663–7.

[107] Yuce E., Minaei S. 'A new active network suitable for realizing ladder filters and transformer simulator'. Journal of Circuits, Systems and Computers. 2007; **16**(1):29–41.

[108] Yuce E., Minaei S., Ibrahim M. A. 'A new simulation of mutually coupled circuit based on CCIIs'. International Journal of Electronics. 2007; **94**(4): 367–72.

[109] Saad R. A., Soliman A. M. 'Generation, modelling and analysis of CCII-based gyrators using the generalized symbolic frame work for linear active circuits'. International Journal of Circuit Theory and Applications. 2008; **36**(3):289–309.

[110] Riewruja V., Petchmaneelumka W. 'Floating current-controlled resistance converters using OTAs'. AEU – International Journal of Electronics and Communications. 2008; **62**(10):725–31.

[111] Minaei S., Yuce E. 'Realization of tunable active floating inductance simulators'. International Journal of Electronics. 2008; **95**(1):27–37.

[112] Yuce E. 'Grounded inductor simulators with improved low-frequency performances'. IEEE Transactions on Instrumentation and Measurement. 2008; **57**(5):1079–84.

[113] Ferri G., Guerrini N., Silverii E., Tatone A. 'Vibration damping using CCII-based inductance simulators'. IEEE Transactions on Instrumentation and Measurement. 2008; **57**(5):907–14.

[114] Maundy B., Gift S., Aronhime P. 'Practical voltage/current-controlled grounded resistor with dynamic range extension'. IET Circuits, Devices and Systems. 2008; **2**(2):201–6.

[115] Sagbas M., Ayten U. E., Sedef H., Koksal M. 'Floating immittance function simulator and its applications'. Circuits, Systems and Signal Processing. 2009; **28**(1):55–63.

[116] Sagbas M., Ayten U. E., Sedef H., Koksal M. 'Electronically tunable floating inductance simulator'. AEU – International Journal of Electronics and Communications. 2009; **63**(5):423–7.

[117] Yuce E., Minaei S. 'On the realization of simulated inductors with reduced parasitic impedance effects'. Circuits, Systems and Signal Processing. 2009; **28**(3):451–65.

[118] Marcellis A. D., Ferri G., Guerrini N. C., Scotti G., Stornelli V., Trifiletti A. 'The VCG-CCII: a novel building block and its application to capacitance multiplication'. Analog Integrated Circuits and Signal Processing. 2009; **58**(1): 55–9.

[119] Khan I. A., Afzal N., Khan M. R. 'Digitally programmable impedance multiplier using CCIIs with high resolution capability'. Journal of Active and Passive Electronic Devices. 2009; **8**(3):247–57.

[120] Lahiri A. 'Comment on electronically tunable floating inductance simulator'. AEU – International Journal of Electronics and Communications. 2009; **63**(10):878.

[121] Saad R. A., Soliman A. M. 'On the systematic synthesis of CCII-based floating simulators'. International Journal of Circuit Theory and Applications. 2010; **38**(9):935–67.

[122] Arslan E., Metin B., Herencsar N., Koton J., Morgul A., Cicekoglu O. 'High performance wideband CMOS CCI and its application in inductance simulator design'. Advances in Electrical and Computer Engineering. 2012; **12**(3):21–6.

[123] Soliman A. M. 'New CCII and ICCII based realizations of L-C and L-R mutators'. Circuits, Systems and Signal Processing. 2010; **29**(6):1181–91.

[124] Horng J. W., Hou C. L., Chun L., Chang C. M., Yang H., Shyu W. T. 'Higher order immittance functions using current conveyors'. Analog Integrated Circuits and Signal Processing. 2009; **61**(2):205–9.

[125] Swamy M. N. S. 'Mutators, generalized impedance converters and inverters and their realization using generalized current conveyors'. Circuits, Systems and Signal Processing. 2011; **30**(1):209–32.

[126] Hou C. L., Wang W. Y. 'Realization of floating immittance function simulators using CCII+'. Microelectronics Journal. 1998; **29**(1–2):59–63.

[127] Pal K. 'Novel floating inductance using current conveyors'. Electronics Letters. 1981; **17**(18):638.

[128] Horng J. W. 'General high-order grounded and floating immittance structures using current conveyors'. Analog Integrated Circuits and Signal Processing. 2012; **71**(2):265–74.

[129] Cabeza R., Carlosena A. 'Analog universal active device: theory, design and applications'. Analog Integrated Circuits and Signal Processing. 1997; **12**(2):153–68.

[130] Yuce E. 'Negative impedance converter with reduced nonideal gain and parasitic impedance effects'. IEEE Transactions on Circuits and Systems-I. 2008; **55**(1):276–83.
[131] Senani R., Bhaskar D. R. 'Versatile voltage-controlled impedance configuration'. IEE Proceedings Circuits, Devices and Systems. 1994; **141**(5): 414–16.
[132] Abuelma'atti M. T. 'Comment on current-mode simulation of lossless floating inductance', International Journal of Electronics. 1999; **86**(13): 321–2.
[133] Zhang Q., Wang C., Sun J., Du S. 'A new type of current conveyor and its application in fully balanced differential current mode elliptic filter design'. Journal of Electrical Engineering. 2011; **62**(3):126–33.
[134] Kapur G., Mittal S., Markan C. M., Pyara V. P. 'Design of analog field programmable CMOS current conveyor'. Science Journal of Circuits, Systems and Signal Processing. 2012; **1**(1):9–21.
[135] Minaei S. 'A new high performance CMOS third generation current conveyor (CCIII) and its application'. Electrical Engineering. 2003; **85**(3): 147–53.
[136] Kacar F., Metin B., Kuntman H. 'A new CMOS dual-X second generation current conveyor (DXCCII) with an FDNR circuit application'. AEU – International Journal of Electronics and Communications. 2010; **64**(8): 774–8.
[137] Piovaccari A. 'CMOS integrated third generation current conveyor'. Electronics Letters. 1995; **31**(15):1228–9.
[138] Elwan H. O., Soliman A. M. 'A novel CMOS differential voltage current conveyor and its applications'. IEE Proceedings Circuits, Devices and Systems. 1997; **144**(3):195–200.
[139] Senani R., Bhaskar D. R., Singh A. K., Singh V. K. 'Current Feedback Operational Amplifiers and Their Applications'. London: Springer Science +Business Media; 2013. p. 227–31.
[140] Kacar F., Metin B., Kuntman H., Cicekoglu O. 'A new high-performance CMOS fully differential second-generation current conveyor with application example of biquad filter realization'. International Journal of Electronics. 2009; **97**(5):499–510.
[141] Chiu W., Liu S. I., Tsao H. W., Chen J. J. 'CMOS differential difference current conveyors and their applications'. IEE Proceedings Circuits, Devices and Systems. 1996; **143**(2):91–6.
[142] Sedef H., Acar C. 'A new floating inductor circuit using differential voltage current conveyors'. Frequenz. 2000; **54**(5–6):123–5.
[143] El-Adawy A. A., Soliman A. M., Elwan H. O. 'A novel fully differential current conveyor and applications for Analog VLSI'. IEEE Transactions on Circuits and Systems-II. 2010; **47**(4):306–13.
[144] Alzaher H. A., Elwan H. O., Ismail M. 'CMOS fully differential second-generation current conveyor'. Electronics Letters. 2000; **36**(13):1095–6.

[145] Sobhy E. A., Soliman A. M. 'Realizations of fully differential voltage second generation current conveyor with an application'. International Journal of Circuit Theory and Applications. 2010; **38**(5):441–52.
[146] Chang C. M., Soliman A. M., Swamy M. N. S. 'Analytical synthesis of low-sensitivity high order voltage-mode DDCC and FDCCII-grounded R and C all-pass filter structures'. IEEE Transactions on Circuits and Systems-I. 2007; **54**(7):1430–43.
[147] Mahmoud S. A. 'New fully-differential CMOS second-generation current conveyor'. ETRI Journal. 2006; **28**(4):495–501.
[148] Alzaher H. A. 'CMOS highly linear fully differential current conveyor'. Electronics Letters. 2004; **40**(4):214–16.
[149] El-Adawy A. A., Soliman A. M., Elwan H. O. 'Low voltage digitally controlled CMOS current conveyor'. AEU – International Journal of Electronics and Communications. 2002; **56**(3):137–44.
[150] Antoniou A. 'Realisation of gyrators using operational amplifiers and their use in RC-active network synthesis'. Proceedings of the IEE. 1969; **116**(11): 1838–50.
[151] Dutta Roy S. C. 'A circuit for floating inductance simulation'. Proceedings of the IEEE. 1974; **62**(4):521–3.
[152] Wise D. R. 'Active simulation of floating lossy inductances'. Proceedings of the IEE. 1974; **121**(2):85–7.
[153] Glover J. 'Lossless ungrounded inductor realisation'. Electronics Letters. 1976; **12**(25):679–81.
[154] The L. Q., Yanagisawa T. 'Some new lossless floating inductance circuits'. Proceedings of the IEEE. 1977; **65**(7):1071–2.
[155] Sudo S., Teramoto M. 'Constitution of floating inductance using operational amplifiers'. IEICE Transactions. 1977; **E60**(4):185–6.
[156] Senani R., Tiwari R. N. 1978 'New canonic active RC realizations of grounded and floating inductors'. Proceedings of the IEEE. **66**(7):803–4.
[157] Patranabis D., Tripathi M. P., Roy S. B. 'A new approach for lossless floating inductor'. IEEE Transactions on Circuits and Systems. 1979; **26**(10): 892–3.
[158] Von Grunigen D. C., Ramseir D., Moschytz G. S. 'Simulation of floating impedances for low-frequency active filter design'. Proceedings of the IEEE. 1986; **74**(2):366–8.
[159] Ishida M., Hirokazu A. J. I., Fukui Y., Ebisutani K. 'Grounded immittance element simulator using current conveyors'. The Transactions of the IEICE. 1988; **J71-A**(8):1644–6.
[160] Toumazou C., Lidgey F. J. 'Novel current-mode instrumentation amplifier'. Electronics Letters. 1989; **25**(3):228–30.
[161] Wilson B. 'Universal conveyor instrumentation amplifier'. Electronics Letters. 1989; **25**(7):470–1.
[162] Bruun E., Olesen O. H. 'Conveyor implementations of generic current mode circuits'. International Journal of Electronics. 1992; **73**(1):129–40.

[163] Lee J. Y., Tsao H. W. 'True RC integrators based on current conveyors with tunable time constants using active control and modified loop technique'. IEEE Transactions on Instrumentation and Measurement. 1992; **41**(5): 709–14.

[164] Hou C.-L., Wu Y.-P., Lu, F.-C. 'Synthetic methods for floating immittances of one-ports and z and y parameters of multiports using CCII'. International Journal of Electronics. 1993; **74**(4):577–86.

[165] Soliman A. M. 'A linear transconductor-multiplier using a matched pair of MOS transistors and a current conveyor'. Frequenz. 1996; **50**(11–12): 292–3.

[166] Gift S. G. J. 'An enhanced current-mode instrumentation amplifier'. IEEE Transactions on Instrumentation and Measurement. 2001; **50**(1):85–8.

[167] Ghallab Y. H., Badawy W., Kaler K. V. I. S., Maundy B. J. 'A novel current-mode instrumentation amplifier based on operational floating current conveyor'. IEEE Transactions on Instrumentation and Measurement. 2005; **54**(5):1941–9.

[168] Galanis C., Haritantis I. 'An improved instrumentation amplifier'. Proceedings of Third International Conference on Electronics, Circuits, and Systems, IEEE. 1996; **1**:65–8.

[169] Khan A. A., Al-Turaigi M. A., El-Ela M. A. 'An improved current-mode instrumentation amplifier with bandwidth independent of gain'. IEEE Transactions on Instrumentation and Measurement. 1995; **44**(4):887–91.

[170] Bruun E. 'Constant bandwidth current-mode operational amplifier'. Electronics Letters.1991; **27**(18):1673–4.

[171] De Jager W., Smit J. 'Design and Symbolic Analysis of Current-Mode Analog Circuits'. Afdelingder Elektrotechniek, 1978.

[172] Senani R. 'Generation of new two-amplifier synthetic floating inductors'. Electronics Letters. 1987; **23**(22):1202–3.

[173] Senani R. 'A novel application of four terminal floating nullors'. Proceedings of the IEEE. 1987; **75**(11):1544–6.

[174] Prescott A. J. 'Loss compensated active gyrator using differential input operational amplifiers'. Electronics Letters. 1966; **2**(7):283–4.

[175] Ford R. L., Girling F. E. J. 'Active filters and oscillators using simulated inductance'. Electronics Letters. 1966; **2**(2):52.

[176] Berndt D. F., Dutta Roy S. C. 'Inductor simulation using a single unity gain amplifier'. IEEE Journal of Solid-State Circuits. 1969; **4**(3):161–2.

[177] Cabeza R., Carlosena A., Serrano L. 'Unified approach to the implementations of universal active devices'. Electronics Letters. 1994; **30**(8): 618–20.

[178] Patranabis D., Ghosh D. K. 'Integrators and differentiators with current conveyor'. IEEE Transactions on Circuits and Systems. 1984; **31**(6):567–9.

[179] Moreira J. P., Silva M. M. 'Limits to the dynamic range of low-power continuous-time integrators'. IEEE Transactions on Circuit and Systems-I. 2001; **48**(7):805–17.

[180] Abuelma'atti M. T. 'On the realization of current-mode integrator using current conveyors'. Active and Passive Electronic Components. 1994; **17**(2): 79–82.

[181] Nandi R., Ray S. B. 'Precise realisation of current mode integrators using current conveyor'. Electronics Letters. 1993; **29**(13):1152–3.

[182] Liu S-I., Kuo J-H., Tsao H-W., Wu J., Tsay, J-H. 'New CCII-based differentiator and its applications'. International Journal of Electronics. 1991; **71**(4):645–52.

[183] Liu S-I., Hwang, Y-S. 'Dual-input differentiators and integrators with tunable time constants using current conveyors'. IEEE Transactions on Instrumentation and Measurement. 1994; **43**(4):650–4.

[184] Bruun E. 'High speed, current conveyor based voltage mode operational amplifier'. Electronics Letters. 1992; **28**(8):742–4.

[185] Gift S. J. G., Maundy B., Muddeen F. 'High-performance current-mode instrumentation amplifier circuit'. International Journal of Electronics. 2007; **94**(11):1015–24.

[186] Wilson B. 'Constant bandwidth voltage amplification using current conveyors'. International Journal of Electronics. 1988; **65**(5):983–8.

[187] Deprettere E. 'On the minimal realisation of the gyrator by means of nullors and resistors I'. International Journal of Circuit Theory. 1975; **3**(4): 383–90.

[188] Deprettere E. 'On the minimal realisation of the gyrator by means of nullors and resistors II'. International Journal of Circuit Theory. 1976; **4**(3): 285–97.

[189] Daniels R. W. 'Gyrators, negative impedance converters and related circuits'. IEEE Transactions on Circuit Theory. 1969; **16**(2):261–2.

[190] Senani R., Bhaskar D. R. 'Comment on Practical voltage/current-controlled grounded resistor with dynamic range extension'. IET Circuits, Devices and Systems. 2008; **2**(5):465–6.

[191] Senani R., Bhaskar D. R. 'Realization of voltage-controlled impedances'. IEEE Transactions on Circuits and Systems. 1991; **38**(9):1081–6; also see ibid, 1992; **39**(2):162.

[192] Senani R., Bhaskar D. R. 'A simple configuration for realizing voltage-controlled impedances'. IEEE Transactions on Circuits and Systems. 1992; **39**(1):52–9.

[193] Fabre A., Saaid O., Wiest F., Boucheron C. 'Current controlled bandpass filter based on translinear conveyor'. Electronics Letters. 1995; **31**(20):1727–8.

[194] Fabre A., Saaid O., Wiest F., Boucheron C. 'High frequency applications based on a new current controlled conveyor'. IEEE Transactions on Circuits and Systems-I. 1996; **43**(2):82–91.

[195] Saaid O., Fabre A. 'Class AB current-controlled resistor for high performance current-mode applications'. Electronics Letters. 1996; **32**(1):4–5.

[196] Chaisricharoen R., Chipipop B., Sirinaovakul B. 'CMOS CCCII: structures, characteristics, and considerations'. AEU – International Journal of Electronics and Communications. 2010; **64**(6):540–57.

[197] Chunhua W., Qiujing Z., Haiguang L. 'CMOS current controlled fully balanced current conveyor'. Journal of Semiconductors. 2009; **30**(7): 075009-1–6.

[198] Abbas Z., Scotti G., Olivieri M. 'Current controlled current conveyor (CCCII) and application using 65 nm CMOS technology'. International Journal of Electronics and Communication Engineering. 2011; **5**(7): 870–4.

[199] Seguin F., Godara B., Alicalapa F., Fabre A. '2.2 GHz all n-p-n second-generation controlled conveyor in pseudo class AB using 0.8-µm BiCMOS technology'. IEEE Transactions on Circuits and Systems-II. 2004; **51**(7): 369–73.

[200] Senani R., Singh A. K., Singh V. K. 'A new floating current-controlled positive resistance using mixed translinear cells'. IEEE Transactions on Circuits and Systems – II. 2004; **51**(7):374–7.

[201] Pawarangkoon P., Kiranon W. 'Electronically-tunable floating resistor'. International Journal of Electronics. 2004; **91**(11):665–73.

[202] Maheshwari S. 'Additional summing/difference amplifiers using CCCIIs'. Journal of Active and Passive Electron Devices. 2005; **1**(2):159–62.

[203] Maheshwari S. 'High CMRR wide bandwidth instrumentation amplifier using current controlled conveyors'. International Journal of Electronics. 2003; **89**(12):889–96.

[204] Ercan H., Tekin S. A., Alci M. 'Voltage and current controlled high CMRR instrumentation amplifier using CMOS current conveyors'. Turkish Journal of Electrical Engineering & Computer Sciences. 2012; **20**(4):547–56.

[205] Kiranon W., Pawarangkoon P. 'Floating inductance simulation based on current conveyors'. Electronics Letters. 1997; **33**(21):1748–9.

[206] Abuelma'atti M. T. 'Comment on Floating inductance simulation based on current conveyors'. Electronics Letters. 1998; **34**(11):1037.

[207] Yuce E., Minaei S., Cicekoglu O. 'Resistorless floating immittance function simulators employing current controlled conveyors and a grounded capacitor'. Electrical Engineering. 2006; **88**(6):519–25.

[208] Metin B. 'On the advantage of floating series RL simulators in ladder filter implementations'. Electronics World. 2008; **114**(1863):46–7.

[209] Sedef H., Sagbas M., Acar C. 'Current-controllable fully-integrated inductor simulator using CCCIIs'. International Journal of Electronics. 2008; **95**(5):425–9.

[210] Minaei S., Yuce E. 'Realization of tunable active floating inductance simulators'. International Journal of Electronics. 2008; **95**(1):27–37.

[211] Abuelma'atti M. T., Tasadduq N. A. 'Electronically tunable capacitance multiplier and frequency-dependent negative-resistor simulator using the current-controlled current conveyor'. Microelectronics Journal. 1999; **30**(9): 869–73.

[212] Yuce E., Minaei S. 'Electronically tunable simulated transformer and its application to stagger-tuned filter'. IEEE Transactions on Instrumentation and Measurement. 2008; **57**(9):2083–8.

[213] Parveen T., Ahmed M. T., Khan I. A. 'A canonical voltage-mode universal CCCII-C filter'. Journal of Active and Passive Electron Devices. 2009; **4**(1–2):7–12.

[214] Ranjan A., Paul S. K. 'Voltage mode universal Biquad using CCCII'. Active and Passive Electronic Components. 2011; **2011**:5 pages; ID 439052.

[215] Altuntas E., Toker A. 'Realization of voltage and current mode KHN biquads using CCCIIs'. AEU – International Journal of Electronics and Communications. 2002; **56**(1):45–9.

[216] Wang C., Liu H., Zhao Y. 'A new current mode current controlled universal filter based on CCCII (±)'. Circuits, Systems and Signal Processing. 2008; **27**(5):673–82.

[217] Senani R., Singh V. K., Singh A. K., Bhaskar D. R. 'Novel electronically controllable current-mode universal biquad filter'. IEICE Electronics Express. 2004; **1**(14):410–15.

[218] Chen H. P., Chu P. L. 'Versatile universal electronically tunable current-mode filter using CCCIIs'. IEICE Electronics Express. 2009; **6**(2):122–8.

[219] Maheshwari S., Khan I. A. 'Novel cascadable current-mode translinear-C universal filter'. Active and Passive Electronic Components. 2004; **27**(4): 215–18.

[220] Yuce E. 'Current-mode electronically tunable biquadratic filters consisting of only CCCIIs and grounded capacitors'. Microelectronics Journal. 2009; **40**(12):1719–25.

[221] Abuelma'atti M. T. 'A novel mixed mode current controlled current conveyor based filter'. Active and Passive Electronic Components. 2003; **26**(3):185–91.

[222] Ozoguz S., Toker A. 'Tunable ladder-type realization of current mode elliptic filters'. AEU – International Journal of Electronics and Communications. 2002; **56**(3):193–9.

[223] Jiraseree-amornkun A., Surakampontorn W. 'Efficient implementation of tunable ladder filters using multi-output current controlled conveyors'. AEU – International Journal of Electronics and Communications. 2008; **62**(1):11–23.

[224] Yuce E., Minaei S. 'On the realization of high-order current mode filter employing current controlled conveyors'. Computers & Electrical Engineering. 2008; **34**(3):165–72.

[225] Koksal M., Sagbas M. 'A versatile signal flow graph realization of a general current transfer function'. AEU – International Journal of Electronics and Communications 2008; **62**(1):33–40.

[226] Ranjan A., Paul S. K. 'Nth order voltage mode active-C filter employing current controlled current conveyors'. Circuits and Systems. 2011; **2**(2):85–90.

[227] Liu S. I., Yang C. Y. 'Higher-order immittance function synthesis using CCIIIs'. Electronics Letters. 1996; **32**(25):2295–6.

[228] Horng J. W. 'A sinusoidal oscillator uses current-controlled current conveyors'. International Journal of Electronics. 2001; **88**(6):659–64.

[229] Fongsamut C., Anuntahirunrat K., Kumwachara K., Surakampontorn W. 'Current conveyor-based single element controlled and current controlled sinusoidal oscillators'. International Journal of Electronics. 2006; **93**(7): 467–78.

[230] Kiranon W., Kesorn J., Wardkein P. 'Current controlled oscillator based on translinear conveyors'. Electronics Letters. 1996; **32**(15):1330–1.

[231] Abuelma'atti M. T., Tasadduq N. A. 'A novel current controlled oscillator using translinear current conveyor'. Frequenz. 1998; **52**(5–6):123–4.

[232] Maheshwari S. 'Current mode third order quadrature oscillator'. IET Circuits, Devices and Systems. 2010; **4**(3):188–95.

[233] Erdal C., Kuntman H., Kafali S. 'A current controlled conveyor based proportional-integral-derivative (PID) controller'. Journal of Electrical & Electronics Engineering. 2004; **4**(2):1243–8.

[234] Maheshwari S., Khan I. A. 'An integrable precision rectifier with current controlled output'. Journal of Active and Passive Electron Devices. 2005; **1**(2): 177–82.

[235] Anuntahirunrat K., Tangsrirat W., Riewruja V., Surakampontorn W. 'Sinusoidal frequency doubler and full-wave rectifier based on translinear current-controlled current conveyors'. International Journal of Electronics. 2004; **91**(4):227–39.

[236] Siripruchyanun M., Jaikla W. 'A novel precision current mode full wave rectifier and class B push-pull current amplifier using BiCMOS current controlled current conveyors'. The Journal of KMITNB. 2006; **16**(1):1–7.

[237] Abuelma'atti M. T., Al-Qahtani M. A. 'A current-mode current-controlled current-conveyor-based analogue multiplier/divider'. International Journal of Electronics. 1998; **85**(1):71–7.

[238] Yuce E. 'Design of a simple current-mode multiplier topology using a single CCCII+'. IEEE Transactions on Instrumentation and Measurement. 2008; **57**(3):631–7.

[239] Petchakit W., Lorsawatsiri A., Kiranon W., Wongtaychathum., Wardkein P. 'Current mode squaring, square-rooting and vector summation circuits'. AEU – International Journal of Electronics and Communications. 2010; **64**(5):443–9.

[240] Dejhan K., Netbut C. 'New simple square-rooting circuits based on translinear current conveyors'. International Journal of Electronics. 2007; **94**(7):707–23.

[241] Netbut C., Kumngern M., Prommee P., Dejhan K. 'New simple square-rooting circuits based on translinear current conveyors'. ECTI Transactions on Electrical Engineering, Electronics and Communications. 2007; **5**(1):10–17.

[242] Abuelma'atti M. T. 'New ASK/FSK/PSK/QAM wave generator using a single current controlled multiple output current conveyors'. International Journal of Electronics. 2002; **89**(1):35–43.

[243] Barthelemy H., Fabre A. 'A second generation current-controlled conveyor with negative intrinsic resistance'. IEEE Transactions on Circuits and Systems-I. 2002; **49**(1):63–5.

[244] Minaei S., Kaymak D., Kuntman H. 'New high performance realization for current-controlled conveyor (CCCII)'. Journal of Electrical & Electronics. 2002; **2**(2):547–53.

[245] Chunhua W., Qiujing Z., Wei Y. 'A second current controlled current conveyor realization using Wilson current mirror'. International Journal of Electronics. 2007; **94**(7):699–706.

[246] Ercan H., Alci M. 'A new design for BiCMOS controlled current conveyor'. Elektronika IR Elektrotechnika. 2013; **19**(1):56–60.

[247] Fabre A., Saaid O., Wiest F., Boucheron C. 'Low power current-mode second-order band pass IF filter'. IEEE Transactions on Circuits and Systems-II. 1997; **44**(6):436–46.

[248] Abuelma'atti M. T., Al-Zaher H. A. 'Nonlinear performance of the mixed translinear loop'. International Journal of Electronics. 1997; **83**(4):467–71.

[249] Kiranon W., Kesorn J., Sangpisit W., Kamprasert N. 'Electronically tunable multifunctional translinear-C filter and oscillator'. Electronics Letters. 1997; **33**(7):573–4.

[250] Abuelma'atti M. T., Abed S. M. 'A new current controlled multiple output current conveyor'. Proceedings of National Science Council ROC (A). 1997; **21**:616–22.

[251] Abuelma'atti M. T., Tasadduq N. A. 'Universal current-controlled current mode filters using the current controlled conveyor'. Proceedings of National Science Council ROC (A)). 1998; **22**:358–61.

[252] Abuelma'atti M. T., Tasadduq N. A. 'Universal current-controlled current mode filters using the multiple-output translinear current conveyor'. Frequenz. 1998; **52**(11–12):252–4.

[253] Abuelma'atti M. T., Al Qahtani M. 'Universal current-controlled current mode filters with three inputs and one output using the current controlled conveyor'. Active and Passive Electronic Components. 1998; **21**(1):33–41.

[254] Abuelma'atti M. T., Tasadduq N. A. 'A novel single-input multiple-output current mode current controlled universal filter'. Microelectronics Journal. 1998; **29**(11):901–5.

[255] Abuelma'atti M. T. 'Comment on Floating inductance simulation based on current conveyors'. Electronics Letters. 1998; **34**(11):1037–8.

[256] Abuelma'atti M. T., Al-Qahtani M. A. 'On the realization of the current controlled current-mode amplifier using the current controlled conveyor'. International Journal of Electronics. 1999; **86**(11):1333–40.

[257] Acar C., Ozoguz S. 'A new versatile building block: current differencing buffered amplifier suitable for analog signal processing filters'. Microelectronics Journal. 1999; **30**(2):157–60.

[258] Khan I. A., Zaidi M. H. 'Multifunctional translinear-C current mode filter'. International Journal of Electronics. 2000; **87**(9):1047–51.

[259] Saaid O., Fabre A. 'Class AB current controlled resistor for high performance current-mode applications'. Electronics Letters. 1996; **32**(1):4–5.

[260] Barthelemy H., Fabre A. 'A new floating controlled resistance operating in Class AB'. IEEE Transactions on Circuits and Systems-I. 2000; **47**(1):67–72.

[261] Seguin F., Fabre A. '2 GHz controlled current conveyor in standard 0.8 μm BiCMOS technology'. Electronics Letters. 2001; **37**(6):329–30.

[262] Minaei S., Cicekoglu O., Kuntman H., Turkoz S. 'High output impedance current-mode low pass, band pass and high pass filters using current controlled conveyors'. International Journal of Electronics. 2001; **88**(8):915–22.

[263] Minaei S., Turkoz S. 'Erratum new current mode current controlled universal filter with single input and three outputs'. International Journal of Electronics. 2001; **88**:955.

[264] Minaei S., Cicekoglu O., Kuntman H., Turkoz S. 'Electronically tunable, active only floating inductance simulation'. International Journal of Electronics. 2003; **89**(12):905–12.

[265] Maheshwari S. 'Active-only current controlled summing/difference amplifiers using CCCIIs'. Active and Passive Electronic Components. 2003; **26**(4):231–4.

[266] Maheshwari S., Khan I. A. 'Simple first-order translinear-C current mode all passes sections'. International Journal of Electronics. 2003; **90**(2):79–85.

[267] Shah N. A., Iqbal S. Z. 'Current mode active only universal filter. International Journal of Electronics. 2003; **90**(6):407–11.

[268] Maheshwari S. 'Electronically-tunable quadrature oscillator using translinear conveyors and grounded capacitors'. Active and Passive Electronic Components. 2003; **26**(3):193–6.

[269] Senani R., Bhaskar D. R., Singh A. K. 'Current Conveyors: Variants, Applications and Hardware Implementations'. London: Springer International Publishing; 2015. p. 33–78.

[270] Maheshwari S. 'New voltage and current mode APS using current controlled conveyor'. International Journal of Electronics. 2004; **91**(12): 735–43.

[271] Minaei S., Turkoz S. 'Current mode electronically tunable universal filter using only plus-type current controlled conveyors and grounded capacitors'. ETRI Journal. 2004; **26**(4):292–6.

[272] Sagbas M., Fidanboylu K. 'Electronically-tunable current-mode second-order universal filter using minimum elements'. Electronics Letters. 2004; **40**(1):2–4.

[273] Chang C. M., Al-Hashimi B. M., Ross J. N. 'Unified active filter biquad structures'. IEE Proceedings Circuits, Devices and Systems. 2004; **151**(4): 273–7.

[274] Shah N. A., Rather M. F., Iqbal S. Z. 'Electronically-tunable high output impedance current mode universal filter'. Journal of Active and Passive Electronic Devices. 2005; **1**:163–9.

[275] Pandey N., Paul S. K, Bhattacharyya A., Jain S. B. 'A novel current controlled current mode universal filter: SITO approach'. IEICE Electronics Express. 2005; **2**(17):451–7.

[276] Yuce E., Minaei S., Metin B. 'Comment on electronically tunable current mode second order universal filters using minimum elements'. Electronics Letters. 2005; **41**(8):453.

[277] Senani R., Singh V. K., Singh A. K., Bhaskar D. R. 'Tunable current-mode universal biquads employing only three MOCCs and all grounded passive elements: additional new realizations'. Frequenz. 2005; **59**(9–10):220–4.

[278] Maheshwari S., Khan I. A. 'High performance versatile translinear-C universal filter'. Journal of Active and Passive Electronic Devices. 2005; **1**(1):41–51.

[279] Zouaoui-Abouda H., Fabre A. 'New high-value floating controlled resistor in CMOS technology'. IEEE Transactions on Instrumentation and Measurement. 2006; **55**(3):1017–20.

[280] Tangsrirat W., Surakampontorn W. 'Electronically-tunable current-mode universal filter employing only plus-type current-controlled conveyors and grounded capacitors'. Circuits, Systems and Signal Processing. 2006; **25**(6):701–13.

[281] Shah N. A., Rather M. F., Iqbal S. Z. 'SIFO electronically tunable current-mode cascadable active only universal filter. Journal of Active and Passive Electronic Devices. 2006; **8**(3–4):327–34.

[282] Shah N. A., Rather M. F., Iqbal S. Z. 'SITO electronically tunable high output impedance current mode universal filter'. Analog Integrated Circuits and Signal Processing. 2006; **47**(3):335–8.

[283] Yuce E., Minaei S., Cicekoglu O. 'Universal current-mode active-C filters employing minimum number of passive elements'. Analog Integrated Circuits and Signal Processing. 2006; **46**(2):169–71.

[284] Metin B., Minaei S., Cicekoglu O. 'Enhanced dynamic range analog filter topologies with a notch/all pass circuit example'. Analog Integrated Circuits and Signal Processing. 2007; **51**(3):181–9.

[285] Siripruchyanun M., Jaikla W. 'Three input single-output electronically controllable dual-mode universal biquad filter using DO-CCCIIs'. Active and Passive Electronic Components. 2007; Article ID 36849.

[286] Chang C. M., Huang T. H., Tu S. H., Hou C. L., Horng J. W. 'Universal active current filter using single second-generation current controlled conveyor'. International Journal of Circuits, Systems and Signal Processing. 2007; **1**(1):194–8.

[287] Minaei S., Yuce E. 'Current-mode active-C filter employing reduced number of CCCII+s'. Journal of Circuits, Systems and Computers. 2007; **16**(7):507–15.

[288] Maheshwari S. 'A new current-mode current-controlled all-pass section'. Journal of Circuits, Systems and Computers. 2007; **16**(2):181–9.

[289] Yuce E. 'Comment on SITO electronically tunable high output impedance current-mode universal filter'. Analog Integrated Circuits and Signal Processing. 2007; **50**(3):271–2.

[290] Keskin A. U., Cam U. 'Insensitive high-output impedance minimum configuration SITO type current mode biquad using dual-output current conveyors and grounded passive components'. AEU – International Journal of Electronics and Communications. 2007; **61**(5):341–4.

[291] Yuce E., Tokat S., Minaei S., Cicekoglu O. 'Stability problems in universal current mode filters'. AEU – International Journal of Electronics and Communications. 2007; **61**(9):580–8.

[292] Tangsrirat W. 'Current-tunable current mode multifunction filter based on dual output current controlled conveyors'. AEU – International Journal of Electronics and Communications. 2007; **61**(8):528–33.

[293] Tangsrirat W., Surakampontorn W. 'High output impedance current mode universal filter employing dual-output current controlled conveyors and grounded capacitors'. AEU – International Journal of Electronics and Communications. 2007; **61**(2):127–31.

[294] Tsukutani T., Sumi Y., Yabuki N. 'Versatile current-mode biquadratic circuit using only plus type CCCIIs and grounded capacitors'. International Journal of Electronics. 2007; **94**(12):1147–56.

[295] Pandey N., Paul S. K., Bhattacharyya A. 'Realization of generalized mixed mode universal filter using CCCIIs', Journal of Active and Passive Electronic Devices. 2010; **5**(3–4), 279–93

[296] Maheshwari S., Khan I. A. 'Novel voltage/current mode translinear-C quadrature oscillator'. Journal of Active and Passive Electronic Devices. 2007; **2**(3):235–9.

[297] Tangsrirat W. 'Current-tunable current-mode multifunction filter based on dual-output current-controlled conveyors'. AEU – International Journal of Electronics and Communications. 2007; **61**(8):528–33.

[298] Pandey N., Paul S. K., Jain S. B. 'Voltage mode universal filter using two plus type CCIIs'. Journal of Active and Passive Electronic Devices. 2008; **3**(2):165–73.

[299] Pandey N., Paul S. K., Jain S. B. 'New high-input impedance voltage-mode universal biquad: multi input multi output'. Journal of Active and Passive Electronic Devices. 2008; **3**(2):93–100.

[300] Yuce E., Kircay A., Tokat S. 'Universal resistorless current mode filters employing CCCIIs'. International Journal of Circuit Theory and Applications. 2008; **36**(5–6):739–55.

[301] Oztayfun S., Kilinc S., Celebi A., Cam U. 'A new electronically tunable phase shifter employing current controlled current conveyors'. AEU – International Journal of Electronics and Communications 2008; **62**(3): 228–31.

[302] Yuce E., Minaei S. 'Universal current-mode filters and parasitic impedance effects on the filter performances'. International Journal of Circuit Theory and Applications. 2008; **36**(2):161–71.

[303] Pandey N., Paul S. K. 'A novel electronically tunable sinusoidal oscillator based on CCCII (-IR)'. Journal of Active and Passive Electronic Devices. 2008; **3**(2):135–41.

[304] Biolek D., Senani R., Biolkova V., Kolka Z. 'Active elements for analog signal processing: classification, review and new proposals'. Radioengineering. 2008; **17**(4):15–32.

[305] Siripruchyanun M., Chanapromma Ch., Silapan Ph., Jaikla W. 'BiCMOS current-controlled current feedback amplifier (CC-CFA) and its applications'. WSEAS Transactions on Electronics. 2008; **5**(6):203–19.

[306] Siripruchyanun M. 'A current-mode analog multiplier/divider based on CCCDTA'. AEU – International Journal of Electronics and Communications. 2008; **62**(3):223–7.

[307] Soliman A. M. 'Comment on: Realization of voltage and current mode KHN biquads using CCCII'. AEU – International Journal of Electronics and Communications. 2009; **63**(10):877.

[308] Siripruchyanun M., Jaikla W. 'Cascadable current-mode biquad filter and quadrature oscillator using DO-CCCIIs and OTA'. Circuits, Systems and Signal Processing, 2009; **28**(1):99–110.

[309] Tangsrirat W., Surakampontorn W. 'Low component current mode universal filter using current controlled conveyors and grounded capacitors'. Journal of Active and Passive Electronic Devices. 2009; **4**(3):259–64.

[310] Shah N. A., Rather M. F. 'Electronically tunable current mode/mixed-mode/voltage mode multifunction active only biquads'. Journal of Active and Passive Electronic Devices. 2009; **4**(3):223–35.

[311] Lahiri A. 'Current controlled inductor simulation using dual-output current controlled conveyors'. Journal of Active and Passive Electronic Devices. 2010; **5**(3–4):191–5.

[312] Pandey N., Paul S. K. 'Digitally switched current mode universal filter'. Journal of Active and Passive Electronic Devices. 2010; **5**(3–4):197–208.

[313] Kumngern M., Jongchanachavawat W., Dejhan K. 'New electronically tunable current mode universal biquad filter using translinear current conveyors'. International Journal of Electronics. 2010; **97**(5):511–23.

[314] Sotner R., Slezak J., Dostal T., Petrzela J. 'Universal tunable current mode biquad employing distributed feedback structure with MO-CCCII'. Journal of Electrical Engineering. 2010; **61**(1):52–6.

[315] Sagbas M., Ayten U. E., Sedef H., Koksal M. 'Reply to comment on "Electronically tunable floating inductance simulator"'. AEU – International Journal of Electronics and Communications. 2012; **66**(1):86–8.

[316] Tangsrirat W., Channumsin O. 'Minimum-component current-mode universal filter'. Indian Journal of Pure and Applied Physics. 2011; **49**: 137–41.

[317] Wang C., Xu J., Keskin A. U., Du S., Zhang Q. 'A new current mode current controlled SIMO type universal filter'. AEU – International Journal of Electronics and Communications. 2011; **65**(3):231–4.

[318] Lahiri A. 'Deriving (MO) (I) CCCII based second order sinusoidal oscillators with non-interactive tuning laws using state variable method'. Radioengineering. 2011; **20**(1):349–53.

[319] Saied A. B., Salem S. B., Masmoudi D. S. 'A new CMOS current controlled quadrature oscillator based on a MCCII'. Circuits and Systems. 2011; **2**(4):269–73.

[320] Yasin M. Y., Gopal B. 'High frequency oscillator design using a single 45 nm CMOS current controlled current conveyor (CCCII+) with minimum passive components'. Circuits and Systems. 2011; **2**(2):53–9.

[321] Fakhfakh M., Pierzchala M., Rodanski B. 'On the design of active inductors with current-controlled voltage sources'. Analog Integrated Circuits and Signal Processing. 2012; **73**(1):89–98.

[322] Metin B., Pal K., Cicekoglu O. 'A new approach for high-input impedance in voltage mode filters using first-generation current conveyor in place of second-generation current conveyor'. International Journal of Electronics. 2012; **99**(1):131–9.

[323] Yildiz H. A., Toker A., Ozoguz S. 'Biquadratic filter applications using a fully differential active-only integrator'. Radioengineering. 2013; **22**(1):3–13.

[324] Srisakultiew S., Lawanwisut S., Siripruchyanun M. 'A current mode electronically controllable multifunction biquadratic filter using CCCIIs'. International Journal of Advances in Telecommunications, Electronics, Signals and Systems. 2013; **2**(2):45–50.

[325] Pandey N., Paul S. K. 'Mixed mode universal filter'. Journal of Circuits, Systems and Computers. 2013; **22**(1):1250064-1–10.

[326] Fabre A., Saaid O. 'Novel translinear impedance convertor and band pass filter applications'. Electronics Letters. 1993; **29**(9):746–7.

[327] Liu S. I., Yang C. Y. 'Higher-order immittance function synthesis using CCIIIs'. Electronics Letters. 1996; **32**(25):2295–6.

[328] Ananda Mohan P. V. 'Grounded capacitor based grounded and floating inductance simulation using current conveyors'. Electronics Letters. 1998; **34**(11):1037–8.

[329] Sedef H., Acar C. 'A new floating FDNR circuit using differential voltage current conveyors'. AEU – International Journal of Electronics and Communications. 2000; **54**:297–301.

[330] Kuntman H., Gulsoy M., Cicekoglu O. 'Actively-simulated grounded lossy inductors using third generation current conveyors'. Microelectronics Journal. 2000; **31**(4):245–50.

[331] Wang H. Y., Lee C. T. 'Systematic synthesis of R-L and C-D immittances using single CCIII'. International Journal of Electronics. 2000; **87**(3): 293–301.

[332] Minaei S., Cicekoglu O., Kuntman H., Turkoz S. 'Electronically tunable, active only floating inductance simulation'. International Journal of Electronics. 2003; **89**(12):905–12.

[333] Khan I. A., Zaidi M. H. 'A novel ideal floating inductor using translinear conveyors'. Active and Passive Electronic Components. 2003; **26**(2):87–9.

[334] Incekaraoglu M., Cam U. 'Realization of series and parallel R-L and C-D Impedances using single differential voltage current conveyor'. Analog Integrated Circuits and Signal Processing. 2005; **43**(1):101–4.

[335] Gift S. J. G. 'New simulated inductor using operational conveyors'. International Journal of Electronics. 2004; **91**(8):477–83.

[336] Ananda Mohan P. V. 'Floating capacitance simulation using current conveyors'. Journal of Circuits, Systems and Computers. 2005; **14**(1):123–8.

[337] Zeki A., Toker A. 'DXCCII-based tunable gyrator'. AEU – International Journal of Electronics and Communications. 2005; **59**(1):59–62.

[338] Yuce E. 'Comment on Realization of series and parallel R-L and C-D impedances using single differential voltage current conveyor'. Analog Integrated Circuits and Signal Processing. 2006; **49**(1):91–2.

[339] Yuce E. 'On the realization of the floating simulators using only grounded passive components'. Analog Integrated Circuits and Signal Processing. 2006; **49**(2):161–6.

[340] Yuce E., Minaei S., Cicekoglu O. 'A novel grounded inductor realization using a minimum number of active and passive components'. ETRI Journal. 2005; **27**(4):427–32.

[341] Yuce E., Cicekoglu O., Minaei S. 'CCII-based grounded to floating immittance converter and a floating inductance simulator'. Analog Integrated Circuits and Signal Processing. 2006; **46**(3):287–91.

[342] Minaei S., Yuce E., Cicekoglu O. 'A versatile active circuit for realizing floating inductance, capacitance, FDNR and admittance converter'. Analog Integrated Circuits and Signal Processing. 2006; **47**(2):199–202.

[343] Yuce E., Minaei S., Cicekoglu O. 'Limitations of the simulated inductors based on a single current conveyor'. IEEE Transactions on Circuits and Systems-I. 2006; **53**(12):2860–7.

[344] Yuce E., Minaei S. 'A new active network suitable for realizing ladder filters and transformer simulator'. Journal of Circuits, Systems and Computers. 2007; **16**(1):29–41.

[345] Minaei S., Yuce E. 'Realization of tunable active floating inductance simulators'. International Journal of Electronics. 2008; **95**(1):27–37.

[346] Yuce E. 'Negative impedance converter with reduced nonideal gain and parasitic impedance effects'. IEEE Transactions on Circuits and Systems-I. 2008; **55**(1):276–83.

[347] Metin B., Herencsar N., Horng J. W. 'DCCII-based novel lossless grounded inductance simulators with no element matching constrains'. Radioengineering. 2014; **23**(1):532–8.

[348] Horng J. W, Hou C. L, Chang C. M., Yang H., Shyu W. T. 'Higher-order immittance functions using current conveyors'. Analog Integrated Circuits and Signal Processing. 2009; **61**(2):205–9.

[349] Sagbas M., Ayten U. E., Sedef H., Koksal M. 'Floating immittance function simulator and its applications'. Circuits, Systems and Signal Processing. 2009; **28**(1):55–63.

[350] Yuce E., Minaei S. 'On the realization of simulated inductors with reduced parasitic impedance effects'. Circuits, Systems and Signal Processing. 2009; **28**:451–65.

[351] Yuce E., Minaei S. 'Novel floating simulated inductors with wider operating-frequency ranges'. Microelectronics Journal. 2009; **40**(6): 928–38.

[352] Sagbas M., Ayten U. E., Sedef H., Koksal M. 'Electronically tunable floating inductance simulator'. AEU – International Journal of Electronics and Communications. 2009; **63**(5):423–7.

[353] Lahiri A. 'Comment on Electronically tunable floating inductance simulator'. AEU – International Journal of Electronics and Communications. 2009; **63**(10):878.

[354] Kacar F., Yesil A. 'Novel grounded parallel inductances simulators realization using a minimum number of active and passive components'. Microelectronics Journal. 2010; **41**(10):632–8.

[355] Soliman A. M. 'On the realization of floating inductors'. Nature and Science. 2010; **8**(5):167–80.

[356] Soliman A. M. 'New CCII and ICCII based realizations of L-C and L-R mutators'. Circuits, Systems and Signal Processing. 2010; **29**(6):1181–91.

[357] Metin B., Minaei S. 'Parasitic compensation in CC-I based circuits for reduced power consumption'. Analog Integrated Circuits and Signal Processing. 2010; **65**(1):157–62.

[358] Soliman A. M. 'Comments on Active simulation of grounded inductors using single current conveyor', IEEE Transactions on Circuits and Systems. 1981; **28**(10):1024.

[359] Horng J. W. 'Lossless inductance simulation and voltage-mode universal biquadratic filter with one input and five outputs using DVCCs'. Analog Integrated Circuits and Signal Processing. 2010; **62**(3):407–13.

[360] Yuce E. 'A novel floating simulation topology composed of only grounded passive components'. International Journal of Electronics. 2010; **97**(3):249–62.

[361] Kacar F. 'New lossless inductance simulators realization using a minimum active and passive components'. Microelectronics Journal. 2010; **41**(2–3): 109–13.

[362] Kacar F., Yesil A. 'FDCCII-based FDNR simulator topologies'. International Journal of Electronics. 2011; **99**(2):285–93.

[363] Pal K. 'Modified current conveyors and their applications'. Microelectronics Journal. 1989; **20**(4):37–40.

[364] Swamy M. N. S. 'Mutators, generalized impedance converters and inverters, and their realization using generalized current conveyors'. Circuits, Systems and Signal Processing. 2011; **30**(1):209–32.

[365] Metin B. 'Supplementary inductance simulator topologies employing single DXCCII'. Radioengineering. 2011; **20**(3):614–18.

[366] Myderrizi I., Minaei S., Yuce E. 'DXCCII-based grounded inductances simulators and filter applications'. Microelectronics Journal. 2011; **42**(9): 1074–81.

[367] Prommee P., Somdunyakanok M. 'CMOS-based current-controlled DDCC and its applications to capacitance multiplier and universal filter'. AEU – International Journal of Electronics and Communications. 2011; **65**(1):1–8.

[368] Ibrahim M. A., Minaei S., Yuce E., Herencsar N., Koton J. 'Lossy/lossless floating/grounded inductance simulation using one DDCC'. Radioengineering. 2012; **21**(1):3–10.

[369] Sagbas M., Ayten U. E., Sedef H., Koksal M. 'Reply to comment on electronically tunable floating inductance simulator'. AEU – International Journal of Electronics and Communications. 2012; **66**(1):86–8.

[370] Yuce E. 'Floating inductance, FDNR, and capacitance simulation circuit employing only grounded passive elements', International Journal of Electronics. 2006; **93**(10):679–88.

[371] Yuce E. 'Inductor implementation using a canonic number of active and passive elements', International Journal of Electronics. 2007; **94**(4):317–26.

[372] Sagbas M. 'Electronically tunable mutually coupled circuit using only two active components'. International Journal of Electronics. 2014; **101**(3):364–74.

[373] Toker A., Cicekoglue O., Kuntman H. 'New active gyrator circuit suitable for frequency-dependent negative resistance implementation', Microelectronics Journal. 1999; **30**(1):59–62.

[374] Soliman A. M. 'Realization of frequency dependent negative resistance circuits using two capacitors and single current conveyor', Proceedings of the IEEE. 1978; **125**(12):1326–37.

[375] Adams K. M., Deprettere E. 'On the realization of gyrators by nullors and resistors'. International Journal of Circuit Theory and Applications. 1974; **2**(3):287–90.

[376] Wang Z. 'Novel voltage-controlled grounded resistor'. Electronics Letters. 1990; **26**(20):1711–12.

[377] Zeki A., Toker A., Cicekoglu O. 'The dual-X current conveyor (DXCCII): a new active device for tunable continuous-time filters'. International Journal of Electronics. 2002; **89**(12):913–23.

[378] Prommee P., Angkeaw K., Somdunyakanok M., Dejhan K. 'CMOS-based near zero-offset multiple input max-min circuits and its applications'. Analog Integrated Circuits and Signal Processing. 2009; **61**(1):93–105.

[379] Laoudias C., Psychalinos C. 'Differential voltage current controlled current conveyor with low-voltage operation capability'. International Journal of Electronics. 2014; **101**(7):939–49.

[380] Myderrizi I., Zeki A. 'Electronically tunable DXCCII-based grounded capacitance multiplier'. AEU – International Journal of Electronics and Communications. 2014; **68**(9):899–906.

[381] De Marcellis A., Ferri G., Guerrini N. C., Scotti G., Stornelli V., Trifiletti A. 'A novel low-voltage low-power fully differential voltage and current gained CCII for floating impedance simulations'. Microelectronics Journal. 2009; **40**(1):20–5.

[382] Jaikla W., Siripruchyanun M. 'Realization of current conveyors-based floating simulator employing grounded passive elements'. 2007 4th International Conference on Electrical Engineering/Electronics, Computer, telecommunications and Information Technology; Thailand: ECTI Association Thailand, May 9–12, 2007:89–92.

[383] Jongkunstidchai C., Fongsamut C., Kumwachara K., Surakampontorn W. 'Full-wave rectifiers based on operational transconductance amplifiers'. AEU – International Journal of Electronics and Communications. 2007; **61**(3):195–201.

[384] Fabre A. 'Third-generation current conveyor: a new helpful active element'. Electronics Letters. 1995; **31**(5):338–9.

[385] Pal K. 'New inductance and capacitor floatation schemes using current conveyors'. Electronics Letters. 1981; **17**(21):807–8.

[386] Sobhy E. A., Soliman A. M. 'Novel CMOS realizations of the inverting second-generation current conveyor and applications'. Analog Integrated Circuits and Signal Processing. 2007; **52**(1–2):57–64.

[387] Soliman A. M. 'Generation and classification of CCII and ICCII based negative impedance converter circuits using NAM expansion'. International Journal of Circuit theory and Applications. 2011; **39**(8):835–47.

[388] Awad I. A., Soliman A. M. 'Inverting second generation current conveyors: the missing building blocks, CMOS realizations and applications'. International Journal of Electronics. 1999; **86**(4):413–32.

[389] Ananda Mohan P. V. 'Grounded capacitor based grounded immittance and floating inductance simulation using current conveyors'. Electronics Letters. 1998; **34**(11):1037–8.

[390] Yuce E., Minaei S., Cicekoglu O. 'Resistorless floating immittance function simulators employing current controlled conveyors and a grounded capacitor'. Electrical Engineering. 2006; **88**(6):519–26.

[391] Saad R. A., Soliman A. M. 'Generation, modelling, and analysis of CCII-based gyrators using the generalized symbolic framework for linear active circuits'. International Journal of Circuit theory and Applications. 2008; **36**(3):289–309.

[392] Soliman A. M. 'On the transformation of grounded inductors to floating inductors using OFA and FCCII'. Journal of Circuits, Systems, and Computers. 2011; **20**(2):243–62.

[393] Senani R. 'On the transformation of grounded inductors to floating inductors using OFA and FCCII'. Journal of Circuits, Systems, and Computers. 2012; **21**(5):3 pages.

[394] Soliman A. M. 'A note on the transformation of grounded inductors to floating inductors using OFA and FCCII'. Journal of Circuits, Systems, and Computers. 2013; **22**(2):3 pages.

[395] Elwan H. O., Soliman A. M. 'A novel CMOS differential voltage current conveyor and its applications'. IEE Proceedings Circuits, Devices and Systems. 2007; **144**(3):195–200.

[396] Braun J. 'Equivalent NIC networks with nullators and norators'. IEEE Transactions on Circuit Theory. 1965; **12**(3):441–2.

[397] Soliman A. M. 'New generalized immittance converter circuits obtained by using the current conveyors'. International Journal of Electronics. 1972; **32**(6):673–9.

[398] Saad R. A., Soliman A. M. 'On the systematic synthesis of CCII based floating simulators'. International Journal of Circuit Theory and Applications. 2010; **38**(9):935–67.

[399] Soliman A. M. 'The inverting second generation current conveyors as universal building blocks'. AEU – International Journal of Electronics and Communications. 2008; **62**(2):114–21.

[400] Carlin H. J. 'Singular network elements'. IEEE Transactions on Circuit Theory. 1964; **11**(1):67–72.

[401] Svoboda J. A. 'Current conveyors, operational amplifiers and nullors'. IEE Proceedings-Circuits, Devices and Systems. 1989; **136**(6):317–22.

[402] Toumazou C., Payne A. 'Operational floating conveyor'. Electronics Letters. 1991; **27**(8):651–2.

[403] Trimmel H. R., Heinlein W. E. 'Fully floating chain-type gyrator circuit using operational trans-conductance amplifiers'. IEEE Transactions on Circuit Theory. 1971; **18**(6):719–22.

[404] Soliman A. M., Saad R. A. 'On the introduction of new floating current conveyors'. Journal of Circuits, Systems, and Computers. 2009; **18**(6):1005–16.

[405] Chong C. P., Smith K. C. 'Biquadratic filter sections employing a single current conveyor'. Electronics Letters. 1986; **22**(22):1162–4.

[406] Patranabis D., Paul A. N., Dey A. N., Ghosh D. K. 'A floating NIC and floating ideal inductors'. IETE Journal of Research. 1983; **29**(1):1–3.

[407] Paul A. N., Dey A. N., Patranabis D. 'A floating NIC without constraints and its applications'. IEEE Transactions on Circuits and Systems. 1983; **CAS-30**(3):181–3.

[408] Silva M., Saraga W. 'On the classification of active-RC circuits simulating of floating inductors'. Proceedings of the 3rd International Symposium on Network Theory, 1975, 489–96.

[409] Senani R. 'Novel Analog Circuits Using Op-amps and Current Conveyors'. Germany: LAP LAMBERT Academic Publishing; 2014. p. 123–5.

[410] Herencsar N., Lahiri A., Koton J., Vrba K., Metin B. 'Realization of resistor less lossless positive and negative grounded inductor simulators using single ZC-CCCITA'. Radioengineering. 2012; **21**(1):264–72.

[411] Metin B. 'Canonical inductor simulators with grounded capacitors using DCCII'. International Journal of Electronics. 2012; **99**(7):1027–35.

[412] Prasad D., Ahmad J. 'New electronically-controllable lossless synthetic floating inductance circuit using single VDCC'. Circuits and Systems. 2014; **5**(1):13–15.

[413] Kacar F., Yesil A., Minaei S., Kuntman H. 'Positive/negative lossy/lossless grounded inductance simulators employing single VDCC and only two passive elements'. AEU – International Journal of Electronics and Communications. 2014; **68**(1):73–8.

Chapter 5

Current feedback-op-amp-based synthetic impedances

Abstract

Besides the op-amp based and operational transconductance amplifier–based gyrators and impedance simulators, the current-feedback operational amplifier (CFOA)-based circuits of these elements are attractive because of the commercial availability of the CFOA AD844 as an off-the-shelf IC. In this chapter, we consider the prominent circuits for realizing gyrators, lossless grounded inductance (GI)/ lossless floating inductance (FI) circuits as well as the more generalized generalized positive impedance converter/inverter and generalized negative impedance converter/inverter elements using CFOAs. It is found that a class of economic impedance simulation circuits based on the use of single CFOA offer the attractive features of single resistance tunability which is not available in the well-known single op-amp-based simulators of the same kind of impedances. A number of inductors and resonators using CFOA poles and a class of GI/FI simulators using a modified CFOA have also been included.

5.1 Introduction

From the available accounts, it turns out that the first current-feedback operational amplifier (CFOA) based on a bipolar current conveyor (CC) topology was designed by Barrie Gilbert in 1967 or so but was finally produced as an off-the-shelf IC by Analog Devices Inc. much later (around 1984 or so) after the *true complementary bipolar process* was available to integrate the circuit architecture of AD844 which required close matching between $p-n-p$ and $n-p-n$ transistors to offer the performance benefits which a CFOA offered. This architecture offered a number of potential advantages in analog circuit design as compared to the traditional bipolar voltage mode op-amp (VOA) which was widely used till then as the main work-horse of analog circuit design.

Internally, the CFOA AD844 employed a bipolar CC – a building block which, as per the published open literature is credited to Smith and Sedra [1,2], though there is enough evidence [3,4] to acknowledge that it was Gilbert who first coined the term *current conveyor* and also invented the translinear CC architecture and that he had made the term *current conveyor* known to people working in industry by word of mouth [3] earlier than the publication of [1,2].

The internal architecture of AD844 CFOA comprises a translinear CC (which, as per [2], is actually the so-called second-generation CC (CCII)) followed by a translinear voltage follower (VF). Gilbert invented this *CC architecture* in 1967 driven by his quest to make a different kind of op-amp with one input having ideally infinite input impedance (thereby being suitable to be driven by an input voltage source) and the other input having ideally zero input impedance (thereby being suitable to be derived by an input current source).

A traditional VOA has two high input impedance input terminals and low output impedance output terminal and is essentially a differential-input–single-output voltage-controlled voltage source (VCVS). By contrast, the CFOA has one input designated as Y having high input impedance, while the other input designated as X has low input impedance. Then there is a current output terminal designated as Z such that the current inputted at the X-terminal is *conveyed* to the Z-terminal (along with the inversion of its direction) which has very high output impedance. If Z-terminal is terminated into a load Z_p then the voltage across it is buffered by the VF and is made available at the voltage output terminal W which has very low output impedance. With no external termination at terminal Z, the output impedance looking into this terminal is found to be composed of a resistance R_p in parallel with a capacitance C_p where R_p represents the parallel combination of the output resistance of the CCII+ and the input resistance of the VF. Since both these parasitic resistances are quite high, their parallel combination, for AD844 CFOA, is around 3 MΩ. On the other hand, the parasitic capacitance C_p is a parallel combination of the output capacitance of the CCII+ in parallel with the input capacitance of the VF and is estimated to be around 3–5 pF. Thus, $Z_p = R_p//(1/sC_p)$ is the parasitic termination at port Z in the absence of any external termination therein.

An ideal characterization of the CFOA can be represented by the instantaneous terminal equations: $i_y = 0$, $v_x = v_y$ and $i_z = +i_x$ (representing the CCII+ part) and in addition $v_w = v_z$ (representing the VF part). It therefore follows that the voltage output at the W-terminal designated as V_{out} can be written as $V_{out} = V_w = V_z = -i_z Z_p = -i_x Z_p = -i_{in} Z_p$. Thus, the gain of the CFOA between the output voltage and one of the inputs, namely the current i_{in} applied to terminal-X is actually a *transimpedance* (Z_p) which is *ideally* infinite! Due to this reason, the Z-terminal is also sometimes referred to as the *transimpedance* (*TZ*) terminal and the CFOA is, thus, an operational trans-impedance amplifier. Some authors also refer to this terminal as the *compensation pin*, because it is here that an external capacitor could be connected to ground to tailor the characteristics of the CFOA as per the required frequency compensation, whenever needed.

A very common topology using the CFOA is the noninverting amplifier which is made just like the conventional VOA, i.e. connecting a potential divider made by connecting a resistor R_b between X and W terminals and another resistor R_a from X to ground and then applying an input V_{in} at terminal Y and taking output V_{out} from terminal W with terminal Z left open. Note that still the parasitic impedance Z_p is in picture. Furthermore, in contrast to the noninverting amplifier made from a VOA which has a *voltage feedback* through the potential divider, this CFOA-based noninverting amplifier has a *current feedback* via the current i_x!

It is this *current feedback* which makes the behaviour of the CFOA-based noninverting amplifier different from the VOA-based noninverting amplifier. In particular, it is found that [5], while in the conventional circuit, the maximum gain (at zero frequency) and the 3-dB bandwidth are *coupled*, in the CFOA-based amplifier, they become *decoupled* in the sense that in the latter, it is possible to have *constant-bandwidth* (dependent only on the feedback resistor R_b) but *variable-gain* (which can be continuously varied by adjusting the resistor R_a). This is a potential advantage with the CFOA vis-à-vis the conventional VOA.

Another significant advantage of the CFOA is its exceptionally high (theoretically infinite!) slew rate (*SR*) which stems from the nature of the output current i_z of the *translinear* CCII+ which is found to be a sinh hyperbolic function of the differential input $(v_y - v_x)$. In fact, $i_z = 2I_B\sinh[(v_y - v_x)/V_T]$. It is readily seen that for large differential inputs, i_z tends to become infinite. Since the *SR* is defined as *SR* = maximum charging current/value of the compensation capacitor, for the CFOA this equals $SR = i_z/C_p = \infty$! Practically, the value of *SR* for AD844 is around 2,000 V/μs. Thus, two other significant advantages of the CFOA-based circuits would be their functionality over a much wider frequency range and increased input signal handling capability than their VOA-based counterparts. Besides these, yet another advantage of CFOAs is that circuits made from them invariably need a minimum number of external passive components *without requiring any* component-matching conditions or equality constraints.

Because of the numerous advantages, the CFOAs have received considerable attention in the past three decades as attractive alternative building blocks for analog circuit design. As a consequence, their various hardware implementations and numerous applications in linear and nonlinear signal processing/generation have been widely investigated. Interested readers may see [1–46] and others cited therein. In this chapter, we will focus upon the use of CFOAs in the realization of gyrators, inductors and a variety of related impedances.

5.2 The IC CFOA AD844

A wide variety of CFOAs are currently available from various analog IC manufacturers; however, the AD844 from Analog Devices Inc. is the only one which has the compensation pin no. 5 (*TZ*-terminal or the *Z*-terminal of the internal CCII+) accessible from outside. For the details of many alternatives CFOA architecture of various CFOAs, the reader is referred to [6]; however, for convenience, the simplified block diagram of the AD844 is shown here in Figure 5.1(a) whereas as its block representation is shown in Figure 5.1(b) and its accepted symbolic notation in Figure 5.1(c).

In the circuit of Figure 5.1(a) the transistors Q_1–Q_4 constitute the ‘mixed translinear cell’ wherein the transistors Q_1–Q_2 each get the dc bias current I_B which is forced by two constant current sources, thereby forcing $i_y = 0$. When $i_x \neq 0$, the

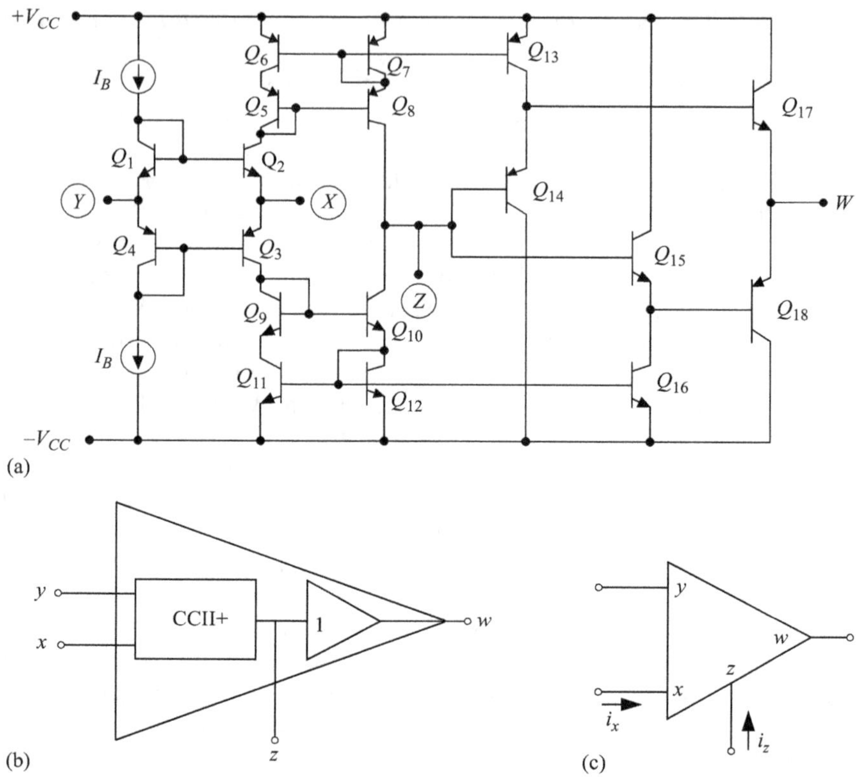

Figure 5.1 The AD844 CFOA: (a) simplified equivalent of the internal circuit architecture, (b) internal block diagram, (c) symbolic notation

modified Wilson current mirrors composed of Q_5–Q_8 and Q_9–Q_{12} ensure $i_z = i_x$. It has been found [5] that for $i_x \ll 2I_B$, $v_x \cong v_y + R_x i_x$ where $R_x \cong V_T/2I_B$. Thus, for $R_x \to 0$, $v_x \cong v_y$. The last stage consisting of transistors $Q_{13} - Q_{16}$ is another 'mixed translinear cell' which functions as a VF.

A more accurate relation between v_y, v_x and i_x or i_z is i_z (flowing out) $= 2I_B\, sinh\big((v_y - v_x)/V_T\big)$ whose plot is shown in Figure 5.2(a) from where it is clear that for large differential inputs $(v_y - v_x)$, $i_z \to \infty$! Since i_z is the charging current for the compensating capacitor C_p (at Z-terminal), this gives rise to infinite SR (theoretically). The specified SR which is practically attainable for AD844 is about 2,000 V/μs, whereas the resistive and capacitive components of the parasitic output impedance looking into terminal Z of AD844 are typically $R_p = 3$ MΩ and $C_p = 4$–5 pF.

Furthermore, it is found that input impedance at terminal-Y has a resistive and capacitive part with typical values being $R_y = 2$ MΩ and $C_y = 2$ pF. The last stage (VF) also has finite low output impedance R_w which is around 50 Ω. Thus, nonideal model of the CFOA, taking the various parasitics into account, can be shown as in Figure 5.2(b).

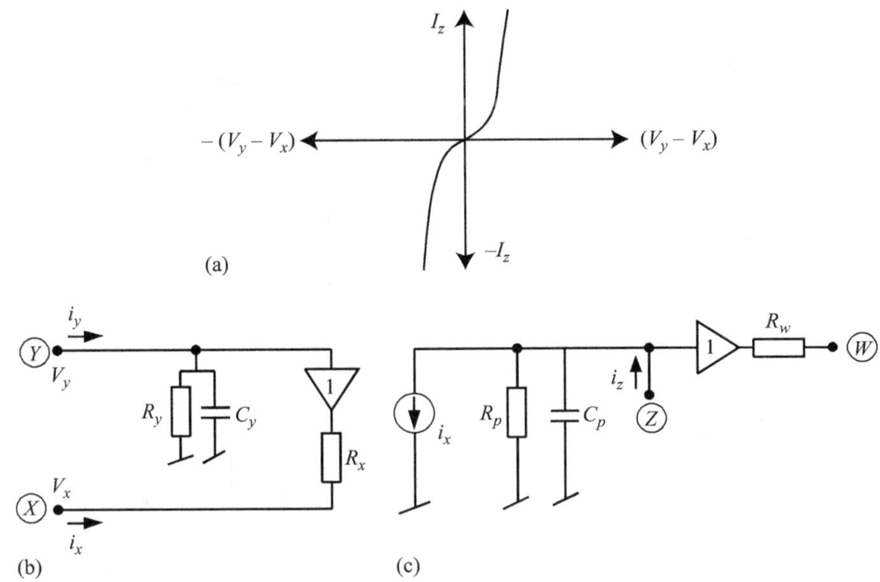

Figure 5.2 Transfer characteristics and nonideal parameters: (a) transfer characteristics of the CFOA, (b) various parasitic impedances of the CFOA

5.3 Systematic synthesis of gyrators/grounded inductance simulators

It is well known that to make a gyrator or a positive impedance inverter (PII), one needs voltage-controlled current sources (VCCS), both inverting and noninverting types. Figure 5.3(a)–(f) shows a variety of VCCS configurations which can be made from CFOAs along with a bare minimum, only one resistor. These circuits can be readily employed to synthesize a gyrator/PII or a negative impedance inverter (NII) (for the simulation of negative elements). The circuit of Figure 5.3(a) is an inverting VCCS giving $I_0 = -(V_{in}/R_1)$ while that of Figure 5.3(b) is a noninverting VCCS providing $I_0 = +(V_{in}/R_1)$. The third circuit of Figure 5.3(c) realizes a differential VCCS characterized by $I_0 = (V_1 - V_2)/R_1$, but it may be noted that both the inputs *do not* have infinite input impedances. The circuits of Figure 5.3(d) and (e) overcome this limitation and provide ideally infinite input impedances for both the inputs and give $I_0 = (V_1 - V_2)/R_1$ (going in) or $I_0 = (V_1 - V_2)/R_1$ (going out). The circuit of Figure 5.3(f) gives two complementary outputs $I_0 = +(V_1/R_1)$ and $I_0 = -(V_1/R_1)$ while the circuit of Figure 5.3(g) accepts differential input ($V_1 - V_2$) and delivers two complementary output currents $I_0 = \pm(V_1 - V_2)/R_1$. These circuits would be seen to be useful building blocks for realizing a variety of impedance simulation circuits.

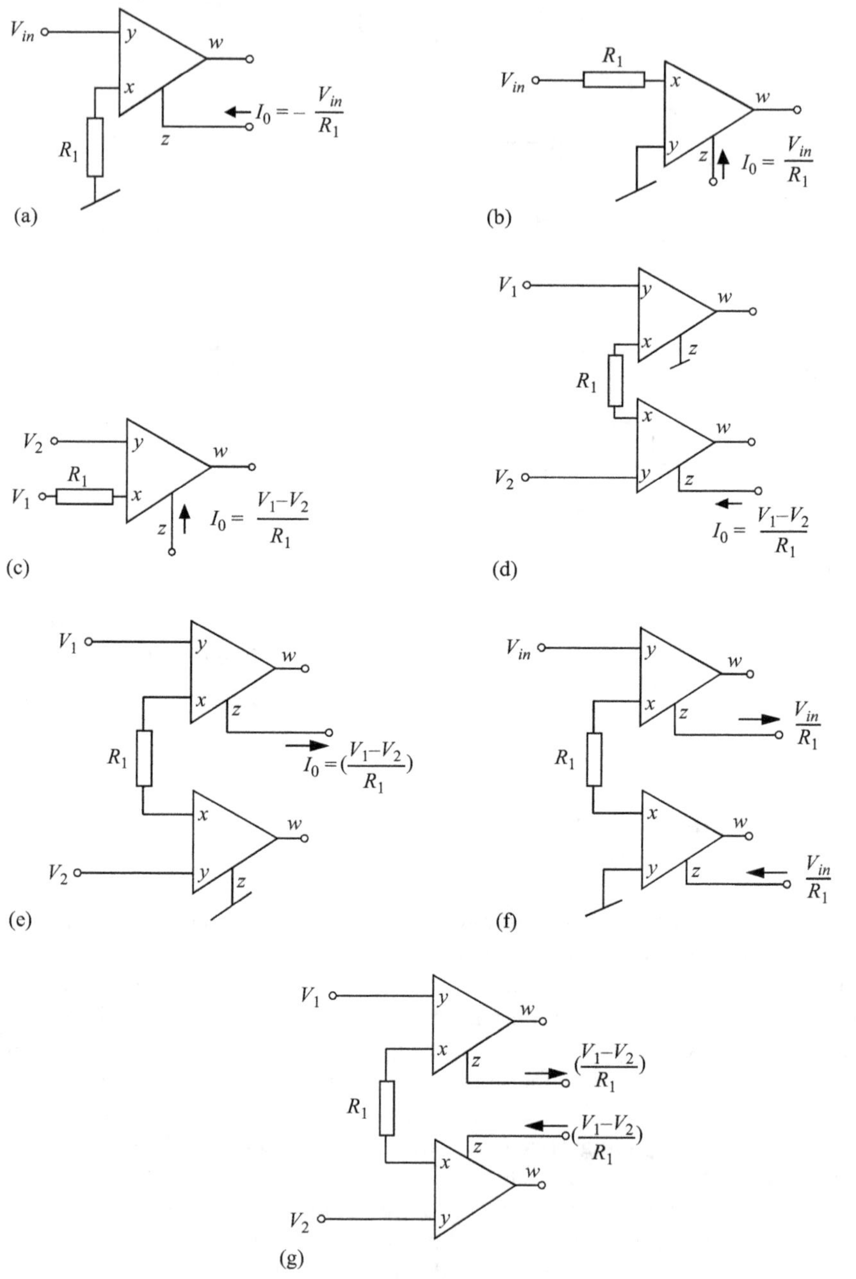

Figure 5.3 Seven possible realizations (a)-(g) of VCCS elements using CFOAs

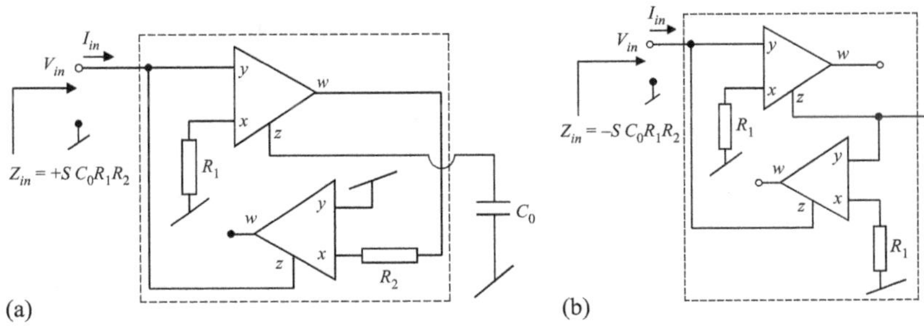

Figure 5.4 Gyrator/PII and NII realizations using CFOAs: (a) gyrator/PII, (b) NII

Let us first start with a gyrator/PII from its [Y] matrix:

$$[Y] = \begin{bmatrix} 0 & \dfrac{1}{R_1} \\ \pm\dfrac{1}{R_2} & 0 \end{bmatrix} \tag{5.1}$$

It is seen that we need a parallel back-to-back connection of two VCCSs of opposite polarities to realize the [Y] matrix of (5.1). Hence, the circuit shown in Figure 5.4(a) is obtained by appropriate connections of an inverting and other noninverting VCCS structures.

When terminated into a capacitor C_0 at port 2, the input impedance looking into port 1 yields $Z_{in}(s) = sC_0R_1R_2$. Likewise an NII can be synthesized by a parallel back-to-back connection of two VCCS structures of same polarity. One such NII is shown in Figure 5.4(b) which, when terminated into a capacitor C_0, leads to the input impedance $Z_{in}(s) = -sC_0R_1R_2$. Note that in the circuit of Figure 5.4(a), the upper CFOA along with C_0 makes a noninverting integrator giving the voltage across C_0 as $V_{in}(s)/sC_0R_1$ and the lower CFOA along with R_2 taps this voltage from the *W*-terminal of this CFOA and converts this into a current $V_{in}(s)/sC_0R_1R_2$ (note that there is a virtual ground at terminal-*X* of the lower CFOA). Thus, the input current becomes $I_{in}(s) = V_{in}(s)/sC_0R_1R_2$, thereby leading to $Z_{in}(s) = sC_0R_1R_2$. Similarly, in the NII circuit, the integrator part is the same as in the PII but the VCCS made from lower CFOA and R_2 is of opposite polarity thereby leading to $I_{in}(s) = -V_{in}(s)/sC_0R_1R_2$ and consequently, $Z_{in}(s) = -sC_0R_1R_2$.

5.4 Lossless FI simulators using CFOAs

We now show how, by employing the same methodology, lossless floating positive inductor and floating negative inductor can be created. In the first case, we use the VCCS circuit of Figure 5.3(c) which gives a current $((V_1 - V_2)/R_1)$ coming out of *Z*-terminal of $CFOA_1$ which flows into C_0 and creates a voltage $((V_1 - V_2)/sC_0R_1)$. Now the remaining part of the circuit is again a VCCS of type of Figure 5.3(g) which creates two complementary currents $\pm((V_1 - V_2)/sC_0R_1)$ which are appropriately connected to ports 1 and 2, thereby leading to

$$I_1(s) = \frac{V_1 - V_2}{sC_0R_1R_2} = -I_2(s) \tag{5.2}$$

which represents a floating lossless inductance of value

$$L_{eq} = sC_0R_1R_2 \tag{5.3}$$

between ports 1 and 2. Following the same procedure, one can see the formulation of the lossless floating negative inductance circuit of Figure 5.3(b) in which the basic blocks are the same and the only difference is the manner in which the signs of the currents $I_1(s)$ and $I_2(s)$ are reversed at the rightmost end of the circuit to enable

$$I_1(s) = \frac{V_2 - V_1}{sC_0R_1R_2} = -I_2(s) = -\frac{V_1 - V_2}{sC_0R_1R_2} \tag{5.4}$$

which therefore realizes a floating negative inductance of value

$$L_{eq} = -sC_0R_1R_2 \tag{5.5}$$

We now show that using the 'same integrator followed by VCCS' approach applied on the previous circuits, we can easily synthesize lossless floating inductors – both positive and negative. Thus, to create a lossless FI, first we need a differential integrator which can easily be obtained by using the two-CFOA VCCS of Figure 5.3(g) and then putting a capacitor C_0 at Z-terminal of the upper CFOA and grounding the Z-terminal of the lower CFOA. This yields a transfer function:

$$V_0(s) = \frac{V_1 - V_2}{sC_0R_1} \tag{5.6}$$

Next we need to convert the voltage $V_0(s)$ into two currents: $(V_1 - V_2)/sC_0R_1R_2$ and $-(V_1 - V_2)/sC_0R_1R_2$ for which we employ the VCCS of Figure 5.3(f) or (g) and then connect the Z-output of the two CFOAs therein as shown in Figure 5.5(a) which finally yields

$$\begin{bmatrix} I_1(s) \\ I_2(s) \end{bmatrix} = \frac{1}{sC_0R_1R_2}\begin{bmatrix} 1 & -1 \\ -1 & 1 \end{bmatrix}\begin{bmatrix} V_1(s) \\ V_2(s) \end{bmatrix} \tag{5.7}$$

Thus, the circuit realizes a lossless FI of value $L_{eq} = sC_0R_1R_2$.

On the other hand, the synthesis of a negative FI follows the same steps except the last one, in which case the current outputs of the differential VCCS are connected to ports 1 and 2 as shown in Figure 5.5(b). The resulting circuit simulates a negative inductance of value $L_{eq} = -sC_0R_1R_2$.

The authors of this monograph have noticed with great surprise and curiosity that many researchers (including even the more experienced ones) have discussed the realization frequency-dependent negative resistances (FDNRs) (both lossy and

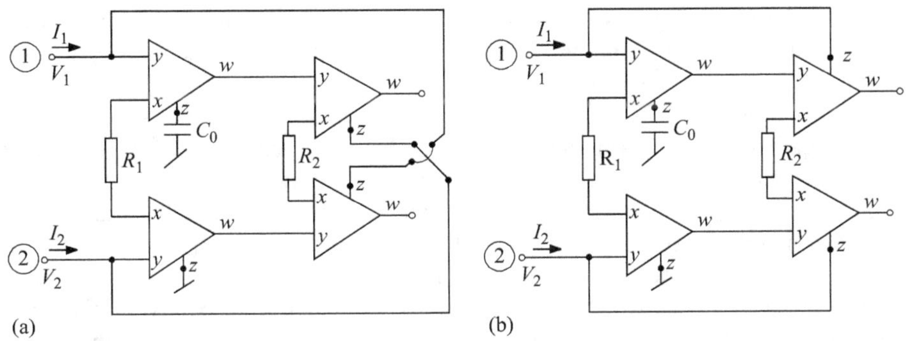

Figure 5.5 Floating inductor realizations: (a) floating lossless positive inductance, (b) floating lossless negative inductance

lossless) separately projecting as if it is something entirely different than inductance simulation (both lossless and lossy) and as if it is not connected with the latter! In view of this, it appears appropriate that it should be clarified once for all that inductance simulation circuits and FDNRs are closely related by the classical *RC* – *CR* transformation which is as follows (see [47], pp. 87–88).

Theorem 1: *Let a given* RC *network be characterized by a driving-point (transfer) impedance function* $Z_{RC}(s)$. *The corresponding network function* $Z'_{RC}(s)$ *of the transformed* RC *network, obtained by* RC:CR *transformation would be given by*

$$Z'_{RC}(s) = \frac{1}{s} Z_{RC}\left(\frac{1}{s}\right) \tag{5.8}$$

It has a companion part which is as follows:

Theorem 2: *The transfer voltage (current) ratio* $T_{RC}(s)$ *of an* RC *network and the corresponding transfer function* $T'_{RC}(s)$ *of the transformed* RC *network, obtained by an* RC:CR *transformation of the original* RC *network, are related as*

$$T'_{RC}(s) = T_{RC}\left(\frac{1}{s}\right) \tag{5.9}$$

For a topological proof of the above theorems, the reader is referred to a paper by Hakimi and Cruz.[1] It should be pointed out that the above theorems are applicable to active RC networks as well, provided they can be modelled by VCVS and VCCS which have dimensionless gain functions. Note that active *RC* circuits containing unity gain VFs, unity gain CFs, op-amps, CCs (which can be modelled by unity

[1]Hakimi S. L., Cruz Jr. J. B. 'On minimal realisation of RC two-ports'. Proceeding of NEC. 1960; 16:258–67.

gain VCVS and unity gain CCCS), as well as CFOAs which also can be modelled by two unity gain VCVS and one unity gain CCCS, are all valid candidates for the application of these transformations.

Furthermore, if an active-*RC* circuit contains a large number of resistors and/or capacitors then resistors (or capacitors) which appear as ratios in the relevant function of interest, need not be transformed into capacitors (or resistors).

Theoretically, the *RC*:*CR* transformation can also be applied to active *RC* circuits containing more complex active building blocks, but it is obvious that the models of these also require VCCS or CCVS, then in applying the transformation, the gain factors of these which are dimensionally $1/R$ and R, respectively, will also need to be transformed into sC and $1/sC$, respectively, in which case the transformed circuits may not be realizable as simply as in the cases pointed out earlier.

Thus, if there is an *RC* active one having $Z(s) = sL$, application of *RC*:*CR* transformation converts it into another network where input impedance function would be given by $Z'(s) = (1/s)(L/s) = L/s^2$ which represents an FDNR. Likewise, if the chosen active-*RC* network has $Z(s) = R + sL$, it can be easily converted into a network having $Z'(s) = (1/sC + 1/s^2)$, which represents a series combination of a C and D element (FDNR). Lastly, if there is an active-*RC* circuit having $Y(s) = ((1/R) + (1/sL))$, the application of *RC*:*CR* transformation would transform this into $Y'(s) = (sC + (s^2/L))$, thereby representing a parallel $C - D$ network.

In view of the above appraisals, it follows that if *RC*:*CR* transformation is applied on the circuit of Figure 5.4(a), then the resulting circuit would have

$$Z'(s) = \frac{1}{s^2 C_1 C_2 R_0} = \frac{1}{Ds^2} \tag{5.10}$$

where $D = C_1 C_2 R_0$, which represents an FDNR.

On the other hand, the application of *RC*:*CR* transformation on the circuit of Figure 5.4(b) will result in a circuit which will have

$$Z'(s) = -\frac{1}{s^2 C_1 C_2 R_0} = -\frac{1}{Ds^2} \tag{5.11}$$

For sinusoidal excitation and for steady-state behaviour, we put $s = j\omega$ which leads to $Z'(j\omega) = 1/\omega^2 C_1 C_2 R_0$; hence, the realized impedance can be called as *frequency-dependent positive resistance* (FDPR).

We now turn our attention to the problem whether a lossless FI can be realized by less than three CFOAs? It was first shown by Chang and Hwang [10] that a three-CFOA-based FI simulator is indeed possible. Their circuit from [10] is shown here in Figure 5.6 and is characterized by

$$\begin{bmatrix} I_1(s) \\ I_2(s) \end{bmatrix} = \frac{1}{sC_0 R_1 R_2} \begin{bmatrix} 1 & -1 \\ -1 & 1 \end{bmatrix} \begin{bmatrix} V_1(s) \\ V_2(s) \end{bmatrix} \tag{5.12}$$

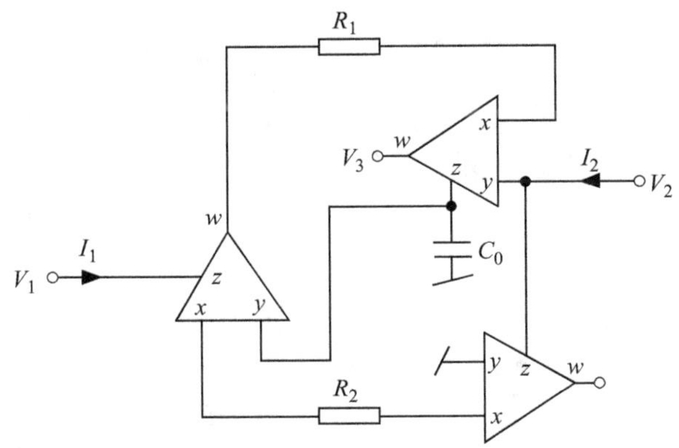

Figure 5.6 Realization of lossless FI using only three CFOAs due to Chang and Hwang [10]

and simulates a lossless FI of value $L_{eq} = sC_0R_1R_2$ without any constraints and without requiring any components-matching conditions or equality constraints. This circuit can be interpreted in two different ways:

By simple analysis, we can find

$$I_1(s) = -\frac{V_3(s)}{R_2} \quad \text{and} \quad I_2(s) = \frac{V_3(s)}{R_2} = -I_1(s) \tag{5.13}$$

where

$$V_3(s) = \frac{1}{sC_0}\left\{\frac{V_2(s) - V_1(s)}{R_1}\right\} \tag{5.14}$$

which gives

$$I_1(s) = \left(\frac{V_1 - V_2}{sC_0R_1R_2}\right) = -I_2(s) \tag{5.15}$$

and the circuit, thus, realizes a lossless FI of value $L_{eq} = sC_0R_1R_2$.

It is interesting to point out that Chang and Hwang [10] had extracted this circuit from their earlier published multifunction active filter realized with three CFOAs [9], by deleting a series resistor connected at the input and a shunt capacitor connected at one of the outputs of the filter.

An alternative three-CFOA-based lossless FI circuit was independently evolved, at about the same time, by Senani [16] which is shown in Figure 5.7. A straightforward analysis of this circuit reveals the following equations:

$$\frac{V_1 - V_2}{R_1} = -sC_0V_3, \quad I_1(s) = -\frac{V_3}{R_2}; \quad I_2(s) = \frac{V_3}{R_2} \tag{5.16}$$

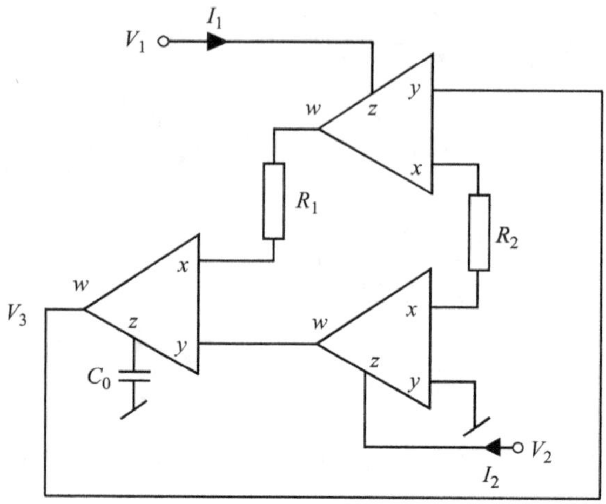

Figure 5.7 An alternative realization of lossless FI using only three CFOAs due to Senani [16]

which yields

$$I_1(s) = \left(\frac{V_1 - V_2}{sC_0R_1R_2}\right) = -I_2(s) \tag{5.17}$$

thereby confirming that this circuit too simulates a lossless FI of value $L_{eq} = sC_0R_1R_2$. The inductance value can be changed easily by varying R_1 and/or R_2.

In retrospection, it is interesting to find that if the capacitor C_0 is taken out from both the circuits, and the port thus created is called port 3, the resulting three-port circuit has the following $[Y]$ matrix:

$$\begin{bmatrix} I_1 \\ I_2 \\ I_3 \end{bmatrix} = \begin{bmatrix} 0 & 0 & \dfrac{-1}{R_2} \\ 0 & 0 & \dfrac{1}{R_2} \\ \dfrac{1}{R_1} & \dfrac{-1}{R_1} & 0 \end{bmatrix} \begin{bmatrix} V_1 \\ V_2 \\ V_3 \end{bmatrix} \tag{5.18}$$

which represents a three-port gyrator.

Yet another three-CFOA-based lossless FI circuit has been proposed by Abuelma'atti and Dhar [43] which is reproduced in Figure 5.8. A straightforward analysis of this circuit reveals that

$$I_1(s) = \left(\frac{V_1 - V_3}{Z_1}\right) = -I_2(s) \quad \text{and} \quad \left(\frac{V_2 - V_1}{Z_3}\right) = \frac{V_3 - V_1}{Z_2} \tag{5.19}$$

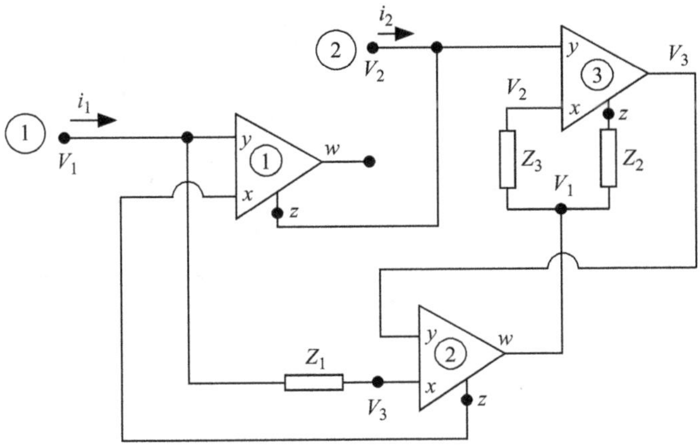

Figure 5.8 Floating admittance converter/inverter using only three CFOAs [43]

which can be simplified as

$$V_3 = V_1 + \left(\frac{Z_2}{Z_3}\right)(V_2 - V_1) \tag{5.20}$$

thus, the substitution of the value of V_3 in the equation of $I_1(s)$ and $I_2(s)$ finally results in

$$I_1(s) = \left(\frac{Z_2}{Z_1 Z_3}\right)(V_1 - V_2) = -I_2(s) \tag{5.21}$$

so that with $Z_2 = (1/sC_2)$, $Z_1 = R_1$ and $Z_3 = R_3$, the circuit simulates a lossless FI with $L_{eq} = sC_2R_1R_3$.

Two interesting questions come to mind after witnessing three three-CFOA-based lossless FIs of Figures 5.6–5.8. First, given three CFOAs and only three passive elements (two resistors and a capacitor which may or may not be grounded), how many three-CFOA lossless FIs can be formulated? Second, can we realize a lossless FI using less than three CFOAs? While as of now, there is no answer to the first query and this, therefore, constitutes an interesting problem for further research, the second question was addressed in [35] by Senani and Bhaskar who came with a two-CFOA-based circuit for FI simulation reproduced in Figure 5.9. Analysis of this circuit reveals its [Y] matrix as

$$[Y] = \left(\frac{1}{R_1} - \frac{1}{R_2} + \frac{1}{sC_1R_1R_2}\right)\begin{bmatrix} 1 & -1 \\ -1 & 1 \end{bmatrix} \tag{5.22}$$

Thus, it can be seen that for $R_1 < R_2$, the circuit simulates a lossy parallel RL admittance with parameters given by

$$\frac{1}{R_{eq}} = \left(\frac{1}{R_1} - \frac{1}{R_2}\right); \quad L_{eq} = C_1R_1R_2 \tag{5.23}$$

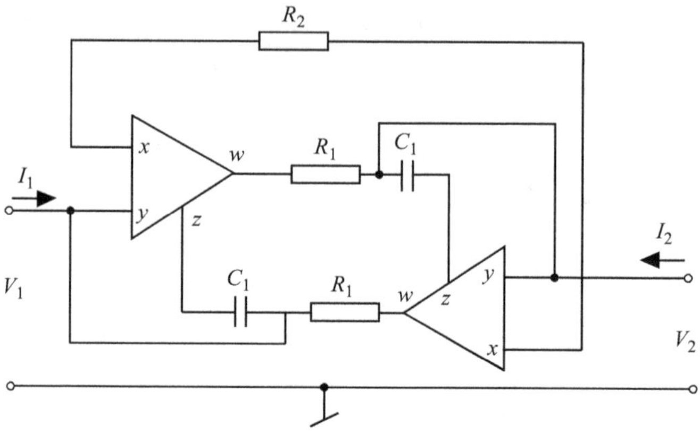

Figure 5.9 Lossy/lossless FI simulator using only two CFOAs proposed by Senani and Bhaskar [35]

However, if one chooses $R_1 = R_2 = R_0$, a lossless FI is realized with equivalent inductance given by $L_{eq} = C_1 R_0^2$. A nonideal analysis of this circuit, taking into account various parasitics, has been carried out in [35]; some results from the same are reproduced here. The various parasitics considered are the finite nonzero input resistance R_x looking into terminal-X and the finite output impedance looking into terminal-Z which comprises a resistance R_p in parallel with a capacitance C_p. Since the circuit is symmetrical in appearance, the nonideal parameters are too found to retain this similarity in the sense that $Y'_{11} = Y'_{22}$ and $Y'_{12} = Y'_{21}$ with values given by

$$Y'_{11} = Y'_{22}$$

$$= \frac{1}{R_1} + \frac{sC_1}{(1 + sC_1Z_p)} - \frac{sC_1Z_p}{(1 + sC_1Z_p)(R_2 + 2R_x)} + \frac{Z_p}{(1 + sC_1Z_p)(R_1R_2 + 2R_1R_x)} \tag{5.24}$$

$$Y'_{12} = Y'_{21}$$

$$= -\left[\frac{sC_1Z_p}{R_1(1 + sC_1Z_p)} - \frac{sC_1Z_p}{(1 + sC_1Z_p)(R_2 + 2R_x)} + \frac{Z_p}{(1 + sC_1Z_p)(R_1R_2 + 2R_1R_x)}\right] \tag{5.25}$$

It is easy to see that with $R_x \to 0$, $R_p \to \infty$ and $C_p \to 0$, the above nonideal values reduce to their ideal counter parts of (5.22).

Looking at the nonideal expressions, it is easy to visualize that the various parasitics are likely to restrict the operational frequency range of the circuit at higher frequencies. It is also apparent that the resistive and inductive parts of all the four nonideal Y-parameters would be both frequency dependent. With $R_x = 50\ \Omega$,

$R_p = 3\ M\Omega, C_p = 4.5$ pF and the circuit designed to realize a lossless FI of value $L = 1$ mH by choosing $C_1 = 1$ nF, $R_0 = 1$ kΩ, the MATLAB® plots of frequency responses of $|Y_{11}| = |Y_{22}|$ and $|Y_{12}| = |Y_{21}|$ were obtained from which it was estimated that the circuit has a usable frequency range of about 10 MHz which was also substantiated by the SPICE simulations of the circuit for the same component values.

Application: This circuit has been successfully used in the design of a fourth-order low-pass Butterworth filter design having cut-off frequency of 500 kHz with component values $R_L = 1$ kΩ, $L_1 = 0.2437$ mH, $C_1 = 0.5884$ nF, $L_2 = 0.5884$ mH, $C_2 = 0.2437$ nF. AD844 CFOAs were biased with ±12 V. The experimental frequency response was found to agree well with the one obtained from SPICE simulations, thereby confirming the workability of the FI configuration. A limitation of the circuit of Figure 5.9 is that it employs a more number of passive components, namely three resistors and two capacitors than the minimum necessary, i.e. only two resistors and a capacitor.

A question which, at the moment, is unanswered is, can one realize a lossless floating inductance using only two CFOAs and three passive elements (namely only two resistors and a single capacitor which may or may not be grounded)? This is open to research.

However, given only two CFOAs and three passive elements (three impedances), a lossy FI (be either series RL or parallel RL) could be surely realized and this will be demonstrated in a subsequent section. In addition, given the same number of CFOAs and passive components, a generalized negative impedance converter/inverter can also be realized which will be discussed in the next section.

5.5 Grounded/floating generalized positive impedance converters/inverters (GPIC/GPII) and generalized negative impedance converters/inverters (GNIC/GNII)

If a circuit has grounded/floating impedance realized as either as a grounded one-port or as a floating two-port given by

$$Z_{in}(s) = \frac{Z_1 Z_3}{Z_2} \quad \text{or} \quad Z_{in}(s) = \frac{Z_1 Z_3 Z_5}{Z_2 Z_4},$$

then the circuit would be called generalized positive impedance converter/inverter (GPIC/GPII) or generalized negative impedance converter/inverter (GNIC/GNII) depending upon the sign (+ or −) involved in the equation. Such circuits are more versatile as apart from their use in grounded/floating inductance simulation they can also be used to realize other related and useful impedances. In this context, it may be noted that the circuit employing five impedances is, obviously, more versatile. For example, an FDNR having $Z(s) = 1/Ds^2$ can be realized by selecting Z_1 and Z_3 as capacitors; a resistively variable capacitance by choosing Z_3 as a capacitor and a frequency-dependent negative conductance (FDNC) having $Z(s) = Ms^2$ by choosing Z_2 and Z_4 both as capacitors.

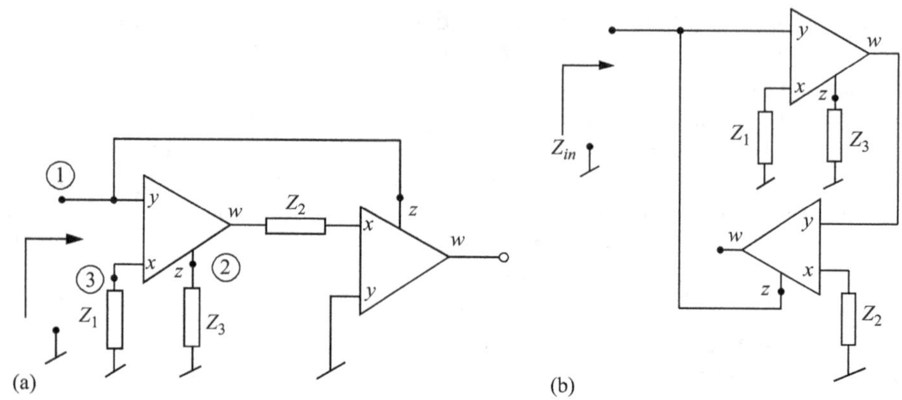

Figure 5.10 Realization of impedance converters/inverters: (a) generalized positive impedance converter/inverter, (b) generalized negative impedance converter/inverter

Two obvious circuits emanate from those of Figure 5.4 by replacing the three passive components shown therein and are shown in Figure 5.10. The input impedance is

$$Z_{in}(s) = \frac{Z_1 Z_2}{Z_3} \tag{5.26}$$

Thus, impedance connected at port 2 is *inverted* while that connected at port 3 (i.e. Z_1) is converted by a factor (Z_2/Z_3). Note that with $Z_1 = 1/sC_1, \; Z_2 = 1/sC_2$ and $Z_3 = R_3, \; Z_{in}(s) = 1/s^2 C_1 C_2 R_3$, and the circuit, thus, realizes an FDNR with $D = C_1 C_2 R_3$.

The other circuit of Figure 5.10(b) is a generalization of that of Figure 5.4(b) and has

$$Z_{in}(s) = -\frac{Z_1 Z_2}{Z_3} \tag{5.27}$$

This circuit, therefore, represents a generalized negative impedance converter/inverter.

For obtaining GPIC/GPII and GNIC/GNII circuits, we would need to embed one more CFOA along with two more impedances. This could be done in a number of ways.[2] We show here two circuits which employ a maximum possible number of grounded impedances which has its own merits. These are shown in Figure 5.11. It is apparent that because of employing five impedances, the circuits of Figure 5.11 are more versatile, for instance, by choosing Z_2 and Z_4 both as capacitors they realize, respectively, $Z(s) = s^2 C_2 C_4 R_1 R_3 R_5$ and $Z(s) = -s^2 C_2 C_4 R_1 R_3 R_5$ which

[2] Senani R. 'Universal floating positive VCZ structure evolution'. Research Note 1700; 1996.

represent an FDNC and an FDPR which, however, cannot be realized from the circuits of Figure 5.10.

We now show how to realize floating GPIC/GPII and floating GNIC/GNII. A floating version of the circuit of Figure 5.10(a) was derived by Senani [16] by ungrounding the impedances Z_1 and Z_2 tying them together and connecting a voltage V_2 obtained by inserting a third CFOA appropriately, which also makes the port-2 current i_2 equal to $-i_1$. The resulting circuit is shown in Figure 5.12.

The circuit of Figure 5.12 is characterized by the following equations:

$$\frac{V_2 - V_1}{Z_1} = \frac{V_2 - e}{Z_3} \tag{5.28}$$

$$i_1 = \frac{e - V_2}{Z_2} = \frac{1}{Z_2}\left\{\frac{Z_3}{Z_1}(V_1 - V_2)\right\} \tag{5.29}$$

and

$$i_2 = \frac{V_2 - e}{Z_2} \tag{5.30}$$

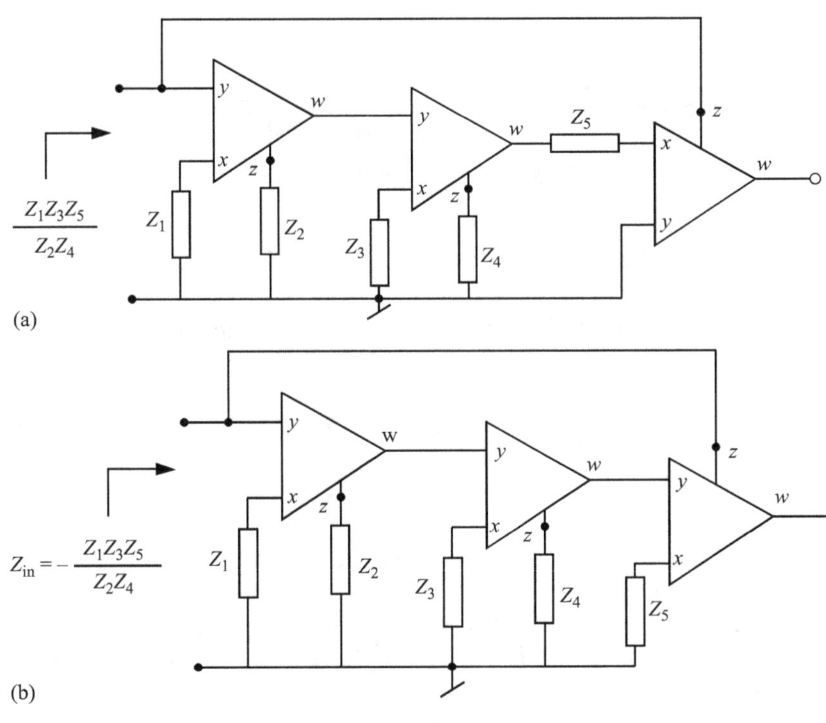

Figure 5.11 Extended impedance converters/inverters: (a) GPIC/GPII, (b) GNIC/GNII

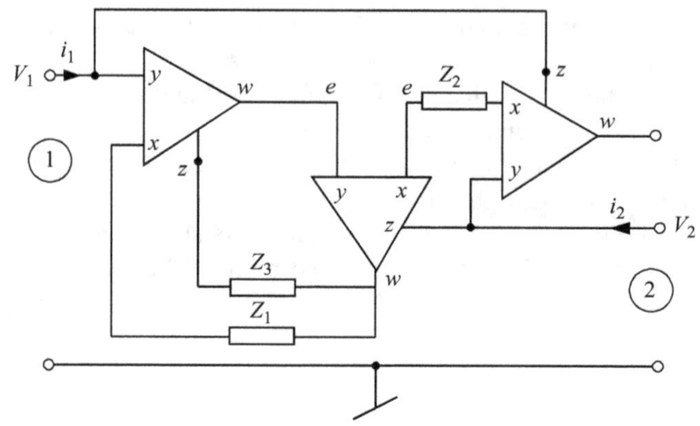

Figure 5.12 Floating GPIC/GPII configurations due to Senani [16]

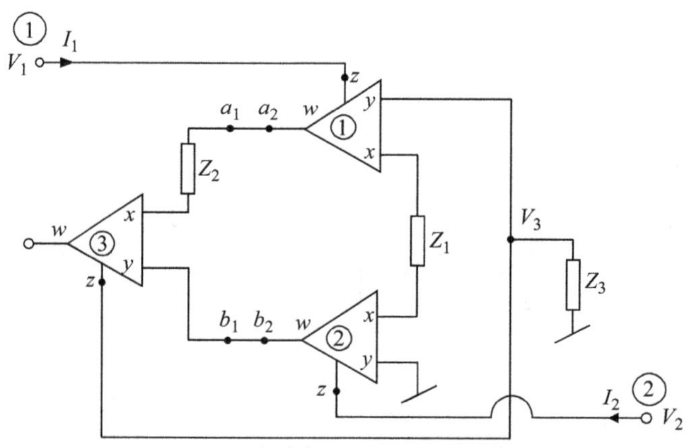

Figure 5.13 An alternate floating GPIC/GNII configuration [16]

By solving these equations, it is found that

$$-i_2 = i_1 = \frac{Z_3}{Z_1 Z_2}(V_1 - V_2) \tag{5.31}$$

and thus, the circuit realizes a floating impedance of value Z_1Z_2/Z_3.

An alternative three-CFOA GPIC/GPII structure with having one of the three impedances connected to ground was also presented in [16] which is shown in Figure 5.13. In this circuit, CFOA_1 and CFOA_2 realize a VCCS of type of Figure 5.3(f), providing

$$I_1 = -\frac{V_3}{Z_1}; \quad I_2 = \frac{V_3}{Z_1} \tag{5.32}$$

Now by using the W-terminal of $CFOA_1$ and $CFOA_2$, in conjunction with impedances Z_2 and Z_3 and the third CFOA, to create the node equation

$$-\frac{V_3}{Z_3} = \frac{V_2 - V_1}{Z_2} \tag{5.33}$$

The combination of the above equations yields

$$I_1 = -I_2 = \frac{Z_3}{Z_1 Z_2}(V_1 - V_2) \tag{5.34}$$

An interesting feature of this circuit is that if we change the connections ($a_1 - a_2$, $b_1 - b_2$) to ($a_1 - b_2$, $a_2 - b_1$), the modified circuit would realize a floating GNIC/GNII circuit.

Another interesting modification is to introduce an additional CFOA along with impedances Z_4 and Z_5, configured as a noninverting amplifier, the resulting circuit takes the form as shown in Figure 5.14. This circuit has been extracted from a voltage-controlled floating GIC presented in [28]. The analysis of Figure 5.14 yields

$$\frac{V_2 - V_1}{Z_3} = \frac{V_3}{Z_2}, \quad I_1 = -\left(\frac{Z_4}{Z_1 Z_5}\right) V_3 = -I_2 \tag{5.35}$$

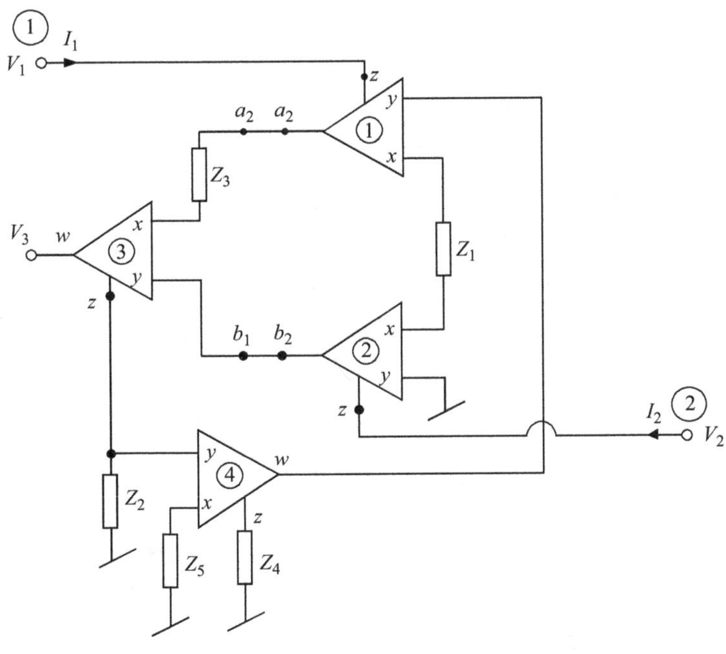

Figure 5.14 A more generalized floating GPIC/GPII which can also realize GNIC/GNII

The combination of these equations leads to

$$I_1 = -I_2 = \frac{Z_2Z_4}{Z_1Z_3Z_5}(V_1 - V_2) \tag{5.36}$$

Resulting in the floating impedance realized by this circuit as

$$Z_{1-2}(s) = \frac{Z_1Z_3Z_5}{Z_2Z_4} \tag{5.37}$$

Like the circuit of Figure 5.13, here also negative-valued impedance can be realized by the simple artifice of interchanging the connections $(a_1 - a_2, b_1 - b_2)$ to $(a_1 - b_2, a_2 - b_1)$, leading to

$$Z_{1-2}(s) = -\frac{Z_1Z_3Z_5}{Z_2Z_4} \tag{5.38}$$

It, may, however, be pointed out that the negative floating elements realizable from this alternative configuration still need to be explored for their possible applications.

Another four-CFOA-based floating generalized impedance converter was presented by Psychalinos–Pal–Vlassis [26] which is shown in Figure 5.15. This circuit which has been reproduced here with impedances renumbered also realizes the equivalent impedance between ports 1 and 2 given by $Z_{1-2}(s) = Z_1Z_3Z_5/Z_2Z_4$. It, may, however, be mentioned that no claim has been made by the proposers of this circuit regarding the usability of this circuit to also realize a floating generalized negative impedance converter too!

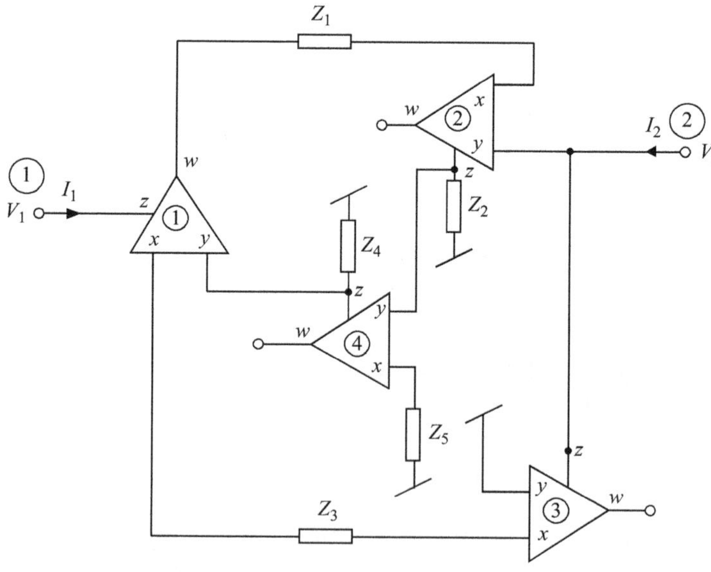

Figure 5.15 Floating GPIC/GPII of [26]

Four special cases of these circuits appear to be of practical interest and the same are, therefore, elaborated here:

Case 1 Capacitance-floatator cum multiplier:

Choosing Z_5 as capacitors and remaining impedances as resistors, which represents a floating capacitance and has a multiplication factor of (R_2R_4/R_1R_3) which may enable one to realize a high valued floating capacitance while employing reasonable values of resistors. For example, with $C_5 = 100$ pF, $R_1 = R_3 = 1$ kΩ and $R_2 = R_4 = 100$ kΩ, the realized capacitance would be 1 μF!

Case 2 Tunable floating inductance employing a grounded capacitor:

By choosing $Z_4 = 1/sC_4$ with remaining impedances as resistors, we obtain

$$Z_{1-2}(s) = \frac{sC_4R_1R_3R_5}{R_2} \tag{5.39}$$

The realized inductance can be varied by having any one resistor being variable.

Case 3 Single-resistance tunable floating FDNR:

By choosing $Z_1 = 1/sC_1$ and $Z_3 = 1/sC_3$ with remaining impedances as resistors, we obtain

$$Z_{1-2}(s) = \frac{R_5}{s^2C_1C_3R_2R_4} \tag{5.40}$$

which is a single-resistance-tunable floating FDNR.

Case 4 Single-resistance-tunable floating FDNC:

If we choose Z_2 and Z_4 as capacitors and the remaining impedances as resistors, the realized impedance takes the form

$$Z_{1-2}(s) = s^2C_2C_4R_1R_3R_5 = Ms^2 \quad \text{with} \quad M = C_2C_4R_1R_3R_5 \tag{5.41}$$

which represents a single-resistance-tunable floating FDNC element which, however, cannot be realized by the three-CFOA-based floating GPIC/GPII of Figures 5.12 and 5.13.

5.6 Economic simulation of lossy grounded inductors

As it happens, whenever a new analog building block gets introduced, quite often people like to see if any previously known circuit based on prevalent building blocks could be realized with the new one possibly with some advantage(s). Apparently with this motivation, Soliman [13] derived two lossy inductor simulation circuits which are shown in Figure 5.16. The first circuit is clearly a

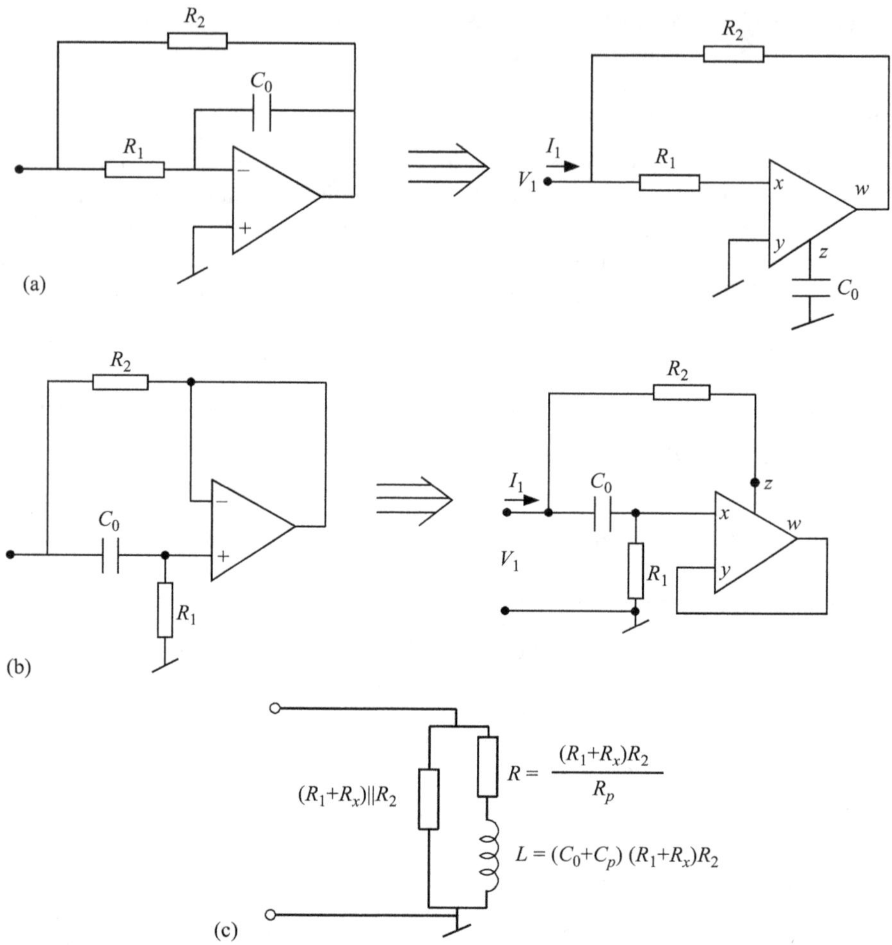

Figure 5.16 CFOA versions of two well-known op-amp-based economic GIs: (a) Ford–Girling circuit and its CFOA analog, (b) Berndt–Dutta Roy circuit and its CFOA analog, (c) nonideal equivalent of the single-CFOA parallel RL simulator of (a)

CFOA-analog of the Ford–Girling (FG) circuit (detailed in Chapter 2). Note that the FG circuit is obtained by putting an additional resistive feedback around an inverting integrator formed by the op-amp along with R_1 and C_0. In the CFOA version, the inverting integrator has the capacitor C_0 grounded which is an advantage not only from the viewpoint of IC implementation but also from the viewpoint of absorption of the parasitic capacitance C_p of the Z-terminal in the main capacitance C_0. Of course, the presence of parasitic resistance R_p will also require to be taken into account. Furthermore, the parasitic finite input resistance

R_x of X-terminal gets added up in R_1. Thus, taking these nonideal parameters into account, the modified expression for the input admittance $Y(s)$ becomes

$$Y(s) = \frac{1}{R_1 + R_x} + \frac{1}{R_2} + \frac{1}{\{s(C_0 + C_p) + (1/R_p)\}(R_1 + R_x)R_2} \tag{5.42}$$

A nonideal equivalent of this circuit would look like as shown in Figure 5.16(c).

On the other hand, the circuit of Figure 5.16(b) is loosely inspired by the Berndt–Dutta Roy circuit [30] (see Chapter 2). The only difference is that the unity gain amplifier made from the op-amp is replaced by a unity conversion constant negative impedance converter made from the CFOA by shorting its Y- and W-terminals and providing a feedback from the Z-terminal to the input of the circuit. A straightforward analysis of the circuit reveals the input admittance to be given by

$$Y(s) = \frac{sC_0R_2 + 1}{sC_0R_1R_2 + (R_2 - R_1)} \tag{5.43}$$

Now, if $R_2 = R_1 = R_0$ then

$$Y(s) = \frac{1}{R_0} + \frac{1}{sC_0R_0^2} \tag{5.44}$$

and the circuit simulates a parallel RL admittance with $R_{eq} = R_0$ and $L_{eq} = C_0R_0^2$. Thus, as compared to the Berndt–Dutta Roy's circuit which simulates bilinear RL grounded impedance, this CFOA analog simulates a clear parallel RL impedance.

5.7 Low-component-count lossy FI simulation

In Section 5.3, it has been discussed that lossless FIs typically require three CFOAs but may not need more than a minimum possible number of passive elements, namely only two resistors and a capacitor to create an impedance having dimensions of CR^2 which represents an equivalent inductance value. However, if the constraint on 'lossless-ness' is dropped, one can expect lossy FIs in the form of series RL or parallel RL impedances to be realizable with less than three CFOAs. That it is indeed possible was first demonstrated by Senani *et al.* [5] and simultaneously, by Bhaskar and Senani [36,38] who presented new lossy FI simulators which required only two CFOAs along with only three passive components and had the other novel features of (i) not requiring any component-matching condition(s) or cancellation constraint(s) to realize the intended type of impedances and (ii) provided single-resistance control of the realized inductance value.

Consider the circuits from [38] first which are shown in Figure 5.17. With standard characterization of CFOAs, it is easy to verify that the circuit of Figure 5.16(a) is characterized by the $[Y]$ matrix:

$$[Y] = \frac{1}{R_1 + sC_0R_1R_2}\begin{bmatrix} 1 & -1 \\ -1 & 1 \end{bmatrix} \tag{5.45}$$

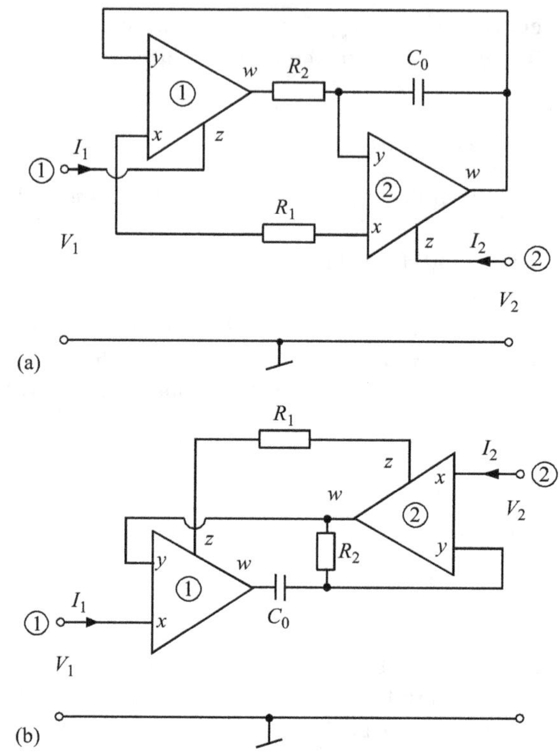

Figure 5.17 The canonic floating inductance simulators: (a) series-RL FI simulator, (b) parallel-RL FI simulator

while the circuit of Figure 5.17(b) has the $[Y]$ matrix given by

$$[Y] = \left(\frac{1}{R_1} + \frac{1}{sC_0R_1R_2}\right)\begin{bmatrix} 1 & -1 \\ -1 & 1 \end{bmatrix} \tag{5.46}$$

Thus, both the circuits realize floating RL impedances, series and parallel, respectively, such that the realized inductance in both the cases has a value $L_{eq} = C_0R_1R_2$ which is single-resistance tunable through R_2 while on the other hand, the series resistance is given by $R_s = R_1$ in first case while the parallel resistance R_p, in the second case, is given by $R_p = R_1$. When these circuits are, however, analysed by taking cognisance of the various parasitic impedances, the four y-parameters no longer remain identical and equal (see [38]). In the first case, it turns out that

nonideally $y'_{11} = -y'_{21} = 1/D_1(s)$ and that

$$y'_{12} = -y'_{22} = \frac{-(1 + (R_2/Z_y))}{D_1(s)} \tag{5.47}$$

where

$$D_1(s) = (R_1 + 2R_x)\left(1 + sC_0R_2 + \frac{R_2}{Z_y}\right) \tag{5.48}$$

However, in the second case, the various nonidealities of the CFOAs manifest to make all the four y-parameters to have entirely different expressions. The net effect of the various nonidealities has been found to be eventually limiting the operational frequency range of these simulated inductors which is normally taken to be the minimum of the operational frequency range over which $|y_{ij}(\omega)|_{\text{nonideal}}$ is closer to the $|y_{ij}(\omega)|_{\text{ideal}}$ within a specified acceptable error (say 5%). Applications given in [38] have demonstrated that with AD844 types of CFOAs, both the circuits can be satisfactorily employed to realize filters with resonant frequencies in the range of a few kHz up to which the deviations between the measured frequency responses and theoretical frequency responses remain within the acceptable limits.

After the publication of the lossy FI circuits of Figure 5.17(a) and (b) in [38], Abuelma'atti and Dhar [39], [43] and Abuelma'atti *et al.* [44] came up with alternative configurations for the same. Although in the latter publication, the authors have presented a variety of CFOA-based circuits for realizing floating negative impedances in different forms too, several of the circuits suffer from the drawback of requiring more than the minimum number of components. In view of this, the problem of devising all possible series-*RL*/parallel-*RL* FIs using no more than two CFOAs along with only three passive elements does not appear to have been fully resolved yet and is thus open to investigation.

5.8 Single resistance-controllable single CFOA simulators

In Chapters 2 and 4, it has been observed that as compared to lossless inductors/FDNRs in both grounded and floating forms, it is possible to realize their nonideal/lossy counterparts using a relatively smaller number of active and passive components. The same has also been found to be true in the case of lossy/nonideal impedance simulators using CFOAs. In this section, we will highlight some significant configurations made for inductor/impedance simulation in both grounded and floating forms.

Liu and Hwang [8] considered a general configuration using a single CFOA along with four admittances from which they derived two interesting economic impedance simulation circuits which are shown in Figure 5.18.

The circuit of Figure 5.18(a) simulates series-*RL* impedance subjected to taking the capacitor values equal. The advantages of this circuit are (i) use of two equal-valued grounded capacitors, as preferred for IC implementation; (ii) feasibility of absorbing the parasitic impedance $(R_p//(1/sC_p))$ as part of external *RC* components connected to *Z*-terminal and (iii) single-resistance tunability of the realized inductance value.

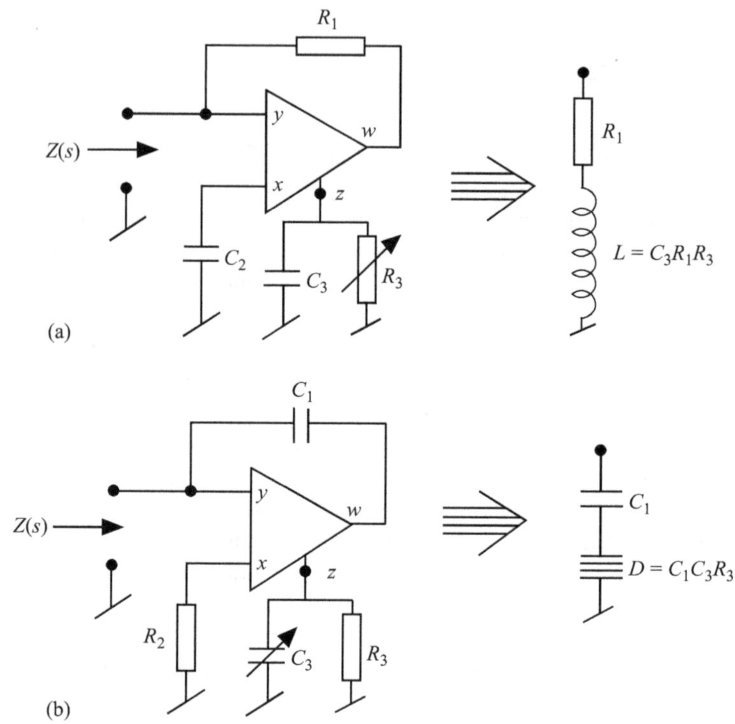

Figure 5.18 Economic series RL/CD simulators with single-element control [8], (a) Series RL, (b) Series CD

The second circuit of Figure 5.18(b) is actually the *RC*:*CR* transformed version of the first circuit and thus, as explained earlier, simulates a series combination of a capacitance and a *D*-element as shown. Here, also the parasitics of the *Z*-terminal of the CFOA can be absorbed in the external passive elements and the value of *D* is single-element controllable through a variable capacitance C_3.

Abuelma'atti [33] derived a number of immittance function simulator circuits using a single CFOA, based on two generalized circuit configurations involving four to five admittances. Out of the various special cases devised therein, we present here two special cases each of which realize a series *RL*-type GI and are distinctly different from the similar circuits published earlier. These circuits are shown in Figure 5.19 along with their equivalent circuits and parameter values.

Both the circuits employ a bare minimum number of only three passive components and realize inductance values (lossy) which are single-resistance tunable in each case.

Kacar and Kuntman [31] proposed four single-CFOA inductance simulators; however, only one of these realizes a positive lossy inductor and hence is worth considering here. This circuit is reproduced in Figure 5.20. The circuit realizes series *RL* impedance with

$$R_{eq} = (R_1 + R_2) \quad \text{and} \quad L_{eq} = CR_1R_2 \tag{5.49}$$

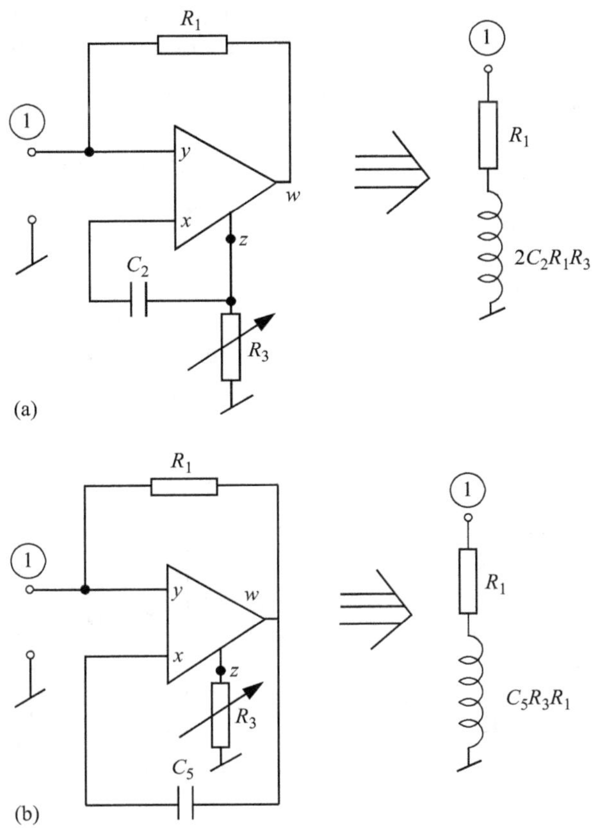

Figure 5.19 Two Single CFOA-based impedance simulators derived in [33], (a) series RL simulator, (b) an alternative series RL simulator

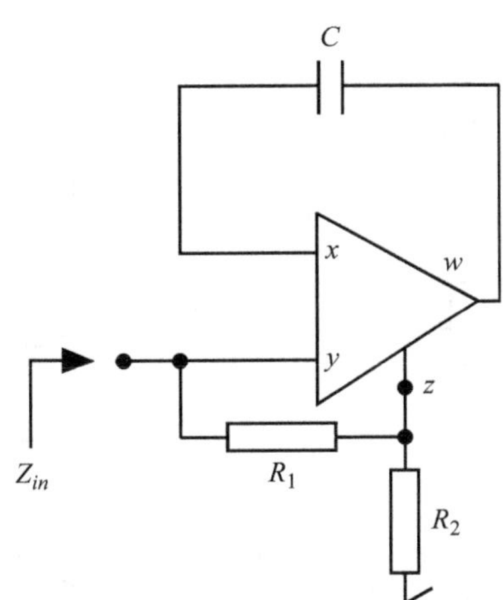

Figure 5.20 A series-RL simulator from [31]

Yuce [29] proposed four new single-CFOA inductance simulators which are shown in Figure 5.21. All the circuits simulate a series-*RL*-type GI with equivalent parameters given by for $Z_{eq} = R_{eq} + sL_{eq}$:

From Figure 5.21 we obtain

$$R_{eq1} = R_1; \quad L_{eq1} = C_0R_1R_2 \qquad \text{and} \qquad R_{eq2} = R_1; \quad L_{eq2} = C_0R_1R_2 \tag{5.50}$$

$$R_{eq3} = R_1; \quad L_{eq3} = 2C_0R_1R_2 \qquad \text{and} \qquad R_{eq4} = \frac{R_1}{2}; \quad L_{eq4} = \frac{C_0R_1R_2}{2} \tag{5.51}$$

It may be noted that all the circuits have the provision of single-resistance-control of L_{eq} value through a single variable resistance R_2.

Application: A novel application of these circuits was presented in realizing a low-pass filter by ungrounding R_2 and applying V_{in} on the input terminal and then

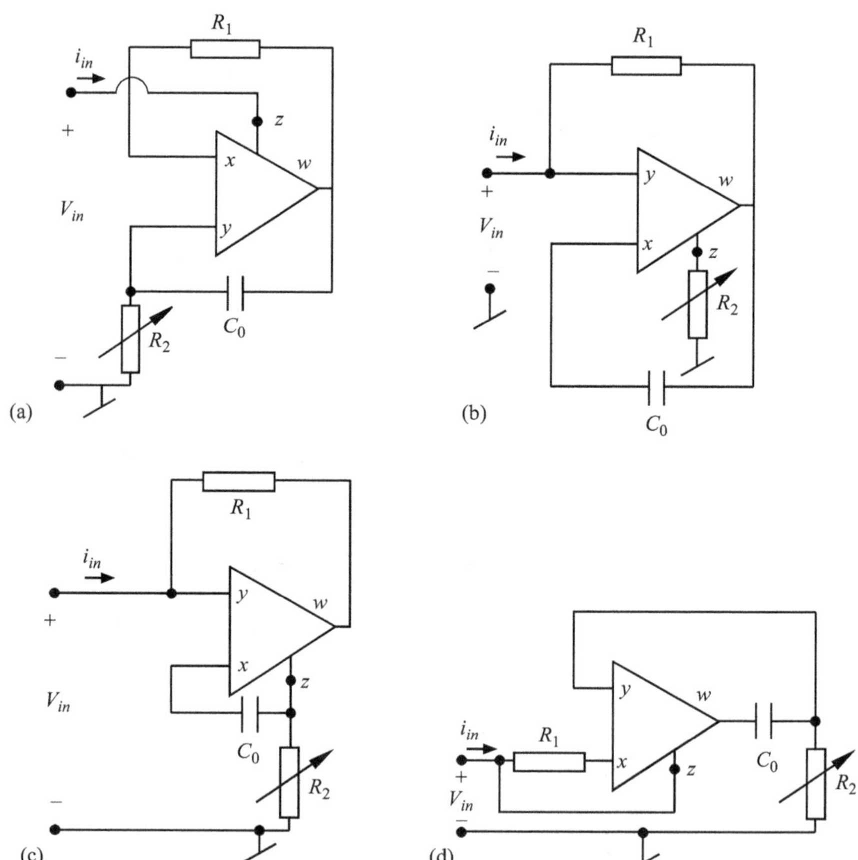

Figure 5.21 Some single-resistance-tunable series RL-type GIs [29]: (a) $R_{eq1} = R_1$; $L_{eq1} = C_0R_1R_2$; (b) $R_{eq2} = R_1$; $L_{eq2} = C_0R_1R_2$; (c) $R_{eq3} = R_1$; $L_{eq3} = 2C_0R_1R_2$; (d) $R_{eq4} = R_1/2$; $L_{eq4} = C_0R_1R_2/2$

connecting a capacitor across port 1 and connecting a capacitor C_1 there and taking output across the grounded capacitor C_1. These four filters, resulting from the use of each of the four inductance simulators of Figure 5.21, have been dealt with in [29]; however, here we present two of them which are shown in Figure 5.22.

At the first glance, this method may appear to be erroneous [29]; however, it can be proved to be correct as follows (see Figure 5.22(c)):

Let N be characterized by

$$\begin{bmatrix} i_2 \\ i_A \end{bmatrix} = \begin{bmatrix} y_{11} & y_{12} \\ y_{21} & y_{22} \end{bmatrix} \begin{bmatrix} v_{LP} \\ v_A \end{bmatrix} \tag{5.52}$$

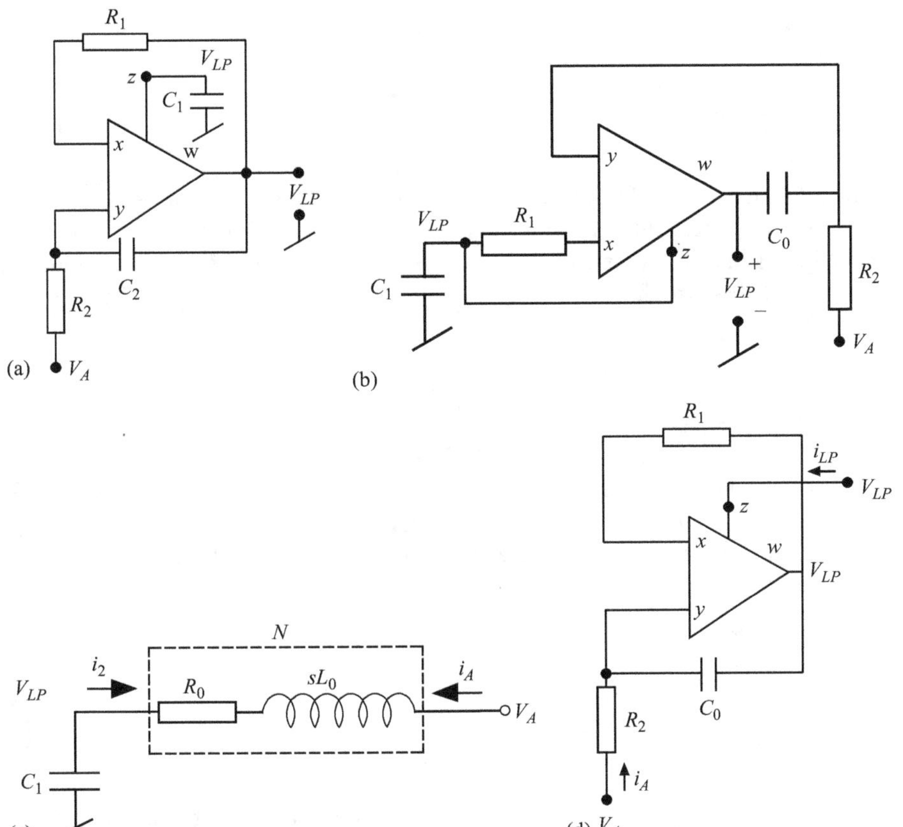

Figure 5.22 LPF application of the lossy inductors presented in Figure 5.21: (a) and (b) two specific simulated inductors, (c) LP filter realization using a unilateral floating RL impedance, (d) a two-port derived from the GI of (a)

Therefore,

$$i_2 = y_{11} v_{LP} + y_{12} v_A \tag{5.53}$$

and

$$V_{LP}\, sC_1 = -i_2 \tag{5.54}$$

If we assume

$$y_{11} = -y_{12} = \frac{1}{R_{eq} + sL_{eq}} \tag{5.55}$$

we get

$$(-sC_1 V_{LP}) = \frac{1}{R_{eq} + sL_{eq}}(V_{LP} - V_A) \tag{5.56}$$

which leads to

$$T(s) = \frac{V_{LP}(s)}{V_A(s)} = \frac{1}{s^2 C_1 L_{eq} + sC_1 R_{eq} + 1} \tag{5.57}$$

From (5.57) it is clear that in this case parameters Y_{21} and Y_{22} of network N are inconsequential! Consider now the two-port circuit of Figure 5.22(a), without connecting capacitor C_1, as shown in Figure 5.22(d). A routine analysis yields the $[Y]$ matrix of this circuit as

$$\begin{bmatrix} I_{LP} \\ I_A \end{bmatrix} = \begin{bmatrix} \dfrac{1}{R_1 + sC_0 R_1 R_2} & \dfrac{-1}{R_1 + sC_0 R_1 R_2} \\ \dfrac{-sC_0}{1 + sC_0 R_2} & \dfrac{sC_0}{1 + sC_0 R_2} \end{bmatrix} \begin{bmatrix} V_{LP} \\ V_A \end{bmatrix} \tag{5.58}$$

It is, thus, found that indeed Y_{11} and Y_{12} are both inductive: $(R_1 + sC_0 R_1 R_2)$ but Y_{21} and Y_{22} are not. Hence, as derived earlier, this circuit would become a low-pass filter when a capacitor is connected at V_{LP}. The resulting transfer function is given by

$$\begin{aligned} T_1(s) = \frac{V_{LP}(s)}{V_A(s)} &= \frac{1}{s^2 C_1 C_0 R_1 R_2 + sC_1 R_1 + 1} \\ &= \frac{(1/C_0 C_1 R_1 R_2)}{s^2 + (s/C_0 R_2) + (1/C_0 C_1 R_1 R_2)} \end{aligned} \tag{5.59}$$

The circuit of Figure 5.22(b) similarly would realize a transfer function of type (Equation (5.59)) with $R_{eq} = R_1/2$; $L_{eq} = C_0 R_1 R_2/2$. An interesting point to be noticed is that in both these circuits the output need not be taken across capacitor C_1; in fact, a better output terminal will be the W-terminal where $V_w = V_z = V_{LP}$ with the advantage of having low output impedance.

5.9 Inductors and resonators using CFOA poles

In the 1970s and 1980s, a variety of analog circuit designs based on the so-called Active-R network was investigated by several researchers in which the otherwise *parasitic* op-amp pole of the open-loop gain of internally compensated type op-amps characterized by $A(s) = (A_0\omega_p)/(s+\omega_p)$ was used as an useful parameter (typically expressing $A(s) \cong (A_0\omega_p/s)$, for $\omega \gg \omega_p$) to devise circuits which employed only op-amps and resistors without requiring any external capacitors (thereby the name 'Active-R') to design Active-R filters, oscillators and immittance simulators. The advantages of such circuits were claimed to be suitability for IC implementation due to the elimination of lossy external capacitors, relatively higher operational frequency range, etc.

This Active-R approach has been attempted even with CFOA-based circuits, and a number of researchers have presented Active-R filters and oscillators using CFOA poles. Note that the transimpedance gain of the CFOA can be expressed as

$$Z_p = \left(R_p // \frac{1}{sC_p}\right) = \frac{(R_P/sC_p)}{R_p + (1/sC_p)} = R_p\left(\frac{(1/R_pC_p)}{s + (1/C_pR_p)}\right) \tag{5.60}$$

which can be written as

$$Z_p = R_p\left(\frac{\omega_0}{s+\omega_0}\right) \tag{5.61}$$

where $\omega_0 = (1/C_pR_p)$ is the CFOA pole which is almost analogous to the open-loop gain function of a VOA.

Singh and Senani [19] presented a class of three Active-R resonators using CFOA poles which are shown in Figure 5.23, a straightforward analysis reveals that the nonideal expression for the input admittance of the circuits of Figure 5.23(a), and Figure 5.23(b) turns out to be the same and is given by:

$$Y(s) = \frac{1}{(R_1'R_2'/R_{p1}) + sC_{p1}R_1'R_2'} - \frac{1}{R_2'} + \frac{1}{R_{p2}'} + sC_{p2}' \tag{5.62}$$

where $R_i' = (R_i + R_{xi});\ \ i = 1, 2$

$$\left.\begin{aligned} R_{p2}' &= (R_{p2}//R_{y1}//R_{y2}) \\ C_{p2}' &= C_{p2} + C_{y2} + C_{y1} \end{aligned}\right\} \text{for the circuit of Figure 5.23}(a) \tag{5.63}$$

and

$$\left.\begin{aligned} R_{p2}' &= (R_{p2}//R_{y2}) \\ C_{p2}' &= C_{p2} + C_{y2} \end{aligned}\right\} \text{for the circuit of Figure 5.23}(b) \tag{5.64}$$

It is easy to see that both the circuits realize a parallel RLC resonator for which the values of the various parameters are given by

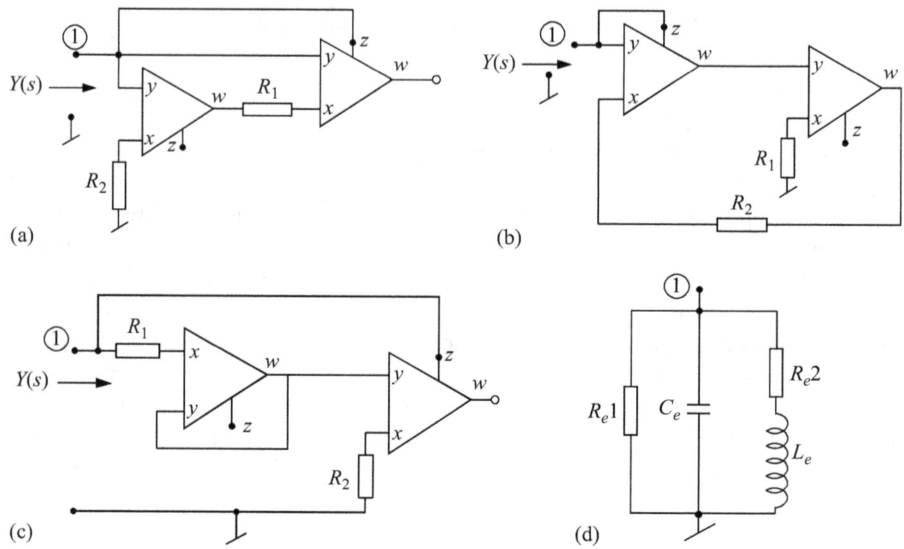

Figure 5.23 Three Active-R resonators using CFOA poles [19], (a) first proposed resonator, (b) second proposed resonator, (c) third proposed resonator, (d) general equivalent circuit

$$\left.\begin{aligned} \frac{1}{R_{e1}} &= \frac{1}{R'_{p2}} - \frac{1}{R'_2} \\ R_{e2} &= \frac{R'_1 R'_2}{R_{p1}} \\ C_e &= C'_{p2} \\ L_e &= C_{p1} R'_1 R'_2 \end{aligned}\right\} \tag{5.65}$$

On the other hand, the input admittance $Y(s)$ of the circuit of Figure 5.23(c) is given by

$$Y(s) = \frac{1}{R'_1} + \frac{1}{R_{p2}} + sC_{p2} + \frac{\left(1 + \left(R'_2/R'_1\right)\right)}{sC_{p1}R'_1R'_2 + R'_2\left(\left(R'_1/R_{p1}\right) - 1\right)} \tag{5.66}$$

From the above, the parameters of this resonator can be expressed as

$$\left.\begin{aligned} \frac{1}{R_{e1}} &= \frac{1}{R_{p2}} + \frac{1}{R'_1} \\ R_{e2} &= R'_2\left(\frac{R'_1}{R_{p1}} - 1\right) \Big/ \left(1 + \frac{R'_2}{R'_1}\right) \\ C_e &= C_{p2} \\ L_e &= C_{p1}R'_1R'_2 \Big/ \left(1 + \frac{R'_2}{R'_1}\right) \end{aligned}\right\} \tag{5.67}$$

A specific case of this for $R'_1 = R_{p1}$ is interesting as this eliminates R_{e2} (shorted), thereby leaving a parallel-*RLC* resonator.

If we consider ideal CFOAs, then the resonators of Figure 5.23(a) and (b) yield two interesting active gyrators which are shown in Figure 5.24, which are distinctly different than Fabre's gyrator described earlier in this Chapter. If port 2 of these gyrators is terminated into a capacitor C_1, the input impedance looking into port 1 is found to be $Z(s) = (R_0 - R_2) + sC_1R_1R_2$ and thus, both the circuits function as *active gyrators* with respect to ports 1 and 2 (simulating lossless inductance) with $R_0 = R_2$.

Application: With both the ports terminated into two grounded capacitors C_1 and C_2, these two active gyrators yield two novel single-resistance-controlled oscillators (SRCOs). The condition of oscillation (CO) and frequency of oscillation (FO) of these SRCOs are found to be

$$\frac{1}{R'_0} = \frac{\left(1 - \left(C'_2R'_2/C'_1R_{p1}\right)\right)}{R_2} \quad \text{and} \quad f_0 = \frac{1}{2\pi\sqrt{C'_1C'_2R'_1R'_2}}\left(1 - \left(C'_2R'_1R'_2/C'_1R^2_{p1}\right)\right) \tag{5.68}$$

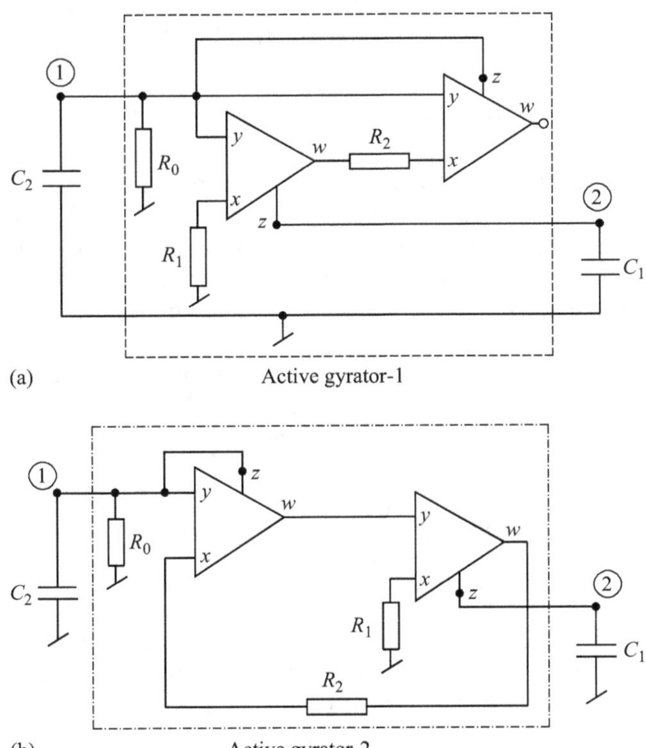

Figure 5.24 Two interesting SRCOs based on the new Active-R gyrators [19], (a) the first proposed active-gyrator, (b) the second proposed active-gyrator

where

$$\left.\begin{aligned} C_1' &= (C_1 + C_{p1}) \\ C_2' &= (C_2 + C_{p2} + C_{y1} + C_{y2}) \\ R_1' &= (R_1 + R_{x1})\,; R_2' = (R_2 + R_{x2}) \\ \frac{1}{R_0'} &= \left(\frac{1}{R_0} + \frac{1}{R_{p2}} + \frac{1}{R_{y1}} + \frac{1}{R_{y2}}\right) \end{aligned}\right\} \text{for the SRCO of Figure 5.24}(a) \quad (5.69)$$

and

$$\left.\begin{aligned} C_1' &= (C_1 + C_{p1}) \\ C_2' &= (C_2 + C_{p2} + C_{y2}) \\ R_1' &= (R_1 + R_{x1})\,; R_2' = (R_2 + R_{x2}) \\ \frac{1}{R_0'} &= \left(\frac{1}{R_0} + \frac{1}{R_{p2}} + \frac{1}{R_{y2}}\right) \end{aligned}\right\} \text{for the SRCO of Figure 5.24}(b) \quad (5.70)$$

A novel feature of these SRCOs is that the independent single-resistance controllability of the CO and FO remains intact even under the influence of various CFOA parasitics, the former by R_0 and the latter by R_1. In addition, these SRCOs are able to absorb all the parasitic impedances of both the CFOAs.

5.10 GI/FI simulators using modified CFOAs

Over the years, a number of variants of CFOAs have been proposed in the literature from time to time among which the current-controlled CFOA (CC-CFOA) [25], the inverting CFOA [40], the dual output CFOA and the so-called modified CFOA (MCFOA) [22] appear to be the most significant ones. Use of these MCFOAs has been prominently made in the realization of synthetic impedances which is elaborated in this section.

The CC-CFOA is basically a building block containing a current-controlled (second generation) conveyor II followed by a voltage buffer and is, hence, characterized by the instantaneous terminal equations:

$$\begin{bmatrix} I_y \\ V_x \\ I_z \\ V_w \end{bmatrix} = \begin{bmatrix} 0 & 0 & 0 & 0 \\ 1 & R_x & 0 & 0 \\ 0 & 1 & 0 & 0 \\ 0 & 0 & 1 & 0 \end{bmatrix} \begin{bmatrix} V_y \\ I_x \\ V_z \\ I_w \end{bmatrix} \quad (5.71)$$

where $R_x \cong V_T/2I_B$ and is, therefore, current-controllable through an external dc bias current. Of course, the Z-port parasitics (comprising $R_p//(1/sC_p)$) and Y-port parasitics $(R_y//(1/sC_y))$ of the normal CFOA are also present in the CC-CFOA and must be considered for a realistic design.

Note that although a CC-CFOA is not commercially available as an off-the-shelf IC yet but architecturally, it does not require any additional hardware than that of a normal CFOA; the only change required is that the two internal current mirrors/repeaters be derived from an external dc bias current source for which an extra terminal (into which this current would be fed from outside) needs to be brought out as an externally accessible pin (as is available in several commercially available IC operational transconductance amplifiers such as LM3080/ LM13600, etc.).

Second, a CC-CFOA can also be realized by CMOS hardware almost in a way analogous to the one employed in a bipolar CC-CFOA; only difference will be the expression for R_x which, instead of depending upon I_B linearly, would become a function of $\sqrt{I_B}$. Furthermore, BiCMOS CC-CFOA design is also feasible as demonstrated in [25].

An interesting lossless GI simulator using CC-CFOA was proposed in [25] which is reproduced in Figure 5.25(a).

Analysis of this circuit yields its input impedance as $Z_{in}(s) = sC_0R_{x1}R_{x2}$, leading to the simulated inductance value as $L_{eq} = \left(C_0V_T^2/4I_{B1}I_{B2}\right)$, which can be electronically controlled by I_{B1} and/or I_{B2}.

It is not difficult to decipher that with capacitor C_0 taken out and terminal 2 treated as port 2, the Y-parameters of the left-out circuit (see Figure 5.25(b)) are actually given by

$$\begin{bmatrix} i_1 \\ i_2 \end{bmatrix} = \begin{bmatrix} 0 & \dfrac{1}{R_{x2}} \\ -\dfrac{1}{R_{x1}} & 0 \end{bmatrix} \begin{bmatrix} v_1 \\ v_2 \end{bmatrix} \tag{5.72}$$

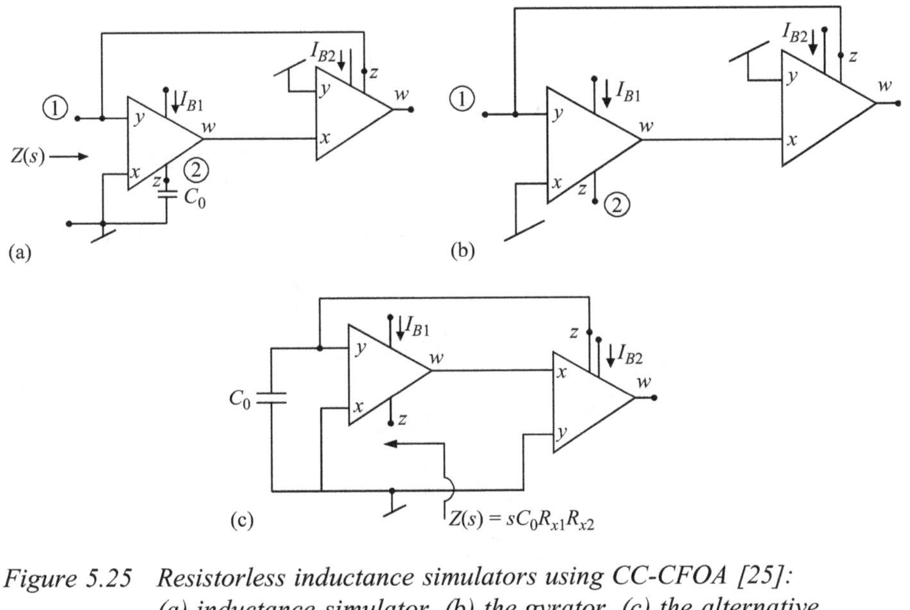

Figure 5.25 Resistorless inductance simulators using CC-CFOA [25]: (a) inductance simulator, (b) the gyrator, (c) the alternative inductance simulator

The two-port thus formed, therefore, represents an ideal gyrator with both the gyration resistances being electronically controllable with external dc bias currents I_{B1} and I_{B2}, respectively. Hence, one other variant of this circuit is also possible (although not explicitly spelt out in [25]) which is shown in Figure 5.25(c).

Alpaslan and Yuce [40] utilized an inverting CFOA to evolve three interesting GI circuits which are shown in Figure 5.26. Here, the inverting CFOA, designated as CFOA−, is characterized by

$$\begin{bmatrix} I_y \\ I_{z-} \\ V_x \\ V_w \end{bmatrix} = \begin{bmatrix} 0 & 0 & 0 & 0 \\ 0 & 0 & -\gamma & 0 \\ \beta_1 & 0 & R_x & 0 \\ 0 & \beta_2 & 0 & R_x \end{bmatrix} \begin{bmatrix} V_y \\ V_{z-} \\ I_x \\ I_w \end{bmatrix} \tag{5.73}$$

By straightforward analysis, the input impedances/admittances of the circuits of Figure 5.26 are found to be

$$Z_{in}(s) = R_1 + sC_0R_1R_2 \quad \text{for the circuit of Figure 5.26}(a) \tag{5.74}$$

$$Y_{in}(s) = \frac{1}{R_2} + \frac{1}{sC_0R_1R_2} \quad \text{for the circuit of Figure 5.26}(b) \tag{5.75}$$

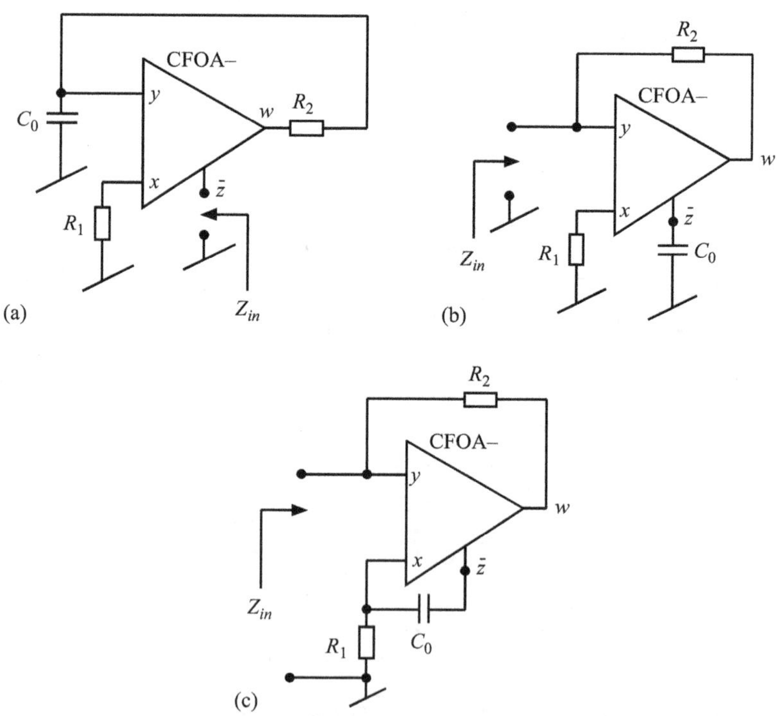

Figure 5.26 Some GI circuits using CFOA− [40], (a) series RL, (b) parallel RL, (c) lossless L

$$Z_{in}(s) = sC_0R_1R_2 \quad \text{for the circuit of Figure 5.26}(c) \tag{5.76}$$

Thus, it may be noted that two circuits realize lossy GIs but have the advantage of (i) single-resistance controllability of the realized inductance value and (ii) use of grounded capacitor. On the other hand, the third circuit has the drawback of employing a floating capacitor but has the advantage of (i) realizing a lossless GI and (ii) possessing single resistance tunability of the realized inductance value. No component matching is required in any of the circuits of Figure 5.26.

If we assume the gains λ, β_1 and β_2 as unity but consider the parasitic output impedances looking into terminal-X and terminal-W (both are taken as R_x because identical subcircuits employed for creating a voltage buffer between Y- and X-terminal and another between Z-terminal and W-terminal), the nonideal expressions turn out to be as follows:

$$Y_{in}(s) = \frac{1}{(R_1 + R_x) + sC_0(R_1 + R_x)(R_2 + R_x)} \quad \text{for the circuit of Figure 5.26}(a) \tag{5.77}$$

$$Y_{in}(s) = \frac{1}{(R_2 + R_x)} + \frac{1}{sC_0(R_1 + R_x)(R_2 + R_x)} \quad \text{for the circuit of Figure 5.26}(b) \tag{5.78}$$

$$Y_{in}(s) = \frac{1}{\{(R_1 + R_x)(R_2 + 2R_x)/2R_x\}} + \frac{1}{sC_0(R_1 + 2R_x)(R_2 + R_x)} \quad \text{for the circuit of Figure 5.26}(c) \tag{5.79}$$

The workability of these propositions has been verified by SPICE simulations using two normal kinds of CFOAs (CFOA+: AD844) and the circuits have been found to work as predicted by theory.

However, in a subsequent publication, Yuce and Minaei [24] employed a more elaborate four-port MCFOA characterized by the following equations:

$$\left.\begin{aligned} i_y &= -i_w \\ v_x &= v_y \\ i_z &= i_x \\ v_w &= v_z \end{aligned}\right\} \tag{5.80}$$

Thus, the terminal equations of the MCFOA differ from those of the normal CFOA only in terms of i_y which is $i_y = 0$ for a normal CFOA but $i_y = -i_w$ for the MCFOA. In the third paragraph of the introduction [24], the authors have admitted that 'the proposed MCFOA is based upon the composite CC reported in [2]'. However, in fact, the MCFOA is, indeed, *exactly* the same as the composite CC of [2] which is reproduced in Figure 5.27.

Since it contains a CCII+ and a CCII−, it follows by inspection that $i_y = -i_w$ (due to CCII−), $v_x = v_y$ (due to CCII+), $i_z = i_x$ (due to CCII+) and finally, $v_w = v_z$

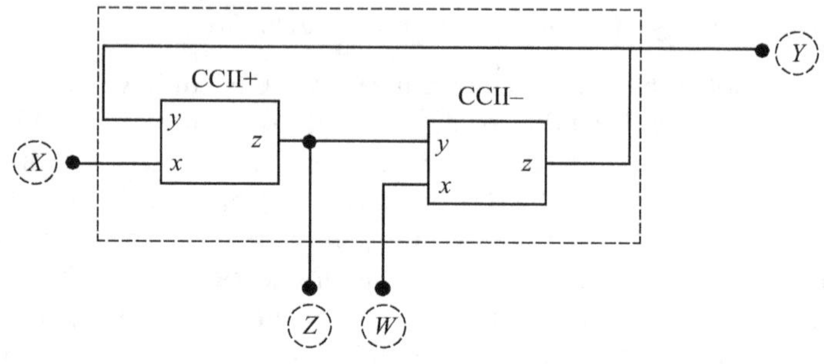

Figure 5.27 The modified CFOA is the same as the composite current conveyor [2]

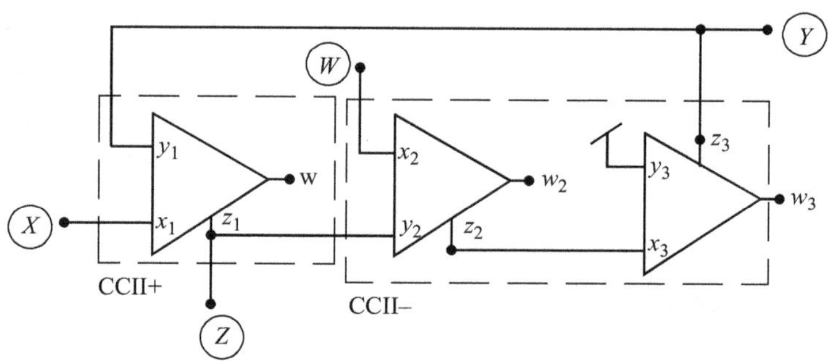

Figure 5.28 The AD844-based implementation of MCFOA presented in [24] which is the same as composite connection of CCII+ and CCII− of [2]

(see Figure 5.27). Thus, MCFOA has *exactly* the same characterization as that of a composite CC.

Furthermore, Yuce and Minaei [24] have presented a CMOS circuit for the implementation of the MCFOA which can be readily decoded as a CMOS CCII+ and another CMOS CCII− connected together exactly according to the schematic of the composite CC of Figure 5.27.

If one considers the CFOA (AD844)-based implementation of the MCFOA reproduced in Figure 5.28, it becomes absolutely clear that MCFOA of [24] is *exactly the* same as the composite CC of [2].

It must, however, be admitted that in spite of this, the MCFOA, if made available as an off-the-shelf IC, can be put to use quite effectively in many applications. Yuce and Minaei [24] have successfully demonstrated its use in devising a variety of very compact realizations of a number of grounded and floating inductance circuits which we will elaborate now.

Realizing that inside the MCFOA reside two CCIIs of opposite polarity, the grounded impedance simulation circuits shown in Figure 5.29 are rather apparent.

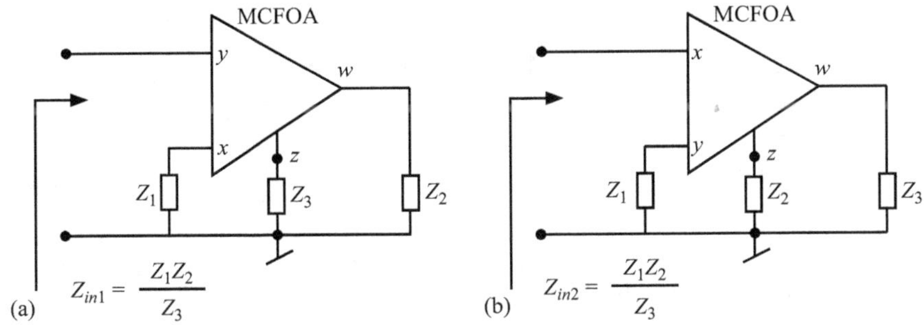

Figure 5.29 Two GIs using MCFOA [24], (a) the first proposition, (b) the second proposition

A nonideal characterization of the MCFOA can be expressed by the following equation:

$$\begin{bmatrix} I_y \\ V_x \\ I_z \\ V_w \end{bmatrix} = \begin{bmatrix} 0 & 0 & 0 & -\alpha_2 \\ \beta_1 & 0 & 0 & 0 \\ 0 & \alpha_1 & 0 & 0 \\ 0 & 0 & \beta_2 & 0 \end{bmatrix} \begin{bmatrix} V_y \\ I_x \\ V_z \\ I_w \end{bmatrix} \tag{5.81}$$

Under the above characterization, the input admittances of the circuits of Figure 5.29 get modified as

$$Y_{in1} = \frac{y_1y_2}{y_3}(\alpha_1\alpha_2\beta_1\beta_2) \quad \text{and} \quad Y_{in1} = \left(\frac{y_1y_2}{y_3}\right)\frac{1}{(\alpha_1\alpha_2\beta_1\beta_2)} \tag{5.82}$$

Thus, when coefficients α_1, α_2 ,β_1 ,β_2 assume their ideal values of unity; both the circuits simulate exactly the same input admittance. These circuits are useful for simulating the following type of synthetic impedances:

1. Simulated grounded inductor is obtained for $y_1 = 1/R_1$, $y_2 = 1/R_2$ and $y_3 = sC_3$ so that $Z_{in1}(s) = Z_{in2}(s) = sC_3R_1R_2$ leading to $L_{eq} = C_3R_1R_2$.
2. Capacitance multiplier is realized when $y_1 = sC_1$, $y_2 = 1/R_2$ and $y_3 = 1/R_3$, which leads to $Z_{in1}(s) = Z_{in2}(s) = (R_2/sC_1R_3)$ which realizes $C_{eq} = C_1R_3/R_2$.
3. FDNR simulator is obtained when $y_1 = sC_1$, $y_2 = sC_2$ and $y_3 = (1/R_3)$, which gives $Z_{in1}(s) = Z_{in2}(s) = (1/s^2C_1C_2R_3) = (1/Ds^2)$ with $D = C_1C_2R_3$.

For a detailed nonideal analysis, these circuits can be seen in [24]. In particular, the nonideal expression for the input impedance of the first circuit (for low to medium frequencies) is found to be

$$Z_{in1}(s) \cong (R_1 + R_x)(R_2 + R_w)\left\{sC_3 + sC_z + \frac{1}{R_z}\right\} \tag{5.83}$$

provided the effect of $R_y//(1/sC_y)$ is ignored. The quality factor is given by

$$Q = \omega R_z(C_3 + C_z) \approx \omega R_z C_3 \tag{5.84}$$

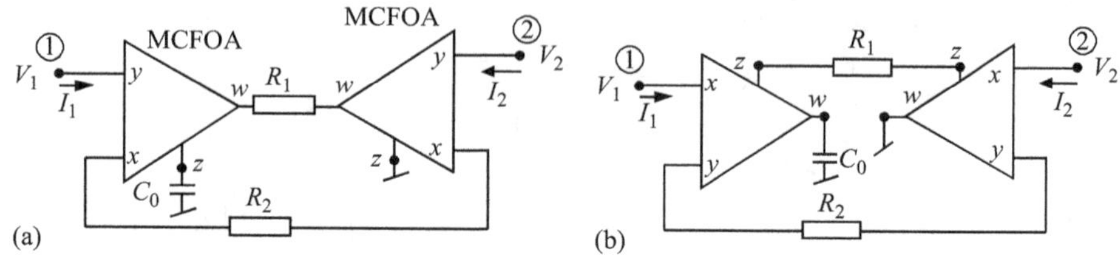

Figure 5.30 Two circuits for the simulation of floating inductors [24], (a) the first proposition, (b) the second proposition

Since $C_z \ll C_3$, for $R_z = 100$ kΩ, $C_3 = 0.1$ nF, Q is found to be 200 at $f = 3.183$ MHz. A nonideal analysis of these circuits reveals that both the circuits are open circuit stable as well as short circuit stable.

Yuce and Minaei [24] also presented two circuits for simulating lossless floating inductance, which are shown in Figure 5.30.

Analysis of both the circuits reveals that they are characterized by the [Y] matrix,

$$[Y] = \frac{1}{sC_0R_1R_2}\begin{bmatrix} 1 & -1 \\ -1 & 1 \end{bmatrix} \tag{5.85}$$

and thus, simulates a lossless floating inductance $L_{eq} = C_0R_1R_2$.

Yuce [22] defined another kind of MCFOA which is a six-port analog building block which is characterized by the following [Y] matrix:

$$\begin{bmatrix} I_y \\ I_{z1} \\ I_{z2} \\ I_{z3} \\ V_x \\ V_w \end{bmatrix} = \begin{bmatrix} 0 & 0 & 0 & 0 & 0 & -\alpha_1 \\ 0 & 0 & 0 & 0 & \alpha_2 & 0 \\ 0 & 0 & 0 & 0 & -\alpha_3 & 0 \\ 0 & 0 & 0 & 0 & 0 & \alpha_4 \\ \beta_1 & 0 & 0 & 0 & 0 & 0 \\ 0 & \beta_2 & 0 & 0 & 0 & 0 \end{bmatrix}\begin{bmatrix} V_y \\ V_{z1} \\ V_{z2} \\ V_{z3} \\ I_x \\ I_w \end{bmatrix} \tag{5.86}$$

Using this version of MCFOA, it has been shown in [22] that a lossless floating inductance can be realized using only one MCFOA along with three passive elements as shown in Figure 5.31.

A routine analysis of both the circuits of Figure 5.31 reveals that both are characterized by the [Y] matrix (assuming MCFOA to be ideal, i.e. $\alpha_i = 1;\ i = 1-4$ and $\beta_j = 1;\ j = 1, 2$):

$$[Y] = \left(\frac{y_1y_2}{y_3}\right)\begin{bmatrix} 1 & -1 \\ -1 & 1 \end{bmatrix} \tag{5.87}$$

Thus, equivalent admittance realized between ports 1 and 2 is given by $y_{eq} = y_1y_2/y_3$.

1. Capacitance multiplier: with y_2 as a capacitor and other admittances as resistors we get $y_{eq}(s) = (sC_2R_3/R_1)$, thereby representing realized floating capacitance as $C_{eq} = (C_2R_3/R_1)$.

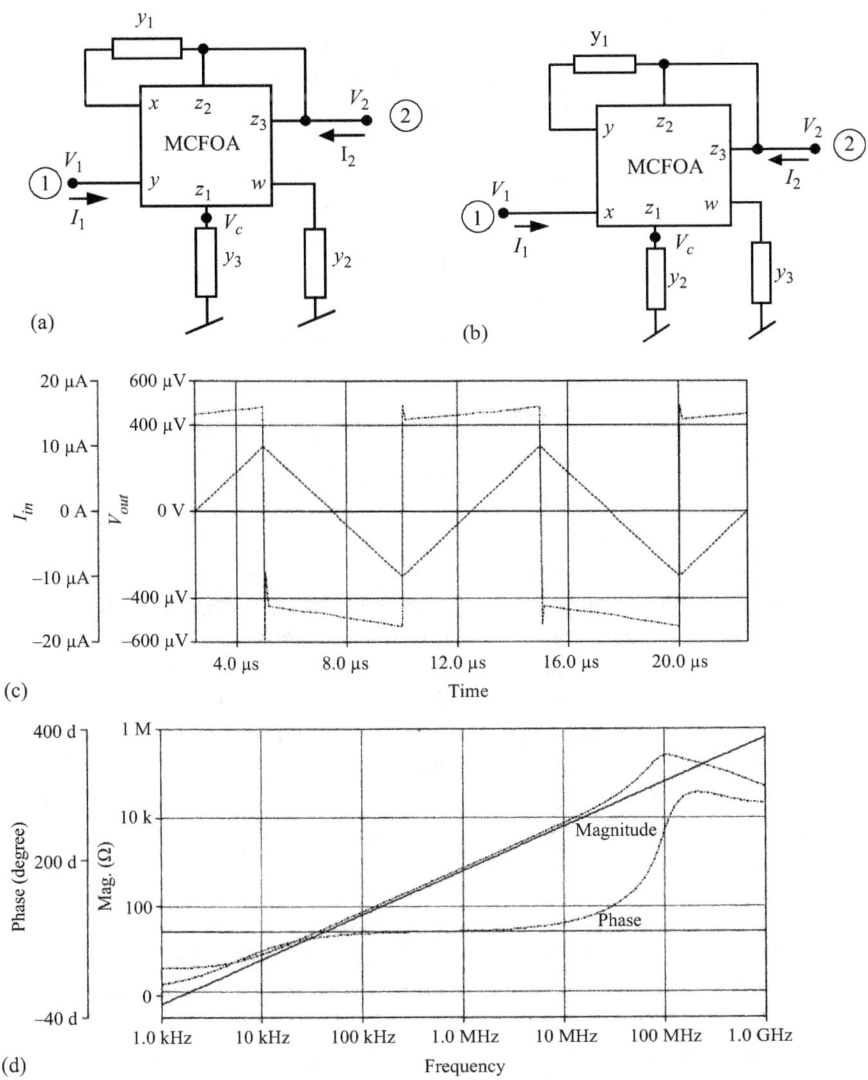

Figure 5.31 Lossless FI realizations using MCFOA: (a) and (b) two FI configurations using MCFOA, (c) transient response with input current, (d) magnitude and phase responses of the impedance. Reprinted with permission from [22] © 2006 Elsevier GmbH

2. Simulated inductor: with $y_3 = sC_3$, $y_1 = (1/R_1)$ and $y_2 = (1/R_2)$, the circuit realizes $Z_{eq}(s) = sC_3R_1R_2$ representing $L_{eq} = C_3R_1R_2$.
3. FDNR realization: this requires y_1 and y_2 as capacitors and y_3 as a resistor, we get $Z_{eq}(s) = (1/s^2C_1C_2R_3)$ representing an FDNR with $D_{eq}(s) = C_1C_2R_3$.

A nonideal analysis of the two circuits taking various parasitic impedances of the MCFOA reveals [22] that both the circuits can satisfactorily realize floating inductance and floating FDNR at higher frequencies. This has been demonstrated in [22] using a CMOS MCFOA using 0.35 μm TSMC CMOS technology parameters. The circuits were simulated in SPICE to determine its time response by applying two triangular wave input currents I_{in} of magnitude 0.01 mA and frequency 100 kHz. As expected, the voltage across inductor was found to be a square wave (see Figure 5.31(c)). In another test for frequency response analysis, both magnitude and phase of the simulated impedance were obtained from SPICE simulations and plotted against ideal ones, which are shown in Figure 5.31(d). These results confirm the workability of these circuits.

5.11 Concluding remarks

This chapter has elaborated a number of prominent CFOA-based circuits developed during the last three decades for the realization of both synthetic inductors and other impedances like FDNRs and capacitive multipliers. The realization of gyrators and grounded/floating synthetic impedances was considered.

A majority of the circuits have been evolved around the commercially available and popular CFOA, namely the AD844. Of course, other varieties of IC CFOAs are available from several IC manufacturers which, in contrast to AD844, do not have external access to the compensation/*TZ* pin and one or two authors have also investigated the use of such CFOAs in impedance simulation, for instance, see [14,21]; however, the results obtained with them are not very encouraging.

Besides the traditional CFOAs, several variants of the CFOAs, namely the inverting CFOA, the CC-CFOA and two types of MCFOAs have also been employed for impedance simulation, and several interesting circuits using these elements have also been included. However, unlike the traditional CFOA AD844, only CFOA− and the MCFOA type I can be implemented using the commercially available CFOAs AD844. Others have only been shown to be realizable using CMOS hardware and the circuits employing them should, therefore, be seen from the perspective of IC implementation in CMOS technology. By comparison, the AD844-based circuits can also be used in discrete applications for which a number of alternatives to choose from have been described in this chapter.

References

[1] Smith K. C., Sedra A. S. 'The current conveyor—a new circuit building block'. Proceedings of the IEEE. 1968; **56**(8):1368–9.

[2] Smith K. C., Sedra A. S. 'Realization of the Chua family of new nonlinear network elements using the current conveyor'. IEEE Transactions on Circuit Theory. 1970; **CT-17**(1):137–9.

[3] Gilbert B., 'About CFA etc.' Private communication with R. Senani, dated March 8, 2015.

[4] Gilbert B. 'The current-mode muddle'. International Journal of Electronics and Communications (AEU). 2018; **83**(1):398–406.

[5] Senani R., Bhaskar D. R., Singh A. K., Singh V. K. 'Current Feedback Operational Amplifiers and Their Applications'. 1st Edition. New York, NY: Springer Science + Business Media; 2013. p. 249.

[6] 'Analog Devices Linear Products Data Book'. Norwood, MA: Analog Devices Inc.; 1990.

[7] Fabre A. 'Gyrator implementation from commercially available transimpedance operational amplifiers'. Electronics Letters. 1992; **28**(3):263–4.

[8] Liu S. I., Hwang Y. S. 'Realisation of R-L and C-D impedances using a current feedback amplifier and its applications'. Electronics Letters. 1994; **30**(5):280–1.

[9] Chang C. M., Hwang C. S., Tu S. H. 'Voltage-mode notch, lowpass and bandpass filter using current-feedback amplifiers'. Electronics Letters. 1994; **30**(24):2022–3.

[10] Chang C. M., Hwang C. S. 'Comment: Voltage-mode notch, lowpass and bandpass filter using current-feedback amplifiers'. Electronics Letters. 1995; **31**(4):246.

[11] Higashimura M. 'Filters and immittances using current-feedback amplifiers'. Proc. 20th International Conference on Microelectronics (MIEL'95); Niš, Serbia, September 12–14, 1995.

[12] Abdullah Al-Walaie S., Alturaig M. A. 'Current mode simulation of lossless floating inductance'. International Journal of Electronics. 1997; **83**(6):825–9.

[13] Soliman A. M. 'Applications of the current feedback operational amplifiers'. Analog Integrated Circuits and Signal Processing. 1996; **11**(3):265–302.

[14] Serrano L., Carlosena A. 'Active RC impedances revisited'. International Journal of Circuit Theory and Applications. 1997; **25**(4):289–305.

[15] Deng J., Aronhime P., Maundy B. 'Simulation of coupled tuned circuits using CFOAs'. Proceedings of the 1998 IEEE International Symposium on Circuits and Systems; Monterey, CA, 31 May–3 June, 1998.

[16] Senani R. 'Realization of a class of analog signal processing/signal generation circuits: novel configurations using current feedback op-amps'. Frequenz. 1998; **52**(9–10):196–206.

[17] Singh A. K., Senani R., Tripathi M. P. 'Low-component-count, high-frequency resonators and their applications, using op-amp compensation poles'. Frequenz. 1999; **53**(7–8):161–9.

[18] Weng R.-M., Lai J.-R., Lee M.-H.'Realization of nth-order series impedance function using only (n-1) current-feedback amplifiers'. International Journal of Electronics. 2000; **87**(1):63–9.

[19] Singh A. K., Senani R. 'Active-R design using CFOA-poles: new resonators, filters, and oscillators'. IEEE Transactions on Circuits and Systems-II: Analog and Digital Signal Processing. 2001; **48**(5):504–11.

[20] Soliman A. M., Awad I. A. 'Fully differential CMOS current feedback operational amplifier'. Analog Integrated Circuits and Signal Processing. 2005; **43**(1):61–9.

[21] Natarajan S. 'Inductance simulation using modern current feedback amplifiers (CFAs)'. Proceedings of the Thirty Seventh South-Eastern Symposium on System Theory 2005; Tuskegee, AL, March 20–22, 2005.

[22] Yuce E. 'On the implementation of the floating simulators employing a single active device'. International Journal of Electronics and Communications (AEU). 2007; **61**(7):453–8.

[23] Yuce E., Minaei S. 'Realisation of various active devices using commercially available AD844s and external resistors'. Electronics World. 2007; **113**(1857):46–9.

[24] Yuce E., Minaei S. 'A modified CFOA and its applications to simulated inductors, capacitance multipliers, and analog filters'. IEEE Transactions on Circuits and Systems-I: Regular Papers. 2008; **55**(1):266–75.

[25] Siripruchyanun M., Chanapromma C., Silapan P., Jaikla W. 'BiCMOS current-controlled current feedback amplifier (CC-CFA) and its applications'. WSEAS Transactions on Electronics. 2008; **6**(5):203–19.

[26] Psychalinos C., Pal K., Vlassis S. 'A floating generalized impedance converter with current feedback amplifiers.' International Journal of Electronics and Communications (AEU). 2008; **62**(2):81–5.

[27] Kacar F., Kuntman H. 'On the realization of the FDNR simulators using only a single current feedback operational amplifier'. International Conference on Electrical and Electronics Engineering-ELECO 2009; Bursa, Turkey, November 5–8, 2009.

[28] Senani R., Bhaskar D. R., Gupta S. S., Singh V. K. 'A configuration for realizing floating, linear, voltage-controlled resistance, inductance and FDNC elements'. International Journal of Circuit Theory and Applications. 2009; **37**(5):709–19.

[29] Yuce E. 'Novel lossless and lossy grounded inductor simulators consisting of a canonical number of components'. Analog Integrated Circuit and Signal Processing. 2009; **59**(1):77–82.

[30] Abuelma'atti M. T. 'Comment on "Novel lossless and lossy grounded inductor simulators consisting of a canonical number of components"'. Analog Integrated Circuit and Signal Processing. 2011; **68**(1):139–41.

[31] Kacar F., Kuntman H. 'CFOA-based lossless and lossy inductance simulators'. Radioengineering. 2011; **20**(3):627–31.

[32] Lahiri A., Gupta M. 'Realizations of grounded negative capacitance using CFOAs'. Circuits, Systems and Signal Processing. 2011; **30**(1):143–55.

[33] Abuelma'atti M. T. 'New grounded immittance function simulators using single current feedback amplifier'. Analog Integrated Circuit and Signal Processing. 2012; **71**(1):95–100.

[34] Yuce E. 'Reply to comment on: "Novel lossless and lossy grounded inductor simulators consisting of a canonical number of components"'. Analog Integrated Circuit and Signal Processing. 2012; **72**(2):505–7.

[35] Senani R., Bhaskar D. R. 'New lossy/loss-less synthetic floating inductance configuration realized with only two CFOAs'. Analog Integrated Circuit and Signal Processing. 2012; **73**(3):981–7.

[36] Bhaskar D. R., Senani R. 'Simulation of a floating inductance: a new two-CFOA-based configuration'. 2013 Fifth International Conference on Computational Intelligence, Modelling and Simulation; Seoul, South Korea, September 24–25, 2013.

[37] Jothimurugan R., Suresh K., Ezhilarasu P. M., Thamilmaran K. 'Improved realization of canonical Chua's circuit with synthetic inductor using current feedback operational amplifiers'. International Journal of Electronics and Communications (AEU). 2014; **68**(5):413–21.

[38] Bhaskar D. R., Senani R. 'Synthetic floating inductors realized with only two current feedback op-amps'. American Journal of Electrical and Electronics Engineering. 2015; **3**(4):88–92.

[39] Abuelma'atti M. T., Dhar S. K. 'New CFOA-based floating lossless negative immittance function emulators'. TENCON 2015-2015 IEEE Region 10 Conference; Macao, China, November 1–4, 2015.

[40] Alpaslan H., Yuce E. 'Inverting CFOA based lossless and lossy grounded inductor simulators'. Circuits, Systems and Signal Processing. 2015; **34** (10):3081–100.

[41] Swamy M. N. S. 'Modified CFOA, its transpose, and applications'. International Journal of Circuit Theory and Applications. 2016; **44**(2):514–26.

[42] Abuelma'atti M. T., Dhar S. K. 'A new CFOA-based low frequency lowpass filter for biomedical applications'. 2016 IEEE EMBS Conference on Biomedical Engineering and Sciences (IECBES); Kuala Lumpur, Malaysia, December 4–8, 2016.

[43] Abuelma'atti M. T., Dhar S. K. 'New CFOA-based floating immittance emulators'. International Journal of Electronics. 2016; **103**(12):1984–97.

[44] Abuelma'atti M. T., Dhar S. K., Khalifa Z. J. 'New two-CFOA-based floating immittance simulators'. Analog Integrated Circuits and Signal Processing. 2017; **91**(3):479–89.

[45] Yuce E., Minaei S. 'Commercially available active device based grounded inductor simulator and universal filter with improved low frequency performances'. Journal of Circuits, Systems, and Computers. 2017; **26** (4):1750052-1–13.

[46] Keskin A. E. 'Single CFA-based NICs with impedance scaling properties'. Journal of Circuits, Systems, and Computers. 2005; **14**(2):195–203.

[47] Mitra S. K. 'Analysis and Synthesis of Linear Active Networks'. London: John Wiley & Sons, Inc; 1968. p. 86–90.

Chapter 6
Applications of FTFN/OFA and OMAs in impedance synthesis

Abstract

This chapter presents an overview of the significant contributions made on the realization of gyrators and various other kinds of impedances with particular emphasis on floating inductance simulation for which a specific form of *nullor* namely the so-called four-terminal-floating-nullor (FTFN) and the operational floating amplifier and somewhat similar appearing element, namely the *operational mirrored amplifiers* (OMAs) have been found to be particularly versatile. Most of the FTFN-based circuits described in this chapter can be practically implemented using commercially available integrated circuit (IC) current feedback op-amps (CFOAs; such as AD844 from Analog Devices Inc.), while the OMA-based circuits can be implemented with commercially available IC op-amps such as μA741 or LF356 in conjunction with mixed transistor arrays such as CA3096/LM3096.

6.1 Introduction

Before the introduction of the *gyrator* as a new network element by Tellegen in 1948, the resistor, the capacitor and the inductor as two-terminal one-port elements and the transformer as the four-terminal two-port element were the only circuit elements used in practical electronic circuit designs and dealt with in the Circuit theory and Network synthesis books. When Tellegen introduced the *gyrator* as the fifth basic passive but non-reciprocal element, the domain of passive linear network elements was supposed to be completed. However, in 1961 Carlin and Youla [1] introduced two *pathological* network elements, namely the *nullator* and *norator* (see Figure 6.1) into the domain of circuit theory in the context of their notion of an *ideal amplifier*. The nullator was defined as a degenerate one-port network element characterized by the terminal equations $v = 0$ and $i = 0$, while the norator was defined as another degenerate element having the characterization $v =$ arbitrary, $i =$ arbitrary. The unifying role of nullors in representing a variety of network elements and active circuit building blocks and their use in generating equivalent realizations from a given one have both been well understood in the literature, have been put to use by many researchers from time to time and have been well

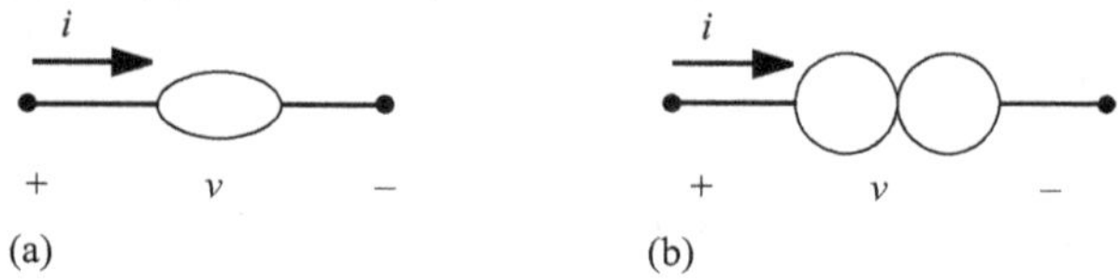

Figure 6.1 The two pathological elements [1] (a) the nullator and (b) the norator

documented in analog circuit literature (see [1–109] and the references cited therein).

In this chapter, we shall focus on the use of nullors in the realization of gyrators and other related impedances and will demonstrate that a fully floating version of the nullor known as four-terminal-floating-nullor (FTFN), the operational floating amplifier (OFA) and the closely related OMAs have been particularly useful in the realization of floating impedances of various kinds.

6.2 An overview of nullors, FTFN/OFA and OMAs

It is obvious from the formal definitions of the nullator and norator that none of these elements is physically realizable in isolated forms by any physical devices. When these elements were first introduced, they were welcomed as well as opposed! In fact, Tellegen was the most vociferous critic of the very basic idea of these elements and argued that these elements are merely mathematical abstractions with no physical realizations and no real practical applications [8].

Soon, it was demonstrated [17] that the transmission matrix of an ideal op-amp with infinite input impedance, infinite voltage gain and zero output impedance is a *null* matrix. If we consider the equivalent model of a non-ideal op-amp (see Figure 6.2), the transmission matrix is found to be

$$\begin{bmatrix} V_1 \\ I_1 \end{bmatrix} = \begin{bmatrix} \dfrac{1}{A} & \dfrac{R_0}{A} \\ \dfrac{1}{AR_{in}} & \dfrac{R_0}{AR_{in}} \end{bmatrix} \begin{bmatrix} V_2 \\ -I_2 \end{bmatrix} \tag{6.1}$$

With ideal values of these parameters, i.e. $R_{in} = \infty$, $R_0 = 0$ and $A = \infty$, the above transmission matrix $[T]$ reduces to the following:

$$[T] = \begin{bmatrix} 0 & 0 \\ 0 & 0 \end{bmatrix} \tag{6.2}$$

which is a null matrix. The transmission matrix of the ideal op-amp, thus, implies that the differential input voltage and input current of an ideal op-amp are both zero regardless of the output voltage and output current, both of which could assume any arbitrary values. Therefore, the input terminals of an ideal op-amp behave like a

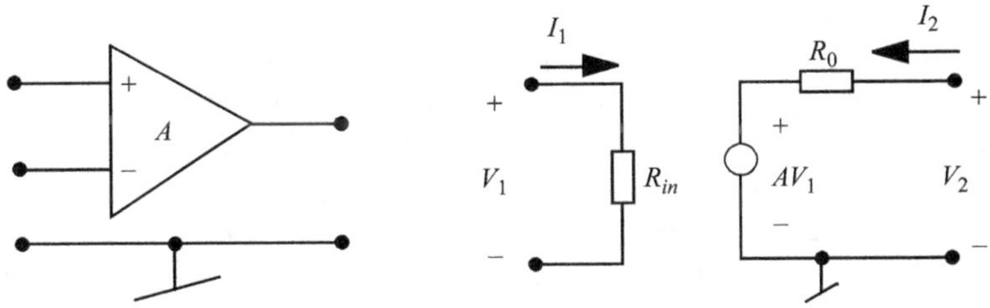

Figure 6.2 The non-ideal op-amp and its equivalent circuit [17]

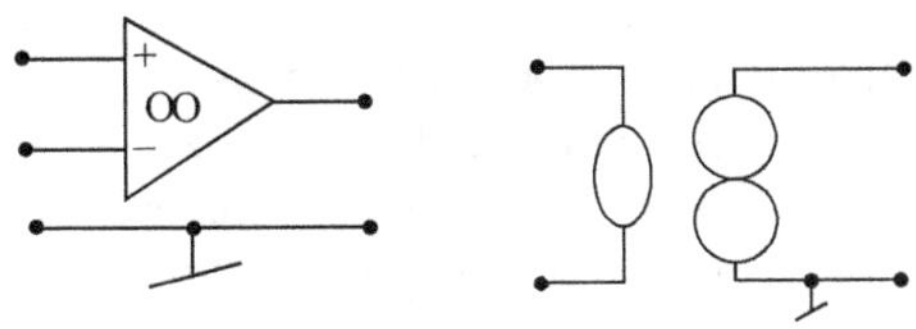

Figure 6.3 Ideal op-amp as a nullor

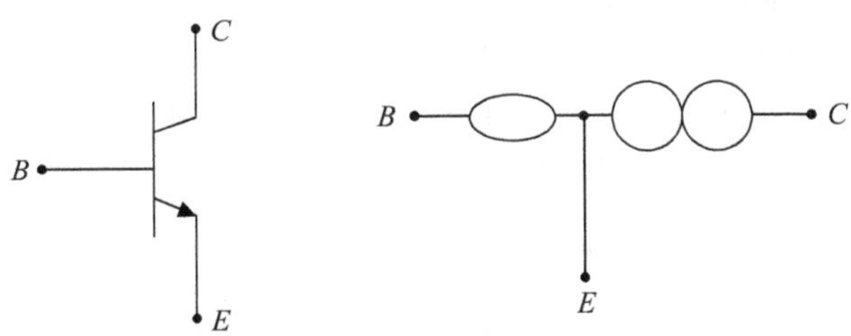

Figure 6.4 The equivalence of a BJT as a three-terminal nullor [3]

nullator while the output terminal *with respect to the ground* behaves as a norator. Hence, an ideal op-amp is a nullor as shown in Figure 6.3.[1]

Subsequently, it was proved [3] that an ideal bipolar junction transistor (BJT) is equivalent to a three-terminal nullor – having a nullator and norator connected in series with their junction forming the third terminal such that the free end of the nullator represents the base terminal, the free end of the norator representing the collector terminal and the junction representing the emitter terminal (see Figure 6.4).

After these concepts were well understood, the maximum use of the nullator and norator elements happened in deriving new equivalent circuits of the transistorized gyrators as well as op-amp gyrators. Recalling that it was Tellegen who had opposed the nullator and norator by labelling them as practically useless and abstract mathematical entities only with no practical use, it was *ironic* that these

[1]Antoniou, A. 'New gyrator circuits obtained by using nullors'. Electronics Letters. 1968; **4**(5):87–8.

elements, on the contrary, were shown to be quite useful in deriving new circuits for the *gyrator* – an element which was invented by none other than Tellegen!

As a further development, nullator and norators were also found to be quite useful in representing and devising a variety of circuit models (nullor models) of the four controlled sources, namely the voltage-controlled-voltage-source (VCVS), voltage-controlled-current-source (VCCS), current-controlled-current-source (CCCS) and the current-controlled-voltage-source (CCVS), some of which are summarized in Figure 6.5 (for others, the readers are referred to [17,30]).

It is not difficult to comprehend that, like the BJT, an ideal MOSFET with infinite g_m and infinite r_o can also be represented as a three-terminal nullor. Furthermore, an OTA with a finite transconductance gain g_m can be represented by two pairs of nullators and norators along with a resistance R taken equal to $1/g_m$.

An important result in this contest is that an ideal controlled source (ideal in the sense of R_{in}, R_{out} and the gain) is exactly equivalent to a nullor (a pair of nullator and norator). Thus, an ideal OTA (with infinite R_{in}, infinite R_{out} and infinite transconductance gain g_m) can also be represented by a pair of nullator and norator. As an alternative to the rigorous mathematical proof, this can be more intuitively proved using the nullor models of the controlled sources. Figure 6.7 shows some nullor identities which are useful in the analysis and understanding of nullor-RC networks.

We can now take note of the following:

1. In the four controlled sources of Figure 6.5, if we make the concerned gain as infinite, it can be easily proved would that all of them converge to a pair of nullator or norator.
2. In the OTA models of Figure 6.6, if the transconductance gain is made infinite (i.e. the resistor R shorted), then using the identities of Figure 6.7 it can be easily checked that both the nullor model of the OTA would reduce to a pair of nullator and norator.

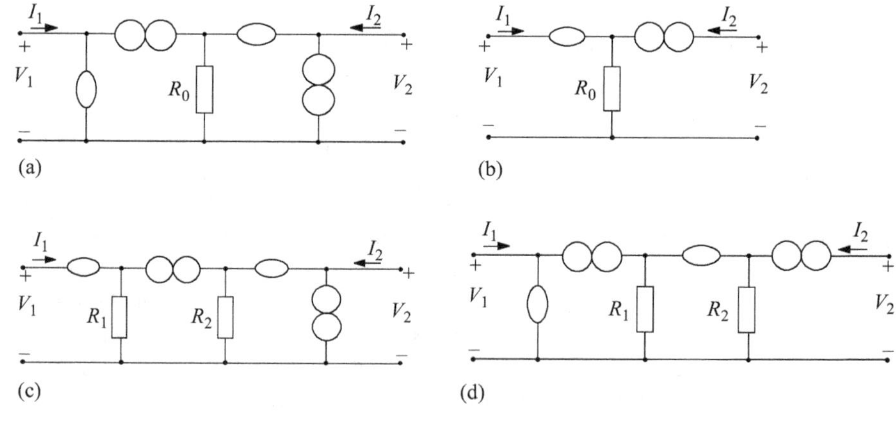

Figure 6.5 Some exemplary nullor models of the four controlled sources: (a) CCVS, (b) VCCS, (c) VCVS and (d) CCCS

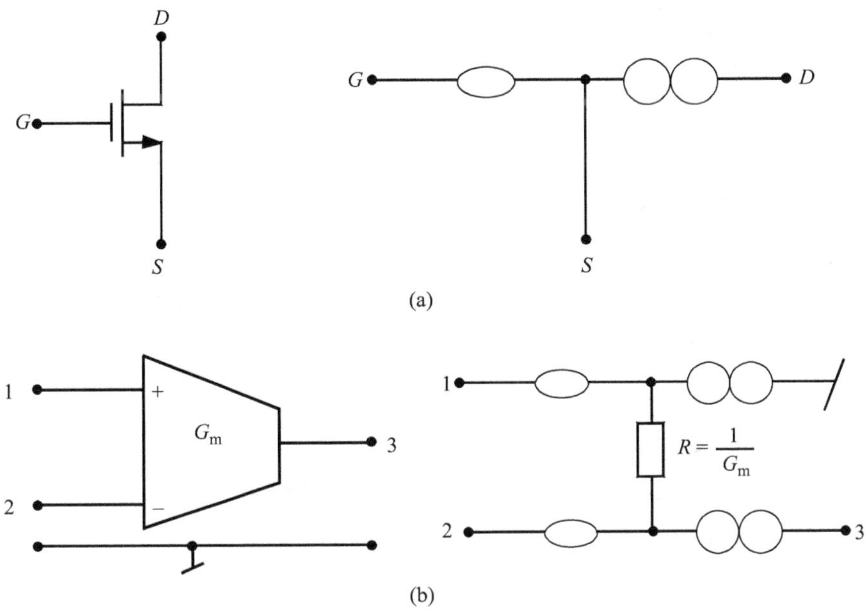

Figure 6.6 Nullor models of the (a) MOSFET and (b) the OTA

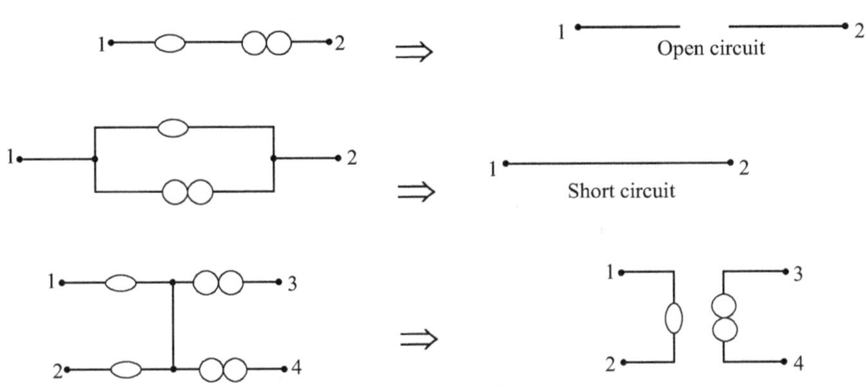

Figure 6.7 Some nullor identities useful for simplifying networks containing nullators and norators

To sum up, it is now well understood that an ideal op-amp, an ideal BJT/MOSFET, an OTA, a voltage follower, a Current follower, a CCII−, a CCII+, a negative impedance converter and a generalized impedance converter, all can be represented by nullor models (see Figure 6.8).

The real impetus came to this area with the introduction of the three-terminal monolithic nullor in [28] and, subsequently, the introduction of the OFA by Huijsing in [46], and many similar ideas by a number of other researchers such as

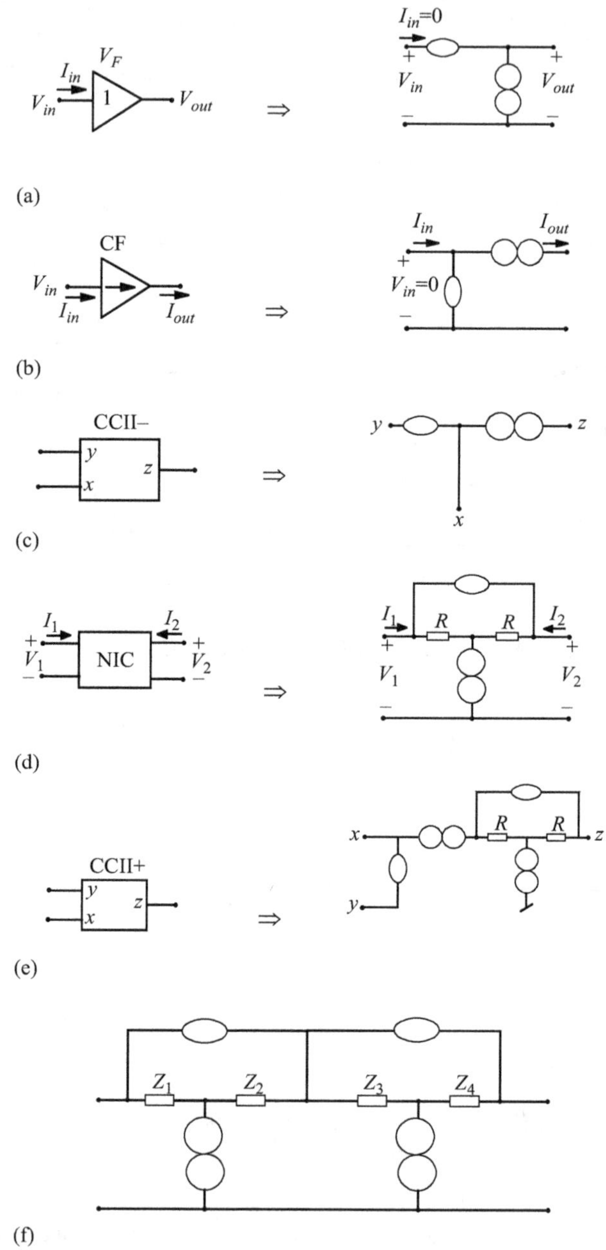

Figure 6.8 Nullor models of some other active elements: (a) VF, (b) CF, (c) CCII−, (d) NIC, (e) CCII+ and (f) GIC

Nordholt [31], Stevenson [33] and Senani [38,39] which lead to the notion of a *FTFN* ([38,39] where the acronym FTFN appears to have been explicitly coined for the first time). Thus, the FTFN is a four-terminal element (with the ground being the fifth external terminal) which is symbolically shown in Figure 6.9 and is characterized by the terminal equations: $i_y = 0$, $i_x = 0$, $v_y = v_x$; $i_w = -i_z = arbitrary$, $v_w = arbitrary$, $v_z = arbitrary$.

An FTFN can be realized in a number of ways. Stevenson presented an entirely op-amp version in [33], while Senani proposed an op-amp-OTA based circuit for the same in [39]; these are shown in Figure 6.10.

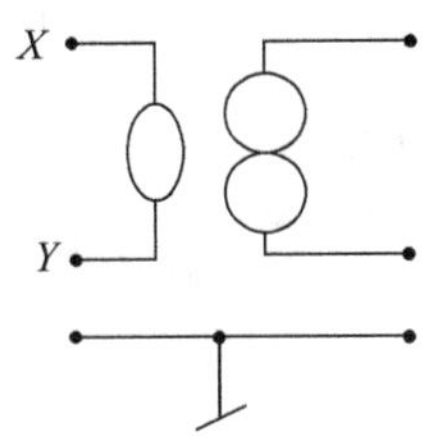

Figure 6.9 The four-terminal-floating-nullor

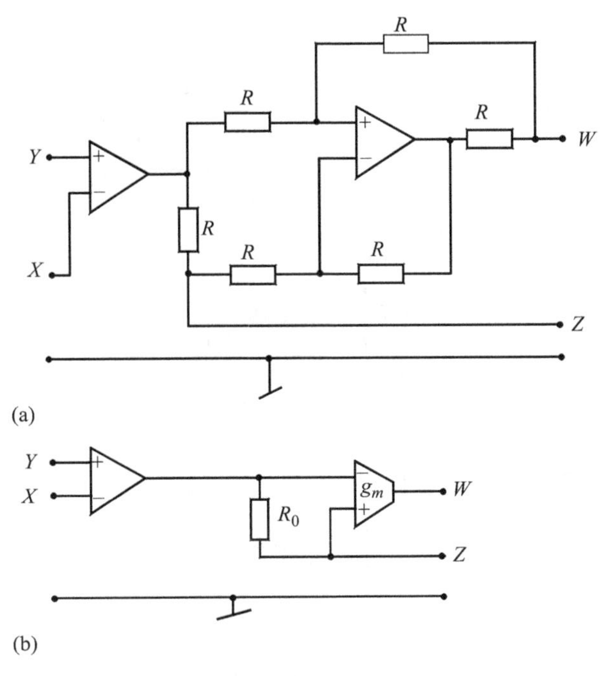

Figure 6.10 FTFN implementation using commercially available ICs: (a) an entirely op-amp-based implementation [33] and (b) an op-amp-OTA-based implementation [39]

It was pointed out in a footnote of [39] that an FTFN can be realized by a composite connection of two current conveyors, and this has been widely recognized and utilized in the published literature, for instance see [73,74]. Thus, an FTFN can be realized using two CCII− elements, or even using two CCII+ elements as well as two commercially available integrated circuit (IC) current feedback op-amps (CFOAs) such as AD844 (see Figure 6.11). In fact, the FTFN has been found to be a quite versatile element, particularly in the synthesis of floating impedances and was once considered a potential candidate for being called a *universal* active element [51,54,58,60,68,102].

Two other circuit elements which are closely related to the FTFN are the so-called OMAs [35] which have two distinct varieties, namely the OMA+ and the OMA− (see Figure 6.12).

Consider first the OMA+. Assuming that the op-amp employed is of a type which does not have any components from the internal circuit to the external world except the two input terminals, two DC power supply terminals and the output

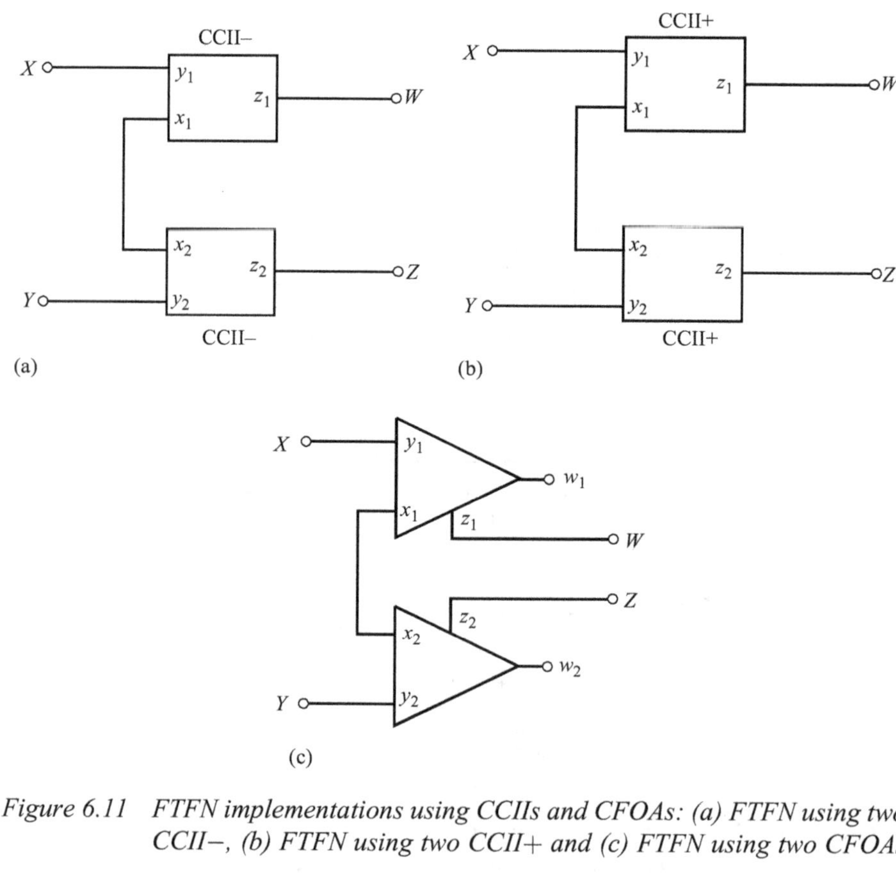

Figure 6.11 FTFN implementations using CCIIs and CFOAs: (a) FTFN using two CCII−, (b) FTFN using two CCII+ and (c) FTFN using two CFOAs

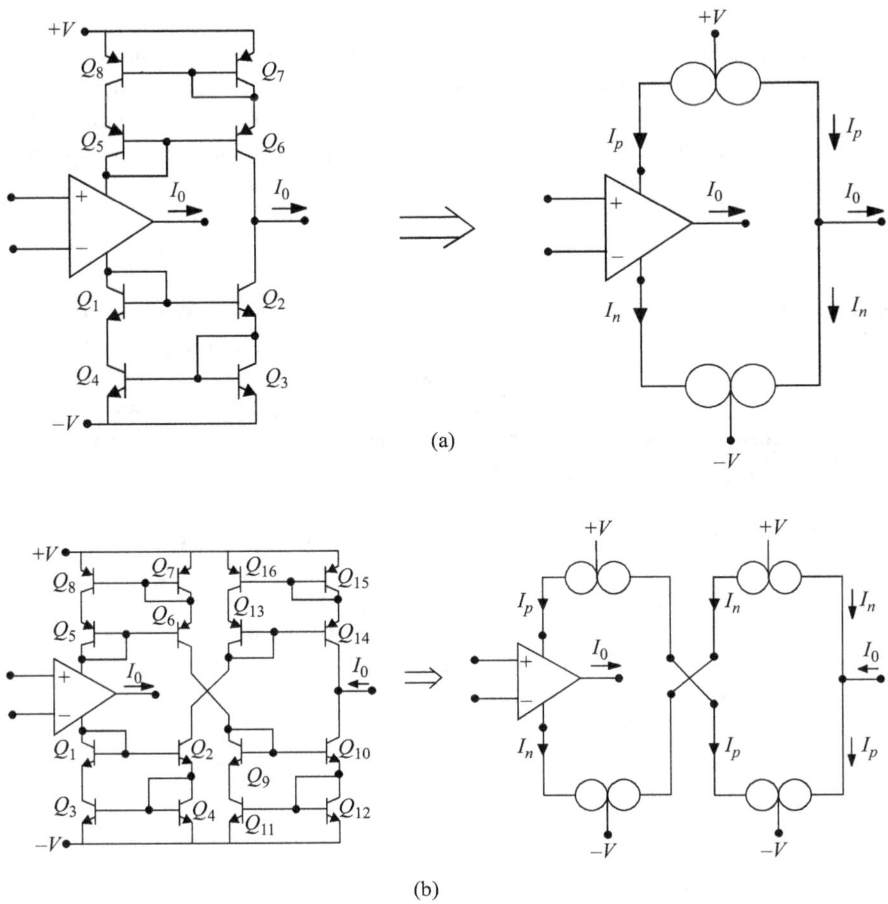

Figure 6.12 Operational mirrored amplifier (OMA) formulations [35]: (a) the OMA+ and (b) the OMA−

terminal. It, therefore, follows that the total current going into the op-amp should be equal to the total current going out. Therefore, the five currents in the terminals mentioned with I_1, I_2 as the input currents are related by the following equation:

$$I_1 + I_2 + I_p = I_n + I_0 \tag{6.3}$$

Since the two input currents are almost zero, the output current is, therefore, given by $I_0 = I_p - I_n$. Now, if the power supply currents I_p and I_n are sensed by a *pnp* current mirror and *npn* current mirror, respectively, and if the output terminal of these mirrors are joined to create a current output terminal, it is easy to see that the output current going out of this terminal will also be equal to I_0. This formulation is referred to as an OMA. In this formulation, however, it is preferred to use either a modified Wilson current mirror or a four transistor cascode current mirror because of two reasons: (i) both these two alternative four-transistor current mirrors have reduced error in the current gain as compared to the two-transistor simple current mirror and (ii) both these mirrors have higher output impedance of the order of $\beta r_o/2$ as compared to the output resistance of the simple current mirror which is r_o only. Furthermore, it should be observed that by inserting two current mirrors in the power supply lines, the basic nature of the DC bias of the op-amp does not change; the only thing different is that the DC bias voltage, instead of being $\pm V_{cc}$, now becomes slightly reduced (by an amount equal to $2V_{BE}$). This OMA formulation is normally called the OMA+. Thus, an OMA+ has a low output impedance voltage terminal and another high output impedance current output terminal with the property that either the current flowing *out* of the voltage terminal and that going *out* of the current output terminal both are equal or conversely, the current going *into* the voltage output terminal and the one going *into* the current output terminal are equal.

A different variant of the OMA, known as OMA−, is obtained by using one more pair of current mirrors (as shown in Figure 6.12(b)), thereby reversing the direction of I_p and I_n. In this case, the output current going out of the current terminal is negative of the current going out of the voltage output terminal. In other words, whatever current enters the voltage output terminal, the same is current is retrieved and flows out of the current output terminal.

A number of authors have defined a variant of the FTFN which they called PFTFN or FTFN$^+$ characterized by the following equations:

$$\left.\begin{aligned} &I_1 = 0, \quad I_2 = 0, \quad V_1 - V_2 = 0, \\ &I_3 = I_4 = arbitrary, \quad V_3 - V_4 = arbitrary \end{aligned}\right\} \tag{6.4}$$

The initial definition of this PFTFN or FTFN$^+$ was based upon the conception of a norator$^+$ [41]. It may be pointed out that even at that time, the first author of this monograph had questioned in [40] that while a normal *norator* is a two-terminal network element characterized by $V =$ arbitrary and $I =$ arbitrary, it is not clear how the *norator*$^+$ could be similarly defined as a two-terminal element? Later on, it was shown in [94] (and later in [102]) that a PFTFN is actually not a *floating* element at all and should be better called an OMA+.

On the other hand, the properties of an OMA− and those of FTFN or the OFA are exactly the same. It may, however, be noted that although the two output terminals of an ideal FTFN should have identical properties, the one made from OMA− normally has the two output terminals having different kinds of output impedances (low output impedance looking into the W-terminal but high output impedance looking into the Z-terminal). However, with some modification in the circuit, an FTFN with two symmetrical output terminals can also be made (see Figure 6.13), although to the best knowledge of the authors, such a modified FTFN, based on an OMA, has never been perceived explicitly or utilized by anybody in the literature so far.

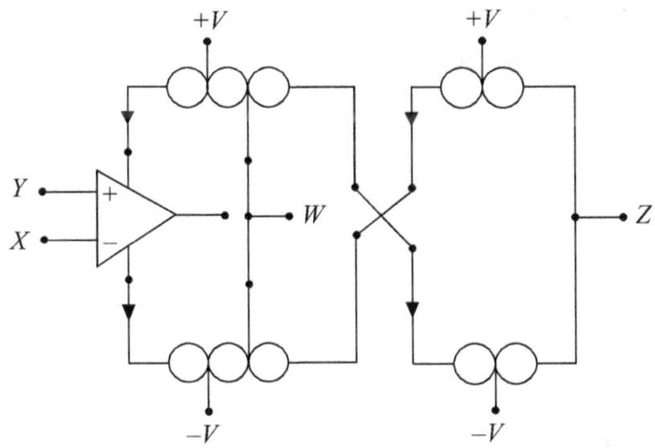

Figure 6.13 FTFN with two symmetrical outputs made from OMA technique

6.3 Generation of FTFN-based floating immittances

Simulated inductors are important elements in a number of methods of designing both active filters and sinusoidal oscillators. Undoubtedly, one of the most well-recognized and widely utilized inductance simulation circuits in many applications continues to be the Antoniou's generalized impedance converter (GIC) [19] based grounded lossless inductance simulation circuit. However, a floating inductance simulation circuit using the same GIC requires double the number of active and passive components requiring a perfect matching between the two pairs therein and, hence, is not practically appealing. Before the advent of FTFN, because of the requirement of a larger number of component as well as component-matching requirements, the problem of synthesizing floating immittances was generally regarded to be a relatively more difficult than their grounded counterparts. All circuits/techniques known earlier required two identical op-amp RC sub-circuits in some form or the other which lead to the requirement of double the number of op-amps and RC elements (see the relevant references in Chapter 2 of this monograph) or required a large number of resistors with matching constraints and/or cancellation conditions when only two op-amps were employed.[2]

It was shown by Senani in [39] that FTFNs provide a novel solution to floating inductor (FI) simulation overcoming the quoted limitations and difficulties. In this section, we present a number of techniques of realizing floating impedances of various kinds using the concept of nullors/FTFNs. Due to this reason, in this chapter, we would be mainly focusing on the circuits and techniques of employing FTFN for realizing FI circuits only. In fact, FTFNs have been used to realize grounded impedances of the various kinds only occasionally, such as in [59].

[2]A classic example of such a realization is the circuits of The–Yanagisawa [103] for lossless FI simulation.

6.3.1 Realization of floating generalized impedance converters/inverters

We now present a novel method of realizing an floating impedance from a given op-amp-based grounded impedance circuit, based upon the following theorem [38,102].

Theorem 6.1: *Suppose that a one-port active network* N, *containing grounded and floating, two-terminal passive elements, nullators and norators realizes a grounded driving point impedance (GDPI)* $Y_{11}(s) = y(s)$. *If a two-port network* N^* *be formulated from* N *by disconnecting from ground all grounded passive elements and terminals (of grounded nullators and norators) and joining them together to form port 2, then* N^* *will have short circuit admittance matrix* $[Y]^* = [Y]$ *(with* $i_1 = -i_2 = y(s)\,[v_1 - v_2]$*).*

The well-known Antoniou's GIC is shown in Figure 6.14(a), while in Figure 6.14(b), we show its nullor model. In Figure 6.14(c), we show the two-port obtained by *un-grounding* both the norators as well as the impedance Z_5. Now from Theorem 6.1 it follows that the circuit of Figure 6.14(c) and its FTFN implementation of Figure 6.14(d) both would realize the same impedance in floating form as the grounded impedance realized by the GIC-based circuit of Figure 6.14(a).

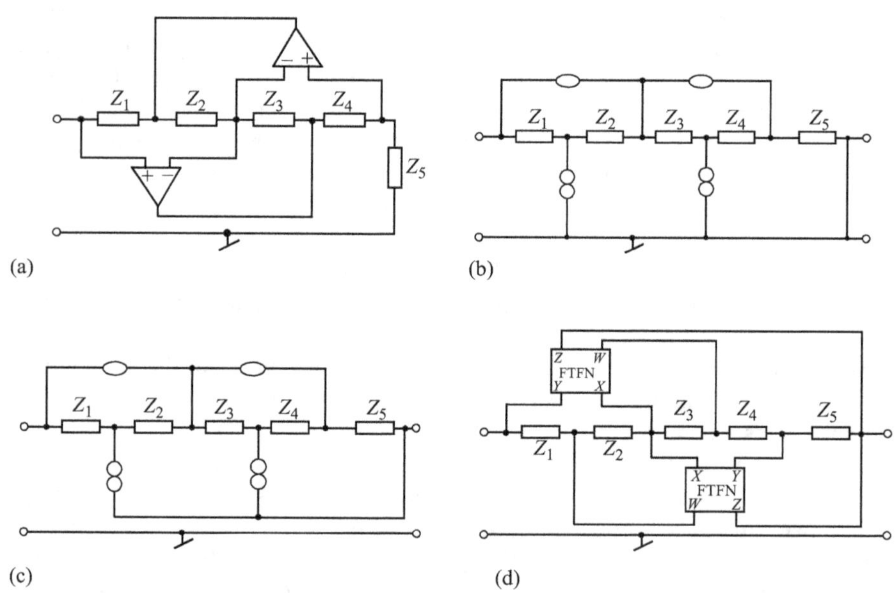

Figure 6.14 Conversion of GIC-based grounded impedance into FTFN-based floating impedance: (a) Antoniou's GIC-based grounded impedance [19], (b) the nullor model of the circuit of Figure 14(a), (c) a floating version of the GIC resulting from the application of Theorem 6.1 [39] and (d) an FTFN-based floating generalized impedance as per the method first introduced explicitly in 1987 in [38,39]

Thus, the two-port networks of Figure 6.14(c) and (d) both are characterized by the following [*Y*]-matrix:

$$[Y] = \frac{Z_2 Z_4}{Z_1 Z_3 Z_5} \begin{bmatrix} 1 & -1 \\ -1 & 1 \end{bmatrix} \tag{6.5}$$

which represents a floating impedance between ports 1 and 2 having the value $Z(s) = Z_1 Z_3 Z_5 / Z_2 Z_4$.

An alternative floating structure of GIC-based generalized floating impedance proposed by Cabeza and Carlosena in 1997 [58] is shown here in Figure 6.15(a), which also realizes a floating impedance of the same value. Yet another circuit having the same characterization was proposed by Cam *et al.* in 2001 [71] (also see [72]) as shown here in Figure 6.15(b).

It is easy to figure out that the configuration of Figure 6.14(d) as well as those of Figure 6.15, all the three can be used to realize a number of useful floating impedances by appropriate selection of the five impedances in these circuits. In fact, all the three quoted circuits can be considered to be generalized positive impedance converters (GPICs)/generalized positive impedance inverters (GPIIs). The three most useful and commonly needed circuit elements, from the point of view of various active network synthesis applications, are the synthetic floating *L*, floating FDNR (an element having $Z(s) = 1/Ds^2$) and floating FDNC (an element

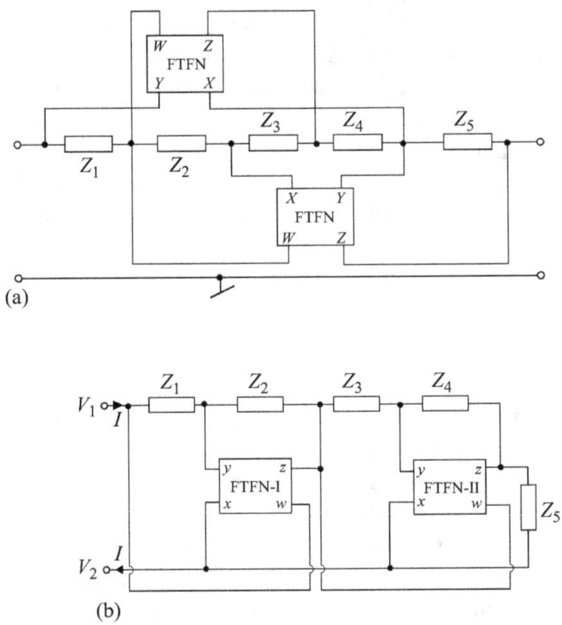

Figure 6.15 Two alternative floating GICs: (a) an alternative generalized floating impedance using FTFNs [58] and (b) another FTFN-based generalized floating impedance simulation [71]

having $Z(s) = Ms^2$). With these GPIC/GPII elements, these are realizable more efficiently than the op-amp-based simulations previously known, as follows:

1. Floating inductance (FI): choosing Z_2 as a capacitor with the remaining impedances chosen as resistors, the simulated L is given by

$$L_{1-2} = \frac{C_2 R_1 R_3 R_5}{R_4} \tag{6.6}$$

2. Floating FDNR: choosing Z_1 and Z_5 as capacitors, with the remaining impedances selected as resistors, the circuit simulates a floating FDNR having value

$$D_{1-2} = \frac{C_1 C_5 R_2 R_4}{R_3} \tag{6.7}$$

3. Floating FDNC: choosing Z_2 and Z_4 as capacitors with all other impedances taken as resistors, the circuit simulates a floating FDNC having value

$$M_{1-2} = R_1 R_3 R_5 C_2 C_4 \tag{6.8}$$

It may be noted that the use of the networks of Figures 6.14(d) and 6.15(a) and (b) makes it possible to realize these elements with only two FTFNs, while providing the following merits simultaneously: (i) reduced number of active and passive elements, (ii) complete lack of any passive component-matching requirements, (iii) low sensitivities to element value changes and (iv) single-resistance-tunability of the realized immittances in all the cases.

Finally, in the context of the material presented in this section, it may be pointed out that Palomera-Garcia in [79] has presented an FTFN re-location techniques by which as many as nine distinctly different lossless FI circuits based upon the GIC have been derived from each of which, as many as 18 circuits can be further evolved by invoking nullator–norator pairings. However, to the best knowledge of the authors, any practical evaluation of these 72 possible lossless FI circuits, with a view to find the best ones, has never been carried out in open literature yet and, thus, constitutes an interesting problem for research.

6.3.2 Generation of lossless FI circuits using a single FTFN

Till date, *only* four single-op-amp circuits have been published in the open literature which can realize lossless grounded inductance.[3] However, all these circuits have a capacitor which is floating. Thus, by deleting this capacitor although these circuits would be all active gyrators but such –gyrators have one port (the input port) grounded but the other port (the output port) floating. Obviously, these single-op-amp active gyrators cannot be used as two-port grounded gyrators, but nevertheless, these circuits are interesting and important in their own right at least in applications which require economical lossless GIs realizable with a single op-amp.

[3]These four circuits have been proposed by Orchard–Wilson in 1974 [104], Schmidt–Lee in 1975 [105], Ramsey in 1978 [106] and Horn–Moschytz in 1979 [107].

In the following, it will be shown that all these four circuits can be transformed as per the techniques of [38,39] into as many as *16*[*] lossless floating inductance circuits using Theorem 6.1 in conjunction with Theorem 6.2 (to be presented subsequently).

Consider the grounded inductance simulation circuit of Horn–Moschytz [107] reproduced here in Figure 6.16(a) which realizes a lossless grounded inductance of value

$$\mathrm{L} = CR_3R_6\left(1+\frac{R_2}{R_1}\right)\left(1+\frac{R_5}{R_4}\right)^{-1} \text{provided } \frac{R_2}{R_4} = \frac{R_1+R_6}{R_5} + \frac{R_1}{R_3}\left(1+\frac{R_6}{R_5}\right) \tag{6.9}$$

The process of deriving an FTFN based floating inductance circuit corresponding to this circuit is shown in Figure 6.16(b)–(c) and follows the same sequence of steps as explained in the previous sub-section.

We will now elaborate as to how, starting from the FI circuit of Figure 6.16(c), three additional FI circuits can be derived:

1. An equivalent of the circuit of Figure 6.16(c) can be generated by invoking the following theorem.

Theorem 6.2: *Let there be a two-port active-RC network* N_1 *containing floating two-terminal passive elements, floating nullators and floating norators has short circuit admittance matrix:*

$$[Y_1] = y(s)\begin{bmatrix} 1 & -1 \\ -1 & 1 \end{bmatrix} \tag{6.10}$$

If N_1 *is transformed into another network* N_2 *by interchanging all nullators and norators, the transformed network* N_2 *has the same* Y*-matrix.*

Thus, if we apply Theorem 6.2 to the FI circuit of Figure 6.16(c), the resulting second equivalent FI turns out to be as shown in Figure 6.16(d).

2. Since the two-port circuit of Figure 6.16(c) realizes a lossless inductance employing a single floating capacitor, the circuit can be looked upon as a floating gyrator with terminals 1 and 2 constituting the input port and terminals 3 and 4 (between which the capacitor is connected) constituting the output port. Thus, gyrator action is available even if the input and the output ports are transposed, i.e. the capacitor is connected between terminals 1 and 2 and the floating impedance is looked between 3 and 4. This results in the third equivalent FI shown in Figure 6.16(e).
3. Finally, the fourth and the last additional equivalent FI of Figure 6.16(f) is obtained by swapping the nullator and norator in the circuit of Figure 6.16(e) in accordance with Theorem 6.2.

Lastly, it must be pointed out that twelve additional lossless FI circuits, each employing only a single FTFN, are similarly realizable starting from the lossless grounded inductance circuits of Orchard–Wilson [104], Schmidt–Lee [105] and Ramsey [106] by following the above-outlined procedure.

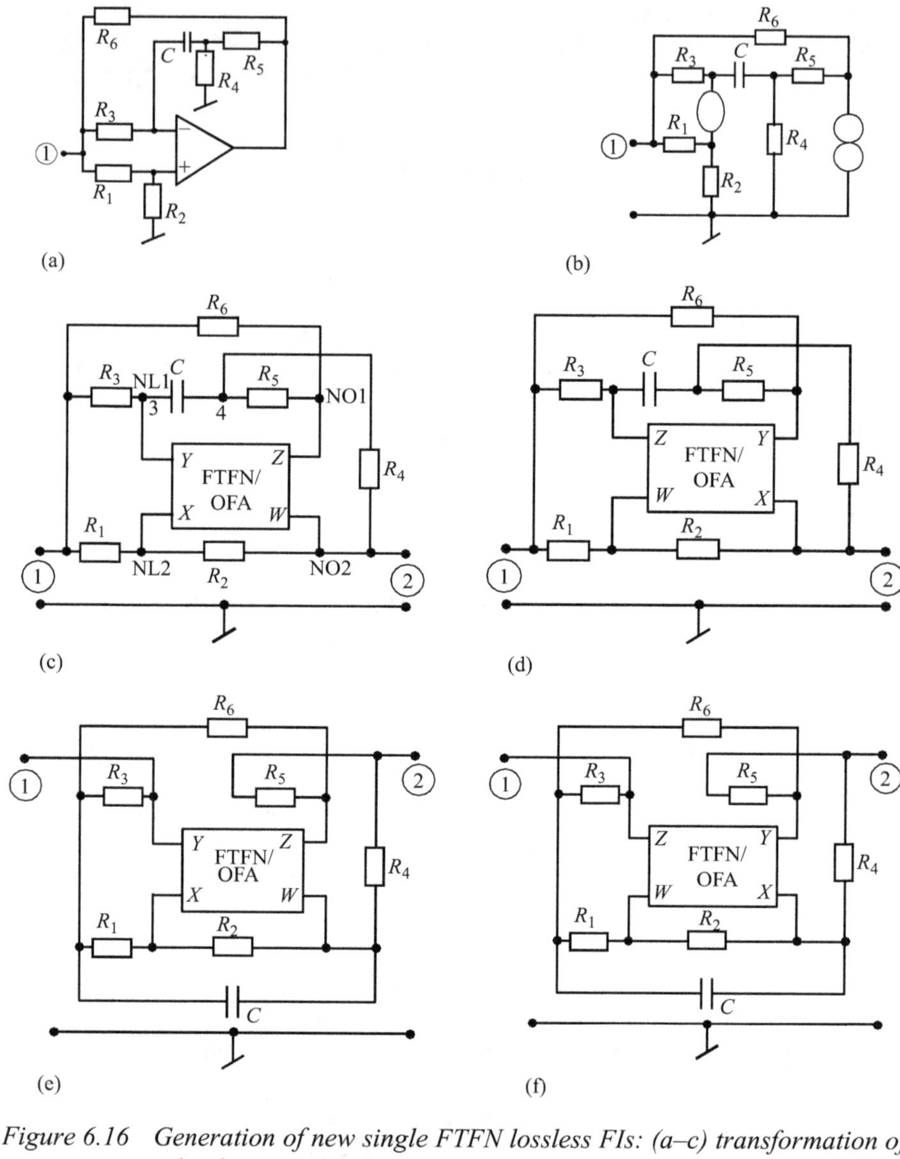

Figure 6.16 Generation of new single FTFN lossless FIs: (a–c) transformation of a lossless grounded inductor into an FI using an FTFN or OFA [38,39] and (d–f) three other equivalent FIs having the same value of realized inductance under the same condition of realization

6.3.3 Single-resistance-tunable lossy FI simulation using only a single FTFN

The techniques explained in the previous sub-sections suggest that a large number of floating impedance circuits can be evolved starting from various known grounded impedance simulation circuits. Figure 6.17 shows the conversion of a single-op-amp simulated GI into a floating inductance circuit. Starting from the grounded inductance circuit of [108], which simulates a single-resistance-tunable lossy GI, we first draw its nullor model. From this nullor model, we then create a two-port by un-grounding all grounded elements/terminals where from finally,

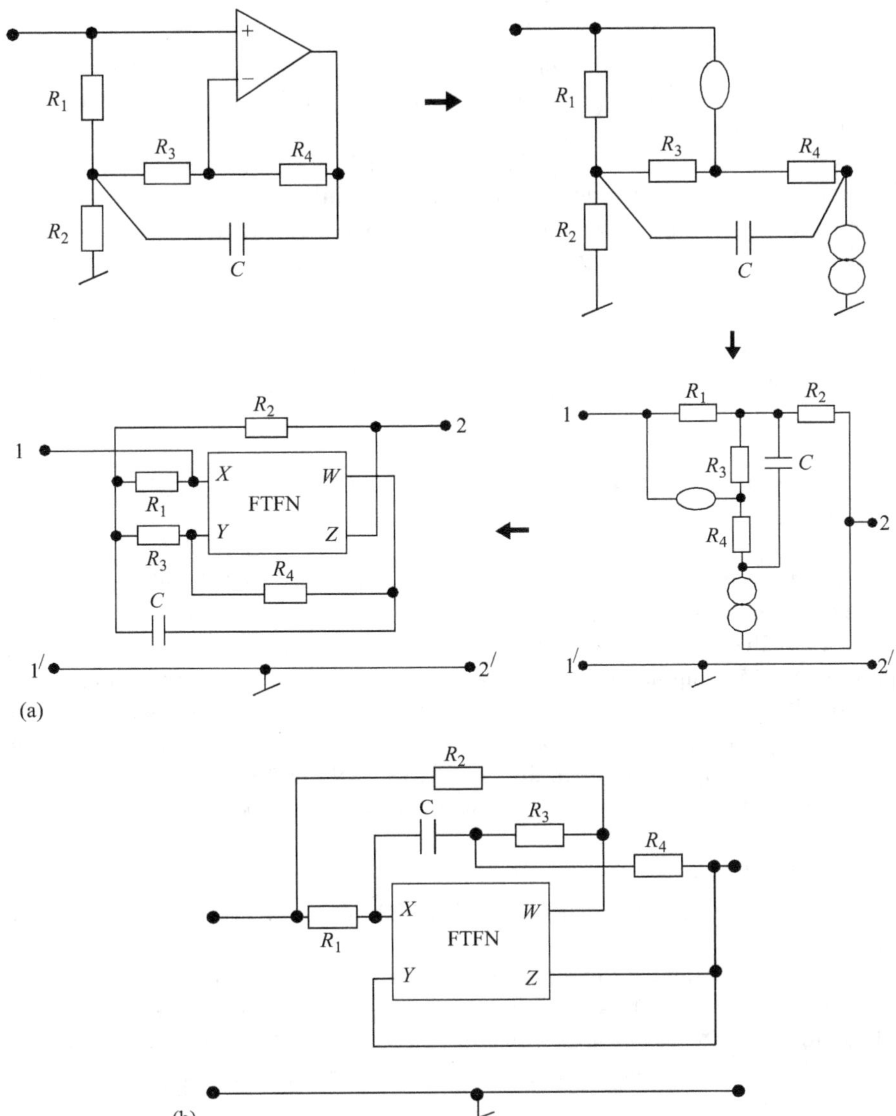

Figure 6.17 Conversion of an op-amp-based GI into FTFN-based FIs: (a) floating series RL and (b) floating parallel RL

a circuit simulating the same kind of impedance in floating form using a single FTFN is derivable as shown in Figure 6.17.

Thus, starting GI circuit as well as the finally derived single-FTFN based FI circuit both simulate a series RL impedance with equivalent resistance and inductance values given by

$$R_s = R_1 + R_2 + \frac{R_1 R_2}{R_3}; L_s = CR_1 R_2 \left(1 + \frac{R_4}{R_3}\right) \tag{6.11}$$

From the above, it is seen that in both GI as well as the derived FI, the simulated inductance L_s is independently adjustable by R_4 which does not appear in the expression for R_s.

It is interesting to mention that a single FTFN-based floating parallel RL impedance possessing single-resistance controllability of inductance value is similarly derivable from the tunable grounded parallel RL impedance circuit of [108] which is shown in Figure 6.17(d) for which equivalent resistance R_p and inductance L_p are given by

$$\frac{1}{R_p} = \left(\frac{1}{R_1} + \frac{1}{R_2} + \frac{R_3}{R_1 R_2}\right); \quad L_p = (CR_1 R_2)\left(1 + \frac{R_3}{R_4}\right)^{-1} \tag{6.12}$$

6.4 Operational mirrored amplifiers (OMA)–based simulators

From the discussion of the previous section, we note that an IC FTFN as an off-the-shelf component has not yet been produced by any IC manufacturer, though it can be physically realized using two commercially available AD844 CFOA ICs for discrete applications. In a similar way, an OMA too can be realized from commercially available IC op-amps of the type uA741 or LF356 in conjunction with CA3096/LM3096 type mixed transistor arrays and that is how OMA-based circuits can be put to use in discrete applications. In this section, we will discuss how OMAs could be usefully employed in impedance synthesis-particularly the *floating* ones.

6.4.1 OMA-based floating GIC

Using OMA one can construct a differential VCCS (DVCCS) as shown in Figure 6.18.[4]

In this circuit, the two op-amps are both configured as OMA+. It may also be noted that within the OMA+ the internal op-amp is connected as a unity gain voltage follower thereby ensuring maximum possible 3-dB bandwidth (equal to ω_t the gain-bandwidth-product) of the op-amp employed). From a straightforward

[4]Toumazou C., Lidgey F. J., Makris, C. A. 'Extending voltage-mode op amps to current-mode performance'. IEE Proceedings Pt. G. 1990; **137**(2):116–30.

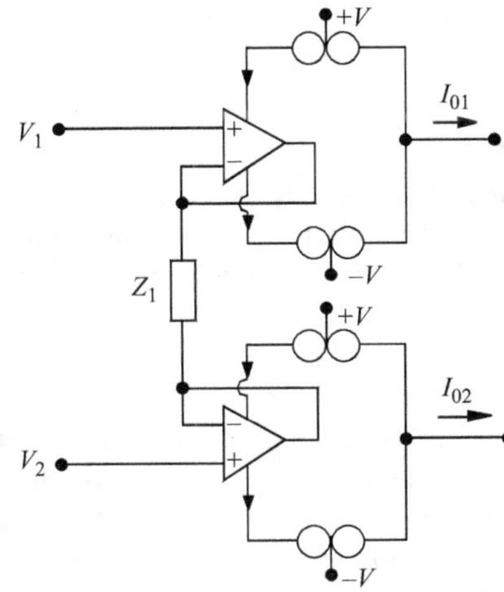

Figure 6.18 Complementary output differential voltage-controlled current source

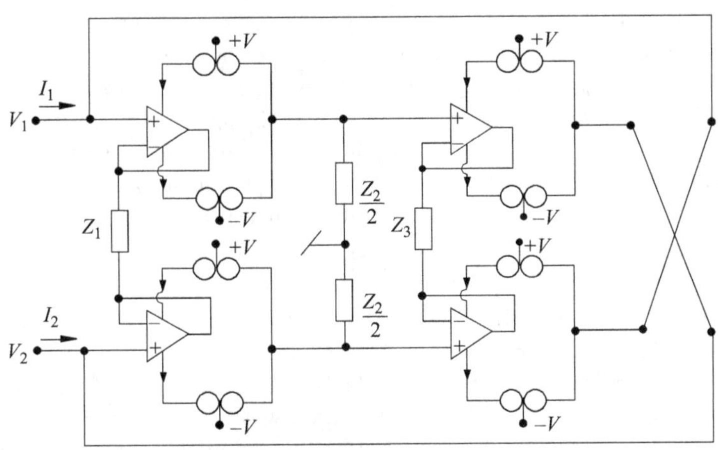

Figure 6.19 Floating GIC using OMA+. Reprinted, with permission, from [45]

analysis of this circuit, it can be easily determined that the two output currents are given by

$$I_{01} = -I_{02} = \frac{V_1 - V_2}{Z_1} \tag{6.13}$$

Using two such DVCCS structures along with two more impedances as shown in Figure 6.19, Toumazou and Lidgey [45] obtained a configuration which can act as a floating GIC.

This can be ascertained from the [Y]-matrix of the circuit which is found to be:

$$\begin{bmatrix} I_1 \\ I_2 \end{bmatrix} = \left(\frac{Z_2}{Z_1 Z_3}\right) \begin{bmatrix} 1 & -1 \\ -1 & 1 \end{bmatrix} \begin{bmatrix} V_1 \\ V_2 \end{bmatrix} \tag{6.14}$$

From the above, the realized floating impedance between port 1 and 2 is given by $Z_{eq} = Z_1Z_3/Z_2$. Thus, for simulating a floating inductance we can choose $Z_1 = R_1$,$Z_3 = R_3$ and $Z_2 = 1/sC_2$, thereby leading to the realized inductance value as $L_{eq} = C_2R_1R_3$.

One might wonder as to why the impedance Z_2 has been split into two half parts each equal to $Z_2/2$ with their junction connected to ground? Particularly in the event when if a single Z_2 is connected between the two current output terminals of the first DVCCS, by routine analysis it can be confirmed that even then the circuit under consideration will have exactly the same [*Y*]-matrix. It has been found that with the alternative allocation of Z_2 the entire circuit does not have even a single element connected to ground. To verify this, one can see that there is no connection from the internal op-amp circuit to the ground; all *pnp* and *npn* current mirrors used to make the OMA also do not have any connection to ground except through the two power supplies employed to bias current mirrors; lastly, the three external impedances are also all floating. Although there is no rigorous mathematical proof but practical experience with such OMA-based circuits shows that if there is not even a single terminal/component connected to ground then such a circuit does not work in practice. *It is, therefore, conjectured that for this 'OMA−' based FI to work satisfactorily, at least two ground elements/terminals are necessary.* Thus, by splitting Z_2 into two equal parts and then *grounding* their junction, one is able to get exactly two impedances connected to ground and therefore, this version of the circuit is practically viable and works well. This has been found to be practically verifiable in hardware[5] implementations.

6.4.2 A floating GIC using OMA

In Chapter 4 it was shown that two CCII− are sufficient to realize a floating generalized impedance converter (GIC). A special case of this, namely the floating FDNR circuit, was also shown to be feasible using only two CCII− elements. However, this development at that point of time was only a theoretical proposition since neither any current conveyor IC was available nor any of the available hardware implementation of the CCs available till then was either popular or convenient to use. Wilson demonstrated in [34,44] that an OMA+ can be made from a commercially available IC op-amp such as μA741/LF356/LM 301 and a pair of *pnp* and *npn* current mirrors both of which could be conveniently implemented using commercially available mixed transistor arrays such as CA3096/LM3096.

[5]It should be pointed out that a software implementation of an OMA− based FI of type shown in Fig. 6.19 in SPICE may employ a *macro-model* of the op-amps which would normally have several components connected to *ground*, and hence, such OMA− based FIs may actually work in simulations even with two-grounded $Z_2/2$ impedances replaced by a floating one, but such a circuit would not work in hardware using the same type of *real* op-amps (which do not have any elements of the internal circuit connected to ground) as explained in Section 6.4.1.

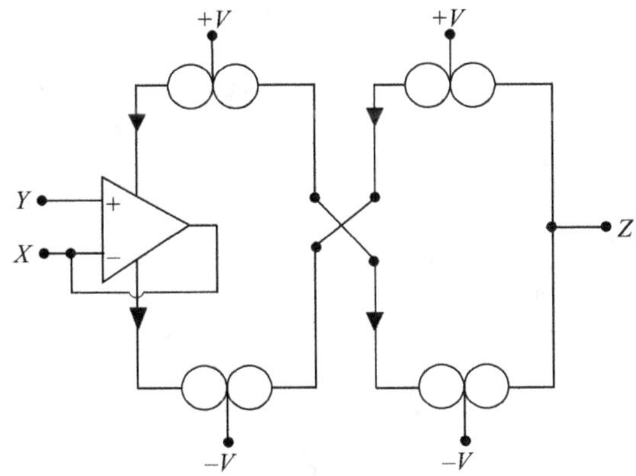

Figure 6.20 CCII− formulation employing an OMA− [34]

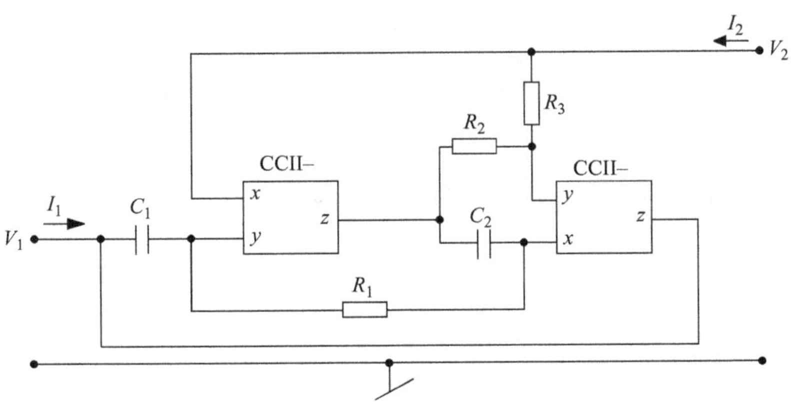

Figure 6.21 Floating FDNR [32]; $Z(s) = R_3/s^2C_1C_2R_1R_2$

He subsequently demonstrated that using two pairs of current mirrors with output of the first pair cross coupled with input of the second pair (see Figure 6.20) one can obtain a realization of CCII−.

It is interesting to note that although a CCII+ can be considered to be made from OMA+ likewise a CCII− could be considered to be made from OMA−. Thus, the previously known CCII− based floating GIC can also be made from two OMA− elements. A floating FDNR realizable from one of these circuits, whose experimental results were published by Wilson [34] himself, is shown in Figure 6.21.

6.4.3 Floating impedance realization using a dual OMA

An alternative use of OMAs in FI realization was advanced by Normand in [35]. He introduced the notion of inverting dual OMA, which can be understood from the formulation shown in Figure 6.22.

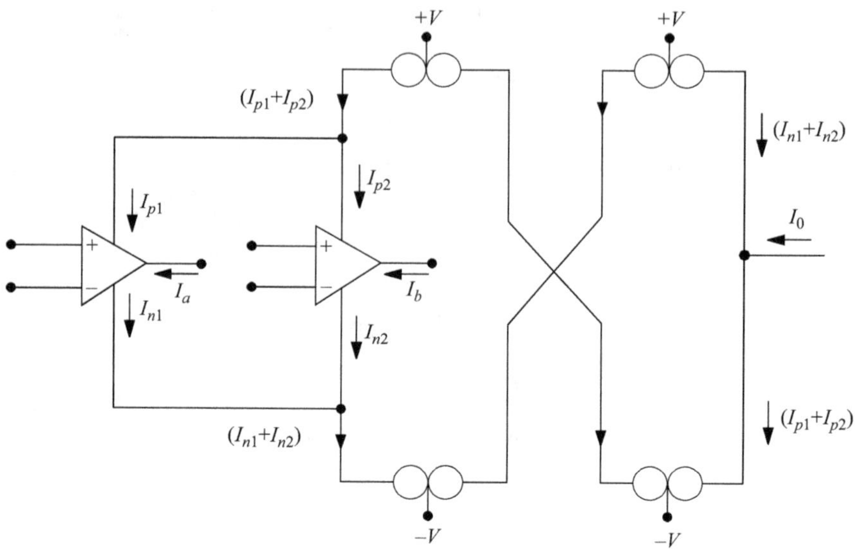

Figure 6.22 Dual OMA formulation proposed by Normand [35]

Analysis of this circuit reveals the following:

$$\left.\begin{aligned} &I_{p1} + I_a = I_{n1} \\ &I_{p2} + I_b = I_{n2} \\ &I_0 + (I_{n1} + I_{n2}) = (I_{p1} + I_{p2}) \\ &I_0 = (I_{p1} - I_{n1}) + (I_{p2} - I_{n2}) = -(I_a + I_b) \end{aligned}\right\} \tag{6.15}$$

The above equation implies that the sum of the two output currents going into the voltage output terminals of the op-amps can be retrieved using power supply current sensing technique and using current mirrors, from the output current I_0 of the current output terminal as I_0. This can readily be used to create a floating GIC as discussed next.

In Figure 6.23, we show that if the only grounded impedance Z_5 of the GIC simulated impedance circuit, is grounded, the flow of the various currents in the circuit shows that $I_2 \neq -I_1$, and hence, the circuit does not simulate an FI as is revealed from its $[Y]$ matrix which is given by

$$\begin{bmatrix} I_1 \\ I_2 \end{bmatrix} = \begin{bmatrix} \dfrac{Z_2Z_4}{Z_1Z_3Z_5} & -\dfrac{Z_2Z_4}{Z_1Z_3Z_5} \\ -\dfrac{1}{Z_5} & \dfrac{1}{Z_5} \end{bmatrix} \begin{bmatrix} V_1 \\ V_2 \end{bmatrix} \tag{6.16}$$

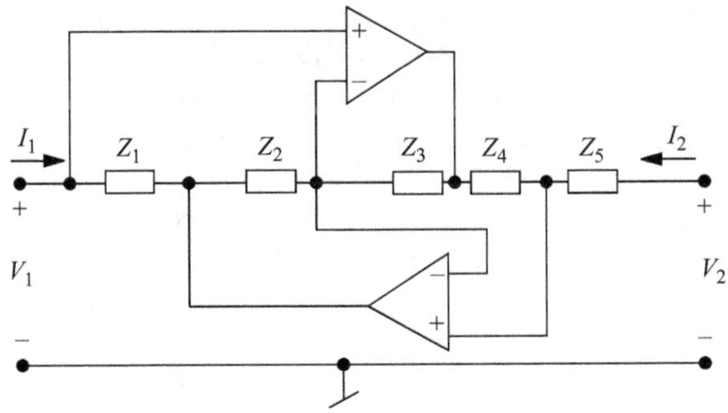

Figure 6.23 A two-port obtained from the classical GIC circuit

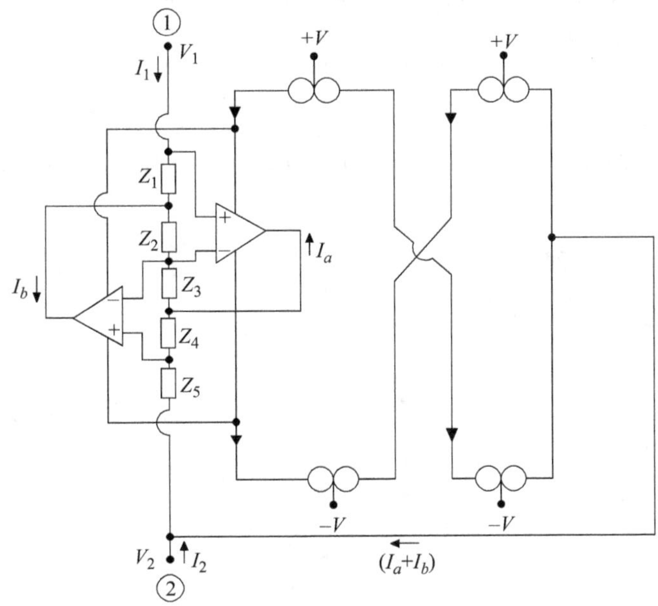

Figure 6.24 A floating GIC formulation as per the method of Normand [35]

However, when the two op-amps are replaced by a dual OMA as explained above, the resulting circuit, as shown in Figure 6.24, would clearly have $I_2 = -I_1$ due to which the $[Y]$ matrix would then become

$$\begin{bmatrix} I_1 \\ I_2 \end{bmatrix} = \frac{Z_2 Z_4}{Z_1 Z_3 Z_5} \begin{bmatrix} 1 & -1 \\ -1 & 1 \end{bmatrix} \begin{bmatrix} V_1 \\ V_2 \end{bmatrix} \tag{6.17}$$

6.4.4 Three OMA-based floating impedance simulators

In Section 6.4.1, we have shown that with four OMAs and a few external impedances, it is possible to realize a fully floating gyrator by using a parallel back-to-back connection of two dual complementary output VCCS structures. A generalized version of this circuit with the two resistors and the capacitor replaced by general impedance, the circuit can be looked upon as a floating GPIC (FGPIC) and with some interconnections changed, the topology can be converted into a floating generalized negative impedance converter (FGNIC) too.

On the other hand, using two CCII− and not three but five impedances, one can simulate a floating FDNR [32] and can also realize FGPIC elements, three circuits for which were reported in [36]. Note that, since a CCII− is realizable from an OMA−, these circuits can also be looked upon as two OMA− based FGPICs using five general impedances. A natural question now arises is: given three impedances, how many OMAs are needed to create a FPGIC? It was found that the minimum number of OMAs required is three. Consequently, Senani and Malhotra in [55] presented four such *minimal* realizations, each of which realizes FGPICs using only three OMAs and only three external impedances. These minimal realizations of FGPICs are shown in Figure 6.26.

In order to simplify the circuit diagrams, we have employed the following simplified notations of the OMA− and OMA+ as shown in Figure 6.25.

In the circuit of Figure 6.26(a) the current entering in port 2 is given by $I_2 = (e_x - V_2)/Z_2$, and from the directions of currents shown on the diagrams,

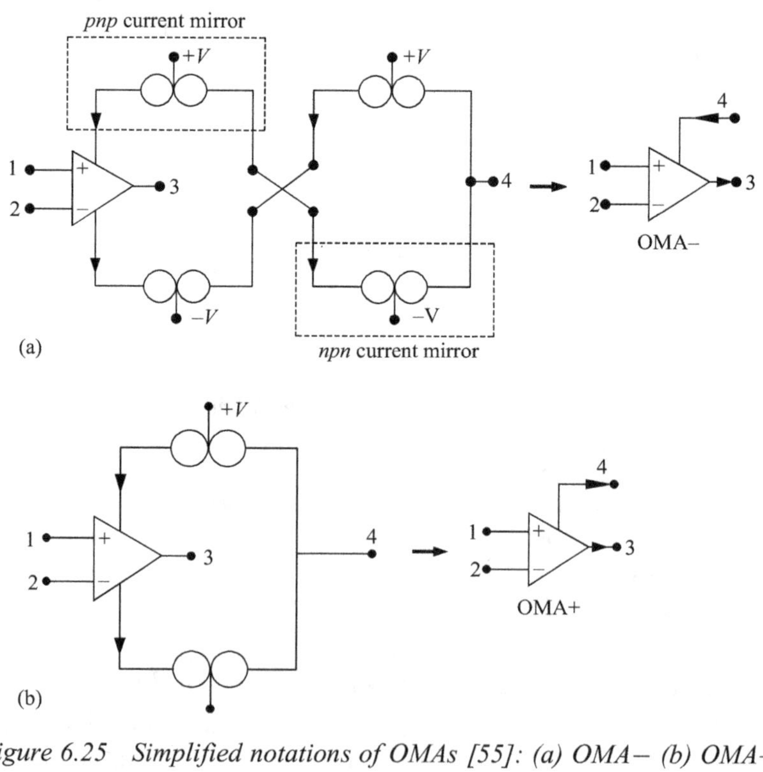

Figure 6.25 Simplified notations of OMAs [55]: (a) OMA− (b) OMA+

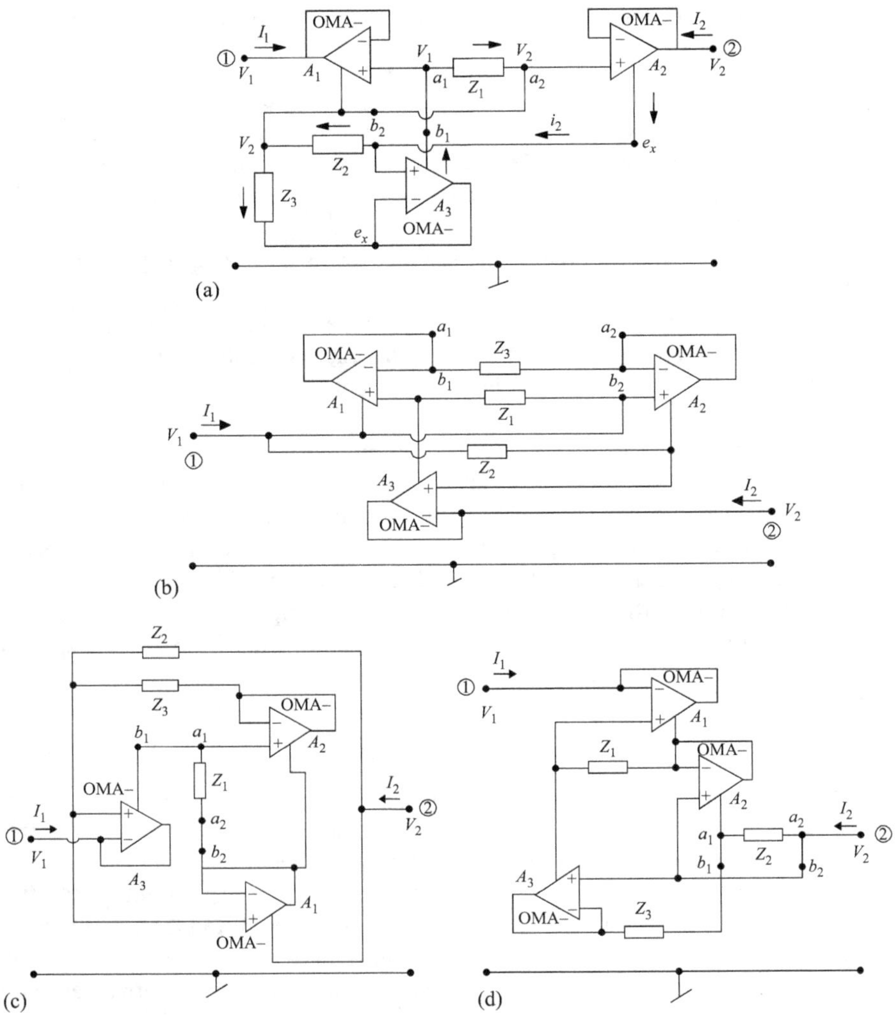

Figure 6.26 Four minimal realizations (a)-(d) of FGPIC [55]

it may also be noted that $(V_2 - e_x)/Z_3 = (V_1 - V_2)/Z_1$. Combining these equations, it turns out that $I_2 = (Z_3/Z_1Z_2)(V_2 - V_1)$. Furthermore, utilizing the properties of OMA– it can be easily verified by inspection that $I_1 = -I_2$. Thus, the [*Y*]-matrix of the circuit would be given by

$$[Y] = \begin{bmatrix} I_1 \\ I_2 \end{bmatrix} = \frac{Z_3}{Z_1Z_2}\begin{bmatrix} 1 & -1 \\ -1 & 1 \end{bmatrix}\begin{bmatrix} V_1 \\ V_2 \end{bmatrix} \tag{6.18}$$

From the directions of currents shown in the remaining three circuits it can be easily verified that their synthesis is also dependent upon the same type of equations as elaborated above. Therefore, all of them are characterized by the same matrix as in (6.18). The following features are noteworthy in the circuits of Figure 6.26:

1. Changing the connections $a_1 - b_1$, $a_2 - b_2$ to $a_1 - b_2$, $a_2 - b_1$ facilitates the realization of FGNIC from the same circuits.
2. By proper selection (resistive/capacitive) of the three impedances the circuits can realize floating lossless inductance ($Z_1 = R_1$, $Z_2 = R_2$ and $Z_3 = 1/sC_3$; $L_{eq} = C_3R_1R_2$), floating FDNR ($Z_1 = 1/sC_1$, $Z_2 = 1/sC_2$ and $Z_3 = R_3$; $D = C_1C_2R_3$) and resistively variable floating capacitance (with Z_1 or Z_2 as capacitor and Z_3 as resistor).
3. In all the cases, the realized elements are single resistance controllable.
4. All the circuits are minimal in the sense that they not only require a minimum number of passive components but also a minimum number of OMAs (given only three passive elements).
5. The most interesting feature is that in all the circuits OMA$-s$ A_1 and A_2 both can be replaced by OMA$+s$ without any change in the realized floating impedance values. Since, an OMA+ requires only a pair of current mirrors as opposed to OMA− which needs two pairs of current mirrors, these alternative versions obviously would require less hardware than the circuits of Figure 6.26.
6. It may be noticed that in all the circuits, the op-amp is configured as a unity gain voltage follower; hence, they offer the maximum possible frequency range of operation and ensure stable operation.

6.4.5 OMA-based FI using op-amp pole

In all the applications of the OMAs described in this chapter, the op-amp employed within has been considered to be ideal having (infinite gain) and therefore, external capacitors need to be employed to create different frequency selective networks. In [89], a new application of OMAs in which a realization of a floating inductance (FI) was achieved by incorporating the op-amp pole in a composite-OMA formulation (combining OMA+ and OMA− in the same block) was reported. The circuit from [89] is reproduced here in Figure 6.27, which does not require an external capacitor and has only one external resistor, which can also be simulated electronically, thereby, facilitating the realization of an electronically controllable FI.

By straightforward analysis, the [Y]-matrix of this circuit is found to be

$$[Y] = \left(\frac{\omega_t}{R_0 s}\right)\begin{bmatrix} 1 & -1 \\ -1 & 1 \end{bmatrix} \tag{6.19}$$

where it has been assumed that for $\omega \gg \omega_p$, the frequency-dependent gain of the op-amp can be modelled as an integrator, i.e.

$$A(s) = \frac{A_0\omega_p}{s + \omega_p} \approx \frac{A_0\omega_p}{s} = \frac{\omega_t}{s} \tag{6.20}$$

where $\omega_t = A_0\omega_p$ represents the gain bandwidth product, A_0 is the DC gain and ω_p is the 3-dB frequency of the open-loop gain.

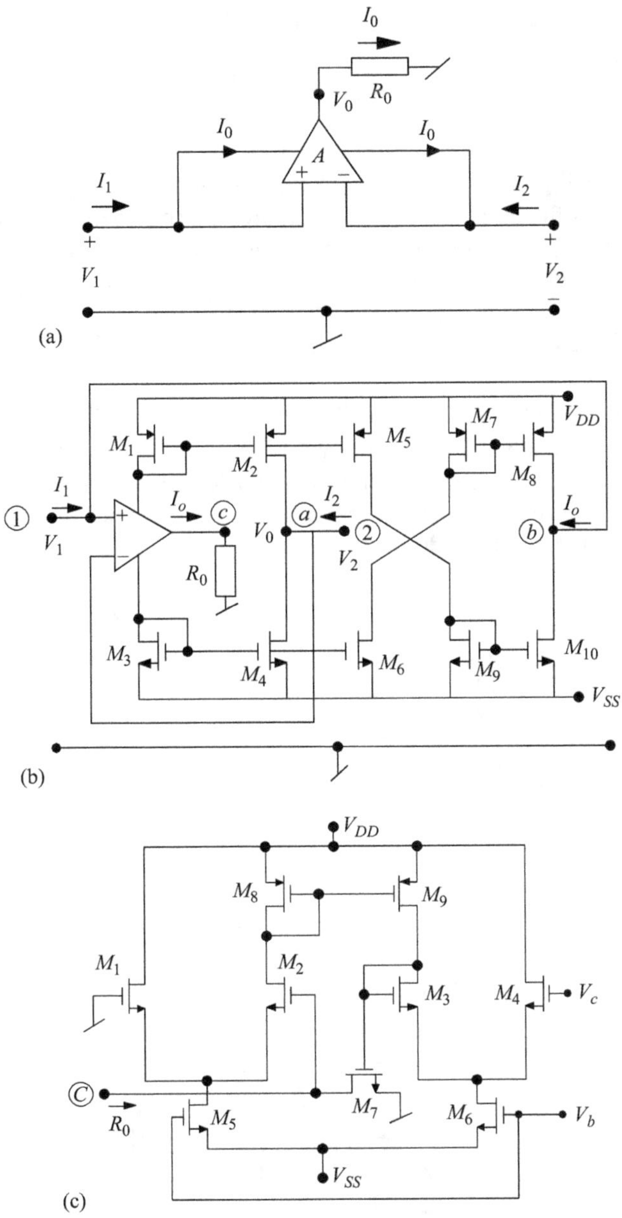

Figure 6.27 Active-R OMA-based FI [89]: (a) the basic configuration, (b) the transistor-level implementation and (c) the CMOS VCR employed in place of R_0. Reprinted, with permission, from [109]

Thus, floating inductance realized by the circuit by the circuit is given by

$$L_{eq} = \frac{R_0}{\omega_t} \tag{6.21}$$

and would be electronically controllable if resistor R_0 is replaced by a linear voltage-controlled resistance (VCR). Assuming a CMOS op-amp, an entirely CMOS version of the circuit is shown in Figure 6.27(b) where the only resistor R_0 can be replaced by a CMOS VCR for which an exemplary circuit is shown in Figure 6.27(c) [109]. The realizability and the variability of the inductance with external control voltage V_c have been well corroborated by the SPICE simulation results given in [89].

6.4.6 OMA-based FI with extended frequency range

In the previous section, we presented an FI formulation, the CMOS version of which was completely resistor-less and capacitor-less. An improved FI formulation which offers an extended dynamic range over the previous FI formulation was presented in [91] and is reproduced here in Figure 6.28.

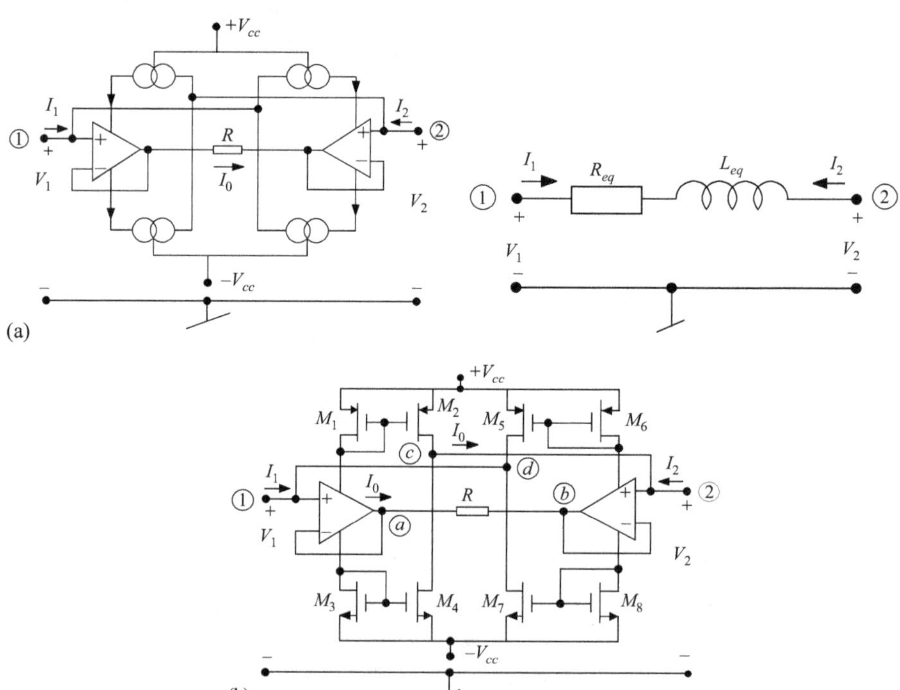

Figure 6.28 Improved OMA-based FI [91]: (a) the basic configuration and its equivalent and (b) CMOS implementation of the associated circuit

The proposed configuration has been detailed out in Figure 6.28 where Figure 6.28(a) shows the basic schematic using a composite-OMA and its equivalent and Figure 6.28(b) shows its complete transistor-level realization. In the circuit of Figure 6.28(b), the op-amp employed in the circuit senses the voltages V_1 and V_2 and creates a current I_0 flowing into R proportional to $(V_1{-}V_2)$. The current mirrors M_1–M_8 create a replica of this output current between the junctions '1' and '2'.

By straightforward analysis, the currents I_1 and I_2 are related to I_0 and are given as

$$I_1 = -I_2 = I_0 = \frac{V_1 - V_2}{R(1 + (s/\omega))} \tag{6.22}$$

and this circuit, thus, simulates a floating series RL impedance with their values given by

$$R_{eq} = R; \quad L_{eq} = \frac{R}{\omega_t} \tag{6.23}$$

By SPICE simulation results, it has been confirmed that this circuit offers the advantages of increased signal handling capability, larger operational frequency range and electronic tunability of the realized floating inductance value.

6.5 Operational floating amplifiers and their use in floating gyrator and FI realization

Huijsing [46,51] proposed the OFA as the most general nullor implementation which is symbolically shown in Figure 6.29 which is characterized by $V_1 = V_2$, $I_1 = 0 = I_2$ and $I_{01} = -I_{02}$. It was also shown that like a conventional op-amp the OFA accurately equates two voltages at the input port, but, in addition, an OFA also equates two currents connected to the output terminals along with a sign reversal.

In this section, we would highlight the applications of OFAs and their variants in the realization of floating gyrator and various kinds of floating impedances.

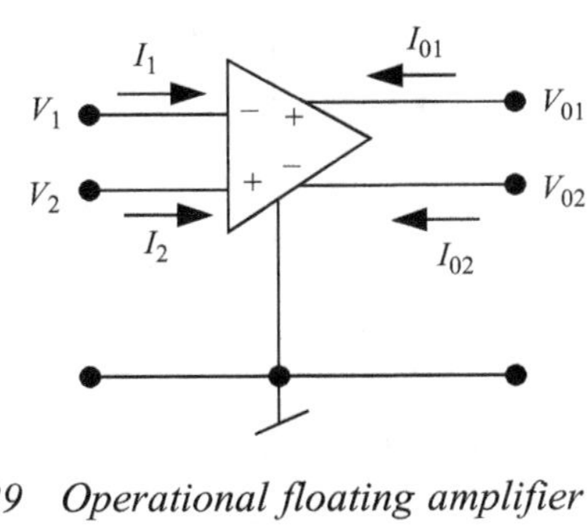

Figure 6.29 Operational floating amplifier (OFA) [46]

6.5.1 The OFA-based floating gyrator

Among other applications, Huijsing [46] demonstrated that OFA is particularly useful in realizing a voltage to current converter. In particular, it was shown that whereas a good V to I converter using a conventional op-amp require typically three op-amps and five external resistors along with perfect matching between four of them, an OFA can realize the same function with only a single resistor. Furthermore, a DVCCS with two complementary current outputs can be made from two OFAs along with a single resistor. Thus, two such DVCCS in a back-to-back cross-coupled connection as shown in Figure 6.30 can realize a fully floating four-port gyrator. By straightforward analysis, the [Y]-matrix of this four-port gyrator can be found to be

$$\begin{bmatrix} I_1 \\ I_2 \\ I_3 \\ I_4 \end{bmatrix} = \begin{bmatrix} 0 & 0 & -1/R_2 & 1/R_2 \\ 0 & 0 & 1/R_2 & -1/R_2 \\ 1/R_1 & -1/R_1 & 0 & 0 \\ -1/R_1 & 1/R_1 & 0 & 0 \end{bmatrix} \begin{bmatrix} V_1 \\ V_2 \\ V_3 \\ V_4 \end{bmatrix} \tag{6.24}$$

The above Y-matrix characterizes a fully floating gyrator whose input port is constituted from terminal 1 and 2 and the output port is composed of terminals 3 and 4. Thus, if a capacitor (C) is connected between port 3 and 4, it can be easily shown that with respect to ports 1 and 2, the circuit will have a [Y]-matrix given by

$$\begin{bmatrix} I_1 \\ I_2 \end{bmatrix} = \frac{1}{sCR_1R_2} \begin{bmatrix} 1 & -1 \\ -1 & 1 \end{bmatrix} \begin{bmatrix} V_1 \\ V_2 \end{bmatrix} \tag{6.25}$$

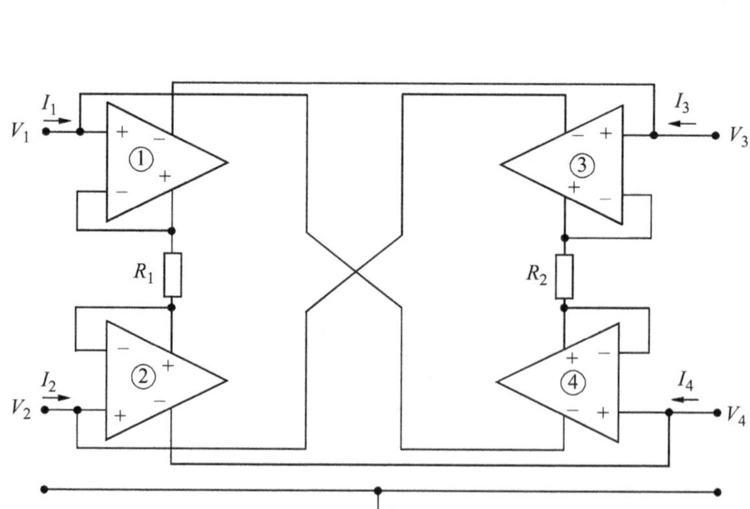

Figure 6.30 Floating gyrator with two balanced voltage to current converters. Reprinted, with permission, from [46]

This OFA based circuit, therefore, realizes a lossless floating inductor using only two resistors and a single capacitor – a task which is impossible to be achieved by a conventional op-amp!

6.5.2 The DD-OFA and its use in FI synthesis

The differential difference amplifier (DDA) has been known in the analog circuit literature since long [37]. Motivated by the concepts of DDA on one hand and that of OFA on the other hand, Mahmoud and Soliman in [61] introduced a wide range differential difference OFA (DDOFA) which is symbolically shown in Figure 6.31.

Two output currents of the DDOFA can be expressed as

$$I_{0+} = -I_{0-} = G_0[(V_2 - V_1) - (V_4 - V_3)] \tag{6.26}$$

where G_0 is the open-loop transconductance gain. With a negative feedback and infinite transconductance gain, it turns out that

$$(V_2 - V_1) = (V_4 - V_3) \tag{6.27}$$

Mahmoud and Soliman [61] proposed CMOS architecture for the DDOFA which consisted of a high gain differential difference transconductor with large signal handling capability along with a single input and differential output current op-amp. In the context of this chapter, the applications of DDOFA in realizing grounded and floating electronically controlled resistors and MOS-C floating inductor are of interest and these are discussed next.

In Figure 6.32, we show an electronically controlled grounded resistor. It may be observed that, in this circuit, V_2 and V_3 are grounded; therefore, it follows that

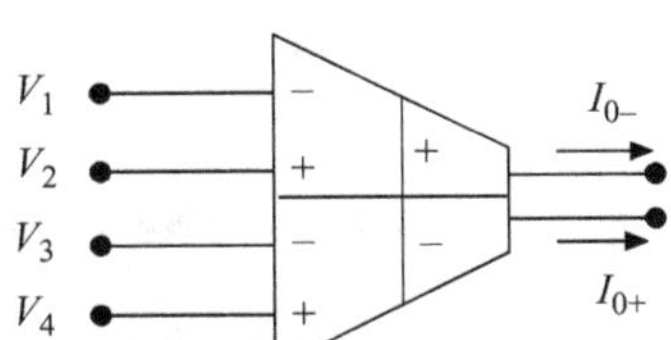

Figure 6.31 Proposed symbol of DDOFA

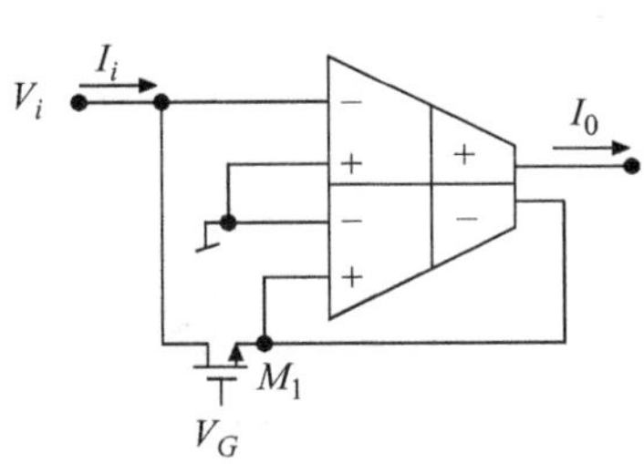

Figure 6.32 DDOFA-based grounded resistor [61]

$V_4 = -V_1$. Assuming that MOSFET M_1 is operated in triode region, the input current of the circuit is given by

$$\left.\begin{aligned} I_i = I_D &= 2K\left(V_{GS} - V_{th} - \frac{V_{DS}}{2}\right)V_{DS} \\ &= 2K\left(V_G + V_1 - V_{th} - \frac{2V_1}{2}\right)V_{DS} \\ &= 2K(V_G - V_{th})V_{DS} \end{aligned}\right\} \tag{6.28}$$

where $K = (\mu_n C_{ox}(W/L))/2$.

Therefore, the square nonlinearity of the MOSFET is *elegantly* cancelled out and the circuit realizes a linear voltage-controlled grounded resistor, with equivalent resistance given by $R_{eq} = 1/2K(V_G - V_{th})$.

In Figure 6.33, we show the voltage-controlled floating resistor devised in [61] which uses two DDOFAs and two identical (matched) MOSFETs operating in the triode region. The expression for the input and output currents of the realized floating resistor is given by

$$I_i = I_0 = (I_1 - I_2) = 2K(V_G - V_{th})(V_1 - V_2) \tag{6.29}$$

Thus, the value of the equivalent floating resistance realized is given by

$$R = \frac{1}{2K(V_G - V_{th})} \tag{6.30}$$

In Figure 6.34, a MOS-C electronically controllable floating inductor employing four DDOFAs, four MOSFETs and a grounded capacitor is shown. Assuming MOSFETs to be operated in triode region, a routine circuit analysis yields

$$I_i = I_0 = \frac{4K_1K_2(V_{G1} - V_{th})(V_{G2} - V_{th})}{sC}(V_1 - V_2) \tag{6.31}$$

From (6.31), the value of floating voltage-controlled inductor (L_{eq}) is given by

$$\left.\begin{aligned} L_{eq} &= \frac{C}{4K_1K_2(V_{G1} - V_{th})(V_{G2} - V_{th})}; \quad \text{where} \\ K_i &= \frac{\mu_n C_{ox}(W/L)_i}{2}; \quad i = 1 - 2 \end{aligned}\right. \tag{6.32}$$

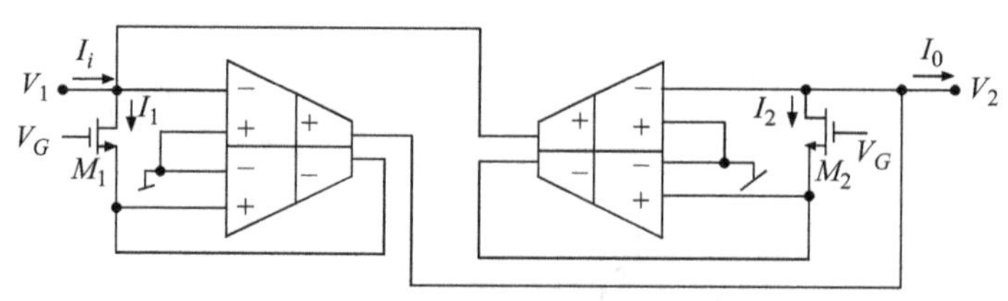

Figure 6.33 DDOFA-based floating resistor [61]

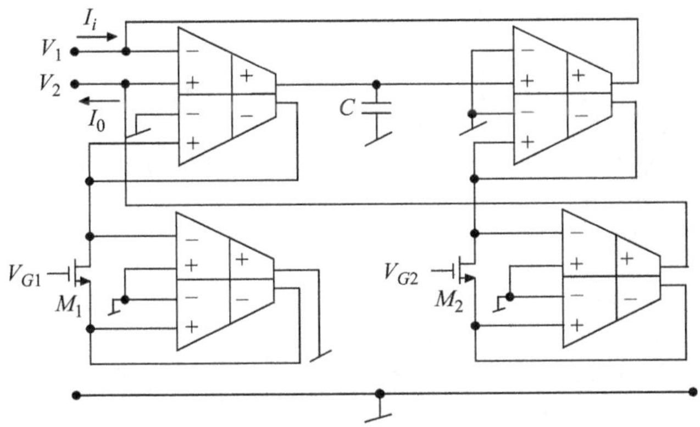

Figure 6.34 DDOFA-based MOS-C floating inductor [61]

6.6 Concluding remarks

In this chapter, we had considered the synthesis of grounded and floating inductors using the FTFNs as active elements. It was highlighted how the notion of fully floating FTFN provided a novel solution to the relatively difficult problem of realizing a lossless floating inductance which previously required interconnection of two identical RC-subcircuits and therefore suffered from the drawback of requiring two capacitors and compulsorily requiring matching between the two sub-sections. With the advent of a number of FTFN implementations using two commercially available IC current conveyors or CFOAs such as AD844, the FTFN, got going as a versatile active element. As a consequence, FTFNs were widely investigated for their possible use in numerous areas. In this chapter, a number of prominent contributions made in the area of impedance realization using FTFN have been highlighted.

The concept of FTFN is very close to the notion of the so-called OFA. In fact, FTFN and OFA both to be particularly useful in realizing fully floating gyrators and floating impedances which have been highlighted in this chapter.

Floating impedance circuits using OMA have also been considered. It must be reiterated that whereas an OMA− is the same thing as an FTFN but an OMA+ is distinctly different than an FTFN or OFA.

Although several of the circuits described in this chapter can be practically implemented using commercially available ICs some of them may need additional transistor arrays, which are fortunately also available in IC form. On the other side, along with the Bipolar and CMOS hardware of OMAs, OFAs and FTFNs, several of the circuits appear to be suitable to be carried over for complete IC implementation in Bipolar/CMOS technologies.

References

[1] Carlin H. J., Youla, D. C. 'Network synthesis with negative resistors'. Proceedings of the IRE. 1961; **49**(5):907–20.

[2] Carlin, J. 'Singular network elements'. IEEE Transactions on Circuit Theory. 1964; **11**(1):67–72.

[3] Martinelli, G. 'On the nullor'. Proceedings of the IEEE'. 1965; **53**(3):332.

[4] Myers B. R., Martinelli, G. 'Nullor model of the transistor'. Proceedings of the IEEE. 1965; **53**(7):758–9.

[5] Shenoi, B. A. 'Practical realization of a gyrator circuit and RC-gyrator filters'. IEEE Transactions on Circuit Theory. 1965; **CT-12**(3):374–80.

[6] Brownlie, J. D. 'On the stability properties of a negative impedance converter'. IEEE Transactions on Circuit Theory. 1966; **13**(1):98–9.

[7] Davies, A. C. 'Matrix analysis of networks containing nullators and norators'. Electronics Letters. 1966; **2**(2):48–9.

[8] Tellegen, B. D. H. 'On nullators and norators'. IEEE Transactions on Circuit Theory. 1966; **13**(4):466–9.

[9] Bendik, J. 'Equivalent gyrator networks with nullators and norators'. IEEE Transactions on Circuit Theory. 1967; **14**(1):98.

[10] Yanagisawa T., Kawashima, Y. 'Active gyrator'. Electronics Letters. 1967; **3**(3):105–7.

[11] Mitra, S. K. 'Equivalent circuits of gyrators'. Electronics Letters. 1967; **3**(7): 333–4.

[12] Mitra, S. K. 'Nonreciprocal negative-impedance invertor'. Electronics Letters. 1967; **3**(8):388.

[13] Davies, A. C. 'The significance of nullators, norators and nullors in active-network theory'. Radio and Electronic Engineer. 1967; **34**(5):259–67.

[14] Davies, A. C. 'Stability properties of a negative immittance converter'. IEEE Transactions on Circuit Theory. 1968; **15**(1):80–1.

[15] Martinelli G., Di Porto, P. G. 'Minimal number of nullors for realising active gyrators'. Electronics Letters. 1968; **4**(13):273–4.

[16] Bruton, L. T. 'Frequency stability using positive impedance converter-type networks'. Proceedings of the IEEE. 1968; **56**(8):1378–9.

[17] Mitra S. K. 'Analysis and Synthesis of Linear Active Networks'. London: John Wiley & Sons, Inc.; 1968.

[18] Daniels, R. W. 'Gyrators, negative impedance converters, and related circuits'. IEEE Transactions on Circuit Theory. 1969; **16**(2):261–2.

[19] Antoniou, A. 'Realisation of gyrators using operational amplifiers, and their use in RC-active-network synthesis'. Proceedings of the IEE. 1969; **116**(11): 1838–50.

[20] Pauker, V. M. 'Equivalent networks with nullors for positive immittance converters'. IEEE Transactions on Circuit Theory. 1970; **17**(4):642–5.

[21] Bruton, L. T. 'Nonideal performance of two-amplifier positive-impedance converters'. IEEE Transactions on Circuit Theory. 1970; **17**(4):541–9.

[22] Cox N., Su K., Woodward, R. 'A floating three-terminal nullor and the universal impedance converter'. IEEE Transactions on Circuit Theory. 1971; **18**(3):399–400.

[23] Yarlagadda R., Ye, F. 'Active RC networks synthesis using nullators and norators'. IEEE Transactions on Circuit Theory. 1972; **19**(4):317–22.

[24] Adams K. M., Deprettere, E. 'On the realisation of gyrators by nullors and resistors'. International Journal of Circuit Theory and Applications'. 1974; **2**(3):287–90.

[25] Deprettere, E. 'On the minimal realization of the gyrator by means of nullors and resistors. I'. International Journal of Circuit Theory and Applications. 1975; **3**(4): 383–90.

[26] Deprettere, E. 'On the minimal realization of the gyrator by means of nullors and resistors. II'. International Journal of Circuit Theory and Applications. 1976; **4**(3): 285–97.

[27] Moschytz, G. S. 'The morphological approach to network and circuit design'. IEEE Transactions on Circuits and Systems. 1976; **23**(4):239–42.

[28] Huijsing J. H., De Korte, J. 'Monolithic nullor—a universal active network element'. IEEE Journal of Solid State Circuits. 1977; **SC-12**(1):59–64.

[29] Silva, M. D. M. 'On the realization of immittance inverters with a minimum number of active components'. IEEE Transactions on Circuits and Systems. 1979; **26**(11):931–5.

[30] Bruton L. T. 'RC-Active Circuits Theory and Design'. London: Prentice-Hall, Inc.; 1980.

[31] Nordholt, E. 'Extending op amp capability by using a current-source power supply'. IEEE Transactions on Circuits and Systems. 1982; **29**(6):411–14.

[32] Senani, R. 'Floating ideal FDNR using only two current conveyors'. Electronics Letters. 1984; **20**(5):205–6.

[33] Stevenson, J. K. 'Two-way circuits with inverse transmission properties'. Electronics Letters. 1984; **20**(23):965–7.

[34] Wilson, B. 'Floating FDNR employing a new CCII—conveyor implementation'. Electronics Letters. 1985; **21**(21):996–7.

[35] Normand, G. 'Floating-impedance realisation using a dual operational-mirrored amplifier'. Electronics Letters. 1986; **22**(10):521–2.

[36] Senani, R. 'On the realization of floating active elements'. IEEE Transactions on Circuits and Systems. 1986; **33**(3):323–4.

[37] Sackinger E., Guggenbuhl, W. 'A versatile building block: the CMOS differential difference amplifier'. IEEE Journal of Solid-State Circuits. 1987; **22**(2):287–94.

[38] Senani, R. 'Generation of new two-amplifier synthetic floating inductors'. Electronics Letters. 1987; **23**(22):1202–3.

[39] Senani, R. 'A novel application of four-terminal floating nullors'. Proceedings of the IEEE. 1987; **75**(11):1544–6.

[40] Senani, R. 'Floating immittance realisation: nullor approach'. Electronics Letters. 1988; **24**(7):403–5.

[41] Higashimura M., Fukui, Y. 'Novel method for realizing lossless floating immittance using current conveyors'. Electronics Letters. 1987; **23**(10): 498–9.
[42] Higashimura M., Fukui, Y. 'Realisation of impedance function using current conveyors'. International Journal of Electronics. 1988; **65**(2):223–31.
[43] Singh, V. 'Proof of recent conjecture concerning impedance floatation'. IEEE Transactions on Circuits and Systems. 1989; **36**(8):1110–13.
[44] Wilson, B. 'Recent developments in current conveyors and current-mode circuits'. IEE Proceedings Pt. G. 1990; **137**(2):63–77.
[45] Toumazou C., Lidgey, F. J. 'Floating-impedance converters using current conveyors'. Electronics Letters. 1985; **21**(15):640–2.
[46] Huijsing, J. H. 'Operational floating amplifier'. IEE Proceedings Pt. G. 1990; **137**(2):131–6.
[47] Higashimura, M. 'Realisation of current-mode transfer function using four terminal floating nullor'. Electronics Letters. 1991; **27**(2):170–1.
[48] Higashimura M., Fukui, Y. 'Realization of immittance floatator using nullors'. IEICE Transactions on Fundamentals. 1992; **E75-A** (6):644–9.
[49] Higashimura M., Fukui, Y. 'Realization of immittance function using a single FTFN and its application to filters'. Proceedings of the 1992 IEEE International Symposium on Circuits and Systems; San Diego, CA, May 10–13, 1992; 2:851–4.
[50] Serrano L., Carlosena, A. 'Resistively variable capacitors: voltage and current approaches'. Proceedings of the 35th Midwest Symposium on Circuits and Systems; Washington, DC, August 9–12, 1992; 1:314–17.
[51] Huijsing J. H. 'Design and applications of the operational floating amplifier (OFA): the most universal amplifier'. Analog Integrated Circuits and Signal Processing. 1993; **4**(2):115–29.
[52] Carlosena A., Moschytz, G. S. 'Nullators and norators in voltage to current mode transformations'. International Journal of Circuit Theory and Applications. 1993; **21**(4):421–4.
[53] Malhotra J., Senani, R. 'Class of floating, generalised, positive/negative immittance convertors/inverters realised with operational mirrored amplifiers'. Electronics Letters. 1994; **30**(1):3–5.
[54] Cabeza R., Carlosena A., Serrano, L. 'Unified approach to the implementations of universal active devices'. Electronics Letters. 1994; **30**(8):618–20.
[55] Senani R., Malhotra, J. 'Minimal realisations of a class of operational-mirrored-amplifier-based floating impedances'. Electronics Letters. 1994; **30**(14):1113–14.
[56] Senani, R. 'Floating GNIC/GNII configuration realised with only a single OMA'. Electronics Letters. 1995; **31**(6):423–5.
[57] Topor-Kaminski L., Pasko, M. 'Nullator/norator models of active circuits with controlled parameters'. Proceedings of First International Caracas Conference on Devices, Circuits and Systems; Caracas, Venezuela, December 12–14, 1995.
[58] Cabeza R., Carlosena, A. 'Analog universal active device: theory, design and applications'. Analog Integrated Circuits and Signal Processing. 1997; **12**(2): 153–68.

[59] Wang H.-Y., Lee, C.-T. 'Realisation of R-L and C-D immittances using single FTFN'. Electronics Letters. 1998; **34**(6):502–3.

[60] Cabeza R., Carlosena, A. 'Computational synthesis of arbitrary floating impedances'. International Journal of Circuit Theory and Applications. 1998; **26**(5):463–75.

[61] Mahmoud S. A., Soliman, A. M. 'The differential difference operational floating amplifier: a new block for analog signal processing in MOS technology'. IEEE Transactions on Circuits and Systems-II. 1998; **45**(1):148–58.

[62] Awad I. A., Soliman, A. M. 'Inverting second generation current conveyors: the missing building blocks; CMOS realizations and applications'. International Journal of Electronics. 1999; **86**(4):413–32.

[63] Cam U., Kuntman, H. 'CMOS four terminal floating nullor design using a simple approach'. Microelectronics Journal. 1999; **30**(12):1187–97.

[64] Cam U., Toker A., Kuntman, H. 'CMOS FTFN realization based on translinear cells'. Electronics Letters. 2000; **36**(15):1255–6.

[65] Cam U., Kuntman, H. 'A new CMOS realisation of a four terminal floating nullor (FTFN)'. International Journal of Electronics. 2000; **87**(7):809–17.

[66] Cam U., Cicekoglu O., Kuntman, H. 'Universal series and parallel immittance simulators using four terminal floating nullors'. Analog Integrated Circuits and Signal Processing. 2000; **25**(1):59–66.

[67] Cabeza R., Carlosena, A. 'On the use of symbolic analyzers in circuit synthesis'. Analog Integrated Circuits and Signal Processing. 2000; **25**(1): 67–75.

[68] Schmidt, H. 'Approximating the universal active element'. IEEE Transactions on Circuits and Systems-II. 2000; **47**(11):1160–9.

[69] Tangsrirat W., Unhavanich S., Dumawipata T., Surakampontorn, W. 'FTFN with variable current gain'. Proceedings of the IEEE Region 10 International Conference on Electrical and Electronic Technology; Singapore, August 19–22, 2001; 1:209–12.

[70] Jiraseri-amomkun A., Chipipop B., Surakampontorn, W. 'Novel translinear-based multi-output FTFN'. The 2001 International Symposium on Circuits and Systems; Sydney, NSW, May 6–9, 2001; 1:180–3.

[71] Cam U., Cicekoglu O., Kuntman, H. 'Novel lossless floating immittance simulator employing only two FTFNs'. Analog Integrated Circuits and Signal Processing. 2001; **29**(3): 233–5.

[72] Gandhi, G. 'Comment on "Novel lossless floating immittance simulator employing only two FTFNs"'. Analog Integrated Circuits and Signal Processing. 2002; **32**(2):191.

[73] Kumar P., Senani, R. 'Bibliography on nullor and their applications in circuit synthesis and design'. Analog Integrated Circuits and Signal Processing. 2002; **33**(1):65–76.

[74] Alzaher H., Ismail, M. 'A CMOS fully balanced four terminal floating nullor'. IEEE Transactions on Circuits and Systems-I. 2002; **49**(4):413–24.

[75] Awad, I. A., Soliman, A. M. 'On the voltage mirrors and current mirrors'. Analog Integrated Circuits and Signal Processing. 2002; **32**(1):79–81.

[76] Cam U., Cicekoglu O., Kuntman, H. 'Reply to Comment on "Novel lossless floating immittance simulator employing only two FTFNs"'. Analog Integrated Circuits and Signal Processing. 2002; **32**(2):193.

[77] Wang H. Y., Chung H. W., Huang, W. C. 'Realization of an nth-order parallel immittance function employing only $(n-1)$ FTFNs'. International Journal of Electronics. 2002; **89**(8):645–50.

[78] Schmid, H. 'Comments on: A CMOS fully balanced four terminal floating nullor'. IEEE Transactions on Circuits and Systems-I. 2003; **50**(12):1601.

[79] Palomera-Garcia, R. 'Generation of equivalent circuits by FTFN relocation'. 2005 International Symposium on Circuits and Systems; Kobe, Japan, May 23–26, 2005; 1:252–5.

[80] Haigh D. G., Tan F. Q., Papavassiliou, C. 'Systematic synthesis of active-RC circuit building blocks'. Analog Integrated Circuits and Signal Processing. 2005; **43**(3):297–315.

[81] Youssef M. A., Soliman, A. M. 'A novel CMOS realization of the differential input balanced output current operational amplifier and its applications'. Analog Integrated Circuits and Signal Processing. 2005; **44**(1):37–53.

[82] Haig D. G., Clarke T. J. W., Radmore, P. M. 'Symbolic framework for linear active circuits based on port equivalence using limit variables'. IEEE Transactions on Circuits and Systems-II. 2006; **53**(9):2011–24.

[83] Haig D. G., Radmore, P. M. 'Admittance matrix models for the nullor using limit variables and their application to circuit design'. IEEE Transactions on Circuits and Systems. 2006; **53**(10):2214–23.

[84] Haig, D. G. 'A method of transformation from symbolic transfer function to active-RC circuit by admittance matrix expansion'. IEEE Transactions on Circuits and Systems-I. 2006; **53**(12):2715–28.

[85] Grigorescu, L. 'Impedance converters'. Romania Journal of Physics. 2008; **53**(1–2):111–15.

[86] Kumar, P. 'FTFNs and Their Applications in Circuit Synthesis and Design' (Ph.D. thesis). 2003–2008; University of Delhi, India.

[87] Tlelo-Cuautle E., Duarte-Villasenor M. A., Guerra-Gomez, I. 'Automatic synthesis of VFs and VMs by applying genetic algorithms'. Circuits, Systems and Signal Processing. 2008; **27**: 391–403.

[88] Saad R. A., Soliman, A. M. 'Use of mirror elements in the active device synthesis by admittance matrix expansion'. IEEE Transactions on Circuits and Systems-I. 2008; **55**(9):2726–35.

[89] Sharma R. K., Senani R., Bhaskar D. R., Singh A. K., Gupta, S. S. 'Electronically-controllable floating inductor using operational mirrored amplifier'. Journal of Circuits, Systems, and Computers. 2009; **18**(1):59–66.

[90] Soliman, A. M. 'Adjoint network theorem and floating elements in the NAM'. Journal of Circuits, Systems and Computers. 2009; **18**(3):597–616.

[91] Sharma R. K., Senani R., Bhaskar D. R., Singh A. K., Gupta, S. S. 'Electronically-controllable floating inductor using OMA with enhanced input dynamic range'. 2009 International Conference on Electrical and Electronics Engineering-ELECO-2009; Bursa, Turkey, November 5–8, 2009; 1–4.

[92] Hashemian, R. 'Local biasing and the use of nullator-norator pairs in analog circuit designs'. VLSI Design. 2010; 2010 (Article ID 297083): 12 pages.

[93] Soliman A. M., Saad, R. A. 'The voltage-mirror current mirror pair as universal element'. International Journal of Circuit Theory and Applications. 2010; **38**(8):787–95.

[94] Soliman, A. M. 'On the four terminal floating nullor (FTFN) and the operational mirror amplifier (OMA)'. Journal of Active and Passive Electronic Devices. 2010; **5**:209–19.

[95] Sanchez-Lopez C., Fernandez F. V., Tlelo-Cuautle, E. 'Generalized admittance matrix models of OTRAs and COAs'. Microelectronics Journal. 2010; **41**(8):502–5.

[96] Saad R. A., Soliman, A. M. 'A new approach for using the pathological mirror elements in the ideal representation of active devices'. International Journal of Circuit Theory and Applications. 2010; **38**(2):148–78.

[97] Tlelo-Cuautle E., Sanchez-Lopez C., Moro-Frias, D. 'Symbolic analysis of (MO) (I) CCI (II) (III)-based analog circuits'. International Journal of Circuit Theory and Applications. 2010; **38**(6):649–59.

[98] Tlelo-Cuautle E., Guerra-Gomez I., Duarte-Villasenor M. A., *et al.* 'Applications of evolutionary algorithms in the design automation of analog integrated circuits'. Journal of Applied Sciences. 2010; **10**(17): 1859–72.

[99] Vandewalle J., Nossek, J. A. 'Nullators and norators in circuits education: a benefit or an obstacle?' 2011 IEEE International Symposium of Circuits and Systems; Rio de Janeiro, Brazil, May 15–18, 2011; 1–4.

[100] Sanchez-Lopez C., Fernandez F. V., Tlelo-Cuautle E., Sheldon, X-D. T. 'Pathological element-based active device models and their application to symbolic analysis'. IEEE Transactions on Circuits and Systems-I. 2011; **58**(6):1382–95.

[101] Soliman, A. M. 'Nodal admittance matrix and pathological realization of BOOA, DDA, DDOFA and DDOMA'. Singapore Journal of Scientific Research. 2011; **1**(2):149–63.

[102] Senani R., Singh A. K., Kumar P., Sharma, R. K. 'Nullors, their bipolar and CMOS implementations and applications in analog circuit synthesis and design'. In Tlelo-Cuautle E. (Ed.). Integrated Circuits for Analog Signal Processing. New York: Springer-Verlag; 2012. p. 31–59.

[103] The L. Q., Yanagisawa, Y. 'Some new lossless floating inductance circuits'. Proceedings of the IEEE. 1977; **65**(7):1071–2.

[104] Orchard H. J., Wilson, A. N. 'New active-gyrator circuit'. Electronics Letters. 1974; **10**(13):261–2.

[105] Schmidt C. E., Lee, M. S. 'Multipurpose simulation network with a single amplifier'. Electronics Letters. 1975; **11**(1):261–2.

[106] Ramsey, W. T. 'High pass and filter using a new single amplifier simulated inductor'. IEE Journal on Electronic Circuits and Systems. 1978; **2**(3): 79–85.

[107] Horn P., Moschytz, G. H. S. 'Active RC single op-amp design of driving point impedances'. IEEE Transactions on Circuits and Systems. 1979; **26**(1):22–30.
[108] Senani, R. 'Author's reply on comments on "New canonic active RC realizations of grounded and floating inductors"'. Proceedings of IEEE. 1982; **70**(1):101–3.
[109] Wilson G., Chan, P. K. 'Novel voltage-controlled grounded resistor'. Electronics Letters. 1989; **25**(7):1725–6.

Chapter 7
Realization of voltage-controlled impedances

Abstract

This chapter presents a variety of voltage-controlled impedance realizations most of which, apart from dealing with voltage-controlled resistances, also deal with voltage-controlled capacitances, voltage-controlled, inductances and voltage-controlled frequency-dependent negative resistance. Such voltage-controlled, impedances have been realized so far using a variety of active integrated circuit building blocks such as operational amplifiers, operational transconductance amplifiers, second-generation current conveyors and current feedback operational amplifiers. Various configurations for realizing grounded/floating, positive/negative voltage-controlled impedance circuits have been detailed out, and some of their representative applications have been illustrated.

7.1 Introduction

It is well known [1,2] that a junction field-effect transistor (JFET), when operated below pinch-off, and a metal oxide field-effect transistor (MOSFET), when operated in *triode* region, realize a voltage-controlled resistance (VCR) which is, in general, nonlinear due to the presence of square term in their I_D versus V_{DS} characteristics when operated as a linear VCR (see Figure 7.1), the drain current of a JFET is given by

$$I_D = \frac{2I_{DSS}}{V_p^2}\left[V_{GS} - V_p - \frac{V_{DS}}{2}\right]V_{DS} \tag{7.1}$$

where I_{DSS} is the saturation current and V_P is the pinch-off voltage.

When V_{DS} is kept small (typically ≤ 50 mV), the square term V_{DS}^2 is smaller and can be neglected thereby leading to

$$I_D \cong \frac{2I_{DSS}}{V_p^2}\left[V_{GS} - V_p\right]V_{DS};\;\; R_{eq} \cong \frac{V_p^2}{2I_{DSS}\left[V_{GS} - V_p\right]}; \tag{7.2}$$

It has also been known that this linear range can be extended by applying a simple resistive feedback around the FET as shown in Figure 7.2 with the gate voltage modified as

$$V_{GS} = \frac{(V_C + V_{DS})}{2} \tag{7.3}$$

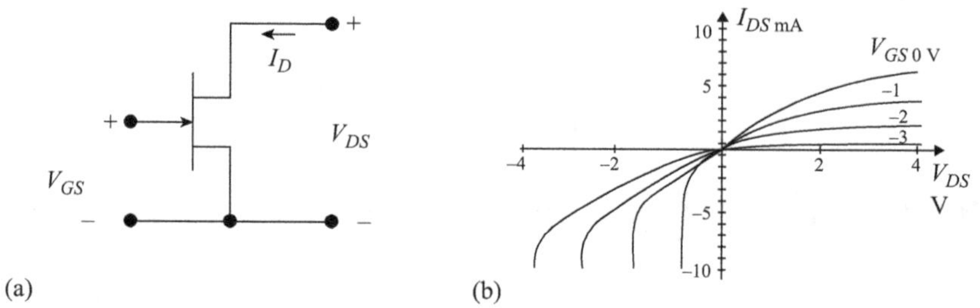

Figure 7.1 FET as a VCR: (a) the basic VCR configuration and (b) a typical $I_D - V_{DS}$ characteristics of a FET

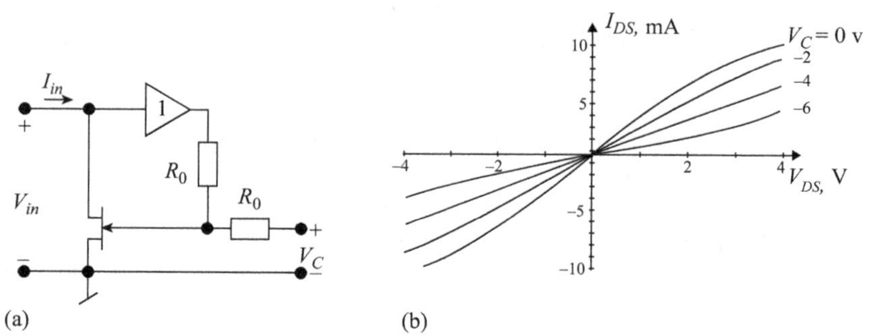

Figure 7.2 Non-linearity cancellation in the FET-VCR [3]: (a) the circuit and (b) the $I_D - V_{DS}$ characteristics showing extended linear range of operation

The unity gain voltage follower in Figure 7.2 can be realized by an op-amp configured as a follower.

I_{in} can now be expressed as

$$I_{in} = I_D = \frac{2I_{DSS}}{V_p^2}\left[\frac{(V_C + V_{DS})}{2} - V_p - \frac{V_{DS}}{2}\right]V_{DS} = \frac{2I_{DSS}}{V_p^2}\left[\frac{V_C}{2} - V_p\right]V_{DS} \tag{7.4}$$

from where it is seen that the square nonlinear term gets cancelled, thereby leading to an extension of the linear operating range of the VCR (Figure 7.2(b)).

If an n-channel enhancement-type MOSFET is used instead of JFET in the schematic of Figure 7.2, (7.1) would be replaced by

$$I_d = \mu_s C_{ox}\left(\frac{W}{L}\right)\left[(V_{GS} - V_{TH}) - \frac{V_{DS}}{2}\right]V_{DS} \tag{7.5}$$

where μ_s is the surface mobility of the majority carriers, C_{ox} is the gate capacitance per unit area, W and L are the width and length of the channel, respectively, (W/L) represents the aspect ratio of the MOSFET and V_{TH} is the threshold voltage.

With the V_{GS} arranged as per (7.3) the square nonlinearity of (7.5) would be cancelled out to yield

$$I_{in} = \mu_s C_{ox}\left(\frac{W}{L}\right)\left(\frac{V_C}{2} - V_{TH}\right)V_{in} \tag{7.6}$$

which implies that the equivalent linear resistance realized by the circuit is given by

$$R_{in} = \frac{1}{\mu_s C_{ox}(W/L)[(V_C/2) - V_{TH}]} \tag{7.7}$$

Thus, all the circuits and techniques which are discussed in this chapter employing a JFET (as they were originally conceived by their proposers) equally apply if JFET in the concerned circuits is replaced by MOSFET.

7.2 Grounded VCZ realization using op-amps

In spite of the emergence of newer analogue circuit building blocks such as operational transconductance amplifiers (OTA) [4,5], second-generation current conveyors (CCII) [6,7] and current feedback operational amplifiers (CFOA) [8–12] – all of which are now also commercially available as off-the-shelf ICs, the traditional voltage mode op-amp (VOA), which is considered to be the workhorse of analogue circuit design is still dominantly employed in practical analog circuit applications. Since a large number of voltage-controlled impedance (VCZ) circuits have been proposed and built around VOAs, we shall discuss in this chapter the various VCZ configurations realizable with VOAs. Subsequently, we shall also discuss VCZ configurations evolved using CFOAs, CCIIs and operational amplifiers.

7.2.1 *Nay–Budak voltage-controlled resistors with extended dynamic range*

Nay and Budak [3] were the first to present an interesting technique of realizing a linear VCR with wide dynamic range using a JFET and IC op-amps. A simplified form of their proposition is shown in Figure 7.3.

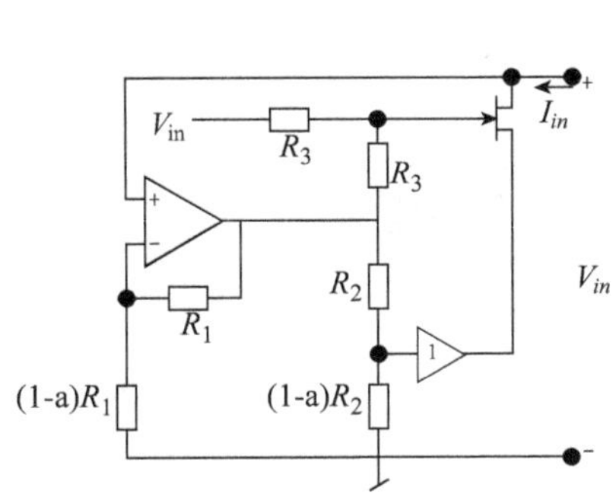

Figure 7.3 VCR with extended dynamic range adapted from Nay and Budak [3]

Assuming ideal op-amps, it can be readily determined by analysis that the op-amp circuit modifies the V_{GS} and the V_{DS} applied to the FET as follows:

$$V_{GS} = \frac{(V_C + aV_{in})}{2}; \quad V_{DS} = aV_{in} \tag{7.8}$$

As a consequence, the expression for the input current (I_{in}) of the circuit can be written as

$$I_{in} = I_{DS} = \frac{2I_{DSS}}{V_p^2}\left[\frac{(V_C + aV_{in})}{2} - V_p - \frac{aV_{in}}{2}\right]aV_{in} = \frac{2I_{DSS}}{V_p^2}\left[\frac{V_C}{2} - V_p\right]aV_{in} \tag{7.9}$$

From the above, the input resistance (R_{in}) of the circuit is given by

$$\begin{aligned} R_{in} &= \frac{V_{in}}{I_{DS}} = \frac{V_p^2}{2I_{DSS}\left[(V_C/2) - V_p\right]a} = \left(\frac{1}{a}\right)R_{DS} \quad \text{where} \\ R_{DS} &= \frac{V_p^2}{2I_{DSS}\left[(V_C/2) - V_p\right]} \end{aligned} \tag{7.10}$$

Nay and Budak in [13] (as a sequel to [3]) demonstrated that the same configuration with parameter $a < 0$ can realize a voltage-controlled linear negative resistance of the same value (see (7.10)). In this circuit, it may be noticed that V_{in} is equal to $(1/a)V_{DS}$. Therefore, smaller the value of the parameter a, wider would be dynamic range for any specified amount of acceptable distortion [3].

It has been experimentally demonstrated in [3] that with $a = 0.1$ and using LF 356 op-amps and VCR4N JFET, the circuit of Figure 7.3(a) exhibits an excellent linearity for input voltage range of ±4 V. The experimental results for the voltage-controlled-negative resistance (VCNR) mode of the circuit are also shown in [13]. The v–i characteristics of the circuit for the positive and negative values of the parameter a are shown in Figure 7.4.

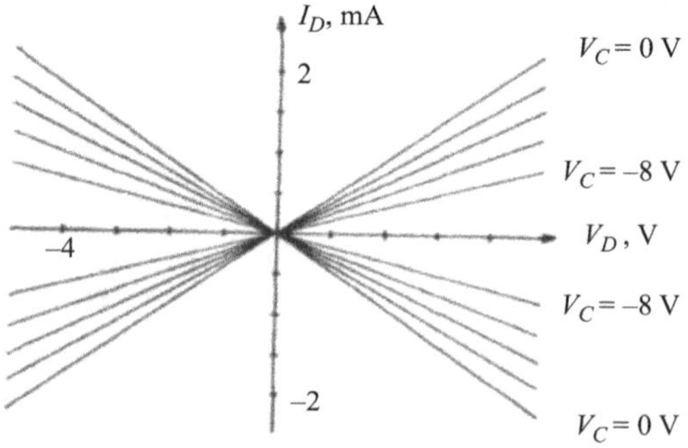

Figure 7.4 v–i characteristics of the circuit of Figure 7.3 for a > 0 (VCPR) as well as a < 0 (VCNR)

It may be pointed out that the dynamic range depends upon the parameter a. However, if the FET is limited to some maximum V_{DS}, say $V_{DS(\max)}$, then the maximum extended range up to which the circuit can operate as intended, would be given by $V_{D(\max)} = V_{DS(\max)}/|a|$. Hence, smaller the value of $|a|$, wider will be the dynamic range. However, to avoid the op-amps from being saturated, V_D is required to be kept less than about half of the DC power supply voltage. But the frequency range of operation of the circuit could be extended by choosing a wider bandwidth FET as well as wider bandwidth op-amps.

7.2.2 Senani–Bhaskar VCZ configurations

Improvizing upon the idea contained in [3,13], it was thought that if it be possible to make the parameter a as the ratio of two impedances (e.g., something like $a = Z_1/Z_2$) and instead of the parameter a appearing at two places as in Nay and Budak [3] circuit (Figure 7.3), if an alternative circuit could be devised in which a could be made to appear only once in the circuit then such a modified circuit would be more versatile than Nay and Budak circuit. Through a systematic synthesis,[1] Senani and Bhaskar in [14] (also see [15]) presented two circuits – one for the realization of positive VCZ and the other for the realization of a negative VCZ. The circuit for positive VCZ from [14] is shown here in Figure 7.5, the analysis of which reveals that the output voltage of the op-amp A_2 is given by

$$V_{02} = (1 - a)V_{in}; \quad \text{where } a = \frac{Z_1}{Z_2} \tag{7.11}$$

On the other hand, the output voltage of the op-amp A_1 is given by

$$V_{01} = \left(2 - \frac{Z_1}{Z_2}\right)V_{in} \tag{7.12}$$

It follows, therefore, that V_{DS} and V_{GS} would be

$$V_{DS} = \left(\frac{Z_1}{Z_2}\right)V_{in}; \quad V_{GS} = \frac{V_C + (Z_1/Z_2)V_{in}}{2} \tag{7.13}$$

Therefore, the input current I_{in} is given by

$$\begin{aligned} I_{in} = I_D &= \frac{2I_{DSS}}{V_p^2}\left[\frac{(V_C + (Z_1/Z_2)V_{in})}{2} - V_p - \frac{V_{in}(Z_1/Z_2)}{2}\right]V_{in}\left(\frac{Z_1}{Z_2}\right) \\ &= \frac{2I_{DSS}}{V_p^2}\left[\frac{V_C}{2} - V_p\right]V_{in}\left(\frac{Z_1}{Z_2}\right) \end{aligned} \tag{7.14}$$

which is simplified to

$$I_D = \left(\frac{Z_1}{Z_2}\right)\left(\frac{1}{R_{DS}}\right)V_{in}, \text{ therefore } Z_{in} = \frac{V_{in}}{I_{in}} = \left(\frac{Z_2}{Z_1}\right)R_{DS} \tag{7.15}$$

[1]Senani R. 'On the realisation of voltage-controlled impedance elements. IEEE Transactions on Circuits and Systems'. 1986; **CASL-321** (unpublished).

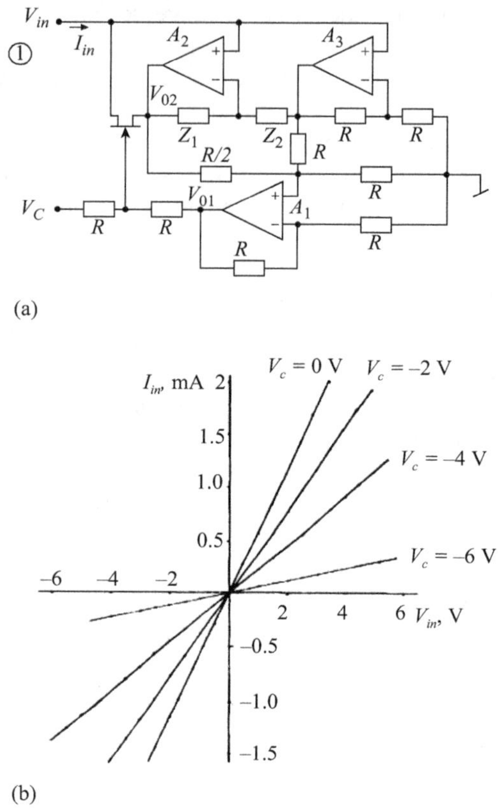

Figure 7.5 VCZ configuration [14]: (a) the circuit and (b) the measured $V_{in} - I_{in}$ characteristics

From (7.15) it can be seen that different kinds of voltage-controlled impedances can be realized by resistive/capacitive choice of impedances Z_1 and Z_2 as follows:

1. a linear VCPR is realized when $Z_1 = R_1$ and $Z_2 = R_2$
2. a linear voltage-controlled positive inductance (VCPL) is realized when $Z_1 = 1/sC_1$ and $Z_2 = R_2$, with the value of the realized inductance given by $L = C_1R_2R_{DS}$
3. a linear voltage-controlled positive capacitance (VCPCC) is realized when $Z_1 = R_1$ and $Z_2 = 1/sC_2$. The value of the realized capacitance (C) is given by

$$C = \frac{C_2R_1}{R_{DS}} \tag{7.16}$$

A nonideal analysis of these cases has been carried out in [14] taking the gain of the three op-amps to be represented as $A_i = \omega_{ti}/s;\ \ i = 1 - 3$ for $\omega \gg \omega_{pi}$, where ω_{ti}

is the gain bandwidth product of the ith op-amp and ω_{pi} is the dominant pole of the open-loop transfer function of the ith op-amp. The nonideal expression for the Y_{in} of the configuration of Figure 7.5(a) has been found [14] to be approximated as

$$Y_{in} \cong \frac{a}{R_{DS}}\left[\frac{1/2+(((1+a)/2A_2a)-(1/A_3))+((1+a)/aA_2A_3)}{1/2+((1/A_3)+((1+a)/2A_2))+((1+a)/A_2A_3)}\right] \tag{7.17}$$

The stability properties of the VCPR, VCPL and VCPC can be ascertained by examining the poles and zeros of the above function, as in the following cases.

Case 1: VCPR (in case $a = R_1/R_2$ and using $a \ll 1$). Under this case, the nonideal Y_{in} becomes

$$Y_{in} \cong \left(\frac{a}{R_{DS}}\right)\left[\frac{(s^2/a)+s((\omega_{t3}/2a)-(\omega_{t2}/(1+a)))+(\omega_{t2}\omega_{t3}/2(1+a))}{s^2+s((\omega_{t3}/2)+(\omega_{t2}/(1+a)))+(\omega_{t2}\omega_{t3}/2(1+a))}\right] \tag{7.18}$$

Since $(\omega_{t3}/2a)-(\omega_{t2}/(1+a))$ is always positive as $a \ll 1$, follows that all the poles and zeros of Y_{in} are located in left half of s-plane, and thus, the VCPR circuit is open circuit stable (OCS) as well as short circuit stable (SCS).

Case 2: VCPL (in this case $a = 1/sC_1R_2 = 1/s\tau_1$). Equation (7.12) becomes

$$Y_{in} \cong \left(\frac{1}{s\tau_1 R_{DS}}\right)\left[\frac{s^3+s^2((1/\tau_1)+(\omega_{t3}/2))+(s/\tau_1)[(\omega_{t3}/2)-\omega_{t2}]+\omega_{t2}\omega_{t3}/2\tau_1}{(s^2/\tau_1)+(s/\tau_1)((\omega_{t3}/2)+\omega_{t2}+(1/\tau_1))+(\omega_{t3}/2\tau_1)(\omega_{t2}+(1/\tau_1))}\right] \tag{7.19}$$

Since the poles of the nonideal admittance function are located in left half of the s-plane, the circuit is SCS. However, since $((\omega_{t3}/2)-\omega_{t2})$ is negative, the circuit is OCS. Experiments have confirmed this. However, in applications such as those of filters/oscillators where input port of the VCPL would be shunted by appropriate impedance, the circuit is still usable in such applications.

Case 3: VCPL ($a = sC_2R_1 = s\tau_1$)

In this case, the applicable nonideal Y_{in} is

$$Y_{in} \cong \left(\frac{s\tau_2}{R_{DS}}\right)\left[\frac{(s^2/\tau_2)+(s/\tau_2)((1/\tau_2)-\omega_{t2}+(\omega_{t3}/2))+(\omega_{t3}/2\tau_2)((1/\tau_2)+\omega_{t2})}{s^3+s^2((\omega_{t3}/2)+(1/\tau_2))+s((\omega_{t2}/\tau_2)+(\omega_{t3}/2\tau_2))+(\omega_{t2}\omega_{t3}/2\tau_2)}\right] \tag{7.20}$$

In this case, all poles of the function are located in left half of s-plane and the circuit is, thus, SCS. However, for the circuit to be OCS too, the following condition needs to be satisfied $((1/\tau_2)-\omega_{t2}+(\omega_{t3}/2))>0$; thus, τ_2 needs to be chosen judiciously which does appear to be too restrictive. Alternatively, if the polarities of both A_2 and A_3 are reversed, the nonideal expression of Y_{in} modifies to

$$Y_{in} \cong \left(\frac{s\tau_2}{R_{DS}}\right)\left[\frac{(s^2/\tau_2) + (s/\tau_2)((1/\tau_2) + (\omega_t/2)) + (\omega_t/2\tau_2)(\omega_t - (1/\tau_2))}{s^3 + s^2((1/\tau_2) - (\omega_t/2)) - s(3\omega_t/2\tau_2) + (\omega_t^2/2\tau_2)}\right] \tag{7.21}$$

With the changes, although the circuit becomes short circuit unstable, but it can be made OCS by choosing τ_2 such that $\tau_2 > 1/\omega_t$, this alternative choice offers more flexibility in choosing τ_2 than the condition $((1/\tau_2) - \omega_{t2} + (\omega_{t3}/2)) > 0$. These assertions about the stability behaviour of the circuit of Figure 7.5, in all its three modes, have been experimentally confirmed in [14].

In [14], the experimental evidence of the workability of the circuit of Figure 7.5(a) has been demonstrated by building the circuit using μA741 type op-amps and BFW10 type JFETs. The practically observed *v–i* characteristics of the VCPR is shown in Figure 7.5(b) which demonstrates excellent linearity with maximum signal handling capability (SHC) of the circuit to be around ±5 V (*p–p*) with op-amps biased with ±12 V dc supply.

Application: The use of VCLC to realize a bandpass filter with voltage-controllable centre frequency was verified with component values $C_0 = 0.012$ μF, $R_0 = 2.7$ kΩ, and $C_1 = 0.12$ μF, $R_2 = 1$ kΩ and a typical frequency response for $V_C = -5.0$ V are shown in Figure 7.6(a). The centre frequency was found to be variable from 12.5 kHz to 5.95 kHz by varying V_c from −1 V to −6 V, while the SHC was found to be of the same order as that of the VCPR.

The use of the VCLC as a voltage-controlled oscillator (VCO) is shown here in Figure 7.6(b). The VCO circuit has been obtained by augmenting the VCLC by resistors R_0, R_0' and capacitor C_0. The variability of the oscillation frequency with V_C was experimentally verified. Over the entire range of frequency controllable by V_C, the circuit exhibited good quality sine waves [14].

Like Nay and Budak's negative resistance configuration, Senani and Bhaskar [14] also proposed a voltage-controlled negative impedance (VCNZ) circuit whose simplified form has been shown here in Figure 7.7(a). By straightforward analysis, it can be easily derived that the equivalent input impedance realized by this circuit is given by:

$$Z_{eq} = -\left(\frac{Z_2}{Z_1}\right)R_{DS} \tag{7.22}$$

where the value of R_{DS} is the same as in (7.10). It is, thus, obvious that this circuit can realize a VCNR, a voltage-controlled negative inductance (VCNL) and voltage-controlled negative capacitance (VCNC) by appropriate choice of impedances Z_1 and Z_2. The experimentally observed $V_{in} - i_{in}$ characteristics of the VCNR are shown in Figure 7.7(b) which shows excellent linearity.

7.2.3 Leuciuc–Goras VCZ configurations based upon GIC

Leuciuc–Goras [16] presented two GIC-based configurations for realizing VCZ whose schematics are shown in Figure 7.8(a) and (b). Under the assumptions of

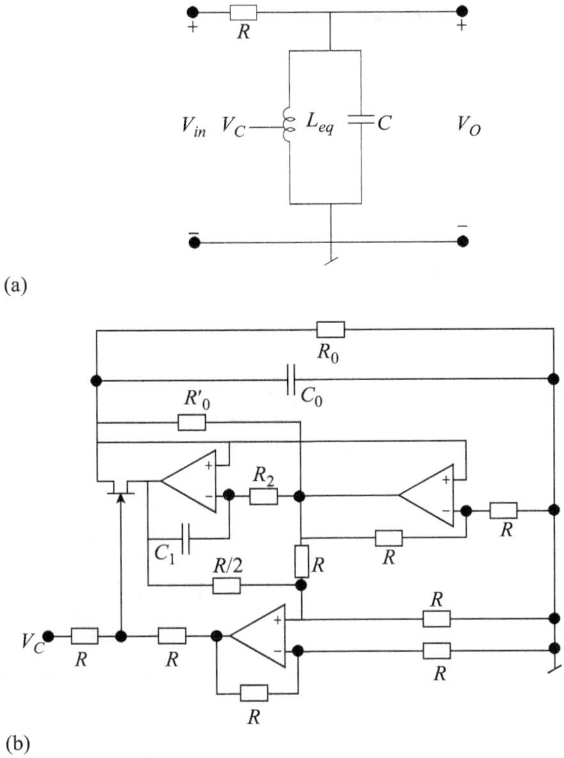

Figure 7.6 Applications of the VCL circuit [14]: (a) a BP Filter with voltage-tuneable centre frequency and (b) a sinusoidal VCO

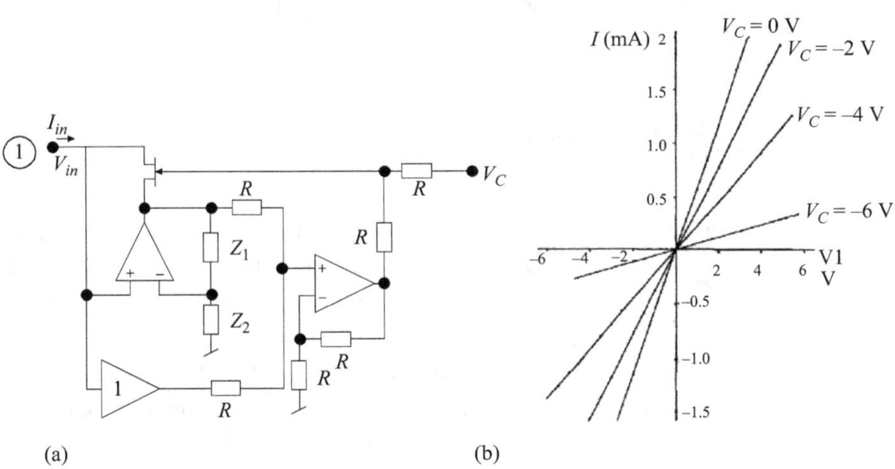

Figure 7.7 The schematic of the voltage-controlled negative impedance realization [14]: (a) the configuration and (b) the measured $V_{in} - I_{in}$ characteristics as a VCNR

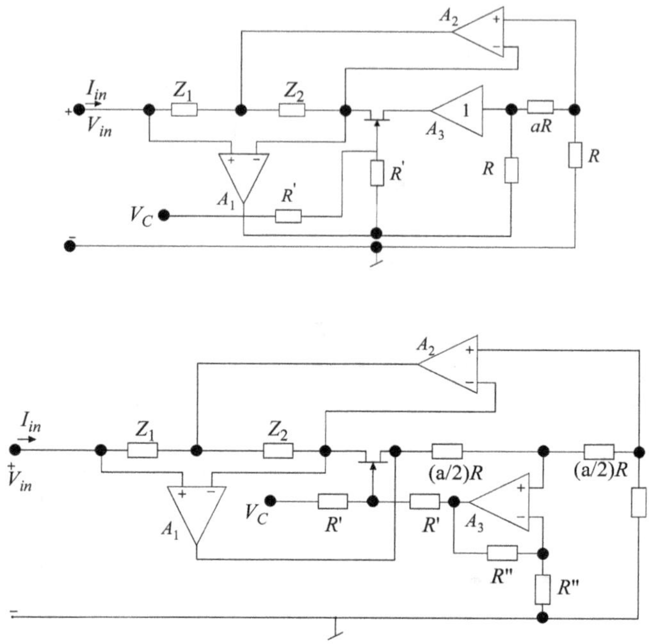

(a)

(b)

Figure 7.8 The GIC-based linear VCZ configurations [16]: (a) the first circuit and (b) the alternative circuit

ideal op-amps and the JFET operating in triode region, the input impedance of the circuit of Figure 7.8 is found to be

$$Z_{eq} = -\left(\frac{Z_1}{aZ_2}\right)R_{DS}, \quad \text{where } R_{DS} = \frac{V_p^2}{I_{DSS}\left[V_C - 2V_p\right]} \tag{7.23}$$

The novelty of this circuit is that the drain to source voltage of the JFET is given by parameter a, and thus, it is independent of Z_1 and Z_2. Thus, as compared to the VCZ configuration of [14], these circuits provide the following advantages:

1. Since V_{DS} of the JFET is independent of Z_1 and Z_2, the linear dynamical range is frequency-independent even for realizing VCPL and VCPC elements.
2. The circuit of [14] requires a large number of resistors (as many as 11), but, by contrast, this circuit employs a reduced number of only 5–7 resistors.
3. The circuit is OCS as well as SCS for all the three cases. Thus, the circuit is unconditionally stable.

A nonideal analysis of the circuit, assuming $A_k \cong \omega_{tk}/s; \quad k = 1 - 3$ reveals [16] the following nonideal expression:

$$Z_{in} = Z_1 \frac{R_{DS}\left[1+\left(\frac{(2+a)}{A_1}\right)\left(1+\left(\frac{1}{A_2}\right)\right)\right]\left(1+\left(\frac{1}{A_3}\right)\right)+Z_2\left[\left(\frac{(2+a)}{A_1A_2}\right)\left(1+\left(\frac{1}{A_3}\right)\right)+\left(\frac{(1+a)}{A_2}\right)\right]}{R_{DS}\left(\frac{(2+a)}{A_1}\right)\left(1+\left(\frac{1}{A_2}\right)\right)\left(1+\left(\frac{1}{A_3}\right)\right)+Z_2\left[a+\left(\frac{(2+a)}{A_1A_2}\right)\left(1+\left(\frac{1}{A_3}\right)\right)+\left(\frac{(1+a)}{A_2}\right)-\left(\frac{1}{A_3}\right)\right]} \tag{7.24}$$

which, with $Z_1 = R_1$ and $Z_2 = 1/sC_2$, can be written as (for $\omega \ll \omega_{tk}$) the following after neglecting the second-order terms in (ω/ω_{tk})

$$Z_{in}(j\omega) \cong (j\omega C_2 R_1 R_3) \frac{\left[1+\left(\dfrac{(1+a)}{\omega_{t2}C_2R_3}\right)+j\omega\left(\left(\dfrac{1}{\omega_{t3}}\right)+\left(\dfrac{(2+a)}{\omega_{t1}}\right)+\left(\dfrac{(2+a)}{\omega_{t1}\omega_{t2}C_2R_3}\right)\right)\right]}{a+j\omega\left[\left(\dfrac{(1+a)}{\omega_{t2}}\right)-\left(\dfrac{1}{\omega_{t3}}\right)\right]} \tag{7.25}$$

Thus, high Q voltage-controlled inductance can be attained under the following conditions:

$$\left(\frac{1+a}{\omega_{t3}}+\frac{a(2+a)}{\omega_{t1}}-\frac{1+a}{\omega_{t2}}\right)=0; \quad C_2R_3 \gg \frac{1}{\omega_{tk}} \tag{7.26}$$

For $\omega_{t1} = \omega_{t2} = \omega_{t3} = \omega_t$, the above implies $\omega_{t2} = ((a+1)/(a^2+3a+1))\omega_t$. For instance, if we choose $a = 0.1$, then we need to select op-amps such that $\omega_{t2} = 0.84\omega_t$.

7.2.4 Three-op-amp-based VCZ structure by Senani–Bhaskar

Yet another circuit capable of realizing linear VCPR, VCPL, VCPC elements was presented by Senani and Bhaskar in [17] and is shown in Figure 7.9. This circuit is so devised that the gate voltage of FET is set to

$$V_G = \frac{V_{in}+V_s}{2}+\frac{V_c}{4} \tag{7.27}$$

with source terminal getting a voltage

$$V_s = (1-a)V_{in}; \quad a = \frac{Z_1Z_3}{Z_2Z_4} \tag{7.28}$$

With the above values substituted in (7.1), the square non-linearity is cancelled out and the input impedance of the circuit is given by

$$Z_{in} = \left(\frac{Z_2Z_4}{Z_1Z_3}\right)R_{DS}, \quad \text{where } R_{DS} = \frac{2V_p^2}{I_{DSS}\left[V_C-4V_p\right]} \tag{7.29}$$

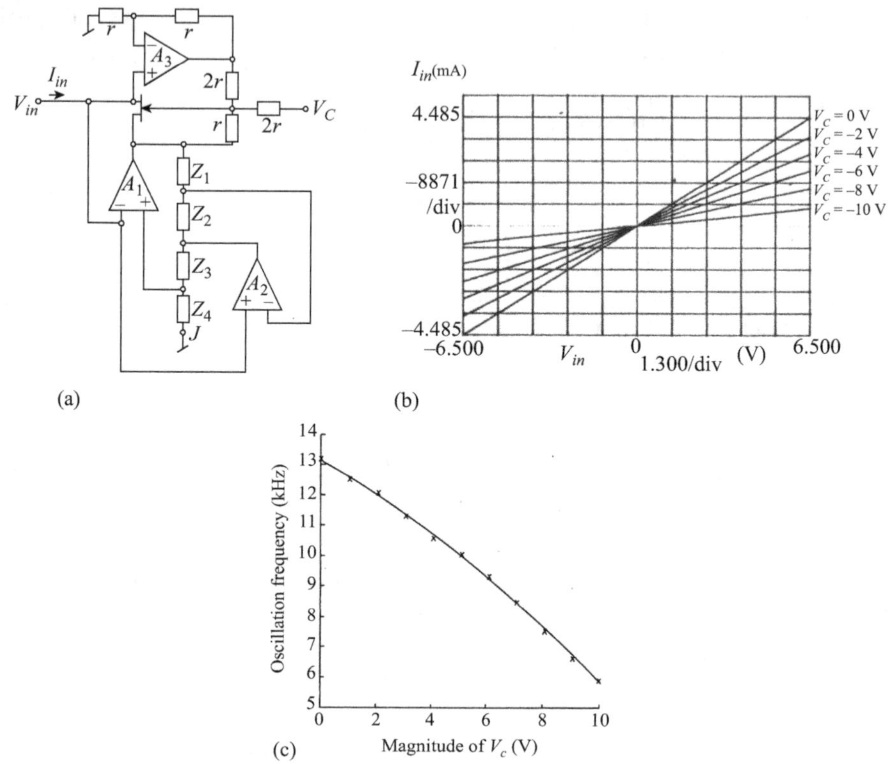

Figure 7.9 Generalized VCZ structure proposed by Senani and Bhaskar [17]: (a) the circuit, (b) the experimentally determined v–i characteristics and (c) variation of oscillation frequency with V_c. © 1994 IEE

From (7.29), it is clear that by appropriate choice of the four impedances, this circuit can not only realize linear VCPR, VCPL and VCPC elements it can also realize VC-FDNR element (by choosing Z_2 and Z_4 both as capacitors) as well as VC-FDNC element (by choosing Z_1 and Z_3 both as capacitors).

Application: The experimentally observed $V_{in} - i_{in}$ characteristics of this circuit realized with BFW10 type JFET, μA741 type op-amps biased with ±15 V dc power supplies using $R_1 = R_2 = R_4 = 10$ kΩ and $R_3 = 1$ kΩ, $r = 10$ kΩ have shown a linear range of about 13 V (*p–p*)! Furthermore, the versatility of the circuit has been confirmed by using it in the same applications as in [14] and in addition, confirming the use of VC-FDNR element in realizing a bandpass filter and a VCO (by shunting the input port of the VC-FDNR by a resistor R_0). The details of these results can be found in [17].

7.2.5 Senani's universal VCZ structure with only two op-amps

A VCZ configuration, which is capable of realizing not only VCPR, VCPL and VCPC elements but also VCNR, VCNL and VCNC elements and, hence, termed as a universal VCZ configuration, was proposed by Senani in [18]. The circuit from [18] is reproduced here in Figure 7.10. Expressing the drain current of the JFET operating in triode region as

$$i_D = \beta\left[V_{GS} - V_p - \frac{V_{DS}}{2}\right]V_{DS} \quad \text{where } \beta = \frac{2I_{DSS}}{V_p^2} \tag{7.30}$$

The circuit has been so devised that the gate to source voltage of the FET is made to be

$$V_{GS} = \left[\frac{a}{2}(V_j - V_i) + \frac{V_c}{4}\right]; \quad \text{where } a = \frac{Z_1}{Z_2} \tag{7.31}$$

On the other hand, the drain to source voltage is made to be $V_{DS} = a(V_j - V_{in})$. It is then found that substituting these values in (7.30), the input current of the circuit is given by

$$i_1 = \beta\left[\frac{V_c}{4} - V_p\right]\left[a(V_j - V_{in})\right] \tag{7.32}$$

The following cases may now be considered.

Case 1: If terminal J is grounded ($V_j = 0$), the input impedance of the circuit is given by

$$Z_{in} = -\left(\frac{Z_2}{Z_1}\right)R_{DS} \tag{7.33}$$

and the circuit, thus, realizes a voltage-controlled generalized negative impedance converter (NIC)/inverter (GNIC/GNII).

Case 2: If terminal J is connected to V_{01} (i.e., $V_j = 2V_{in}$) then

$$Z_{in} = +\left(\frac{Z_2}{Z_1}\right)R_{DS} \tag{7.34}$$

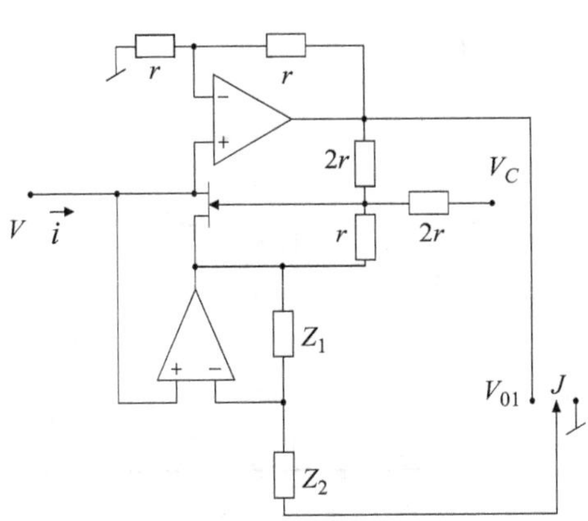

Figure 7.10 Universal VCZ configuration proposed by Senani [18]. © 1994 IEE

and under this condition, the circuit becomes a voltage-controlled generalized positive impedance converter/inverter (GPIC/GPII). When compared with the circuit of [14], this circuit has double the functionality but employs one less op-amp and two less number of resistors. On the other hand, by experiments, it is found [18] that linear range of VCPR and VCNR of this circuit is about two orders of magnitudes larger than that of a conventional FET-based VCR. This circuit too can be used to advantage in all those applications outlined in [14].

In retrospection, it was reported in [18] that the earlier configuration of [17] can also be endowed with these properties, i.e. capability to realize a generalized positive impedance $Z_{in} = +(Z_2Z_4/Z_1Z_3)R_{DS}$ or a generalized negative impedance $Z_{in} = -(Z_2Z_4/Z_1Z_3)R_{DS}$ by similar artifice of appropriately connecting the terminal marked J to either ground or to the output of the first op-amp.

7.2.6 Ndjountche configuration using MOS resistive circuit

A circuit which retains all the functionalities of the universal VCZ configuration of Senani [18] but does not need a switch to alternate between the positive and negative VCZ functions was advanced by Ndjountche [19] which is reproduced here in Figure 7.11. Assuming that all MOSFETs are matched and are operating in triode region, their drain currents can be expressed as

$$I_D = 2K\left[(V_{GS} - V_p)V_{DS} - \frac{V_{DS}^2}{2}\right]; \quad \text{where } K = \mu_s C_{OX}\frac{W}{2L} \tag{7.35}$$

A straightforward analysis of the circuit yields

$$Z_{eq} = 2KR(V_{c1} - V_{c2})\frac{Z_1Z_3}{Z_2} \tag{7.36}$$

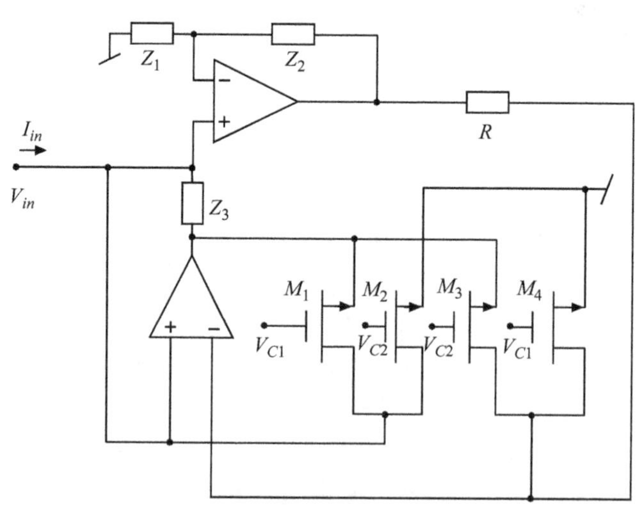

Figure 7.11 A VCZ configuration using MOS-resistive-circuit due to Ndjountche [19]. © 1996 IET

Thus, by appropriate choice of the two control voltages, the circuit can realize a variety of positive as well as negative voltage-controlled impedances, i.e. VCR, VCL, VCC and VC-FDNR in their positive as well negative forms.

Application: It may be mentioned that the circuit functions well for frequency ranges given by $f \ll 0.1f_t$, where f_t is the gain bandwidth product of the op-amps employed. The workability of this circuit was established in [19] from experimental results of a centre frequency tunable bandpass filter using dual op-amps and MOS transistor array. The filter was tunable by varying the control voltages to achieve to f_0 values of 1, 2, 3, 4 and 5 kHz.

7.2.7 Economical VCZ configurations

As compared to the various VCZ configurations described earlier, Senani and Bhaskar in [20] presented a structure employing only two unity gain amplifiers and only four resistors to realize a VCR. This circuit from [20] is reproduced here in Figure 7.12(a). The input resistance of this circuit is given by

$$R_{eq} = R_1\left[1 + \left\{\frac{V_p^2}{I_{DSS}(V_c - 2V_p)R_2}\right\}\right] \tag{7.37}$$

which is controllable by V_c. Since

$$V_{GS} = \left(\frac{V_c + V_{DS}}{2}\right) \tag{7.38}$$

the square nonlinearity is cancelled. Furthermore, since R_2 along with the drain to source resistance of FET constitutes a potential divider. Thus, the mentioned factors make it possible to extend the linear range of the R_{eq} realized by this circuit.

If resistors R_1 and R_2 are replaced by general impedances Z_1 and Z_2, the equivalent impedance realized is given by

$$Z_{eq} = Z_1\left[1 + \left\{\frac{V_p^2}{I_{DSS}}(V_c - 2V_p)Z_2\right\}\right] \tag{7.39}$$

with resistive/capacitive choices of Z_1 and Z_2, the following two special cases are of interest.

Case 1: With $Z_1 = R_1$ and $Z_2 = 1/sC_2$, the circuit realizes a lossy VCL (series RL) for which equivalent parameters are given by

$$L_{eq} = \frac{R_1 C_2 V_p^2}{I_{DSS}(V_c - 2V_p)}; \quad R_{eq} = R_1 \tag{7.40}$$

Case 2: With $Z_1 = 1/sC_1$ and $Z_2 = R_2$, the circuit realizes a lossy VCC

$$C_{eq} = \frac{C_1}{\left[1 + \left\{\left(V_p^2/I_{DSS}\right)(V_c - 2V_p)R_2\right\}\right]} \tag{7.41}$$

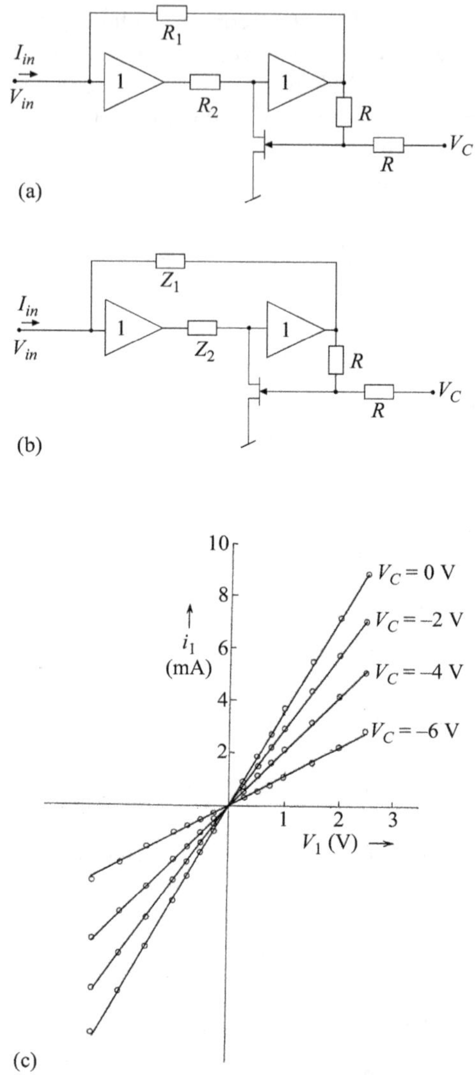

(a) (b) (c)

Figure 7.12 Economical VCZ configurations due to Senani and Bhaskar [20]: (a) VCR, (b) general configuration and (c) experimentally determined v–i characteristics of the circuit of Figure 7.12(a)

It is worth mentioning that the implementation of the VCR can be simplified by removing the output buffer in which case, the realized equivalent resistance assumes the following form:

$$R_{eq} = \frac{R_2 + R_{DS}}{1 + (R_2/R_1)} \cong R_1 + \frac{R_1 R_{DS}}{R_2}; \quad \text{provided } \frac{R_2}{R_1} \gg 1 \tag{7.42}$$

with

$$R_{DS} = \frac{V_p^2}{I_{DSS}[V_C - 2V_p]} \tag{7.43}$$

The experimentally observed $V_{in} - I_{in}$ characteristics where the circuit was realized with μA741 type op-amps biased with ±15 V dc and n-channel JFET BFW10 was employed along with passive components $R_1 = R_2 = 100\ \Omega$, $R = 10\ \text{k}\Omega$ shows excellent linearity and signal handling capacity of several volts. The realized VCR has an applicable frequency range of about 25 kHz with maximum signal handling capacity of around 6 V (*p–p*) for $V_c = -2$ V for maximum allowable distortion of 1%. From a nonideal analysis of the circuit taking into account the op-amp pole, the nonideal equivalent circuits of the realized VCZs, distortion analysis and CMOS-compatible versions of the circuits, the reader is referred to [20]. The nonideal analysis [20] shows, however, that the VCZ of Figure 7.12 is both OCS and SCS under all special cases.

Application: In [21] Yu and Tseng have proposed a sensorless control scheme based on virtual neutral voltage for brushless direct current motor drive. This method requires a practical circuit for commutating signal generation. The economical VCR of Figure 7.12(a) has been employed therein[2] as a part of the feedback system. This application demonstrated that the circuit of the kind described in the chapter has the potential of being incorporated in real-life applications in various areas of engineering.

7.3 Grounded VCZ configurations using CFOAs

When CFOAs are employed as active elements, one should expect a higher frequency operational range and increased signal handling capacity (due to very high slew rate) and employment of a reduced number of external passive components which is so characteristic with both CFOAs and current conveyors (CC). Two low component count VCZ configurations were reported in [22] by Senani which are shown in Figure 7.13. The first circuit realizes

$$Z_{in}(s) = \left(\frac{Z_2}{Z_3}\right) R_{DS} \tag{7.44}$$

whereas the second one realizes

$$Z_{in}(s) = \left(\frac{Z_1 Z_2}{R_{DS}}\right) \tag{7.45}$$

[2]While describing this circuit however, the authors have stated '. . . employs two op-amps (one connected as a noninverting amplifier, the other connected as a unity gain buffer). . . '. This statement is, however, not correct since in the circuit of Figure 7.12(a) (employed in [21]) both the op-amps are configured as unity gain buffers only.

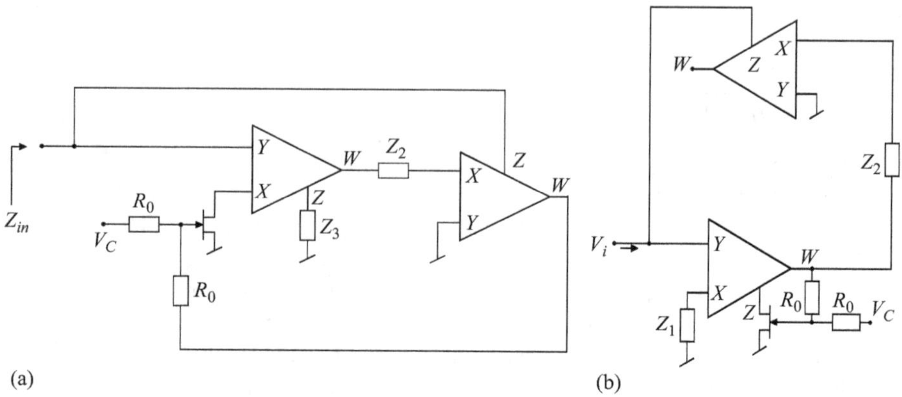

Figure 7.13 Two grounded VCZ configurations (a), (b) using CFOAa [22]

It is, thus, obvious that both the circuits can be used to realize a variety of VCZ like VCR, VCC, VCL and VC-FDNR by appropriate choice of circuit impedances Z_1, Z_2 and Z_3.

7.4 VCZ configurations using current conveyors

Maundy *et al.* [23] presented voltage/current-controlled grounded resistor circuits using current conveyors which were implemented using AD844 and are based upon ideas which are more or less similar to those described in this chapter. One such circuit from [23] is shown here in Figure 7.14 which realizes

$$R_{in} = \frac{V_{in}}{I_{in}} = \frac{V_p^2}{2I_{DSS}} \frac{1}{a(V_c - V_p)} \quad \text{where } a = \frac{R_1}{R_1 + R_2} \tag{7.46}$$

However, curiously, the authors of [23] failed to mention most of the earlier works on such voltage-controlled impedance circuits of the research group to which the authors of this monograph belong, see [24,25].

7.5 The floating VCR

We now present a floating VCR (FVCR) circuit from [26] which was inspired by the ideas contained in [27]. Consider the general scheme of Figure 7.15(a) which is used to generate an appropriate gate voltage as a function of V_1, V_2 and control voltage V_c such that the square nonlinearity is cancelled and a linear voltage-controlled floating resistance is realized between ports 1 and 2. Assuming that $(V_1 - V_2)$ is small enough to let the FET operate in triode region where we use the equation

$$I_D = \beta\left[V_{GS} - V_p - \frac{V_{DS}}{2}\right]V_{DS}; \quad \text{where } \beta = \frac{2I_{DSS}}{V_p^2} \tag{7.47}$$

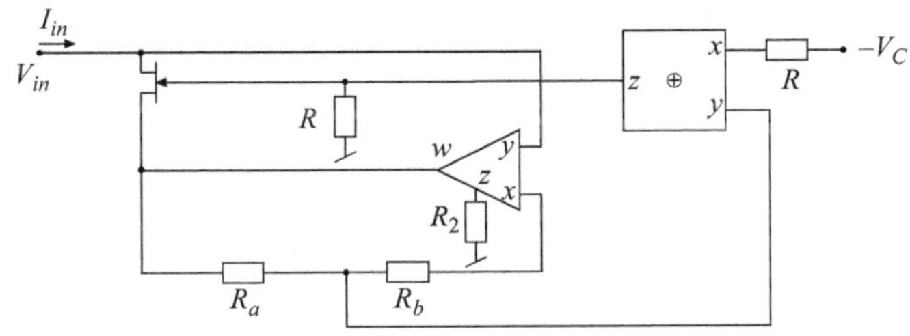

Figure 7.14 Compact representation of VCR configuration using current conveyors by Maundy et al. [23]

By straightforward analysis, it can be verified that the two-port of Figure 7.15(a) is characterized by

$$\begin{aligned} i_1 &= \beta\left\{V_1\left(K_1 - \frac{1}{2}\right) + V_2\left(K_2 - \frac{1}{2}\right) + K_3 V_c - V_p\right\}(V_1 - V_2) \quad \text{and} \\ i_2 &= -i_1 \end{aligned} \tag{7.48}$$

Thus, it turns out that for cancellation of nonlinearity one needs $K_1 = K_2 = 1/2$, whereas K_3 can have any arbitrary value. If we choose K_1 also as 1/2, the resulting circuit can be implemented as shown in Figure 7.15(b). This circuit realizes an FVCR of value

$$R_{eq} = \frac{V_p^2}{I_{DSS}\left[V_C - 2V_p\right]} \tag{7.49}$$

A simplified version of the circuit is obtainable as in Figure 7.15(c) by omitting two voltage followers provided the resistors R used in the summing network and the noninverting amplifier are chosen of large values (say 1 MΩ–10 MΩ).

The experimental results of this FVCR using BFW10 JFET and μA741 IC op-amps biased with ±15 V dc power supplies, with $R = 1$ MΩ as shown in Figure 7.15(d) demonstrate good linearity over extended voltage range of about one order of magnitude larger than applicable to a simple FET-VCR, whereas the operating frequency range of the circuit when realized with μA741 type of op-amps was found to be around 30 kHz [26].

Application: An interesting application of the economical F-VCR of Figure 7.15(c) is in converting the classical single-op-amp-based astable multivibrator into a voltage-controlled square wave generator wherein the floating resistor in the negative feedback path of the multivibrator is replaced by the F-VCR. It was found that the frequency of the output wave could be made voltage controllable through V_C.

It must be mentioned that with either port 1 or port 2 shorted to ground one obtains a grounded VCR which employs exactly the same number of components as the FVCR. Thus, using the circuit of Figure 7.15(a) both grounded and FVCRs are realizable with exactly the same number of active and passive components. This feature is particularly significant in several functional circuits in the context of

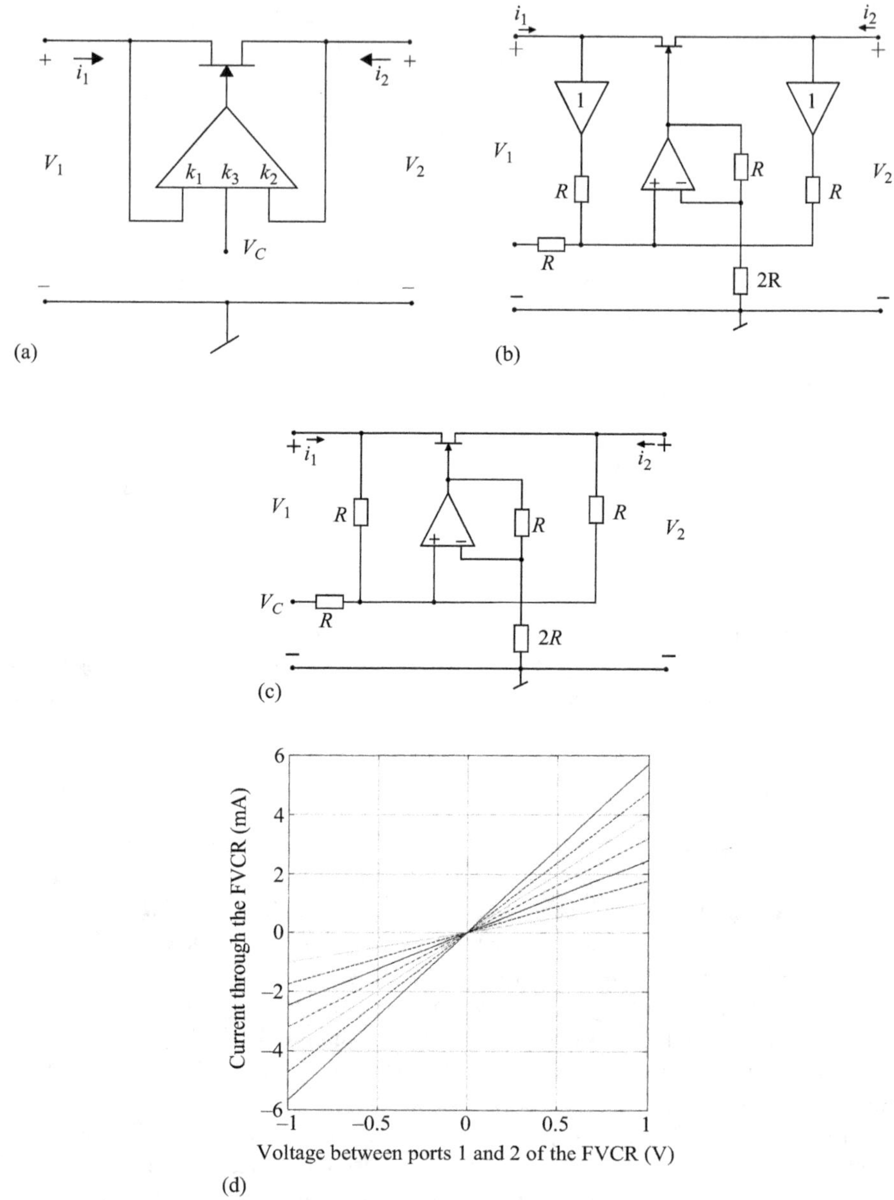

Figure 7.15 Floating voltage-controlled positive resistance due to Senani [26]: (a) the general scheme, (b) circuit implementation, (c) simplified version of the circuit of (b) and (d) experimental results.

voltage-control of intended parameters for which the condition that 'the controlling resistor be grounded' is no longer necessary. In view of this, the VCR structure of Figure 7.15(c) is quite a favourable choice for use in the realization of sinusoidal VCOs, voltage variable capacitors and other voltage-controllable functional circuits.

7.6 The floating/grounded voltage-controlled GIC/GIIs using CFOAs

We now show a floating generalized GPIC/GPII using CFOAs and FET along with its associated nonlinearity cancellation circuitry. One such circuit was first presented by Senani in [22] and is reproduced here in Figure 7.16(a). This circuit is characterized by the $[y]$-matrix:

$$\begin{bmatrix} i_1 \\ i_2 \end{bmatrix} = \left(\frac{R_{DS}}{Z_1 Z_2}\right)\begin{bmatrix} 1 & -1 \\ -1 & 1 \end{bmatrix}\begin{bmatrix} v_1 \\ v_2 \end{bmatrix}; \quad \text{where } R_{DS} = \frac{2V_p^2}{I_{DSS}\left[V_C - 2V_p\right]} \tag{7.50}$$

and realizes an equivalent floating impedance

$$Z_{eq}(s) = \left(\frac{Z_1 Z_2}{R_{DS}}\right) \tag{7.51}$$

between port 1 and port 2 and can, thus, realize a VCR, VCC and VC-FDNR elements. Furthermore, the same kind of elements with negative values can be obtained by the simple artifice of changing the connections [a_1–a_2, b_1–b_2] with [a_1–b_2, b_1–a_2]. Experiments have shown that the $v-i$ characteristics of VCPR and VCNR obtained with $R_1 = R_2 = R_0 = 10\ \text{k}\Omega$ shown excellent linearity over an input range of two orders of magnitude larger than the conventional FET-VCR.

In Figure 7.16(b) [28], we show the most versatile universal floating VCZ configuration which realizes a floating impedance of value

$$Z_{1-2} = \left(\frac{Z_1 Z_3}{Z_2 Z_4}\right) R_{DS} \tag{7.52}$$

from which it follows that linear, floating, positive, VCR, VCL, VCC, VC-FDNR or VC-FDNC all elements can be realized from this circuit by appropriate (resistive/capacitive) choice of circuit impedances Z_1, Z_2, Z_3 and Z_4.

Furthermore, various negative-valued elements corresponding to those mentioned above can be obtained from the circuit of Figure 7.16(b) by the simple artifice of connecting a_1–b_2 and a_2–b_1 in the circuit, due to which the value of realized floating impedance modifies to

$$Z_{1-2} = -\left(\frac{Z_1 Z_3}{Z_2 Z_4}\right) R_{DS} \tag{7.53}$$

It also worth pointing out that grounded form of all the types of impedances described above can be obtained by shorting terminal 1 or terminal 2 to ground. However, with terminal 2 grounded, CFOA$_2$ becomes redundant, whereas

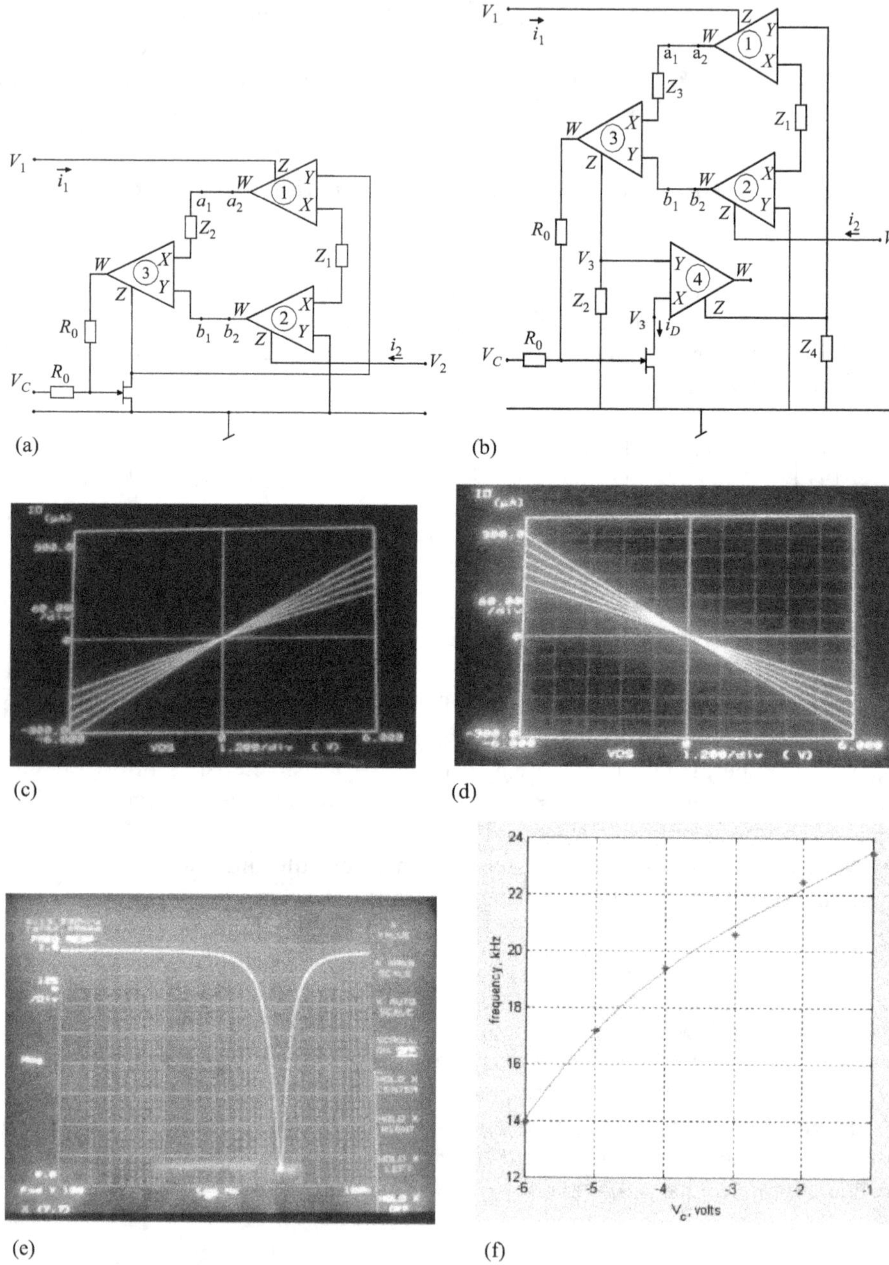

Figure 7.16 *Generalized voltage-controlled floating GIC/GII configurations: (a) Floating GPIC/GPII, (b) a versatile VCZ, (c) v–i characteristic of floating positive VCR, (d) v–i characteristic of floating negative VCR, (e) frequency response of a voltage tunable notch filter and (f) voltage control of centre frequency of a bandpass filter using VC-FDNC. © 2004 Wiley*

y-terminal of $CFOA_3$ and R_1 can be connected to ground directly. Consequently, the circuit gets simplified to have only three CFOAs and is still capable by realizing

$$Z_{in} = \pm\left(\frac{Z_1 Z_3}{Z_2 Z_4}\right) R_{DS} \tag{7.54}$$

In view of its versatility, the circuit of Figure 7.16(b) can be called a universal voltage-controlled grounded/floating, positive/negative GIC/GII configuration.

Experimental results and applications: The experimentally observed *v–i* characteristic of the configuration of the Figure 7.16(b) with $R_1 = R_2 = R_3 = 10\ \text{k}\Omega$, $R_2 = R_4 = 1\ \text{k}\Omega$ with AD844 type CFOA biased with ± 15 V dc power supply in conjunction with BFW11 type JFETs are shown in Figure 7.16(c), which exhibit a linear range of the VCR to be about ± 6 V dc. It is thus seen that this configuration extends the linear range of the VCR by two orders of magnitude larger than conventional FET VCR. Figure 7.16(d) similarly demonstrates the linear range of operation of the negative FVCR also to be around ± 6 V dc. The floating VCL was employed to realize a second-order voltage tunable notch filter having a parallel LC tank circuit in the series arm a resistor in the shunt arm. The filter was designed $f_0 = 15.9$ kHz, $H_0 = 1$ and $Q_0 = 2$ for which resistor values chosen were all 10 kΩ and both capacitors taken as 1 nF. The frequency response of the notch filter for $V_C = -3$ V is shown in Figure 7.16(e). In the last, Figure 7.16(f) shows the voltage controllability of the centre frequency of a bandpass filter designed by using a VC-FDNC.

The reader is referred to [28] for various experimental demonstrations of the VCL and VC-FDNC elements in the design of voltage-controllable notch and bandpass filters. Furthermore, the viability of implementing these configurations in CMOS using CMOS CFOAs has also been demonstrated in [28].

In the last, it must be pointed out that both the circuits discussed in this section although realize floating generalized impedances, they can be readily used to realize various positive/negative impedances of the same kind by shorting port two to ground. This operation, however, makes $CFOA_2$ in both the cases redundant which can be eliminated, thereby reducing one CFOA from each configuration.

7.7 Floating VC-negative-impedance realization using OMAs

Floating negative impedance realization requires normally a cascade of two NICs with impedance embedded in between them, thereby necessitating two op-amps and five impedances with the condition that two pairs of resistances in the two NICs must be matched. Alternatively, one can employ a cascade of two CCII+ elements, both configured as NICs, along with an impedance embedded in between them but such a circuit does not enable the realization of floating negative inductance elements and is limited to the realizing of only floating negative resistors/capacitors. By contrast to these realizations, a circuit for realizing generalized floating VCNZ was proposed by Senani in [29] which is reproduced here in Figure 7.17. The circuit employs two OMA-elements (each of

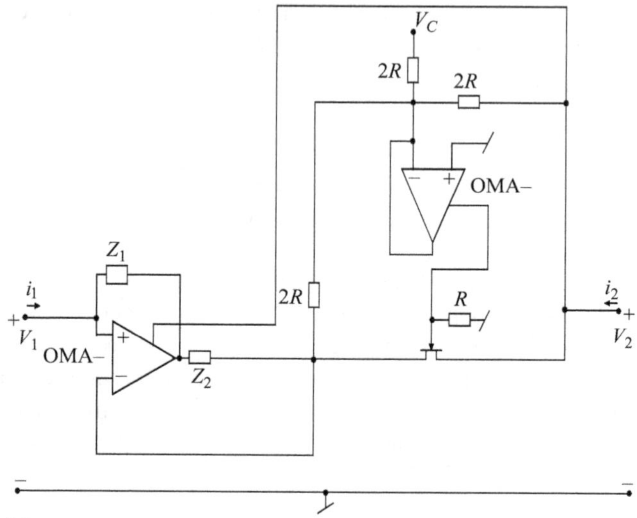

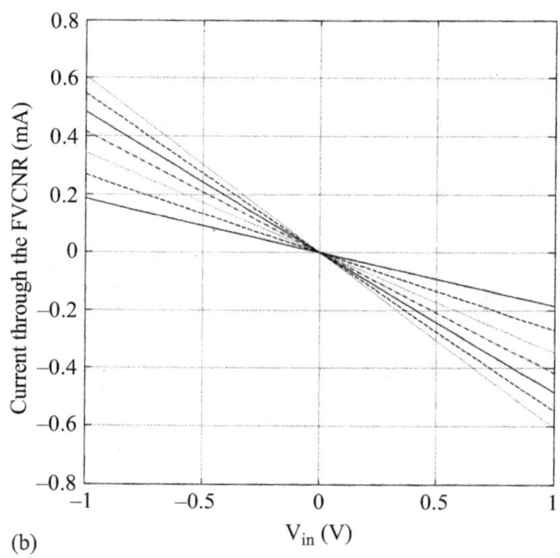

Figure 7.17 A linear VC-floating negative impedance realization due to Senani [29]: (a) the configuration, (b) the v–i characteristics obtained from SPICE simulations. © 1995 IEE

which is equivalent to an FTFN). The floating impedance realized by this circuit is given by

$$Z_{1-2} = -\left(\frac{Z_1}{Z_2}\right) R_{DS}, \quad \text{where } R_{DS} = \frac{V_p^2}{I_{DSS}\left[V_C - 2V_p\right]} \tag{7.55}$$

The workability of the structure of Figure 7.17 was demonstrated by SPICE simulations based upon OMA-implemented from macromodel of AD704 IC op-amps and modified Wilson current mirrors [30].

7.8 Floating/grounded VCZ structures using CFOAs and analogue multipliers

In this section, we present a number of circuit configurations for realizing voltage-controlled impedances using analog multipliers. Two such canonic floating voltage-controlled inductance configurations using only two resistors and a grounded capacitor (as preferred for IC implementation) were presented by Senani *et al.* in [31]. These are shown in Figure 7.18.

Both the circuits of Figure 7.18 are characterized by the [*y*] matrix:

$$\begin{bmatrix} i_1 \\ i_2 \end{bmatrix} = \left(\frac{V_{ref}}{V_c}\right)\left(\frac{1}{sC_0R_1R_2}\right)\begin{bmatrix} 1 & -1 \\ -1 & 1 \end{bmatrix}\begin{bmatrix} v_1 \\ v_2 \end{bmatrix} \tag{7.56}$$

and, thus, realize an equivalent floating inductance (FI) of value

$$L_{eq} = (C_0R_1R_2)\left(\frac{V_c}{V_{ref}}\right) \tag{7.57}$$

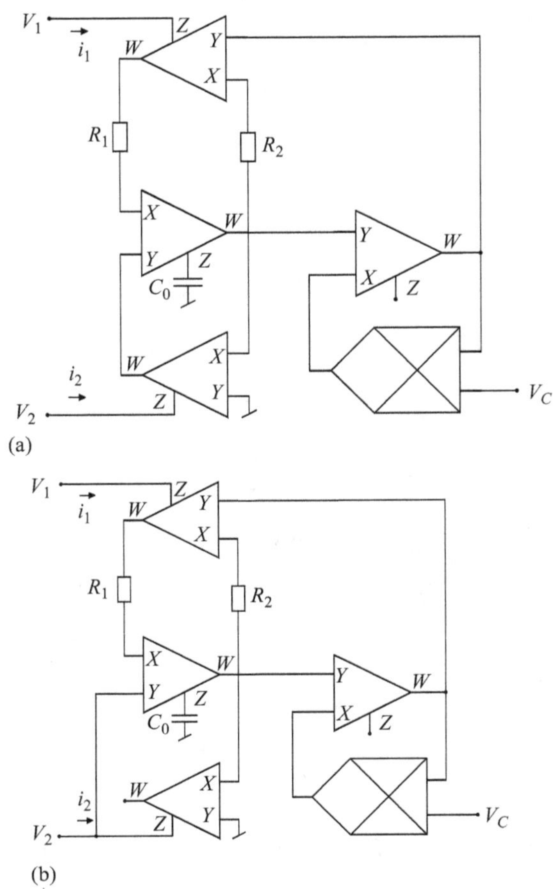

Figure 7.18 Two VC-inductance configurations (a), (b) using analog multipliers [31]

On the other hand, the circuits of Figure 7.19 are characterized by the [y] matrix:

$$\begin{bmatrix} i_1 \\ i_2 \end{bmatrix} = \left(\frac{V_c}{V_{ref}}\right)\left(\frac{1}{sC_0R_1R_2}\right)\begin{bmatrix} 1 & -1 \\ -1 & 1 \end{bmatrix}\begin{bmatrix} v_1 \\ v_2 \end{bmatrix} \tag{7.58}$$

and thus, realize a VC-FI of value

$$L_{eq} = (C_0R_1R_2)\left(\frac{V_{ref}}{V_c}\right) \tag{7.59}$$

In [31], the validity of the configuration of VC-FI of Figure 7.18(b) has been demonstrated by hardware implementation of a voltage-controllable band-reject filter. The circuit was realized using AD844 type CFOAs biased with ± 12 V dc

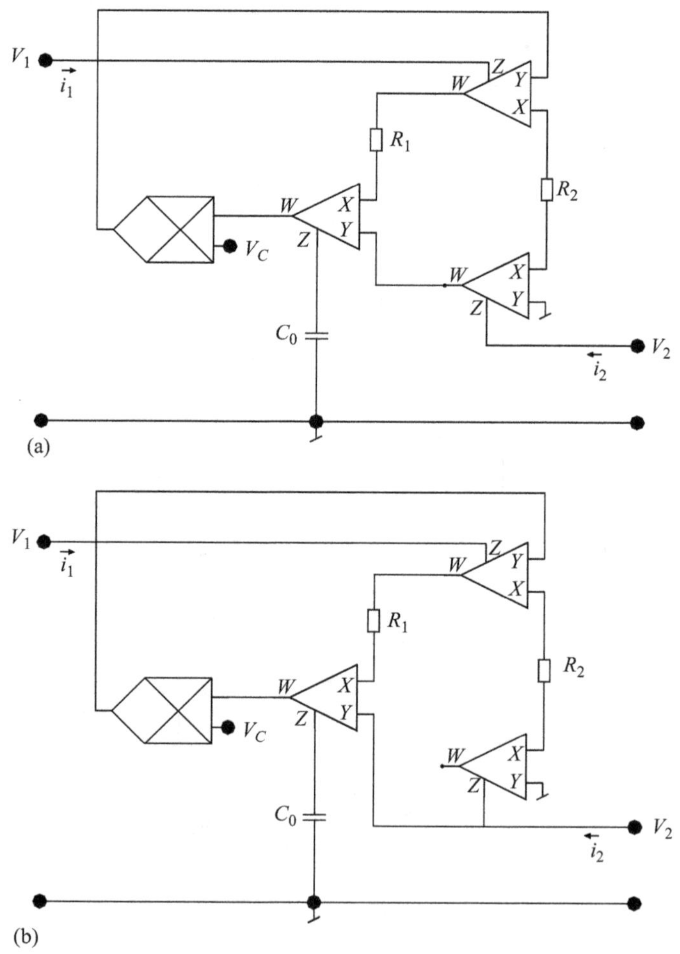

Figure 7.19 Floating VCL circuits using only two CFOAs [31]

power supply and MPY534 type analog multipliers and component values used were $R_1 = R_2 = 1\ \text{k}\Omega$, $C_1 = 10\ \text{nF}$, $R_0 = 680\ \Omega$ to obtain $f_0 = 5.2\ \text{kHz}$ and bandwidth of 5.58 kHz. The control voltage V_c was varied from 1 to 10 V.

Lastly, we show in Figure 7.20 [31] two configurations which can be considered to be VC-floating generalized impedance converters/inverters having equivalent floating impedance values given by

$$Z_{1-2} = \left(\frac{V_{ref}}{V_c}\right)\left(\frac{Z_1 Z_3 Z_5}{Z_2 Z_4}\right); \quad \text{for Figure 7.20}(a) \tag{7.60}$$

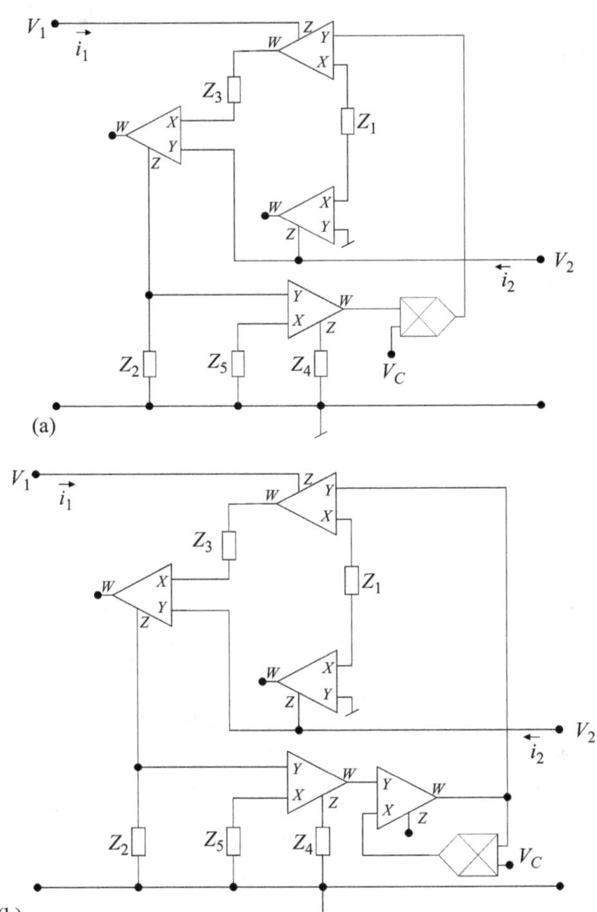

Figure 7.20 Floating VC GIC/GII configurations [31]: (a) circuit providing floating impedance proportional to reciprocal of V_c and (b) circuit providing floating impedance proportional to V_c

and

$$Z_{1-2} = \left(\frac{V_c}{V_{ref}}\right)\left(\frac{Z_1 Z_3 Z_5}{Z_2 Z_4}\right); \text{ for the circuit of Figure 7.20}(b) \tag{7.61}$$

It may be pointed out that the circuits of Figure 7.18 can also be generalized similarly by replacing the elements R_1, R_2, C_0 by general impedances Z_1, Z_2 and Z_3. Finally, negative-valued elements can be realized from all the circuits described in this section by the simple artifice of interchanging certain connections as outlined in an earlier section of this chapter.

The workability of the circuits of Figures 7.19 and 7.20 has been verified by employing them in the design of filters with voltage controllable centre/cut-off frequency. In Figures 7.21 and 7.22, we demonstrate the results of the design of voltage-controlled band-reject filters based on the VC-FI of Figure 7.19(a) using AD 844 type CFOAs MPY534 type analog multipliers both biased with ± 12 V dc power supply. Taking the component values as $R_1 = R_2 = 1$ kΩ, $C_1 = 1$ nF with V_c varied from 1 V to 10 V to vary the centre frequency.

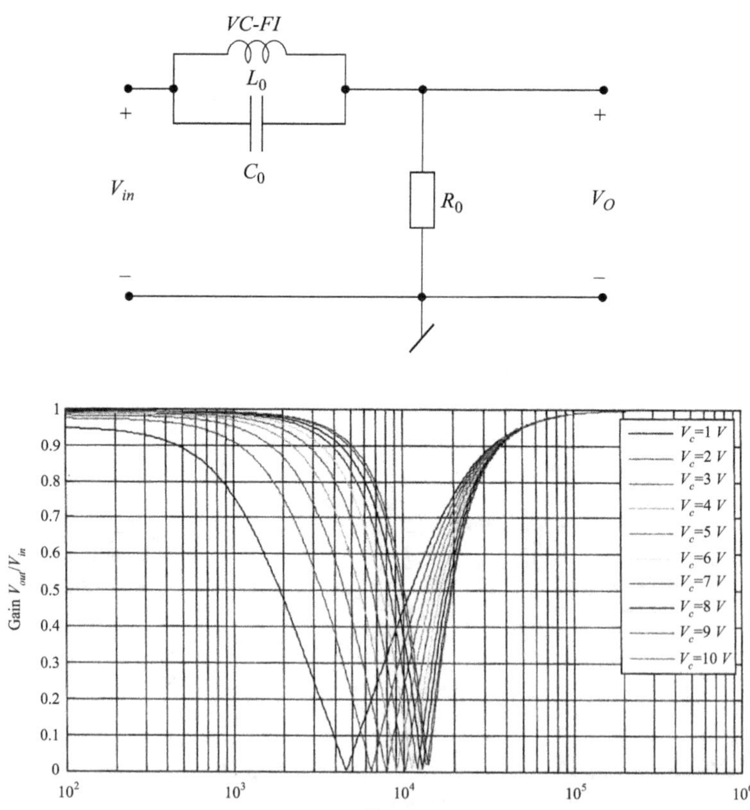

Figure 7.21 Notch filter design using VC-FI: (a) LC prototype filter using VC-FI and (b) controllability of the centre frequency through V_c

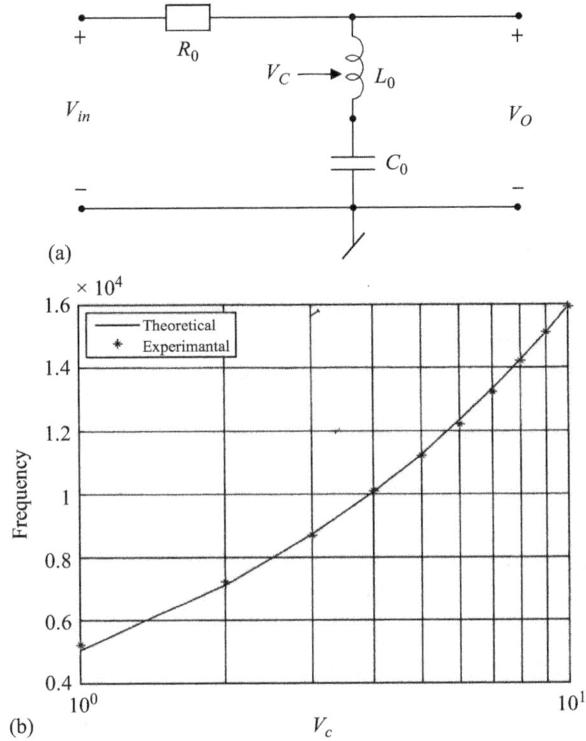

Figure 7.22 Another voltage-controlled notch filter: (a) the circuit and (b) the variation of the centre frequency through V_c

7.9 Concluding remarks

This chapter has presented a variety of configurations for realizing voltage-controlled impedances in both grounded and floating forms using voltage-mode op-amps, current conveyors, operational mirrored amplifiers and current feedback amplifiers as active building blocks along with the FET or AM as the nonlinear elements. Several of the circuits are universal in the sense that the same circuit can be used to realize linear VCR, VCC and VCL (and in some cases VC-FDNR VC-FDNC elements too) in both positive and negative forms from the same circuits. These voltage-controlled impedances have been found to be useful elements in the realization of linear VCRs and in the design of voltage-controlled filters, voltage-controlled oscillators/waveform generators.

The circuits presented here have, however, not exhausted all possibilities, and there appears to be ample scope of improvizing upon the described structures with a view to reduce the number of component-count and/or improving the performance. The authors are of the opinion that there is ample scope[3] of devising new and

[3]For example, see [32–34] for some ideas, different than those described here, for realizing voltage-controlled elements.

improved VCZ structures using CCs/CFOAs as active elements and MOSFETs or AMs as nonlinear elements facilitating voltage-control aspect.

References

[1] Bilotti A. 'Operation of a MOS transistor as a variable resistor'. Proceedings of the IEEE. 1966; **54**(8):1093–5.

[2] Sugita E., Yasuda T., Matsumoto T. 'A solid-state variable resistor using a junction FET'. IEEE Transactions on Parts, Hybrids, and Packaging. 1976; **PHP-12**(3):260–4.

[3] Nay K., Budak A. 'A voltage-controlled resistance with wide dynamic range and low distortion'. IEEE Transactions on Circuits and Systems. 1983; **30**(10):770–2.

[4] Wheatley C. F., Whittlinger H. A., 'OTA obsoletes op-amp'. Proceedings of the Electronics Conference; 1969:152–7.

[5] Bialko M., Newcomb R. W. 'All finite linear circuits using the integrated DVCCS/DVCVS'. IEEE Transactions on Circuit Theory. 1971; **18**(6):719–21.

[6] Sedra A. S, Smith K. C. 'A second generation current conveyor and its applications'. IEEE Transactions on Circuit Theory. 1970; **17**(2):132–4.

[7] Senani R., Bhaskar D. R., Singh A. K. 'Current conveyors: varieties, applications and hardware implementations'. Springer; 2014.

[8] Toumazou C., Lidgey F.J. 'Current feedback op-amps: a blessing in disguise?'. IEEE Circuits and Devices Magazine. 1994; **10**(1):34–37.

[9] Lidge F. J., Hayatleh K. 'Current feedback amplifiers and their applications'. Electronic Communication Engineering Journal. 1997; **9**:176–82.

[10] Senani R. 'A class analog signal processing/signal generation circuits: novel configurations using current feedback op-amps'. Frequenz. 1998; **52**(9–10): 196–206.

[11] Soliman A. M. 'Applications of current feedback operational amplifiers'. Analog Integrated Circuits and Signal Processing. 1997; **11**(3):265–302.

[12] Senani R., Bhaskar D. R., Singh A. K., Singh V. K. 'Current Feedback Operational Amplifiers and Their Applications'. Springer; 2013.

[13] Nay K., Budak A. 'A variable negative resistance'. IEEE Transactions on Circuits and Systems. 1985; **32**(11):1193–4.

[14] Senani R., Bhaskar D. R. 'Realization of voltage-controlled impedances'. IEEE Transactions on Circuits and Systems. 1991; **38**(9):1081–6.

[15] Senani R., Bhaskar D. R. 'Correction to realization of voltage-controlled impedances'. IEEE Transactions on Circuits and Systems. 1992; **39**(2):162.

[16] Leuciuc A., Goras L. 'New general immittance converter JFET voltage-controlled impedances and their applications to controlled biquad synthesis'. IEEE Transactions on Circuits and Systems-I. 1998; **45**(6):678–82.

[17] Senani R., Bhaskar D. R. 'Versatile voltage-controlled impedance configuration'. IEE Proceedings—Circuits, Devices and Systems. 1994; **141**(5):414–16.

[18] Senani R. 'Universal linear voltage-controlled-impedance configuration'. IEE Proceedings—Circuits, Devices and Systems. 1995; **142**(3):208.

[19] Ndjountche T. 'Linear voltage-controlled-impedance architecture'. Electronics Letters. 1996; **32**(17):1528–9.

[20] Senani R., Bhaskar D. R. 'A simple configuration for realizing voltage-controlled impedances'. IEEE Transactions on Circuits and Systems. 1992; **39**(1):52–9.

[21] Yu C. H., Tseng C. Y. 'Sensor-less control based on virtual neutral voltage for BLDC motor drive'. Applied Mechanics and Material, 2015; **789–790**:962–6.

[22] Senani R. 'Realization of a class of analog signal processing/signal generation circuits: novel configurations using current feedback op-amps'. Frequenz: Journal of Telecommunications. 1998; **52**(9/10):196–206.

[23] Maundy B., Gift S., Aronhime P. 'Practical/current controlled grounded resistor with wide dynamic range extension'. IET Circuits, Devices and Systems. 2008; **2**(2):201–6.

[24] Senani R., Bhaskar D. R. 'Comment practical voltage/current-controlled grounded resistor with dynamic range extension'. IET Circuits, Devices and Systems. 2008; **2**(5):465–6.

[25] Maundy B., Gift S., Aronhime P. 'Reply to: Practical voltage/current-controlled grounded resistor with dynamic range extension'. IET Circuits, Devices and Systems. 2008; **2**(5):467.

[26] Senani R. 'Realization of linear voltage-controlled-resistance in floating form'. Electronics Letters. 1994; **30**(23):1909–11.

[27] Tsividis Y. P., Banu M., Khoury J. 'Continuous-time MOSFET-C filters in VLSI'. IEEE Transactions on Circuits and Systems, 1986; **33**(2):125–40.

[28] Senani R., Bhaskar D. R., Gupta S. S., Singh V. K. 'A configuration for realizing floating, linear voltage-controlled resistance, inductance and FDNC elements'. International Journal of Circuit Theory and Applications. 2009; **37**(5):709–19.

[29] Senani R. 'Floating GNIC/GNII configuration realized with only a single OMA'. Electronics Letters. 1995; **31**(6):423–5.

[30] Spencer R. G. 'Analysis of the modified MOS Wilson current mirror: a pedagogical exercise in signal flow graphs, mason's gain rule, and driving-point impedance techniques'. IEEE Transaction on Education. 2001; **44**(4): 322–8.

[31] Senani R., Bhaskar D. R., Tripathi M. P., Jain M. K. 'Canonic realizations of voltage-controlled floating inductors using CFOAs and analog multipliers'. Circuits and Systems. 2016; **7**(11):3617–25.

[32] Yang X., Shiyan Z. 'Design of FDNR continuous time MOSFET-C band-pass filters with low sensitivity and fewer number of op-amps'. International Symposium on Circuits and Systems; Singapore, June 11–14, 1991.

[33] Qiu D. 'New simulated inductor using four-quadrant analogue multipliers'. International Journal of Electronics. 1993; **75**(6):1177–84.

[34] Sato T., Takagi S., Fujii N. 'A design of MOSFET-C impedance simulation circuits based on a GIC'. Transactions on Electrical and Electronic Engineering. IEEJ. 2007; **2**(5):547–55.

Chapter 8

Impedance synthesis using modern active building blocks

Abstract

In this chapter, we present some prominent contributions made in the realization of gyrators and related impedances using electronic circuit building blocks of relatively more recent origin. Thus, we discuss the impedance synthesis using electronic circuit building blocks such as operational transresistance amplifier, Unity gain voltage follower/current follower, current-differencing buffered amplifier, voltage-differencing current conveyor (CC), current-differencing CC, current-differencing transconductance amplifier, current-follower transconductance amplifier, CC transconductance amplifier, voltage-differencing differential input buffered amplifier, voltage-differencing buffered amplifier and many others. Since the work on the use of all these building blocks done so far is by no means complete, we would also highlight what has not been attempted so far.

8.1 Introduction

In the earlier chapters of this monograph, we have dealt with gyrators, simulated inductors and related impedances realized with integrated circuit (IC) op-amps, operational transconductance amplifier (OTA), current conveyors (CCs), current feedback op-amps (CFOAs) and four-terminal-floating nullors (FTFNs). However, over the years, a large number of other building blocks have been proposed in the analog circuit literature, a comprehensive review and classification of which was given by Biolek *et al.* in [1] wherein a number of new composite building blocks were also proposed. These new composite building blocks have been very actively pursued by a number of researchers around the world during the past one decade [1–4]. Out of these, we have already discussed a number of variants of the CCs and their use in impedance simulation circuits in Chapter 4 of this monograph.

In this chapter, therefore, after giving an overview of the various newer building blocks, we shall elaborate upon their use in the realization of grounded and floating impedances (FIs) of various kinds, which are currently receiving wide interest in the technical literature [5–113].

8.2 An overview of modern electronic circuit building blocks

Among the various new building blocks, other than the commercially available ones like op-amps, OTAs, CCs, CFOAs, the operational transresistance amplifier (OTRA) is one of the earlier propositions in this regard. The bipolar and CMOS implementations of the OTRAs and their applications in linear and nonlinear circuit design have been widely investigated over the last two decades, and a comprehensive *state-of-the-art* review of the OTRAs, their implementations and applications have now been made available recently in [15].

Besides this, some of the prominent new building blocks which were postulated in [1] and which have been extensively used in synthetic impedance realizations include unity gain voltage followers (VFs) and unity gain current followers (CFs), differential difference amplifiers (DDAs), current-differencing buffered amplifiers (CDBAs), current-differencing transconductance amplifiers (CDTAs), current-follower transconductance amplifiers (CFTAs), CC transconductance amplifiers (CCTA), voltage-differencing CCs (VDCCs), current differencing CCs, voltage-differencing transconductance amplifiers (VDTAs), voltage-differencing buffered amplifiers (VDBAs), voltage-differencing inverting buffered amplifiers (VDIBAs), voltage-differencing differential input buffered amplifiers (VD-DIBAs), CFCCs, current backwards transconductance amplifier (CBTAs) and several other miscellaneous building blocks.

In the following sections, we will highlight the most significant contributions made in the area of impedance simulation using the above-mentioned new building blocks.

8.3 Grounded impedance and floating impedance synthesis using modern building blocks

We now discuss some of the prominent impedance simulation networks from a large number of such circuits [1–109] proposed during the last decade which make use of the newly introduced analog circuit building blocks.

8.3.1 Unity gain VF/CF-based circuits

The use of unity gain VFs and unity gain CFs has attracted attention of circuit designers from time to time due to two main reasons: (i) wider bandwidth of VF/CF as compared to the traditional building blocks like op-amps which have been the most dominating building blocks employed in analog signal processing/generation and (ii) simpler internal circuit architecture of the VFs and CFs as compared to IC op-amp. Three exemplary implementations to substantiate the first assertion are shown in Figure 8.1.

An overview of the use of VFs and CFs in analog signal processing/generation has been presented in [15]. In this section, we would present some prominent circuits for realizing gyrator and the various kinds of synthetic impedances using VFs and CFs.

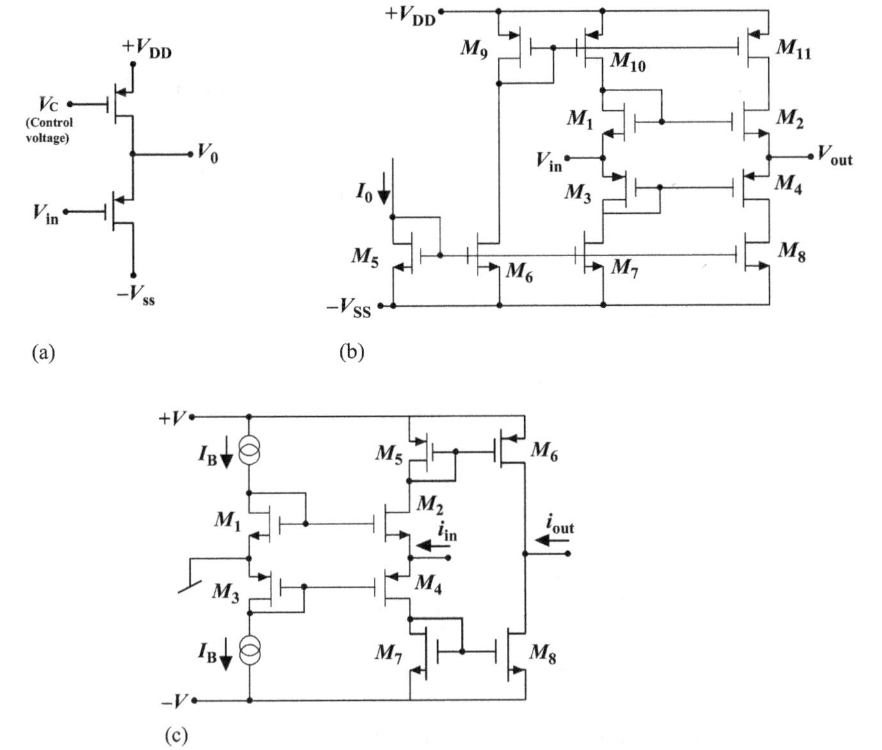

Figure 8.1 Exemplary unity gain VF and CF structures: (a) a simple-voltage-controlled voltage follower [5], (b) a translinear VF [6] and (c) a translinear-voltage-controlled single output current follower [7]

Although a number of researchers have proposed impedance simulation networks using VFs and/or CFs, for instance see [5–15] and references cited therein, one of the earliest contributions in this area appears to be that of Linggard [8] who proposed that the well-known concept of constructing an ideal gyrator using parallel back to back connection of two VCCS of opposite polarities can be readily implemented from only VFs and CF as per the schematics of Figure 8.2.

It can be easily found by employing ideal characterization of VF by $i_{in} = 0$, $V_{out} = V_{in}$, $r_{out} = 0$ and that of CF as $V_{in} = 0$, $i_{out} = \pm i_{in}$ and $r_{out} = \infty$ (with CF^+ defined by $i_{out} = +i_{in}$ and CF^- defined by $i_{out} = -i_{in}$), that the active gyrator of Figure 8.2(b) is characterized by

$$\begin{bmatrix} i_1 \\ i_2 \end{bmatrix} = \begin{bmatrix} 0 & +\dfrac{1}{R_2} \\ -\dfrac{1}{R_1} & 0 \end{bmatrix} \begin{bmatrix} V_1 \\ V_2 \end{bmatrix} \tag{8.1}$$

Based upon this gyrator, a lossless simulated inductor is obtained by loading the output port into a capacitor C_0 as shown in Figure 8.3.

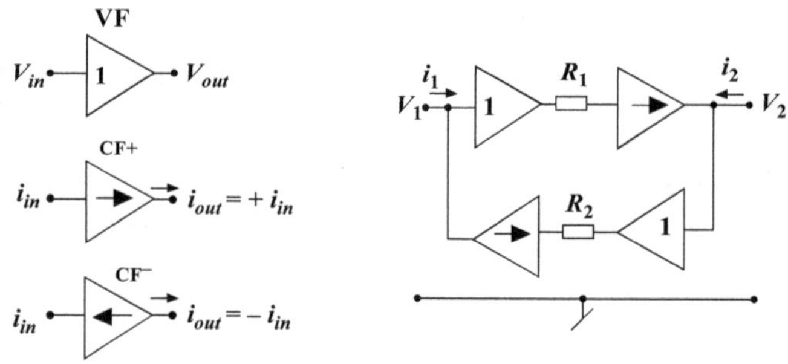

(a) (b)

Figure 8.2 Realization of active gyrator as per the method of Linggard [8]: (a) symbols of VF and CFs and (b) active gyrator [13]

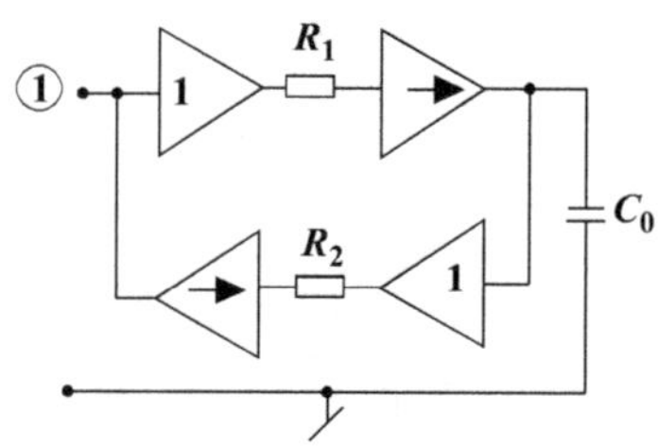

Figure 8.3 Lossless grounded inductor having $L_{eq} = C_0 R_1 R_2$

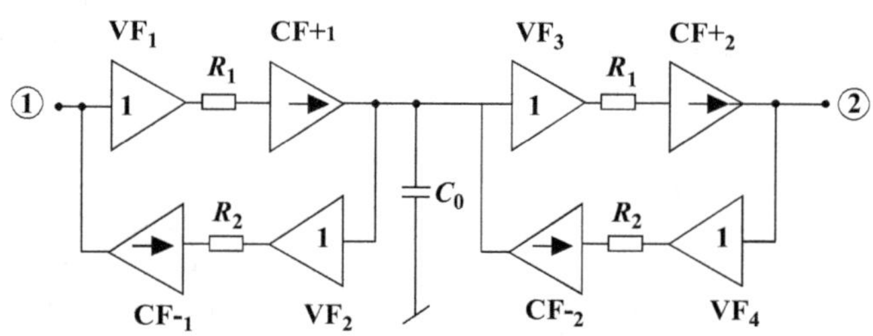

Figure 8.4 A lossless FI circuit employing two identical active gyrators

It is not difficult to devise a floating lossless inductor by employing two identical gyrators with a grounded capacitor embedded between them and such a formulation leads to the floating inductor (FI) circuit of Figure 8.4.

A straightforward analysis of this circuit yields the following transmission matrix:

$$[T] = \begin{bmatrix} 0 & R_1 \\ \dfrac{1}{R_2} & 0 \end{bmatrix} \begin{bmatrix} 1 & 0 \\ sC_0 & 1 \end{bmatrix} \begin{bmatrix} 0 & R_1 \\ \dfrac{1}{R_2} & 0 \end{bmatrix} = \begin{bmatrix} 1 & sC_0R_1R_2 \\ 0 & 1 \end{bmatrix} \tag{8.2}$$

which represents the floating inductance of the value $L_{eq} = C_0R_1R_2$ between ports 1 and 2.

However, a careful inspection of the circuit of Figure 8.4 reveals that VF_2 and VF_3 do the same job; hence, one of these is redundant and can be deleted. This leads to two simplifications one of which is shown in Figure 8.5.

The other alternative circuit is obtainable by retaining VF_2 and omitting VF_3. However, such a circuit would be identical to that of Figure 8.5 with ports 1 and 2 interchanged and, hence, is not repeated here. Both the circuits, however, suffer from the drawbacks of requiring close matching between the resistor pairs of the two gyrators involved.

Yet another gyrator circuit can be obtained without using VFs, if only CFs with complementary outputs are employed. One such circuit, using dual output CFs (DOCF) and complementary output CFs (COCF) is shown here in Figure 8.6.

By straightforward analysis of the circuit, its characterizing equations are found to be

$$i_1 = \frac{V_2}{R_2} \text{ and } i_2 = -\frac{V_1}{R_1} \tag{8.3}$$

Using the active gyrator of Figure 8.6, the circuit for realizing lossless grounded inductor (GI) would be as shown in Figure 8.7. A floating version of this circuit using two identical gyrators with a grounded capacitor embedded between them is obvious and has the drawbacks of requiring matching between identical set of passive components and, hence, is not shown explicitly here.

Psychalinos and Spanidou [10] have presented circuits and techniques of realizing grounded and floating inductance simulations using multiple output current amplifiers (CAs) (MOCAs), the underlying principles of which are quite similar to those contained in the circuit of Figure 8.7. The grounded inductor circuit using two MOCA is shown in Figure 8.8.

In the circuit of Figure 8.8, the current entering CA_1 is given by $i_1 = g_{m1}V$ where g_{m1} is the small signal transconductance parameter of the diode-connected transistor at the input of CA_1. Thus, the voltage across capacitor C_0 is $g_{m1}V/sC_0$. Now CA_2 converts this voltage into a current $i_2 = g_{m1}g_{m2}V/sC_0$ and one gets

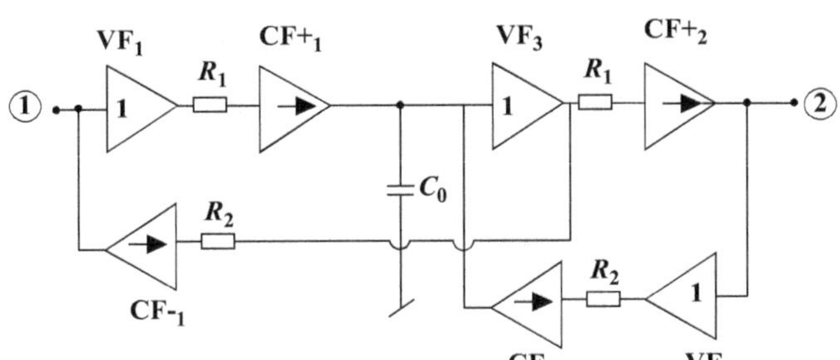

Figure 8.5 Simplified version of the circuit of Figure 8.4 with one reduced VF

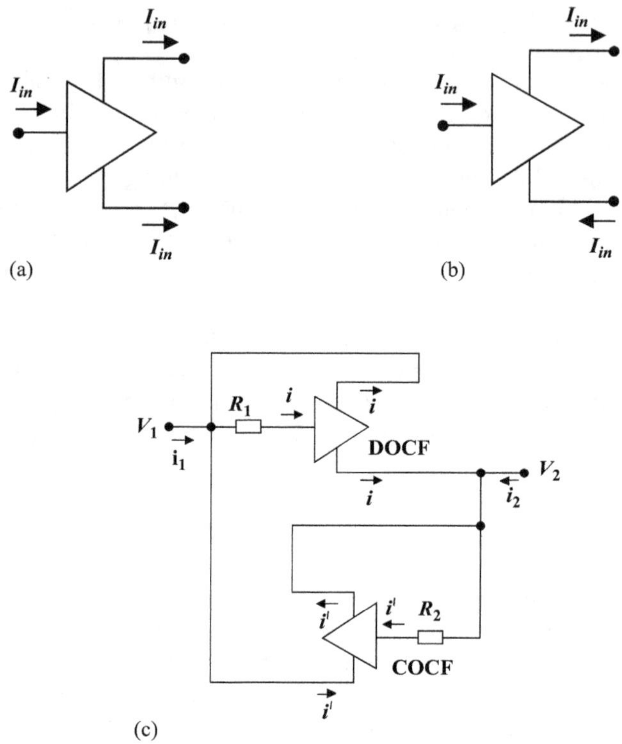

Figure 8.6 (a) Symbol of DOCF, (b) Symbol of COCF, (c) An Active gyrator using DOCF and COCF

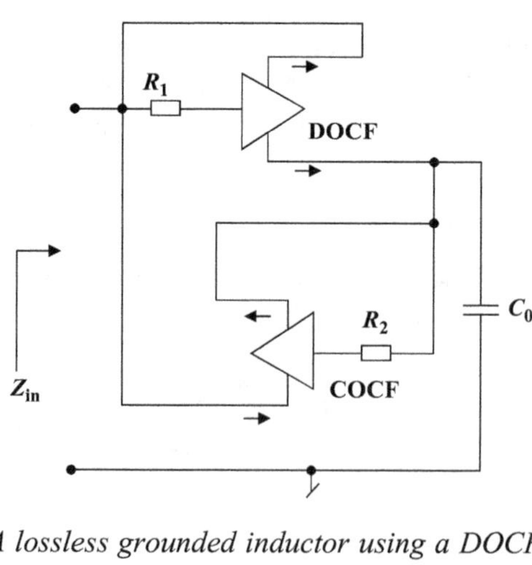

Figure 8.7 A lossless grounded inductor using a DOCF and a COCF

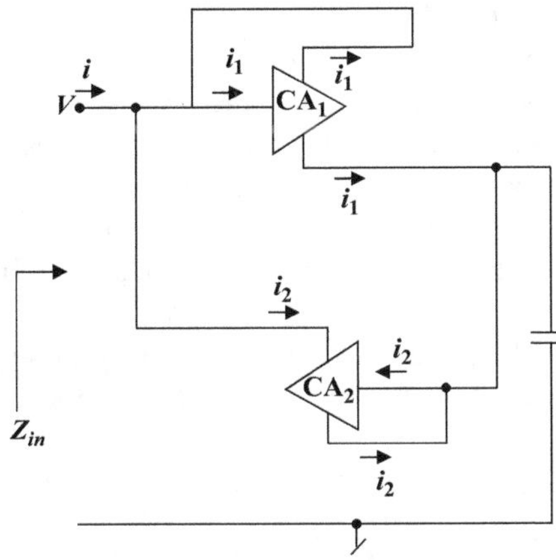

Figure 8.8 Grounded inductance simulation using two current amplifiers [10]

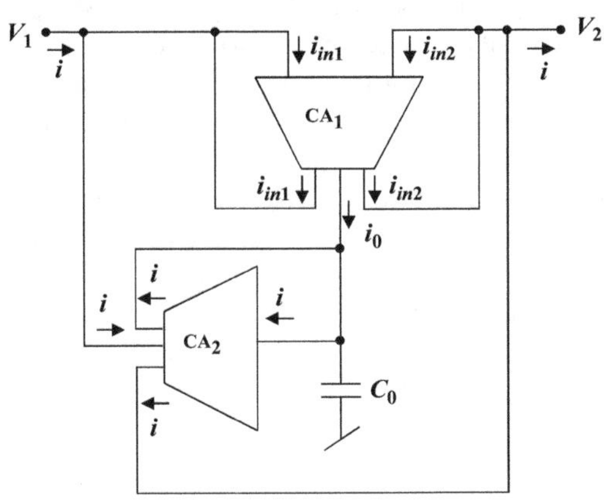

Figure 8.9 Lossless floating inductance using two CAs [10]

$Z_{in} = sC_0/g_{m1}g_{m2}$. In retrospect, it can be seen that the essence of this circuit technique employed in the circuit is analogous to that implicit in the circuit of Figure 8.7, while on the other side, it is also analogous to the two-OTA-based grounded inductance simulation circuit elaborated in Chapter 3.

An extension of this technique to create a lossless floating inductance too was proposed by Psychalinos and Spanidou in [10] which is shown here in Figure 8.9. Here, CA building blocks employed are extended from the basic CA to have two inputs and three outputs as well as a CA with single input and three outputs.

This circuit has an underlying circuit principle similar to the one employed in the three-OTA lossless FI circuit explained in Chapter 3. Here CA$_1$ provides an output current i_0, which is equal to $i_0 = (i_{in1} - i_{in2})$, thereby creating a voltage $(g_{m1}V_1 - g_{m2}V_2)/sC_0$, which is converted into two currents that create two identical currents i given by $(g_m(V_1 - V_2)g_{m3})/sC_0$, where $g_{m1} = g_{m2} = g_m$, so that the floating inductance realized by this circuit is given by $L_{eq} = C_0/g_m g_{m3}$.

From a review of the work done on impedance synthesis using VFs and CFs, it is found that although a lossless grounded inductor can be realized using only two terminal VFs and CFs as shown here, a floating lossless inductor cannot be realized with only VFs and CFs unless two terminal NICs are also incorporated as done by Alpaslan *et al.* in [12], but the resulting circuit, while using a grounded capacitor, still requires matching between two pairs of resistors. Hence, the problem of matching two terminal VFs and CFs using only three passive components and without requiring any component matching constraints is not resolved and, hence, is open to investigation.

8.3.2 OTRA-based circuits

Other than op-amps, OTAs, CCs, CFOAs, FTFNs and VF/CF elements, another building block which has been extensively employed in analog signal processing/generation is the so-called OTRA. The use of OTRA in impedance synthesis has been explored by several authors such as [16–26]; however, the work done so far does not suggest OTRA to be really very useful for this task as we shall demonstrate here.

The OTRA is essentially a differential current-controlled voltage source having two inputs as p and n receiving two current inputs i_p and i_n at the input nodes both of which offer ideally zero input impedance, thereby making both the input nodes as the 'virtual grounds'. The output is a voltage given by $v_0 = R_m(i_p - i_n)$ where negative feedback and infinite $R_m(R_m \to \infty)$ force the two input currents to be equal. An extensive survey [26] of the literature on OTRAs reveals that perhaps the best impedance simulation circuit which could be evolved using OTRA is the generalized impedance converter (GIC) [21] shown here in Figure 8.10, which requires two OTRAs and an additional unity gain VF but has six impedances (instead of five) to produce an impedance Z_{in} which is given by

$$Z_{in} = \frac{Z_1 Z_3 Z_5}{Z_2 Z_4} \tag{8.4}$$

The above expression is although exactly same as obtainable with the classical Antoniou's GIC based on op-amps, and this circuit has all the functionality of the latter circuit, but the penalty to be paid is an extra VF (besides two OTRAs as active elements) and one matching condition (not needed in Antoniou's GIC).

In spite of the deficiencies of the configuration of Figure 8.10(a) as mentioned, a strikingly unique feature of this circuit is that if the p-terminal of the OTRA_1 is left open with Z_5 is connected between the VF and the n-terminal of OTRA_1 and with impedance Z_4 used in its feedback path, then the input impedance modifies to

$$Z_{in} = -\frac{Z_1 Z_3 Z_5}{Z_2 Z_4} \tag{8.5}$$

In this mode, therefore, the circuit becomes a generalized negative impedance converter (GNIC). Note that such a circuit cannot be made from a conventional op-amp-based GIC.

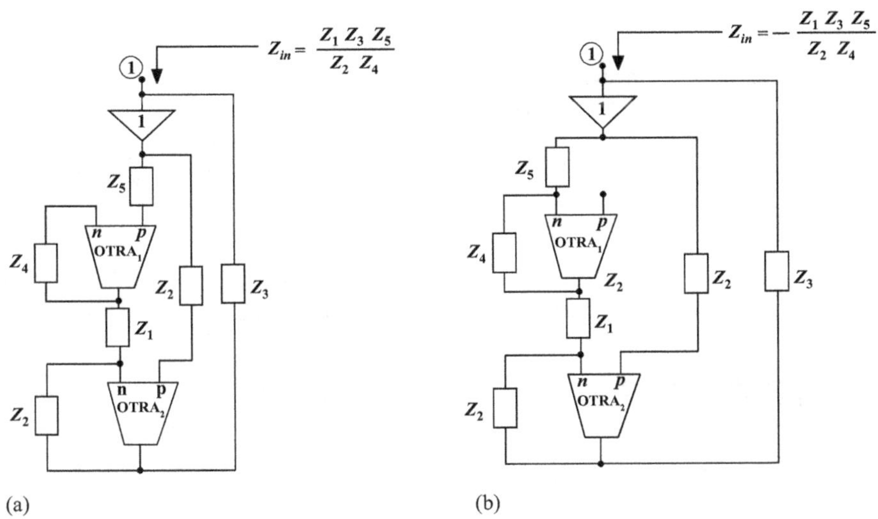

Figure 8.10 OTRA-based impedance converters [21]: (a) generalized positive impedance converter and (b) generalized negative impedance converter

Beyond the above circuits using OTRA, the survey presented in [26] reveals the following in the context of impedance simulation using OTRAs: (i) no single OTRA-based canonic realizations of lossless grounded inductor or ideal FDNR have been discovered so far; (ii) given even two OTRAs, no circuit has so far been discovered to enable realization of a lossless gyrator or lossless inductor using a single capacitor; (iii) any lossless gyrator realization with both ports grounded has not been discovered given any number of OTRAs and resistors and (iv) not even a single circuit has so far been reported to simulate any kind of floating inductor (lossy series /parallel RL or lossless) using OTRAs.

The above, therefore, constitute some open problems worthy of further investigation.

8.3.3 DDA-based circuits

The DDA was first introduced in the area of analog circuits in 1987 by Sackinger and Guggenbuhl [27]. A prominent advantage of this concept is that an infinite gain DDA with feedback makes it possible to design a number of useful circuits such as adder, subtractor and an inverting amplifier without using any external resistors in contrast to the corresponding op-amp-based circuits for realizing these functions which compulsorily require external resistors.

The symbolic notation of the DDA is shown in Figure 8.11. A DDA has two differential input ports having input voltages $(V_{pp} - V_{pn})$ and $(V_{np} - V_{nn})$ and its output is given by

$$V_0 = A_0\big((V_{pp} - V_{pn}) - (V_{np} - V_{nn})\big) \tag{8.6}$$

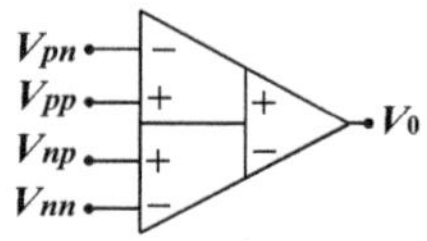

Figure 8.11 The symbolic representation of the DDA [27]

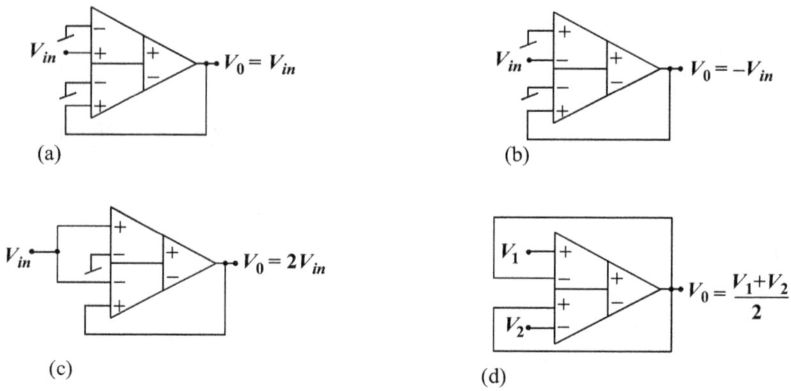

Figure 8.12 Some basic circuits using DDAs without requiring any external resistors [27]: (a) voltage follower, (b) voltage inverter, (c) voltage doubler and (d) average circuit

where A_0 is the open-loop gain of the DDA. If a negative feedback is applied by connecting V_0 to V_{np}, (8.6) forces the two differential inputs to be equal:

$$V_{pp} - V_{pn} = V_{np} - V_{nn} \quad (\text{since } A_0 \to \infty) \tag{8.7}$$

This can be alternatively written as

$$V_0 = V_{pp} - V_{pn} + V_{nn} \tag{8.8}$$

Using (8.8), a number of basic circuits are conceivable (which are shown in Figure 8.12) where no external components are required for the various functions performed.

It is observed from the above that, contrary to the conventional op-amp, a DDA makes it possible to carry out the operation of addition, subtraction, sign inversion (inverting amplifier) and averaging just by application of appropriate signals on various input terminals and/or grounding of appropriate input terminal(s), without requiring any resistors!

It was demonstrated by the Singh *et al.* [28] for the first time that the commercially available video amplifier AD830 from Analog Devices Inc. is a good candidate for implementing DDA-based circuits. The general block diagram of the internal circuit of the AD830 and its pin configuration is shown in Figures 8.13 and

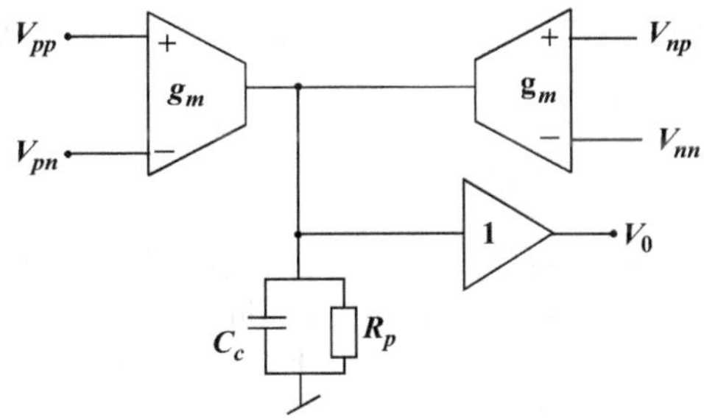

Figure 8.13 The block diagram of AD830

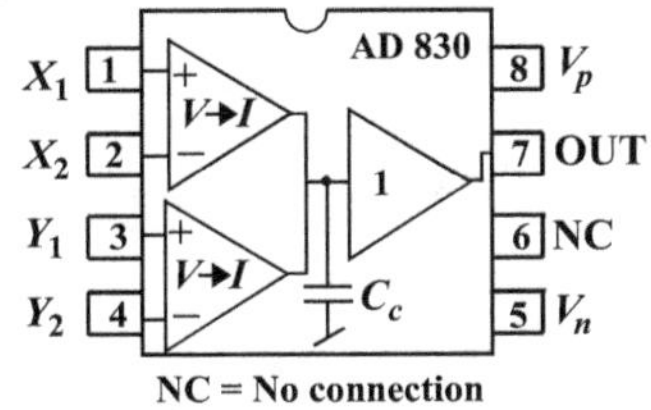

Figure 8.14 Pin configuration of AD830

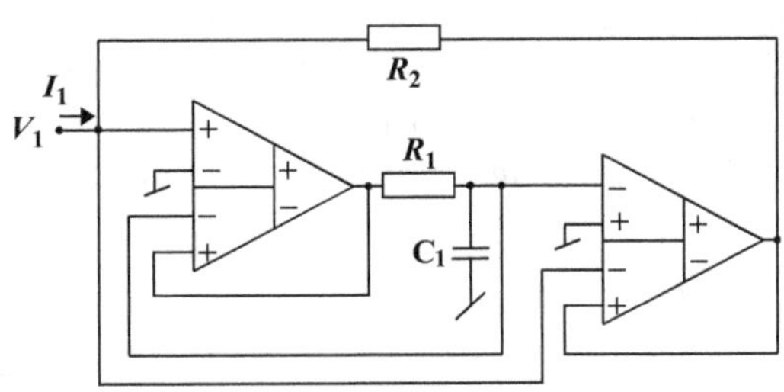

Figure 8.15 DDA-based grounded lossless inductor [29]

8.14, respectively.[1] The ideal characterizing equation for this building block, as per (8.7), with negative feedback turns out to be (with an appropriate change in input terminals) same as (8.8).

To the best knowledge of the authors, not much work on inductance simulation circuits using DDAs has been reported in the open literature, so far and some circuits included here are, in fact, being presented for the first time in this monograph only. A grounded lossless inductor simulation circuit can be devised using only two DDAs along with only three passive components-two resistors and a grounded capacitor as shown in Figure 8.15.

[1]Datasheet of AD830, high speed, video differential amplifier (Analog Devices).

A straightforward analysis of this circuit gives its input impedance Z_{in} as

$$Z_{in} = sC_1R_1R_2 \tag{8.9}$$

where it is seen that the value of inductance can be varied by changing either R_1 or R_2.

From the discussions in various chapters of this monograph it is well known and established that as compared to lossless inductor, a lossy one can be realized with a smaller number of active elements. Hence, it is not surprising that lossy inductors in the form of a series RL impedance or a parallel RL admittance can be realized using only a single DDA and two resistors along with a grounded capacitor as shown in Figure 8.17.[2]

The input impedance of the circuit of Figure 8.16(a) is given by

$$Z_{in} = 2R_2 + sC_1R_1R_2 \tag{8.10}$$

whereas the input admittance of the circuit of Figure 8.16(b) is given by

$$Y_{in} = \frac{2}{R_2} + \frac{1}{sC_1R_1R_2} \tag{8.11}$$

It may be noted that in both the cases, the simulated inductance value is single resistance controllable, by a variable resistance R_1 in the former and through a variable resistance R_2 in the latter.

Finally, a circuit for realizing a lossless floating inductance can be devised using only three DDAs, three resistors (two of which are required to be matched) and again a single grounded capacitor. This circuit, in fact, is a *DDA-analogue* of the op-amp-based lossless floating inductor circuit proposed earlier by Senani in [30].

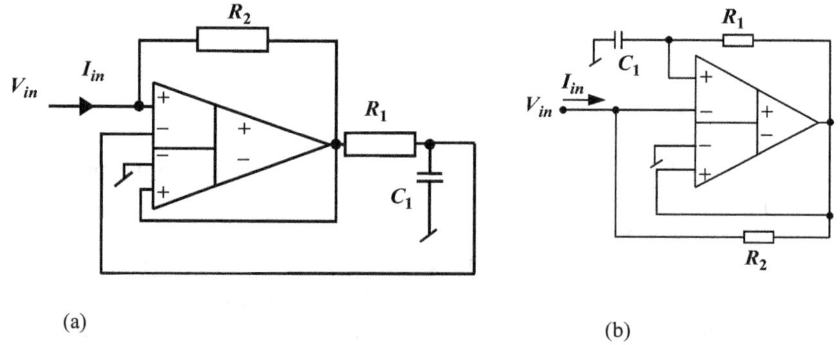

Figure 8.16 Lossy grounded inductance simulators: (a) series RL and (b) parallel RL

[2]Senani R., Singh A. K., Singh B. 'Novel synthetic inductors using DDAs employing a grounded capacitor'. Analog Integrated Circuits and Signal Processing. 2019 (communicated).

Note that this DDA-version does not require any matched resistors to realize the three unity-gain summers as required in the synthesis of lossless floating inductance presented in [30].

In the circuit of Figure 8.17 if one takes $R_1 = R_2 = R$, then it realizes a lossless floating inductor having equivalent inductance.

$$L_{eqv} = C_0 R_0 R \tag{8.12}$$

which is single-resistance controllable through a single variable resistance R_0.

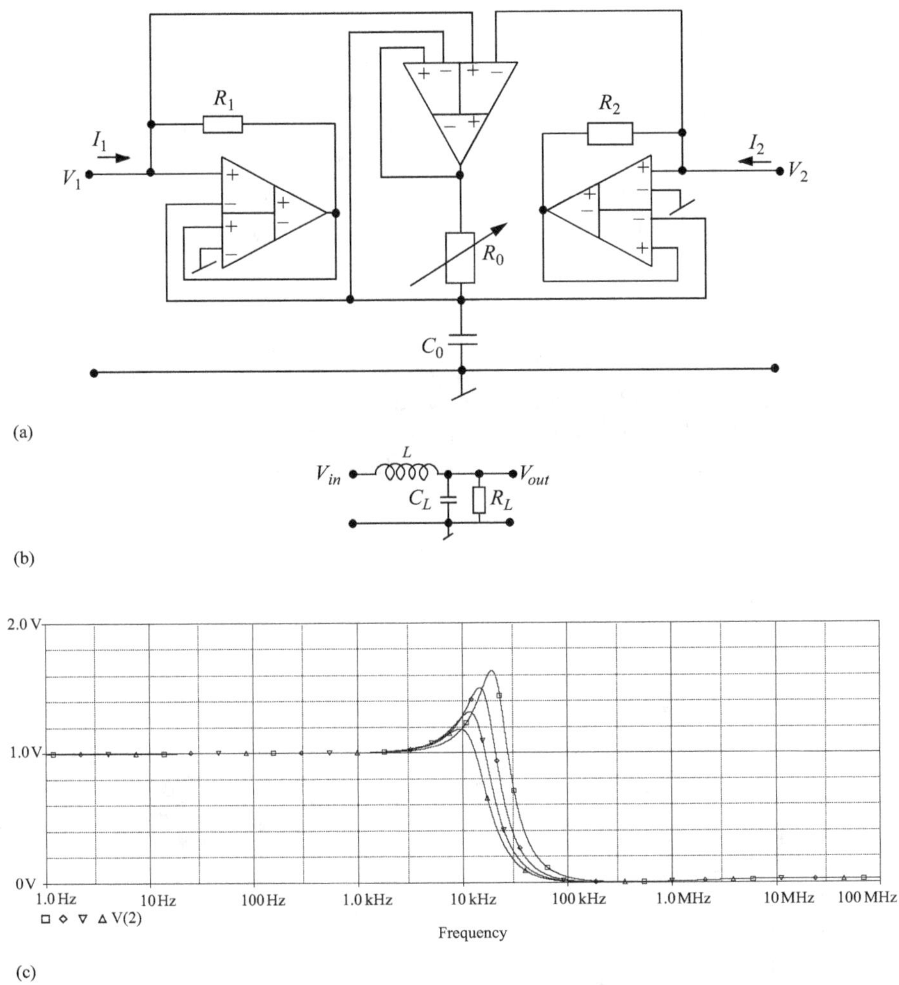

Figure 8.17 Realization of floating inductance using DDA: (a) lossless floating inductor analogous to the op-amp-based FI of [30], (b) second-order low pass filter used to check the validity of the DDA-based lossless FI [29] and (c) frequency response of the low pass filter using the DDA-simulated FI [29]

To verify the workability of the DDA-based lossless floating inductor circuit of Figure 8.17(a), PSPICE simulations have been carried out using a CMOS DDA. The values of the resistors and the capacitor were taken as $R_1 = R_2 = 10$K and $C_0 = .01$ μf. The value of R_0 has been changed to 400 Ω, 700 Ω and 1 kΩ when the floating inductor is employed to realize the LC low-pass filter as shown in Figure 8.17(b) where the resistor R_L and the capacitor C_L were set to 10 kΩ and 10 pF, respectively. Figure 8.17(c) shows the frequency responses of the low-pass filter for different inductor values which prove the variability of the inductance through the resistor R_0 as outlined above.

8.3.4 Translinear operational current amplifier

The translinear operational CA (TOCA) was introduced as a new building block by Fabre in [31] and its bipolar implementation together with a variety of its applications were elaborated in [31,32]. It is basically a differential CCCS having single/dual outputs with its symbolic notation as shown in Figure 8.18.

However, as far as is known, any circuit for realizing gyrator or any synthetic impedances using TOCAs has not been reported so far and this constitutes an open problem.

8.3.5 CDTA-based circuits

The CDTA as a new building block for current mode analog signal processing was introduced by Biolek in [33]. It is symbolic notation and circuit realization in terms of two second-generation CCs (CCIIs) and an OTA are shown in Figure 8.19.

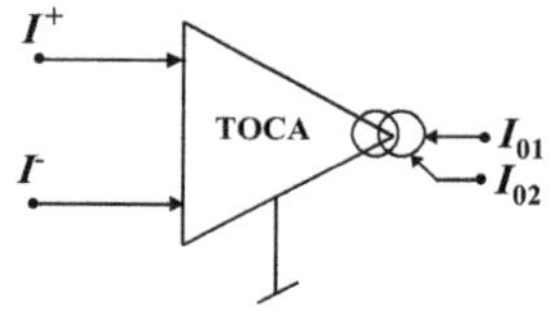

Figure 8.18 Symbolic notation of differential CCCS [31]

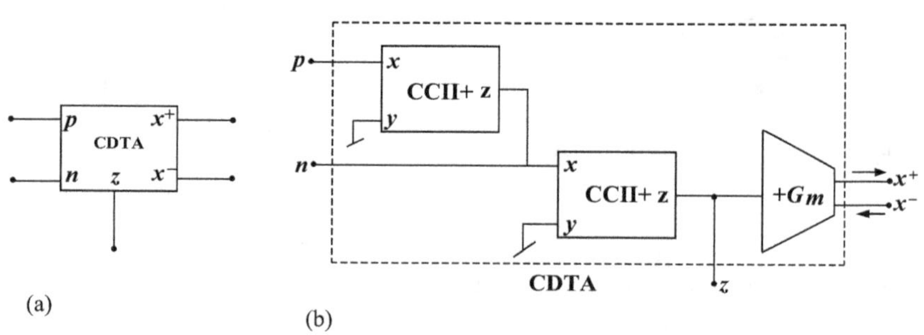

Figure 8.19 CDTA representations [33]: (a) symbolic notation and (b) implementation using CCIIs and OTA

A CDTA is characterized by $i_z = i_p - i_n$, $i_x^+ = G_m V_z$ and $i_x^- = -G_m V_z$.

Apart from its various applications demonstrated in [1,2,33–40], Biolek demonstrated that with two identical CDTAs one can easily realize a positive impedance inverter (an active gyrator) as per the schematic of Figure 8.20.

The input impedance of this circuit is found to be $Z_{in} = Z/G_{m1}G_{m2}$, hence with Z taken as a capacitor C_0, the circuit simulates a lossless grounded impedance of value $L_{eq} = C_0/G_{m1}G_{m2}$.

A slightly different circuit arrangement for simulating a lossless GI using two CDTAs was reported by Prasad *et al.* [34] which is shown in Figure 8.21(a).

The equivalent Z_{in} for this circuit is given by $Z_{in} = C_0/G_{m1}G_{m2}$ which is electronically controllable through I_{B1} and/or I_{B2}. A floating version of this circuit is shown in Figure 8.21(b) for which straightforward analysis gives the same value of equivalent realized inductance as $L_{eq} = C_0/G_{m3}G_m$ where $G_{m1} = G_{m2} = G_m$.

In [35] Siripruchyanun and Jaikla presented a current-controlled CDTA (CC-CDTA) characterized by $V_p = I_pR_p$, $V_n = R_nI_n$, $I_z = I_p - I_n$ $I_x^+ = +G_mV_z$ and

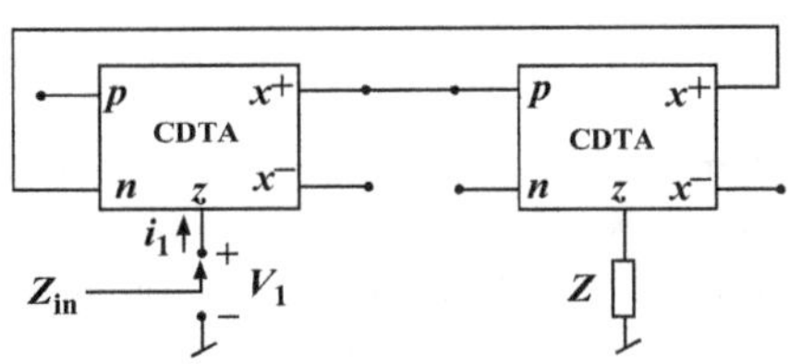

Figure 8.20 A positive impedance inverter (an active gyrator) using CDTAs [33]

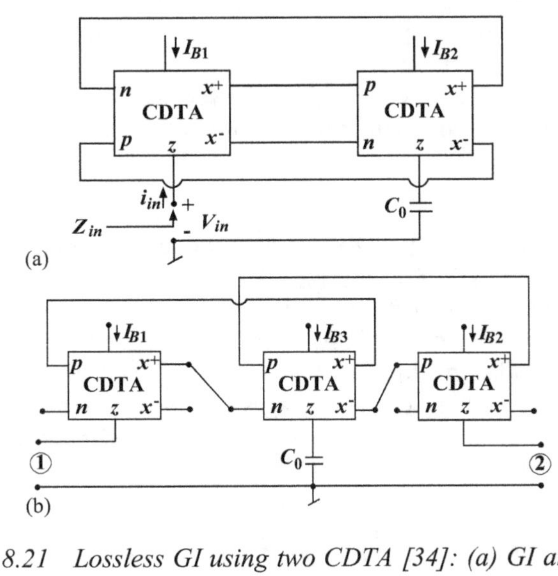

Figure 8.21 Lossless GI using two CDTA [34]: (a) GI and (b) FI

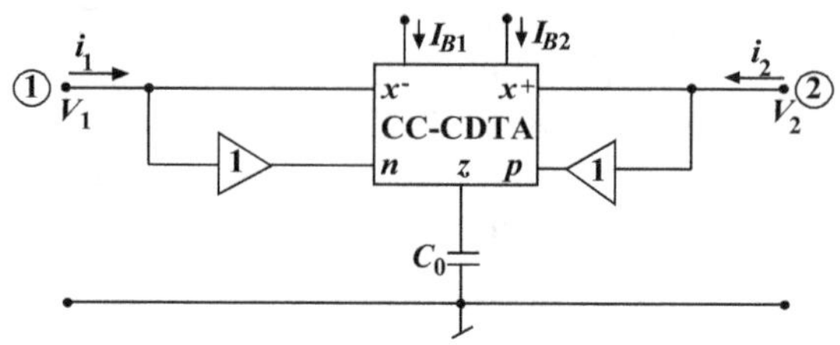

Figure 8.22 A lossless floating inductor using a CC-CDTA [35]

$I_x^- = -G_m V_z$ where $R_p = R_n = V_T/2I_{B1}$ and $G_m = I_{B2}/2V_T$ with I_{B1} and I_{B2} being external DC bias currents for the current differencing and the transconductance amplifier, respectively. They demonstrated that a CC-CDTA can be deployed to realize a floating inductance as per the circuit configuration shown in Figure 8.22.

An analysis of this circuit shows that the two-port currents are related by the following equation:

$$i_1 = -i_2 = \frac{G_m}{sC_0 R}(V_1 - V_2) \tag{8.13}$$

The circuit, therefore, simulates a lossless FI of value $L_{eq} = C_0 R/G_m$ which is electronically controllable through G_m and/or R where $R_p = R_n = R = V_T/2I_{B1}$ and $G_m = I_{B2}/2V_T$. This circuit has the attractive features of electronic tunability, use of grounded capacitor and being free from component-matching.

The same authors in [36] proposed a CC-CDTA implementable in CMOS technology and included the circuit of Figure 8.22 as an example of FI simulation with CMOS CC-CDTA. In CMOS version, the simulated inductance value is $L_{eq} = (C_0 R/G_m) = C_0/\left(\beta_n \sqrt{8 I_{B1} I_{B2}}\right)$, where I_{B1} is the external dc bias current of the CMOS current differencing unit (CDU) of the CDTA architecture and I_{B2} is the external DC bias current of the CMOS transconductance amplifier part.

Thus, the circuit of Figure 8.22 has been verified to be a good FI simulator suitable for implementation in either bipolar or CMOS technology.

8.3.6 CDBA-based circuits

A new building block, namely CDBA, was introduced by Acar and Ozoguz [41] which is characterized by $i_z = i_p - i_n$, $v_p = 0$, $v_n = 0$ and $v_w = v_z$. Though a number of researchers have proposed its integratable circuit architecture, it can also be realized using two AD844 type CFOAs as per the schematic of Figure 8.23(b).

A voltage-controlled gyrator was shown to be realizable using two CDBAs and two pairs of MOSFETs having control voltages V_{G1} and V_{G2} as shown in Figure 8.24.

When port 2 is terminated into a capacitance C_0 the input impedance of the circuit is given by

$$Z_{in} = \frac{C_0}{\left[\mu_s C_{ox}(W/L)_1 (V_{G1} - V_{G2})\right]\left[\mu_s C_{ox}(W/L)_2 (V_{G1} - V_{G2})\right]} \tag{8.14}$$

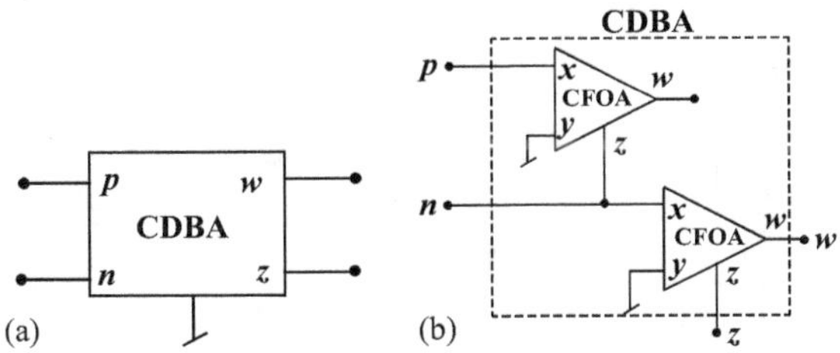

Figure 8.23 CDBA characterization and realization [41]: (a) circuit symbol and (b) implementation using two CFOAs

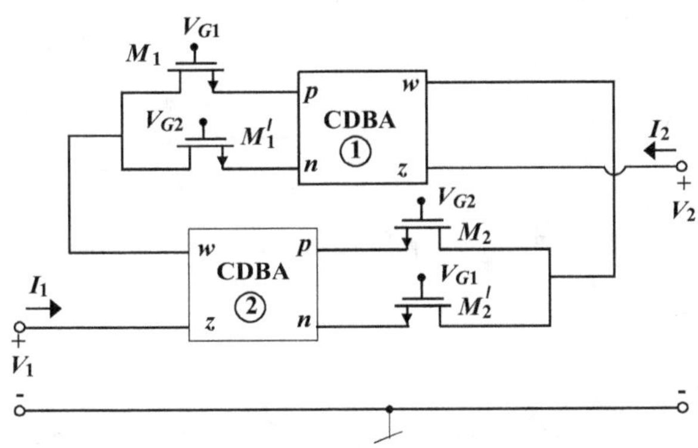

Figure 8.24 Voltage-controlled active gyrator realization [42]

Thus, the simulated inductance can be tuned/varied by V_{G1} and/or V_{G2} and can also be made either positive or negative. In [42], it has been shown that this CDBA-based gyrator can be used to realize a fully integratable Tow–Thomas equivalent biquad filter.

Biolek *et al.* [43] proposed a circuit for the simulation of a lossless floating inductance which is shown in Figure 8.25.

The circuit simulates a lossless FI of value $L_{eq} = C_0 r^2$ but needs four-matched resistors each equal to r.

Keskin and Hancioglu [44] presented another circuit for FI simulation which, however, needs four CDBAs along with four MOS resistive circuits.

It is by now well understood that as compared to an ideal gyrator or lossless GI simulation, a nonideal gyrator/lossy inductor can be realized more economically. This is also true in the context of CDBA-based gyrator/GIs. Consider now the parallel-RL simulator proposed by Pathak *et al.* [45] shown in Figure 8.26(a).

The input admittance of this circuit is given by

$$Y_{in} = \left(\frac{1}{R_1} + \frac{1}{R_2}\right) + \frac{1}{sC_0R_1R_2} \tag{8.15}$$

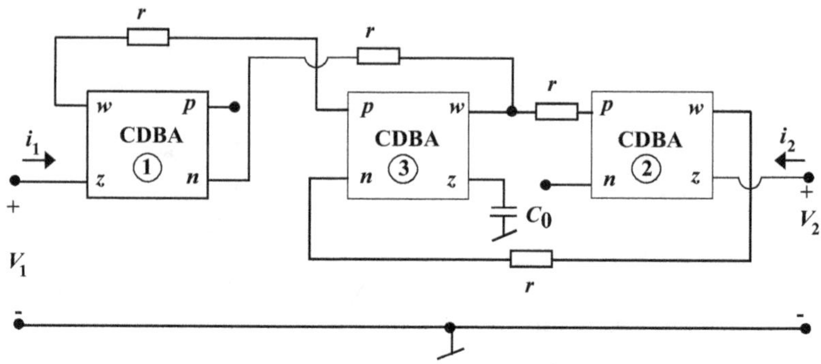

Figure 8.25 CDBA-based lossless FI simulation [43]

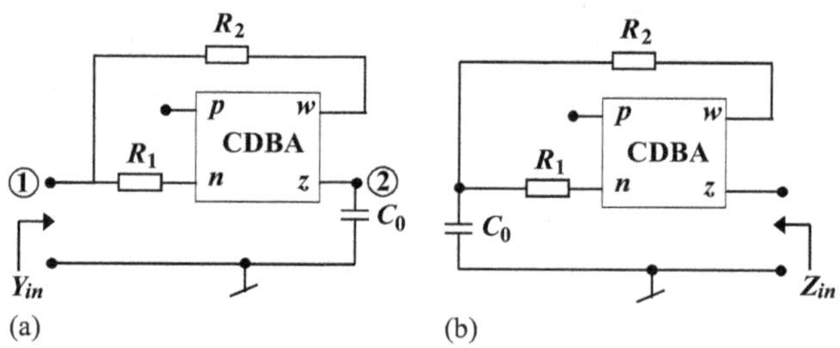

Figure 8.26 Lossy GI simulators [45]: (a) parallel RL and (b) series RL

which represents a grounded parallel RL admittance. The circuit can be looked upon as a nonideal gyrator with respect to port 1 and port 2 created by removing the grounded capacitor C_0. Hence, if the capacitor is connected at port 1, the input impedance looking into port 2 is found to be $Z_{in} = (R_1 + R_2) + sC_0R_1R_2$, thereby simulating a series RL type grounded inductor. This circuit was published earlier in [106].

Analogues to the controlled CCs dealt in Chapter 4, Tangsrirat and Surakampontorn [46] presented an electronically tunable floating inductance simulation which employed a current-controlled CDBA (CC-CDBA). The circuit from [46] is shown here in Figure 8.27.

An analysis of this circuit shows that the circuit is characterized by the following equation:

$$i_1 = -i_2 = \frac{(V_1 - V_2)}{sC_0R_{x1}R_{x2}} \tag{8.16}$$

and thus simulates a lossless FI of value $C_0R_{x1}R_{x2}$. It may, however, be noted that this structure appears somewhat similar to the CDTA-based structure of Figure 8.21.

Biolek *et al.* [47] further modified the basic CDBA and proposed Z-copy (ZC)-controlled CDBA; however, its use in immittance simulation has not been explored so far. On the other hand, Jaikla and Siripruchyanun [48] have demonstrated that

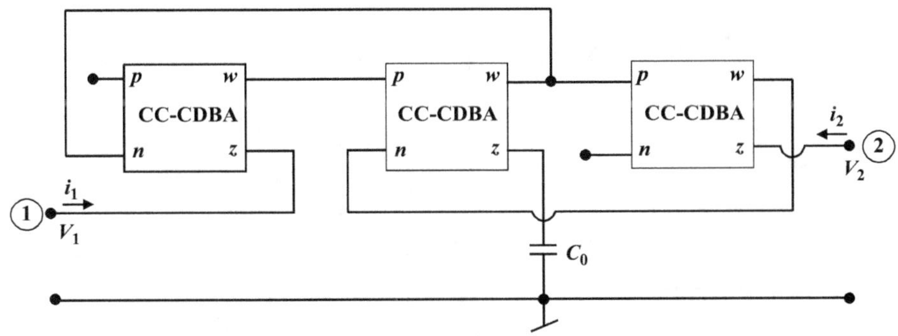

Figure 8.27 Lossless floating inductance using CC-CDBA [46]

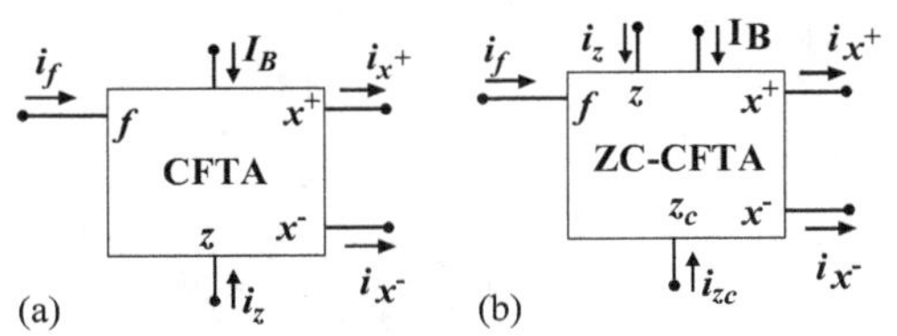

Figure 8.28 CFTA and its variant [49]: (a) symbolic notation of CFTA and (b) symbolic notation of ZC-CFTA

both grounded and floating negative inductor can be simulated using two CC-CDBAs only.

8.3.7 CFTA-based circuits

CFTA is a simplified form of CDBA in which instead of CDU a CF has been employed. The CFTA was introduced by Herencsar *et al.* in [49]. This simplification was introduced by observing that in quite a few application circuits using CDTAs, either ‘p’ or ‘n’ terminal is left open which may cause the same noise injection into the monolithic circuit. The CFTA not only overcomes this limitation, but the simplified building block also has reduced transistor count. Another variant of this is current-inverting transconductance amplifier (CIFTA) which is also termed inverting CFTA. In the following, we present some interesting impedance simulation circuits using CFTAs as active elements. The schematic symbol of the CFTA is shown in Figure 8.28 and its characterization is $V_f = 0$, $i_z = i_f$, $i_x^+ = g_m V_z$ and $i_x^- = -g_m V_z$ where g_m is given by $(I_B/2V_T)$. On the other hand, a ZC-CFTA has one additional terminal having $i_{zc} = i_f$.

Although at least three CFTAs of the normal kind are needed to simulate an FI or generalized floating admittance converter, two ZC-CFTAs suffice to create a floating lossless FI simulator as demonstrated by Herencsar *et al.* in [50] (Figure 8.29).

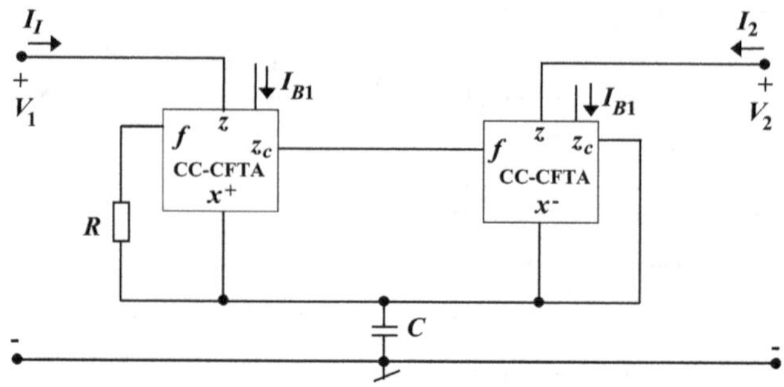

Figure 8.29 Lossless FI using Z-copy CFTAs [50]

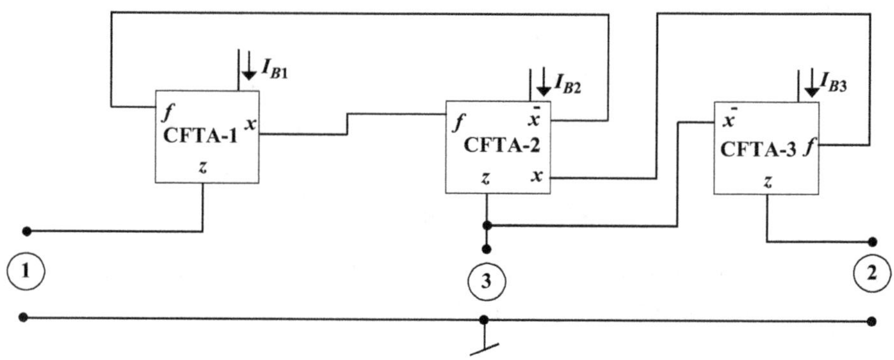

Figure 8.30 Three-port gyrator using CFTAs [51]

An analysis of this circuit reveals that the value of the FI is given by $L_{eq} = \mathrm{CR}/g_m$ where for a CMOS implementation of the CFTA, g_m is given by

$$g_m = \sqrt{kI_B}\ ;\ k = \mu_s C_{ox}\left(\frac{W}{L}\right) \tag{8.17}$$

where (W/L) is the aspect ratio of the relevant MOSFETs in the CMOS structure of ZC-CFTA of [50] (see Figure 2 therein).

Li [51] demonstrated that with three CFTAs a three-port gyrator can be realized without requiring any external passive elements as per the schematic of Figure 8.30.

With port 3 terminated into an impedance Z_L, the FI realized by this gyrator is given by

$$Z_{eq} = \frac{1}{g_m g_{m3} Z_L} \quad \text{where } g_{m1} = g_{m2} = g_m \tag{8.18}$$

The circuits for floating capacitance multiplier given in [51] require four CFTAs while those for floating FDNR require as many as five CFTAs [51].

Koton *et al.* [52], while reviewing history, progress and new results on synthetic passive element design employing CFTAs, have established that the circuit of Figure 8.31 is the optimum solution of the floating inductor simulation using CFTAs.

It is worth pointing out that the optimum circuit for floating capacitor presented in [52] still requires four CFTAs though it simulated a lossless floating capacitance multiplier employing a grounded capacitor.

An electronically controllable version of the CFTA called current-controlled CFTA (CC-CFTA) was proposed by Siriphot *et al.* in [53] along with its CMOS implementation. It was shown that a grounded inductance simulator is realizable with only one ZC-CC-CFTA as per Figure 8.32. A straightforward analysis of the circuit of Figure 8.32 yields

$$L_{eq} = \frac{C_0}{\sqrt{8K_{RF}K_{gm}I_0I_B}} \tag{8.19}$$

where K_{RF} and K_{gm} are transconductance parameters of the internal structure of CMOS CC-CFTA of [53].

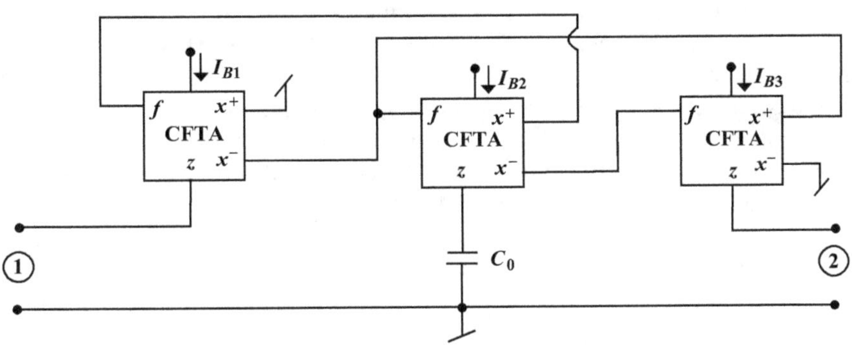

Figure 8.31 Optimized lossless FI circuit using CFTAs [52]

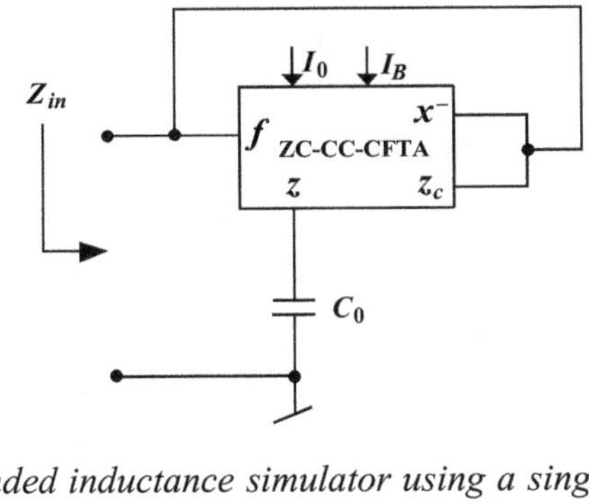

Figure 8.32 Grounded inductance simulator using a single ZC-CC-CFTA [53]

8.3.8 VDTA-based impedance simulators

Analogous to the CDTA, another building block called VDTA was also proposed by Biolek *et al.* in [1] which is symbolically shown here in Figure 8.33.

The VDTA is characterized by the following equations: $i_z = G_{mF}(V_p - V_n)$ (with both p and n input terminals assumed to have ideally infinite input impedance), $i_x^+ = G_{ms}V_z$ and $i_x^- = -G_{ms}V_z$. In terms of OTAs, it can be represented as shown in Figure 8.33(b). Thus, it is clear that a single VDTA is actually equivalent to two OTAs – one with a differential input and a single output and the other with a single input and dual complementary outputs. Depending upon the realization, both transconductances can be either current-controllable (in case of bipolar implementation of OTAs) or voltage-controllable (in case of CMOS implementation of OTAs).

From Chapter 3 of this monograph, it is very well established that a lossless grounded L can be realized using only two OTAs, a lossless FI can be realized with three OTAs of the single-input type, and finally, if dual output OTAs are considered then both lossless GI and lossless FI can be realized with only two OTAs. These basic circuits here are summarized in Figure 8.34 for ready reference.

It is found that besides other applications, several researchers have proposed VDTA-based grounded and FIs of various kinds. Let us consider them one by one.

First, we take the case of lossless grounded inductance simulation. Circuits using a single VDTA for this purpose have been proposed by Srivastava and Prasad [55] and Srivastava *et al.* [56]. These are shown in Figure 8.35 with internal OTA-based implementation of VDTA shown inside the CDTA block in each case to correlate how they relate to the basic two-OTA-based GI simulators of Figure 8.34.

Analysis of the circuit of Figure 8.35(a) gives

$$V_3 = [G_{m1}(V_{in} - V_3) + G_{m2}V_3]\frac{1}{sC_0} \tag{8.20}$$

Equation (8.20) can be simplified as

$$V_3[sC_0 + (G_{m1} - G_{m2})] = G_{m1}V_{in} \tag{8.21}$$

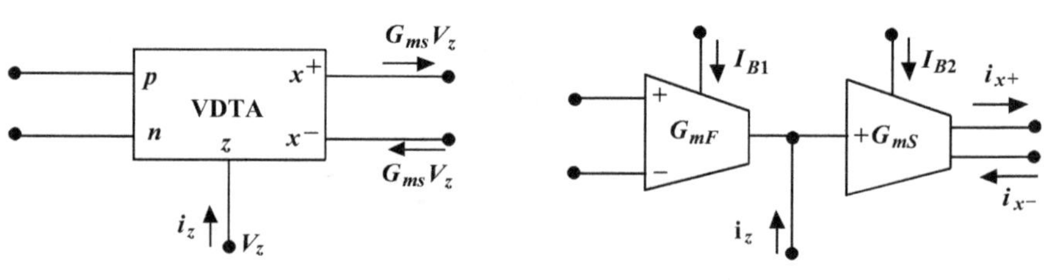

(a) (b)

Figure 8.33 The VDTA: (a) symbolic notation and (b) OTA implementation of the VDTA [1]

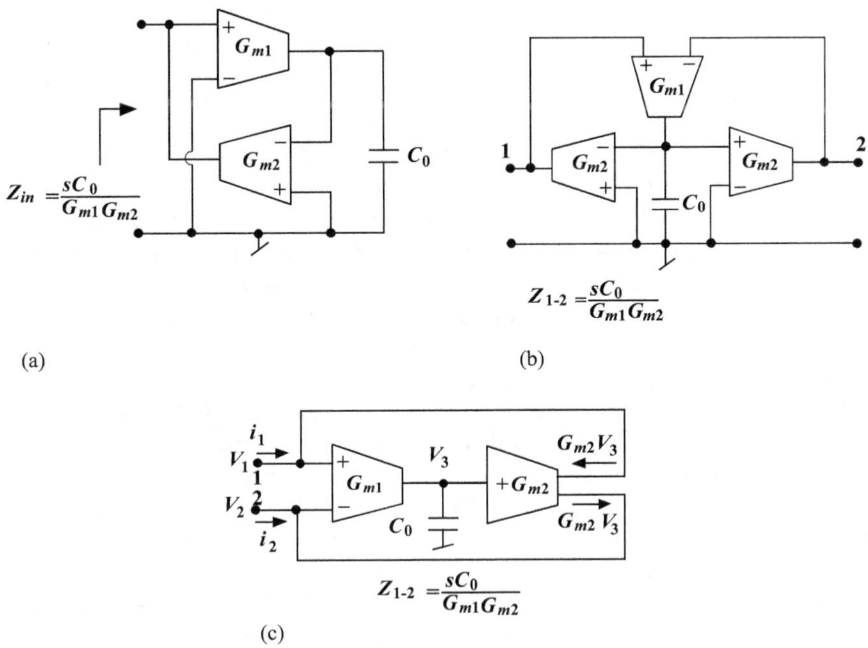

Figure 8.34 *Basic OTA-based simulated inductors: (a) lossless GI and (b, c) lossless FIs*

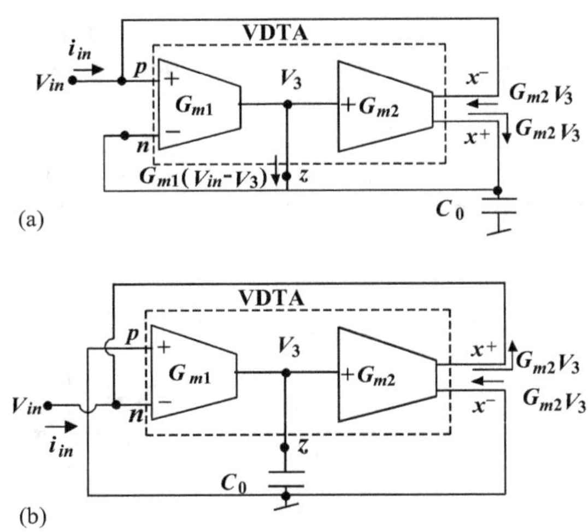

Figure 8.35 *Lossless GIs using single VDTA: (a) Srivastava and Prasad circuit [55] and (b) Srivastava, Prasad and Bhaskar circuit [56]*

and

$$i_{in} = G_{m2}V_3 = \frac{G_{m2}G_{m1}V_{in}}{sC_0 + (G_{m1} - G_{m2})} \tag{8.22}$$

Thus, with $G_{m1} = G_{m2} = G_m$ does one obtain $Z_{in} = G_m^2/sC_0$ representing a grounded inductance $L_{eq} = G_m^2/C_0$.

In the other circuit of Figure 8.35(b), with terminal p and x^- grounded, the resulting circuit essentially follows the mechanism of the circuit of Figure 8.34(a) with polarities of G_{m1} and G_{m2} reversed! In conclusion, the circuit of Figure 8.35(b) here is similar to the circuit of Figure 8.34(a) whereas that of Figure 8.35(a) is although different but suffers from the drawback that it compulsorily requires an equality constraint $G_{m1} = G_{m2} = G_m$ to realize a lossless GI.

Consider now the circuit proposed by Bang and Lee [57] shown here in Figure 8.36. It must be mentioned that the circuit mentioned in [56] is somewhat erroneous, as although unused terminal x^- can be left open, the unused terminal n cannot be left open and needs to be grounded (the same, however, has been *corrected* here). Then, one can see that the resulting structure is conceptually the same as that of the two OTA GI of Figure 8.34(a).

On the similar lines, from a re-examination of the floating active inductor of Bang and Lee [57], it can be readily deciphered that the internal mechanism of this circuit is no different than the two-gyrator-grounded capacitor–based FI wherein each two-OTA-based gyrator is implemented by one VDTA.

Let us now turn our attention to other lossless FI simulators using VDTAs. In this context comes the single VDTA-based FI simulator presented by Tangsrirat and Unhavanich [58] which is reproduced here in Figure 8.37 with the OTA-based VDTA implementation expanded inside. It is clear that the basic scheme of this circuit is the same as that of the lossless FI of Figure 8.34(c).

Let us now look into the two-VDTA-based floating capacitance multiplier circuit proposed by Tangsrirat and Unhavanich in [58] which is shown here in Figure 8.38, in a similarly expanded form.

It is readily seen that $VDTA_1$ makes a three-port gyrator at whose port 3 (similar to the one shown in Figure 8.34(c)) the second VDTA-simulated lossless

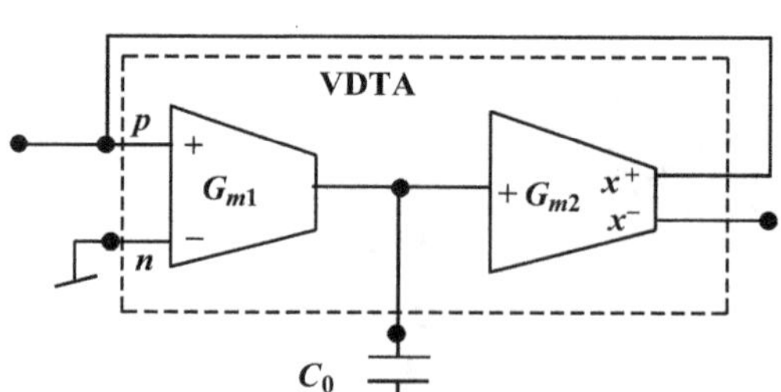

Figure 8.36 The expanded form of the single CDTA GI simulator of Bang and Lee [57]

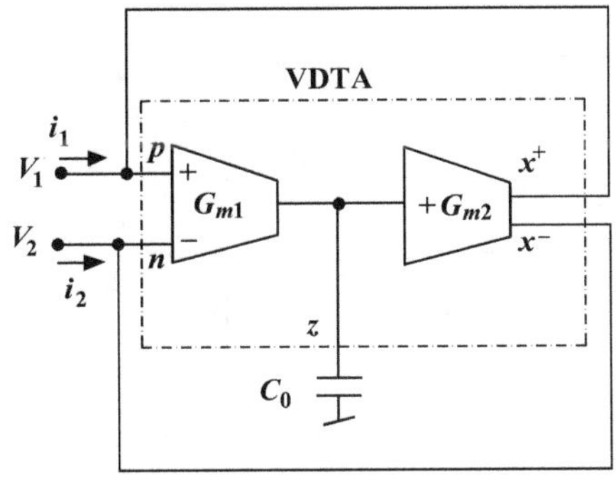

Figure 8.37 Expanded form of a lossless FI circuit due to Tangsrirat and Unhavanich [58]

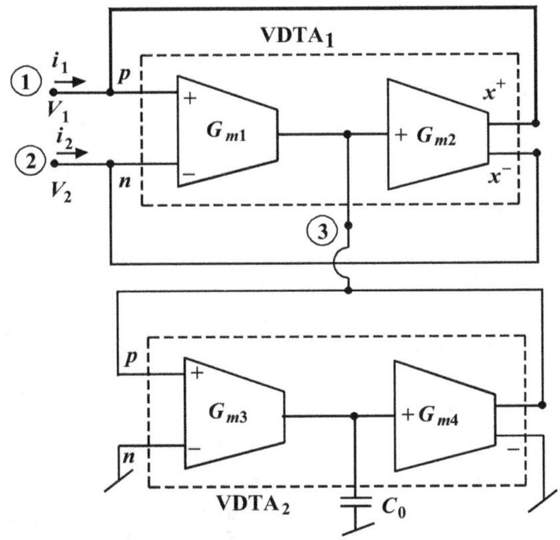

Figure 8.38 Expended form the lossless floating-capacitance multiplier circuit [58]

GI has been terminated as a load. The resulting circuit, thus, would simulate a lossless floating capacitance of value.

$$C_{eq} = \frac{G_{m1}G_{m2}C_0}{G_{m3}G_{m4}} \tag{8.23}$$

Lastly, let us consider the lossless FDNR circuit proposed by Tangsrirat and Mongkolwai [59] shown here in Figure 8.39.

It is obvious that $VDTA_1$ constitutes a three-port gyrator on port 3 of which $VDTA_2$ and $VDTA_3$ along with capacitors C_1 and C_2 realize a grounded FDNC

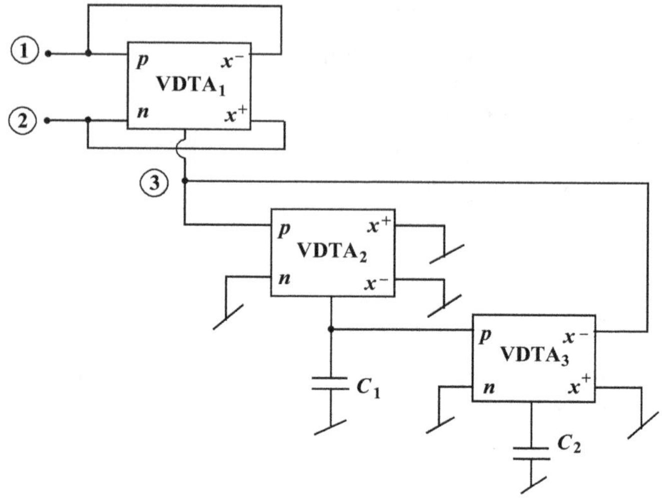

Figure 8.39 The floating FDNR circuit proposed by Tangsrirat and Mongkolwai [59]

having $Z_{eq} = s^2C_1C_2/G_{m3}G_{m5}G_{m6}$ so that the overall circuit simulates a floating FDNR with its impedance given by

$$Z_{1-2} = \frac{G_{m3}G_{m5}G_{m6}}{s^2C_1C_2G_{m1}G_{m2}} \tag{8.24}$$

Next, we take up some so-called active-only circuits made from VDTAs. The three-VDTA-based floating resistor presented by Srivastava and Prasad [60] requires matching of six transconductances. However, a much simpler two OTA-based floating resistor requiring matching of only two OTAs is well known. An alternative circuit can be made with only a single-dual complementary output-based OTA (with no such matching requirements). Both of these circuits have been detailed out in Chapter 3 of this monograph. The same comment applies to the other two-VDTA-based active floating resistors presented by the same authors in [61].

From these expositions, it is apparent that many of the VDTA-based simulation circuits for lossless GI, lossless FI, lossless floating capacitance and lossless FDNR are conceptually quite similar to their OTA counterparts if each VDTA is considered to be a composite connection of two OTAs. However, when VDTA is considered to be a single block with its internal circuit implemented in terms of bipolar or CMOS transistors, then only these circuits can be considered to be different than OTA circuits.

8.3.9 CCTA-based circuits

The CCTA as a new composite analog circuit building block appears to have been proposed first by Propkop and Musil [66]. Since it is a composite of a CCII and an OTA with complementary outputs, a number of basic circuits including those of simulated impedances can be readily formulated based upon those realizable with CCIIs or those realizable with OTAs. However, the extra degrees of freedom available due to either CCII or the OTA can be clearly exploited to either overcome some deficiency of the parent CCII- or OTA-based circuit or to facilitate the introduction of one or more of the various novel features into the corresponding circuit made from CFTA such as better (say electronic) tunability, reduction in the number of external passive components or elimination of component-matching requirement.

However, what appears to be more interesting is the consideration of several interesting improvizations of the basic CCTA such as current-controlled CCTA (CCCCTA) [70–72], differential voltage CCTA (DVCCTA) [68,69], ZC current-controlled CIFTA (CCCITA) [74] and dual-X current-controlled differential input transconductance amplifier [75].

In the following, we discuss some prominent results on immittance simulation using CCTA and their above-mentioned variants.

The behavioural model of the CCTA is shown in Figure 8.40(a), and it is characterized by $i_y = 0$, $v_x = v_z$, $i_z = i_x$, $i_O = \pm G_m v_z$. In Figure 8.40(b)–(d), it is shown how a single CCTA can be employed to realize a lossless grounded FDNR, capacitance multipliers.

Note that a VCCS can be easily realized by another CCII along with a resistor; hence, all the above circuits, in essence, are the same that would be realizable with

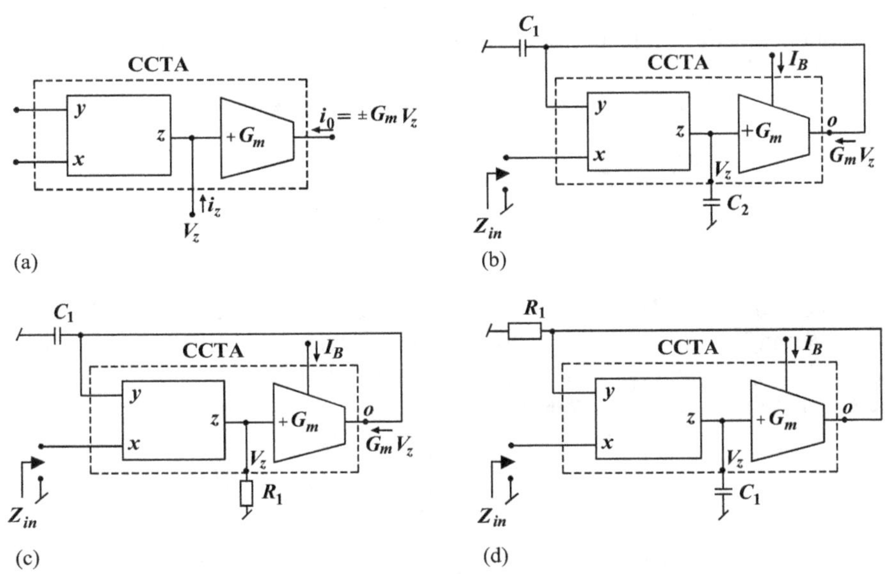

Figure 8.40 Single CCTA simulators based upon the propositions of Jantakun [67]: (a) behavioural model of CCTA, (b) grounded FDNR realization; $Z_{in} = G_m/s^2C_1C_2$, (c) capacitance multiplier; $Z_{in} = G_mR_1/sC_1$ and (d) another grounded capacitance multiplier giving $Z_{in} = G_mR_1/sC_1$

two CCIIs of opposite polarities with two capacitors and a resistor used for FDNR realizations and one capacitor and two resistors (in two different ways) used to realize a grounded simulated capacitor.

From the published works on basic CCTAs, it is found that the most interesting applications of the CCTAs are in the realization of a grounded FDNR and grounded capacitance multipliers, since all of these can be realized using a single CCTA and grounded capacitors as demonstrated by Jantakun [67].

On the other hand, if the normal CCII is replaced by a differential voltage CCII (DVCC) to construct the so-called DVCCTA (as shown in Figure 8.41) characterized by $V_x = V_{y1} - V_{y2}$, $i_z = i_x$, $i_0 = G_m V_z$ with $i_{y1} = 0$, $i_{y2} = 0$, this provides somewhat more capability in realizing synthetic impedances of various types.

Several researchers have demonstrated that it is relatively easier to devise circuits for simulating inductors and other kinds of FIs. However, simulating such impedances using two DVCCTAs is not surprising since it has already been shown in Chapters 3 and 4 that such impedances are realizable with two CCIIs and in some limiting cases (i.e. that of lossless FIs) even with two OTAs with dual complementary outputs. The only additional benefits of devising FI circuits using two DVCCTAs is the possibility of using all grounded components/capacitors in addition to the electronic tunability of the parameters of the realized impedances which would be clear from the DVCCTA-based circuits which have been discussed in the following.

The first example of the kind anticipated above is the floating lossless immittance function simulator proposed by Nandi *et al.* [68] which is shown here in Figure 8.42.

This circuit of Figure 8.42 realizes an admittance matrix

$$[Y] = \begin{bmatrix} \dfrac{Y_1 Y_{m1}}{Y_0} & -\dfrac{Y_{m1} Y_{m2}}{Y_0} \\ -\dfrac{Y_1 Y_2}{Y_0} & \dfrac{Y_{m2} Y_2}{Y_0} \end{bmatrix} \tag{8.25}$$

From the above, it follows that an FI is realizable if $Y_{m1} = Y_{m2} = g_m = 1/R$, $Y_1 = Y_2 = 1/R$ and $Y_0 = sC_0$, thereby realizing an equivalent floating inductance of value $L_{eq} = C_0 R/g_m$. Thus, although the circuit employs all grounded passive elements, the matching requirement is a deficiency of this circuit.

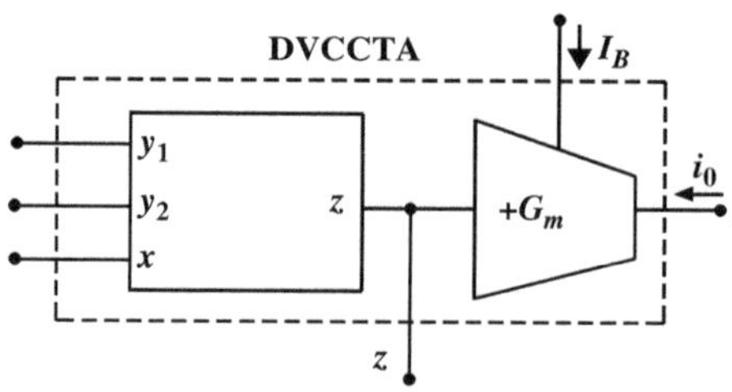

Figure 8.41 Symbolic notation of the DVCCTA [68]

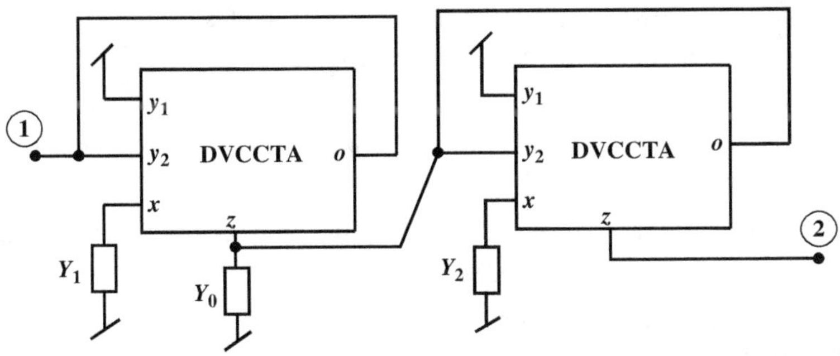

Figure 8.42 Simulation of generalized floating impedance using DVCCTA as proposed by Nandi et al. [68]

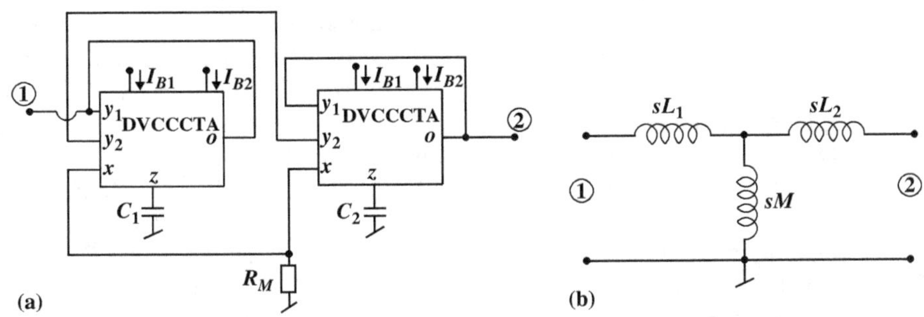

Figure 8.43 Simulation of a mutually coupled circuit using DVCCCTA as per the proposal of Pandey et al. [73]: (a) circuit and (b) equivalent

It has been found that as compared to the CCTA, its electronically controllable version, namely the CCCCTA, is more versatile for impedance simulation. This has been amply demonstrated by several researchers such as Siripruchyanun and Jaikla [70], Siripruchyanun *et al.* [71] and Li [72].

It is interesting to note that Pandey *et al.* [73] demonstrated that given only two DVCCCTAs and only a few passive components, one can even simulate a mutually coupled circuit as per the schematic of Figure 8.43.

Assuming DVCCCTA characterization as $i_{y1} = 0$, $i_{y2} = 0$, $V_x = (V_{y1} - V_{y2}) + R_x i_x$, $i_z = i_x$, $i_0 = g_m V_z$, whereas in case of a CMOS implementation, R_x and g_m are given by

$$R_x \cong \frac{1}{\sqrt{8\mu_s C_{ox}(w/L)I_{B1}}} \tag{8.26}$$

and

$$g_m \cong \sqrt{\mu_s C_{ox}\left(\frac{W}{L}\right)I_{B2}} \tag{8.27}$$

where I_{B1} and I_{B2} are the external DC bias currents of the DVCCII and OTA, respectively. The parameters of the mutually coupled circuit of Figure 8.43 are given by

$$L_1 = \frac{CR_{x1}}{g_m}, \quad L_2 = \frac{CR_{x2}}{g_m} \quad \text{and} \quad M = \frac{CR_m}{g_m} \tag{8.28}$$

where $g_{m1} = g_{m2} = g_m$.

In this circuit, the primary and secondary impedances are independently controllable by I_{B1} and I_{B2}, respectively, whereas the mutual coupling is adjustable through the grounded resistor R_m.

8.3.10 VDBA-based circuits

While the previously described immittance simulators employed composite building blocks many of which had a CDU at the front end, such as CDTA and CDBA, we now consider some interesting impedance simulators which employ composite building blocks which have a voltage difference unit at the front end. On one hand, there are building blocks like VDBA (and its fully balanced version called FB-VDBA), VD-DIBA and VDIBA. On the other hand, the so-called VDCC also belong to the same class.

A VDBA is characterized by ideally infinite input impedance at both its input terminals, thereby implying $I_p = 0$, $I_n = 0$, while it has a current output terminal z (with ideally infinite output impedance) providing $I_z = G_m\,(V_p - V_n)$ and has a voltage output terminal w (which has ideally zero output impedance) having $V_w = V_z$. Thus, internally, it could be thought of to be consisting of an OTA with transconductance G_m, followed by a unity gain buffer (see Figure 8.44). An exemplary transistor-level schematic suitable for bipolar technology was presented by Unhavanich *et al.* [76]. A simplified representation of a CMOS structure, presented by Tangsrirat [80], is shown here in Figure 8.45.

Unhavanich *et al.* [76] presented a single-capacitor single VDBA-based capacitance multiplier which is reproduced here in Figure 8.46(a), whereas its expanded form is shown in Figure 8.46(b).

It is easy to derive that $Z_{in} = [1/(sC_0(1 + G_mR_0))]$ which also follows from the elaborated diagram from which it is readily seen that $V_0/V_{in} = -G_mR_0$, so that using Miller's theorem, Z_{in} is given by

$$Z_{in} = \frac{1/sC_0}{1 - (V_0/V_{in})} = \frac{1}{sC_0(1 + G_mR_0)} \tag{8.29}$$

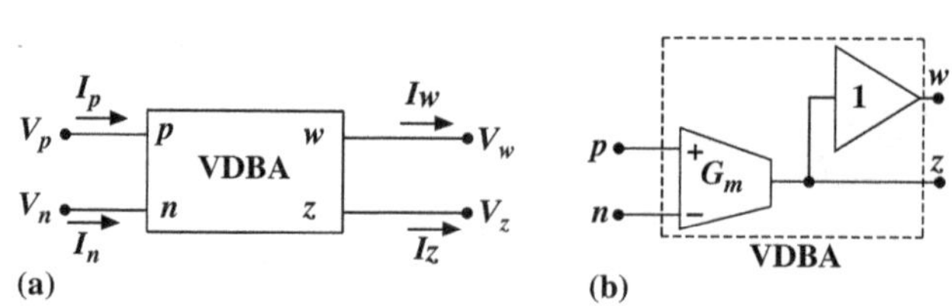

Figure 8.44 The VDBA: (a) symbolic notation and (b) OTA and buffer-based implementation of the VDBA

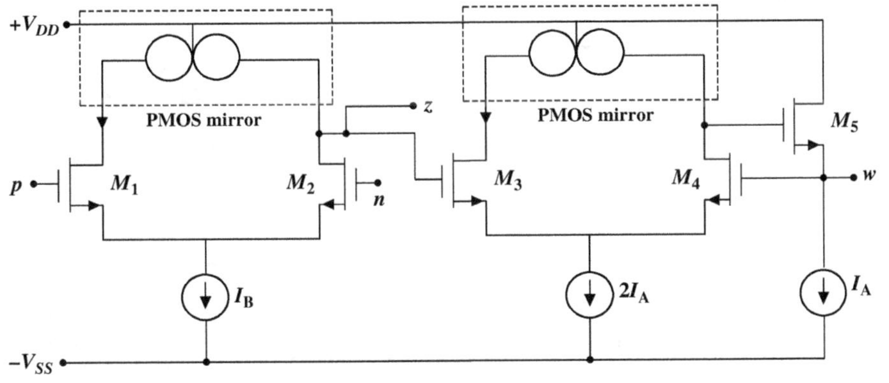

Figure 8.45 A simplified form of the CMOS VDBA architecture proposed by Tangsrirat in [80]

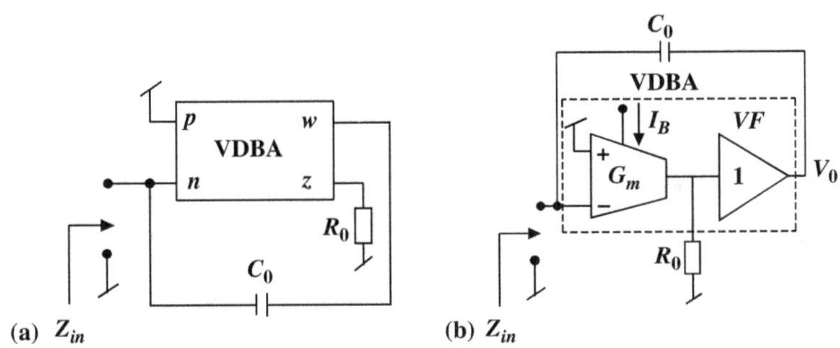

Figure 8.46 A VDBA-based capacitance multiplier [76]: (a) the circuit and (b) elaborated diagram with internal architecture of VDBA split out

and thus, the capacitance multiplication factor can be changed through a variable resistance R_0 or can be electronically controlled by changing G_m through I_{B1} since $G_m = I_{B1}/2V_T$.

Yesil *et al.* [77] postulated a ZC-VDBA whose symbolic notation is shown in Figure 8.47(a), the behavioural model is shown in Figure 8.47(b), while its implementation using OPA860 is detailed in Figure 8.47(c); its application in inductance simulation is shown in Figure 8.47(d).

From the circuit of Figure 8.47(d), it is easy to decipher that Z_{in} of the circuit would be given by $Z_{in} = sC_0R_0\,R_{01}$ which represents an inductive $L_{eq} = C_0R_0R_{01}$ which is controllable by R_0 and/or R_{01}.

However, in comparison to these circuits, a generalized impedance simulator proposed by Mongkolwai and Tangsrirat [78] is comparatively more interesting and versatile proposition which is reproduced here in Figure 8.48.

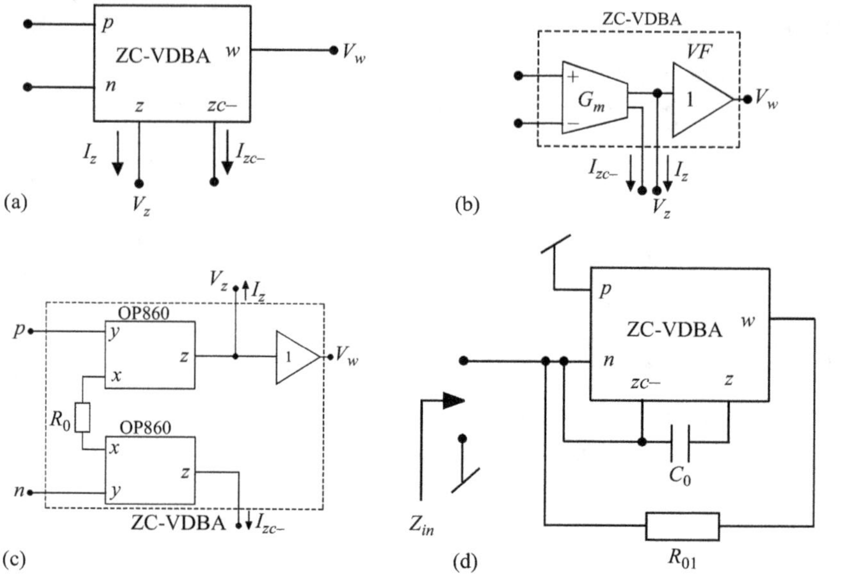

Figure 8.47 ZC-VDBA, its model, implementation and application [77]: (a) symbolic notation of ZC-VDBA, (b) OTA-buffer implementation, (c) OP860 implementation and (d) ZC-VDBA-based inductance simulator

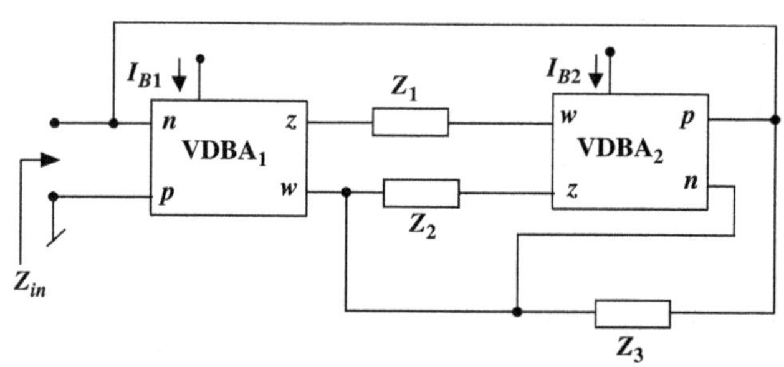

Figure 8.48 Generalized impedance simulator proposed by Mongkolwai and Tangsrirat [78]

The input impedance of this circuit is found to be

$$Z_{in} = \left(\frac{g_{m2}}{g_{m1}}\right)\frac{Z_2 Z_3}{Z_1} \tag{8.30}$$

From which it is clear that with an appropriate choice of circuit impedances (resistive/capacitive), the same circuit can be used to realize an electronically tunable (through g_{m1} or g_{m2}, therefore by I_{B1} or I_{B2}) grounded inductor, capacitance multiplier or FDNR as follows:

1. Grounded inductor: $Z_1 = 1/sC_1$, $Z_2 = R_2$, $Z_3 = R_3$ yielding

$$L_{eq} = \frac{g_{m2}C_1R_2R_3}{g_{m1}} \tag{8.31}$$

2. Capacitance multiplier: $Z_1 = R_1$, $Z_2 = 1/sC_2$, $Z_3 = R_3$ giving

$$C_{eq} = \frac{g_{m1}R_1C_2}{g_{m2}R_3} \tag{8.32}$$

3. FDNR: $Z_1 = R_1$, $Z_2 = 1/sC_2$ and $Z_3 = 1/sC_3$ yielding

$$Z(s) = \frac{g_{m2}}{s^2 g_{m1}R_1C_2C_3} \tag{8.33}$$

A notable feature of this circuit is electronic tunability of the realized grounded impedance in all the cases in a *temperature-insensitive* manner since the impedance scaling factor (g_{m2}/g_{m1}) is equal to (I_{B2}/I_{B1}) wherein the temperature-dependent parameter V_T has been cancelled out.

A number of authors have reported single VDBA-based lossy inductor simulation circuits which, however, are hardly surprising given that a VDBA is equivalent to an OTA and VF.

A number of researchers have presented two VDBA and grounded capacitor-based lossless/lossy inductor simulation circuits [80,81] or floating capacitance multiplier circuits using FB-VDBA [79]. In some cases, such structures have even been realized by BIMOS FB-VDBA, see [79]. The readers are referred to [76–81] for details.

8.3.11 VD-DIBA-based circuit

An active building block somewhat similar to the VDBA is the so-called VD-DIBA [1] which has two voltage input terminals V^+ and V^- in which input currents $I_V{}^+$ and $I_V{}^-$ both are zero. However, current at Z is given by $I_Z = g_m(V_V{}^+ - V_V{}^-)$ but the buffered amplifier with inputs V_V and V_Z and output V_w is related by $V_w = (V_Z - V_V)$ with I_W being arbitrary. This building block is equivalent to a composite connection of an OTA and unity gain differential amplifier as shown in Figure 8.49.

From the internal structure of VDTA, it is clear that simulating a lossless GI using two VD-DIBAs, as in [83], is straight forward since this task can be easily done by two OTAs as explained in Chapter 3 of this monograph. Also, for the simulation of a lossless FI using three VD-DIBAs, the same comment applies, i.e. if a lossless FI can be simulated using three OTAs, it can surely be done using three VD-DIBAs as done in [83]. However, lossless GI using only one VD-DIBA(−) and lossless FI using only two VD-DIBAs(−) both employing a single grounded capacitor are of interest and the circuits presented by Bhaskar *et al.* in [82] which

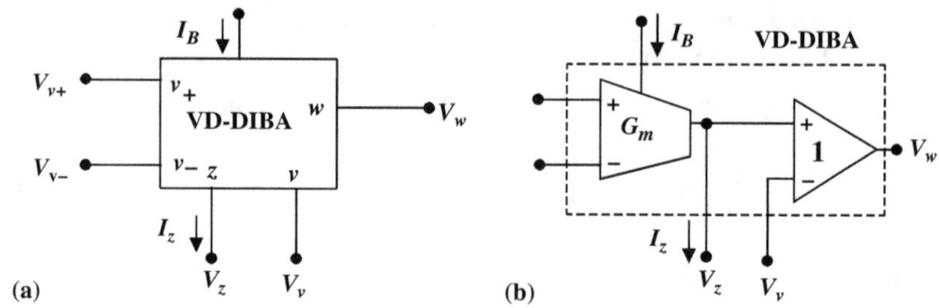

Figure 8.49 The VD-DIBA: (a) symbolic notation and (b) behavioural model of a VD-DIBA

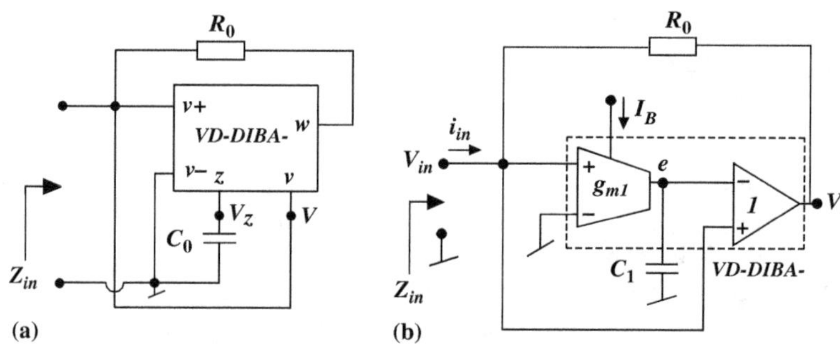

Figure 8.50 VD-DIBA-based inductance simulation [82]: (a) VD-DIBA-based lossless inductor and (b) expanded version

accomplish exactly this are reproduced here in Figure 8.50. Note that in contrast to VD-DIBA defined earlier, a VD-DIBA(−) has somewhat different characterization: $I_V{}^+ = 0$, $I_V{}^- = 0$, $I_Z = g_m\,(V_V{}^+ - V_V{}^-)$, $I_{V+} = 0$ and $V_w = (V_V - V_Z)$.

When viewed it, its expanded form as in Figure 8.50(b), the realization of Figure 8.50(a) has simple internal mechanism. Note that V_W is given by

$$V_w = V_{in} - \frac{g_m}{sC_i} V_{in} \tag{8.34}$$

so that with a resistive feedback input current is given by

$$i_{in} = V_{in} - \left[V_{in} - \frac{g_m}{sC_1} V_{in} \right] = + \frac{g_m V_{in}}{sC_1 R_0} \tag{8.35}$$

thereby yielding

$$Z_{in} = \frac{sC_1 R_0}{g_m} \tag{8.36}$$

which is controllable by I_B.

Now the GI circuit Figure 8.50(a) can be converted into an FI in two different ways:

1. Unground V^- and apply port-2 voltage there so that I_{in} gets modified to

$$i_{in1} = \frac{\{V_{in1} - [V_{in1} - (V_{in1} - V_{in2})(g_m/sC_1)]\}}{R_0} = \frac{g_m}{sC_1R_0}(V_{in1} - V_{in2}) \qquad (8.37)$$

   as required.
2. To get $i_{in2} = i_{in1}$ one can either convert the internal voltage e into a current through another VD-DIBA containing OTA with transconductance g_{m2} and applying its output current at port 2 to get $i_{in2} = -i_{in1}$ with $g_{m2} = 1/R_1$ or putting this OTA across R_1 and making its output current applied at port 2 to yield $i_{in2} = -i_{in1}$; subject to $g_{m2}R_0 = 1$.

These two alternatives are shown in Figure 8.51.

It may, however, be observed that since in neither of the cases, the UGDA and its w-terminal are engaged in FI formulation, there is no need to connect V-terminal of the second VD-DIBA to be connected anywhere, it can also be left open (like the w-terminal of the second VD-DIBA(−) which is kept open). However, the

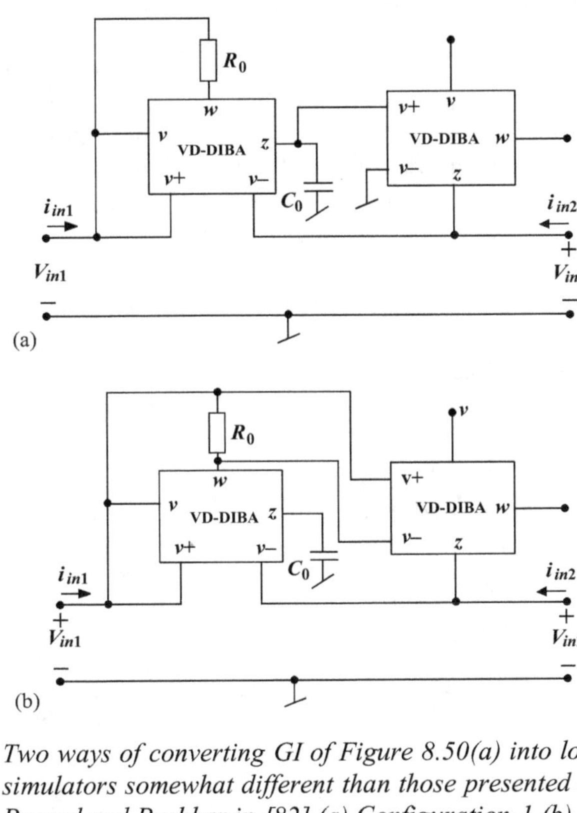

Figure 8.51 Two ways of converting GI of Figure 8.50(a) into lossless FI simulators somewhat different than those presented by Bhaskar, Prasad and Pushkar in [82] (a) Configuration-1 (b) Configuration-2

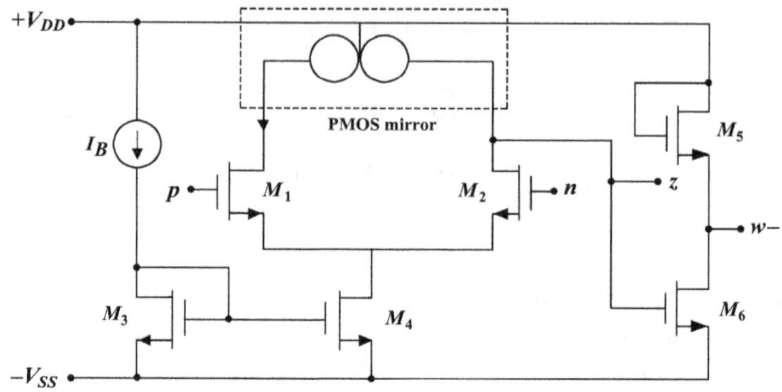

Figure 8.52 Simplified representation of the CMOS realization of VDIBA proposed by Tangsrirat [85]

determination of any advantages of the circuits of Figure 8.51 over those of [82] appears to be an interesting problem for further work.

Finally, we may comment that a single-VDBA is sufficient to simulate lossy (series RL or parallel RL) inductors as shown in [77].

A number of authors have also dealt with a slightly modified form of VDBA which they call VDIBA which has possibly the simplest CMOS architecture as shown in Figure 8.52. Thus, a VDIBA differs from VDBA only in the relation between V_Z and V_w which is $V_w = -V_Z$ in case of the VDIBA. With identical MOSFETs M_5 and M_6, the inverting amplifier formed by $M_5 - M_6$ has the distinct advantage of having zero output offset voltage.

8.3.12 CFCC-based circuits

The current follower CC (CFCC) [86] or ZC-CFCC [87] is an active building block whose symbolic notation is shown in Figure 8.53, and it is characterized by the following matrix equation:

$$\begin{bmatrix} V_p \\ I_z \\ V_i \\ I_{x+} \\ I_{x-} \\ I_{zc} \end{bmatrix} = \begin{bmatrix} R_p & 0 & 0 \\ 1 & 0 & 0 \\ 0 & 1 & -R_i \\ 0 & 0 & 1 \\ 0 & 0 & -1 \\ 1 & 0 & 0 \end{bmatrix} \begin{bmatrix} I_p \\ V_z \\ I_i \\ V_{x+} \\ V_{x-} \\ V_{zc} \end{bmatrix} \tag{8.38}$$

Use of CFCCs and ZC-CFCC in evolving a class of novel lossy/lossless GI circuits has recently been demonstrated by Singh *et al.* [87]. Here, we present some representative circuits from those presented in [87] in Figure 8.55.

The circuits of Figure 8.55(a) and (b) simulate lossy grounded inductors: parallel RL and series RL, respectively; that of Figure 8.55(c) simulates a lossless

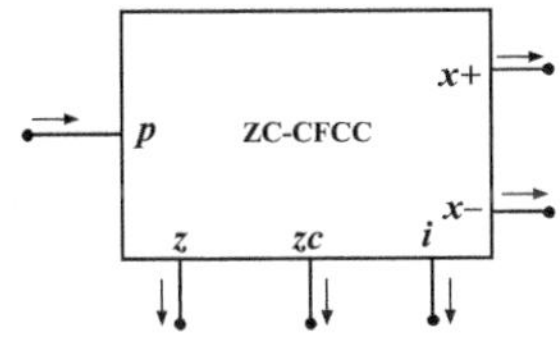

Figure 8.53 The symbolic notation of the ZC-CFCC [86]

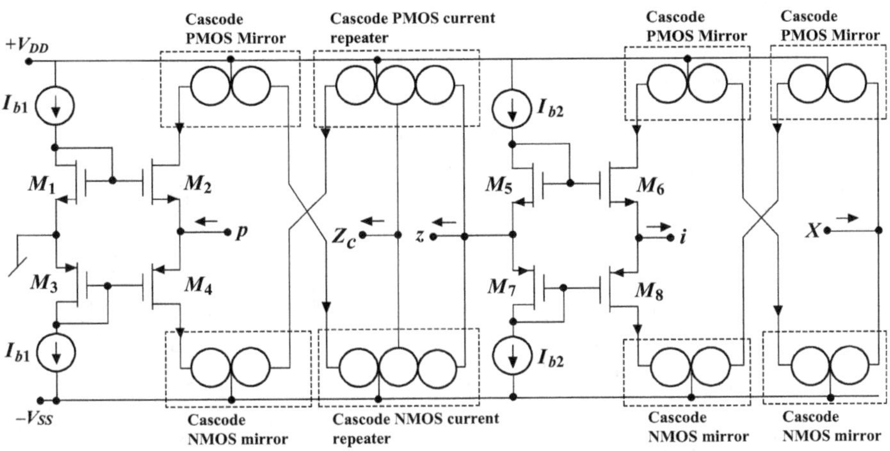

Figure 8.54 Simplified form of an exemplary CMOS implementation of the ZC-CFCC [87]

grounded inductor, whereas the last one simulates an ideal FDNR. The notable features of these circuits are as follows: (i) use of only one active element, (ii) deployment of a canonical number (only three) of passive elements, (iii) employment of single grounded capacitor in case of inductance simulation circuits and (iv) single resistance controllability of the realized impedances in all the cases.

Singh *et al.* [87] have demonstrated that if the intrinsic p- and i-port input/output resistances which are external current-controllable are taken into account as a useful design parameter, rather than being treated as a parasitic influence, the ZC-CFCC can be employed to devise a novel active-only lossless gyrator which leads to a single-ZC-CFCC-based lossless grounded inductor which is shown here in Figure 8.56.

A straightforward analysis of the circuit of Figure 8.56 gives its Z_{in} as

$$Z_{in}(s) = sCR_pR_i \tag{8.39}$$

where

$$R_p = \frac{1}{\sqrt{8\mu_s C_{ox}(W/L)I_{b1}}} \quad \text{and} \quad R_i = \frac{1}{\sqrt{8\mu_s C_{ox}(W/L)I_{b2}}} \tag{8.40}$$

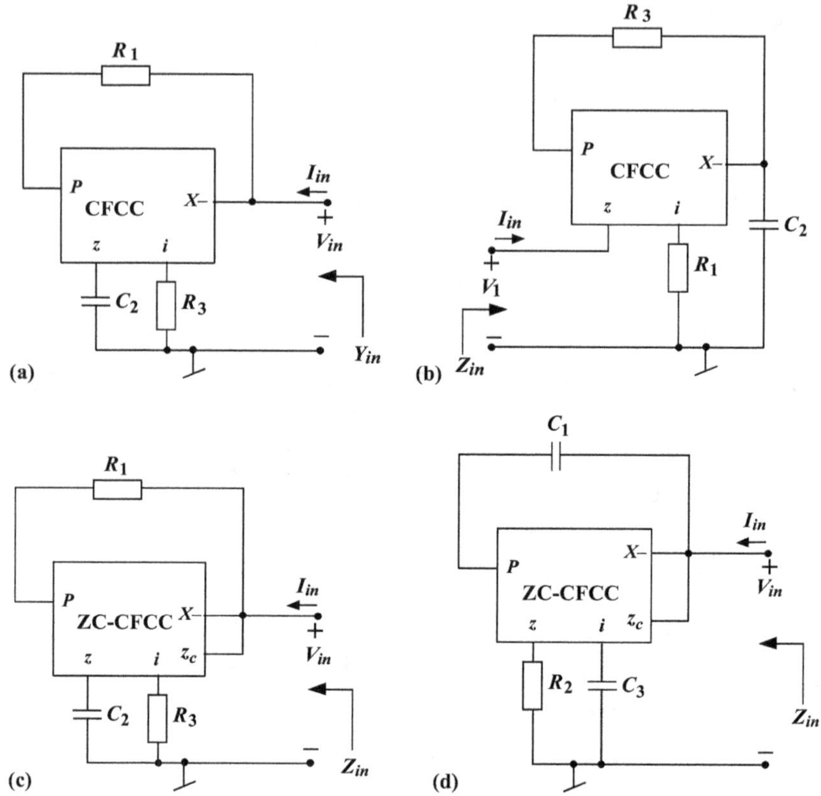

Figure 8.55 *Some inductance/FDNR simulators from the catalogue of single CFCC impedance simulators presented by Singh* et al. *[86]: (a) parallel RL: $R_{eq} = R_1$ and $L_{eq} = C_2R_1R_3$, (b) series RL: $R_{eq} = R_1$ and $L_{eq} = C_2R_1R_3$, (c) grounded Inductance: $L_{eq} = R_1R_3C_2$ and (d) grounded FDNR $D_{eq} = R_2C_1C_3$*

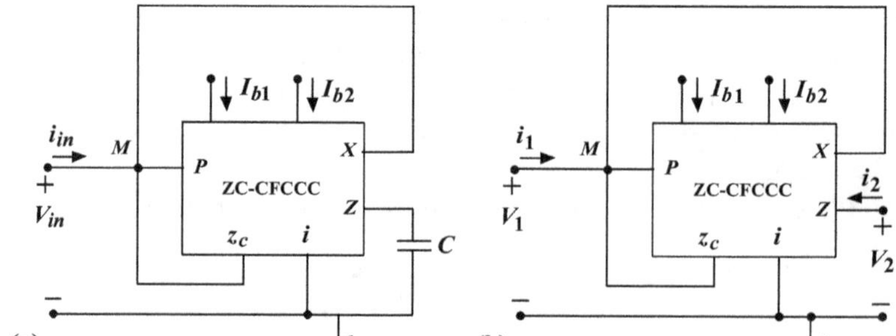

Figure 8.56 *Electronically tunable circuits using ZC-CFCC [87]: (a) grounded inductance and (b) an active gyrator*

The circuit, thus, simulates a lossless grounded inductor whose value can be electronically controlled by varying any or both of the external DC bias currents I_{b1} and I_{b2}.

It is interesting to note that with the grounded capacitor C deleted, the resulting two-port is an excellent gyrator with its [Y]-matrix given by

$$[Y] = \begin{bmatrix} 0 & +\sqrt{8\mu_s C_{ox}\left(\frac{W}{L}\right)I_{b2}} \\ -\sqrt{8\mu_s C_{ox}\left(\frac{W}{L}\right)I_{b1}} & 0 \end{bmatrix} \tag{8.41}$$

A novel feature of this gyrator is that both the gyration resistances can be independently controlled through external control currents I_{b2} and I_{b1}, respectively.

Having identified the gyrator, the creation of a lossless floating inductor using two such gyrators with a grounded capacitor embedded between them as shown in Figure 8.57 is straightforward. Analysis of the circuit of Figure 8.57 proposed by Singh *et al.* [87] yields its short circuits admittance matrix is found to be (with $R_p = R_i = R$)

$$[Y] = \frac{1}{sCR^2}\begin{bmatrix} 1 & -1 \\ -1 & 1 \end{bmatrix} \tag{8.42}$$

Application: An example of the application of the electronically tunable, lossless grounded inductor in the realization of a fourth-order Butterworth high pass filter was demonstrated in [87]. Starting from a prototype fourth-order high pass filter; the normalized filter for cut-off frequency of 1 Hz was designed for which the component values were $R_s = R_L = 1\ \Omega$, $L_1 = L_2 = 0.1217$ H, $C_1 = 0.2768$ F and $C_3 = 0.0780$ F [20]. After performing appropriate frequency and impedance scaling, a passive filter circuit was designed for achieving the cut-off frequency of 1.59 MHz for which the final values of components were $R_s = R_L = 1\ \mathrm{k}\Omega$, $L_1 = L_2 = 76.54\ \mu$H, $C_1 = 1.74.1$ pF and $C_3 = 49.1$ pF. The inductor of 76.54 μH was then simulated using the circuit of Figure 8.56(a). The cut-off frequency obtained from SPICE simulations was found to be $f_0 = 1.5952$ MHz.

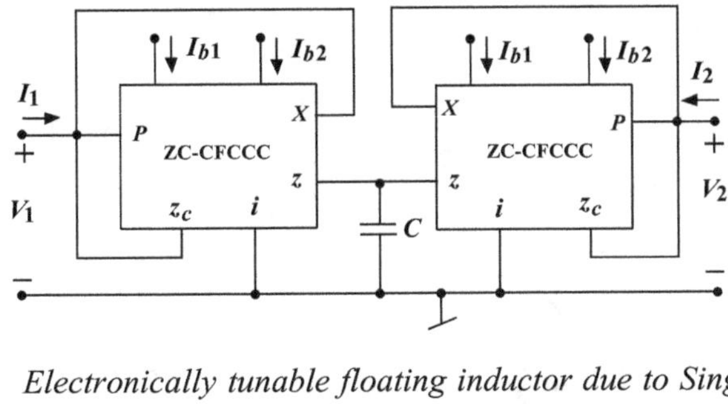

Figure 8.57 Electronically tunable floating inductor due to Singh et al. [87]

8.3.13 Inductance simulation using OTA–COA

Yamacli *et al.* [88] proposed an interesting idea to employ the composite connection of an OTA and a current operational amplifier (COA) to realize an active-only grounded inductance simulator which is shown in Figure 8.58.

In this circuit, the COA gain is supposed to be characterized by a one-pole model approximated by

$$\frac{I_{out}(s)}{I_{in1}(s) - I_{in2}(s)} = \frac{B}{s} \tag{8.43}$$

Open-loop gain of COA is as an integrator resulting from (8.43), where B is the gain bandwidth product of the COA. By a routine analysis, the input impedance of this circuit is found to be

$$Z_{in}(s) = \frac{s}{g_m B} \tag{8.44}$$

With the OTA and COA both realized in CMOS technology, this inductance simulation circuit is suitable for CMOS implementation. Furthermore, the inductance value can be controlled by varying the transconductance g_m of the OTA through its external DC bias current.

The proposers of the circuit of Figure 8.58 have demonstrated that when OTA and COA were implemented in 0.5-μm CMOS technology prevalent, the designed circuit could be operated up to 30 MHz and the inductance value could be adjusted electronically between 3.9 μH and 37 μH.

8.3.14 FDNR realization using capacitive gyrators

A *capacitive gyrator* is a two-port active element quite similar to the conventional two-port gyrator, the only difference being that the two gyration conductances are replaced by capacitive admittances due to which its $[Y]$-matrix can be written as

$$[Y] = \begin{bmatrix} 0 & \pm sC \\ \mp sC & 0 \end{bmatrix} \tag{8.45}$$

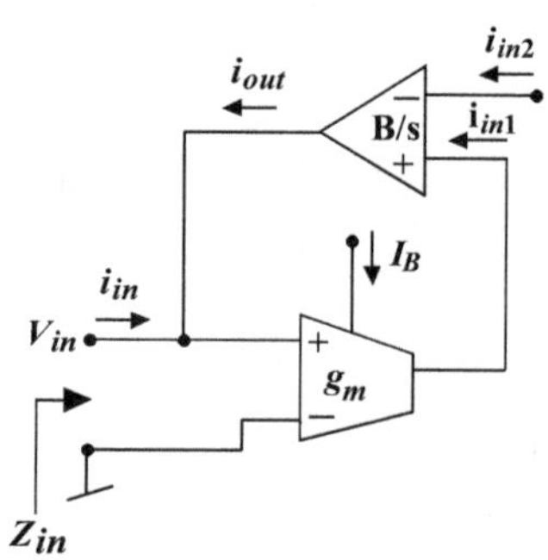

Figure 8.58 An inductance simulator using OTA–COA combination [88]

where sC represents the transadmittance of the gyrator which is capacitive.

The use of such a gyrator was proposed by Kacar in [89] to realize a grounded FDNR using two trans-capacitance amplifiers of opposite polarity, connected back-of-back with a grounded resistor embedded between them; see Figure 8.59. Thus, a completely CMOS circuit is obtainable by replacing the passive resistor by a grounded MOS resistor comprising two diode-connected MOSFETs operating in saturation region [107].

Using the same idea, a scheme of realizing floating FDNR was also given (see Figure 8.60) which requires two *trans-capacitive* gyrators with a grounded resistor embedded between them.

In both the cases, the entire circuit would be resistor-less (if the R is simulated by two-MOSFET circuit of [107]) and, therefore, suitable for implementation in CMOS technology.

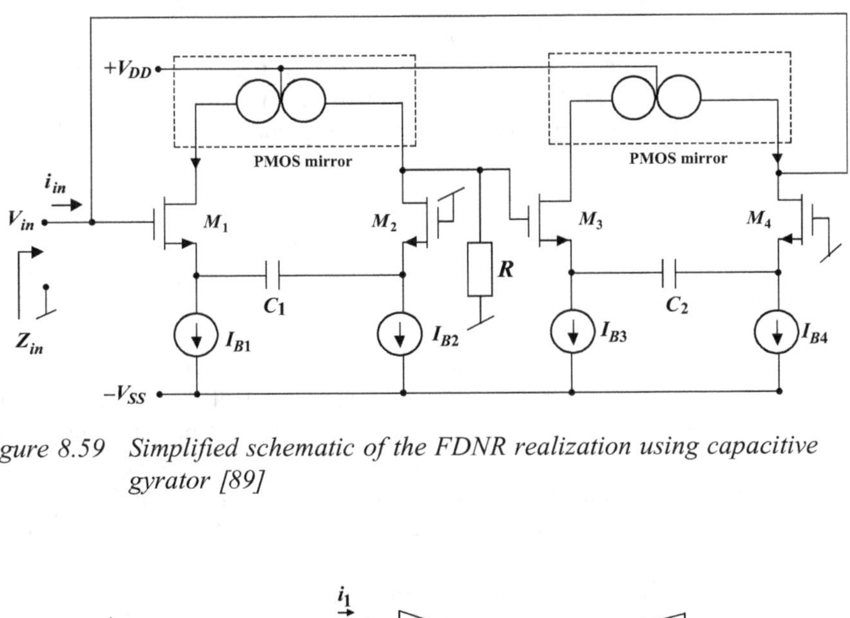

Figure 8.59 Simplified schematic of the FDNR realization using capacitive gyrator [89]

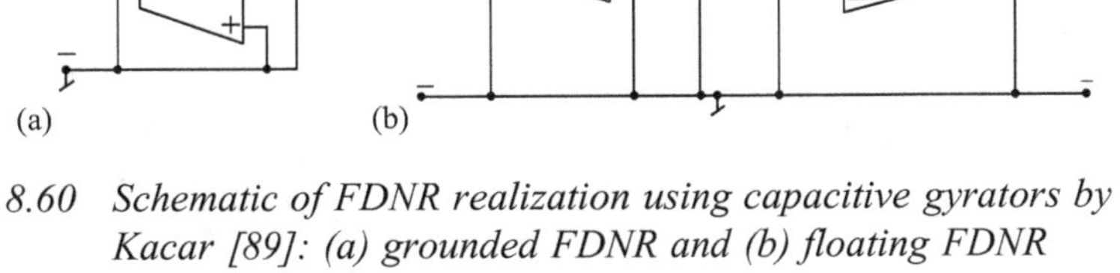

Figure 8.60 Schematic of FDNR realization using capacitive gyrators by Kacar [89]: (a) grounded FDNR and (b) floating FDNR

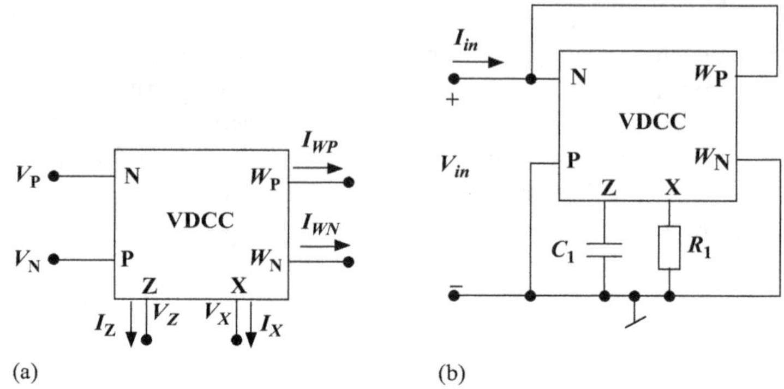

Figure 8.61 Lossless grounded inductor using a single VDIBA [90]: (a) symbolic notation and (b) the GI circuit

8.3.15 Lossless grounded inductance simulator using VDCC

Kacar *et al.* [90] demonstrated that using a single VDCC and two passive elements only, a variety of grounded impedances and FI [91] can be realized. Out of their catalogue of six single-VDCC-based circuits, we present here one exemplary realization in Figure 8.61. The VDCC, as an active element, is characterized by

$$I_N = 0, \quad I_p = 0, \quad I_Z = g_m(V_p - V_n), \quad V_x = V_z, \quad I_{WP} = I_X, \quad I_{WN} = -I_X$$

This circuit simulates a lossless grounding inductance of value $L_{eq} = sC_1R_1/g_m$, and therefore, the value of the realized inductance is tunable through g_m. It is obvious that with capacitor C deleted and port 2 created therein, the resulting circuit would turn out to be a lossless two-port gyrator.

An exemplary CMOS implementation of the VDCC is shown in Figure 8.62.

Application: The utility of this circuit was demonstrated in [90] in the design of a third-order Butterworth filter containing a T-network, between the source resistance and load resistance, comprising two capacitors in the series branches and an inductor in the shunt branch. This resulted in a very economical design of the third-order filter requiring a single VDCC which simulated the grounded inductance part. It was established that when the DC bias current of the VDCC was varied to 10 μA, 30 μA and 60 μA, correspondingly, the realized inductance value was observed to be 0.64 mH, 0.36 mH and 0.27 mH, respectively. The circuit was implemented in 0.18μm CMOS technology [109].

8.3.16 Inductance simulation using VDIBA

The VDIBA is a four terminal active building block characterized by the following matrix equation:

$$\begin{bmatrix} i_p \\ i_n \\ i_z \\ v_{w-} \end{bmatrix} = \begin{bmatrix} 0 & 0 & 0 & 0 \\ 0 & 0 & 0 & 0 \\ g_m & -g_m & 0 & 0 \\ 0 & 0 & -\beta & 0 \end{bmatrix} \begin{bmatrix} v_p \\ v_n \\ v_z \\ i_{w-} \end{bmatrix} \tag{8.46}$$

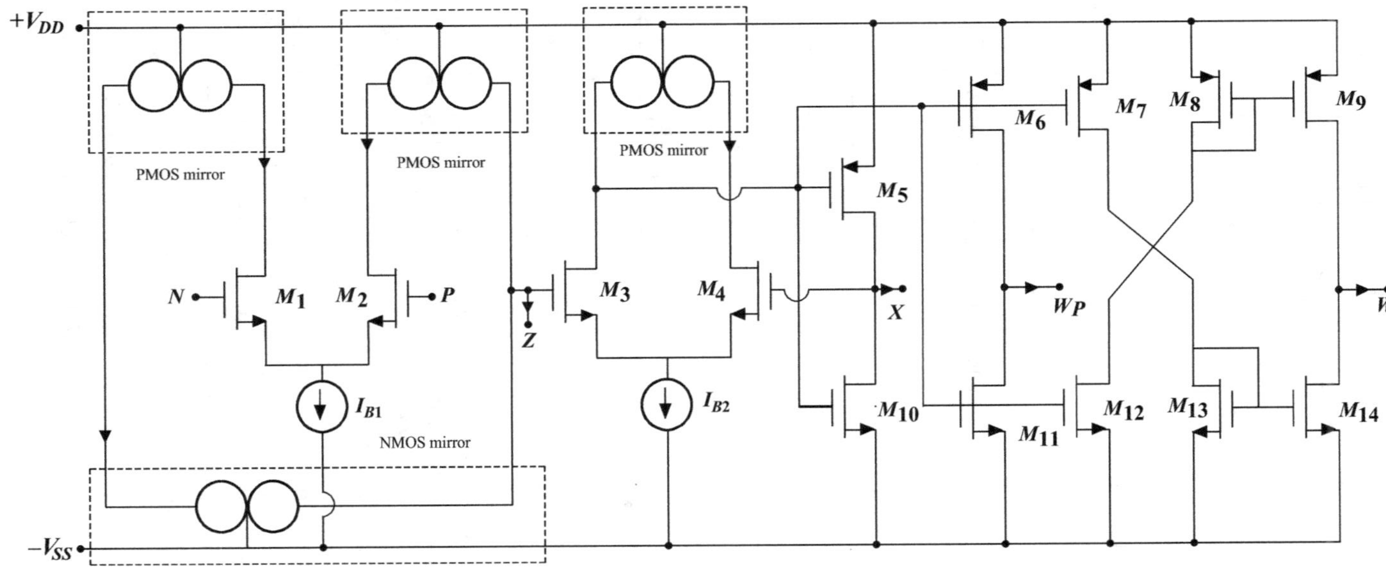

Figure 8.62 Simplified representation of an exemplary CMOS implementation of the VDCC [90]

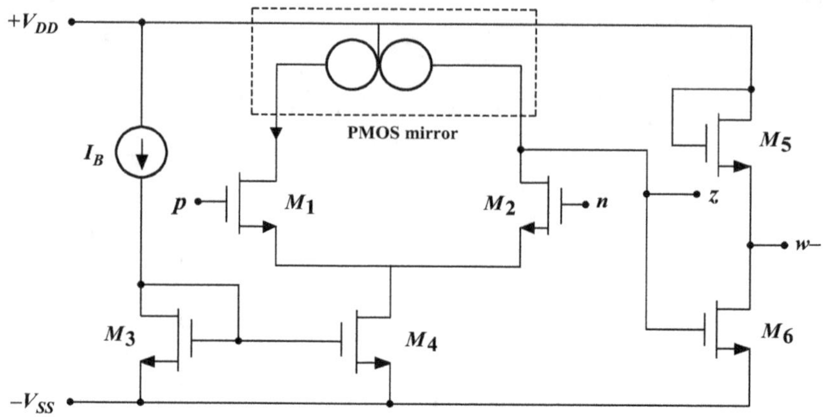

Figure 8.63 A simplified representation of the CMOS VDIBA [85]

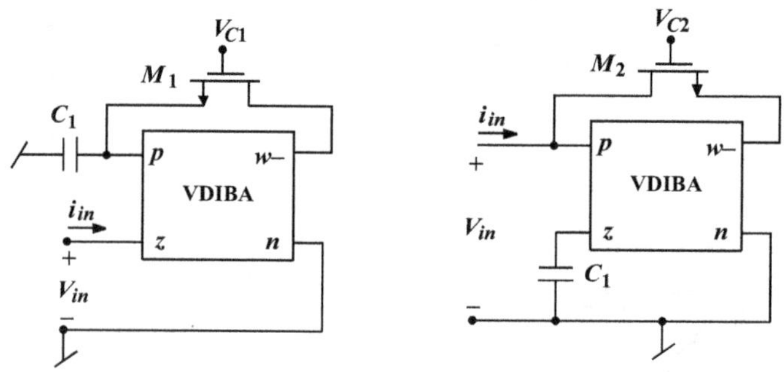

Figure 8.64 Electronically tunable lossy inductors using a single VDIBA proposed by Tangsrirat [85]

In the above characterization, g_m represents the transconductance of the differential transconductance amplifier, whereas β is the voltage gain of the inverting amplifier. An exemplary CMOS implementation is shown in Figure 8.63 from where it is seen that this is possibly one of the simplest looking active building blocks from the viewpoint of internal circuit architecture.

Tangsrirat [85] employed this building block to devise two single-VDIBA-based circuits (Figure 8.64) each employing only one additional grounded capacitor and one MOSFET to realize series RL and parallel RL type grounded impedances, both of which are electronically controllable. Both the circuits were shown to realize second-order resonant circuits with the resonance frequency being electronically tunable.

8.3.17 General floating immittance simulator using CBTA

CBTA is another newly proposed building block which is characterized by the following equation:

$$I_z = g_m(s)(V_p - V_n); \quad V_W = \mu_W(s)V_z, \quad I_p = \alpha_p(s)I_w; \quad I_n = -\alpha_n(s)I_W \qquad (8.47)$$

In the above characterization, except g_m, the nominal values of all the other parameters are unity.

Ayten *et al.* [92] presented a configuration employing two CBTA elements and three grounded admittances (Figure 8.65) which realized [*Y*]-matrix as given in the following equation:

$$[Y] = \left(\frac{\alpha_1 \mu_{w1}}{\alpha_2 \mu_{w2}}\right)\left(\frac{g_{m1} Y_1 Y_3}{g_{m2} Y_2}\right)\begin{bmatrix} +1 & -1 \\ -1 & +1 \end{bmatrix} \tag{8.48}$$

From the above equation, it is obvious that by proper selection of different admittances, this circuit can realize a floating inductance, a floating resistance or a floating capacitance – all elements being electronically controllable.

Later, Sagbas in [94] presented a number of CBTA-based circuits, all using only a single CBTA to realize a variety of impedances which included $\pm L$, $\pm C$ and $\pm R$ circuits from the same structure.

Koksal *et al.* [93] presented a three CC-CBTA-based circuit for the simulation of an active transformer which is shown in Figure 8.66.

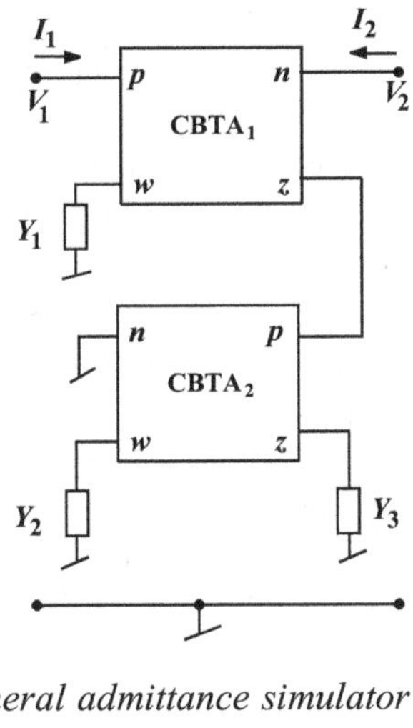

Figure 8.65 General admittance simulator using CBTAs [92]

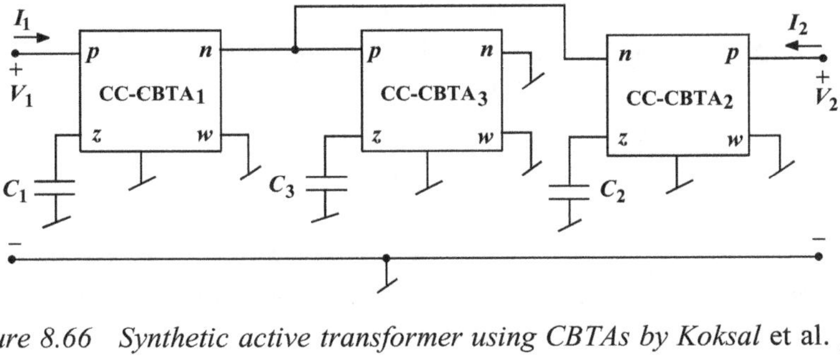

Figure 8.66 Synthetic active transformer using CBTAs by Koksal et al. *[93]*

The circuit could realize a synthetic active transformer with several advantages such as employment of all grounded capacitors, capability to implement both positive and negative couplings, not requiring any component matching and suitability for IC implementation.

8.3.18 Simulation of inductor using current differential amplifiers (CDA)

Current differential amplifier (CDA) was introduced by Souliotis *et al.* [95] which is characterized by $i_{out} = K(I^+ - I^-)$. Among the various applications of the CDA and its dual output version, it was also used to simulate an FI as shown in Figure 8.67. With $K_1 = 1$, $K_3 = 3$, the circuit simulates an FI of value $L = 3sCR^2$. Contrary to numerous circuits for FI realization described earlier, the circuit of Figure 8.67 is deficient due to the requirement of equal-valued resistors and very specific values of gains K_1 and K_2 needed for lossless FI realization, though the authors did demonstrate successfully that even CDAs could be used to realize a lossless FI!

A digitally programmable lossless FI based upon this scheme was subsequently devised by Said *et al.* in [96].

8.3.19 GI/FI using other miscellaneous active elements

Although we have described in this chapter a number of prominent circuits for realizing inductors, gyrators and even synthetic transformers using the most popularly studied building blocks, our compilation would be by no means complete until we mention the following additional efforts made by researchers for evolving interesting impedance simulators:

- Zhou and Zeng [97] proposed a configuration for FI simulation using a dual complementary output DVCC and a CCCII along with an external resistor R and a capacitor C that simulates a lossless FI. In this circuit, the DVCC along with the resistor R creates a differential voltage-controlled current source whose output currents are flown into the capacitor C, the voltage across which is again converted into a current which constitutes the port 2 current. (This circuit, however, needs to be corrected by having the CCCII with complementary outputs so

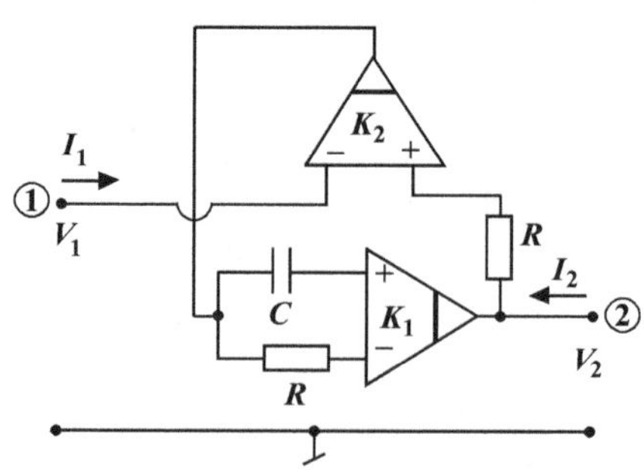

Figure 8.67 A floating inductor simulation using CDAs [95]

that the input current is derived from the other output current of the CCCII to ensure that this makes $I_1 = -I_2$.)

- While re-surveying the literature in connection with the writing of this monograph we were reminded of an interesting paper by Riordan [98] who had proposed a new active element termed *tri-amplifier* as early as in 1975 with the sole objective of creating a configuration for realizing a gyrator. In fact, more appropriately, it should better be called a GIC because it actually realizes an equivalent FI given by $Z_1Z_3Z_5/Z_2Z_4$ which is more akin to a GIC than a gyrator. However, the bipolar implementation proposed therein, apart from using 20 BJTs, required as many as 18 resistors and 3 capacitors, was perhaps not so encouraging to stimulate further interest in the tri-amplifier. The authors of this monograph believe that the development of a better bipolar (perhaps translinear) and/or CMOS implementation of the internal circuit architecture of the tri-amplifier is a good area for further research.
- Alzaher [99] proposed an interesting element called digitally controlled CF (DCCF) with two complementary outputs and programmable gain. Besides other applications, it was demonstrated that with two VFs and two DCCFs along with two resistors and a capacitor, a lossless inductor can be realized with digital tuning of its value.
- De Marcellis *et al.* [100] came up with another new active element which they called voltage and current gained CCII and demonstrated its application in realizing a capacitance multiplier which was shown to multiply a given capacitance of value 10 pF by a factor tunable from 1 to 3,100, thereby achieving a capacitance multiplication for more than six decades frequency range from 0.15 to 865 kHz. The designed circuits were verified in 0.35-μm CMOS technology.
- A symmetry type FI scaling circuit concept was advanced by Matsumoto *et al.* in [101] which was made from CMOS voltage followers and CFs. The circuit could multiply a given capacitance by 50 times which was claimed to be suitable for applications in low frequency, low power application in biomedical devices.
- Yet another new active element called *current differencing transresistance amplifier* (CDTRA) was postulated by Siripruchyanun *et al.* [102] which has four external terminals labelled *p*, *n*, *z* and *o* with the terminal characteristics represented by the following matrix equation:

$$\begin{bmatrix} V_p \\ V_n \\ I_z \\ V_o \end{bmatrix} = \begin{bmatrix} 0 & 0 & 0 & 0 \\ 0 & 0 & 0 & 0 \\ 1 & -1 & 0 & 0 \\ R_m & -R_m & 0 & 0 \end{bmatrix} \begin{bmatrix} I_p \\ I_n \\ V_z \\ I_o \end{bmatrix} \tag{8.49}$$

A bipolar implementation and its application in realizing a grounded inductor was given but the circuit, besides a CDTRA, also needs a unity gain VF and suffers from the drawback of requiring two equal-valued capacitors to realize a lossless inductor.

- Yesil *et al.* [103] presented two configurations for lossless grounded inductance (GI) realization using a CCII and two inverting unity gain amplifiers (each realizable from a very simple implementation requiring only two MOS transistors operating in saturation). Two circuits were proposed which require only two resistors and a capacitor (though floating) but do not require any component matching constraints or cancellation conditions. One of the circuits, when realized using a CMOS CCII with very small intrinsic x-port input resistance allows the GI to be operated over a wide range from 200 Hz to 10 MHz.
- Yesil *et al.* [104], in another publication, employed the combination of an ICCII and inverting/noninverting unity gain VF to realize a capacitance multiplier with the advantage that the capacitance to be multiplied is connected to ground.
- Pandey and Paul in [105] presented the so-called differential difference CC transconductance amplifier (DDCCTA) as a new analog building block for signal processing and demonstrated its use in various analog signal processing/generation circuits. It was also shown in [105] that a single DDCCTA along with a grounded resistor and a grounded capacitor can yield a lossless GI circuit. Although not so recognized therein by the authors of [105], it is interesting to point out that with the grounded capacitor removed from the quoted inductance simulation circuit, the remaining two-port circuit would function as a lossless gyrator!

8.4 Concluding remarks

In this chapter, we have discussed some prominent gyrator circuits, grounded impedance circuits and FI circuits using new analog circuit building blocks of more recent origin. These composite building blocks are generally active elements having four or more external terminals, the internal structure of which is normally containing two *primary* active elements. A large number of new active composite building blocks have been proposed during the last two decades which, apart from their other applications in analog signal processing and signal generation, have also found potential applications in impedance synthesis [1–4]. Most of the authors who have employed such building blocks have also come up with their possible hardware implementation suitable for bipolar or CMOS technology. Thus, it is anticipated that the real potential of the circuits which can be made from these building blocks will be actually realized in practice when integratable versions of these elements are produced and are made available as off-the-shelf ICs. On the other hand, in most of the cases, the realization of these composite building blocks has also been suggested in terms of the various commercially available ICs. Although such implementations are mainly suitable for discrete applications, the ideas utilized in making the internal circuit architecture in this manner could also be carried over to IC implementations.

Thus, a large number of circuits have been developed so far [5–113] out of which we have deliberated upon only those which are considered more significant than the others. However, in doing so, apart from bringing out the salient features of the considered circuits, an attempt has also been made to explain the underlying common principles which not only give an insight into the synthesis of these circuits but is also useful from the viewpoint of correlating the various circuits made from seemingly different devices. In view of the vast amount of literature [5–113] on the theme of this chapter, a task which continues to remain un-attempted is to figure out as to which composite building block can be considered to be the best from the point view of realizing various impedance simulation circuits. It is hoped that ongoing/ future research in this direction might be able to answer this important question.

References

[1] Biolek D., Senani R., Biolkova V., Kolka Z. 'Active elements for analog signal processing: classification, review, and new proposals'. Radioengineering. 2008; **17**(4):15–31.

[2] Biolek D., Biolkova V. 'Implementation of active elements for analog signal processing by Diamond transistors'. International Conference in EDS'09, IMAPSCS, Brno, Czech Republic, 2009:304–09.

[3] Kuntman H. K. 'New advances and possibilities provided by alternative active elements in analogue circuit design'. Elektrik-Elektronik ve Bilgisayar Sempozyumu 2011.

[4] Abdalla K. K., Bhaskar D. R., Senani R. 'A review of the evolution of current-mode circuits and techniques and various modern analog circuit building blocks'. Nature and Science. 2012; **10**(10):1–13.

[5] Alpaslan H., Yuce E. 'New CMOS based current follower and its applications to inductor simulator and band-pass filter'. Indian Journal of Pure & Applied Physics. 2016; **54**(8):511–16.

[6] Tlelo-Cuautle E., Duarte-Villasenor M. A., Reyes-Garcia C. A., *et al.* 'Designing VFs by applying genetic algorithms from nullator-based descriptions'. 18th European Conference on Circuit Theory and Design. 2007; 555–8.

[7] Chaisricharoen R., Chipipop B., Sirinaovakul B. 'CMOS CCCII: structures, characteristics, and considerations'. International Journal of Electronics and Communications (AEU). 2010; **64**(6):540–57.

[8] Linggard R. 'High-*Q* gyrator circuit'. Proceedings of the IEEE. 1970; **58**(11): 1871–2.

[9] Keskin A. U., Toker A. 'A NIC with impedance scaling properties using unity gain cells'. Analog Integrated Circuits and Signal Processing'. 2004; **41**(1):85–7.

[10] Psychalinos C., Spanidou A. 'Current amplifier based grounded and floating inductance simulators'. International Journal of Electronics and Communications (AEU). 2006; **60**(2):168–71.

[11] Fakhfakh M., Pierzchala M. 'On the design of active inductors with current-controlled voltage sources'. Analog Integrated Circuits and Signal Processing. 2012; **73**(1):89–98.

[12] Alpaslan H., Yuce E., Tokat S. 'A new lossless floating inductor simulator employing only two-terminal active devices'. Indian Journal of Engineering & Material Sciences'. 2013; **20**(1):35–41.

[13] Senani R., Bhaskar D. R., Singh A. K., Sharma R. K. 'Synthesis of analog circuits using only voltage and current followers as active elements'. In M. Fakhfakh, E. Tlelo-Cuautle, R. Castro-Lopez (Eds.). Analog/RF and Mixed-Signal Circuit Systematic Design. London, Berlin, Heidelberg: Springer-Verlag; 2013. p. 289–315.

[14] Alzaher H. 'Current follower based reconfigurable integrator/differentiator circuits with passive and active components'. Microelectronics Journal. 2015; **46**(2):135–42.

[15] Fakhfakh M., Pierzchala M. 'Synthesis of active circuits by combining UGCs'. International Journal of Electronics. 2015; **103**(3):455–67.

[16] Kacar F., Cam U., Cicekoglu O., Kuntman H., Kuntman A. 'New parallel immittance simulator realizations employing a single OTRA'. The 2002 45th Midwest Symposium on Circuits and Systems, MWSCAS-2002; Tulsa, OK, USA, August 4–7, 2002.

[17] Cam U., Kacar F., Cicekoglu O., Kuntman H., Kuntman A. 'Novel two OTRA-based grounded immittance simulator topologies'. Analog Integrated Circuits and Signal Processing. 2004; **39**(2):169–75.

[18] Kilinc S., Salama K. N., Cam U. 'Realization of fully controllable negative inductance with single operational transresistance amplifier'. Circuits, Systems and Signal Processing. 2006; **25**(1):47–57.

[19] Kafrawy A. K., Soliman A. M. 'A modified CMOS differential operational transresistance amplifier (OTRA)'. International Journal of Electronics and Communications (AEU). 2009; **63**(12):1067–71.

[20] Gupta A., Senani R., Bhaskar D. R., Singh A. K. 'OTRA-based grounded-FDNR and grounded-inductance simulators and their applications'. Circuits, Systems and Signal Processing. 2012; **31**(2):489–99.

[21] Gupta A., Senani R., Bhaskar D. R., Singh A. K. 'New OTRA-based generalized impedance simulator'. ISRN Electronics. 2013; **2013**:10 pages.

[22] Pandey R., Pandey N., Paul S. K., Singh A., Sriram B., Trivedi K. 'Novel grounded inductance simulator using single OTRA'. International Journal of Circuit Theory and Applications. 2014; **42**(10):1069–79.

[23] Ghosh M., Paul S. K. 'Design of lossless grounded negative inductance simulator using single operational transresistance amplifier'. Revue Roumaine des Sciences Techniques. 2014; **59**(4):381–90.

[24] Senani R., Singh A. K., Gupta A., Bhaskar D. R. 'Simple simulated inductor, low-pass/band-pass filter and sinusoidal oscillator using OTRA'. Circuits and Systems. 2016; **7**(3):83–99.

[25] Nagar B. C., Paul S. K. 'Lossless grounded FDNR simulator and its applications using OTRA'. Analog Integrated Circuits and Signal Processing. 2017; **92**(3):507–17.

[26] Singh A. K., Senani R., Gupta A. 'OTRA, its implementations and applications: a state-of-the-art review'. Analog Integrated Circuits and Signal Processing. 2018; **97**(2):281–311.

[27] Sackinger E., Guggenbuhl W. 'A versatile building block: the CMOS differential difference amplifier'. IEEE Journal of Solid-State Circuits. 1987; **SC-22**(2):287–94.

[28] Singh B., Singh A. K., Senani R. 'A new universal biquad filter using differential difference amplifiers and its practical realization'. Analog Integrated Circuits and Signal processing. 2013; **75**(2):293–7.

[29] Singh B. 'A Class of Analog Signal Processing Circuits for Bipolar/CMOS Implementation' (Ph.D. thesis). 2017; AKTU Lucknow (supervised by Senani R. and Singh A. K.).

[30] Senani R. 'Realization of single-resistance-controlled lossless floating inductance'. Electronics Letters. 1978; **14**(25):828–9.

[31] Fabre A. 'The translinear operational current amplifier: a new building block'. International Journal of Electronics. 1986; **60**(2):275–9.

[32] Fabre A., Rochegude P. 'Current processing circuits with translinear operational current amplifiers'. International Journal of Electronics. 1987; **63**(1): 9–28.

[33] Biolek D. 'CDTA-building block for current-mode analog signal processing'. ECCTD'03; Krakow, Poland, 2003. p. 397–400.

[34] Prasad D., Bhaskar D. R., Singh A. K. 'New grounded and floating simulated inductance circuits using current differencing transconductance amplifiers'. Radioengineering. 2010; **19**(1):194–8.

[35] Siripruchyanun M., Jaikla W. 'CMOS current-controlled current differencing amplifier and applications to analog signal processing'. International Journal of Electronics and communications (AEU). 2008; **62**(4):277–87.

[36] Siripruchyanun M., Jaikla W. 'Realization of current controlled current differencing transconductance amplifier (CCCDTA) and its applications'. ECTI Transactions on Electrical Engineering, Electronics and Communication. 2007; **5**(1):41–50.

[37] Siripruchyanun M., Jaikla W. 'Current-controlled current differencing transconductance amplifier and applications in continuous-time signal processing circuits'. Analog Integrated Circuits and Signal Processing. 2009; **61**(3): 247–57.

[38] Biolek D., Vavra J., Keskin A. U. 'CDTA-based capacitance multipliers'. Circuits, Systems and Signal Processing. 2019; **38**(4):1466–81.

[39] Chaichana A., Jaikla W. 'Electronically tunable inductance simulator for monitoring biomedical signals'. International Journal of Bioscience, Biochemistry, and Bioinformatics. 2014; **4**(1):45–8.

[40] Prasad D. Haseeb Z., Mainuddin, Akram Md W. 'Realization of resistorless floating inductor using modified CDTA'. Indian Journal of Pure & Applied Physics. 2019; **57**(1):29–32.

[41] Acar C., Ozoguz S. 'A new versatile building block: current differencing buffered amplifier suitable for analog signal-processing filters'. Microelectronics Journal. 1999; **30**(2):157–60.

[42] Toker A., Ozoguz S., Acar C. 'CDBA-based fully-integrated gyrator circuit suitable for electronically tunable inductance simulation'. International Journal of Electronics and Communications. 2000; **54**(5):293–6.

[43] Biolek D., Olsak M., Biolkova V. 'Optimization of elliptic leap-frog CDBA-based filters'. Computational Methods in Circuits and Systems Applications. 2003; **1**:221–5.

[44] Keskin A. U., Hancioglu E. 'CDBA-based synthetic floating inductance circuits with electronic tuning properties'. ETRI Journal. 2005; **27**(2):239–42.

[45] Pathak J. K., Singh A. K., Senani R. 'New canonic lossy inductor using a single CDBA and its application'. International Journal of Electronics. 2015; **103**(1):1–13.

[46] Tangsrirat W., Surakampontorn W. 'Electronically tunable floating inductance simulation based on current-controlled current differencing buffered amplifiers'. Thammasat International Journal of Science and Technology 2006; **11**(1):60–5.

[47] Biolek D., Bajer J., Biolkova V., Kolka Z., Kubicek M. 'Z-copy controlled gain-current differencing buffered amplifier and its applications'. International Journal of Circuit Theory and Applications. 2011; **39**(3):257–74.

[48] Jaikla W., Siripruchyanun M. 'Current controlled CDBAs based–novel floating and grounded negative inductance simulators'. The Proceedings of the International Technical Conference on Circuits/Systems, Computers, and Communications (ITC-CSCC 2006); Chiang Mai, Thailand, 2006:**3**:701–4.

[49] Herencsar N., Koton J., Lattenberg I., Vibra K. 'Signal-flow graphs for current-mode universal filter design using current follower transconductance amplifiers (CFTAs)'. Proceeding International Conference Applied Electronics (APPEI); Pilsen, Czech Republic, 2008; 69–72.

[50] Herencsar N., Lahri A., Koton J., Vibra K., Sotner R. 'New floating lossless inductance simulator using *Z*-copy current follower transconductance amplifiers'. Proceedings of 22nd International Conference Radioelektronika, 2012; 1–4.

[51] Li Y. 'A series of new circuits based on CFTAs'. International Journal of Electronics and Communications (AEU). 2012; **66**(7):587–92.

[52] Koton J., Herencsar N., Venclovsky M. 'History, progress and new results in synthetic passive element design employing CFTAs'. International Journal of Advances in Telecommunications, Electronics, Signals and Systems. 2015; **4**(1):15–26.

[53] Siriphot D, Maneewan S, Jaikla W. 'Single active element based electronically controllable grounded inductance simulator'. The 6th 2013 Biomedical Engineering International Conference, 2013.

[54] Herencsar N., Koton J., Vibra K., Lahri A. 'Floating simulators based on current followers transconductance amplifiers (CFTAs)'. Proceedings of the European Conference of Circuit Technology and Devices-ECCD, 2010:23–6.

[55] Srivastava M., Prasad D. 'Minimum component, electronically tunable simulator for grounded inductor with parasitic effects'. Journal of Telecommunication, Electronic and Computer Engineering. 2017; **9**(2):41–6.

[56] Srivastava M., Prasad D., Bhaskar D. R. 'New electronically tunable grounded inductor simulator employing single VDTA and one grounded capacitor'. Journal of Engineering Science and Technology. 2017; **12**(1):113–26.

[57] Bang J., Lee, J-Y. 'A new floating inductor using a voltage differencing transconductance amplifier'. The Transactions of the Korean Institute of Electrical Engineers. 2015; **64**(1):143–8.

[58] Tangsrirat W., Unhavanich S. 'Voltage differencing transconductance amplifier-based floating simulators with a single grounded capacitor'. Indian Journal of Pure & Applied Physics. 2014; **52**(6):423–8.

[59] Tangsrirat W., Mongkolwai P. 'VDTA-based floating FDNR simulator topology'. KMITL Science and Technology Journal. 2013; **13**(1):17–21.

[60] Srivastava M., Prasad D. 'VDTA based electronically tunable purely active simulator circuit for realizing floating resistance'. Journal of Engineering Science and Technology Review. 2015; **8**(3):112–16.

[61] Srivastava M., Prasad D. 'A novel purely active electronically controllable configuration for simulating resistance in floating form'. Electronics. 2016; **20**(2):90–4.

[62] Prasad D., Bhaskar D. R. 'Grounded and floating inductance simulation circuits using VDTAs'. Circuits and Systems. 2012; **3**(4):342–7.

[63] Mehra R., Kumar V., Islam A. 'Floating active inductor based class-C VCO with 8 digitally tuned sub-bands'. International Journal of Electronics and Communications (AEU). 2018; **8**:1–10.

[64] Mehra R., Kumar V., Islam A., Kaushik B. K. 'Variation-aware widely nanoscale design of CMOS active inductor-based RF bandpass filter'. International Journal of Circuit Theory and Applications. 2017; **45**(12): 2181–200.

[65] Mehra R., Kumar V., Islam A. 'Reliable and *Q*-enhanced floating active inductors and their application in RF bandpass filters'. IEEE Access. 2018; **6**:48181–94.

[66] Propkop R., Musil V. 'New modern circuit block CCTA and some its applications'. The Fourteenth International Scientific and Applied Science Conference Electronics ET, 2005; **Book 5**:93–8.

[67] Jantakun A. 'A simple grounded FDNR and capacitance simulator based-on CCTA'. International Journal of Electronics and Communications (AEU). 2015; **69**(6):950–7.

[68] Nandi R., Das S., Venkateswaran P. 'Floating lossless immittance functions using DVCCTA'. International Journal of Electronics Letters. 2014; **4**(1): 117–26.

[69] Nandi R., Kar M., Das S. 'Synthetic inductor based resonators using DVCCTA'. 2011 IEEE EUROCON—International Conference on Computer as a Tool, 2011:1–3.

[70] Siripruchyanun M., Jaikla W. 'Current controlled current conveyor transconductance amplifier (CCCCTA): a building block for analog signal processing'. Electrical Engineering. 2008; **90**(6):443–53.

[71] Siripruchyanun M., Silapan P., Jaikla W. 'Realization of CMOS current controlled current conveyor transconductance amplifier (CCCCTA) and its applications'. Journal of Active and Passive Electronic Devices. 2009; **4**: 35–53.

[72] Li Y. 'NAM expansion method for systematic synthesis of floating gyrators using CCCCTAs'. Analog Integrated Circuits and Signal Processing. 2015; **82**(3):733–43.

[73] Pandey N., Arora S., Takkar R., Pandey R. 'DVCCCTA-based implementation of mutually coupled circuit'. ISRN Electronics. 2012; Article ID 303191: 6 pages.

[74] Herencsar N., Lahri A., Koton J., Vibra K., Metin B. 'Realization of resistorless lossless positive and negative grounded inductor simulators using single ZC-CCCITA'. Radioengineering. 2012; **21**(1):264–72.

[75] Mohammad F., Sampre J., Shireen S., Ali S. H. M. 'Minimum passive components based lossy and lossless simulators employing a new active block'. International Journal of Electronics and Communications (AEU). 2017; **82**:226–40.

[76] Unhavanich S., Onjan O., Tangsrirat W. 'Tunable capacitance multiplier with a single voltage differencing buffered amplifier'. Proceeding of the International Multiconference of Engineers and Computer Scientists. 2016:**11**:568–71.

[77] Yesil A., Kacar F., Gurkan K. 'Lossless grounded inductance simulator employing single VDBA and its experimental band-pass filter application'. International Journal of Electronics and Communications (AEU). 2013:**68**(2): 143–50.

[78] Tangsrirat W., Mongkolwai P. 'Generalized impedance function simulator using voltage differencing buffered amplifiers (VDBAs)'. Proceedings of the International Multiconference of Engineers and Computer Scientist; Hong Kong, 2016; 11.

[79] Channumsin O., Tangsrirat W. 'Electronically tunable floating capacitance multiplier using FB-VDBAs'. Engineering Letters. 2016; **24**(3):1–5.

[80] Tangsrirat W. 'Actively floating lossy inductance simulators using voltage differencing buffered amplifiers'. IETE Journal of Research. 2019; **65**(4): 446–59.

[81] Channumsin O., Tangsrirat W. 'Actively tunable lossless floating inductance simulator using voltage differencing buffered amplifiers'. IAENG International Journal of Computer Science. 2016; **43**(4):469–73.

[82] Bhaskar D. R., Prasad D., Pushkar K. L. 'Electronically-controllable grounded-capacitor-based grounded and floating inductance simulated circuits using VD-DIBAs'. Circuits and Systems. 2013; **4**(5):422–30.

[83] Prasad D., Bhaskar D. R., Pushkar K. L. 'Realization of new electronically controllable grounded and floating inductance circuits using voltage differencing differential input buffered amplifiers'. Active and Passive Electronic Components. 2011; Article ID 101432: 1–8.

[84] Srivastava M., Bhardwaj K. 'Compact lossy inductance simulators with electronic control'. Iranian Journal of Electrical and Electronic Engineering. 2019; **15**(3):343–51.

[85] Tangsrirat W. 'Synthetic grounded lossy inductance simulators using single VDIBA'. IETE Journal of Research. 2017; **63**(1):134–41.

[86] Singh A. K., Kumar P., Senani R. 'New grounded immittance simulators employing a single CFCC'. The Journal of Engineering. 2017; **8**:435–47.

[87] Singh A. K., Kumar P., Senani R. 'Electronically tunable grounded /floating inductance simulators using Z-copy CFCCC'. Turkish Journal of Electrical Engineering & Computer Sciences. 2018; **26**(2):1041–55.

[88] Yamacli S., Ozcan S., Kuntman H. 'New active-only grounded inductance simulator employing current-mode approach suitable for wide band operation'. International Journal of Electronics. 2011; **98**(8):1–14.

[89] Kacar F. 'A new tunable floating CMOS FDNR and elliptic filter applications'. Journal of Circuits, Systems, and Commuters. 2010; **19**(8):1641–50.

[90] Kacar F., Yesil A., Minaei S., Kuntman H. 'Positive/negative lossy/lossless grounded inductance simulators employing single VDCC and only two passive elements'. International Journal of Electronics and Communications (AEU). 2014; **68**(1):73–8.

[91] Srivastava M., Bhanja P., Mir S. F. 'A new configuration for simulating passive elements in floating state employing VDCCs and grounded passive elements'. 2016 IEEE 1st International Conference on Power Electronics, Intelligent Control and Energy Systems; Delhi, India, July 4–6, 2016.

[92] Ayten U. E., Sagbas M., Herencsar N., Koton J. 'Novel floating general element simulators using CBTA'. Radioengineering. 2012; **21**(1):11–19.

[93] Koksal M., Ayten U. E., Sagbas M. 'Realization of new mutually coupled circuit using CC-CBTAs'. Circuits, Systems, and Signal Processing. 2012; **31**(2):435–46.

[94] Sagbas M. 'Component reduced floating ±L, ±C and ±R simulators with grounded passive components'. International Journal of Electronics and Communications (AEU). 2011; **65**(10):794–8.

[95] Souliotis G., Chrisanthopoulos A., Haritantis I. 'Current differential amplifiers: new circuits and applications'. International Journal of Circuit Theory and Applications. 2001; **29**(6):553–74.

[96] Said L. A., Madian A., Ismail M. H., Soliman A. M. 'Digitally programmable lossless floating inductor realization using current differential amplifier (CDA)'. 16th IEEE Mediterranean Electrotechnical Conference, Tunisia. 2012; 840–3.

[97] Zhou X., Zeng R. 'Floating inductance simulators based on active components on active components'. Asia-Pacific Engineering and Technology Conference. 2017; 1904–8.

[98] Riordan R. H. 'The triamplifier, a new circuit element and its use in active filters'. IEEE Journal of Solid-State Circuits. 1975; **10**(6):505–8.

[99] Alzaher H. L. 'CMOS digitally programmable quadrature oscillators'. International Journal of Circuit Theory and Applications. 2008:**36**(8):953–66.

[100] De Marcellis A., Ferri G., Guerrini N. C., Scotti G., Stornelli V., Trifiletti A. 'The-VCG-CCII: a novel building block and its application to capacitance multiplication'. Analog Integrated Circuits for Signal Processing. 2009; **58**(1):55–9.

[101] Matsumoto F., Nishioka S., Ohbuchi T., Fujii T. 'Design of a symmetry-type floating impedance scaling circuits for a fully differential filter'. Analog Integrated Circuits and Signal Processing. 2015: **85**(2):253–61.

[102] Siripruchyanun M., Poonnoy N., Upathamkuekool Ch., Payakkakul K. 'Current differencing transresistance amplifier (CDTRA) and its application for analog signal processing'. Procedia Computer Science. 2016; **86**:184–7.

[103] Yesil A., Yuce E., Minaei A. 'Grounded capacitance multipliers based on active elements'. International Journal of Electronics and Communications (AEU). 2017; **79**:243–9.

[104] Yesil A., Yuce E., Minaei A. 'Inverting voltage buffer based lossless grounded inductor simulators'. International Journal of Electronics and Communications (AEU). 2018; **83**:131–7.

[105] Pandey N., Paul S. K. 'Differential difference current conveyor transconductance amplifier: A new analog building block for signal processing'. Journal of Electrical and Computer Engineering. 2011; Article ID 361384: 10 pages.

[106] Gulsoy M., Cicekoglu O. 'Lossless and lossy synthetic inductor employing single current differencing buffered amplifier'. IEICE Transactions on Communications. 2005; **E88-B**(5):2152–5.

[107] Wang Z. '2-MOSFET transistors with extremely low distortion for output reaching supply voltage'. Electronics Letters. 1990; **26**(13):951–52.

[108] Ayten A. E., Sagbas M., Sedef H. 'Current mode leapfrog ladder filters using a new active block'. International Journal of Electronics and Communications (AEU). 2010; **64**(6):503–11.

[109] Pimpol J., Channumsin O., Thongsopa C., Tangsrirat W. 'VDIBA-based floating lossless inductance simulator employing a single grounded capacitor'. Far East Journal of Electronics and Communications. 2016; **16**(3): 615–27.

[110] Herencsar N., Lahiri A., Koton J., Vrba K., Metin B. 'Realization of resistor less lossless positive and negative grounded inductor simulators using single ZC-CCCITA'. Radioengineering. 2012; **21**(1):264–72.

[111] Metin B. 'Canonical inductor simulators with grounded capacitors using DCCII'. International Journal of Electronics. 2012; **99**(7):1027–35.

[112] Prasad D., Ahmad J. 'New electronically-controllable lossless synthetic floating inductance circuit using single VDCC'. Circuits and Systems. 2014; **5**(1):13–15.

[113] Kacar F., Yesil A., Minaei S., Kuntman H. 'Positive/negative lossy/lossless grounded inductance simulators employing single VDCC and only two passive elements'. AEU-International Journal of Electronics and Communications. 2014; **68**(1):73–8.

Chapter 9

Transistor-level realization of electronically controllable grounded and floating resistors

Abstract

Electronically controllable resistors find numerous applications in the realization of electronically controlled amplifiers, integrators, filters, oscillators and in a number of other linear/nonlinear functional circuits. This chapter presents the significant developments made in the realization of positive and negative, grounded and floating electronically variable resistors implementable in both bipolar and CMOS technology. Salient features of the various circuits have been pointed out and some applications have been outlined.

9.1 Introduction

Integrated passive resistors can be implemented in silicon technology using polysilicon or diffusion areas in monolithic integrated circuits. However, such resistors occupy a large chip area, are restricted to the realization of moderately low values of resistances (typically less than 10 kΩ), have large tolerances and their values cannot be varied. This leads to a lot of research in realizing on-chip variable resistors implemented from bipolar and MOS transistors only. These actively realized resistances can easily realize electronically tunable resistances.

The electronically variable resistors find numerous applications in the realization of electronically controlled amplifiers, integrators, filters, oscillators and in a number of other linear/nonlinear functional circuits.

It may be mentioned that the biploar junction transistor (BJT) based electronically controllable resistors which are mainly based upon translinear (TL) circuits, realize linear resistance subject to constraints on the input signal of the type $V_{in} \ll 2V_T$ while, by contrast, the MOSFET-based circuits achieve a linear resistance by means of some kind of nonlinearity cancellation employed due to which they can handle relatively large input signals and have a wider input dynamic range.

This chapter presents some of the prominent circuits from the various electronically controllable *linear* resistance configurations evolved so far. The coverage includes positive and negative as well as grounded and floating electronically controllable resistance circuits realized for implementation in bipolar as well as CMOS technology [1–117].

Before outlining the ley result in this area, we may mention that extensive research on realizing linear resistors in bipolar and MOS technology has led to considerable research effort in many diverse areas of analog circuit design which have employed some kind of linearization techniques in devising the intended circuit functions such as realizing linear voltage-controlled resistances (VCR) as in [3–9,11,13–16,18,20,22,23,25–29,31,35,38,40–52,54–57,60,62,63,66,70,71,73–76, 80–82,88–92,95–108,113,114,116,117]; linear negative VCRs such as those of [2,10,21,32,33,61,68,79,83,84,86,110]; linear transconductors [12,17,19,53,59]; linear trans-resistance elements such as those in [24,64]. Effort has also been put up on applying temperature compensation to such circuits [1], creating a composite MOSFETs with in-built nonlinearity cancellation (popularly known as COMFET) [30,37,39] and several others aspects as in [34,36,58,65,67,69,72,78,85,93,109, 111,112].

9.2 BJT-based translinear current-controlled resistors

In this section we describe a number of interesting bipolar circuit arrangement which has been devised to realize linear current-controlled resistors in both grounded and floating forms using BJTs only. The basic idea appears to stem from the observation that a BJT-based current conveyor (CC) which employs a four-transistor mixed translinear cell (MTC) exhibits a sine hyperbolic relation between the differential input voltage $V_y - V_x$ and the current entering terminal-X of the CC which implies that the small signal input resistance looking into terminal-X is finite and nonzero and is not fixed but rather is variable through the external dc bias current applied to the CC.

9.2.1 Current-controllable grounded/floating resistors

A current-controllable grounded resistor implementable in bipolar technology was *first* proposed by Saaid and Fabre [49] (also see [48,50,62]) which is based upon an MTC [49] (see Figure 9.1).

Assuming all matched transistors having large β (hence, negligible base currents), a straightforward analysis of the circuit of Figure 9.1 yields the following equations:

Considering the closed loop of the voltages V_{BE1}, V_{BE3} and V_1, one can write

$$V_1 = V_{BE1} - V_{BE3} \quad \text{which leads to } V_1 = V_T \ln\left(\frac{I_B}{I_{C3}}\right) \tag{9.1}$$

Similarly, considering the closed loop containing voltages V_{EB2}, V_{EB4} and V_1, one can also write

$$V_1 = (V_{EB4} - V_{EB2}) \quad \text{which leads to } V_1 = V_T \ln\left(\frac{I_{C3} + i_1}{I_B}\right) \tag{9.2}$$

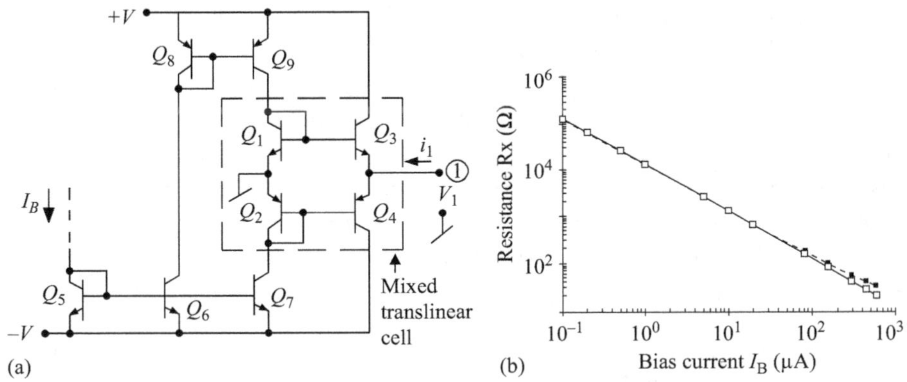

Figure 9.1 BJT-based electronically controllable grounded resistor: (a) expanded form of the grounded current-controllable resistor proposed by Saaid and Fabre [49], (b) theoretical and measured variation of R_{in} with respect to external bias current I_B. © 1996 IET

From (9.1) and (9.2), one can write

$$I_{C3} = I_B e^{-(V_1/V_T)} \quad \text{and} \quad I_{C3} + i_1 = I_B\, e^{(V_1/V_T)} \tag{9.3}$$

From (9.3) we obtain

$$i_1 = 2I_B\left(\frac{e^{(V_1/V_T)} - e^{-(V_1/V_T)}}{2}\right) = 2I_B \sinh\left(\frac{V_1}{V_T}\right) \tag{9.4}$$

which can be rewritten as

$$V_1 = V_T \sinh^{-1}\left(\frac{i_1}{2I_B}\right) \approx \left(\frac{V_T}{2I_B}\right) i_1; \quad \text{provided } i_1 \ll 2I_B \tag{9.5}$$

Thus, subject to the constraint given in (9.5), the input resistance realized by this circuit is given by

$$R_{in} \simeq \left(\frac{V_T}{2I_B}\right) \tag{9.6}$$

which is *electronically controllable* through the external dc bias current I_B.

From the SPICE simulations using bipolar transistor parameters of the complementary high performance HF3 CMOS process from SGS Thomson, Saaid and Fabre [49] demonstrated (see Figure 9.1(b)) that for a dc bias of ±2.5 V, the circuit realizes controllable resistance for over three decades of the external bias current I_B.

The frequency response of this circuit showed that the realized resistance value remains nearly constant up to around 100 MHz. The circuit was found to work satisfactorily in the realization of a gain-variable current amplifier using a CCII+.

9.2.2 A translinear current-controlled floating resistor due to Barthelemy and Fabre

Another circuit capable of realizing the same type of electronically controllable resistance, but in floating form, was subsequently proposed by Barthelemy and Fabre [62] and is shown in Figure 9.2. In this circuit a new eight-transistor MTC was utilized which comprises transistors Q_1–Q_8. The circuit incorporates a number of current mirrors which create copies of the external dc bias current I_0 in such a manner that the currents I_Y and I_X become equal and opposite, thereby leading to

$$I_Y = -I_X \tag{9.7}$$

Furthermore, the three (TL) loops in the circuit, namely Q_1–Q_3–Q_A–Q_C, Q_5–Q_7–Q_B–Q_D and Q_A–Q_2–Q_4–Q_D–Q_B–Q_6–Q_8–Q_C, enforce the following relations between the various collectors currents of the transistors involved in these TL loops [62]:

$$I_{C1}I_{C3} = I_A^2 \tag{9.8}$$

$$I_{C5}I_{C7} = I_B^2 \tag{9.9}$$

and

$$I_{C2}I_{C6} = I_AI_B \tag{9.10}$$

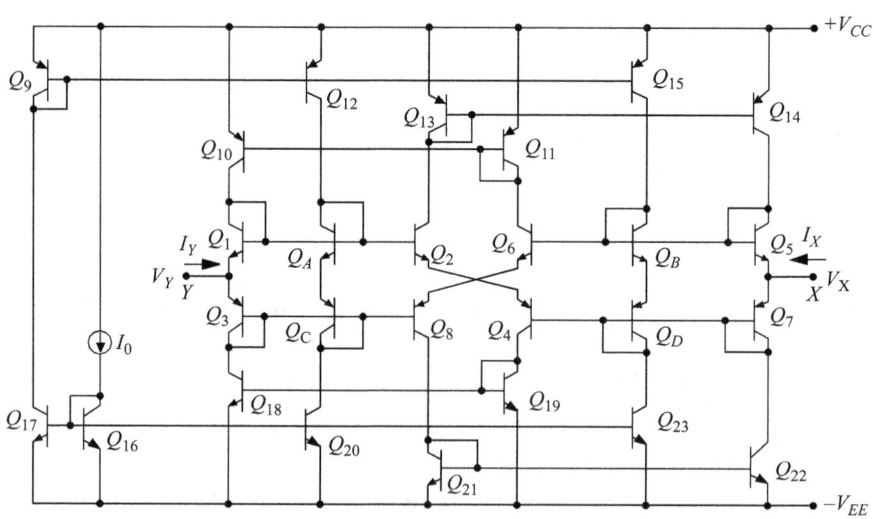

Figure 9.2 The electronically controlled linear floating resistance [62]

By solving (9.8)–(9.10), one finds that $I_A = I_B = I_0$. This means that with $I_X = 0$, all the currents are equal to I_0, as expected. Finally, the solution to the various currents is found to be [62]

$$I_{C1} = I_{C6} = I_{C7} = I_{C8} = \frac{I_Y}{2} + \sqrt{\frac{I_Y^2 + 4I_0^2}{2}} \tag{9.11}$$

$$I_{C2} = I_{C3} = I_{C4} = I_{C5} = -\frac{I_Y}{2} + \sqrt{\frac{I_Y^2 + 4I_0^2}{2}} \tag{9.12}$$

Since all the collector currents are always positive, the operation of the circuit in class AB is, thus, confirmed. Finally, the following result is obtained for the port currents in terms of the differential input:

$$V_{x-y} = 4V_T \sinh^{-1}\left(\frac{I_X}{2I_0}\right) \tag{9.13}$$

For $I_X \ll 2I_0$, the above equation yields the floating resistance realized between ports X and Y as

$$R_{eq} \cong \frac{2V_T}{I_0} \tag{9.14}$$

which is electronically controllable through I_0.

Application: An interesting application of this circuit was demonstrated [62] in a single current feedback op-amp (CFOA)-based single resistance controlled oscillator (SRCO) shown in Figure 9.3.

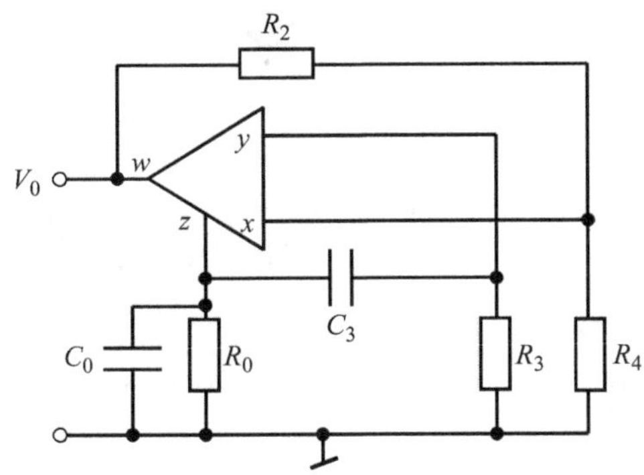

Figure 9.3 Application of the floating CCR [62] in realization of a current-controlled sinusoidal oscillator based upon the SRCO of Senani and Singh [115]

This SRCO from [115] is characterized by the following condition of oscillation (CO) and frequency of oscillation [62]:

$$\frac{R_3'}{R_4} \geq 1 + \frac{C_0'}{C_6} + \frac{R_3'}{R_0'} + R_x\left(\frac{1}{R_4} + \frac{1}{R_2'}\right)\left(1 + \frac{C_0'}{C_6} + \frac{R_3'}{R_0'}\frac{C_y}{C_6}\right) + \frac{R_3'}{R_2'}\frac{C_y}{C_3} \tag{9.15}$$

$$\omega_0 = \sqrt{\frac{(1/R_0') + (1/R_2')\{1/(1 + R_x((1/R_4) + (1/R_2')))\}}{R_3'C_0'C_3 + R_3'C_y(C_0' + C_3)}} \tag{9.16}$$

In the above expressions, $R_0' = (R_0||R_z)$, $C_0' = (C_0 + C_z)$, $R_3' = (R_3||R_y)$ and $R_2' = (R_2 + R_w)$, where R_x represents the X-port parasitic input resistance, R_y represents the nonideal input resistance of terminal Y, R_z is the output resistance looking into terminal Z and R_w is the output resistance of terminal W. Similarly, C_y and C_z represent the input and output capacitance of the Y and Z ports, respectively.

Application: A current-controlled oscillator was realized by replacing the floating resistance R_2 by the circuit of Figure 9.2 which was varied by adjusting the external bias current from 50 μA to 450 μA, thereby resulting in the variation of oscillation frequency from 4.5 MHz to 8.5 MHz. It was anticipated that this floating VCR which uses bipolar transistors only is eminently suitable for IC realization and in this way, many improved structures for filters, oscillators, amplifiers and programmable analog circuits could be designed using the floating VCR of Figure 9.2.

9.2.3 A translinear current-controllable floating negative resistor

A circuit for realizing floating negative resistance was proposed by Kiranon *et al.* [61] using two four-transistor MTCs as shown in Figure 9.4(a), where the symbolic notations of the current mirrors have been employed, which have been explained in Figure 9.4(b) and (c). Using straightforward analysis, the expression for the current flowing from node A to B is found to be

$$I_1 = -2I_0 \sinh\left(\frac{V_1 - V_2}{2V_T}\right) = -I_2 \tag{9.17}$$

Expanding the sinh term in (9.17) with Taylor's series and assuming $(V_1 - V_2) \ll 2V_T$, (9.17) can be approximated as

$$I_1 \cong -I_0\left(\frac{V_1 - V_2}{V_T}\right) = -I_2 \tag{9.18}$$

The equivalent negative resistance between nodes A and B is, therefore, given by

$$R_{eq} = -\left(\frac{V_T}{I_0}\right) \tag{9.19}$$

A drawback of this circuit is that realized resistance is temperature sensitive. To circumvent this problem Kiranon *et al.* devised a translinear-compensated circuit

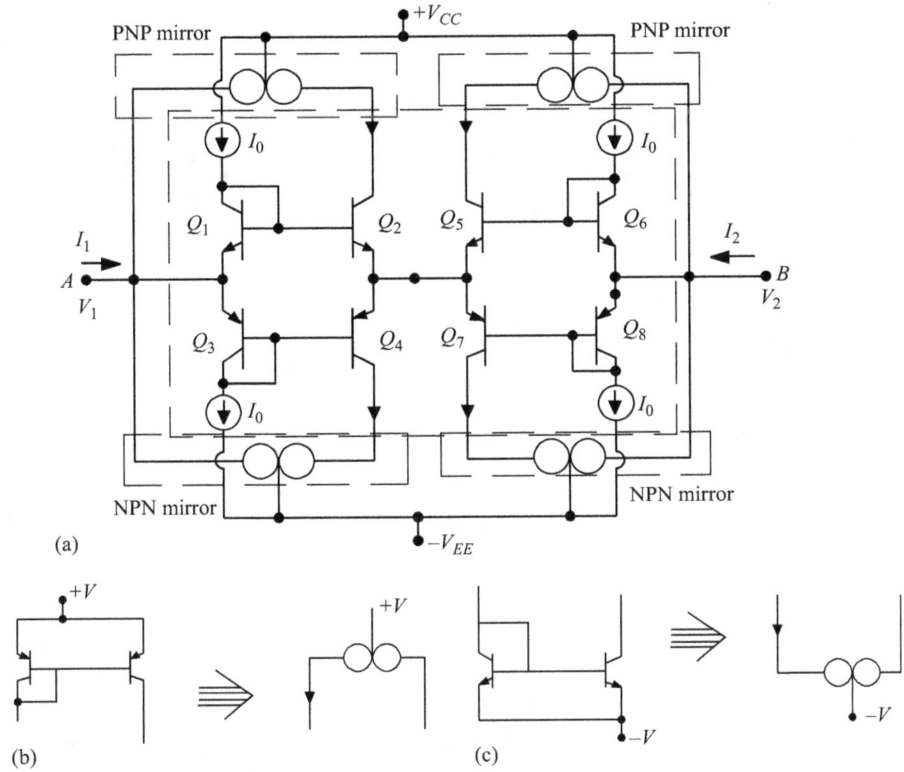

Figure 9.4 (a) Simplified schematic of the circuit for the realization of floating negative resistor [61]: (b) symbolic notation of PNP mirror, (c) symbolic notation of NPN mirror

(see Figure 9.5) which, subject to the assumption $V_s \ll V_T$, generates a dc bias current:

$$I_0 = \frac{I_C I_D V_T}{2 I_R V_S} \tag{9.20}$$

Therefore, connecting the current I_0 generated by the circuit of Figure 9.5 into the circuit of Figure 9.4 leads to

$$R_{AB} = -2\frac{I_R V_S}{I_C I_D} \tag{9.21}$$

Finally, if one takes $I_C = I_D = I_R$, the resistance realized by the modified circuit reduces to

$$R_{AB} = -2\left(\frac{V_S}{I_R}\right) \tag{9.22}$$

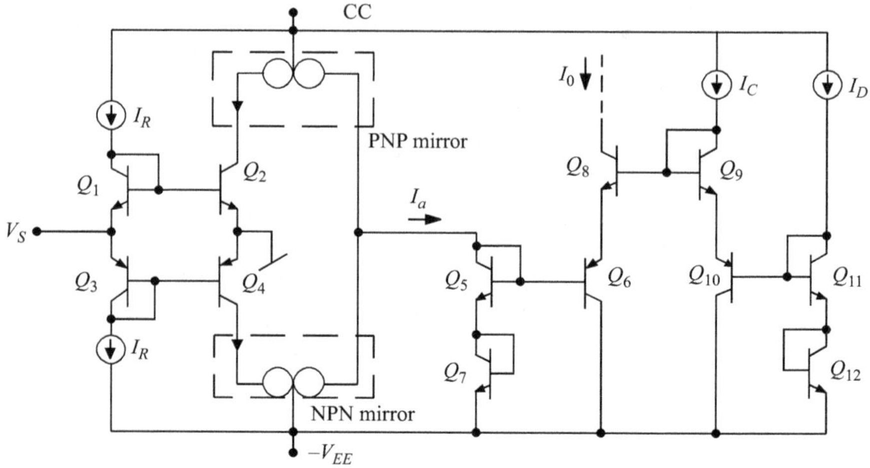

Figure 9.5 Simplified representation of the translinear temperature-compensated circuit [61]

It is, thus, seen that the realized negative resistance value has now become temperature insensitive. The validity of this circuit was verified [61] by realizing the temperature-compensated negative resistance circuit using PR100N and NR100N transistors, $V_S = 3$ mV, $I_R = 50$ μA and dc bias supply voltage as ±2.5 V. A good correspondence was achieved between theoretical and experimental results with realized resistance value remaining unaffected for temperature variation from 27°C to 100°C.

Application: Kiranon *et al.* [61] demonstrated that this floating negative resistance circuit can be effectively used to improve the quality factor of series resonant *RLC* circuit by inserting the negative resistance circuit of appropriate value in series with the resonant circuit. The measured value of the quality factor confirmed the effectiveness of this approach.

9.2.4 Translinear floating current-controlled positive resistance due to Senani, Singh and Singh

A circuit for realizing floating current-controlled positive resistance (FCCPR) was proposed by Senani *et al.* [70], which is shown in Figure 9.6. This circuit employs two matched MTCs, one composed of bipolar transistors Q_1, Q_2, Q_3, Q_4 and the other composed of Q_5, Q_6, Q_7, Q_8 with the other transistors configured as current mirrors/repeaters for providing the dc bias current I_0 to various parts of the circuits. An analysis of the circuit using translinear principle reveals that

The ports currents I_1 and I_2 can be expressed as

$$I_1 = -I_2 = 2I_0 \sinh\left(\frac{V_1 - V_2}{V_T}\right) \tag{9.23}$$

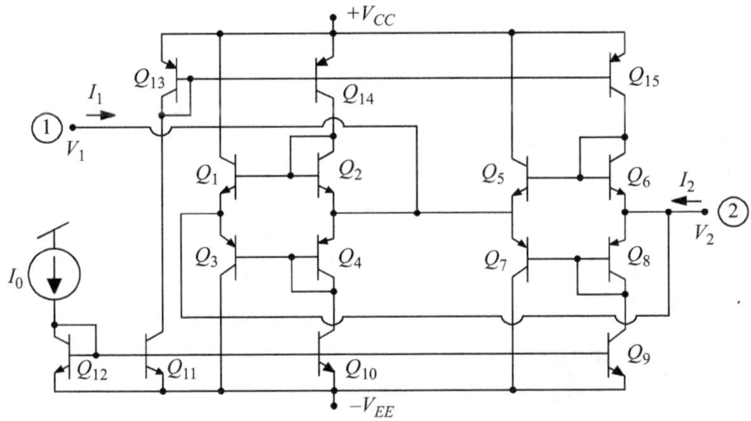

Figure 9.6 Floating current-controlled positive resistance [70]

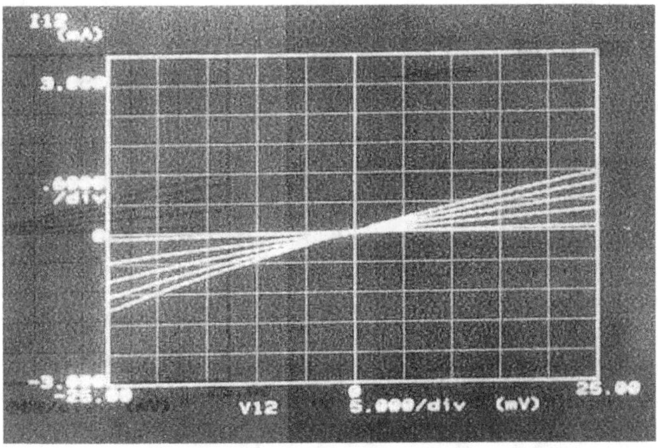

Figure 9.7 V–I characteristics of the circuit of Figure 9.6 [70]

For $|V_1 - V_2| \ll V_T$, (9.23) can be approximated as

$$I_1 = -I_2 \cong \left(\frac{2I_0}{V_T}\right)(V_1 - V_2) \tag{9.24}$$

which represents a floating positive resistance between ports 1 and 2 having value $R_{eq} = (V_T/2I_0)$. The experimentally observed V–I characteristics of this circuit, realized by using CA3096 transistor arrays with dc power supply voltage of ±2.5 V and then external dc bias current varied from 50 μA to 250 μA, are shown in Figure 9.7, which exhibits a linear range of about 50 mV peak to peak.

When compared to the FCCPR circuit of Barthelemy and Fabre [62], which employed 27 bipolar transistors, the circuit of Figure 9.6 requires only 15. On the other hand, it has been demonstrated in [70] that performance-wise the circuit is comparable to the circuit of Barthelemy and Fabre [62] with a slight edge in terms of better bilaterality.

Applications: An interesting application of the new FCCPR was demonstrated in [70] wherein a new frequency shift keying (FSK) generator was realized as shown in Figure 9.8. This circuit involved a CFOA-based SRCO comprising $CFOA_2$ in which, on the external dc bias current terminal of the FCCPR, a square wave signal current was superimposed through a voltage-controlled current source (VCCS) realized by $CFOA_1$ and resistor R_3. The oscillation frequency of the SRCO and the CO are given by:

$$f_0 = \frac{1}{2\pi}\sqrt{\frac{1}{C_1C_2R_0R_1}} \tag{9.25}$$

and

$$\left(1 + \frac{C_2}{C_1}\right) \geq \frac{R_2}{R_1} \tag{9.26}$$

where R_0 is an FCCPR realizing an equivalent resistance $R_0 = (V_T/2I_B)$. Since I_B is composed of a fixed dc current I_0 and a slowly varying square wave current signal i_s obtained from VCCS realized from $CFOA_2$ and R_3, it was shown that by alternating i_s between 0 μA and 100 μA with I_0 taken as 10 μA, the total dc bias current was alternated between 10 μA (LOW) and 110 μA (HIGH). With component values taken as $R_1 = 1$ kΩ, $R_2 = 2.013$ kΩ, $R_3 = 50$ Ω, and $C_1 = C_2 = 0.01$ μF, the circuit was found to generate FSK signal with frequency shifting from $f_1 = 46.2$ kHz to $f_2 = 14$ kHz when input signal V_s alternated between LOW and HIGH.

Another application of the FCCPR was demonstrated in [70] in the realization of an electronically controllable band-pass filter (BPF) which is shown in Figure 9.9(a). In this circuit, which is based upon a CFOA-based-gyrator simulated inductance comprising the two CFOAs along with resistors R_1, R_2 and capacitor C_2

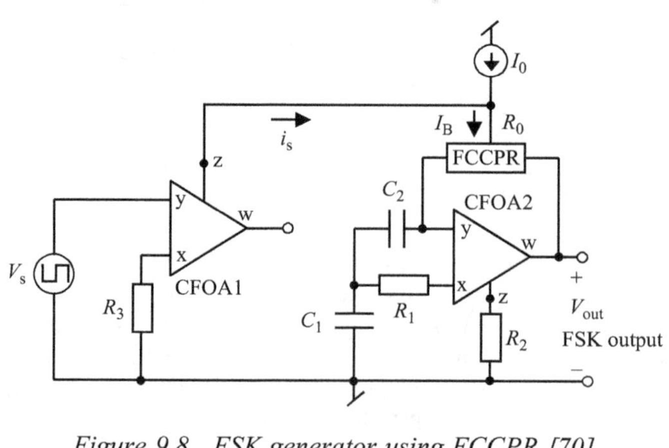

Figure 9.8 FSK generator using FCCPR [70]

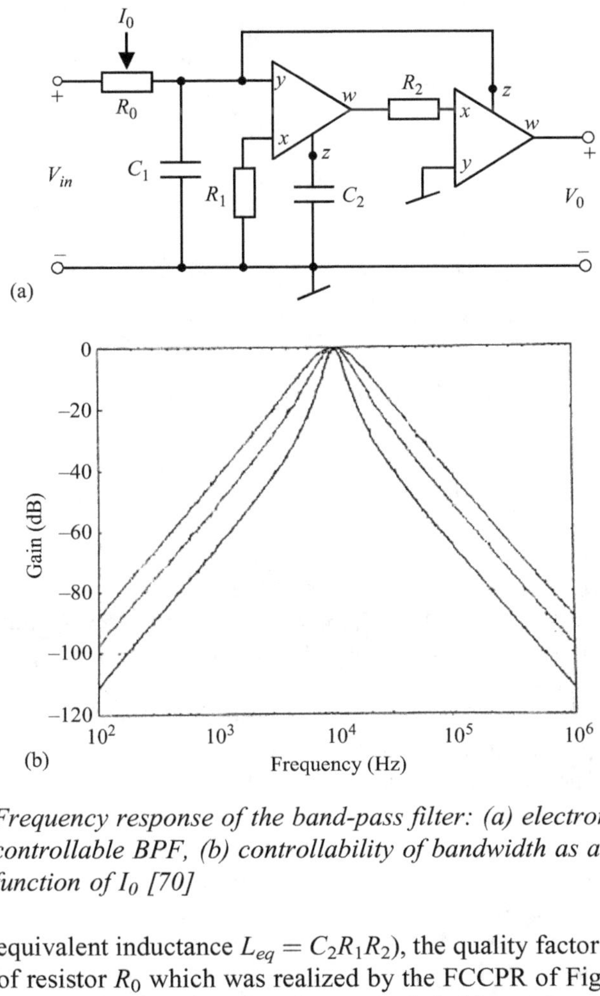

Figure 9.9 Frequency response of the band-pass filter: (a) electronically controllable BPF, (b) controllability of bandwidth as a function of I_0 [70]

(realizing an equivalent inductance $L_{eq} = C_2R_1R_2$), the quality factor Q_0 of the BPF is a function of resistor R_0 which was realized by the FCCPR of Figure 9.6. It was demonstrated in [70] that by changing the external dc bias current of the FCCPR from 7.85 μA to 25 μA, the bandwidth of the BPF was found to be controllable from 3.812 kHz to 12 kHz as shown in Figure 9.9(b).

9.2.5 A circuit to realize current-controllable floating positive/negative resistance

In [71] Pawarangkoon and Kiranon presented another electronically tunable floating resistance circuit which was implemented using a BJT-based multiple output second-generation current-controlled conveyors (CCCIIs) with appropriate connections of the three Z-outputs (Figure 9.10) to yield a two-port network which realizes an equivalent floating resistance of value:

$$R_{eq} = \left(\frac{V_T}{2I_0}\right)\left[\frac{1}{1-K}\right] \tag{9.27}$$

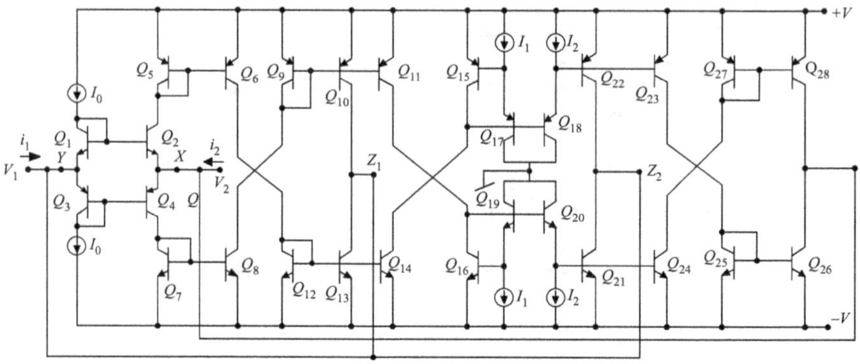

Figure 9.10 Current-controlled floating resistor [71]

It may be pointed out that the circuit of Figure 9.10 contains a CCCII with inputs X and Y and three Z outputs with X shorted to Z_3 and Y shorted to Z_1 and Z_2, where $V_x \cong V_y + R_x i_x$, $i_y = 0$, $i_{z1} = -i_x$, $i_{z2} = Ki_x$, $i_{z3} = -Ki_x$, where $K = (I_1/I_2)$ and $R_x = V_T/2I_0$.

Thus, the same circuit can realize a floating positive resistance for $K < 1$ and a floating negative resistance by setting $K > 1$. In both cases, the realized resistance is electronically controllable through the external bias current I_0. Furthermore, the same circuit can realize a current-controlled grounded positive/negative resistor of the same value, by grounding port 2.

The workability of this circuit was demonstrated by realizing the circuit using bipolar transistors NR100N and PR100N with CCCII circuit biased from ± 2.5 V dc power supply. The simulation results were found to be in good agreement with the theory [71].

Application: To demonstrate the applicability of their floating resistance circuit, Pawarangkoon and Kiranon [71] realized an oscillator using this negative resistor, a positive resistor, an inductor and a capacitor.

As compared to other circuits, the capability of creating either a positive or negative grounded or floating resistors from the same topology constitutes a significant advantage of the circuit of Figure 9.10.

9.2.6 A low transistor count current-controlled grounded/floating, positive/negative resistor due to Arslanalp, Yuce and Tola

Arslanalp *et al.* presented low-power low-component count grounded and floating current-controllable resistance circuits in [97], out of which their proposition for realizing a current-controllable grounded positive resistor is shown in Figure 9.11. By straightforward analysis as demonstrated earlier in this chapter, for the input voltage V_1 much smaller than thermal voltage, i.e. $|V_1| \ll V_T$, the input current of the circuit is given by

$$i_1 \cong \frac{V_1 I_0}{V_T} \tag{9.28}$$

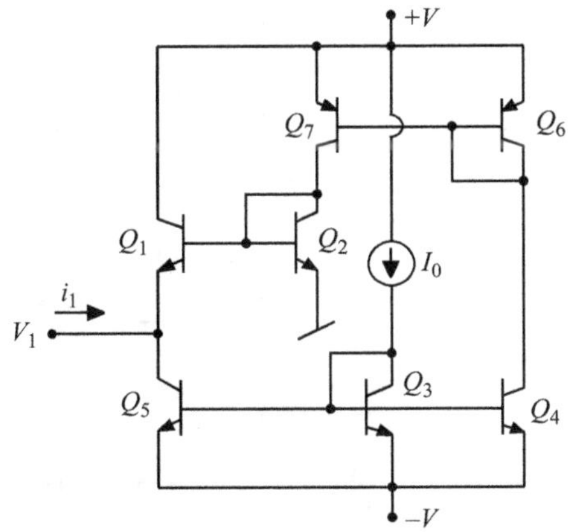

Figure 9.11 Current-controlled grounded positive resistance [97]

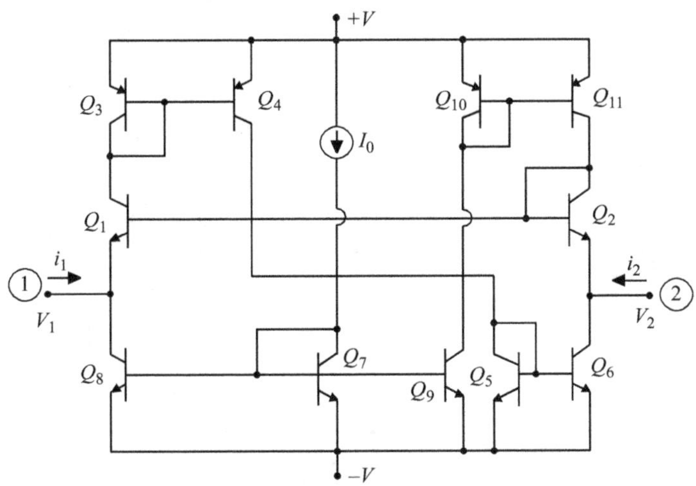

Figure 9.12 Current-controlled floating positive resistance [97]

which gives the input resistance as

$$R_{in} \cong \frac{V_T}{I_0} \tag{9.29}$$

A floating version of this circuit, realizing exactly the same value of equivalent floating resistance between ports 1 and 2 is shown in Figure 9.12.

Employing a four-transistor-based negative impedance converter (NIC), a circuit capable of realizing a current-controlled grounded negative resistor evolved in [97], is shown in Figure 9.13. It can be readily seen that in this circuit, the part of

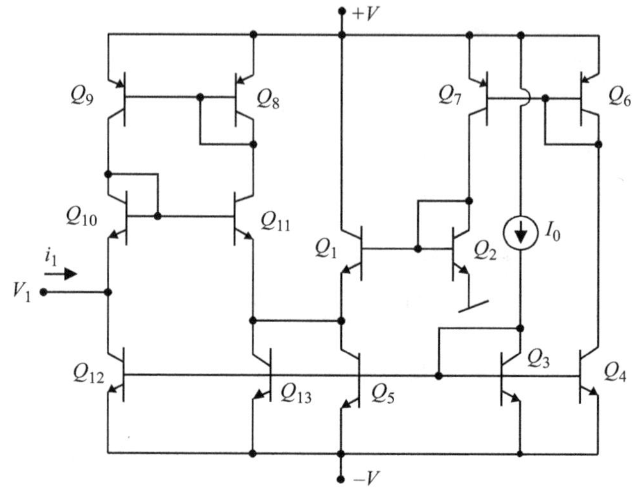

Figure 9.13 Current-controlled grounded negative resistance [97]

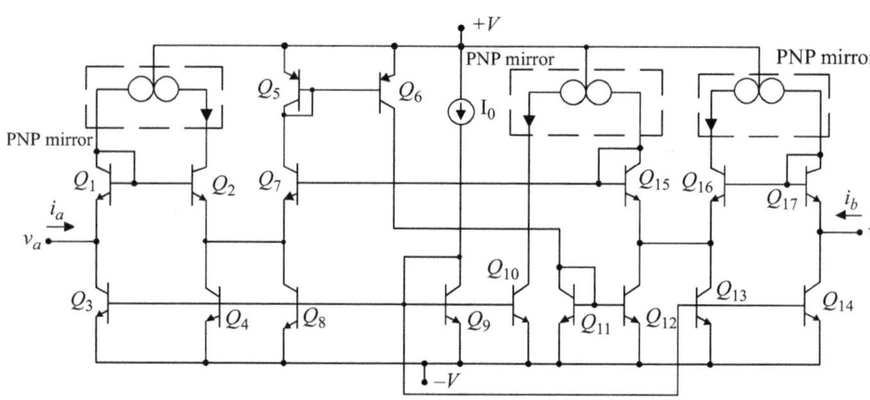

Figure 9.14 Current-controlled floating negative resistance [97]

the circuit comprising transistors Q_1–Q_7 is the same as the grounded positive resistor of Figure 9.11, while the remaining circuit comprising BJTs Q_8–Q_{13} is a simple NIC. The equivalent resistance realized by this circuit is

$$R_{eq} \cong -\frac{V_T}{I_0} \tag{9.30}$$

Finally, using two BJT-based NICs, the circuit of the floating current-controlled negative resistance is shown in Figure 9.14. By realizing all the circuits using transistor arrays and all the simulated resistors biased with ±1.5 V bias current taken as 10 μA, by SPICE simulations carried out with 27°C temperature, it has been found that all the circuits have indeed low power consumption varying between 87.8 μW and 190 μW. For the current-controlled positive grounded/floating resistors, by changing I_0 from 100 nA to 1 mA, the resistance value was found to vary over a range of nearly four decades (from 34 Ω to 285 Ω). Frequency responses of the circuits confirmed their operational bandwidth to be around 10 MHz.

Applications: In [97], the above-mentioned circuits were shown to be useful in two applications. In the first application, two current-controlled positive resistors (one floating and the other grounded) were shown to be useful in converting a simple *RC* second-order BP filter into a current-controllable BPF with current-controlled variable centre frequency. The second example was that of an *RC* circuit containing a capacitor and three negative resistors which were realized by employing two grounded negative resistance circuits and one floating negative resistor – all having the control of resistance value through respective external dc bias currents. The practical viability of this circuit which simulated a negative inductance was demonstrated by confirming the magnitude and phase responses.

9.2.7 *Current-controlled-resistor based upon a new eight-transistor mixed-translinear-cell (MTC)*

In [95], Merz *et al.* presented a new eight-transistor MTC with an increased input linear range as compared to the traditional four-transistor MTC proposed by Fabre [49]. They demonstrated the usefulness of this circuit in realizing a current-controlled resistor and controlled CC; see Figure 9.15.

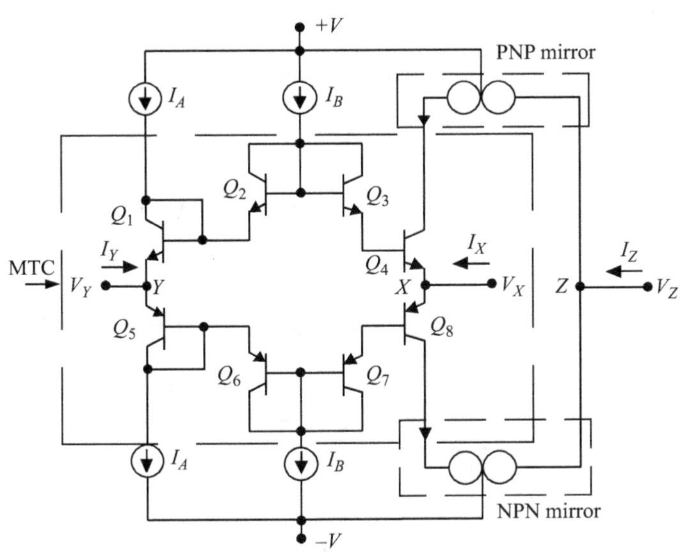

Figure 9.15 CCCII architecture based on a new mixed translinear cell [95]

9.2.8 Electronically tunable active resistance circuits based upon differential amplifiers

Tekin and Alci [100] introduced two electronically tunable active resistance circuits based upon differential amplifiers. The floating positive resistance circuit incorporating two differential transconductance amplifiers is shown in Figure 9.16. It is well known that the traditional differential transconductance amplifier implements a tanh function between its output current (I_0) and its differential input voltage ($V_{id} = V_1 - V_2$), i.e.

$$I_0 = I_B \tanh\left(\frac{V_1 - V_2}{2V_T}\right) \tag{9.31}$$

For $(V_1 - V_2) \ll 2V_T$, the above expression can be approximated as

$$I_0 \cong I_B\left(\frac{V_1 - V_2}{2V_T}\right) = G_m(V_1 - V_2), \quad \text{where } G_m = I_B/2V_T \tag{9.32}$$

Thus, two such differential transconductance stages with their output currents appropriately connected to ports 1 and 2 lead to the circuit of Figure 9.16 for realizing a floating positive resistor and the circuit of Figure 9.17 for the realization of floating negative resistance, respectively. The magnitude of the value of the floating resistance realized in each case is given by $R_{eq} = 2V_T/I_B$.

A nonideal analysis of these circuits, considering the effect of finite β, indicated that there would be some error in the dc offset current of each differential transconductance stage which can, however, be reduced by employing MOS current mirrors with perfectly matched MOSFETs.

The validity of these two configurations has been demonstrated in [100] using NR100N and PR100N BJTs, with the circuits biased from ±1.5 V dc and the external dc bias currents varied from 0.1 μA to 10 mA. From simulation results, the linear range of positive and negative floating resistors has been found to be ±30 mV, whereas the resistance value was found to be variable between ±43 Ω and ±516 kΩ with I_B varied from 0.1 μA to 1.2 mA.

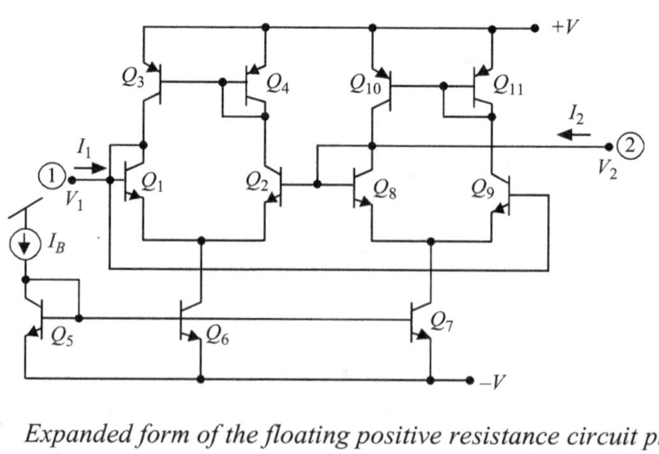

Figure 9.16 Expanded form of the floating positive resistance circuit proposed by Tekin and Alci [100]

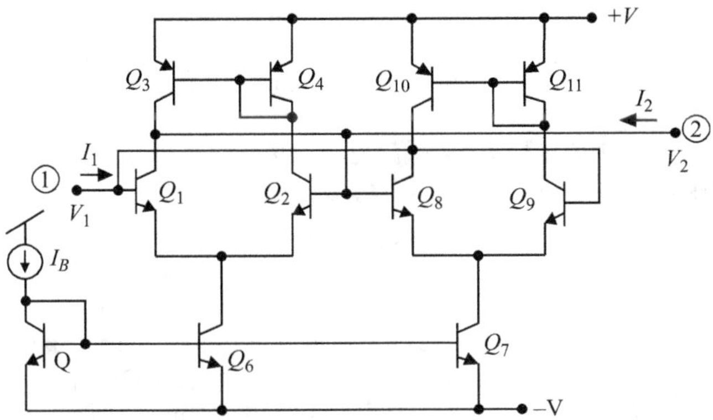

Figure 9.17 Expanded form of the floating negative resistance proposed by Tekin and Alci [100]

Applications: Two applications were demonstrated in [100] for the circuits of Figures 9.16 and 9.17. In the first application the negative resistance was used in constructing an oscillator in conjunction with a positive resistor, a capacitor and an inductor. On the other hand, the current-controlled positive resistor was used to realize a temperature-insensitive current-controlled current amplifier. In this case, the intended current amplifier was constructed using CCCII which is known to have a nonzero current-controlled input resistance looking into its X-terminal therefore, with the terminal-X directly connected to ground and the current-controlled resistance shown here in Figure 9.17 connected from Y-terminal to ground, when output is taken from Z-terminal, the current gain of the circuit is given by

$$A_I = \frac{I_{out}}{I_{in}} = \frac{R_{eq}}{R_x} = \frac{2V_T/I_{B1}}{V_T/2I_{B2}} = 4\left(\frac{I_{B2}}{I_{B1}}\right) \tag{9.33}$$

where I_{B1} is the external dc bias current of the current-controlled circuit of Figure 9.17 and I_{B2} is the external dc bias current of the CCCII. Thus, the temperature-dependent terms were cancelled out and the gain of this current amplifier is controllable by the ratio of two bias currents. SPICE simulations confirmed that the current gain of this amplifier remains nearly constant when temperature was changed from 0° to 90°.

9.3 CMOS linear voltage/current-controlled grounded/ floating resistors

With the advent of continuous time MOSFET-C filters pioneered by Tsividis [14], Czarnul [16], Ismail [11] and others as continuous-time alternatives to discrete time MOS switched-filters, there was spurt of studies on the methods of realizing voltage/current-controlled linear resistive circuits for implementation in CMOS technology. Thus, commencing from 1984, when Han and Park [7] demonstrated a simple connection of two MOS transistors to realize a voltage-controlled resistor

(VCR), a number of researchers proposed from time to time a variety of circuit configurations of varying complexity and features. In this section, we highlight some prominent circuit configurations for realizing CMOS VCRs evolved till date.

9.3.1 A two-MOSFET-based linear voltage-controlled resistor devised by Han and Park

The circuit[1] proposed by Han and Park [7] is shown in Figure 9.18, which uses two depletion mode MOSFETs. This circuit can be analysed as follows: when the applied voltage is positive, i.e. $V_{in} = V_{12} > 0$, terminal 1 acts as drain and terminal 2 as the source for both the MOSFETs and hence, due to the virtual ground at the input terminal of the op-amp (which is assumed to be CMOS op-amp), it follows that $V_{GS} = V_G$. If $V_{12} < (V_G + |V_{thn}|)$, the MOSFETs operate in triode region and hence, the current flowing from nodes 1 to 2 is given by

$$I_{in} = I_{d1} + I_{d2} = K(V_G - 2V_{th})V_{12} \tag{9.34}$$

On the other hand, when the applied voltage is negative, i.e. $V_{in} < 0$, terminal 2 is regarded as drain while terminal 1 as the source (recalling the symmetry of the MOSFET structure). It, therefore, follows that in this case, V_{GS} of M_1 would be $(V_G - V_{in})$ while V_{GS} of M_2 would be zero. Thus, in this case too, adding the two drain currents the nonlinear terms cancel out and one obtains

$$I_{21} = K(V_G - 2V_{th})V_{21} \tag{9.35}$$

Thus, a combination of the two equations proves that the composite connection of the two MOSFETs behaves as an equivalent resistance of value:

$$R_{eq} = \frac{1}{K(V_G - 2V_{th})} \tag{9.36}$$

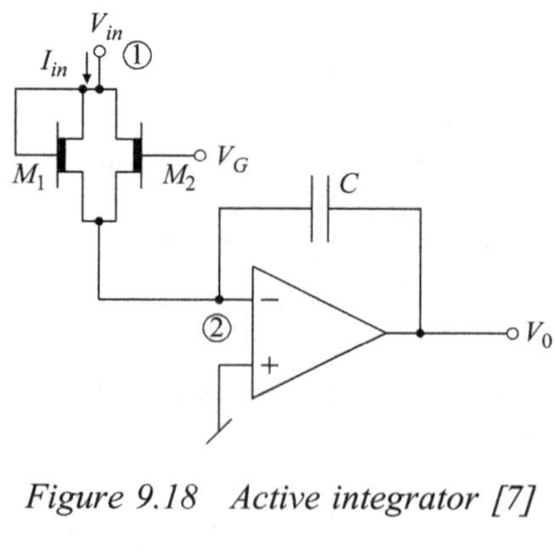

Figure 9.18 Active integrator [7]

[1]The same basic interconnection of two MOSFETs as linear voltage-controlled resistor has also been discussed and analysed in detail in [27].

9.3.2 Some general techniques of realizing linear MOS-resistive circuits

Based upon their extensive research on MOSFET-C networks and study of nonlinearity cancellation mechanism in one, two and four MOSFETs, Tsividis *et al.* [14] summed up several generalized techniques of nonlinearity cancellation, some of which are shown here in Figure 9.19. In all these circuits, the MOSFETs are

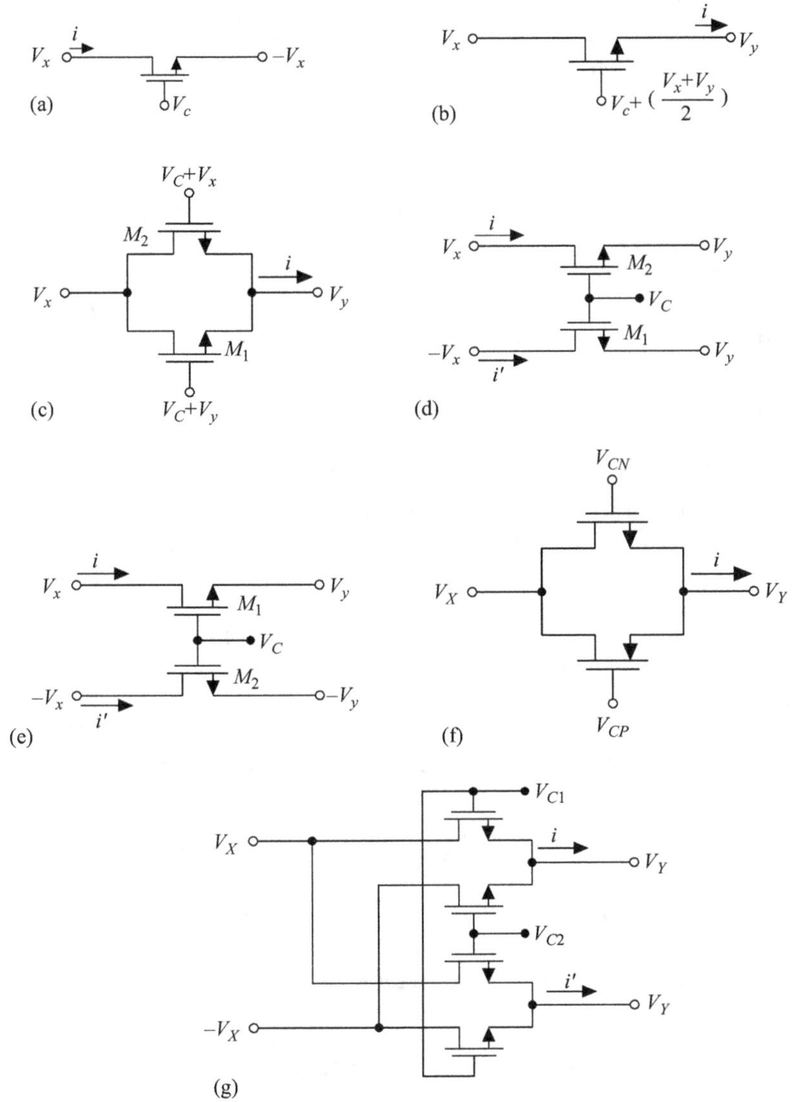

Figure 9.19 Seven general techniques (a)-(g) of nonlinearity cancellation in MOSFETs [14]

operated in triode region by keeping $V_{DS} < (V_{GS} - V_{th})$ so that the drain current I_d is given by

$$I_d = 2K\left[(V_{GS} - V_{th})V_{DS} - \frac{V_{DS}^2}{2}\right] \tag{9.37}$$

where $K = 1/2\{\mu_s C_{ox}(W/L)\}$ and the symbols have their usual meanings.

In the circuit of Figure 9.19(a), the square nonlinearity of the MOSFETs operated in triode/VCR region is cancelled out by applying negative of the drain voltage (V_x) on the source terminal, thereby leading to $V_{GS} = (V_C + V_x)$ and $V_{DS} = 2V_x$, which upon substituting in (9.36) leads to

$$i = I_d = \frac{2V_x}{R_{eq}}, \quad \text{where} \quad R_{eq} = \frac{1}{2K(V_C - V_{th})} \tag{9.38}$$

In the circuit of Figure 9.19(b), the drain and the source terminals are kept at different potentials (V_x and V_y, respectively), thereby making $V_{DS} = (V_x - V_y)$, but the gate voltage is modified to $V_G = \left(V_C + \left((V_x + V_y)/2\right)\right)$, thereby yielding $V_{GS} = (V_G - V_y) = \left(V_C + \left((V_x - V_y)/2\right)\right)$ so that the substitution of these values in the equation for I_d (operating in triode region) gives

$$i = I_d = 2K\left[(V_{GS} - V_{th}) - \frac{V_{DS}}{2}\right]V_{DS} = 2K(V_C - V_{th})(V_x - V_y) \tag{9.39}$$

thereby realizing a linear resistance $R_{eq} = 1/(2K(V_C - V_{th}))$.

The next four circuits employ two matched MOSFETs, both operating in triode region. In the case of the circuit of Figure 9.19(c), the two MOSFETs are connected in parallel, their drains and sources are at different potentials V_x and V_y, respectively, but the gate voltages are made $(V_C + V_x)$ and $(V_C + V_y)$, respectively, while both have their $V_{DS} = (V_x - V_y)$. Consequently, $V_{GS1} = V_C + V_y - V_y = V_C$ while $V_{GS2} = V_C + (V_x - V_y)$ due to which the current i, flowing from one end to the other is now sum of the two drain currents:

$$\begin{aligned} i = I_{d1} + I_{d2} &= 2K\left[(V_C - V_{th}) - \frac{(V_x - V_y)}{2}\right](V_x - V_y) \\ &\quad +2K\left[V_C + (V_x - V_y) - V_{th} - \frac{(V_x - V_y)}{2}\right](V_x - V_y) \\ &= 4K[(V_C - V_{th})](V_x - V_y) = \frac{2(V_x - V_y)}{R_{eq}} \end{aligned} \tag{9.40}$$

thereby representing an equivalent resistance

$$R_{eq} = \frac{1}{2K(V_C - V_{th})}. \tag{9.41}$$

In the next configuration of Figure 9.19(d), the drain terminals of the two MOSFETs are arranged to have complementary voltage V_x and $-V_x$, respectively, while their source terminals are kept at the same potential V_y. This makes $V_{GS1} = (V_C - V_y) = V_{GS2}$

while $V_{DS1} = (V_x - V_y)$ but $V_{DS2} = (-V_x - V_y)$. As a consequence, the differential output current for this configuration is given by

$$\begin{aligned} i &= I_{d1} - I_{d2} \\ &= 2K\left[(V_{GS1} - V_{th}) - \frac{V_{DS1}}{2}\right]V_{DS1} - 2K\left[(V_{GS2} - V_{th}) - \frac{V_{DS2}}{2}\right]V_{DS2} \end{aligned} \tag{9.42}$$

Substituting the mentioned values, it is found that $i = 2K(V_C - V_{th})(2V_x)$, thereby yielding $R_{eq} = 1/(2K(V_C - V_{th}))$. The next configuration of Figure 9.19(e) is quite similar to that of Figure 9.19(d) with only difference being that the voltages on the source terminals of the two MOSFETs too are complementary: V_y and $-V_y$, respectively. As a consequence, $V_{GS1} = (V_C - V_y)$ but $V_{GS2} = (V_C + V_y)$; also while $V_{DS1} = (V_x - V_y)$, $V_{DS2} = (-V_x + V_y)$.

When these values are substituted in the expressions for the drain currents of the two MOSFETs, one finds

$$I_{d1} = 2K\left[(V_C - V_y - V_{th}) - \frac{(V_x - V_y)}{2}\right](V_x - V_y) \tag{9.43}$$

$$I_{d2} = 2K\left[(V_C + V_y - V_{th}) - \frac{(-V_x + V_y)}{2}\right](-V_x + V_y) \tag{9.44}$$

Consequently, in the differential current $i = I_{d1} - I_{d2}$, the nonlinear terms cancel out and what is left is

$$i = 2\frac{(V_x - V_y)}{R_{eq}}, \quad \text{where} \quad R_{eq} = \frac{1}{2K(V_C - V_{th})}.$$

While the circuits described so far were using all identical enhancement mode NMOS transistors only, the circuit of Figure 9.19(f) uses one PMOS and one NMOS transistor. Consequently, the equivalent *floating* resistance realized by this circuit is given by

$$i = (V_x - V_y)\left[2K_n(V_{CN} - V_{Tn}) + 2K_p\left(V_{CP} - |V_{Tp}|\right)\right] \tag{9.45}$$

Lastly, the arrangement of Figure 9.19(g), which was originally proposed by Czarnul [16], employs four matched NMOS transistors and has two control voltages V_{C1} and V_{C2} and realizes the differential output current as

$$(i - i') = 2V_x[2K(V_{C1} - V_{Tn}) - 2K(V_{C2} - V_{Tn})] \tag{9.46}$$

The following features of these arrangements are worth mentioning – (i) only the circuits of Figure 9.19(b)–(e) can be used as 'stand-alone' two terminal 'floating' voltage-controlled resistors (VCR), whereas that of Figure 9.19(a) can be employed only as a grounded VCR. Some of these circuits have been actually employed to realize novel VCRs in conjunction with various kinds of active elements in the literature. (ii) Out of the floating VCRs mentioned above, all would need additional active elements to generate the control voltages required for their implementation. (iii) The circuit of Figure 9.19(d) and (g) which need equal voltages V_y at the source terminals of the MOSFETs can be used in configuration with active elements like op-amps, CCs or CFOA – all of which have their two input voltages at the same potential and thereby can lead to fully differential or single-ended biquad active filter topologies. On the other hand, the use of the schematic of Figure 9.19(e) which has complementary voltages at the source ends appears to have been employed in [116].

9.3.3 The two MOSFET transresistor due to Wang

Undoubtedly, one of the simplest MOS-grounded resistors is the one proposed by Wang [24], which uses only two diode-connected NMOS transistors as shown in Figure 9.20. Since both MOSFETs are diode-connected, obviously both operate in saturation with their drain currents given by

$$I_{d1} = K(V_{DD} - V_{in} - V_{th})^2 \quad \text{and} \quad I_{d2} = K(V_{in} + V_{SS} - V_{th})^2 \tag{9.47}$$

Since $I_{in} + I_{d1} = I_{d2}$, the substitution of the values leads to (assuming $|V_{SS}| = |V_{DD}|$)

$$\frac{V_{in}}{I_{in}} = \frac{1}{4K(V_{DD} - V_{th})} \tag{9.48}$$

It is surprising why Wang [24] chose to call the circuit as 'transresistor' when I_{in} and V_{in} are both applied on the same common node. Moreover, true transresistor element should have ideally zero-input impedance at the input node and should offer low-output impedance at the node from where output V_0 is taken. Curiously, neither of the requirements is met by this circuit. Nevertheless, as a 'resistor' with V_{in} and I_{in} referred to the same node, this circuit is perfectly fine and hence, it is not surprising that as a MOS-grounded resistor, this circuit has been extensively

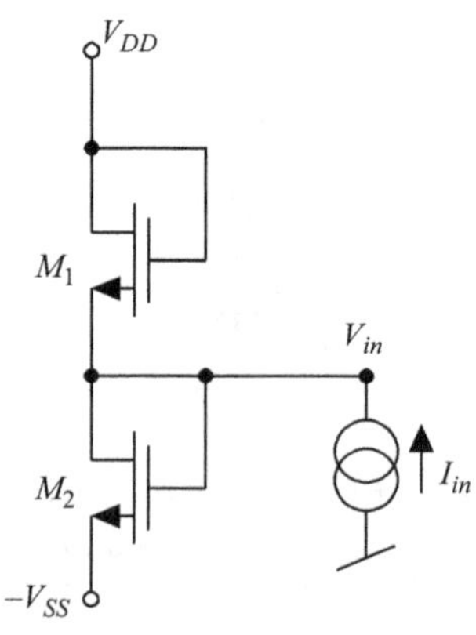

Figure 9.20 A simple linear MOS resistor [24]

employed by numerous researchers to replace all passive resistors with MOS resistors for instance [117]. The attractive feature of this circuit is its apparent simplicity though a limitation is that the value of the realized resistance is not electronically tunable; one has to alter the *W*/*L* ratio of the MOSFETs to realize different values of resistance.

9.3.4 Banu–Tsividis linear voltage-controlled floating linear resistor

Probably, one of the earliest circuits to realize a CMOS 'floating' VCR, exclusively using MOSFETs only, was the one proposed by Banu and Tsividis [5] which is shown in Figure 9.21. The basic core of the circuit is a parallel connection of two PMOS transistors with additional MOS transistors M_3–M_6 arranged to ensure that the gate voltages of MOSFET1 and MOSFET2 are made equal to $(V_C + V_1)$ and $(V_C + V_2)$, respectively. With these values of gate voltages, it is easy to figure out that the equivalent resistance realized between ports 1 and 2 is given by

$$R_{1-2} = \frac{1}{2\mu_s C_{ox}(W/L)[V_c - V_{th}]} \tag{9.49}$$

The relative derivation from linearity for this MOS VCR is shown in Figure 9.21(b).

9.3.5 Linear transconductor due to Park and Schaumann

Another circuit, which was originally introduced as a single-ended transconductor element but was employed as a resistor by many researchers, is the four-MOSFET circuit advanced by Park and Schaumann [17] and is shown in Figure 9.22.

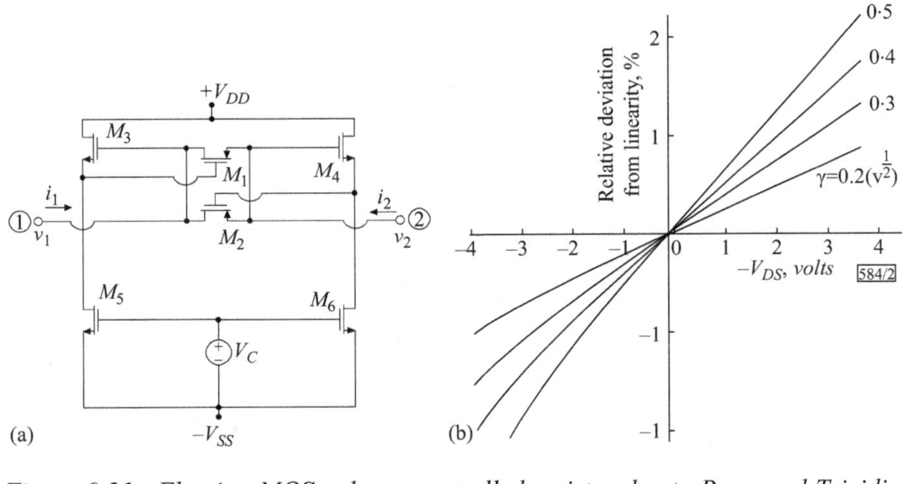

Figure 9.21 Floating MOS voltage-controlled resistor due to Banu and Tsividis [5]: (a) the simplified form of the VCR circuit, (b) relative deviation from linearity for this MOS VCR. © 1982 IEE

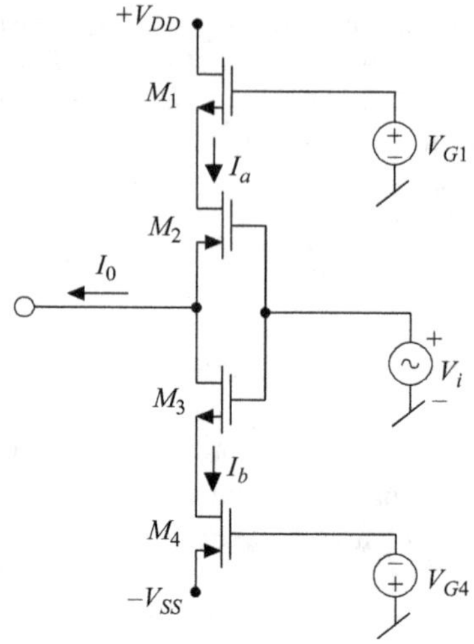

Figure 9.22 Linear tunable transconductance [17]

The circuit employs two matched PMOS transistors and two matched NMOS transistors. A straightforward analysis from [17] reveals the following.

The two drain currents are given by

$$I_a = K_{eff}(V_{G1} - V_i - V_{Tn1} - |V_{TP2}|)^2 \tag{9.50}$$

$$I_b = K_{eff}(V_i + V_{G4} - V_{Tn3} - |V_{TP4}|)^2 \tag{9.51}$$

where

$$K_{eff} = \frac{K_n K_p}{\left(\sqrt{K_n} + \sqrt{K_p}\right)^2} \quad \text{and} \quad K_{n,p} = \left[\frac{\mu_{eff} C_{ox}(W/L)}{2}\right]_{n,p} \tag{9.52}$$

Since $I_0 = (I_a - I_b)$, it turns out that [17]

$$I_0 = -2K_{eff}[V_{G1} + V_{G4} - \Sigma V_T]V_i + K_{eff}[V_{G1} + V_{G4} - \Sigma V_T]\Delta V_T \tag{9.53}$$

where

$$\begin{aligned} \Sigma V_T &= V_{Tn1} + V_{Tn3} + |V_{Tp2}| + |V_{Tp4}| \quad \text{and} \\ \Delta V_T &= (V_{Tn3} - V_{Tn1}) + (|V_{Tp4}| - |V_{Tp2}|) + (V_{G1} - V_{G4}) \end{aligned} \tag{9.54}$$

If one takes $V_{G1} = V_{G4} = V_G$ and MOSFETs are matched ΔV_T becomes zero and one obtains [17]

$$I_0 = -2K_{eff}(2V_G - \Sigma V_T)V_i \tag{9.55}$$

so that the transconductance is electronically tunable with V_G. If output is returned to the input, the circuit would realize a grounded VCR with

$$R_{eq} = \frac{1}{2K_{eff}(2V_G - \Sigma V_T)} \tag{9.56}$$

9.3.6 Linear floating VCR due to Nagaraj

After the floating VCR of Banu and Tsividis [5], another floating VCR circuit was proposed by Nagaraj [13] which is shown in Figure 9.23. The core of the circuit is a two-MOSFET transconductor comprising MOSFETs M_5 and M_6. A straightforward analysis of this circuit can be carried out as follows:

$$\begin{aligned} I_{D5} &= 2K\left[(V_{C2} - V_x - V_{th}) - \frac{(V_y - V_x)}{2}\right](V_y - V_x); \quad K = \frac{\mu_s C_{ox}(W/L)}{2} \\ I_{D6} &= 2K\left[(V_{C1} - V_x - V_{th}) - \frac{(V_y - V_x)}{2}\right](V_y - V_x) \end{aligned} \tag{9.57}$$

Since MOSFETs M_1–M_4 have the same aspect ratios and carry the same bias currents, it can be deduced that $V_x - V_y = V_1 - V_2$. Furthermore, by writing node equations it can also be determined that $I_1 = (I_{D5} - I_{D6})$ and $I_2 = (I_{D6} - I_{D5})$, respectively. Therefore, taking the difference of the two currents, one obtains

$$I_1 = 2K(V_{C1} - V_{C2})(V_1 - V_2) = -I_2 \tag{9.58}$$

The realized equivalent floating resistance is given by

$$R_{eq} = \frac{1}{2K(V_{C1} - V_{C2})} \tag{9.59}$$

Thus, the circuit realizes a floating VCR, the value of which can be made either positive or negative by judicious choice of V_{C1} and V_{C2}. It may be mentioned that in this circuit, operation of M_5 and M_6 is ensured in triode region by ensuring that

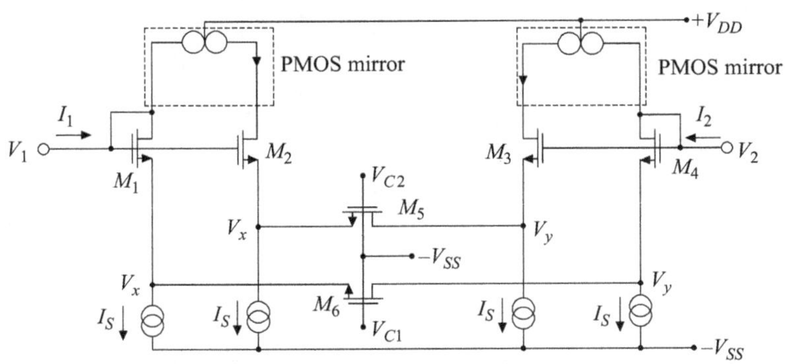

Figure 9.23 Simplified form of the floating VCR proposed by Nagaraj [13]

both V_{C1} and V_{C2} are set sufficiently high while the desired value of resistance can be set by varying difference between them [13]. With the W/L ratio of MOSFETs M_5 and M_6 as 0.1 and that of remaining MOSFETs as 40, SPICE simulations using the then-prevalent 5 μm CMOS technology demonstrated a working linear range of the circuit as ±2.5 V with control voltages varied between 3 V and 4 V. The worst case linearity error was found to be less than 2%.

9.3.7 Wilson and Chan grounded VCR

Wilson and Chan [23] came up with a CMOS-grounded VCR which is shown in Figure 9.24. The basic principle of this circuit is the idea that if an NMOS transistor has its source connected to ground with its drain connected to V_{in} then if the gate voltage is modified to become $V_G = V_C + (V_{in}/2)$, the input current I_{in} (which is the same as the drain current I_d) becomes

$$\begin{aligned} I_{in} = I_d &= 2K\left[(V_{GS} - V_{th}) - \frac{V_{DS}}{2}\right]V_{DS} = 2K\left[\left(V_C + \frac{V_{in}}{2} - V_{th}\right) - \frac{V_{in}}{2}\right]V_{in} \\ &= 2K(V_C - V_{th})V_{in} \end{aligned} \tag{9.60}$$

and the circuit would realize an equivalent resistance

$$R_{eq} = \frac{V_{in}}{I_{in}} = \frac{1}{2K(V_C - V_{th})}; \quad K = \frac{\mu_s C_{ox}(W/L)}{2} \tag{9.61}$$

In the circuit of Figure 9.24, there are two source-coupled differential pairs (M_4–M_5 and M_6–M_7) along with a PMOS current mirror (shown in dotted box) and the two other MOSFETs M_1 and M_3 realizing constant current sources $2I$ each, through a constant gate bias voltage V_b. Thus, the voltage V_{in} is input to the first source-coupled

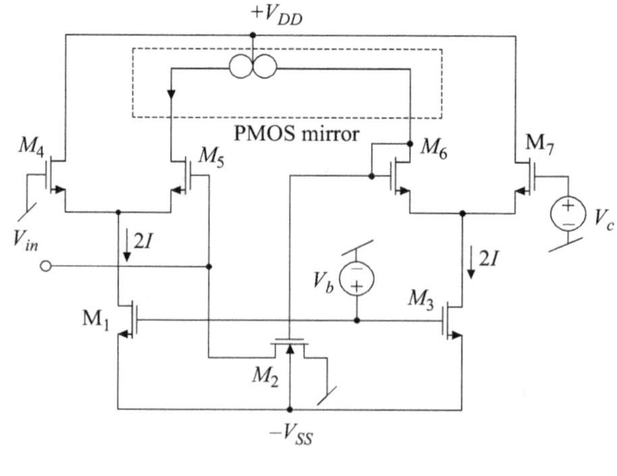

Figure 9.24 Simplified form of the CMOS-grounded VCR due to Wilson and Chan [23]

differential amplifier pair so that through the PMOS current mirror and second differential pair, the voltage applied on the gate of M_2 becomes exactly equal to $(V_C + (V_{in}/2))$ as required for nonlinearity cancellation. Thus, equivalent resistance realized is the same as given in (9.60).

9.3.8 Wang's grounded linear VCR

An interesting technique of eliminating the square nonlinearity of a single MOSFET operated in triode region was proposed by Wang [26]. His proposition is shown in Figure 9.25. The method of cancelling the nonlinearity as advanced by Wang [26] can be explained as follows: The drain current of M_1 operating in triode region is given by

$$I_{d1} = 2K\left[(V_C - V_{th}) - \frac{V_{in}}{2}\right]V_{in} \tag{9.62}$$

with drain current of diode-connected M_2 added to it, the total input current becomes

$$I_{in} = I_{d1} + I_{d2} = 2K[V_C - 2V_{th} + V_{SS}]V_{in} + K(V_{SS} - V_{th})^2 \tag{9.63}$$

The second term on the right-hand side of the above equation can be considered to be an offset current which needs to be removed. Wang [26] added a third transistor M_3 with a gate connected to ground and the source connected to $-V_{SS}$ such that its drain current I_{d3}, through the PMOS current M_4–M_5, is carried to the input node, thereby modifying the input current as

$$I_{in} = (I_{d1} + I_{d2}) - I_{d3} = (I_{d1} + I_{d2}) - K(V_{SS} - V_{th})^2 \tag{9.64}$$

It is, thus, seen that with the subtraction of I_{d3} from $(I_{d1}+I_{d2})$, the unwanted offset current term is completely cancelled out, thereby rendering the input current I_{in} as

$$I_{in} = 2K[V_C - 2V_{th} + V_{SS}]V_{in} \tag{9.65}$$

resulting in an equivalent resistance of value

$$R_{eq} = \frac{1}{2K(V_C - 2V_{th} + V_{SS})} \tag{9.66}$$

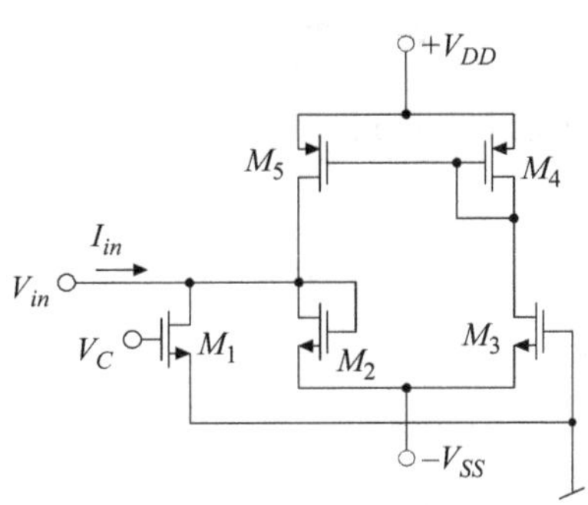

Figure 9.25 Grounded VCR proposed [26]

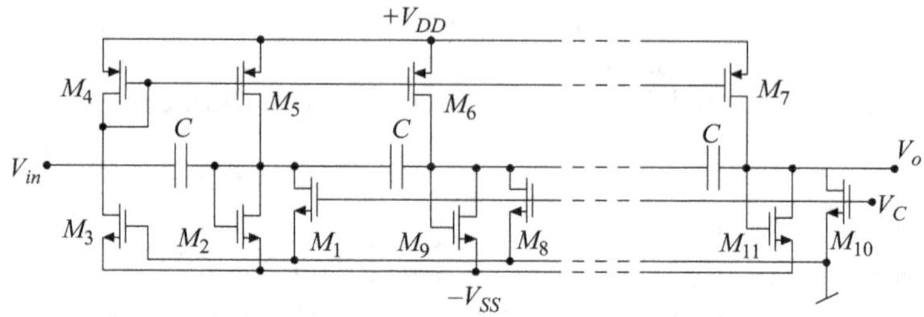

Figure 9.26 Realization of third-order electronically controllable high-pass filter [26]

Application: Wang [26] demonstrated an interesting application of the grounded MOS VCR circuit of Figure 9.25 in realizing an electronically tunable third-order high-pass filter in which all the three equal resistors are to be replaced by three identical VCRs controlled by a common control voltage V_C. The interesting point to be observed is that while the MOSFETs M_1–M_2 are used in all the three VCRs, the remaining three transistors need not to be repeated thrice. In fact, the 'offset-current' cancelling transistor M_3 can be shared with all the three VCRs; the replicas of I_{d3} can be added at the input nodes of the remaining two VCRs by simply modifying the current mirror M_4–M_5 to become a current repeater by adding PMOS transistors M_6 and M_7, thereby delivering copies of I_{d3} at the required nodes as shown in Figure 9.26. SPICE simulations have demonstrated that using ±5 V dc supply and varying input voltages up to 4 V, a linearity with errors not more than 0.6% was observable for the grounded VCR of Figure 9.25 [26].

9.3.9 Positive/negative linear grounded VCRs due to Wang

Wang [33] introduced another grounded MOS VCR circuit which was, in fact, a simple modification of his two-MOSFET trans-resistor of Figure 9.20. This alternative circuit, in contrast to the circuit of [24] which used two diode-connected transistors, instead employed one PMOS and one NMOS transistor – still both of them diode-connected. This alternative circuit is shown in Figure 9.27. Straightforward analysis of this circuit reveals that the input current is given by

$$I_{in} = \left(\frac{V_{in}}{R_{eq}} + I_{off}\right), \text{ where } I_{off} = K\left[(V_{SS} + V_{thn})^2 - \left(V_{DD} - \left|V_{thp}\right|\right)^2\right] \text{ and}$$
$$K = \frac{\mu_s C_{ox}(W/L)}{2} \tag{9.67}$$

The offset in the above equation is cancelled out by an additional offset-compensation network (OCN) realized from one more pair of PMOS and NMOS

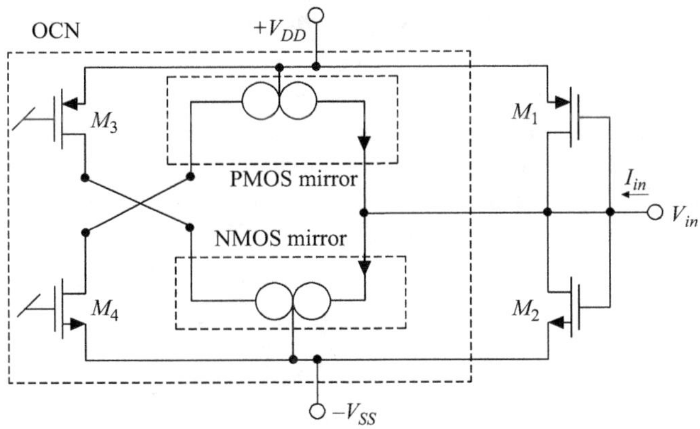

Figure 9.27 Simplified form of the grounded positive VCR with OCN due to Wang [33]

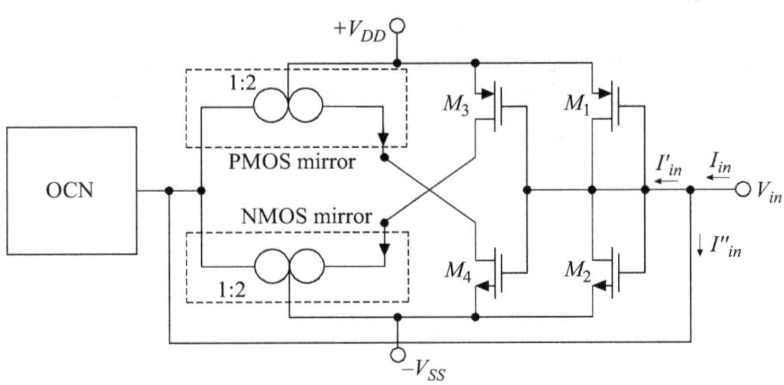

Figure 9.28 A simplified form of the grounded negative VCR with OCN due to Wang [33]

transistors along with a pair of PMOS and NMOS current mirrors. As a consequence, the equivalent resistance realized by this circuit is found to be [33]

$$R_{eq} = \frac{1}{2K\left(V_{DD} - V_{SS} - V_{thn} - \left|V_{thp}\right|\right)} \tag{9.68}$$

whereas to keep all the MOSFETs to be operating in saturation region, the input voltage is constrained by the following condition: $\left(V_{DD} - \left|V_{thp}\right|\right) \leq V_{in} \leq \left(V_{SS} + V_{thn}\right)$.

Using the same basic idea, a circuit to realize a negative VCR was also proposed by Wang [33], which is shown in Figure 9.28, where the input current was made to be negative of that of the circuit of Figure 9.27 by adding two more MOSFETs and with cross-coupling the inputs of two current mirrors. In this circuit also, the offset cancellation circuit is the same as the one in the earlier circuit.

9.3.10 Positive/negative linear grounded VCRs and voltage-controlled gyrator using NICs

In [29], Tabei *et al.* presented a technique of cancelling the square nonlinearities of two matched MOSFETs operating in triode region by arranging the input current to be the difference of two drain currents through an NIC. The scheme of grounded VCR proposed in [29,49] is shown in Figure 9.29(a). A straightforward analysis of this circuit reveals that

$$\begin{aligned} I_{in} = (I_{d1} - I_{d2}) &= 2K\left[V_{C1} - V_{th} - \frac{V_{in}}{2}\right]V_{in} - 2K\left[V_{C2} - V_{th} - \frac{V_{in}}{2}\right]V_{in} \\ &= 2K(V_{C1} - V_{C2})V_{in} \end{aligned} \tag{9.69}$$

thereby yielding,

$$R_{eq} = \frac{1}{2K(V_{C1} - V_{C2})} \tag{9.70}$$

Thus, the novel features of this circuit are (i) the realized resistance value being independent of V_{th}, (ii) feasibility of realizing either positive or negative valued VCRs by a judicious choice of the control voltages V_{C1} and V_{C2}.

A floating version of this circuit is obtained by using only two matched MOSFETs but instead of one NIC, using two NICs as shown in Figure 9.29(b).

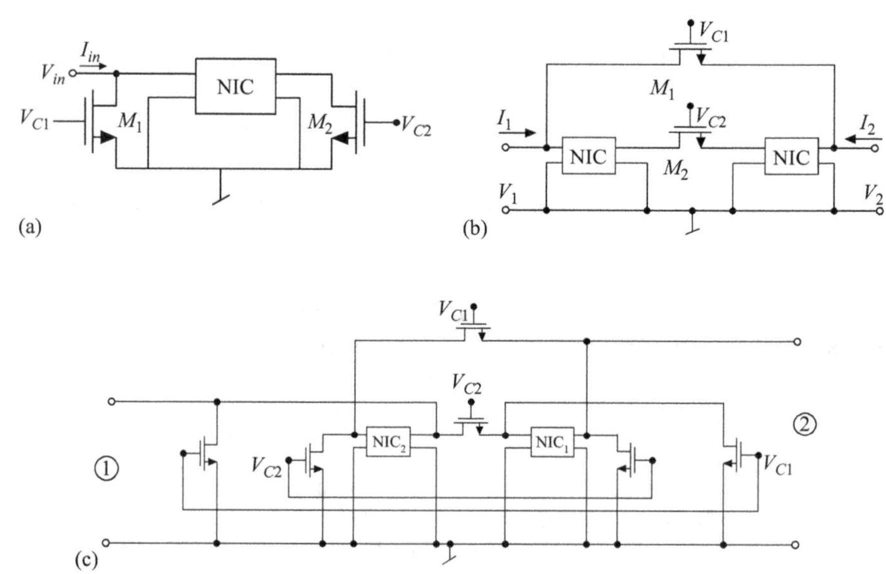

Figure 9.29 Various linearized voltage-controllable circuits proposed by Tabei et al. [29]: (a) grounded VCR circuit, (b) floating VCR, (c) voltage-controlled gyrators.

A straightforward analysis of the circuit, following the various currents in the circuit, reveals

$$I_1 = (I_{d1} - I_{d2}) = -I_2 = 2K(V_{C1} - V_{C2})(V_1 - V_2) \tag{9.71}$$

which represents a floating VCR of the same value as that of the circuit of Figure 9.29(a).

The same idea was extended to realize a MOS gyrator which was well elaborated in detail in [43] which is shown in Figure 9.29(c). The *Y*-matrix of this gyrator circuit is found to be

$$[Y] = \begin{bmatrix} 0 & 2K(V_{C1} - V_{C2}) \\ -2K(V_{C1} - V_{C2}) & 0 \end{bmatrix} \tag{9.72}$$

With port 2 terminated into a capacitor C_0, the input impedance would be given by

$$Z_{in} = \frac{sC_0}{4K^2(V_{C1} - V_{C2})^2} \tag{9.73}$$

Thus, the circuit realizes a lossless inductance controllable by V_{C1} and V_{C2}.

It may be mentioned that Tabei *et al.* [29] had envisaged an op-amp-based NIC which requires two matched passive resistors also. Subsequently, Soliman [53] demonstrated that with NICs realized by a CCII+ the requirement of matched resistors can be completely dispensed with. Thus, if these circuits are realized using CMOS CCII+ based NICs, completely CMOS-based grounded and floating VCRs and a CMOS gyrator would be realizable.

Application: The authors of [43] demonstrated that the devised MOS gyrator could be advantageously employed to design an electronically controllable third-order low-pass filter based upon an *RLC* prototype in which the floating inductor and the two resistors (one at the input signal source end and the other at the load end) encountered therein can be efficiently realized using no more than three NICs, eight MOSFETs along with three grounded capacitors, as preferred for IC implementation. The validity of this design was substantiated using op-amp-based NICs and 2 μm CMOS technology prevalent then with satisfactory performance. Since the issue of the design of a high-frequency CMOS NIC was not resolved till then, from the latter work of Soliman [53], the possibility of obtaining better performance using any new and improved CMOS NICs (either CCII+ based or otherwise) is still worth investigating.

Figure 9.30 shows the CMOS floating resistor circuit proposed by Elvan *et al.* [51]. In this circuit, it is easy to figure out that $I_{in} = I_{out} = I_2 - I_1$, where the individual values of the two currents are given by

$$\left.\begin{aligned} I_1 &= K(V_1 - V_x - V_{th})^2 \\ I_2 &= K(V_2 - V_y - V_{th})^2 \end{aligned}\right\}; \quad \text{with } K = \frac{\mu_s C_{ox}(W/L)}{2} \tag{9.74}$$

From the biasing circuit comprising MOSFETs M_5 and M_9, since $I_{D5} = I_{D9}$, therefore, it follows that

$$K(V_C - V_{DD} - V_{thp})^2 = K(V_x - V_2 - V_{thp})^2.$$

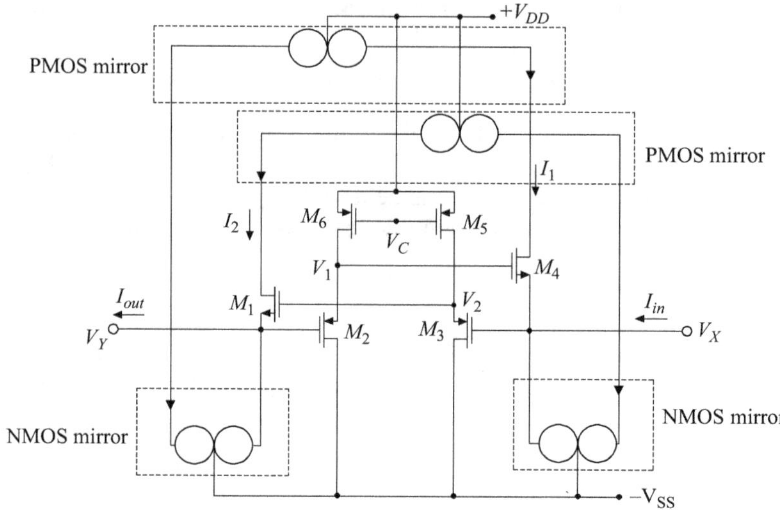

Figure 9.30 Simplified representation of the floating resistor proposed by Elwan et al. [51]

Hence, it is found that the voltages V_1 and V_2 can be expressed as

$$V_2 = V_x - V_C + V_{DD} \quad \text{and} \quad V_1 = V_y - V_C + V_{DD} \tag{9.75}$$

so that we can finally write $I_{in} = I_{out} = 4K(V_{DD} - V_C - V_{th})(V_x - V_y)$ for $|V_{thn}| = |V_{thp}| = V_{th}$, which represents a floating resistor of equivalent value

$$R_{eq} = \frac{1}{4K(V_{DD} - V_C - V_{th})}. \tag{9.76}$$

9.3.11 Floating linear resistor proposed by Elwan, Mahmoud and Soliman

Since the realized resistance is dependent upon the value of the process parameter V_{th}, Elvan *et al.* [51] came up with a modified circuit in which a diode-connected NMOS transistor was added to shift the external control voltage V_C by an amount V_{th} before applying to the biasing circuit. This modifies the expression for the realized resistance value as

$$R'_{eq} = \frac{1}{4K(V_{DD} - V_C)} \tag{9.77}$$

An interesting floating VCR circuit which can realize either a positive or a negative resistor from the same configuration by a judicious choice of two control voltages V_{C1} and V_{C2} and can operate from a low supply voltage of ± 1.5 V only was evolved by Tantry *et al.* [68], which is shown here in Figure 9.31. By following the

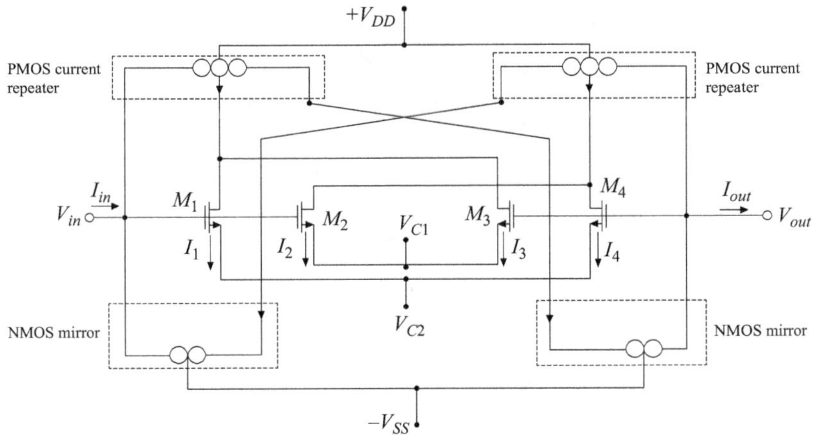

Figure 9.31 Simplified form of the floating resistor proposed by Tantry et al. [68]

circuit diagram, it can be readily seen that the various current mirrors/repeaters realized from matched MOSFETs ensure that the currents I_{in} and I_{out} are equal to $(I_2+I_4-I_1-I_3)$. Furthermore, assuming that the two MOSFETs forming the central core are also matched, it is found that

$$I_{in} = I_{out} = \frac{(V_{in} - V_{out})}{R_{eq}}, \quad \text{where} \quad R_{eq} = \frac{1}{2K(V_{C2} - V_{C1})} \tag{9.78}$$

Thus, it is clear that the realized floating resistance value can be made either positive or negative by arranging $V_{C2} > V_{C1}$ or $V_{C1} > V_{C2}$, respectively. Furthermore, from the circuit configuration it is apparent that the number of MOSFETs directly coming between the two power supply rails is only two, which facilitates the operation of the circuit with low supply voltages.

Yet another circuit capable of realizing either a positive or negative voltage-controlled floating resistance was proposed by Tantry *et al.* [66], which is shown here in Figure 9.32. A straightforward analysis of this circuit reveals that the currents I_1 and I_2 are given by

$$I_1 = I_2 = 2K(V_{C2} - V_{C1})(V_1 - V_2) \tag{9.79}$$

Hence, the realized resistance is independent of the threshold voltage of the matched MOSFETs.

With the circuits biased from ±5 V dc power supply and control voltage V_{C2} varied from 4.5 V to 4.9 V with V_{C1} fixed at 4.7 V, the equivalent positive resistance was found to vary from 1.17 MΩ to ∞, while it varied from −1.21 MΩ to −∞ in the case of negative resistance. In both the cases, the bandwidth was found to be around 100 kHz.

Al-Sarawi [67] presented a topology to convert grounded resistors to equivalent floating resistors such that the value of the floating resistor was equal to sum of

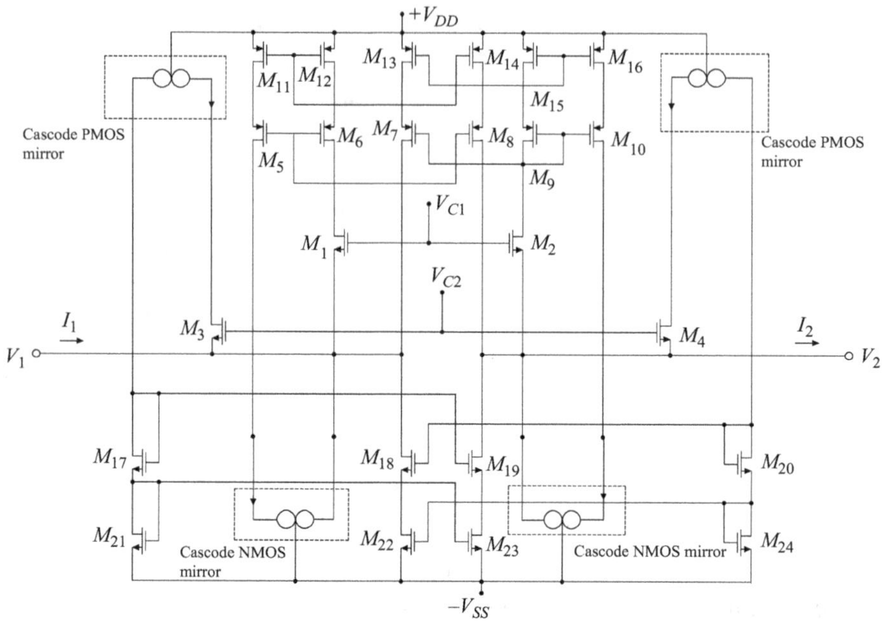

Figure 9.32 Simplified schematic of the bilateral floating resistor due to Oura et al. [66]

the two grounded resistors. Based upon this idea, Al-Sarawi [67] presented a configuration for realizing high-value floating resistor in the GΩ range based upon the utilization of the high-output resistance of the two matched MOSFETs operating in sub-threshold region.

Tekin *et al.* [99] presented a floating active resistor based upon two CMOS differential pairs which could achieve either a positive or negative resistor with a simple modification in the circuit which is shown in Figure 9.33. However, to get the required floating resistor characterized by

$$I_1 = -I_2 = \frac{V_1 - V_2}{R_{eq}}, \quad \text{where } R_{eq} = \frac{1}{\sqrt{2I_0 K_n (W/L)}} \tag{9.80}$$

subject to the fulfilment of the following constraint:

$$4I_0 \gg \left(\frac{1}{2}\right) K_n \left(\frac{W}{L}\right) (V_1 - V_2)^2$$

Ercan [101] introduced a linear current-controlled negative floating resistor based upon a self-cascode composite transistor which is shown in Figure 9.34. Like the earlier circuit, this circuit too realizes a linear floating negative resistance

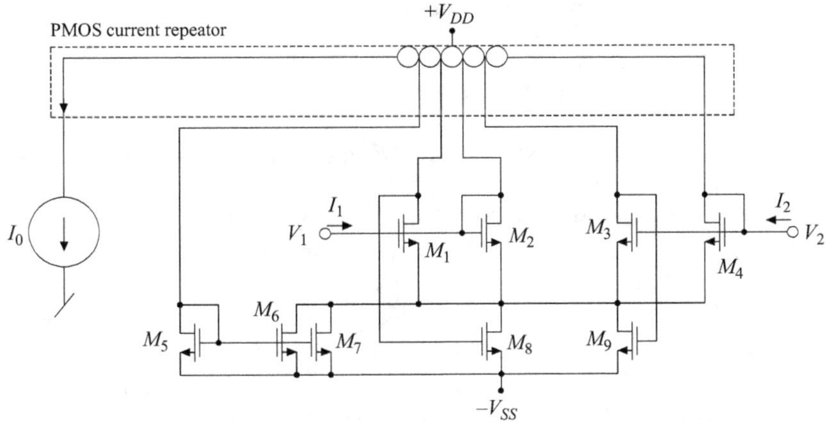

Figure 9.33 Simplified representation of the floating resistor proposed by Tekin et al. [99]

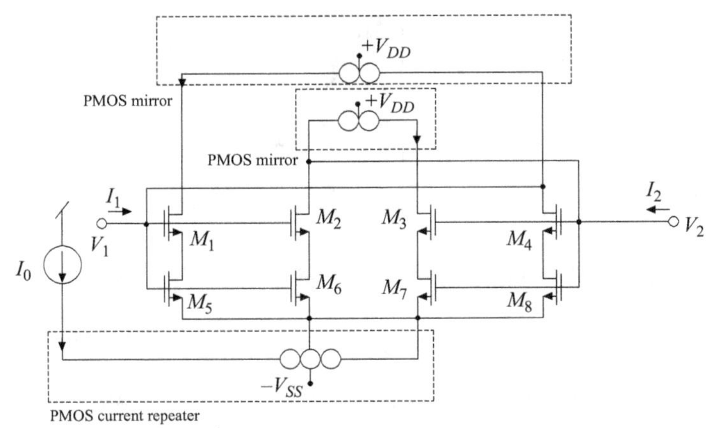

Figure 9.34 Simplified representation of the current-controlled floating negative resistor due to Ercan [101]

subject to a constraint. An analysis of this circuit, as shown in [101], reveals that $I_1 = -I_2$ and that I_1 is given by

$$I_1 \cong -(V_1 - V_2)\left[K_n\left(\frac{a}{a+1}\right)\left(\frac{W}{L}\right)I_0\right]^{1/2} \tag{9.81}$$

provided that $2I_0 \gg (1/2)K_n(a/a+1)(W/L)(V_1 - V_2)^2$.

Thus, the circuit realizes as a negative resistance given by

$$R_{12} \cong -\left[K_n\left(\frac{a}{a+1}\right)\left(\frac{W}{L}\right)I_0\right]^{-1/2} \tag{9.82}$$

where (W/L) ratios of MOSFETs M_1–M_4 are a times the (W/L) ratios of MOSFETs M_5–M_8.

In [104] Yuce *et al.* presented two topologies for realizing grounded voltage-controlled resistor based upon MOSFETs operating in saturation, which are shown in Figures 9.35 and 9.36, respectively. The input current for the first proposed VCR is composed of $I_{in} = -I_{D1} - I_{D2} + I_{D3} + I_{D4}$ which after substituting the value of all the drain currents leads to the equivalent resistance given by

$$R_{eq} = \frac{V_{in}}{I_{in}} = \frac{1}{2K\left(V_{DD} - |V_{tp}| - V_{tn} - V_C\right)}, \quad \text{where } K = \frac{\mu_s C_{ox}(W/L)}{2} \tag{9.83}$$

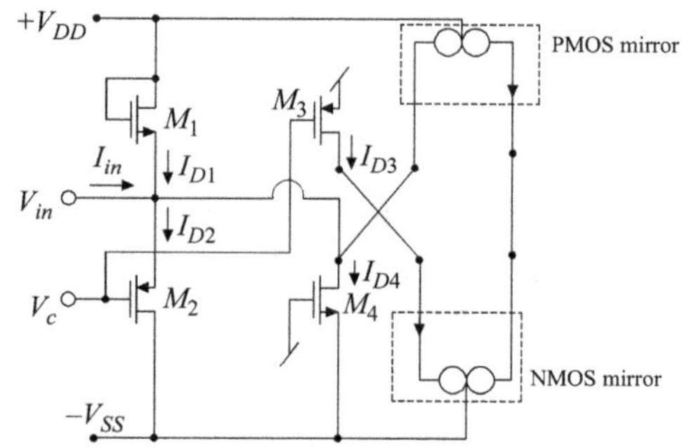

Figure 9.35 Simplified representation of the grounded VCR due to Yuce et al. [104]

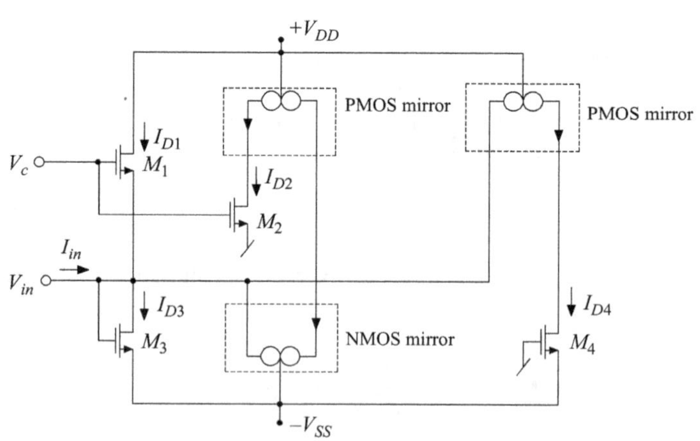

Figure 9.36 Simplified representation of the grounded VCR due to Yuce et al. [104]

however; the controlled voltage of the proposed MOS resistor is required to satisfy the following constraint $V_{SS} - |V_{tp}| \leq V_C \leq -|V_{tp}|$, whereas the input voltage of the circuit is required to satisfy $\text{Max}\{V_C + |V_{tp}|, -V_{tn}\} \leq V_{in} \leq (V_{DD} - V_{tn})$.

In the second configuration, shown here in Figure 9.36, the input current is composed of $I_{in} = -I_{D1} + I_{D2} + I_{D3} - I_{D4}$ and subject to the constraints $(V_{SS} + V_{tn}) \leq V_{in} \leq V_C - V_{tn}$ and $V_{tn} \leq V_C \leq (V_{DD} + V_{tn})$, the equivalent resistance is given by

$$R_{eq} = \frac{V_{in}}{I_{in}} = \frac{1}{2K(V_C - 2V_{tn} - V_{SS})} \tag{9.84}$$

Kushima *et al.* [74] proposed a floating VCR which was based upon the idea that if the gate voltage of an NMOS transistor is organized to be $V_G = V_{bias} + ((V_1 + V_2)/2)$, then the square nonlinearity of the I_D–V_{DS} characteristic equation of the MOSFET is cancelled out, the MOSFET then realizes a floating linear VCR of value $R_{eq} = 1/(2K(V_{bias} - V_{th}))$. Thus, in their circuit shown here in Figure 9.37, one can note that since $I_{D1} = I_{D2}$, we can write [74]

$$K_1(V_1 - V_a - V_{thn})^2 = K_2(V_a - V_{thn})^2 \tag{9.85}$$

Assuming $K_1 = K_2$, it turns out, therefore, that $V_a = (V_1/2)$. From the other side, we similarly note that since $I_{D5} = I_{D6}$, we can write $K_5(V_2 - V_b - V_{thn})^2 = K_6(V_b - V_{thn})^2$. Again, assuming $K_5 = K_6$, it follows that $V_b = (V_2/2)$. Now I_{D7} and I_{D8} are also equal, which leads to $K_7(V_C - V_{DD} - V_{thp})^2 = K_8(V_a - V_G - V_{thp})^2$. With $K_7 = K_8$, this leads to $V_G = (V_a - V_C + V_{DD})$. Lastly, since $I_{D3} = I_{D4}$ gives $K_3(V_{DD} - V_{bias} - V_C - V_{thn})^2 = K_4(V_b - V_{thn})^2$. With $K_3 = K_4$, we get $V_C = (V_{DD} - V_{bias} - V_b)$. From the various expressions given above, it is easily deduced that $V_G = V_{bias} + ((V_1 + V_2)/2)$. Finally, we can write [74]

$$I_1 = -I_2 = 2K_0\left[V_G - V_1 - V_{thp} - \left(\frac{V_1 - V_2}{2}\right)\right](V_1 - V_2) \tag{9.86}$$

Substituting the value of V_G in the above leads to

$$R_{eq} = \frac{1}{2K_0(V_{bias} - V_{thp})} \tag{9.87}$$

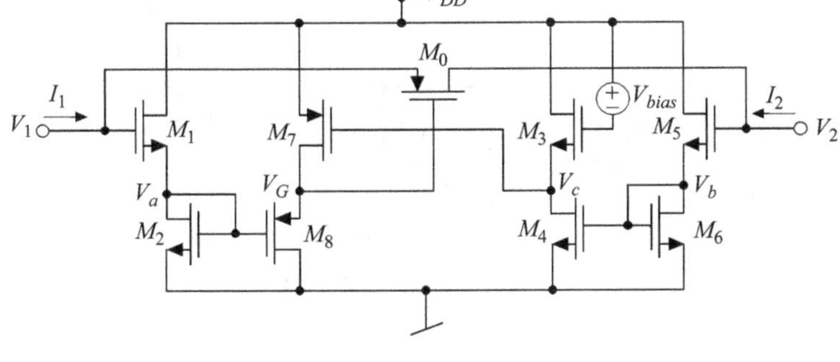

Figure 9.37 Floating voltage-controlled linear resistor [74]

SPICE simulations have shown that by varying V_{bias} from 0 to 1.5 V, a linear floating VCR of value changing from 58 kΩ to 1,600 kΩ was obtainable.

Application: The authors of [74] have demonstrated that an electronically tunable *RC* low-pass filter realized using this floating VCR in conjunction with a grounded capacitor made it possible to realize a low-pass filter with electronically tunable cut-off frequency.

Among the various floating linear VCRs proposed so far, possibly one of the simplest circuit is that proposed by Tajalli *et al.* [80], the other one being the two-MOSFET circuit of [14] (described as the circuit of Figure 9.19(c)). The circuit from [80] employs only three MOSFETs as against dozens of MOSFETs in several circuits described in this chapter earlier. This compact high-valued resistor using PMOS devices is claimed to be highly suitable for either biasing purposes or implementing very low frequency filters.

Yuce *et al.* [103] presented an interesting five-MOSFET-based linear VCR which is shown in Figure 9.38. In this circuit, taking $|V_{DD}| = |V_{tp}|$, $|V_{SS}| = |V_{tn}|$ and assuming all MOSFETs to be matched and considering that M_2 is operated in triode region whereas the remaining ones are in saturation, it can easily be seen that for $V_{in} > 0$, MOSFETs M_2 and M_3 are ON, whereas others are OFF, which leads to $I_{D3} = (\beta/2)V_{in}^2$, but $I_{D4} = 0$. Since $I_{in} = I_{D2} + I_{D3} + I_{D4}$, it follows that

$$I_{in} = \beta\left[(V_C - V_{th})V_{in} - \frac{V_{in}^2}{2}\right] + \frac{\beta}{2}V_{in}^2 = \beta(V_C - V_{th})V_{in}; \quad \beta = \frac{\mu_s C_{ox}(W/L)}{2} \tag{9.88}$$

On the other hand, when $V_{in} < 0$, M_3 is OFF while MOSFETs M_1, M_2, M_4 and M_5 are ON and that $I_{D3} = 0$, but $I_{D4} = (\beta/2)V_{in}^2$. Hence, I_{in} is now given by

$$I_{in} = \beta\left[(V_C - V_{th})V_{in} - \frac{V_{in}^2}{2}\right] + \frac{\beta}{2}V_{in}^2 = \beta(V_C - V_{th})V_{in} \tag{9.89}$$

Thus, it follows that combining both the cases,

$$R_{eq} = \frac{V_{in}}{I_{in}} = \frac{1}{\beta(V_C - V_{th})} \tag{9.90}$$

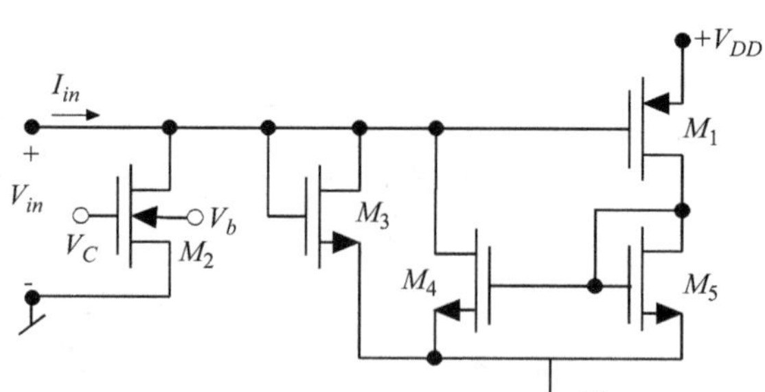

Figure 9.38 CMOS-grounded VCR due to Yuce et al. [103]

However, for proper operation of the circuit, it is imperative that $V_C > V_{thn}$ and $(V_{GS2} - V_{thn}) > |V_{in}|$.

Applications: A significant advantageous feature of this circuit has been claimed to be its ultra-low power consumption. SPICE simulations using IBM 0.13 μm SIGE013 Level-7 CMOS process parameters have been demonstrated to show the utility in the realization of electronically tunable first-order active low-pass filter and a high-Q and high-gain voltage-mode multiple feedback second-order BPF – both realized using ADA 4899-1 op-amp, wherein the proposed circuit of the VCR was used as a replacement of the grounded resistors in both the application examples.

In [107], Yuce *et al.* presented a three-MOSFET circuit for realizing a grounded linear VCR which is shown in Figure 9.39. Assuming that the MOSFET M_2 is operated in triode region and the remaining two in saturation, the input current of the circuit is given by

$$I_{in} = I_{d2} + I_{d3} - I_{d1} \tag{9.91}$$

Assuming all MOSFETs to have the same transconductance parameter K, it follows that

$$\begin{aligned} I_{in} = {} & 2K\left[(V_C - V_{thn})V_{in} - \frac{V_{in}^2}{2}\right] + K(V_{in} - V_{thn} + V_{SS})^2 \\ & -K\left(V_{DD} - |V_{thp}|^2\right) \end{aligned} \tag{9.92}$$

Now, taking $|V_{DD}| = |V_{SS}|$ and $K(V_{DD} - |V_{thp}|)^2 = K(V_{SS} - |V_{thn}|)^2$, it turns out that all square nonlinearities and unwanted terms are cancelled and what is left is given by

$$I_{in} = 2K(V_C - 2V_{thn} + V_{SS})V_{in} \tag{9.93}$$

which represents a linear VCR of value

$$R_{eq} = \frac{V_{in}}{I_{in}} = \frac{1}{2K(V_C + V_{SS} - 2V_{thn})}. \tag{9.94}$$

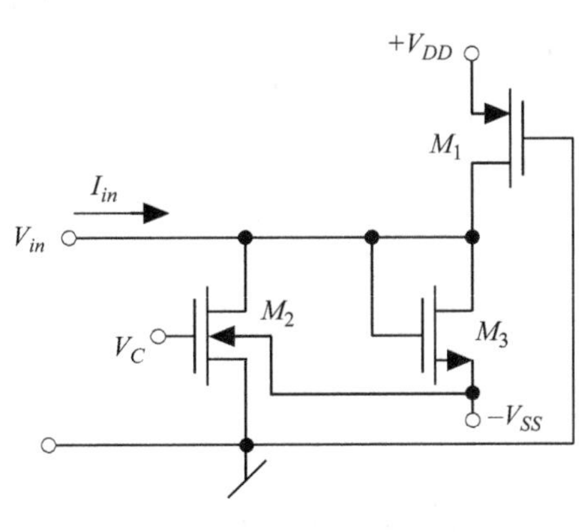

Figure 9.39 Low-component-count grounded VCR [107]

Application: The authors of [107] had devised an interesting application of this circuit in realizing an electronically tunable filter using two DVCCs, two grounded capacitors and three grounded resistors. For electronic tunability, all the three grounded resistors had been replaced by grounded VCRs of [107].

In retrospect, it is interesting to note that, although not explicitly acknowledged in [107], the operating principle as well as the circuit proposed by Yuce *et al.* [107] is quite similar to that of Wang [26]. Note that in both the circuits, the main VCR-creating NMOS transistor operating in triode region and one more NMOS transistor is diode-connected, which is connected between input voltage and negative power supply, thereby creating the sum of the currents as $2K[(V_C - V_{thn})V_{in} - (V_{in}^2/2)] + K(V_{in} - V_{thn} + V_{SS})^2$ which give a term $2K(V_C - 2V_{thn} + V_{SS})V_{in}$ which is proportional to V_{in}, while there is unwanted offset current term equal to $K(-V_{thn} + V_{SS})^2$. In the Wang's circuit [26], this offset current is nullified by the drain current of a third NMOS transistor which has gate connected to ground and source connected to $-V_{SS}$ so that with this transistor operating in saturation region, the drain current of this is given by $K(-V_{thn} + V_{SS})^2$ which is applied to input node such that the input current I_{in} is obtained by subtracting this current from the sum of two drain currents so that the offset current term is exactly cancelled by this third current. By contrast, Yuce *et al.* [107] use PMOS transistor instead whose source is connected to positive supply $+V_{DD}$, with its gate grounded and drain connected to the input junction so that the drain current of this transistor is exactly $K(-V_{thp} + V_{DD})^2$ so that if $|V_{DD}| = |V_{SS}|$, $|V_{thn}| = |V_{thp}|$, this third current cancels the offset current exactly, and the circuit thus realizes a grounded linear VCR.

More recently, a new electronically fine-tunable grounded VCR was introduced by Yucel and Yuce [113], which is shown in Figure 9.40, which consists of six MOSFETs, all operating in saturation.

By a straightforward analysis, it may be easily verified that if $|V_{DD}| = |V_{SS}|$ and all MOSFETs are matched having the same β, it turns out that $I_{in} = [\beta(V_{DD} - V_C - 2V_{th})]V_{in}$ so that circuit realizes a linear grounded VCR with equivalent resistance given by

$$R_{eq} = \frac{1}{\beta(V_{DD} - V_C - 2V_{th})}, \quad \text{where} \quad \beta = \mu_s C_{0x}\left(\frac{W}{L}\right) \tag{9.95}$$

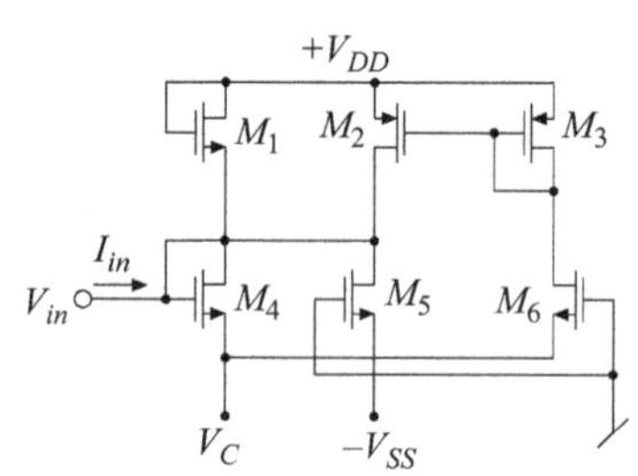

Figure 9.40 Grounded voltage-controlled positive resistor proposed by Yucel and Yuce [113]

For ensuring the MOSFETs to be operating in saturation, the following constraints need to be fulfilled:

$$\left.\begin{array}{l}(V_{DD} - V_{th}) > V_{in} > (V_C + V_{th}) \\ \text{and} \quad V_C < -V_{th}\end{array}\right\} \forall V_{in} \tag{9.96}$$

Furthermore, the parasitic capacitance appearing in parallel with the equivalent R_{eq} is given by

$$C_p \cong C_{gs1} + C_{ds1} + C_{gs4} + C_{ds4} + C_{ds2} + C_{ds5} \tag{9.97}$$

which restricts the operating frequency range of this VCR to be $f \ll f_0$, where $f_0 = \omega_0/2\pi$ and $\omega_0 = 1/R_{eq}C_p$. The recommended operational frequency range of this circuit is $f \leq 0.1f_0$.

SPICE simulations have revealed that the tunable resistance range of this circuit is from about 770 Ω to about 1,370 Ω and the circuit can be satisfactorily operated to a few hundreds of MHz using discrete transistors 2N7000 (NMOS) and BS250 (PMOS).

In the last, it must be mentioned that recently, in view of some interest on FGMOS-based analog circuits, a few attempts have also been made in realizing FGMOS-based VCR circuits but by and large, the topologies evolved and the underlying principles are more or less similar to those of conventional MOSFET-based VCRs. For some recent works on FGMOS-based VCRs, the readers are referred to [87,89,90,98,114].

9.4 Concluding remarks

In this chapter, we have discussed the realization of voltage/current-controlled linear resistance and their applications. The circuit realizations using BJTs were discussed first all of which were based upon the employment of the so-called MTC. The realizations of both grounded and floating current-controlled linear resistors were considered. Although a majority of the circuits included realizing current-controlled positive resistors, some circuits capable of realizing grounded/floating current-controlled negative resistances were also discussed. An attractive feature of these circuits is that all of them use only BJTs and the realized resistance is electrically controllable; hence, they are eminently suitable for IC implementation using bipolar technology. A major drawback of these circuits is the temperature sensitivities of the realized resistances. A method to circumvent this using additional circuitry was also discussed. Another limitation of these BJT-based circuits is that all of them have restriction on the applicable input voltage range which is required to be much smaller than $2V_T$. As far as is known, any remedy to this restriction has so far not been discovered in the open literature yet. In view of this, extending the linear range of operation of BJT-based linearized current-controlled resistor circuits or finding any new technique/circuits with extended range of operation is a challenging problem which is open to investigation.

In the area of linear CMOS resistor realization, pioneering contributions were made by the research group of Tsividis [5,6,12,14,35,40,42,44] and Czarnul [16,18,34,43]. This chapter has given a state-of-the-art survey of prominent works done in this area till date. The central theme of most of the circuits is to employ one of the eight standard methods of the nonlinearity cancellation which represent either a one- (grounded) or a two-port (floating) in which some technique of square nonlinearity cancellation has been put to use. Except the technique of Takagi *et al.* [43] which, apart from matched MOSFETs operating in triode region, also uses NICs (which can be either realized by a CMOS CCII+ or otherwise), all other circuits included here employ exclusively MOSFETs only. Consequently, a number of prominent circuit realizations of linear grounded/floating VCRs have been highlighted. Although a majority of the circuits discussed are based upon operating the MOSFET(s) in triode region, some circuits which realize linear VCR using all MOSFETs operating in saturation have also been described. Furthermore, some circuits for realizing negative linear VCRs have also been discussed.

It is believed that this compendium of linear grounded/floating, positive/negative, bipolar/CMOS VCR circuits would be useful to the designers for incorporating in various analog signal processing/signal generation circuits to make various parameters therein electronically controllable. It is also hoped that this compilation may also serve as a useful reference to researchers interested in working in this area who may find a number of problems to work upon from the various discussions of this chapter.

References

[1] Reddy M. A., Rao Y. G. 'Temperature stabilization of a negative-resistance characteristic'. IEEE Journal of Solid-State Circuits. 1970; **5**(3):124–5.

[2] Pookaiyaudom S., Kuhanont T. 'Integrable accurate linear negative impedance conversion circuits'. Proceedings of the IEEE. 1978; **66**(1):88–9.

[3] Pookaiyaudom S., Surakampontorn W., Kuhanont T. 'Integrable electronically variable general-resistance converter – a versatile active circuit element'. IEEE Transactions on Circuits and Systems. 1978; **25**(6):344–53.

[4] Regtien P. P. L. 'Floating current-controlled resistance'. International Journal of Electronics. 1979; **46**(3):273–80.

[5] Banu M., Tsividis Y. 'Floating voltage-controlled resistors in CMOS technology'. Electronics Letters. 1982; **18**(15):678–9.

[6] Banu M., Tsividis Y. 'Fully integrated active RC filters in MOS technology'. IEEE Journal of Solid-State Circuits. 1983; **SC-18**(6):644–51.

[7] Han I. S., Park B. 'Voltage-controlled linear resistor by two MOS transistors and its application to active RC filter MOS integration'. Proceedings of the IEEE. 1984;**72**(11):1655–7.

[8] Babanezhad J. N., Temes G. C. 'A linear NMOS depletion resistor and its application in an integrated amplifier'. IEEE Journal of Solid-State Circuits. 1984; **19**(6):932–8.

[9] Akiya M., Nakashima S., Kato K. 'A wide range linear variable resistor by buried channel MOS/SIMOX'. IEEE Journal of Solid State Circuits. 1984; **SC-19**(4):526–31.

[10] Surakampontorn W. 'Integrable wide-dynamic-range negative resistance circuits'. Electronics Letters. 1985; **21**(11):506–8.

[11] Ismail M. 'A new MOSFET-capacitor integrator'. IEEE Transactions on Circuits and Systems. 1985; **32**(11):1194–6.

[12] Tsividis Y., Czarnul Z., Fang S. C. 'MOS transconductors and integrators with high linearity'. Electronics Letters. 1986; **22**(5):245–6.

[13] Nagaraj K. 'New CMOS floating voltage-controlled resistor'. Electronics Letters. 1986; **22**(12):667–8.

[14] Tsividis Y., Banu M., Khoury J. 'Continuous-time MOSFET-C filters in VLSI'. IEEE Transactions on Circuits and Systems. 1986; **33**(2):125–40.

[15] Surakampontorn W., Junnapiya S. 'Electronically variable general-resistance circuit'. International Journal of Electronics. 1986; **60**(2):281–5.

[16] Czarnul Z. 'Novel MOS resistive circuit for synthesis of fully integrated continuous-time filters'. IEEE Transactions on Circuits and Systems. 1986; **33**(7):718–21.

[17] Park C. S., Schaumann R. 'A high-frequency CMOS linear transconductance element'. IEEE Transactions on Circuits and Systems. 1986; **33**(11):1132–8.

[18] Czarnul Z. 'Comments on "A linear NMOS depletion resistor and its application in an integrated amplifier"'. IEEE Journal of Solid-State Circuits. 1987; **22**(1):124–7.

[19] Ismail M. 'Four-transistor continuous-time MOS transconductor'. Electronics Letters. 1987; **23**(20):1099–100.

[20] Vanpeteghem P. M., Rice G. L. 'New CMOS resistor implementation for linear IC applications'. Electronics Letters. 1988: **24**(5):288–9.

[21] Brennan R. L., Viswanathan T. R., Hanson J. V. 'The CMOS negative impedance converter'. IEEE Journal of Solid State Circuits. 1988; **23**(5):1272–5.

[22] Singh S. P., Hanson J. V., Vlach J. 'A new floating resistor for CMOS technology'. IEEE Transaction on Circuits and Systems. 1989; **36**(9): 1217–20.

[23] Wilson G., Chan P. K. 'Novel voltage-controlled grounded resistor'. Electronics Letters. 1989; **25**(25):1725–6.

[24] Wang Z. '2-MOSFET transistor with extremely low distortion for output reaching supply voltages'. Electronics Letters. 1990; **26**(13):951–2.

[25] Silva-Martinez J., Steyaert M., Sansen W. 'Very linear CMOS floating resistor'. Electronics Letters. 1990; **26**(19):1610–11.

[26] Wang Z. 'Novel voltage-controlled grounded resistor'. Electronics Letters. 1990; **26**(20):1711–12.

[27] Moon G., Zaghloul M. E., Newcomb R. W. 'An enhancement-mode MOS voltage-controlled linear resistor with wide dynamic range'. IEEE Transactions on Circuits and Systems. 1990; **37**(10):1284–8.

[28] Wang Z. 'Novel Electronically-controlled floating resistors using MOS transistors operating in saturation'. Electronics Letters. 1991; **27**(2):188–9.

[29] Tabei K., Czarnul Z., Takagi S. 'Realisation of highly linear MOS circuits using negative impedance convertors'. Electronics Letters. 1991; **27**(16): 1416–17.

[30] Cheng M. C. H., Toumazou C. 'Linear composite MOSFETs (COMFETs)'. Electronics Letters. 1991; **27**(20):1802–4.
[31] Wilson G., Chan P. K. 'CMOS series/parallel quad resistor'. Electronics Letters. 1992; **28**(3):335–6.
[32] Surakampontorn W. 'CMOS floating voltage-controlled negative resistor'. Electronics Letters. 1992; **28**(15):1457–9.
[33] Wang Z. 'CMOS positive and negative grounded resistors using MOS transistors operating in saturation'. Microelectronics Journal. 1992;**23**(5): 355–7.
[34] Czarnul Z. 'Comments on "Linearization techniques for nth-order sensor models in MOS VLSI technology'. IEEE Transactions on Circuits and Systems – II. 1992; **39**(9):677.
[35] Tsividis Y. P., Vavelidis K. 'Linear, Electronically tunable resistor'. Electronics Letters. 1993; **28**(25):2303–5.
[36] Sakurai S., Ismail M. 'A CMOS square-law programmable floating resistor independent of the threshold voltage'. IEEE Transactions on Circuits and Systems – II. 1992; **39**(8):565–74.
[37] Cheng M. C. H., Toumazou C. 'Linear composite MOSFET: theory, design, and applications'. IEEE Transactions on Circuits and Systems – I. 1993; **40**(5): 297–306.
[38] Wilson G., Chan P. K. 'Floating CMOS resistor'. Electronics Letters. 1993; **29**(3):306–7.
[39] Huang S. C., Ismail M. 'Linear tunable COMFET transconductor'. Electronics Letters. 1993; **29**(5):459–61.
[40] Tsividis Y., Vavelidis K. 'Comment: linear, electronically tunable resistor'. Electronics Letters. 1993; **29**(6):556–7.
[41] Ismail M. 'Reply to "Comments on 'Linearization techniques for nth-order sensor models in MOS VLSI technology'"'. IEEE Transactions on Circuits and Systems – II. 1993; **40**(2):121–2.
[42] Tsividis Y. P. 'Comments on "A nonlinear CMOS analog cell for VLSI signal and information processing"'. IEEE Journal of Solid State Circuits. 1993; **28**(12):1395.
[43] Takagi S., Czarnul Z., Fujii N. 'A synthesis of highly linear MOS circuits and their application to filter realization'. IEICE Transactions on Fundamentals. 1994; **E77-A(2)**:351–5.
[44] Tsividis Y. P., Khoury J. M. 'Reply to recent Comments by M. Ismail'. IEEE Transactions on Circuits and Systems – II. 1994; **41**(3):238–40.
[45] Wilson G., Chan P. K. 'Analysis of nonlinearities in MOS floating resistor networks'. IEE Proceedings Circuits, Devices and Systems. 1994; **141**(2): 82–8.
[46] Al-Ruwaih K. M., Noras J. M. 'A novel linear resistor utilizing MOS transistors with identical sizes and one controlling voltage'. International Journal of Electronics. 1994; **76**(6):1083–98.
[47] Senani R. 'Realisation of linear voltage-controlled resistance in floating form'. Electronics Letters. 1994; **30**(23):1909–11.

[48] Fabre A., Saaid O., Wiest F., Boucheron C. 'Current controlled bandpass filter based on translinear conveyors'. Electronics Letters. 1995; **31**(20): 1727–8.

[49] Saaid O., Fabre A. 'Class AB current-controlled resistor for high performance current-mode applications'. Electronics Letters. 1996; **32**(1):4–5.

[50] Fabre A., Saaid O., Wiest F., Boucheron C. 'High frequency applications based on a new current controlled conveyor'. IEEE Transactions on Circuits and Systems – I. 1996; **43**(2):82–91.

[51] Elwan H. O., Mahmoud S. A., Soliman A. M. 'CMOS voltage controlled floating resistor'. International Journal of Electronics. 1996; **81**(5):571–6.

[52] Urbas A., Galinski B. 'Unbalanced Czarnul resistive MOS circuit in the symmetrical MOSFET-C filters'. Radioengineering. 1996; **5**(3):23–6.

[53] Soliman A. M. 'A linear transconductor/multiplier using a matched pair of MOS transistors and a current conveyor'. Frequenz. 1996; **50**(11–12):292–3.

[54] Surakampontorn W., Kumwachara K., Riewruja V., Surawatpunya C. 'CMOS-based Integrable electronically tunable floating general impedance inverter'. International Journal of Electronics. 1997; **82**(1):33–44.

[55] Teramura M., Miyazaki T., Horie Y., Takeishi T. 'A voltage-controlled resistor for a linear voltage-controlled oscillator'. Electronics and Communication in Japan, Part 2. 1996; **79**(12):24–30.

[56] Al-Ruwaihi K. M. 'A floating voltage-controlled linear resistor and its application to active RC filters'. International Journal of Electronics. 1997; **82**(5):483–98.

[57] Mahmoud S. A., Elwan H. O., Soliman A. M. 'Generation of CMOS voltage-controlled floating resistors'. Microelectronics Journal. 1997; **28**(6–7):627–40.

[58] Abuelma'atti M. T. 'An improved analysis for the harmonic distortion of MOS voltage-controlled-resistors'. Active and Passive Electronic Components. 1997; **19**:253–60.

[59] Kimura K. 'A linear transconductance amplifier obtained by realizing a floating resistor'. IEEE Transactions on Circuits and Systems – I. 1998; **45**(1):108–13.

[60] Tadic N. 'Resistive mirror-based voltage controlled resistor with generalized active devices'. IEEE Transactions on Instrumentation and Measurement. 1998; **47**(2):587–91.

[61] Kiranon W., Loescharataramdee C., Kiatwarin N., Wardkein P. 'Electronically controlled negative resistance based on translinear circuits'. IEEE International Symposium in Intelligent Signal Processing and Communication Systems; Phuket, Thailand,1999; 645–648.

[62] Barthelemy H., Fabre A. 'A new floating controlled resistance operating in class AB'. IEEE Transactions on Circuits and Systems – I. 2000; **47**(1): 67–72.

[63] Tadic N., Gobovic D. 'A voltage-controlled resistor in CMOS technology using bisection of the voltage range'. IEEE Transactions on Measurement and Instrumentation. 2001; **50**(6):1704–10.

[64] Schlarmann M. E., Geiger R. L. 'Simple CMOS transresistor'. Electronics Letters. 2001; **37**(23):1386–7.

[65] Alarcon E., Martinez H., Vidal E., Poveda A. 'Digitally programmable MOS resistive circuit'. Electronics Letters. 2002; **38**(1):11–13.

[66] Oura T., Yoneyama T., Tantry S., Asai H. 'A CMOS floating resistor circuit having both positive and negative resistance values'. IEICE Transactions on Fundamentals. 2002; E85-A(**2**):399–402.

[67] Al-Sarawi S. F. 'A novel topology for grounded-to-floating resistor conversion in CMOS technology'. Microelectronics Journal. 2002; **33**(12):1059–69.

[68] Tantry S., Hiraku Y., Oura T., Yoneyama T., Asai H. 'A low voltage floating resistor circuit having both positive and negative resistance values'. IEICE Transactions on Fundamentals. 2003; **E86-A**(2):335–41.

[69] Lande T. S., Olsen E. S., Toumazou C. 'Resistive equivalents in CMOS'. Electronics Letters. 2003; **39**(17):1233–5.

[70] Senani R., Singh A. K., Singh V. K. 'A new floating current-controlled positive resistance using mixed translinear cells'. IEEE Transactions on Circuits and Systems – II. 2004; **51**(7):374–7.

[71] Pawarangkoon P., Kiranon R. 'Electronically tunable floating resistor'. International Journal of Electronics. 2004; **91**(11):665–73.

[72] Tavakoli M., Sarpeshkar R. 'A sinh resistor and its application to tanh linearization'. IEEE Journal of Solid State Circuits. 2005; **40**(2):536–43.

[73] Hoy A. S., Lin J. L. 'Voltage-controlled isolated linear resistors with wide operation range'. International Journal of Electronics. 2005; **92**(12):697–707.

[74] Kushima M., Inaba M., Tanno K. 'Linear and compact float node voltage-controlled variable resistor circuit'. IEICE Transactions on Fundamentals of Electronics, Communications and Computer Sciences'. 2006; **E89-A**(2): 459–60.

[75] Zouaoui-Abouda H., Fabre A. 'New high-value floating controlled resistor in CMOS technology'. IEEE Transactions on Measurement and Instrumentation. 2006; **55**(3):1017–20.

[76] Han, Il-Song. 'A novel tunable transconductance amplifier based on voltage-controlled resistance by MOS transistors'. IEEE Transactions on Circuits and Systems – II. 2006; **53**(8):662–6.

[77] Worapishet A., Khumsat P. 'Sub-threshold R-MOSFET tunable resistor technique'. Electronics Letters. 2007; **43**(7):390–2.

[78] Al-Shahrani S. M. 'CMOS wideband auto-tuning phase shifter circuit'. Electronics Letters. 2007; **43**(15):804–5.

[79] Jendernalik W. 'A low-voltage CMOS negative impedance converter for analogue filtering applications'. Bulletin of the Polish Academy of Sciences Technical Sciences. 2007; **55**(4):419–23.

[80] Tajalli A., Leblebici Y., Brauer E. J. 'Implementing ultra-high-value floating tunable CMOS resistors. Electronics Letters. 2008; **44**(5):349–50.

[81] Hoong K., Sarpeshkar R. 'An electronically tunable linear or nonlinear MOS resistor'. IEEE Transactions on Circuits and Systems – I. 2008; **55**(9):2573–83.

[82] Torralba A., Lujan-Martinez C., Carvajal R. M., *et al.* 'Tunable linear MOS resistors using quasi-floating-gate techniques'. IEEE Transactions on Circuits and Systems – II. 2009; **56**(1):41–5.

[83] Popa C. 'Negative resistance active resistor with improved linearity and frequency response'. Journal of Circuits, Systems, and Computers. 2009; **18**(1):1–10.

[84] Clark K. D., Comer D. J. 'Tuned negative capacitance circuitry for CMOS amplifier bandwidth extension'. International Journal of Electronics. 2009; **96**(1):1–9.

[85] Bozomitu R. G., Cehan V., Popa V. 'A new linearization technique using "multi-sinh" doublet'. Advances in Electrical and Computer Engineering. 2009; **9**(2):45–57.

[86] Yazgi M., Toker A., Virdee B. S. 'A new negative resistance circuit and an application for loss compensation in a distributed amplifier'. Analog Integrated Circuits and Signal Processing. 2009; **60**(3):215–20.

[87] Gupta M., Pandey R. 'FGMOS based voltage-controlled resistor and its applications'. Microelectronics Journal. 2010; **41**(1):25–32.

[88] Manolescu A., Popa C. 'Low-voltage low-power improved linearity CMOS active resistor circuits'. Analog Integrated Circuits and Signal Processing. 2010; **62**(3):373–87.

[89] Pandey R., Gupta M. 'FGMOS based tunable grounded resistor'. Analog Integrated Circuits and Signal Processing. 2010; **65**(3):437–43.

[90] Pandey R., Gupta M. 'FGMOS based voltage-controlled grounded resistor'. Radioengineering. 2010; **19**(3):455–9.

[91] Yuce E., Minaei S., Alpaslan H. 'Novel CMOS technology-based linear grounded voltage controlled resistor'. Journal of Circuits, Systems, and Computers. 2011; **20**(3):447–55.

[92] Kumngern M., Torteanchai U., Dejhan K. 'Voltage-controlled floating resistor using DDCC'. Radioengineering. 2011; **20**(1):327–33.

[93] Yuce E. 'Multiplier, frequency doubler and squarer circuits based on voltage controlled resistors'. International Journal of Electronics and Communications (AEU). 2011; **65**(3):244–9.

[94] Shiue M. T., Yao K. W., Gong C. S. A. 'Tunable high resistance voltage-controlled pseudo-resistor with wide input voltage swing capability'. Electronics Letters. 2011; **47**(6):377–8.

[95] Merz N., Kiranon W., Wongtachathum C., Pawarangkoon P., Narksarp W. 'A modified bipolar translinear cell with improved linear range and its applications'. Radioengineering. 2012; **21**(2):736–45.

[96] Fard R. A., Pooyan M. 'A low voltage and low power parallel electronically tunable resistor with linear and nonlinear characteristics'. Microelectronics Journal. 2012; **43**(7):492–500.

[97] Arslanalp R., Yuce E., Tola A. T. 'Low-component BJT technology-based current-controlled tunable resistors and their applications'. IET Circuits, Devices and Systems. 2013; **7**(1):21–30.

[98] Gupta R., Sharma S. 'Voltage controlled resistor using quasi-floating-gate MOSFETS'. Maejo International Journal of Science and Technology. 2013; **7**(1):16–25.

[99] Tekin S. A., Ercan H., Alci M. 'Novel low voltage CMOS current controlled floating resistor using differential pair'. Radioengineering. 2013; **22**(2):428–33.

[100] Tekin S. A., Alci M. 'Design and applications of electronically tunable floating resistor using differential pair'. Elektronika IR Elektrotechnika. 2013; **19**(4):41–6.

[101] Ercan H. 'A linear current-controlled floating negative resistor'. Journal of Microelectronics, Electronic Components and Materials. 2013; **43**(4):222–7.

[102] Sitaram R. S., Townsend W. G. 'A voltage-controlled variable-resistance MOSFET. International Journal of Electronics. 2014; **38**(2):253–7.

[103] Yuce E., Minaei S., Herencsar N. 'Grounded voltage controlled positive resistor with ultralow power consumption'. Elektronika IR Elektrotechnika. 2014; **20**(7):45–50.

[104] Yuce E., Minaei S., Alpaslan H. 'Single voltage controlled CMOS grounded resistors and their application to video filter'. Indian Journal of Engineering & Materials Sciences. 2014; **21**(5):501–9.

[105] Holani R., Pandey P. C., Tiwari N. 'A JFET-based circuit for realizing a precision and linear floating voltage-controlled resistance'. Proceedings of the 11th Annual Conference of the IEEE India Council; Pune, India, December 11–13, 2014.

[106] Lopez-Delgadillo E., Diaz-Mendez J. A., Gurrola-Navarro M. A., Garcia-Andrade M. A., Vazquez-Medina R. 'A digitally programmable active resistor in CMOS technology'. IEICE Electronics Express. 2015; **12**(8):1–10.

[107] Yuce E., Tokat S., Yucel F. 'A new wideband electronically tunable grounded resistor employing only three MOS resistors'. Turkish Journal of Electrical Engineering & Computer Science. 2016; **24**(4):2442–53.

[108] Srivastava M., Prasad D. 'A novel purely active electronically controllable configuration for simulating resistance in floating form'. Electronics. 2016; **20**(2):90–4.

[109] Richelli A., Grassi M., Redoute J. M. 'Design of an integrated tunable differential negative resistance in UMC 0.18 μm'. Microelectronics Journal. 2016; **48**(2):1–6.

[110] Tripetch K., Nakano N. 'CMOS positive and negative resistors based on complementary regulated cascode topology with cross-coupled regulated transistors'. International Journal of Electronics and Communication Engineering. 2017; **11**(11):1156–62.

[111] Al-Absi M. A. 'A new CMOS tunable floating capacitance multiplier'. International Journal of Electronics Letters. 2018; **6**(1):48–57.

[112] Polley A., Pandey P., Bloodworth B. E., Cazana C. 'Analog front end for tribo-current-based fly-height sensor for magnetic hard disk drive'. IEEE Transactions on Circuits and Systems – I. 2018; **65**(2):556–66.

[113] Yucel F., Yuce E. 'A new electronically fine tunable grounded voltage controlled positive resistor'. IEEE Transactions on Circuits and Systems – II. 2018; **65**(4):451–5.

[114] Prasad D., Rana C., Afzal N. 'A compact tunable floating gate MOSFET based resistor'. I-manager's Journal on Electronics Engineering. 2018; **8**(3):1–4.

[115] Senani R., Singh V. K. 'Synthesis of canonic single-resistance-controlled oscillators using a single current-feedback-amplifier'. IEE Proceedings Circuits Devices Systems. 1996; **143**(1):71–2.

[116] Czarnul Z., Tsividis Y. P. 'Implementation of MOSFET-C filters based on active RC prototype'. Electronics Letters. 1988; **24**(3):184–5.

[117] Liu S. I., Wei J. F., Sung G. M. 'SPICE macro model for MAGFET and its applications'. IEEE Transactions on Circuits and Systems – II. 1999; **46**(4):370–5.

Chapter 10

Bipolar and CMOS active inductors and transformers

Abstract

This chapter discusses methods and circuits of producing gyrators, simulated grounded and floating inductors and transformers in both bipolar and CMOS technology. While in the case of bipolar circuits, we have included only some of the prominent configurations from amongst the available works, the treatment of CMOS circuits has been restricted to the developments taken place from 2008 onwards. Thus, the circuits and techniques prior to 2008 have been treated in a somewhat concise manner only due to the fact that more detailed treatment of these is already available in some of the recent books on this topic. Wherever possible, the applications suggested by the proposers of the concerned circuits have also been highlighted.

10.1 Introduction

Before the widespread use of the integrated circuit (IC) op-amp as a prominent active circuit building block; the early semiconductor gyrators were designed using bipolar junction transistors (BJTs). Some of the very initial work on simulated inductors too was around the inductive behaviour of bipolar transistors. However, with the impetus on integrating analog circuits along with the digital circuits on the chip using CMOS technology, interest was more diverted to CMOS analog circuits, including CMOS-based gyrators and simulated inductors. Since a MOSFET by itself is a voltage-controlled current source, it soon attracted the attention of researchers and circuit designers to attempt to build a gyrator using MOSFETs. Thus, was born the area of CMOS gyrators and simulated inductors as a consequence of which besides bipolar circuits, a number of CMOS realizations of grounded, floating as well as fully differential inductors and transformers were widely investigated during the past three decades [1–137].

This chapter presents a survey of gyrators, simulated inductors and transformers using bipolar and CMOS technologies. A good source of references and excellent review of work done in these areas till about 2007 is available in the book by Yuan [89]. Hence, instead of duplicating what is already available in [89], we will only very shortly sum-up some prominent CMOS structures developed till 2007 or so and will concentrate more on the developments taken place after the publication of Yuan's book [89].

10.2 BJT-based gyrators and inductance simulators

After the emergence of integrated circuits, it was well understood that since the passive capacitors, even of very small values, occupy much larger chip area as compared to hundreds of transistors, one must find ways of making them using semiconductor approach. Consequently, several ways of creating integratable capacitors of stable values and high-Q suitable for several integrated circuit applications were found. On the other hand, inductors were found to be difficult to be fabricated on the chip since efforts of microminiaturization of inductors by reducing their geometrical dimensions results in not only very low values of inductances but also resulted in highly reduced Q-values which made them impractical and unusable for general applications. Hence, it was natural for the researchers and circuit designers to search ways of 'simulating' inductive behaviour using semiconductor devices.

Thus, there was a very realistic and challenging task of simulating inductive impedances using the then prevalent BJT. One of the first attempts in demonstrating and evaluating various semiconductor inductive elements and to show some of their simple applications were made by Dill [1].

10.2.1 *The early attempts of devising transistor-based gyrators/simulated inductors*

The basic single transistor inductive circuit introduced by Dill [1] was analysed later by Dutta Roy in [4]. The latter author subsequently presented in [4], a novel high-Q inductance circuit comprised of two BJTs and a resistor, which is shown here in Figure 10.1.

Subsequent years witnessed a considerable activity in the proposition of transistorized gyrator circuits using more number of transistors and resistors from the viewpoint of evolving fully integratable IC gyrator circuits from which lossless inductors would be realized. Thus, a number of researches proposed integratable gyrator circuits, for instance, see [5–41].

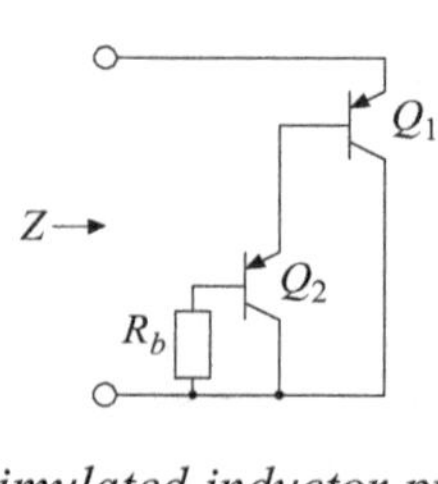

Figure 10.1 A simple simulated inductor proposed by Dutta Roy [4]

In this section, we present some representative BJT-based gyrators including the floating ones. Of course, circuits using BJTs as well as MOS transistors were also advanced. Besides this, method of producing gyrators from two negative impedance converters and a few (three to four) resistors was also proposed (see Harrison [3]).

An integratable time-variable gyrator circuit was proposed by New and Newcomb [6]. It was essentially a three transistor circuit which was an interconnection of a noninverting and an inverting current-controlled voltage source, the former realizable with one transistor and the latter by two transistors. The gyration resistances were proposed to be made time-variable by realizing them by voltage-variable resistors made from FETs operated below pinch-off region. Another three transistor gyrator was advanced by Shenoi [5] which, however, required a large number (as many as 14) resistors along with as many as six capacitors but also required two DC power supplies with supply voltages as large as +150 and +18 V!

10.2.2 A two-transistor semiconductor FI simulator due to Takahashi, Hamada, Watanabe and Miyata

Takahashi *et al.* [38] presented a simple circuit of producing a semiconductor floating inductance using only two BJTs, three resistors and a capacitor, as shown here in Figure 10.2.

Analysis of this circuit shows [38] that the equivalent impedance simulated by this circuit between ports 1 and 2 is given by $Z_{1-2} \cong -K_1R_3 + K_2C_os$; provided $|1/\omega c_o| \gg |R_3(1 + (R_1/R_2))|$ where

$$K_1 = \frac{R_1}{R_2} \quad \text{and} \quad K_2 = K_1(1 + K_1)R_3 \tag{10.1}$$

Since the equivalent impedance contains a negative series resistance, theoretically an infinite Q appears possible but in practice, Q values of the order of 50–100 were achievable. With a constant current source biasing of the transistors, the circuit was successfully employed by the authors [38] in designing a third low pass filter.

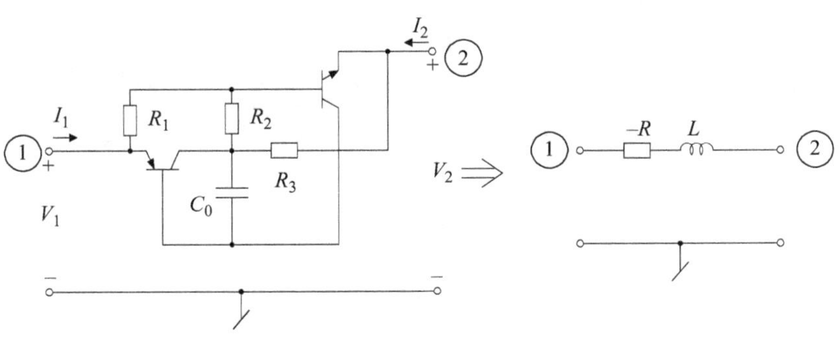

Figure 10.2 A simple two-transistor floating inductance simulator [38]

10.2.3 A direct-coupled fully integratable gyrator due to Chua and Newcomb

A direct-coupled fully integratable gyrator was presented by Chua and Newcomb [16] which is shown here in Figure 10.3. This circuit was obtained by a modification of an earlier gyrator proposed by Rao and Newcomb [7]. The circuit employed 9 BJTs, 12 resistors and 1 diode. Out of the nine transistors, ten were *n–p–n* transistors and remaining two *p–n–p* transistors. The circuit consists of two cross-coupled voltage-controlled current sources, one of which has its output current in the same phase as the input, whereas the other has the output current 180° out of phase with the input voltage. The circuit was designed such that it is well compensated for temperature and DC power supply variation.

According to [16], the connections which are shown by dotted lines were to be connected externally for normal operation of the circuit. On the other hand, for more flexibility 10–11 and 5–12 could facilitate incorporation of variable resistors by opening the connections 1–10 and 4–5.

10.2.4 The Integrated gyrator due to Haykim, Kramer, Shewchun and Treleaven

Subsequently, another fully integratable gyrator was reported by Haykim *et al.* [23]. In this circuit (Figure 10.4), the amplifier consisting of transistor Q_1 connected as emitter follower and the transistor Q_2 connected as common emitter amplifier, the output of which is applied to a Darlington connected transistor made from Q_3 and Q_4. The output signal developed by the transistor Q_4 is, therefore, 180° out of phase with the input signal, such that the transconductance is approximately equal to $1/R_5$. Likewise, the remaining part of the circuit, from the other side, creates another

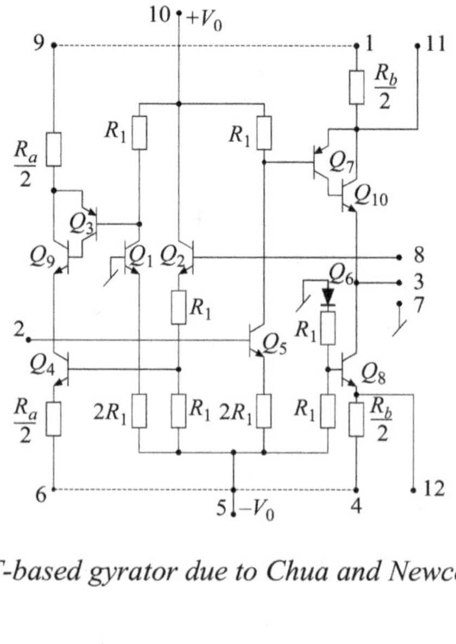

Figure 10.3 The BJT-based gyrator due to Chua and Newcomb [16]. © 1967 IEE

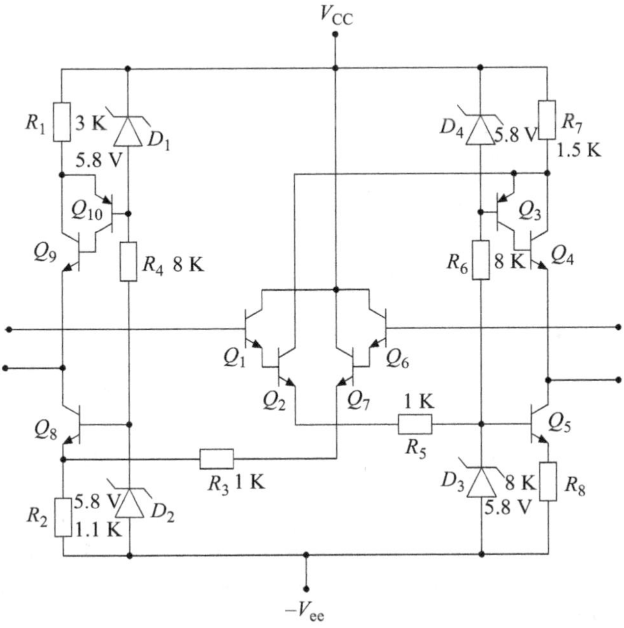

Figure 10.4 A direct-coupled IC gyrator [23]

voltage-controlled current source having transconductance approximately equal to $1/R_3$. This part of the circuit has 0° phase shift as required (Figure 10.4).

10.2.5 Synthesis of three transistor gyrators

Stevenson [41] presented a method of generating new transistor-based gyrator circuits. The key idea of this method is that if there is two-port network or a one-port network consisting of nullators, norators and resistors and if all the nullators are interchanged by norators and vice versa, then the driving point impedance of the one-port network remains unaffected, whereas the short circuit admittance matrix of the two-port network becomes transpose of the *Y*-matrix of the original circuit. It is, therefore, obvious that if the original network happens to be a two-port lossless gyrator then the modified network would also be a lossless gyrator. Stevenson [41] considered the nullor model of the lossless gyrator consisting of three nullors and two resistors and showed that from this model, two different three transistor gyrators can be derived by replacing each combination of a nullator and norator to constitute a transistor one of which is shown in Figure 10.5(a). Now if the quoted transformation is applied on the nullor model of Figure 10.5(a), the transformed nullor will lead to two additional three-transistor gyrators one of which would be as shown in Figure 10.5(b).

It may be pointed out that all the four three-transistor gyrators would require additional circuitry to provide appropriate DC biasing current to all the transistors

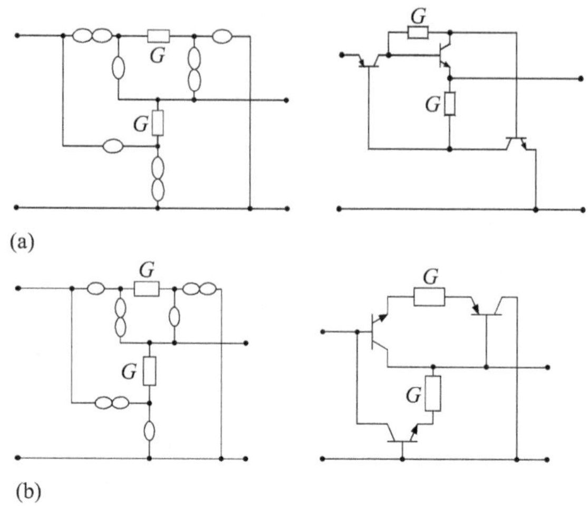

Figure 10.5 *Stevenson's method [41] of generating three-transistor gyrator circuits: (a) nullor model and one of the corresponding transistor circuits and (b) alternate nullor model and one of the corresponding transistor circuits*

which will make the resulting circuit quite complex. Furthermore, the design of additional biasing circuitry is by no means either a straightforward or simple task.

10.2.6 The translinear floating inductance simulator

It is interesting to note that although a number of transistorized gyrators were advanced, they suffered from the drawback of requiring identical number of resistors. With the introduction of the concept of the translinear circuits by Gilbert in 1975, it became possible to think of devising various functional circuits using transistors only without requiring any resistors. Using the concept of translinear circuits, Ngamkham *et al.* [56] presented electronically tunable synthetic floating impedance which is presented here in Figure 10.6. In this circuit, various *npn* current mirrors, *pnp* current mirrors, *npn* current repeaters and *pnp* current repeaters are shown by their symbolic notations as shown in Figure 10.6(b). By straightforward analysis of the translinear cells and following the flow of currents in the circuit, it can be deduced that this circuit simulates a floating inductance of value

$$Z_{1-2} \cong (2CR_xR_{x2})s \text{ provided } V_{1-2} \ll V_T \qquad (10.2)$$

The workability of this circuit was demonstrated in [56] by taking the capacitance value as $C = 3.0$ nF and the bipolar transistors NR100N and PR100N and the circuit biased from the DC power supply of ±2.5 V. By varying the external bias current as 50 μA, 70 μA and 100 μA, accordingly the inductance value was found to be changed to L = 0.4 mH, 0.2 mH, 0.1 mH. The attractive features of this circuit are complete absence of any passive resistor and electronic tunability of the inductance. On the other hand, a limitation of the circuit is its relatively small input

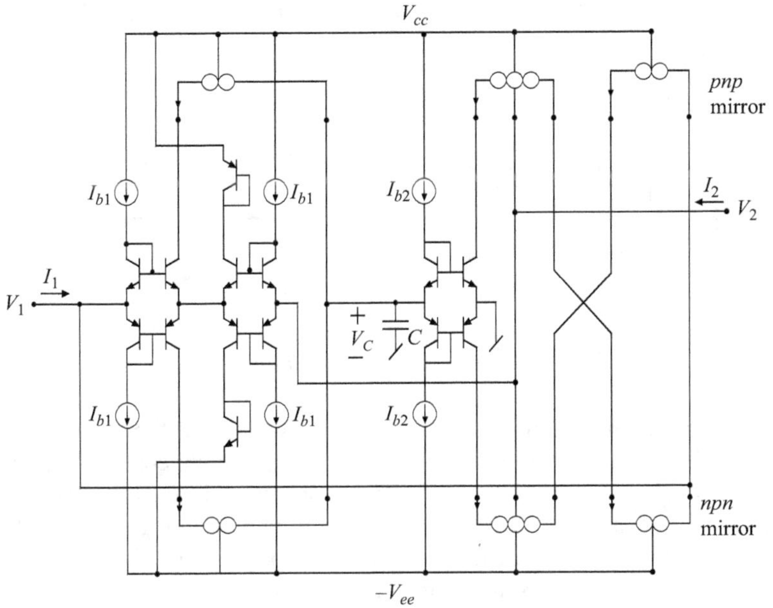

(a)

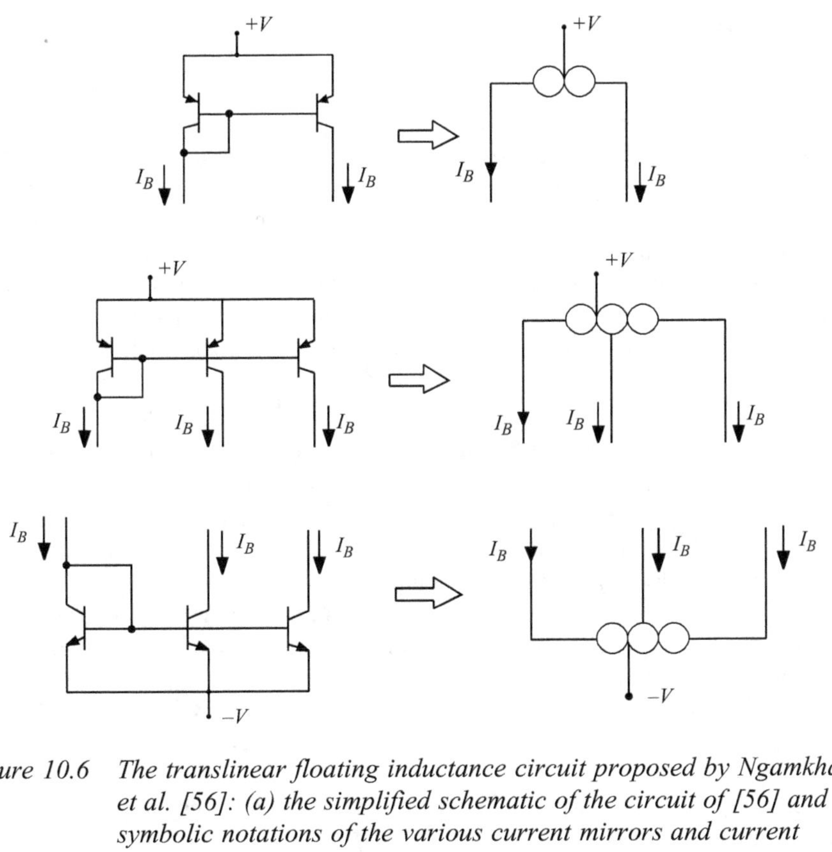

Figure 10.6 The translinear floating inductance circuit proposed by Ngamkham et al. [56]: (a) the simplified schematic of the circuit of [56] and (b) symbolic notations of the various current mirrors and current repeaters

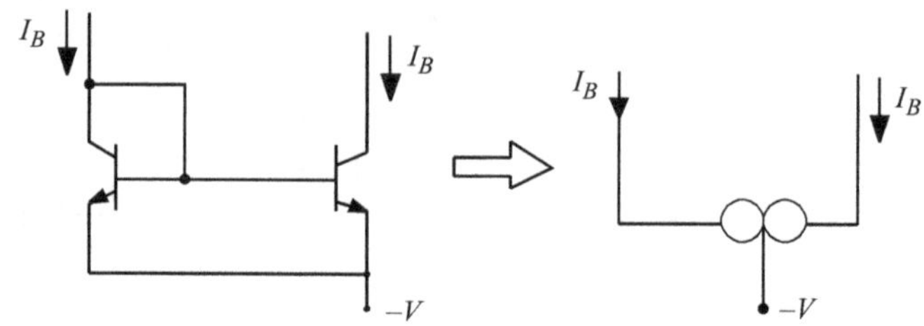

(b)

Figure 10.6 (Continued)

signal handling capability and temperature sensitivity since the parameter R_x is temperature dependent. Fortunately, the temperature compensation techniques which were developed earlier for designing temperature-insensitive current conveyors and OTAs can be applied to this FI circuit as well, but unfortunately, any remedy for the limited input signal handling capability has not been proposed so far, and hence, this constitutes an interesting unresolved problem.

10.3 CMOS active inductors

A popular and widely employed method to synthesize grounded, floating or fully differentially inductors suitable for CMOS implementation is based upon the Gyrator-C approach involving a parallel back to back connection of one inverting and one non-inverting transconductor both of which are realizable by a single or judicious combination of more than one MOSFETs. A comprehensive treatment of such CMOS active inductors developed till 2007 or so has been given in the book by Yuan (see chapter 2 of [89]). In view of this, in order not to duplicate the material already available in [89], in this section, we will discuss some of the prominent works which have appeared 2007 onwards (and not included in [89]).

10.3.1 CMOS inductor proposed by Uyanik and Tarim

Contrary to the utilization of internal parasitics of the MOSFETs, Uyanik and Tarim [76] proposed a circuit which employs as external grounded capacitor (as preferred for IC implementation) along with CMOS circuitry to create an inductor as shown in Figure 10.7. With capacitor C deleted, the rest of the circuit, in fact, realizes an active gyrator so that with output port loaded into port 2, input impedance has a parallel equivalent consisting of a shunt resistor R_p, a shunt capacitor C_p and an inductive shunt branch comprising series impedance (R_s+L_{eq}) with pertinent parameters given by

$$L_{eq} = \frac{C_2}{g_{m1}g_{m2}}; \quad R_s = \frac{G_2}{g_{m1}g_{m2}}; \quad R_p = \frac{1}{G_1}; \quad \text{and} \quad C_p = C_1$$

where G_1, G_2 and C_1, C_2 are respective equivalent conductances and total capacitances at nodes 1 and 2.

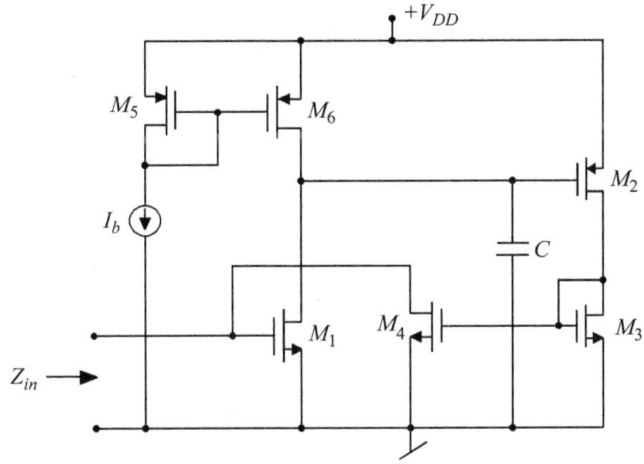

Figure 10.7 Active inductor proposed in [76]

Qualitatively, the role of the various subcircuits can be explained as follows:

MOSFET M_1 in common-source configuration realizes an inverting transconductor, M_2–M_4 form the non-inverting transconductor where the current mirror M_3–M_4 inverts the inverting transconductor of common-source PMOS transistor M_2. Lastly, M_5–M_6 provides the dc bias current. It may be noted that body terminal of all MOSFETs is connected to respective source terminals; hence, the body effect is out of picture. The power consumption is low due to a low number of MOSFETs employed. Moreover, R_s and L_{eq} are independently adjustable through C_2 and G_2 (hence by I_b), respectively.

Simulation of circuit in Cadence UMC 0.13 μm CMOS process with $V_{DD} = 1.2$ V having length of all MOSFETs as 0.13 μm and taking $W_1 = 4$ μm, $W_2 = 14.4$ μm, $W_3 = W_4 = 3.6$ μm, $W_5 = W_6 = 10$ μm, $I_b = 135$ μA and $C = 104$ fF has demonstrated that for input bias voltage of 500 mV the various parameters were found to be $G_{m1} = 1.7$ mS, $G_{m2} = 3.1$ mS, $G_1 = 258$ μS, $G_2 = 304$ μS, $C_1 = 12$ fF. The inductive behaviour of the circuit was found to be prevalent over a wide frequency range spanning from 300 MHz up to the self-resonant frequency of 7.32 GHz with a nominal inductance value of 38 nH and $Q > 100$ (>10) in the 4.8–6.4 GHz (2.35–7 GHz) range.

CMOS active inductors have been found to be very useful in the design of RF/microwave circuit design because of several advantages offered by them such as, high-quality factors, wide tunability, and large inductance values realizable along with small chip area.

10.3.2 CMOS grounded inductor proposed by Reja, Filanovsky and Moez

The CMOS grounded inductor proposed by Reja *et al.* [90] shown in Figure 10.8 is a modified form of the one proposed earlier by Thanachayanont and Payne [48]. In this circuit, the cross-coupled MOSFET pair M_1–M_2 provides a positive feedback

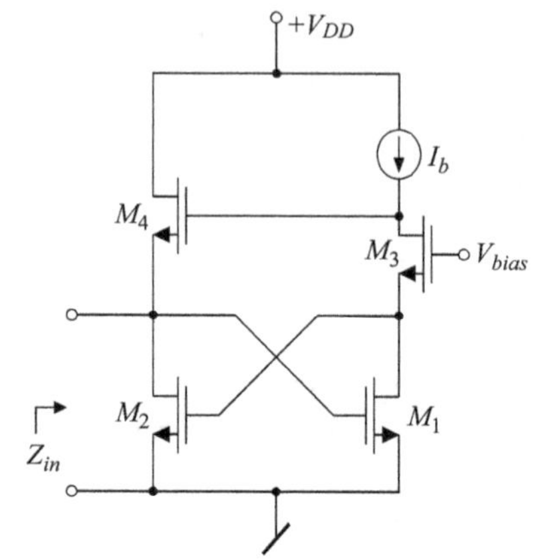

Figure 10.8 Active inductor proposed in [90]

to create inductive impedance. It is readily seen that the DC bias current I_b flows in M_3 and M_1 and that the appropriate bias currents in M_2 and M_4 are set up by appropriately choosing DC bias voltage V_{bias} such that it is more than $(V_{GS3}+V_{GS2})$ which also ensures that M_3 would not entre triode region. Thus, with all the MOSFETs operating in saturation and represented by their transconductances, g_{mi} and gate to source capacitances C_{gsi}; $i = 1$–4 omitting the effect of output resistances r_{dsi} and gate to drain capacitances C_{gdi}, the input impedance of the circuits is found to be [90]

$$Z_{in} \cong \frac{(C_{gs3} + C_{gs4})s}{C_{gs1}(C_{gs3} + C_{gs4})s^2 + C_{gs1}g_{m3}s + g_{m1}(g_{m3} - g_{m4})} \tag{10.3}$$

By simple algebraic calculations [90], the expressions for the self-resonant frequency ω_0 and Q_0 of the essentially bandpass impedance function represented by the above expression, are given by the following:

$$\omega_0^2 = \frac{4(K_3 + K_4)\sqrt{K_1}}{\sqrt{K_3}C_{gs1}(C_{gs3} + C_{gs4})}\sqrt{I_b}\left(\frac{I_b - K_4\sqrt{K_3}(V_{bias} - 2V_{th})}{K_3 + K_4}\right)^{1/2} \tag{10.4}$$

$$Q_0 = \frac{1}{2}\left\{\left[\left(\frac{C_{gs3} + C_{gs4}}{C_{gs1}}\right)\left(1 + \frac{K_4}{K_3}\right)\sqrt{\frac{K_1}{K_3}}\right]\left[1 - \frac{K_4\sqrt{K_3}(V_{bias} - 2V_{th})}{(K_3 + K_4)\sqrt{I_b}}\right]\right\}^{1/2} \tag{10.5}$$

where the symbols have their usual meaning i.e., $K_i = \mu_s C_{ox}/2(W/L)_i$; $i = 1$ to 4.

From the above, it follows that if V_{DD} is chosen close to $2V_{th}$, the quality factor is almost constant and ω_0 is proportional to $\sqrt{I_b}$ which leads to a simple tuning through the external DC bias current I_b. SPICE simulation based on ST Microelectronics 90 nm digital CMOS using the Cadence Spectre simulator have shown [90] that with $W_1 = 2.3$ μm, $W_2 = 5$ μm, $W_3 = 3.4$ μm and $W_4 = 1.5$ μm

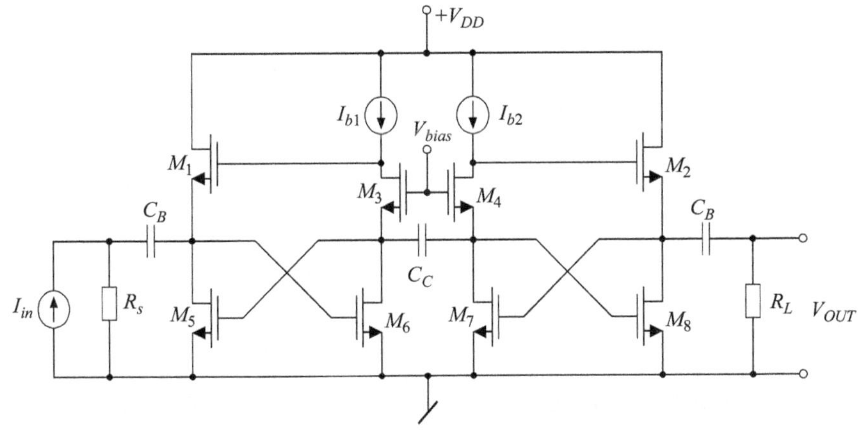

Figure 10.9 Wideband filter/amplifier proposed in [90]

with L_i; $i = 1{,}4$ taken as 0.1 μm inductive impedance was realizable from a few MHz to 12.5 GHz with adequate values of realized Q for $I_B = 60$ μA.

Application: It was demonstrated in [90] that with the two active inductors connected back to back through a coupling capacitor C_C (50 fF) resulted in the wideband active bandpass filter as shown here in Figure 10.9 [90] with $V_{DD} = 0.8$ V, this circuit demonstrated −3 dB bandwidth of 8 GHz with 0.7 dB ripple.

10.3.3 Constant-Q active inductor proposed by Tang, Yuan and Law

Tang *et al.* [91] presented a constant-Q active inductor by improvising upon the classic active inductor of Wu *et al.* [57] which suffers from the basic limitation that both realized inductance and its quality factor are strong functions of the swing of the input current. The proposition of Tang *et al.* [91] is shown in Figure 10.10.

This circuit employs Wu *et al.*'s [57] active inductor comprising M_1, M_2 and current source I_{b1} and additional current feedback network realized by the MOSFETs M_3 to M_7. If the current source I_{b2} is set to maximum input current $I_{in,\max}$, this leads to $i_{D4} = I_{in,\max} - K_1 i_{D2}$ if we take the current gain of the M_2–M_3 current mirror as K_1. Now if we take current gains of mirrors M_4–M_5 and M_6–M_7 as K_2 and K_3, respectively, we can easily see that $i_{D7} = K_2K_3(I_{in,\max} - K_1 i_{D2})$. Thus, MOSFET M_7 injects the negative feedback current i_{D7} in to the input terminal which results in $i_{D2} = i_{D7} + i_{in} + I_{b1} = K_1K_2K_3I_{in,\max} + I_{b1} - K_1K_2K_3i_{D2} + i_{in}$ which can be rearranged to give

$$i_{D2} = \left(\frac{K_1K_2K_3}{1 + K_1K_2K_3}\right) I_{in,\max} + \frac{I_{b1}}{1 + K_1K_2K_3} + \frac{i_{in}}{1 + K_1K_2K_3}$$

If the circuit is so designed that $K_1K_2K_3 \gg 1$, it follows that $i_{D2} \approx I_{in,\max}$. Since the effect of input current i_{in} on the drain current of M_2 and consequently on g_{m2} has been effectively minimized, it is, therefore, ensured that both the realized

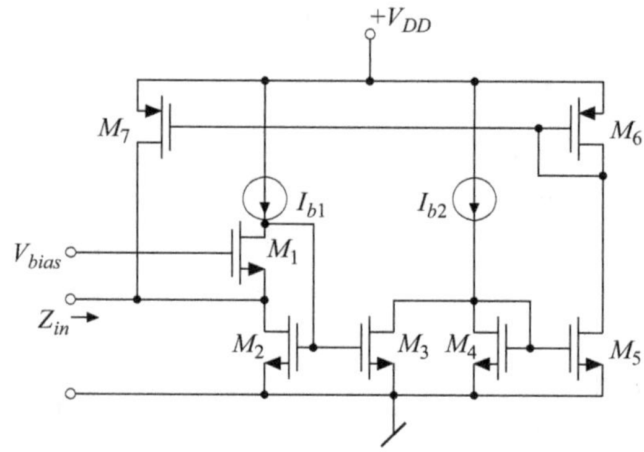

Figure 10.10 Constant-Q active inductor proposed in [91]

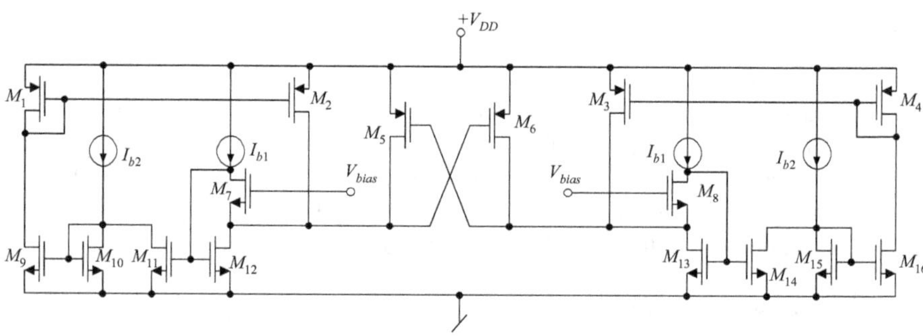

Figure 10.11 LC tank oscillator with constant-Q active inductor [91]

inductance value and its quality factor remains almost unchanged from the influence of widely varying input current.

Application: The proposed inductor of Figure 10.10 is used to realize an LC tank oscillator as in Figure 10.11 whose performance was compared with an LC tank oscillator designed using Wu *et al.*'s active inductor [57] as well as a VCO designed with spiral inductor. The simulations employing UMC-0.13 micrometer 1.2 V CMOS technology demonstrated [91] phase noise improvement (−118 dBc/Hz for the circuit of Figure 10.11 as against −113 dBc/Hz LV tank oscillator based on Wu *et al.*'s [57] active inductor) at the cost of somewhat increased power consumption (mainly due to large current in the feedback network) as compared to LC tank oscillator based on active inductor of [57].

10.3.4 CMOS active inductors due to Krishnamurthy, El-Sankary and El-Masry

It is well recognized that in CMOS active inductors, higher noise, nonlinearity and power consumption are the major problems to be taken care of. Krishnamurthy *et al.* [97] presented a new CMOS active inductor architecture with an inbuilt thermal noise cancellation (NC). From a noise analysis of the circuit of Figure 10.12 they have figured out that the noise contribution of the MOSFET

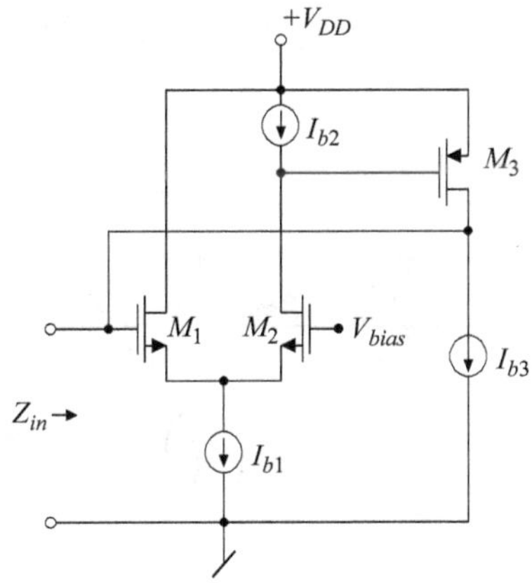

Figure 10.12 Active inductor without NC proposed by Krishnamurthy et al. [97]

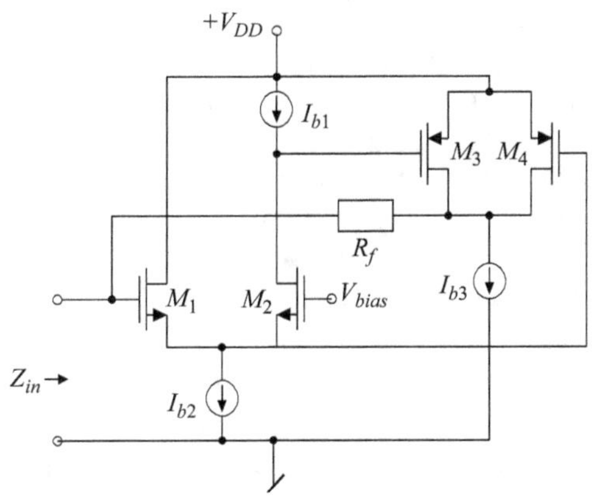

Figure 10.13 Simplified active inductor [97]

appearing in the forward path of the gyrator-C structure is the dominant one which can be cancelled using a feedback path while the noise contribution of the latter is degenerated by inclusion of a shielding resistor in the feedback.

With this strategy, they have come up with the modified CMOS active inductor circuit shown in Figure 10.13.

The equivalent of the input impedance of the circuit represents a parallel combination of a resistor R_P, capacitor C_P and a series-RL branch ($R_s + sL_{eq}$) where the values of the elements associated with RL part are given by

$$L_{eq} = \frac{G_m C_3 \left(1 + (g_4/g_f)\right)}{g_{m1}(g_{m3}g_{m2} + g_{m4}g_3)} \tag{10.6}$$

$$C_P = C_1$$

$$R_s = \frac{G_m g_3}{(g_{m1}g_{m2}g_{m3} + g_{m1}g_{m4}g_3)\left(g_f/(g_4 + g_f)\right)} \tag{10.7}$$

The resonant frequency ω_0 and quality factor are given by

$$\omega_0 = \sqrt{\frac{g_{m1}(g_{m3}g_{m2} + g_{m4}g_3)}{C_1C_3G_m\left(1 + \left(\frac{g_4}{g_f}\right)\right)}};$$

$$Q_0 = \frac{\sqrt{\frac{g_{m1}(g_{m2}g_{m3} + g_{m4}g_3)}{G_mC_1C_3}\left(\frac{g_f}{(g_4 + g_f)}\right)}}{\left(\frac{g_{m4}g_{m1}}{G_mC_1}\right)\left(\frac{g_4g_f}{(g_4 + g_f)}\right) + \left(\frac{g_1}{C_1}\right) + \left(\frac{g_3}{C_3}\right) + \left(\frac{g_1g_3C_2}{G_mC_1C_3}\right) - \omega_0^2\left(\frac{C_2}{G_m}\right)} \tag{10.8}$$

where $G_m = g_{m1} + g_{m2} + g_3$ and $g_f = 1/R_f$.

Application: It is expected that CMOS active inductors with NC could be useful for many applications such as front end active RF filters in receivers and in the design of low-phase noise oscillators.

Simulations were carried out in 90 nm STM CMOS process with aspect ratios (in μm) as M_1 (6/0.18), M_2 (6/0.18), M_3 (11.5/0.18) and M_4 (2.3/0.18) along with $I_{b1} = 280$ μA, $I_{b2} = 400$ μA, $V_{bias} = 0.8$ V and supply voltage taken as $V_{DD} = 1.2$ V. The active inductor was found to have centre frequency of frequency response at $f_0 = 3.8$ GHz with quality factor $Q_0 = 405$. All poles and zeros were found to be in left half of s-plane, thereby indicating stability even while the bias currents were varied from 300 μA to 500 μA to tune the centre frequency and quality factor.

10.3.5 CMOS high-Q active grounded inductor due to Li, Wang and Gong

Li *et al.* in [101] presented a CMOS active grounded inductor with high Q achievable by enhanced feedback while proving high operational frequency (maximum up to 17 GHz, with 1–8 GHz peak operating frequency range) and low power consumption of the order of 0.1 mW.

The circuit technique of Li *et al.* [101] is, in fact, outcome of an improvization applied to the Wu *et al.* [65] circuit in two different ways (see Figure 10.14): (i) in the first case, a feedback is applied through an additional NMOS transistor M_3 and (ii) in the second case, a similar feedback is applied through as additional PMOS transistor M_3. In both cases, the current (I_{b2}) through the feedback transistor must be much smaller than the current I_{b1} flow through the cascaded transistors M_1 and M_2. The input admittances of both of the circuits are found to be crossing of three admittances in parallel as represented by [101].

$$Y_{in}(s) = sC_{eq} + \frac{1}{R_{eq1}} + \frac{1}{R_{eq2} + sL_{eq}} \tag{10.9}$$

where $C_{eq} \approx C_{gs2}$.

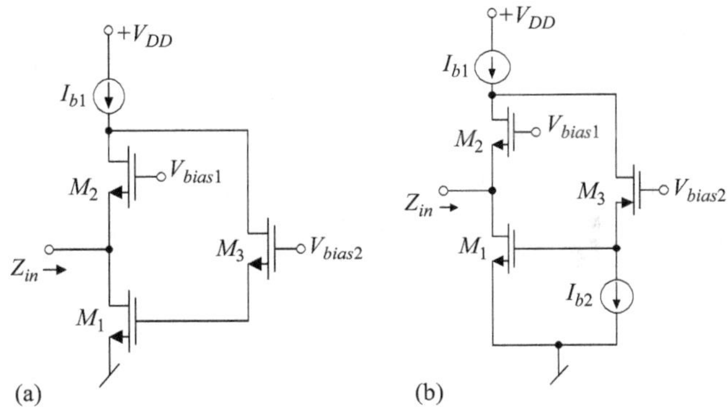

Figure 10.14 Active inductor due to Li et al. [101]: (a) high-Q active inductor with NMOS feedback and (b) inductor with PMOS feedback

$$R_{eq1} \cong \frac{C_A r_{02} + C_B(r_{02} + r_{03} + g_{m3} r_{02} r_{03})}{[g_{m2} + (1/r_{02})](C_A r_{02} + C_B(r_{02} + g_{m3} r_{02} r_{03}))} \tag{10.10}$$

$$R_{eq2} \cong \frac{r_{02} - \omega^2 C_A C_B r_{02}^2 r_{03}}{(g_{m2} r_{02} + 1)\{g_{m1}(g_{m3} r_{03} + 1) r_{02} - 1\}} \tag{10.11}$$

$$L_{eq} \simeq \frac{r_{02}(C_A r_{02} + C_B(r_{02} + r_{03} + g_{m3} r_{02} r_{03}))}{(g_{m2} r_{02} + 1)(g_{m1}(g_{m3} r_{03} + 1) r_{02} - 1)} \tag{10.12}$$

where C_A and C_B are the total parasitic capacitances at nodes A and B of the circuit, respectively, and r_{oi}; $i = 1$ to 3 is the output resistance and the transconductances of the MOSFETs M_i, respectively.

The self-resonance frequency of the circuit has been estimated [101] to be

$$\omega_0 \simeq \sqrt{\frac{(g_{m2} r_{02} + 1)\{g_{m1}(g_{m3} r_{03} + 1) r_{02} - 1\}}{C_{gs2} r_{02}[C_A r_{02} + C_B(r_{02} + r_{03} + g_{m3} r_{02} r_{03})]}} \tag{10.13}$$

Thus, due to the feedback transistor M_3 added, R_{eq1} increases (should be ideally infinity), but there is negative resistance in R_{eq2} such that when $g_{m3} r_{03}$ increases not only equivalent inductance value increases but the negative resistance R_{eq2} also increases, thereby enhancing the quality factor. Thus, by adding M_3, high ω_0 and high Q with low power consumption are achievable but the penalty to be paid is the additional noise introduced by MOSFET M_3.

10.3.6 Tunable CMOS inductor using MOSFETs

Minaei and Yuce [106] proposed a tunable CMOS inductor using MOSFETs as shown in Figure 10.15. In this circuit, a cascade of an inverting amplifier and a two MOSFET transconductor realizes a noninverting transconductor whose output is terminated into a capacitor. The voltage across this capacitor is then applied to an

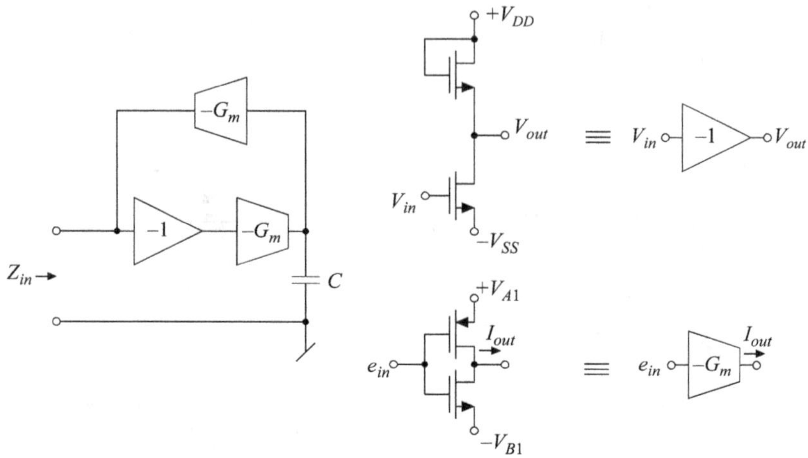

Figure 10.15 Tunable CMOS inductor simulator due to Minaei and Yuce [106]

inverting transconductor whose output current is brought back to the input terminal to constitute the input current.

The circuit, therefore, can be looked upon as a parallel back-to-back connection of (i) a non-inverting transconductor made of a cascade of unity gain inverting amplifier and an inverting transconductor (both of which are realizable with two MOSFETs as shown in the right hand side of Figure 10.15) and (ii) an inverting transconductor. The circuit, obviously, is a gyrator with its second port being the node where the grounded capacitor is connected. An analysis of the inverting amplifier comprising M_1–M_2 reveals that its input–output relation is given by

$$V_0 = V_{DD} - V_{th1} + \sqrt{\frac{K_2}{K_1}}(-V_{SS} + V_{th2}) - \sqrt{\frac{K_2}{K_1}}V_{in} \tag{10.14}$$

from where it turns out that if $K_1 = K_2$, $|V_{DD}| = |V_{SS}|$, $V_{th1} = V_{th2}$ then the circuit realizes $V_0 = -V_{in}$, without any DC offset!

On the other hand, the two MOSFET transconductors composed of M_3–M_4 and M_5–M_6 are both governed by the following output equations:

$$I_{out} \cong \left[-K_P\left(V_A - \left|V_{thp}\right|\right) - K_n(V_B - V_{thn})\right]V_{in}$$

provided $K_P \approx K_n$ and $V_B \approx V_A + V_{thn} - |V_{thp}|$ and the following condition is fulfilled so that transistors operate in saturation as required [106]:

$$V_{D1} + V_{thn} \geq V_{in} \geq V_{D1} - \left|V_{thp}\right| \tag{10.15}$$

Subject to the fulfilment of the above requirements, the input impedance Z_{in} of the circuit of Figure 10.15 is found to be [106]

$$Z_{in}(s) = \frac{sC}{G_{m3}G_{m4}} \tag{10.16}$$

where G_{m3} and G_{m4} are given by

$$G_{m3} = K_{p3}\left[V_{A1} - \left|V_{thp3}\right|\right] + K_{n4}(V_{B1} - V_{thn4}) \tag{10.17}$$

$$G_{m4} = K_{p5}\left[V_{A2} - \left|V_{thp5}\right|\right] + K_{n6}(V_{B2} - V_{thn6}) \tag{10.18}$$

The input impedance of the circuit is found to be composed of three parallel branches having a capacitance C_p, a resistance R_p and the last branch being (r_s+sL_{eq}). A straightforward algebraic calculation reveals [106] that

$$|Z_{in}(j\omega)| = \frac{\omega/C_p}{\sqrt{\left[(1/C_pL_{eq}) - \omega^2\right]^2 + (\omega/C_pR_p)^2}} \tag{10.19}$$

and

$$\phi(\omega) = \frac{\pi}{2} - \tan^{-1}\left\{\frac{\omega/C_pR_p}{(1/L_{eq}C_p) - \omega^2}\right\} \tag{10.20}$$

It must be noted that contrary to the other active inductors described in this chapter which do not require any external capacitor, this one does need an external capacitor. For details of the performance evaluation and applications of this circuit, the readers are referred to [106] which describe techniques of cancellation of unwanted series resistances through two transconductors-based negative resistance circuits made from the same transconductance stage as employed in the main circuit.

10.3.7 CMOS inductor proposed by Sato and Ito

Sato and Ito [108] postulated three basic single MOSFET transconductor cells, namely a common-source transconductor, a common gate transconductor and common drain transconductor. In conjunction with the dominant gate to source parasitic capacitance C_{gs} of the MOSFET these three basic cells realize, respectively, a cascade of a shunt grounded capacitance C_{gs} and an inverting transconductance element ($-g_m$), a shunt capacitance C_{gs} and a noninverting transconductor ($+g_m$) with a capacitance C_{gc} bridging between input and output terminals. They, then demonstrated that two basic CMOS active inductors result by judicious combination of two such cells loosely equivalent to a gyrator-C structure in each case, each of which, with a more exact analysis was actually realizing a parallel RLC resonator (second-order equivalent input impedance) consisting of a resistor R_p, a capacitor C_p and the last parallel branch being a series-RL (R_s+sL_{eq}) such that the values of all these elements were a function of the various parasitic capacitances, transconductances and various output resistances of the MOSFETs employed. For both these circuits, the self-resonant frequency and the Q were found to be [108]

$$\omega_0 = \sqrt{\frac{g_{m1}g_{m2}}{C_{gs1}C_{gs2}}} \quad \text{and} \quad Q_0 = \sqrt{\frac{g_{m2}C_{gs1}}{g_{m1}C_{gs2}}} \tag{10.21}$$

Thus, using these three single-MOSFET transconductor building blocks, the authors formulated a number of active inductors [108] two of which are shown in Figure 10.16(a) and (b).

The same authors then proposed a distortion reduction technique which can be explained as follows.

It was agreed that, in general, the drain current of MOSFET can be expressed as

$$I_D = I_{D0} + g_m V_{in} + \alpha V_{in}^2 \tag{10.22}$$

which shows a square nonlinearity. Now if we add a compensating current I_c which can be done as shown in Figure 10.17(a) and (b), then if this compensating current also has exactly same nonlinear components, the output current I_0 can then be expressed as

$$I_0 = \left(I_{D0} + g_m V_{in} + \alpha V_{in}^2\right) - \left(I_{CO} + \alpha V_{in}^2\right) = (I_{D0} + I_{CO}) g_m V_{in} \tag{10.23}$$

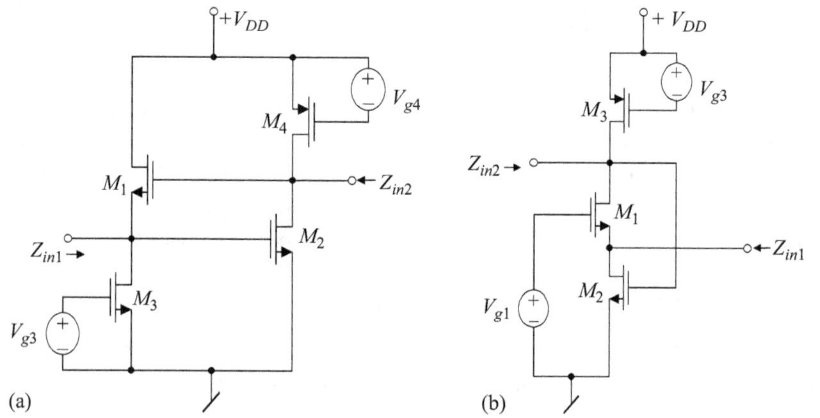

Figure 10.16 *The exemplary active inductors from those presented by Sato and Ito [108]: (a) using a common source and another common drain transconductor and (b) using a common source and another common gate transconductor*

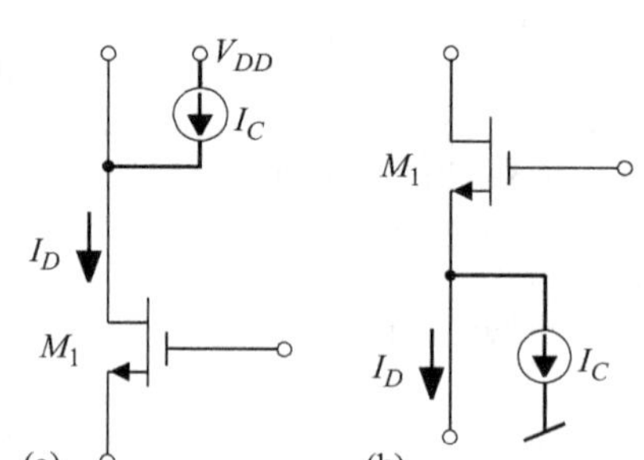

Figure 10.17 *Principle of distortion reduction technique [108]: (a) inserting a compensating current (b) alternative method of inserting compensating current*

and thus, the square nonlinearity has been cancelled out which is, therefore, likely to result in the reduction of distortion.

The authors subsequently applied this technique on previously postulated active inductor circuits resulting in compensated low distortion inductor circuits which are shown in Figure 10.18(a)–(c).

The first low distortion CMOS inductor circuit is shown in Figure 10.18(a). By straightforward analysis [108], the equivalent values of the parameters of this circuit are found to be

$$G_p = g_{m2} + g_{ds1} + g_{ds2} + g_{ds3} + g_{ds4} \quad (10.24)$$

$$C_p = C_{gs2}$$

$$R_s = \frac{g_{ds2} + g_{ds4}}{(2g_{m1} - g_{ds2} - g_{ds4})(g_{m2} + g_{ds2} + g_{ds4})} \quad (10.25)$$

$$L = \frac{C_{gs1}}{(2g_{m1} - g_{ds2} - g_{ds4})(g_{m2} + g_{ds2} + g_{ds4})} \quad (10.26)$$

It is noteworthy that assuming all g_{dsi} to be much smaller than g_{mi}, it is found [108] that L and R_s become half of that of the circuit of Figure 10.16(a) which

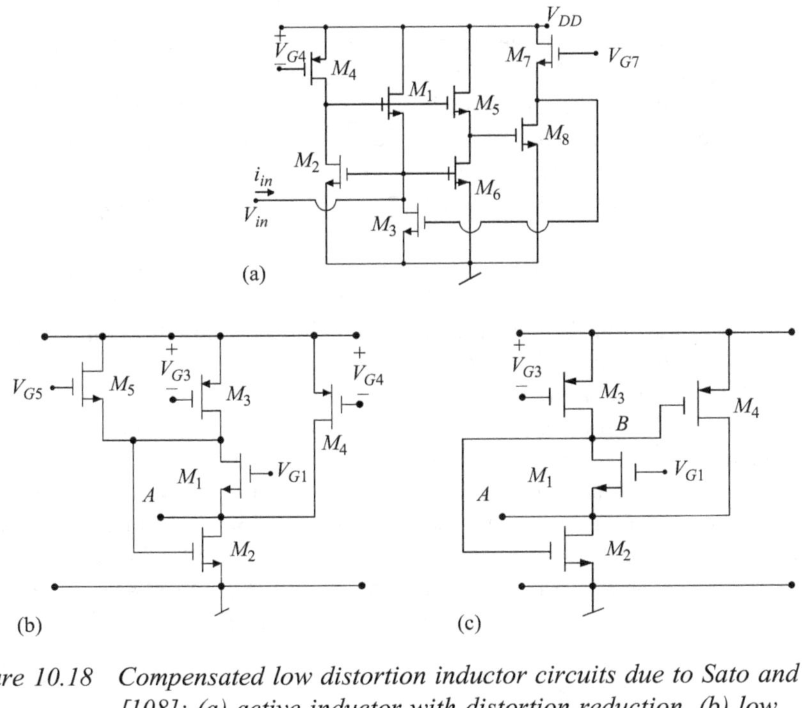

Figure 10.18 Compensated low distortion inductor circuits due to Sato and Ito [108]: (a) active inductor with distortion reduction, (b) low distortion active inductor using M_5 and (c) low distortion active inductor using M_4

means the self-resonant frequency of this circuit would become $\sqrt{2}$ times higher than that of the circuit of Figure 10.16(a). Thus, the applied distortion reduction technique also makes the circuit useful for an extended frequency range of operation.

Figure 10.18(b) shows the second reduced distortion version of the circuit in which the required compensation current is inserted by MOSFET M_5. However, this causes an increase in the equivalent resistance R_s which limits the available frequency range of operation of the circuit.

Lastly, another reduced distortion version is shown here in Figure 10.18(c) where M_4 is the MOSFET creating the compensation current.

Application: Validity of this distortion reduction technique has been verified in [108] by HSPICE simulations using 0.18 μm CMOS process BSIM3 parameters. Applications of these inductors have been verified in the realization of the balanced bandpass filter, and it has been shown that the THD of the realized bandpass filter is reduced to 0.26% at 1 GHz. On the other hand, the VCO using the proposed inductors could be designed to provide oscillations up to 950 MHz with improved phase noise due to incorporation of the suggested distortion reduction technique.

10.4 CMOS active transformers

Like spiral inductors which can be fabricated on the Silicon chip, there have a number of efforts on the design of CMOS on-chip transformer. It is well known that the spiral inductors which are intrinsic to their spiral layout suffer from several drawbacks such as realizability of low Q factor, low self-resonant frequency, non-tunable inductance value and occupying a large area on the chip. Because of these disadvantages, some efforts have been made in literature in realizing transformers using CMOS transistors.

Tang *et al.* [94] introduced a number of active transformers synthesized using MOS transistors only. The active CMOS transformers offer a number of attractive features over spiral transformers such as (i) tunability of coupling ratio, (ii) large and tunable self and mutual inductances, (iii) high and tunable quality factors and (iv) a smaller silicon area on the chip.

Based upon the grounded inductor structures proposed by Wu *et al.* [57,65], two basic active transformer schemes were first proposed in [94] as shown in Figure 10.19. In these circuits, the coupling from primary to secondary is implemented by the MOSFET M_3 configured as a voltage-controlled-current-source. On the other hand, the external DC bias currents I_p and I_s control the inductances of the primary and secondary windings. Apart from the above, Tang *et al.* [94] also proposed two configurations for realizing bidirectional transformers, which are shown here in Figure 10.20.

Yuan [81] presented a systematic approach for synthesizing active transformers based upon the use of gyrators realized by using MOS transistors. The first basic structure of synthesizing a CMOS transformer employs two ideal gyrators,

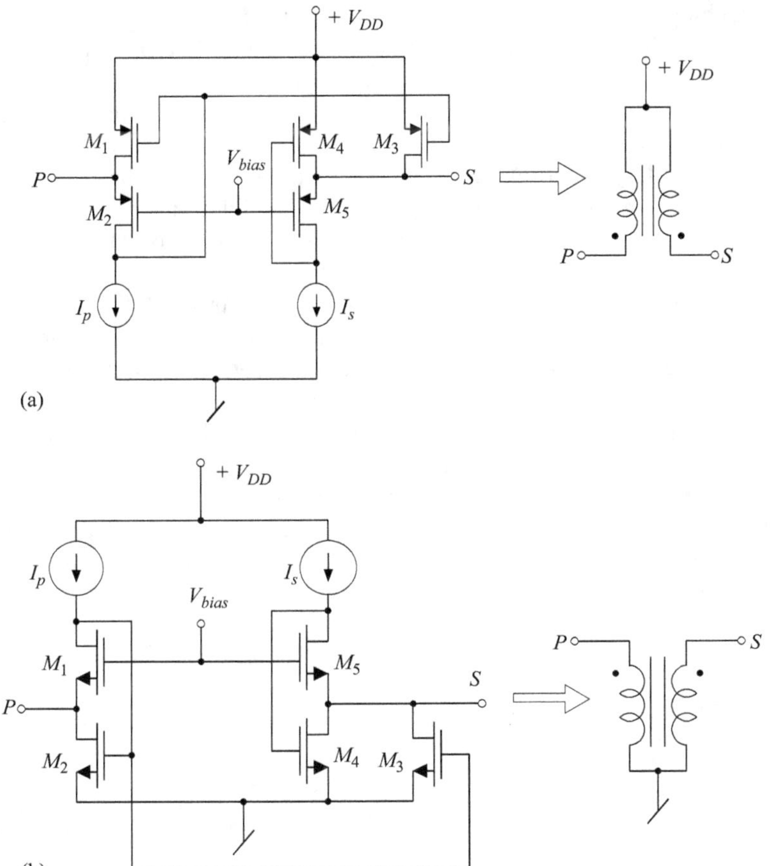

Figure 10.19 CMOS active transformers [94]: (a) PMOS-configured active transformer and (b) NMOS-configured active transformer

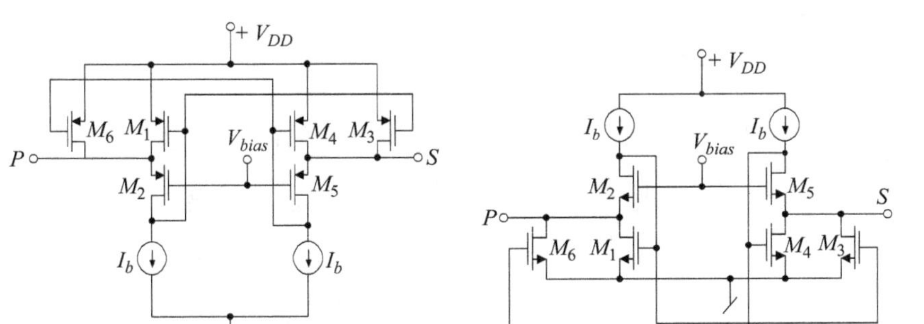

Figure 10.20 Bidirectional active transformers [94]: (a) PMOS-configured active transformer and (b) NMOS-configured active transformer

two transconductance amplifiers and two grounded capacitors are shown in Figure 10.21. The second topology presented by Yuan [81] is structurally quite similar to the first one; the only difference being the interchange of input and output terminals in the two single-ended transconductors employed therein.

The same basic idea was extended to design active transformer with multiple secondary winding. Subsequently, a number of basic transformer structures were elaborated using simple two MOSFET transconductance stages. The interested readers may see the complete catalogue of all CMOS active transformers in [81].

Application: DiClemente, Yuan and Tang [77] utilized the active CMOS transformer in the design of a current mode phase lock loop (PLL) which consists of a phase detector and active transformer, a current controlled oscillator and buffer as per the schematic of Figure 10.22.

Thus, in this architecture, the current-mode loop filter has been constructed using an active transformer with multiple primary winding and a single secondary winding. It was demonstrated that by employing a CMOS active transformer, the proposed PLL architecture provides the significant advantages of small chip area, low sensitivity to changes in the supply voltage, fast locking and good phase noise performance.

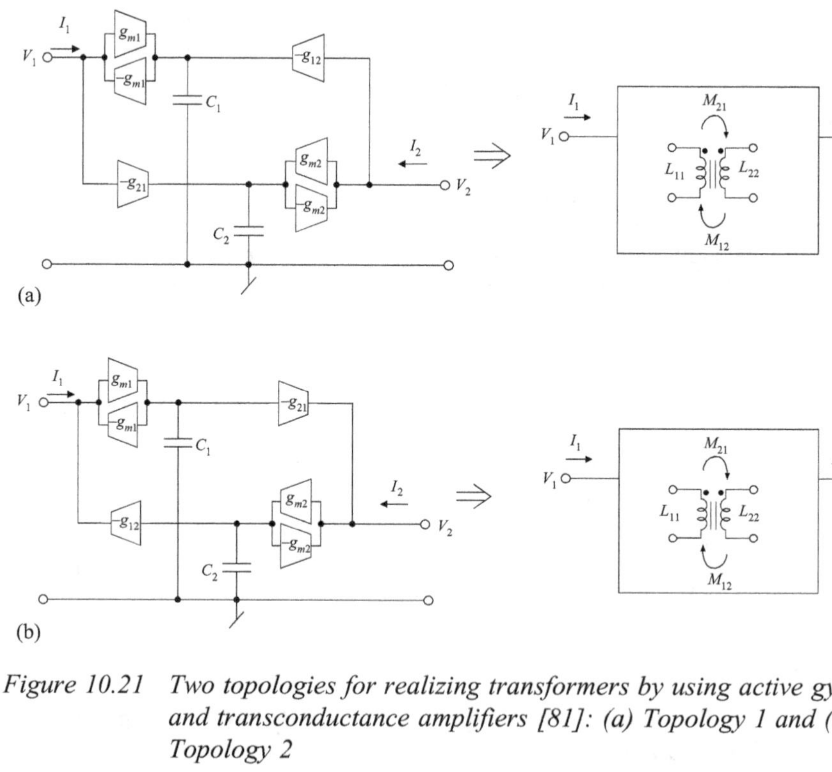

Figure 10.21 Two topologies for realizing transformers by using active gyrators and transconductance amplifiers [81]: (a) Topology 1 and (b) Topology 2

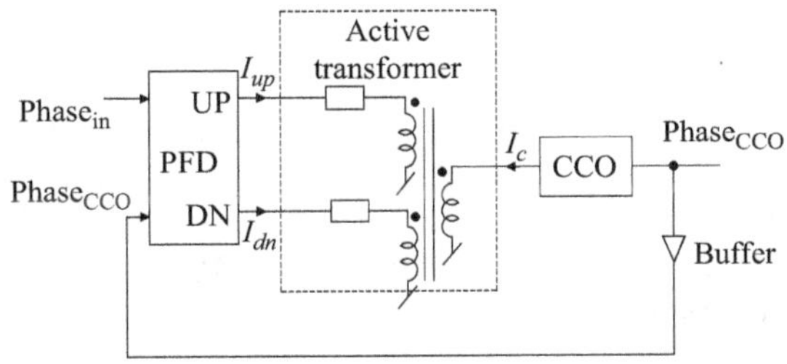

Figure 10.22 Current-mode PLL employing an active transformer [77]

10.5 Concluding remarks

In this chapter, we had focused on circuits and techniques developed to design semiconductor inductors and transformers on Silicon chip. Both bipolar and CMOS-based structure were considered. Historically, the BJT was the first semiconductor device which was considered to devise means for simulating an inductance. Hence, first, we discussed some BJT-based simulated inductors and gyrators. The developments in this area can be broadly divided into two classes – first, the circuits developed using conventional BJTs circuits which employed resistors also along with the BJTs and second, the circuits based upon translinear cells which can be devised without employing any resistors. In the first category, although relatively a larger number of circuits have been developed we have kept the discussion limited to only a few prominent results. On the other hand, to the best of authors' knowledge, there has been only one attempt so far in realizing a translinear electronically tunable floating inductance circuit and the same has been elaborated in Section 10.2.6. Thus, there appears to be enough scope to develop new and improved translinear inductors and gyrators suitable for implementation in bipolar technology. Since translinear concepts are also applicable to MOS circuits, these ideas could also be carried over to MOS-translinear circuits-based gyrators and inductors. This, however, calls for further research.

In the context of VLSI and particularly with reference to the implementation of analog functions in CMOS technology, CMOS-based inductors are natural choice for which the survey of the literature shows that a much larger number of CMOS active inductors have been proposed in the literature as compared to their bipolar counterparts. A very good compendium of CMOS active inductors was made available some time back by Yuan in 2008 [81]; hence, we have limited our discussion to some prominent circuits developed after 2008 only (although a few interesting circuits from the period before 2008 have also been briefly dealt with). The last part of this chapter also discusses some topologies for realizing CMOS active transformers.

Lastly, in view of the continuity of the ongoing work in this area, it is believed that the work done on this topic is by no means complete and it is hoped that many meaningful contributions may still be waiting to be explored in future.

References

[1] Dill H. G. 'Inductive semiconductor elements and their application in bandpass amplifiers'. IRE Transactions on Military Electronics. 1961; **MIL-5**(3):239–50.

[2] Dutta Roy, S. C. 'The inductive transistor'. IEEE Transactions on Circuit Theory. 1963; **10**(1):113–15.

[3] Harrison, T. J. 'A gyrator realization'. IEEE Transactions on Circuit Theory. 1963; **10**(2):303.

[4] Dutta Roy, S. C. 'A novel high-Q inductance and a tuned oscillator for micro-miniature circuits'. Proceedings of the IEEE. 1964; **52**(2):214–15.

[5] Shenoi, B. A. 'Practical realization of a gyrator circuit and RC-gyrator filters'. IEEE Transactions on Circuit Theory. 1965; **CT-12**(3):374–80.

[6] New W., Newcomb, R. W. 'An integratable time-variable gyrator'. Proceedings of the IEEE. 1965; **53**(12):2161–2.

[7] Rao T. N., Newcomb, R. W. 'Direct-coupled gyrator suitable for integrated circuits and time variation'. Electronics Letters. 1966; **2**(7):250–1.

[8] Sheahan D. F., Orchard, H. J. 'High-quality transistorised gyrator'. Electronics Letters. 1966; **2**(7):274–5.

[9] Sheahan D. F., Orchard, H. J. 'Integratable gyrator using M.O.S. and bipolar transistors'. Electronics Letters. 1966; **2**(10):390–1.

[10] Sheahan, D. F. 'Gyrator-floatation circuit'. Electronics Letters. 1967; **3**(1):39–40.

[11] Sheahan D. F., Orchard, H. J. 'Bandpass-filter realisation using gyrators'. Electronics Letters. 1967; **3**(1):40–2.

[12] Holmes W. H., Gruetzmann S., Heinlein, W. E. 'Direct-coupled gyrators with floating ports'. Electronics Letters. 1967; **3**(2):46–7.

[13] Holmes W. H., Gruetzmann S., Heinlein, W. E. 'High-performance direct-coupled gyrators'. Electronics Letters. 1967; **3**(2):45–6.

[14] Holmes W. H. 'Practical realization of an ideal active transformer with dc response'. Proceedings of the IEEE. 1967; **55**(3):471–2.

[15] Rao T. N., Gary P., Newcomb, R. W. 'Equivalent inductance and Q of a capacitor-loaded gyrator'. IEEE Journal of Solid-State Circuits. 1967; **2**(1):32–3.

[16] Chua H. T., Newcomb, R. W. 'Integrated direct-coupled gyrator'. Electronics Letters. 1967; **3**(5):182–4.

[17] Looij, H. TH. Van., Adams, K. M. 'Phase-compensation in electronic gyrator circuits'. Electronics Letters. 1968; **4**(20):430–1.

[18] Looij, H. TH. Van., Adams, K. M. 'Wideband electronic gyrator circuit'. Electronics Letters. 1968; **4**(20):431–2.

[19] Mitra, S. K. 'Synthesizing active filters'. IEEE Spectrum. 1969; **6**(1): 47–63.

[20] McGregor, I. M. 'A fully Integrable gyrator'. IEEE Journal of Solid-State Circuits. 1969; **4**(1):53–4.

[21] Holmes W. H., Heinlein, W. E., Gruetzmann, S. 'Sharp-cutoff low-pass filters using floating gyrators'. IEEE Journal of Solid-State Circuits. 1969; **4**(1):38–50.

[22] Holmes, W. H. 'An improved version of a floating gyrator'. IEEE Journal of Solid-State Circuits. 1969; **4**(3):162–3.

[23] Haykim S. S., Kramer S., Shewchun J., Treleaven, D. H. 'Integrated-circuit implementation of direct-coupled gyrator'. IEEE Journal of Solid-State Circuits. 1969; **4**(3):164–6.

[24] Rao, T. N. 'Readily biased wideband gyrator circuit for floating or earthed inductors'. Electronics Letters. 1969; **5**(14):309–10.

[25] Daniels, R. W. 'Gyrators, negative impedance converters, and related circuits'. IEEE Transactions on Circuit Theory. 1969; **16**(2):261–2.

[26] Daniels, R. W. 'The synthesis of fundamental gyrators'. IEEE Transactions on Circuit Theory. 1969; **16**(4):543–4.

[27] Orchard H. J., Sheahan, D. F. 'Inductorless bandpass filters'. IEEE Journal of Solid-State Circuits. 1970; **5**(3):108–18.

[28] Daniels, R. W. 'Comments on "A fully integrable gyrator"'. IEEE Journal of Solid-State Circuits. 1970; **5**(3):126.

[29] Hazony, D. 'Permutation network operators'. IEEE Transactions on Circuit Theory. 1970; **16**(1):146–8.

[30] Wade J. G., Parker, D. W. 'Stability of a wideband gyrator circuit'. 1971; **7**(9):224–5.

[31] Daniels, R. W. 'A method for generating some active circuits'. IEEE Transactions on Circuit Theory. 1971; **18**(3):397–9.

[32] Trimmel H. R., Heinlein, W. E. 'Fully floating chain-type gyrator circuit using operational transconductance amplifiers'. IEEE Transactions on Circuit Theory. 1971; **18**(6):719–21.

[33] Holmes, W. H. 'On canonical second-order active RC filters'. IEEE Transactions on Circuit Theory. 1971; **18**(6):721–3.

[34] Holmes, W. H. 'On inverters and converters'. IEEE Transactions on Circuit Theory. 1971; **18**(6):723–6.

[35] Muller, J. H. W. 'Hybrid integrated gyrator in dual-in-line package for universal application'. IEEE Transactions on Circuit Theory. 1971; **18**(6):726–8.

[36] Bruton, L. T. 'A transistor realization of the generalized impedance converter'. The Radio and Electronic Engineer. 1972; **42**(3):133–6.

[37] Voorman H. O., Biesheuvel, A. 'An electronic gyrator'. IEEE Journal of Solid-State Circuits. 1972; **7**(6):469–73.

[38] Takahashi T., Hamada S., Watanabe R., Miyata, T. 'A simple method for producing floating semiconductor inductors'. IEEE Journal of Solid-State Circuits. 1975; **10**(1):75–7.

[39] Pookaiyaudom S., Kuhanont, T. 'Integrable accurate linear negative impedance conversion circuits'. Proceedings of the IEEE. 1978; **66**(1):88–9.

[40] Pookaiyaudom S., Surakampontorn, W. 'An accurate Integrable voltage-variable floating gyrator'. IEEE Transactions on Instrumentation and Measurement. 1980; **29**(1):15–19.

[41] Stevenson, J. K. 'Transformation for active-RC networks'. IEE Proceedings Pt. G. 1981; **128**(4):182–3.

[42] Hara S., Tokumitsu T., Tanaka T., Aikawa, M. 'Broad-band monolithic microwave active inductor and its application to miniaturized wide-band amplifiers'. IEEE Transactions on Microwave Theory and Techniques. 1988; **36**(12): 1920–4.

[43] Hara S., Tokumitsu T., Tanaka T., Aikawa, M. 'Lossless broad-band monolithic microwave active inductors'. IEEE Transactions on Microwave Theory and Techniques. 1989; **37**(12): 1979–84.

[44] Qiu, D. 'Circuit design of an integrable simulated inductor and its applications'. IEEE Transactions on Instrumentation and Measurement. 1991; **40**(6):902–7.

[45] Zhang G. F., Ripoll C. S., Gautier J. L., Villegas, M. L. 'Broadband lossless monolithic microwave floating active inductor'. 1993 International Symposium on Circuits and Systems; Chicago, IL, May 3–6, 1993.

[46] Lucyszyn S., Robertson, I. D. 'Monolithic narrow-band filter using ultrahigh-Q tunable active inductors'. IEEE Transactions on Microwave Theory and Techniques. 1994; **42**(12): 2617–22.

[47] Hou C. L., Chen C. C., Hwang C. M., Wu, J. S. 'Switched-OTA technique for floating immittance function simulators'. Microelectronics Journal. 1995; **26**(7):691–5.

[48] Thanachayanont A., Payne, A. 'VHF CMOS integrated active inductor'. Electronics Letters. 1996; **32**(11):999–1000.

[49] Yang S. G., Ryu G. H., Seo, K. S. 'Fully symmetrical, differential-pair type floating active inductors'. 1997 International Symposium on Circuits and Systems; Hong Kong, June 9–12, 1997.

[50] Kaunisto R., Alinikula P., Stadius K., Porra, V. 'A low-power HBT MMIC filter based on tunable active inductors'. IEEE Microwave and Guided Wave Letters. 1997; **7**(8):209–11.

[51] Li W., Ooi B. L., Xu Q. J., Kooi, P. S. 'High Q active inductor with loss compensation by feedback network'. Electronics Letters. 1999; **35**(16): 1328–9.

[52] Carreto-Castro F., Silva-Martinez J., Murphy-Arteaga, R. 'RF low-noise amplifiers in BiCMOS technologies'. IEEE Transactions on Circuits and Systems-II. 1999; **46**(7): 974–7.

[53] Thanachayanont A., Payne, A. 'CMOS floating active inductor and its applications to bandpass filter and oscillator designs'. IEE Proceedings Circuits Devices and Systems. 2000; **147**(1):42–8.

[54] Li D., Tsividis, Y. 'Active LC filters on silicon'. IEE Proceedings Circuits Devices and Systems. 2000; **147**(1):49–56.

[55] Karsilayan A. I., Schaumann, R. 'A high-frequency high-Q CMOS active inductor with DC bias control'. Proceedings of 43rd IEEE Midwest Symposium on Circuits and Systems; Lansing, MI, August 8–11, 2000.

[56] Ngamkham W., Kiranon W., Loescharataramdee C., Sukhwan, T. 'Electronically tunable synthetic floating inductance'. 2000 TENCON Proceedings Intelligent Systems and Technologies for the New Millennium (Cat. No. 00CH37119). 2000; **2**(2):26–9.

[57] Wu Y., Ismail M., Olsson, H. 'CMOS VHF/RF CCO based on active inductors'. Electronics Letters. 2001; **37**(8):472–3.

[58] Wu Y., Ding X., Ismail M., Olsson, H. 'Inductorless CMOS RF bandpass filter'. Electronics Letters. 2001; **37**(16): 1027–9.

[59] Despain E., Choma, Jr., J. 'A new monolithic variable capacitor'. Analog Integrated Circuits and Signal Processing. 2001; **26**(2):89–102.

[60] Pennisi, S. 'CMOS multiplier for grounded capacitors'. Electronics Letters. 2002; **38**(15):765–6.

[61] Pino J. D., Sendra J. R., Hernandez A., *et al.* 'Models and tools for CMOS integrated inductors'. Analog Integrated Circuits and Signal Processing. 2002; **33**(2): 171–8.

[62] Hsiao C. C., Kuo C. W., Ho C. C., Chan, Y. J. 'Improved quality-factor of 0.18 μm CMOS active inductor by a feedback resistance design'. IEEE Microwave and Wireless Components Letters. 2002; **12**(12):467–9.

[63] Yang J. N., Cheng Y. C., Lee, C. Y. 'A design of CMOS broadband amplifier with high-Q active inductor'. Proceedings of the 3rd IEEE International Workshop on System-on-Chip for Real-Time Applications. 2003; 86–9.

[64] Bakken T., Choma, J. 'Gyrator-based synthesis of active on-chip inductances'. Analog Integrated Circuits and Signal Processing. 2003; **34**(3):171–81.

[65] Wu Y., Ding X., Ismail M., Olsson, H. 'RF bandpass filter design based on CMOS active inductors'. IEEE Transactions on Circuits and Systems-II. 2003; **50**(12): 942–9.

[66] Xiao H., Schaumann R., Daasch W. R., Wong P. K., Pejcinovic, B. 'A radio-frequency CMOS active inductor and its application in designing high-Q filters'. IEEE International Symposium on Circuits and Systems (ISCAS 2004); Vancouver, Canada, May 23–26, 2004.

[67] Oh Y. H., Lee, S. G. 'An inductance enhancement technique and its application to a shunt-peaked 2.5 Gb/s transimpedance amplifier design'. IEEE Transactions on Circuits and Systems-II. 2004; **51**(11):624–8.

[68] Mukhopadhyay R., Park Y., Sen P., *et al.* 'Reconfigurable RFICs in Si-based technologies for a compact intelligent RF front-end'. IEEE Transactions on Microwave Theory and Techniques. 2005; **53**(1):81–93.

[69] Lu L. H., Liao Y. T., Wu, C. R. 'A miniaturized Wilkinson power divider with CMOS active inductors'. IEEE Microwave and Wireless Components Letters. 2005; **15**(11):775–7.

[70] Jiang J., Yuan, F. 'New CMOS class AB transmitter for 10 Gb/s serial links'. Analog Integrated Circuits and Signal Processing. 2006; **47** (2):103–2.

[71] Yuan, F. 'A fully differential VCO cell with active inductors for GBPs serial links'. Analog Integrated Circuits and Signal Processing. 2006; **47** (2):213–23.

[72] Yuan, F. 'A modified Park-Kim voltage-controlled ring oscillator for multi-GBPs serial links'. Analog Integrated Circuits and Signal Processing. 2006; **47** (3):345–53.

[73] Mahmoudi F., Salama, C. A. T. '8 GHz 1 V, CMOS quadrature down-converter for wireless applications'. Analog Integrated Circuits and Signal Processing. 2006; **48**(3):185–97.

[74] Lu L. H., Hsich H. H., Liao, Y. T. 'A wide tuning-range CMOS VCO with a differential tunable active inductor'. IEEE Transactions on Microwave Theory and Techniques. 2006; **54**(9):3462–8.

[75] Abdalla M. A. Y., Phang K., Eleftheriades, G. V. 'A 0.13-μm CMOS phase shifter using tunable positive/negative refractive index transmission lines'. IEEE Microwave and Wireless Components Letters. 2006; **16**(12): 705–7.

[76] Uyanik H. U., Tarim, N. 'Compact low voltage high-Q CMOS active inductor suitable for RF applications'. Analog Integrated Circuits and Signal Processing. 2007; **51**(3):191–4.

[77] DiClemente D., Yuan, F., Tang, A. 'Current-mode phase-locked loops-a new architecture'. IEEE Transactions on Circuits and Systems-II. 2007; **54**(4):303–7.

[78] Hsieh H. H., Liao Y. T., Lu, L. H. 'A compact quadrature hybrid MMIC using CMOS active inductors'. IEEE Transactions on Microwave Theory and Techniques. 2007; **55**(6):1098–104.

[79] Chen J., Saibi F., Lin J., Azadet, K. 'Electrical backplane equalization using programmable analog zeros and folded active inductors'. IEEE Transactions on Microwave Theory and Techniques. 2007; **55**(7):1459–66.

[80] Xiao H., Schaumann, R. 'A 5.4-GHz high-Q tunable active-inductor band-pass filter in standard digital CMOS technology'. Analog Integrated Circuits and Signal Processing. 2007; **51**(1):1–9.

[81] Yuan, F. 'CMOS gyrator-C active transformers'. IET Circuits Devices and Systems. 2007; **1**(6):494–508.

[82] Tang A., Yuan F., Law, E. 'A new CMOS active transformer QPSK modulator with optimal bandwidth control'. IEEE Transactions on Circuits and Systems-II. 2008; **55**(1): 11–15.

[83] Ler C. L., A'ain A. K. B., Kordesch, A. V. 'CMOS source degenerated differential active inductor'. Electronics Letters. 2008; **44**(3):196–7.

[84] DiClemente D., Yuan F., Tang, A. 'Current-mode phase-locked loops with CMOS active transformers'. IEEE Transactions on Circuits and Systems-II. 2008; **55**(8):771–5.

[85] Nair M. U., Zheng Y. J., Lian, Y. '1 V, 0.18 μm-area and power efficient UWB LNA utilising active inductors'. Electronics Letters. 2008; **44**(19): 1127–9.

[86] Abbadi M. I., Jaradat, A. R. M. 'Artificial voltage-controlled capacitance and inductance using voltage-controlled transconductance'. International Journal of Electrical and Computer Engineering. 2008; **2**(8):1652–5.

[87] Yuce E., Minaei, S. 'Electronically tunable simulated transformer and its application to stagger-tuned filter'. IEEE Transactions on Instrumentation and Measurement. 2008; **57**(9):2083–8.

[88] Ler C. L., A'ain A. K. B., Kordesch, A. V. 'Compact, high-Q, and low-current dissipation CMOS differential active inductor'. IEEE Microwave and Wireless Components Letters. 2008; **18**(10):683–5.

[89] Yuan F. 'CMOS Active Inductors and Transformers: Principle, Implementation, and Applications'. Boston, MA: Springer; 2008. p. 21–100.

[90] Reja M. M., Filanovsky L. M., Moez, K. 'Wide tunable CMOS active inductor'. Electronics Letters. 2008; **44**(25): 1461–3.

[91] Tang A., Yuan F., Law, E. 'A new constant-Q CMOS active inductor with applications to low-noise oscillators'. Analog Integrated Circuits and Signal Processing. 2009; **58** (1):77–80.

[92] Zito D., Fonte A., Pepe, D. 'Microwave active inductors'. IEEE Microwave and Wireless Components Letters. 2009; **19**(7):461–3.

[93] Ler C. L., A'ain A. K. B., Kordesch A. V. 'CMOS active inductor linearity improvement using feed-forward current source technique'. IEEE Transactions on Microwave Theory and Techniques. 2009; **57**(8):1915–24.

[94] Tang A., Yuan F., Law, E. 'CMOS active transformers and their applications in voltage-controlled quadrature oscillators'. Analog Integrated Circuits and Signal Processing. 2010; **62**(1):83–90.

[95] Li C., Gong F., Wang, P. 'Analysis and design of a high-Q differential active inductor with wide tuning range'. IET Circuits Devices and Systems. 2010; **4**(6):486–95.

[96] Ahmed A., Wight, J. '6.7 GHz high-Q active inductor design using parasitic cancellation with process variation control'. Electronics Letters. 2010; **46**(7):486–7.

[97] Krishnamurthy S. V., El-Sankary K., El-Masry, E. 'Noise-cancelling CMOS active inductor and its application in RF band-pass filter design'. International Journal of Microwave Science and Technology. 2010; Article ID 980957: 8 pages.

[98] Zheng Y., Saavedra, C. E. 'Frequency response comparison of two common active inductors'. Progress in Electromagnetics Research Letters. 2010; **13**:113–19.

[99] Li C., Gong F., Wang, P. 'Analysis and design of a high-Q differential active inductor with wide tuning range'. IET Circuits Devices and Systems. 2010; **4**(6):486–95.

[100] Said L. A., Madian A. H., Ismail M. H., Soliman, A. M. 'CMOS digitally programmable lossless floating inductor'. 2010 International Conference of Electron Devices and Solid-State Circuits (EDSSC); December 15–17, 2010, Hong Kong, China.

[101] Li C., Wang P., Gong, F. 'A low-power active inductor with improved Q-factor and its application to phase shifter'. Analog Integrated Circuits and Signal Processing. 2011; **66**(2):303–7.

[102] Leuzzi G., Stornelli V., Del Re, S. 'A tuneable active inductor with high dynamic range for band-pass filter applications'. IEEE Transactions on Circuits and Systems-II. 2011; **58**(10):647–51.

[103] Ebrahimzadeh M., Rezaei F., Rezaei, S. 'A new active inductor and its application to wide tuning range LC oscillator'. International Journal of Soft Computing and Engineering. 2011; **1**(5):111–14.

[104] Colucci P., Leuzzi G., Pantoli L., Stornelli, V. 'Third order integrable UHF bandpass filter using active inductors'. Microwave and Optical Technology Letters. 2012; **64**(6):1426–9.

[105] Nagel I., Fabre L., Pastre M., Krummenacher F., Cherkaoui R., Kayal, M. 'Tunable floating active inductor with internal offset reduction'. Electronics Letters. 2012; **48**(13):786–8.

[106] Minaei S., Yuce, E. 'A simple CMOS-based inductor simulator and frequency performance improvement techniques'. International Journal of Electronics and Communications (AEU). 2012; **66**(11):884–91.

[107] Shirazi N., Hamzehyan, R. 'Design of VCO with a differential tunable active inductor'. International Journal of Machne Learning and Computing. 2013; **3**(1):13–16.

[108] Sato T., Ito, T. 'Design of low distortion active inductor and its applications'. Analog Integrated Circuits and Signal Processing. 2013; **75**(2):245–55.

[109] Kia H. B., A'ain A. K., Grout I., Kamisian, I. 'A reconfigurable low-noise amplifier using a tunable active inductor for multistandard receivers'. Circuits Systems and Signal Processing. 2013; **32**(3):979–92.

[110] Tripetch, K. 'Symbolic analysis of input impedance of CMOS floating active inductors with application in fully differential bandpass amplifier'. International Journal of Electrical and Computer Engineering. 2013; **7**(12):1612–20.

[111] Pepe D., Zito, D. '50 GHz mm-wave CMOS active inductor'. IEEE Microwave and Wireless Components Letters. 2014; **24**(4):254–6.

[112] Kia H. B., Aain, A. K. 'A high gain low flicker noise CMOS mixer with low flicker noise corner frequency using tunable differential active inductor'. 2014; **79**(1):599–616.

[113] Pantoli L., Stornelli V., Leuzzi, G. 'Class AB tunable active inductor'. Electronics Letters. 2015; **51**(1):65–7.

[114] Leuzzi G., Stornelli V., Pantoli L., Del Re, S. 'Single transistor high linearity and wide dynamic range active inductor'. International Journal of Circuit Theory and Applications. 2015; **43**(3):277–85.

[115] Kao H. L., Lee P. C., Chiu, H. C. 'A wide tuning-range CMOS VCO with a tunable active inductor'. Mathematical Problems in Engineering. 2015; **2015**:7 pages.

[116] Branchi P., Pantoli L., Leuzzi, G. 'RF and microwave high-Q floating active inductor design and implementation'. International Journal of Circuit Theory and Applications. 2015; **43**(8):1095–104.

[117] Matsumoto F., Nishioka S., Ohbuchi T., Fujii, T. 'Design of a symmetry-type floating impedance scaling circuits for a fully differential filter'. Analog Integrated Circuits and Signal Processing. 2015; **85**(2):253–61.

[118] Bhattacharya R., Basu A., Koul, S. K. 'A highly linear CMOS active inductor and its application in filters and power dividers'. IEEE Microwave and Wireless Components Letters. 2015; **25**(11):715–17.

[119] Pantoli L., Stornelli V., Leuzzi, G. 'Low-noise tunable filter design by means of active components'. Electronics Letters. 2016; **52**(1):86–8.

[120] Bhuiyan M. A. S., Zijie Y., Yu J. S., Reaz M. B. I., Kamal N., Chang, T. G. Active inductor based fully integrated CMOS transmit/receive switch for 2.4 GHz RF transceiver'. Anais da Academia Brasileira de Ciencias. 2016; **88**(2):1089–98.

[121] Saberkari A., Ziabakhsh S., Martinez H., Alarcon, E. 'Active inductor-based tunable impedance matching network for RF power amplifier application'. Integration, the VLSI Journal. 2016; **52**:301–8.

[122] Momen H. G., Yazgi M., Kopru R., Saatlo, A. N. 'Design of a new low loss fully CMOS tunable floating active inductor'. Analog Integrated Circuits and Signal Processing. 2016; **89**(3):727–37.

[123] Zhong L., Lai X., Xu D., Short M., Yuan B., Wang, Z. 'An improved CMOS-based inductor simulator with simplified structure for low-frequency applications'. Journal of Computational Electronics. 2016; **15**(3): 1017–22.

[124] Mehra R., Kumar V., Islam A., Kaushik, B. K. 'Variation-aware widely tunable nanoscale design of CMOS active inductor-based RF bandpass filter'. International Journal of Circuit Theory and Applications. 2017; **45**(12): 2181–200.

[125] Kumar V., Mehra R., Islam, A. 'A CMOS active inductor based digital and analog dual tuned voltage-controlled oscillator'. Microsystem Technologies. 2017; https://doi.org/10.1007/s00542-017-3457-y.

[126] Stornelli V., Pantoli L., Leoni A., Leuzzi, G. 'RF and microwave filter design by means of single transistor active inductor: a review'. Microwave Review. 2017; **23**(2):3–7.

[127] M'harzi Z., Alami M., Temcamani, F. 'A new grounded current controlled inductor based on simplified current conveyors'. International Journal of Electronics and Telecommunications. 2017; **63**(2):227–32.

[128] Sagbas M., Ayten U. E., Sedef H., Minaei, S. 'Modified Gorski-Popiel technique and synthetic floating transformer circuit using minimum components'. Journal of Circuits, Systems, and Computers. 2017; **26**(1): 21 pages.

[129] Sotner R., Herencsar N., Jerabek J., Kartci A., Koton J., Dostal, T. 'Pseudo-differential filter design using novel adjustable floating inductance simulator with electronically controllable current conveyors'. Elektronika ir Elektrotechnika. 2017; **23**(2):31–5.

[130] Ou J., Ferreira, P. M. 'Design considerations of CMOS active inductor for low power applications'. Analog Integrated Circuits and Signal Processing. 2018; **94**(3):347–57.

[131] Muhamad M., Soin N., Ramiah, H. 'Design of 2.4 GHz CMOS floating active inductor LNA using 130 nm technology'. International Conference on Applied Electronic and Engineering 2017 (ICAEE2017), IOP Conference Series: Materials Science and Engineering. 2018; **341:** 1–7.

[132] Sachan D., Goswami M., Misra, P. K. 'A high-Q floating active inductor using 130 nm BICMOS technology and its application in IF band pass filter'. Analog Integrated circuits and Signal Processing. 2018; **96**(3):385–93.

[133] Sotner R., Jerabek J., Herencsar N., Langhammer L., Petrzela J., Dostal, T. 'Methods for extension of tunability range in synthetic inductance simulators'. Elektronika ir Elektrotechnika. 2018; **24**(3):41–5.

[134] Patel D. P., Oza-Rahurkar, S. 'Tunable CMOS active inductor using Widlar current source'. Journal of Circuits, Systems, and Computers. 2019; **28**(2):1–12.

[135] D'Angelo G., Fanucci L., Monorchio A., Monterastelli A., Neri, B. 'High-quality active inductors'. Electronics Letters. 199;**35**(20):1727–8.

[136] Li C., Gong F., Wang, P. 'Analysis and design of a high-Q differential active inductor with wide tuning range'. IET Circuits, Devices & Systems. 2010; **4**(6):486–95.

[137] Ranjan R. K., Sagar S., Roushan S., Kumari B., Rani N., Khateb, F. 'High-frequency floating emulator and its experimental results'. IET Circuits, Devices & Systems. 2019; **3**(5):292–302.

Chapter 11
Recent developments and concluding remarks

Abstract

This chapter first presents a retrospection of the developments on gyrators, simulated inductors and other impedances as presented in the preceding ten chapters of this monograph. In doing so, we present several general conclusions about a large number of circuits discussed in this monograph and reiterate a number of unresolved problems which are open to investigation for further research. Besides this, we also outline a number of interesting developments of relatively more recent origin and also give an overview of the impact of the circuits and techniques of impedance simulation on the areas of memristive circuits and fractional-order circuits – both of which are currently being rigorously investigated in the analog circuits' literature.

11.1 Introduction

In this concluding chapter, we present a short summary of the entire content of this monograph wherein we re-iterate a number of problems which have not been attempted so far in the published literature and which, therefore, are open to investigation. We also point out some of the recent developments, along with relevant background [1–69], on the simulated inductor circuits and will demonstrate how the circuits and techniques developed in this area have impacted the research in several other related fields such as devising circuit emulators for the newly researched area of nonlinear network elements – the memristor, memcapacitor and meminductor and the realization of fractional capacitance and fractional inductance elements so vital in the context of modern research on fractional-order circuits and systems.

11.2 Retrospection

In this last chapter of the monograph, we go into a retrospection mode and present an overview about how this area has progressed over the last several decades, what have been the significant developments during this period and what are the most significant results (in the opinion of the authors of this monograph). We also make an attempt to classify the work reported in the earlier ten chapters of this monograph and figure out several general methods and demonstrate how in several cases the circuits and techniques may appear to be different *apparently* but when examined closely, they turn out to be based upon some result known earlier using some other building blocks. However, we may add that this aspect of research is not surprising. The only surprising thing is to observe that more often than not, the proposers of such so-called new circuits have inadvertently or deliberately have never given reference to those earlier works where the ideas responsible for their new propositions were originally published. In the process, we also try to sum up the best known solutions in most of the cases. This part of our discussion may appear to be a repetition, but we believe that the repetition is worth doing since it provides a good insight into what best has already been achieved and what is left out and yet to be achieved on which the current and future researchers may focus in order to bring out significantly useful new results.

Looking back, we can see that the available gyrators/inductance simulation circuits, depending upon the type of active elements deployed, can be broadly categorized into the following categories:

Class A: Circuits using commercially available building blocks – in this category, we can include circuits containing IC op-amps (Chapter 2), IC operational transconductance amplifiers (OTAs) (Chapter 3), IC current conveyors (CCs) (Chapter 4) and IC current feedback op-amps (CFOAs) (Chapter 5); the latter two being popularly implementable using the commercially available AD844-type CFOAs.

Class B: Circuits using building blocks which are although not commercially available as off-the-shelf ICs – in this category, we can include circuits containing four terminal floating nullors (FTFNs), operational transresistance amplifiers (OTRAs), current differencing buffered amplifiers (CDBAs), VF/CF, all of which can be realized using *at the most two* AD844 CFOAs.

Class C: Circuits using controlled CCII, numerous other varieties of CCs (Chapter 4) and circuits using other composite building blocks such as current differencing transconductance amplifier (CDTA), current follower transconductance amplifier (CFTA), current differencing CC (CDCC). voltage differencing CC (VDCC), voltage differencing TA (VDTA), CC transconductance amplifier (CCTA) and numerous others building blocks (Chapter 8) which have been shown to be realisable by various authors using a combination of different ICs such as IC op-amps, IC CFOAs, IC OTAs and others.

Class D: Circuits employing BJTs as active elements.

Class E: Circuits employing MOSFETs as active elements.

In the context of op-amp-based circuits, the following may be noted:

1. As conclusively proved by Willson and Orchard [55], an ideal gyrator with both ports grounded cannot be realized using a single op-amp. As a consequence, all the single op-amp gyrators known so far employ a floating capacitor when used for lossless inductor realization. Moreover, all known single op-amp gyrator/grounded

inductor (GI) circuits necessarily require component-matching or cancellation constraints in terms of resistors employed.

2. Given two or more op-amps, a lossless gyrator/GI with both ports grounded is possible to be realized but still component-matching or cancellation constraints in terms of resistors employed are unavoidable.
3. However, if the capacitor is allowed to be floating, a lossless GI can be realized with only two op-amps and four resistors without requiring any component-matching/cancellation condition requirements using the positive impedance converter (PIC)/generalized impedance converter (GIC)-type circuits. Although six distinctly different circuits of such GIC-based lossless grounded inductors are possible which were reported by different researchers, however, the Antoniou's GIC is widely recognized to be the best circuit for this purpose and the necessary conditions for realizing lossless GI as well as ideal frequency-dependent-negative resistance (FDNR) using this under the influence of finite gain bandwidth product (GBP) of the op-amps employed are well understood and well documented and is available even in several standard texts.
4. It is well known that using IC op-amps, it is impossible to realize any kind of floating inductance or floating FDNR circuits without requiring any component-matching or without fulfilling certain cancellation conditions in terms of the resistors employed. Furthermore, lossless floating inductors (LLFIs) can be realized with four op-amps with a single matching condition but require two floating capacitors. On the other hand, the realizations which can be accomplished with only three op-amps along with a grounded capacitor require more stringent component-matching requirements/cancellation conditions.
5. In contrast to LLFIs, lossy floating inductors (LFIs); series resistance inductance (RL) or parallel RL) can be realized using a smaller number of op-amps (only two) and fewer resistance capacitance (RC) components and need no more than one matching condition but still require two capacitors. Single-capacitor-based LLFI/LFIs possible with only two op-amps, by contrast, suffer from the drawback of requiring more stringent component-matching conditions/cancellation constraints.
6. Curiously, no single op-amp-based FI of any kind (lossy or lossless) has been discovered till date; neither it has been conclusively/mathematically proved anywhere that this cannot be done! This appears to be an interesting problem which is open to investigation.

In the context of OTA-C circuits, the following are worth pointing out:

1. Although there have been circuits which, along with OTAs and capacitors, have also utilized in addition, passive resistors and sometimes other active building blocks such as op-amps and even CCs and their variants, the potential advantages accrue when only OTA-C circuits are employed since such circuits, provide electronic-tunability of parameters of the functional circuits synthesized using them, eliminate the use of resistors, use integrable low-valued capacitors (a few pF) and are hence, suitable for IC implementation in both bipolar (using bipolar OTAs) and CMOS (using CMOS OTAs) technologies.

2. Thus, in the context of OTA-C circuits, it is clear that an ideal gyrator needs two OTAs and a lossless grounded inductor realization too needs two OTAs along with a grounded capacitor. In both the cases only one input terminal of the OTA is employed while the other is grounded and thus, the circuits can be seen to be parallel back-to-back connection of two transconductance elements of opposite polarities to make either a lossless inductor or a two-port gyrator. Surprisingly, if one wants to simulate only a lossy, series RL or parallel RL type of inductor, there is no respite from the number of OTAs required. One still needs two OTAs and a grounded capacitor to make grounded series-RL impedance or grounded parallel-RL admittance. Of course, one can combine one OTA with one unity gain buffer (voltage follower) to create circuits for the same purpose.
3. All OTA-C inductance simulation circuits suffer from temperature-sensitivity in the case of realization using bipolar OTAs since the transconductance of the BJT is temperature-dependent. The only exception is the circuit of a capacitance multiplier which can be realized by using two OTAs and a buffer such that the capacitance multiplying factor being $(1 + (g_{m1}/g_{m2}))$ reduces to $(1 + (I_{B1}/I_{B2}))$ wherein the temperature-dependent term V_T nicely cancels out!
4. Though a number of circuits for simulating floating inductors and other FIs have been presented by various researchers using four to five OTAs, it is well understood that only three OTAs of the commercially available differential-input–single-output (DISO) type would suffice to create a floating inductor using a grounded capacitor. However, in such a circuit, two OTAs are required to have identical g_m, which can be achieved simply by joining the external bias current terminal of these two OTAs together and injecting an external current equal to $2I_B$.
5. On the other hand, if OTAs with differential-input–dual (complementary) outputs (DOTA) are available, an LLFI can be simulated with only two DOTAs whereas the capacitor employed can be either floating or grounded. Obviously, the same circuit can also simulate a lossless grounded inductor by shorting to ground either of the two ports of the quoted FI. Thus, either an electronically controllable LLFI or electronically controllable lossless GI can be realized with only two DOTAs and a grounded capacitor.
6. It is ironical that during the entire research on inductor simulation using OTAs, the work of Sewell[1] went, by and large, unnoticed who had demonstrated as early as in 1969 that using two five terminal voltage-controlled–current-source (VCCS)-type amplifiers, a four-port fully floating gyrator, a three-port grounded gyrator and an LLFI using a floating or grounded capacitor – all can be realized with only two such VCCS elements. Of course, the OTA (which was a VCCS) was not yet produced as an off-the-shelf IC till then and was not in vogue in those days! In fact, an IC OTA CA3080 came into prominence and became widely known much later.

[1]Sewell J. I. 'A simple method for producing floating inductors'. Proceedings of the IEEE. 1969; **57** (12):2155–6.

7. In view of the above, the state-of-the-art of gyrator realization using OTAs can be summed up; thus, a two-port grounded gyrator is realizable with only two commercially available DISO type OTAs. A three-port gyrator can be realized with three DISO-type OTAs. A fully floating four-port gyrator needs either two DOTAs or four DISO-type OTAs. Finally, from all the quoted gyrators which are positive impedance inverters, the corresponding negative impedance inverters can be realized by appropriate changes in the polarities of some OTAs while using exactly the same number and type of OTAs.

Let us now take a look at the gyrators/impedance simulation circuits using CCs next.

1. In the published literature, the CCI was introduced by Smith and Sedra [56] and CCII was proposed by Sedra and Smith [57]. However, as recently pointed out by Gilbert [58], the basic CC topology as well as the name *CC* for it was invented by him as early as in 1967 although its embodiment into a monolithic CFOA architecture devised during the same time was actually brought in the open by Analog Devices Inc. in 1984 when they mass produced the AD844 IC CFOA after a true complementary bipolar IC fabrication process, so vital for the successful integration of AD844, became available to them. In view of the recent disclosures[2] by Gilbert therefore, we believe that the credit for introducing the concept of CCs should be given to Sedra–Smith and Gilbert both.
2. There were several attempts for producing an IC CC by several other IC manufacturers such as CCII01 by LTP Electronics and PA630 by Phototronics Ltd. However, due to the various reasons, as compared to these alternatives, the so-called CFOA AD844 along with OPA860 appear to be the preferred choices for implementing CCII+ or CCII− based circuits. It may be noted that a CCII+ can be realized with only one AD844 while CCII− requires two AD844 ICs.
3. From the summary of op-amp-based impedance simulation as well as OTA-based impedance simulation circuits, it is found that realization of FIs using both of them cannot be done without requiring any component/parameter matching conditions. By contrast, it was shown for the first time by Senani [59] that using CCII− as an active element an FI can be simulated employing a bare minimum number of passive components and without requiring any component-matching conditions! The idea was soon followed up by other researchers who came up with a variety of FI simulators using one to four CCs – all having these two characteristic features, namely employment of minimum number of passive elements and complete lack of any component-matching conditions.
4. An ideal gyrator with both ports grounded can be realized with a CCII+, a CCII− and two grounded resistors only and that two distinctly different circuits are possible.

[2]Gilbert, B. 'About CFA etc.', private communication to Raj Senani of NSIT, March 08, 2015.

5. A three-port gyrator with port 3 grounded can be made from four CCIIs, all of which could be all CCII+ or all CCII−. However, if port 3 is allowed to be floating then such a three-port gyrator can be realized from only three CCIIs.
6. A lossless grounded inductor can be realized with two CCIIs of opposite polarities along with a grounded capacitor. Although single-CC active gyrators are possible and a number of them have been described in Chapter 4, as in the case of single op-amp-based active gyrators, here also, the port 2 cannot be grounded. Thus, a lossless GI is possible to be realized from a single CC (like a single op-amp active gyrator), but it would employ a floating capacitor.
7. A lossless floating inductance, floating FDNR or floating FDNC all can be realized with only two CCII− elements (each representing a three-terminal floating nullor) and only five passive elements (all of which are floating) without requiring any component-matching conditions and three distinctly different circuits of this kind are known.
8. On the other hand, if realizations of LLFI and lossless FDNR elements using a minimum number of only three passive elements are needed then at least three CCIIs (or two CCIIs and an NIC or two CCIIs and a VF) are needed and several such circuits have been described in Chapter 4.
9. If realization of such FIs using a canonic number of passive components without requiring any component-matching conditions along with the employment of a grounded capacitor are needed, then four CCIIs (or three CCIIs and one CCI) are needed and several such circuits have been devised in the literature.
10. If LFIs are needed, these can be readily realized using only two CCs along with only three passive components, again without requiring any component-matching conditions. No such circuits are possible with the conventional op-amps. Moreover, in all these CC-based circuits, the realized inductance value is single-resistance controllable.
11. Although a floating parallel RL, a floating series RL as well as a floating bilinear RL all can be realized using a single CCII− element employing only three passive elements without requiring any component-matching conditions, none of these three circuits provides a single-element-controllability of the realized inductance value which may be needed in some applications. Though such circuits are known,[3] they have so far not been made available in open literature.
12. The discovery of a single-CC LLFI, preferably with single-resistance-control of inductance value, is an unresolved problem which is open to investigation!

[3]A single-CCII-based circuit for single-resistance-controlled floating parallel RL admittance was derived in the Ph. D. thesis of the first author of this monograph: Senani, R. 'Realization of Some Classes of Active Networks'. University of Allahabad; 1987; Chapter 3. p. 143–6 (see [69]), from where the corresponding circuit for series RL FI is derivable by the operation of *port transposition* described in Chapter 4 of this monograph.

The state-of-the-art view of CFOA-based impedance simulation circuits is now in order which is as under:

1. Since a CFOA's internal circuit architecture consists of a CCII followed by a unity gain VF, it provides more capability and flexibility than CCIIs only. It is interesting to point out that AD844 was labelled as '60 MHz high slew rate monolithic op-amp' and thus was *disguised* as an op-amp! It became known much later that internally there is a CCII+ in it. It was demonstrated subsequently that an AD844 could be used as a building block in three different modes: first, as a replacement for a conventional op-amp since AD844 has complete pin compatibility with the popular 741-type op-amp and several others, second, to realize a CCII+ (one AD844) or CCII− (by a cascade of two AD844s) and third, as a four-terminal active building block in its own right.
2. One of the first applications of CFOA has been on gyrator realization wherein Fabre [60] showed that an ideal gyrator can be realized from two CFOAs (also called operational transimpedance amplifiers (TIA) sometimes) using only two resistors without requiring any component-matching condition which can be used to simulate a lossless grounded inductor using a grounded capacitor. Note that because of the CCII+ inside, CFOA-based gyrator too facilitates inductance simulation without requiring any component-matching conditions.
3. It was soon found that three CFOAs are all one needs to simulate an LLFI using only three passive elements (two resistors and a grounded capacitor) without requiring any component-matching conditions. Two such circuits were *independently* discovered at about the same time; one by Senani [61] and Chang [62]. Both the circuits realize three-port gyrators by removing the capacitor and treating the port thus created as port 3.
4. It was quite later found that lossy inductors in the form of series RL or parallel RL could be realized using only two CFOAs using a minimum of only three passive components without requiring any component-matching conditions while also providing single resistance tunability of the realized inductance/FDNR values.
5. In contrast to the op-amp- or CC-based circuits, which can although realize lossy inductors using a canonical number of passive elements along with a single active element but do not provide single resistance control of inductance value, several researchers have demonstrated that single-resistance tunable lossy grounded inductors and other related kind of elements can be realized using a single CFOA.
6. An interesting and unresolved problem in the area of CFOA-based impedance simulation is to devise a single-CFOA-based FI of any kind which does not appear to have been discovered so far!

The state of the art of the circuits of the classes B, C, D and E are well summarized in the respective chapters dealing with the circuits of these classes; hence, it will not be re-iterated here. Furthermore, although the status of impedance simulation circuits using OAs, OTAs, CCIIs and CFOAs is by and large completed, the same is not true about the circuits using other composite devices (on which research is still continuing as will be demonstrated in the next section) or those based on BJTs and MOSFETs as active elements; hence, the time is still not ripe to make any conclusive comments on these classes of circuits.

11.3 Recent developments on inductance simulation and related impedances

In this section, we present some of the more recent research directions concerning the theme of this monograph. In particular, we present the recent innovations of the following:

1. Evolution of single-active-element-based FI configurations.
2. FI configurations which although employ two active elements achieve the novel properties of electronic tunability of the realized impedance value and also *temperature-insensitivity* of the realization.
3. Impact of the circuits and techniques of impedance simulation in the area of memristor, memcapacitance and meminductance simulation.
4. Impact of the circuit techniques of impedance simulation in the area of fractional-order circuits.

These, in our opinion, are the four most significant directions of the current research.

11.3.1 Evolution of single-active-element-based floating impedance configurations

Recently, emphasis is being placed by researchers on evolving LLFI circuits using a single active element along with a minimum possible number of passive components (namely a single capacitor along with only two resistors).[4] Consequently, a number of interesting circuits have recently merged using a variety of building blocks. Of course, the building blocks employed have either more terminals or belong to the category of composite building blocks with enhanced capabilities introduced by Biolek *et al.* [63]. An obvious advantage of such circuits is that the same circuit could be used as either an LLFI or a lossless GI, with exactly the same number of active and passive elements in both the cases. In the following, we discuss various single active element based LFI or LLFI circuits which have so far been discovered in the literature so far.

Yuce [1] presented two circuits employing the so-called[5] modified CFOA which has the following characterization:

$$\begin{aligned} &I_Y = -I_W, \quad I_{Z1} = +I_X, \quad I_{Z2} = -I_X, \\ &I_{Z3} = I_W, \quad V_X = V_Y \quad \text{and} \quad V_W = V_{Z1} \end{aligned} \tag{11.1}$$

[4]One such circuit using a modified CFOA has been described earlier in Chapter 5 of this monograph (see Figure 5.31).

[5]A closer look at this MCFOA reveals that it is structurally same as a 'composite current conveyor' defined by Smith and Sedra [68].

The CMOS implementation presented in [1] (see Fig. 3 therein) clearly reveals that the circuit is made from an appropriate composite connection of two CCIIs having dual complimentary Z-outputs, in view of which, from the considerations made in Chapter 2, it can be appreciated that with such a composite building block, two such elements would suffice to create a floating inductance which can be realized with a bare minimum number of only three circuit admittances. Thus, Yuce [1] came up with two general floating admittance simulation networks, each employing an MCFOA characterized above and only three admittances, as shown here in Figure 11.1.

Both the circuits have exactly the same $[Y]$-matrix characterization:

$$\begin{bmatrix} I_1 \\ I_2 \end{bmatrix} = \frac{y_1 y_2}{y_3} \begin{bmatrix} 1 & -1 \\ -1 & 1 \end{bmatrix} \begin{bmatrix} V_1 \\ V_2 \end{bmatrix} \tag{11.2}$$

From the above, it is obvious that by appropriate selection of the various admittances, these circuits can realize a variety of circuit elements of interest as given below.

Sagbas in [5] proposed that if a current backward transconductance amplifier (CBTA) is employed as an active element, it becomes possible to devise an interesting circuit arrangement which can simulate a class of floating elements of interest using grounded passive elements. The so-called CBTA is a four-port active element which has input terminals labelled as p and n, a current output terminal z and a terminal w and is characterized by the following terminal equations:

$$I_z = g_m(V_p - V_n), \quad V_w = V_z, \quad I_p = I_w \quad \text{and} \quad I_n = -I_w \tag{11.3}$$

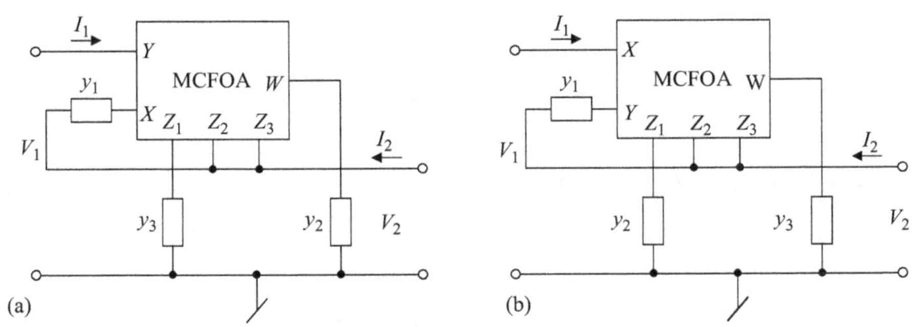

Figure 11.1 *FI, FC and FDNR simulators [1]: (a) lossless floating inductance: results if one selects y_3 as a capacitor with the remaining admittances being both resistors, thereby yielding $Z_{eq} = sC_3R_1R_2$, with the interesting advantage of simulating an LLFI using a grounded capacitor! (b) Floating capacitance simulator employing a grounded capacitor: for this realization one can have a grounded capacitor in place of y_2, thereby obtaining $C_{eq} = (R_3/R_1)C_2$ (c) Floating FDNR: it is obtained by choosing both y_1 and y_2 as capacitors resulting in $Z_{eq} = (1/s^2C_1C_2R_3)$*

The configuration of Figure 11.2 is characterized by the $[Y]$ matrix:

$$\begin{bmatrix} I_1 \\ I_2 \end{bmatrix} = g_m \left(\frac{Z_2}{Z_1} \right) \begin{bmatrix} 1 & -1 \\ -1 & 1 \end{bmatrix} \begin{bmatrix} V_1 \\ V_2 \end{bmatrix} \tag{11.4}$$

It is obvious that the circuit can simulate both an LLFI and lossless floating capacitor using a single grounded capacitor as preferred for IC implementation.

Another circuit exhibiting almost similar properties was presented by Tangsrirat [7] which employed a single active element known as differential voltage CC transconductance amplifier (DVCCTA). A DVCCTA has input terminals Y_1 and Y_2 having ideally infinite input impedance so that it senses a differential voltage and makes the same voltage available at its terminal X whereupon the current flowing into terminal X is conveyed to port Z at an ideally infinite output impedance. Additionally, the voltage at terminal Z is converted into two equal and opposite currents at its current output terminals labelled as O^+ and O^-. The configuration proposed by Tangsrirat [7] is shown in Figure 11.3 and has an equivalent FI given by

$$Z_{eq} = \frac{Z_1}{g_m Z_2} \tag{11.5}$$

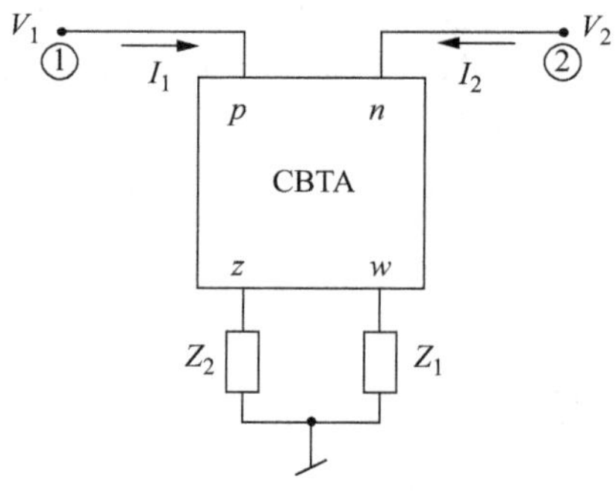

Figure 11.2 An FI configuration using CBTA as proposed by Sagbas [5]

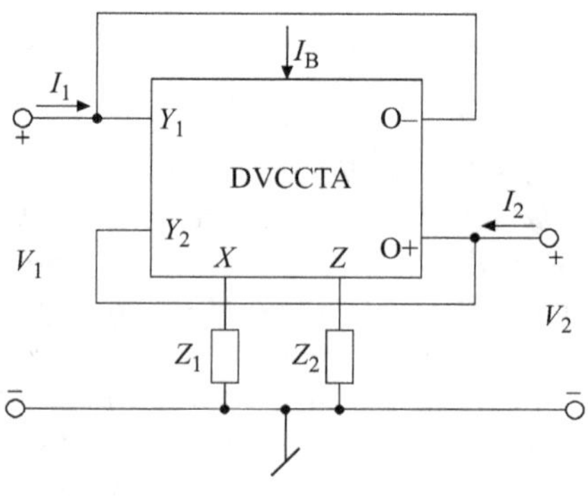

Figure 11.3 Floating simulator using a single DVCCTA [7]

From the above expression, it is not difficult to visualize that by selecting Z_2 as a capacitor and Z_1 as resistor the circuit realizes an electronically controllable FI while by choosing Z_1 as a capacitor and Z_2 as a resistor, the same circuit can realize an electronically controllable floating capacitor from a grounded capacitor.

Ibrahim *et al.* [6] utilized a modified form of DDCC termed a dual output DDCC (DO-DDCC) to evolve an interesting circuit which employed only a single DO-DDCC and only three passive elements to simulate a parallel RL-type FI simulator which is reproduced here in Figure 11.4.

By direct analysis, assuming the active element DO-DDCC to be having terminals Y_1, Y_2, Y_3, X, $Z+$ and $Z-$ with terminal equations given as

$$\begin{aligned} &I_{y1} = 0, \quad I_{y2} = 0, \quad I_{y3} = 0; \\ &V_x = V_{y1} - V_{y2} + V_{y3}; \qquad I_{z+} = +I_z, \quad I_{z-} = -I_z \end{aligned} \tag{11.6}$$

The $[Y]$ matrix of this circuit is found to be

$$[Y] = \left(\frac{1}{R_1} + \frac{1}{R_2} + \frac{1}{sCR_1R_2}\right)\begin{bmatrix} 1 & -1 \\ -1 & 1 \end{bmatrix} \tag{11.7}$$

Thus, the circuit realizes a parallel RL admittance without any passive component-matching or cancellation/equality constraints.

Later on, Abaci and Yuce [14] have presented a number of DDCC-based function circuits employing only grounded passive components, several of which have been described in Chapter 4. Here we highlight a specific circuit which employs only a single DO-DDCC element along with only three passive components, all of which are grounded and realize LFI in the form of a parallel RL admittance. This circuit is shown in Figure 11.5.

By straight-forward analysis, the $[Y]$ matrices of these circuits are found to be

$$[Y] = \left(\frac{1 + sCR_2}{R_1 - R_2 + sCR_1R_2}\right)\begin{bmatrix} 1 & -1 \\ -1 & 1 \end{bmatrix} \tag{11.8}$$

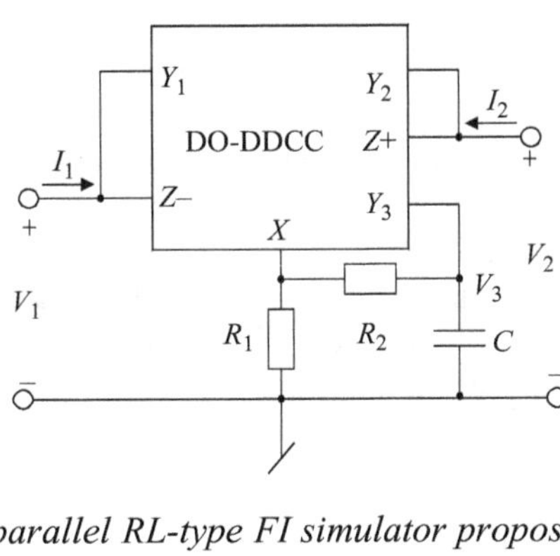

Figure 11.4 A lossy parallel RL-type FI simulator proposed by Ibrahim et al. [6]

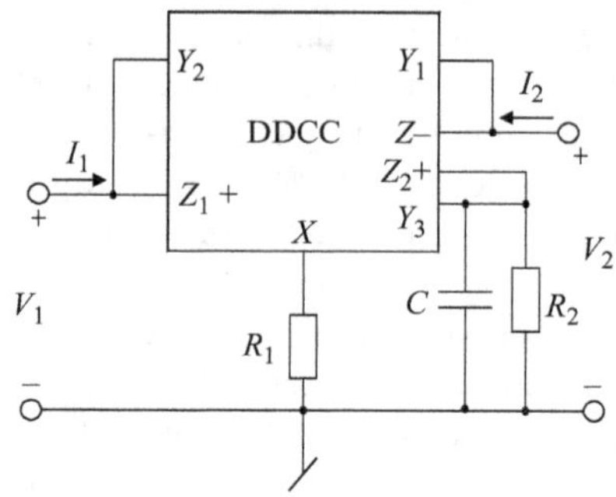

Figure 11.5 LFI realization employing a single DDCC and all grounded passive elements as per Abaci and Yuce [14]

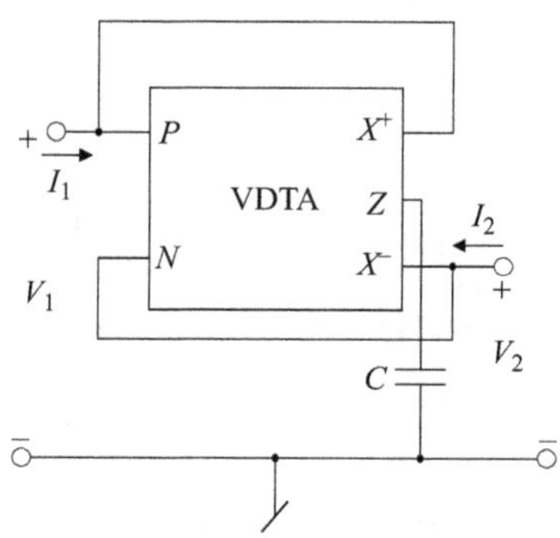

Figure 11.6 Single VDTA-based lossless FI (Shaktour–Biolek [4] and Guney–Kuntman [9]

which, with $R_1 = R_2 = R$ reduces to

$$[Y] = \left(\frac{1}{R} + \frac{1}{sCR^2}\right)\begin{bmatrix} 1 & -1 \\ -1 & 1 \end{bmatrix} \tag{11.9}$$

which represents a floating parallel RL admittance.

An interesting circuit for LLFI simulation was first presented in the Ph.D. thesis of Shaktour[6] [4] which was also presented by Guney and Kuntman [9] in a slightly different form since they called the active element employed as a *Z*-copy VDTA (ZC-VDTA) although this so-called *Z*-copy terminal was not explicitly employed by them in their proposition. Thus, omitting this additional unused terminal, the essential idea of both these circuits is exactly the same. The proposition in question is shown here in Figure 11.6.

[6]Supervised by Professor Dalibor Biolek.

The circuit is characterized by the following $[Y]$-matrix:

$$\begin{bmatrix} I_1 \\ I_2 \end{bmatrix} = \left(\frac{g_{mx}g_{mz}}{sC}\right)\begin{bmatrix} 1 & -1 \\ -1 & 1 \end{bmatrix}\begin{bmatrix} V_1 \\ V_2 \end{bmatrix} \tag{11.10}$$

A number of attempts have been made to employ VDCC as active elements for FI synthesis. In this context it may be noted that what can be achieved by a VDTA can be surely achieved by a VDCC too. The reason for this is as follows. First, it may be recalled that a VDTA is in fact a composite connection of two OTAs – the first one being a differential input single output (DISO)-type OTA while the second OTA being a single input and dual (complementary) output (SIDO)-type OTA. On the other hand, in a VDCC, the first unit is a DISO-type OTA while the second unit is a CCII with two complementary Z-outputs. Thus, in a VDCC, if an external resistor is connected from the X-input of the internal CCII to ground, this part of the circuit would perform exactly the same function as a SIDO-type OTA of the VDTA.

The single VDCC-based LLFI of Figure 11.7(a) is, therefore, analogous to the single VDTA-based FI described earlier while the single VDCC-based floating capacitance multiplier of Figure 11.7(b) is obtainable by interchanging the resistor and capacitor in the former circuit.

The FI simulated by the circuit of Figure 11.7(a) is given by

$$L_{eq} = \frac{CR}{g_m} \tag{11.11}$$

whereas the floating capacitance simulated by the circuit of Figure 11.7(b) is given by

$$C_{eq} = CRg_m \tag{11.12}$$

Obviously, in both the cases the element values simulated are electronically controllable by varying g_m which is controllable by an external DC bias current applied to the DISO-type OTA stage of the VDCC.

We now present one more single active element based LLFIsimulators using a somewhat more complex composite building blocks than those dealt with so far, namely the *dual-X CC differential input transconductance amplifier* (DXCCITA) [13]. The DXCCITA has two X-terminals and one Y-terminal such that a voltage V_y

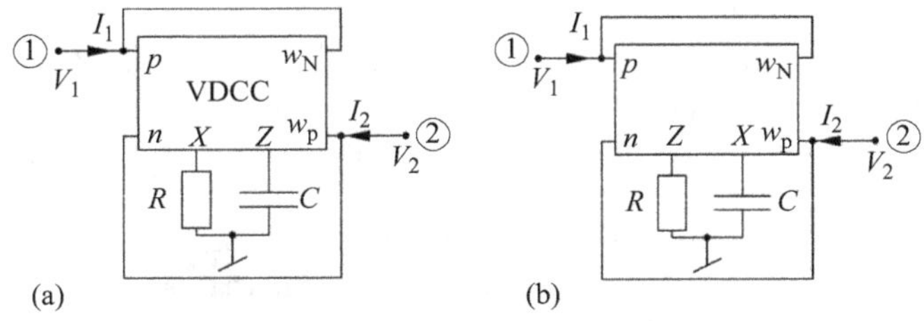

Figure 11.7 Two single-VDCC floating simulators: (a) the lossless FI simulator due to Prasad and Ahmed [8], (b) the floating capacitance multiplier due to Kartci et al. [11]

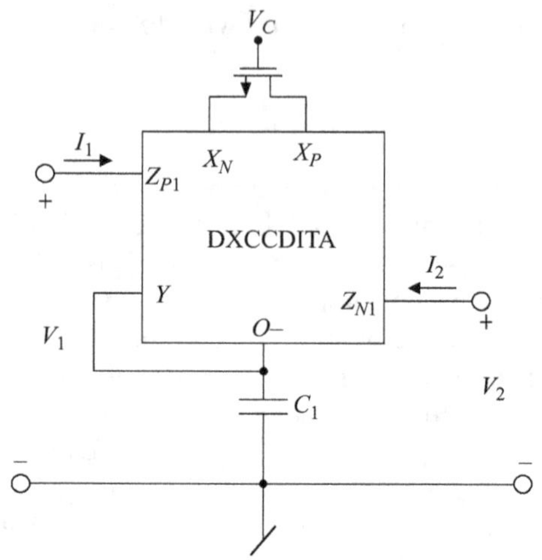

Figure 11.8 Single active element based electronically-controllable MOSFET-C FI due to Mohammad et al. [13]

applied on the high input impedance input terminal Y gets transferred to the two X-terminals as $V_{XP} = +V_Y$ and $V_{XN} = -V_Y$. On the other hand, the currents I_{XP} and I_{XN} are conveyed to ports I_{ZP1}, I_{ZP2} and I_{ZN1}, I_{ZN2} according to $I_{ZP1} = I_{XP}$, $I_{ZP2} = I_{XP}$ and $I_{ZN1} = I_{XN}$, $I_{ZN2} = I_{XN}$. Finally, the two output currents at its current output terminals $O+$ and $O-$ are given by $I_{o+} = -I_{o-} = g_m(V_{ZP1} - V_{ZN1})$. By a straightforward analysis of the circuit of Figure 11.8, the equivalent floating inductance value is given by

$$L_{eq} = \frac{sC_1 R_m}{2g_m} \tag{11.13}$$

where R_m is the equivalent voltage-controlled resistance of the nonlinearity cancelled MOSFET, controllable by the external control voltage V_c. Thus, this simulated FI has dual electronic tunability!

Finally, we present the single active element based FI proposed by Alpaslan [12] who built an FI around a building block called modified voltage differencing voltage transconductance amplifier (MVDVTA) which is nearly as complex as the preceding one. An MVDVTA is, in fact, composed of an OTA and a DVCC. Its various terminals are labelled as P, N, V, YO, O_C, V_C, X, $Z+$ and $Z-$ such that the voltage at the X-terminal is equal to the difference of the voltages on its Yo and V-terminal. O_c terminals are auxiliary terminals which can provide positive or negative output currents. The current entering at the X-terminal is conveyed to $Z+$ and $Z-$ as per usual CC mechanism. The external DC control voltage controls the transconductance of the OTA block at the input. Its internal circuit architecture can be made from appropriate connection of a dual complementary/same polarity output OTA and a DVCC with complementary Z-outputs as shown in Figure 11.9.

For the general configuration of Figure 11.9(a), the floating admittance realized by the circuit for Y_3 chosen as infinity is given by $Y_{eq} = g_m(Y_1/Y_2)$ from

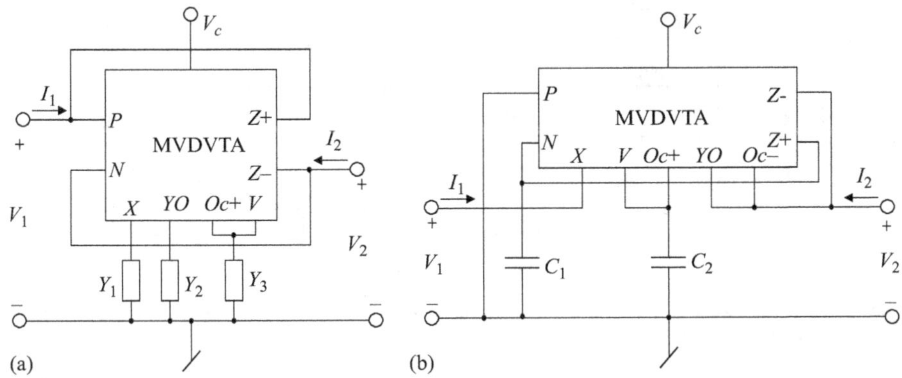

Figure 11.9 A single active element floating impedance simulator due to Alpaslan [12]: (a) generalized FI simulator, (b) a floating FDNR simulator

where it is obvious that by proper selection of the admittances an LLFI and a lossless inductor are realizable. On the other hand, the circuit of Figure 11.9(b) realizes a floating FDNR or *D*-element with $D_{eq} = C_1C_2/g_m$. All realized elements are electronically controllable.

11.3.2 Floating impedance configurations having electronic tunability and temperature-insensitivity

The circuits describe so far were aimed at devising single active element based FI configurations, with a minimum number of passive components without requiring any component-matching conditions with a single grounded capacitor and two/no resistors and if possible with electronic controllability of the realized impedance value.

On the other hand, there have been few interesting attempts in devising circuits which offer electronic tunability/variability of the realized FI but using two instead of one active element but managing to have temperature insensitive performance. We now present two such configurations.

First such configuration was presented by Jantakun *et al.* [3] employing two DVCCTAs and is shown here in Figure 11.10.

The DVCCTA is a composite building block at the front end of this is a DVCC and at the back end is the OTA with dual complementary outputs. It has two *Y*-terminals, an *X*-terminal, a *Z*-terminal and another current output terminal *O* with various terminal equations being

$$I_{Y1} = 0, \quad I_{Y2} = 0; \qquad V_X = V_{Y1} - V_{Y2}; \qquad I_Z = I_X, \quad I_O = \pm g_m V_Z \tag{11.14}$$

An analysis of this circuit reveals that the equivalent FI realized by this circuit, if a bipolar DVCCTA is formulated, is given by

$$Z_{eq} = \frac{g_{m2}Z_1Z_3}{g_{m1}Z_2} = \left(\frac{I_{B2}}{I_{B1}}\right)\frac{Z_1Z_3}{Z_2} \tag{11.15}$$

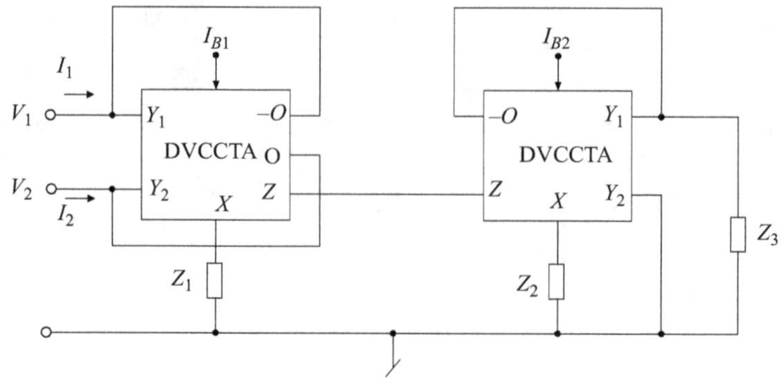

Figure 11.10 Temperature-insensitive FI simulator proposed by Jantakun et al. [3]

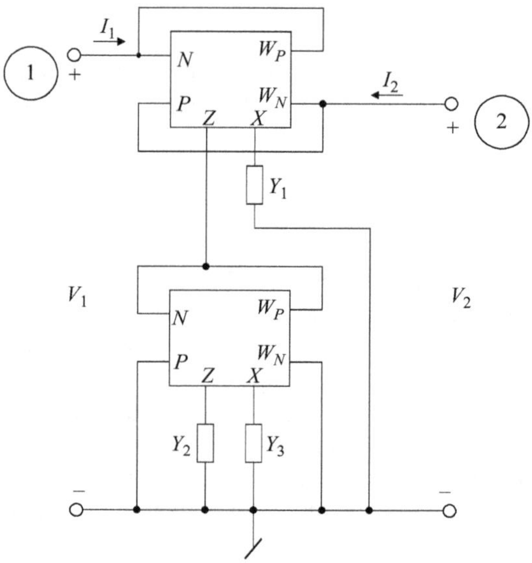

Figure 11.11 General floating impedance simulator using all grounded passive elements due to Kartci et al. [10]

Thus, the thermal voltage V_T is beautifully cancelled out and the realized impedance can be electronically tuned by varying any of the two external DC bias currents I_{B1} and I_{B2}. Note that this circuit can realize a floating inductance, a floating capacitance and a floating FDNR and even a floating (electronically variable) resistance by appropriate selection of circuit admittances and in all cases, the circuit employs grounded capacitor(s).

Second such configuration employing two VDCCs and three admittances was presented by Kartci*et al.* [10] and is shown here in Figure 11.11.

11.3.3 Impact of the circuits and techniques of impedance simulation on the area of memristive circuits

In 1971, Chua [51] postulated that since there are four fundamental variables, namely charge, flux, voltage and current, there should be a fourth fundamentally different basic circuit element having a constitutive relation between flux and charge which in his opinion was missing till then. He named this missing circuit element as the *memristor*. However, at the time of its proposition, no element was known to exhibit the properties of this new element. The world-wide interest in *memristor* was revived when HP Labs in 2008 announced [64] a physical device whose behaviour was similar to that of the memristor and thus, the memristor found life as a physically existing circuit element. This resulted in a spurt of studies on the modelling, simulation and discovery of numerous new applications which could be made using traditional active and passive elements in conjunction with the now available new element – *the memristor*! Since then, a lot of research has undergone in the area of memristive circuits and systems; see, for instance, [15–36,51] and the references cited therein.

To recapitulate, Chua himself wrote an excellent tutorial titled *The Fourth Element* [21] and in conjunction with other authors, highlighted and re-emphasized the importance of studies and further research on memristors by bringing out interesting facts from the history which disclosed several unknown facts such as *the first radios were made using memristors* [24] and went on to point out that even the *human brains are made of memristors* [26]!

As a consequence of the above-mentioned developments, the class of nonlinear circuit elements was further enlarged by putting memory into the capacitors and inductors too, thereby leading to the creation of new nonlinear network elements which were termed *memcapacitors* and *meminductors* [15–20,27,29,30]. To bring these elements to a more practical footing, a number of researchers worked on evolving electronic *emulators* of these elements for possible practical use of these in many new applications of the memristors which were continuously being sought by academicians, circuit designers and researchers. In this context, another class of circuit elements known as *mutators* which were also introduced by Chua in 1968 [50] were found to be particularly helpful in converting memristors into memcapacitors or memristors into meminductors.

In retrospection, it was found that in the design of emulators of memcapacitors and meminductors and more particularly in the synthesis of floating emulators of various kinds, a number of techniques which were evolved by various researches in the context of grounded/FI simulation were found to be particularly helpful.

In this section, therefore, we will highlight several examples from this area to demonstrate the impact of the techniques of gyrator realization or synthesis of grounded and FI simulation on the evolution of a class of memristive circuits.

Before discussing various circuits, it is worthwhile to briefly review the notion of memristors, memcapacitors and meminductors and some mutators relevant to the present discussion.

Memristor: the *memristor* can be looked upon as a circuit element which provides the missing link between flux ϕ and charge q. The *memristance* $M(q)$ can be defined as the derivative of the flux ϕ with respect to charge q, i.e.

$$M(q) = \frac{d\phi(q)}{dq} \tag{11.16}$$

A memristor exhibits a single-valued q–ϕ curve whereas it shows a *pinched i–v hysteresis* loop. Furthermore, as the reciprocal of resistance is a conductance, the reciprocal of *memristance* is called *memductance.*

Memcapacitor: The *memcapacitance* is defined by

$$C(\phi) = \frac{d\sigma(\phi)}{d\phi} \tag{11.17}$$

where $\sigma(\phi)$ is a differentiable function of ϕ with $\sigma = \int_{-t}^{\infty} \phi(\tau)d\tau$. A memcapacitance exhibits a single-valued function in σ–ϕ plane whereas a *pinched hysteresis* loop in q–v plane under periodic excitation.

Meminductance: the *meminductance* can be defined as

$$L(q) = \frac{d\rho(q)}{dq} \tag{11.18}$$

where $\rho(q)$ is a differentiable function of q with $\rho = \int_{-t}^{\infty} q(\tau)d\tau$. A meminductance exhibits q to be a single-valued function of ρ while it exhibits a *pinched hysteresis* loop in i–ϕ plane under periodic excitation.

Mutators: The mutators were introduced by Chua [50] as the basic two port elements (along with *rotators* and *reflectors)* for transforming one kind of nonlinear element into another kind of nonlinear element. Thus, there are L–R mutators, C–R mutators and L–C mutators as the basic elements which convert nonlinear resistors into nonlinear inductors, nonlinear resistors into nonlinear capacitors or nonlinear capacitors into nonlinear inductors, respectively. Of course, these elements can also be used to convert linear element of one kind into linear element of other kind and consequently, in the past, have been utilized to simulate inductors and other elements. Several other kinds of mutators have been defined in the literature subsequently, for instance, R–D mutators which convert a resistance into an FDNR which is an element having $Z(s) = 1/Ds^2$. Furthermore, it has been well understood that several types of impedance converters can also be looked upon as impedance inverters and converters and vice versa.

With the advent of memristor, memcapacitor and meminductor elements, several new kinds of mutators have come into being. The following mutators are of interest in the context of the present discussion:

M–R mutators: M–R mutators transform a resistor with nonlinear i–v characteristic into a memristor with a similar shape of charge–flux (q–ϕ) characteristics. It has been proposed by Biolkova *et al.* [22] that there are fundamentally two types of M–R mutators which can lead to four types of realizations for which they proposed four circuit implementations using operational TIA which is another name for the so-called CFOA.

Thus, if we look upon the two-port mutator as a network on the port 2 of which one connects a nonlinear resistor having terminal voltage and terminal current as v_R and i_R, respectively, with the memristive behaviour then observed looking into its port 1 between the terminal voltage v_M and terminal current i_M, then the two fundamental types of M–R mutators can be defined as follows:

Type-IA M–R mutator is defined as a two-port network element characterized by the following equations:

$$v_R = \frac{1}{k_x}\int v_m dt; \quad i_M = k_y \frac{d}{dt} i_R \tag{11.19}$$

On the other hand, type-IB M–R mutator is characterized by

$$v_M = k_x \frac{d}{dt} v_R; \quad i_R = \frac{1}{k_y}\int i_M dt \tag{11.20}$$

Similarly, type-IIA M–R mutator is defined by the following characterizing equations:

$$v_M = k_y \frac{di_R}{dt}; \quad v_R = \frac{1}{k_x}\int i_M dt \tag{11.21}$$

Lastly, type-IIB M–R mutator is characterized by

$$i_m = k_x \frac{dv_R}{dt}; \quad i_R = \frac{1}{k_y}\int v_m dt \tag{11.22}$$

MR–MC mutator: This kind of mutator is an active-RC two-port network element which converts a *memristor* into a *memcapacitance* and is characterized by

$$i_2 = \frac{1}{k_x}\int i_1 dt; \quad v_1 = k_y v_2 \tag{11.23}$$

MR–ML mutator: This kind of mutator is a two-port active RC network element which transforms a *memristor* into a *meminductor*. It is characterized by

$$v_{ML} = k_x \frac{dv_R}{dl}; \quad i_{ML} = k_y i_R \tag{11.24}$$

Pershin and Ventra [17] presented two simple circuits for converting a memristor into equivalent impedance comprising a resistance in series with a memcapacitance and another one to convert a memristor into an equivalent combination of a resistance in series with meminductance. These circuits are shown in Figure 11.12.

From the economic lossy grounded inductance simulation circuits dealt in Chapter 2 of this monograph, these topologies can be readily identified to be based upon the Prescott lossy inductor (series RL)-type simulator configuration with one of the resistors replaced by a memristor. Curiously, the authors of [17] have not acknowledged so!

Biolek and Biolkova in [18] presented a two CFOA-based MR–MC mutator (shown in Figure 11.13) which is closely related to the Fabre gyrator of [60] using two OTRAs (described in Chapter 5).

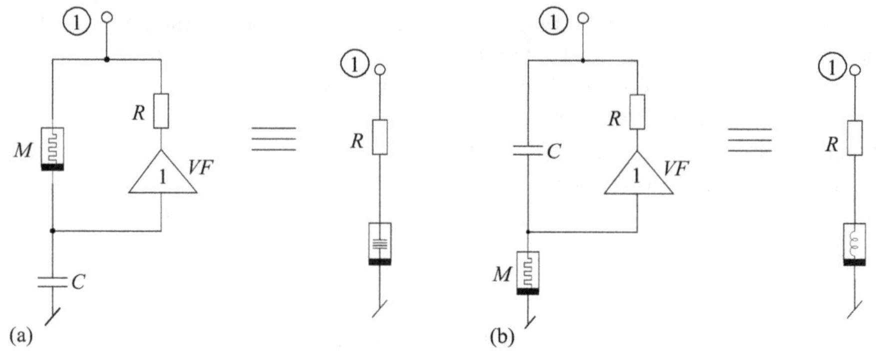

Figure 11.12 Circuits simulating memcapacitor and meminductor [17]: (a) memcapacitor, (b) meminductor

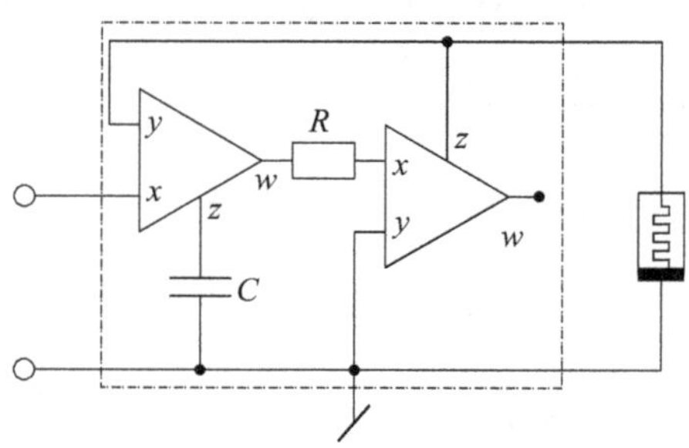

Figure 11.13 An MR–MC mutator proposed by Biolek and Biolkova [18]

Pershin and Ventra in [19] presented a number of CCII+ based circuits that emulate floating memcapacitors and meminductors. Two of these circuits employing four single *Z*-output types CCII+ (which can be implemented from an AD844-type CFOA) are shown here in Figure 11.14. The first of these two circuits realize a floating meminductance assuming that a memristor is available while the other one emulates a floating memcapacitor. It is interesting that these configurations have been derived from the LLFI circuits proposed by Kiranon and Pawarangkoon [65] and it has been duly acknowledged explicitly by Pershin and Ventra in their publication [19].

The second proposition of Pershin and Ventra [19] employs dual complimentary *Z*-output-type CCIIs to carry out the same functions which can then be accomplished by only two dual output CCII (DOCCII) elements as shown in Figure 11.15. The circuits of Figure 11.15 have been based upon the DOCCII-based LLFI circuits originally proposed by Pal [66]; this fact too has been duly acknowledged by Pershin and Ventra [19].

Yu *et al.* [29] presented an universal mutator circuit for transformations among memristor, memcapacitor and meminductor using three CCII+ and three unity gain voltage followers along with five general impedances. When redrawn, in terms of

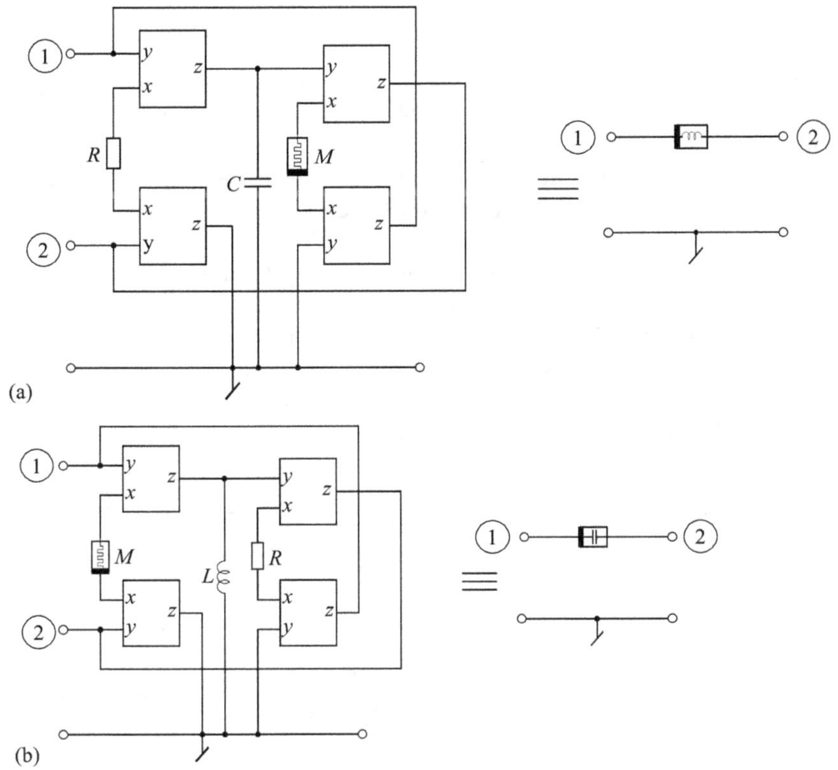

Figure 11.14 Emulators for realizing floating meminductor and memcapacitor as proposed by Pershin and Ventra in [19]: (a) meminductor emulator, (b) memcapacitor emulator

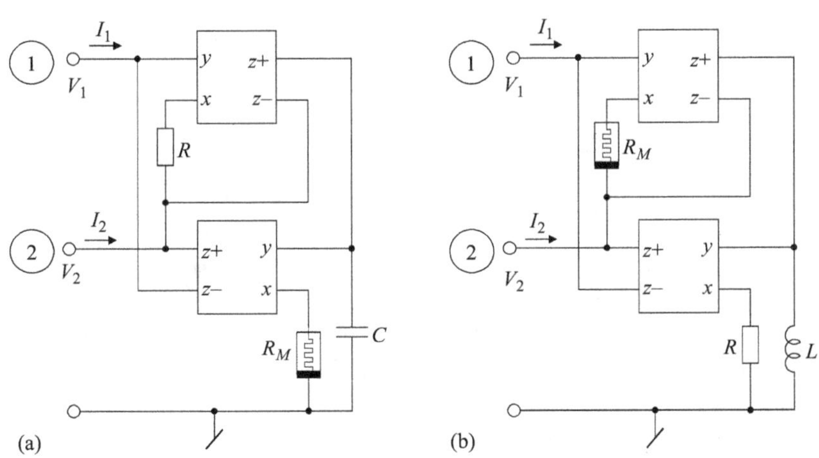

Figure 11.15 Emulators for realizing floating meminductor and memcapacitor using DOCC as proposed by Pershin and Ventra [19]: (a) meminductor emulator, (b) memcapacitor emulator

the CFOA notation, their circuit can be redrawn as shown here in Figure 11.16. These circuits can then be easily identified to be a grounded-GIC circuit realized with three CFOAs as explained in Chapter 5.

We now present two more recent proposals for realizing memristor emulators. The first one is a configuration proposed by Cam and Sedef [33] as shown in Figure 11.17 in which four CCII+ have been employed in an arrangement which, with the analog multiplier and the resistor R removed, quite resembles the four-CCII+ based LLFI configuration (described in Chapter 4).

Lastly, we show a DOTA-based memristive emulator circuit configuration shown in Figure 11.18 [35] which was recently proposed by Taskiran *et al.* [35]. Clearly, with the multiplier and the resistor R removed the circuit is similar to a two DOTA-based LLFI presented in Chapter 3.

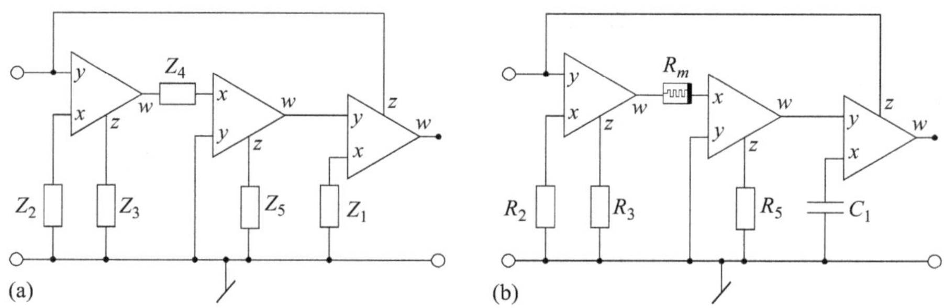

Figure 11.16 CFOA representation of the universal mutator proposed by Yu et al. [29]: (a) the basic universal mutator, (b) special case of the mutator for transforming from an MR into an MC

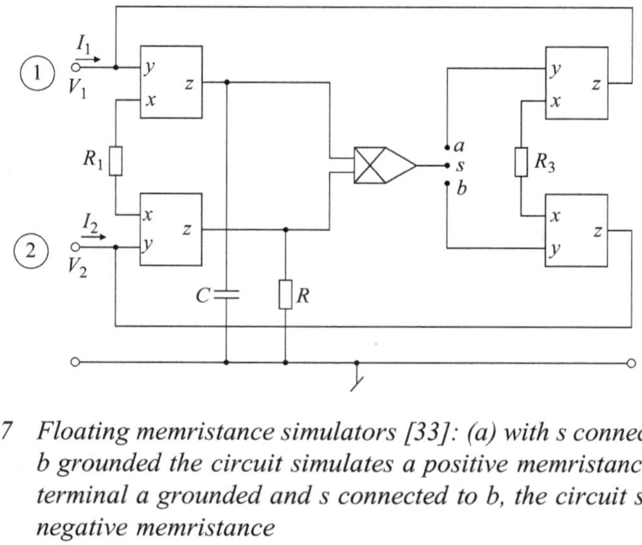

Figure 11.17 Floating memristance simulators [33]: (a) with s connected to a and b grounded the circuit simulates a positive memristance, (b) with terminal a grounded and s connected to b, the circuit simulates a negative memristance

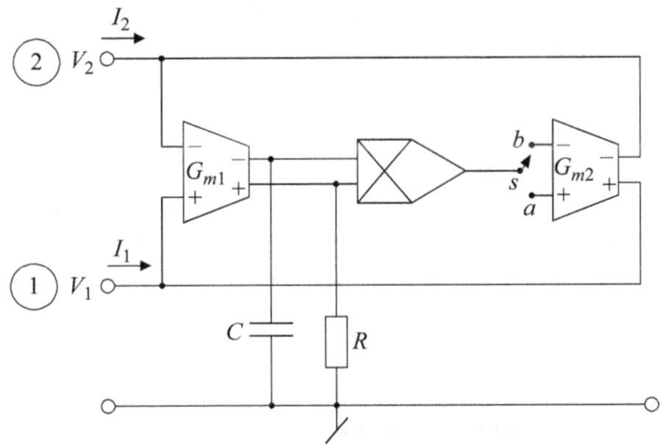

Figure 11.18 Memristance simulation circuits using DOTAs [35]: (a) with terminal a grounded and s connected to b, the circuit simulates a positive memristance; (b) with s connected to a and b grounded, the circuit simulates a negative memristance

These examples are enough testimony to emphasize that the techniques of inductance simulation have been a great influence in the design of a class of non-linear circuit elements known as memristors, memcapacitors and meminductors.

11.3.4 Impact of the circuit techniques of impedance simulation on the area of fractional-order circuits

In this subsection, we will demonstrate that a number of circuits and techniques have also influenced the area of *fractional-order* circuits and systems which is widely investigated recently.

At present, there is growing interest in the synthesis and design of fractional-order circuits and systems, and consequently a number of analog signal processing/signal generation circuits such as fractional-order analog filters, sinusoidal oscillators, PID controllers, etc., using a variety of active elements, have been investigated as these provide additional degrees of freedom/advantages in the design of the quoted circuits, as compared to their integer-order counterparts. Fractional-order integrators and differentiators are generalizations of the integer-order integrators and differentiators and have been known for the past three centuries. However, the fractional-order circuits and systems have attracted attention recently due to their numerous applications in many diverse areas of science and technology. Fractional-order circuits are considered to be a very convenient means of realizing fractional-order dynamics which can also be used for modelling other non-electrical systems and processes.

A fractional-order derivative, as per the definition of Riemann–Liouville [38], is as follows:

$$D_{t_0}^{\alpha} = \frac{1}{\Gamma(m-\alpha)}\frac{d^m}{dt^m}\int_{t_0}^{t} f(u)f(t-u)^{m-\alpha-1}du \tag{11.25}$$

where $m \in R$ such that $(m-1) < \alpha < m$. The *fractance* device is a generalization of the circuit impedance which, in s-domain, can be expressed as

$$Z(s) = K_0 s^{\alpha} \tag{11.26}$$

Thus, with α equal to zero, $Z(s)$ represents a resistor, with $\alpha = 1$ it represents an inductor and with $\alpha = -1$ it represents a capacitor but with $0 < \alpha < 1$, it represents a *fractional*-order inductance. Similarly, if one takes α to be negative, it represents a *fractional* capacitance for which transformed impedance is given by

$$Z(s) = \frac{K}{s^{\alpha}}; \quad 0 < \alpha < 1 \tag{11.27}$$

A fractional-order capacitance is also known as a *constant phase element* (CPE) due to its constant phase angle $\phi = -\alpha\pi/2$.

There have been a number of attempts to create devices which can approximate the behaviour of a CPE using electrochemical materials, polymer composites, and thick film layers of resistive dielectric and insulating materials on silicon, etc.; however, none of these are yet available commercially as off-the-shelf two terminal components. Consequently, a popular and convenient method of realizing CPEs is to employ a rational function approximation of its $Z(s)$ and realize the same by RC networks. A popular method of realizing fractional-order capacitance or inductance employing OTAs is depicted in Figure 11.19.

In this circuit, $H_0(s)$ represents the voltage ratio transfer function corresponding to the rational function approximated $Z(s)$ of the required fractional-order element.

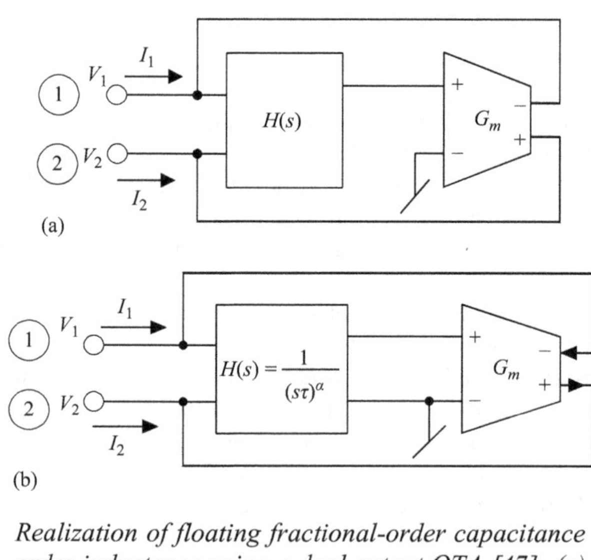

Figure 11.19 Realization of floating fractional-order capacitance or fractional-order inductance using a dual output OTA [47]: (a) generic circuit for realizing an arbitrary floating impedance, (b) a general circuit for realizing a floating fractional-order inductance/capacitance

From a routine analysis, it can be easily derived that the circuit of Figure 11.19(a) realizes FI given by

$$I_1 = -I_2 = G_m H(s)(V_1 - V_2) \tag{11.28}$$

On the other hand, the circuit of Figure 11.19(b) realizes FI given by

$$I_1 = -I_2 = G_m \left(\frac{1}{s\tau}\right)^{\alpha} (V_1 - V_2) \tag{11.29}$$

It is, therefore, obvious that from the circuit of Figure 11.19(a), any arbitrary fractional impedance approximated by the rational function $H(s)$ can be realized while from the circuit of Figure 11.19(b) either a fractional-order inductance or fractional-order inductance can be realized by keeping in mind that in the case of the former, the parameter α can take any value within the range $0 < \alpha < 1$ while in the latter case, the parameter α has to be negative with $0 < |\alpha| < 1$.

In [44], Soltan *et al.* presented fractional-order impedance transformations based on three-port mutators. One such three-port mutator from [44] employing two CCII+ and one DOCCII is shown in Figure 11.20.

The same authors presented two reduced component-count three-port mutators the first using an OTA and a DOCCII while the other one employing a DVCCII and an additional resistor R (implementing the same function as the OTA in the previous circuit). These two circuits are shown in Figure 11.21.

It has been shown that the generic three-port mutators of Figures 11.20 and 11.21 can give rise to a variety of cases depending upon the signs of the various voltage/current gains of the active devices employed and that both positive and negative floating fractional-order impedances can be realized from them. From our point of view, it is interesting to observe that with the impedance Z replaced by a

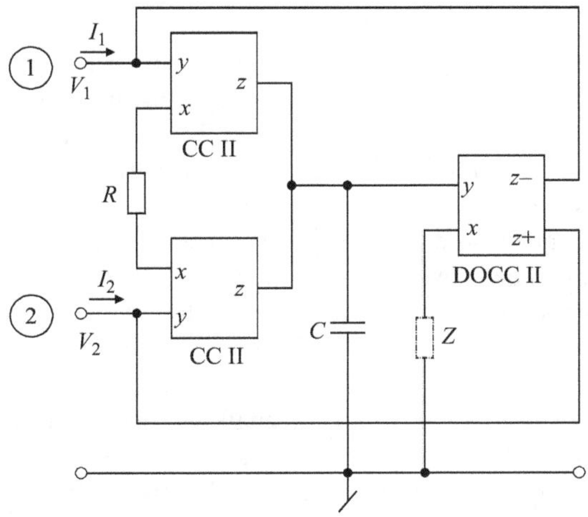

Figure 11.20 A three-port L–R mutator from [44]

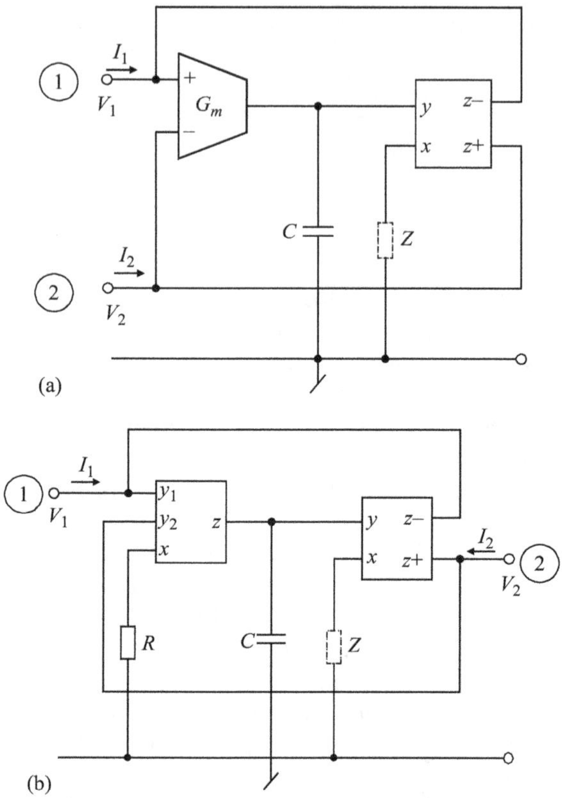

Figure 11.21 Two alternative reduced component-count three-port mutators presented in [44]

resistor, all these circuits are in fact floating inductance circuits the kind of which have been amply discussed in Chapters 4 and 8 of this monograph.

As the last example we show in Figure 11.22, a circuit proposed by Khattab *et al.* [52] for realizing a fractional-order inductance from an available fractional-order capacitance which is explicitly derived from a low-component-count CFOA-based inductance simulation circuit proposed earlier by Kacar and Kuntman [67].

By analysis, it can be shown [67] that the circuit of Figure 11.22(a) simulates a grounded lossy inductor in the form of series RL impedance whereas that of Figure 11.22(b) realizes a fractional-order lossy grounded inductor of the same form [52].

In view of the above few examples, it may be concluded that the circuits and techniques presented in this monograph have ample scope in being advantageously exploited in the synthesis of positive and negative, floating and grounded fractional-order inductance or capacitance elements in a number of ways which are likely to be exploited in future.

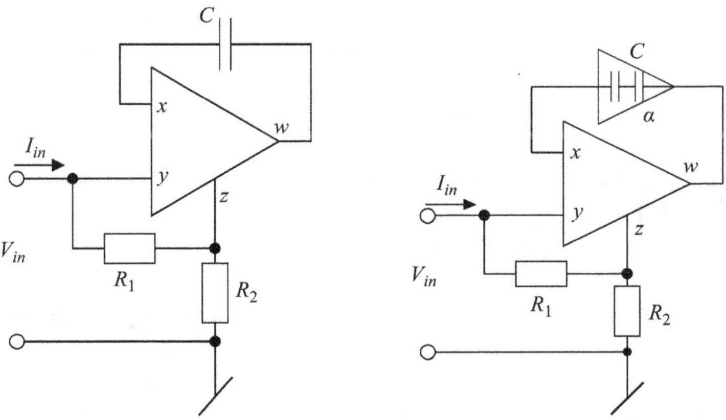

Figure 11.22 Low-component-count fractional inductance simulation: (a) the inductance simulation circuit proposed by Kacar and Kuntman [67], (b) the fractional-order simulated inductor by Khattab et al. [52]

11.4 Concluding remarks

In this chapter, we provided a concise summary of the gyrator realization and impedance simulation networks presented in the various chapters of this monograph. We hope that our exposition provides an appropriate background to the new researchers in this area and also serves to be an up-to-date and comprehensive refresher to the readers who are interested in a detailed overview of this area. Besides this, we hope the compilation of a large number of circuits would also serve as a useful catalogue for the circuit designer who may require a gyrator or an immittance simulation network in any given application. There are a number of unresolved problems which have been identified in each and every chapter of this monograph which could be helpful to the new researchers. Finally, we believe that this compendium on gyrators, simulated inductors and other related immittances, using a variety of active elements, should be useful to anybody interested in electronic circuits, in general and the specific class of circuits considered in this monograph in particular.

11.5 Epilogue

Active realization of gyrators and simulation of both grounded and floating impedances (FI) of various kinds has continued to remain an interesting and exciting area of analog circuits' research for more than four decades due to their potential applications in analog signal processing and signal generation techniques.

This monograph has dealt in detail the research work done on the realization of gyrators, simulated inductors and related immittances using traditional as well as modern electronic circuit building blocks. It may be pointed out that starting from the late eighties; the current conveyors (CCs) started overtaking the ubiquitous IC op-amp as an alternative and attractive building block for analog signal processing, in general and for realizing FIs, in particular. Since then, a large number of new electronic circuit building blocks have been proposed by researchers from time to time which have been employed to realize synthetic grounded and FIs of various kinds.

The building blocks engaged for such purposes include operational transconductance amplifiers (OTAs), four terminal floating nullors (FTFNs), operational transresistance amplifiers (OTRAs), differential difference amplifiers, many different variants of CCs, current differencing buffered amplifiers (CDBAs), current follower transconductance amplifiers (CFTAs), current conveyor transconductance amplifiers (CCTAs) and numerous others.

In this monograph, wherever possible, we have demonstrated the underlying common circuit principles in seemingly different FI realizations and have also highlighted the achievements made in this area. A number of problems which have still remain unattended and hence, unresolved in this area, have also been pointed out at several places. The process of evolution of new gyrator and other impedance simulation circuits using new building blocks (including those which have not been employed so far) does not appear to have been exhausted yet, in view of which, we believe that we may still witness the evolution of a number of new improved gyrator implementations and better immittance simulation circuit configurations in future!

References

[1] Yuce E. 'On the implementation of the floating simulators employing a single active device'. International Journal of Electronics and Communications (AEU). 2007; **61**(7):453–8.

[2] Silapan P., Tanaphatsiri C., Siripruchyanun M. 'Current controlled CCTA based- novel grounded capacitance multiplier with temperature compensation'. 2008 IEEE Asia Pacific Conference on Circuits and Systems; Macao, China, November 30–December 3, 2008; 1:1490–3.

[3] Jantakun A., Pisutthipong N., Siripruchyanun M. 'A synthesis of temperature insensitive/electronically controllable floating simulators based on DV-CCTAs'. 2009 International Conference on Electrical Engineering/Electronics/Computer, Telecommunications and Information Technology; Pattaya, Chonburi, Thailand, May 6–9, 2009; **1**:560–3.

[4] Shaktour M. 'Unconventional Circuit Elements for Ladder Filter Design' (Ph.D. thesis). 2011; BRNO University of Technology, Czech Republic.

[5] Sagbas M. 'Component reduced floating ±L, ±C and ±R simulators with grounded passive components'. International Journal of Electronics and Communications (AEU). 2011; **65**(10):794–8.

[6] Ibrahim M., Minaei S., Yuce E., Herencsar N., Koton J. 'Lossy/lossless floating/grounded inductance simulation using DDCC'. Radioengineering. 2012; **21**(1):3–10.

[7] Tangsrirat W. 'Floating simulator with a single DVCCTA'. Indian Journal of Engineering and Material Science. 2013; **20**(4):79–86.

[8] Prasad D., Ahmad J. 'New electronically controllable lossless synthetic floating inductance circuit using single VDCC'. Circuits and Systems. 2014; **5**(1):13–17.

[9] Guney A., Kuntman H. 'New floating inductance simulator employing a single ZC-VDTA and one grounded capacitor'. 2014 International Conference on Design and Technology of Integrated Systems in Nanoscale Era; Santorini, Greece, May 6–8, 2014; **1**:1–2.

[10] Kartci A., Ayten U. E., Herencsar N., Sotner R., Jerabek J., Vrba K. 'Application possibilities of VDCC in general floating element simulator circuit'. 2015 European Conference on Circuit Theory and Design; Trondheim, Norway, August 24–26, 2015; **1**:1–4.

[11] Kartci A., Ayten U. E., Herencsar N., Sotner R., Jerabek J., Vrba K. 'Floating capacitance multiplier simulator for grounded RC Colpitts oscillator design'. 2015 International Conference on Applied Electronics; Pilsen, Czech Republic, September 8–9, 2015; **1**:1–4.

[12] Alpaslan H. 'A modified VDVTA and its applications to floating simulators and a quadrature oscillator'. Microelectronics Journal. 2016; **51**(1):1–14.

[13] Mohammad F., Sampe J., Shireen S., Ali S. H. M. 'Minimum passive components based lossy and lossless inductor simulators employing a new active block'. International Journal of Electronics and Communications (AEU). 2017; **82**(12):226–40.

[14] Abaci A., Yuce E. 'Single DDCC based new immittance function simulators employing only grounded passive elements and their applications' Microelectronics Journal. 2019; **83**(1):94–103.

[15] Ventra M. D., Pershin Y. V., Chua L. O. 'Putting memory into circuit elements: memristors, memcapacitors and meminductors'. Proceedings of the IEEE. 2009; **97**(8):1371–2.

[16] Ventra M. D., Pershin Y. V., Chua L. O. 'Circuit elements with memory: memristors, memcapacitors and meminductors' Proceedings of the IEEE. 2009; **97**(10):1717–24.

[17] Pershin Y. V., Ventra M. D. 'Memristive circuits simulate memcapacitors and meminductors'. Electronics Letters. 2010; **46**(7):517–18.

[18] Biolek D., Biolkova V. 'Mutator for transforming into memcapacitors'. Electronics Letters. 2010; **46**(21):1428–9.

[19] Pershin Y. V., Ventra M. D. 'Emulation of floating memcapacitors and meminductors using current conveyors'. Electronics Letters. 2011; **47**(4): 243–4.

[20] Biolek D. 'Analog building blocks for emulating memresistors, memcapacitors, and meminductors'. ELECO 2011 7th International Conference on Electrical and Electronics Engineering; Bursa, Turkey, December 1–4, 2011.

[21] Chua L. O. 'The fourth element'. Proceedings of the IEEE. 2012; **100**(6): 1920–6.

[22] Biolkova V., Biolek D., Kolka Z. 'Unified approach to synthesis of mutators employing operational transimpedance amplifiers for memresistor emulation'. Proceedings of 11th WSEAS International Conference on Instrumentation, Measurement, Circuits, and Systems; Rovaniemi, Finland, April 18–20, 2012, 110–15.

[23] Wang H., Wang X., Li C., Chen L. 'SPICE mutator model for transforming memristor into meminductor'. Abstract and Applied Analysis. 2013; **2013**, 5 pages, Article ID 281675.

[24] Gandhi G., Aggarwal V., Chua L. O. 'The first radios were made using memresistors!'. IEEE Circuits and Systems Magazine. 2013; **13**(2):8–16.

[25] Goknar I. C., Minayi E. 'Realizations of mutative 4-ports and their applications to memstor simulations'. Analog Integrated Circuits and Signal Processing. 2014; **81**(1):29–42.

[26] Sah M. P., Kim H., Chua L. O. 'Brains are made of memristors'. IEEE Circuits and Systems Magazine. 2014; **14**(1):12–36.

[27] Dong-Sheng Y., Yan L., Iu H. H. C., Yi-Hua H. 'Mutator for transferring a memristor emulator into meminductive and memcapacitive circuits'. Chinese Physical Society. 2014; **23**(7):070702-1–11.

[28] Sanchez-Lopez C., Mendoza-Lopez J., Carrasco-Aguilar M. A., Muniz-Montero C. 'A floating analog Memristor emulator circuit'. IEEE Transactions on Circuits and Systems-II. 2014; **61**(5):309–13.

[29] Yu D., Liang Y., Iu H. H. C., Chua L. O. 'A universal mutator for transformations among Memristor, memcapacitor, and meminductor'. IEEE Transactions on Circuits and Systems-II. 2014; **61**(10):758–62.

[30] Yin Z. Y., Tian H., Chen G., Chua L. O. 'What are memristor, memcapacitor, and meminductor?'. IEEE Transactions on Circuits and Systems-II. 2015; **62**(4):402–6.

[31] Minaei S., Goknar I. C., Yildiz M., Yuce E. 'Memstor, memstance simulations via a versatile 4-port built with new adder and subtractor circuits'. International Journal of Electronics. 2015; **102**(6):911–31.

[32] Abuelma'atti M. T., Khalifa Z. J. 'A new floating memristor emulator and its application in frequency-to-voltage conversion'. Analog Integrated Circuits and Signal Processing. 2016; **86**(1):141–7.

[33] Cam Z. G., Sedef H. 'A new floating memristance simulator circuit based on second generation current conveyor'. Journal of Circuits, Systems, and Computers. 2017; **26**(2):1750029-1–15.

[34] Abraham I. 'The case for rejecting the memristor as a fundamental circuit element'. Scientific Reports. 2018; **8**:10972.

[35] Taskiran Z. G. C., Ayten U. E., Sedef H. 'Dual-output operational transconductance amplifier-based electronically controllable memristance simulator circuit'. Circuits, Systems and Signal Processing. 2019; **38**(1):26–40.

[36] Xie X., Zou L., Wen S., Zeng Z., Huang T. 'A flux-controlled logarithmic memristor model and emulator'. Circuits, Systems and Signal Processing. 2019; **38**(4):1452–65.

[37] Ahmad W., El-Khazali R., Elwakil A. S. 'Fractional-order Wien-bridge oscillator'. Electronics Letters. 2001; **37**(18):1110–12.

[38] Elwakil A. S. 'Fractional-order circuits and systems: an emerging interdisciplinary research area'. IEEE Circuits and Systems Magazine. 2010; **10**(4):40–50.

[39] Jakubowska-Ciszek A., Jakubowski J. 'Synthesis of electronic fractional-order components of arbitrary fractional-order parameters'. Elektryka. 2016; **2**(238):79–91.

[40] Jakubowska-Ciszek A., Walczak J. 'Electronic realizations of fractional-order elements: I. Synthesis of the arbitrary order elements'. Poznan University of Technology Academic Journals (Electrical Engineering). 2016; **85**:12 pages.

[41] Tsirimokou G., Psychalinos C., Freeborn T. J., Elwakil A. S. 'Emulation of current excited fractional-order capacitors and inductors using OTA topologies'. Microelectronics Journal. 2016; **55**(C):70–81.

[42] Sotner R., Jerabek J., Petrzela J., Dostal T. 'Simple approach for synthesis of fractional-order grounded immittances based on OTAs'. 2016 39th International Conference on Telecommunications and Signal Processing; Vienna, Austria, June 27–29, 2016.

[43] Sotner R., Jerabek J., Petrzela J., Domansky O., Tsirimokou G. 'Synthesis and design of constant phase elements based on the multiplication of electronically controllable bilinear immittances in practice'. International Journal of Electronics and Communications (AEU). 2017; **78**:98–113.

[44] Soltan A., Soliman A. M., Radwan A. G. 'Fractional-order impedance transformation based on three port mutators'. International Journal of Electronics and Communications (AEU). 2017; **81**:12–22.

[45] Bertsias P., Psychalinos C., Radwan A. G., Elwakil A. S. 'High-frequency capacitor-less fractional-order CPE and FI emulator'. Circuits, Systems and Signal Processing. 2018; **37**(7):2694–713.

[46] Tsirimokou G., Psychalinos C., Elwakil A. 'Design of CMOS Analog Integrated Fractional-Order Circuits: Applications in Medicine and Biology'. Switzerland: Springer Briefs in Electrical and Computer Engineering; 2017. p. 114.

[47] Tsirimokou G., Psychalinos C., Elwakil A. S., Salama K. N. 'Electronically tunable fully integrated fractional-order resonator'. IEEE Transactions on Circuits and Systems-II. 2018; **65**(2):166–70.

[48] Verma R., Pandey N., Pandey R. 'Realization of a higher fractional order element based on novel OTA based IIMC and its application in filter'. Analog Integrated Circuits and Signal Processing. 2018; **97**(1):177–91.

[49] Tsirimokou G., Kartci A., Koton J., Herencsar N., Psychalinos C. 'Comparative study of discrete component realizations of fractional-order capacitor and inductor active emulators'. Journal of Circuits, Systems and Computers. 2018; **27**(11):26 pages.

[50] Chua L. O. 'Synthesis of new nonlinear network elements'. Proceedings of the IEEE. 1968; **56**(8):1325–43.

[51] Chua L. O. 'Memristor-The missing circuit element'. IEEE Transactions on Circuit Theory. 1971; **18**(5):507–19.

[52] Khattab K. H., Madian A. H., Radwan A. G. 'CFOA-based fractional order simulated inductor'. 2016 IEEE 59th International Midwest Symposium on Circuits and Systems; Abu Dhabi, United Arab Emirates, October 16–19, 2016.

[53] Khalil N. A., Ahmed R. F., Seoud R. A. A., Soliman A. M. 'An intelligent technique for generating equivalent gyrator circuits using genetic algorithm'. Microelectronics Journal. 2015; **46**(11):1060–8.

[54] Kumar N., Vista J., Ranjan A. 'A tunable active inductor employing DXCCTA: grounded and floating operation'. Microelectronics Journal. 2019; **90**:1–11.

[55] Willson A. N., Orchard H. J. 'Realization of ideal gyrators'. IEEE Transactions on Circuits and Systems. 1974; **21**(6):729–32.

[56] Smith K. C., Sedra A. 'The current conveyor-a new circuit building block'. Proceedings of the IEEE. 1968; **56**(8):1368–9.

[57] Sedra A., Smith K. C. 'A second-generation current conveyor and its applications'. IEEE Transactions on Circuit Theory. 1970; **17**(1):132–4.

[58] Gilbert B. 'The current-mode muddle'. International Journal of Electronics and Communications (AEU). 2018; **83**:398–406.

[59] Senani R. 'Novel active RC circuit for floating-inductor simulation'. Electronics Letters. 1979; **15**(21):679–80.

[60] Fabre A. 'Gyrator implementation from commercially available transimpedance operational amplifiers'. Electronics Letters. 1992; **28**(3):263–4.

[61] Senani R. 'Realization of a class of analog signal processing/signal generation circuits: novel configurations using current feedback op-amps'. Frequenz. 1998; **52**(9–10):196–206.

[62] Chang C. M., Hwang C. S. 'Comment on voltage-mode notch, low pass and band pass filter using current-feedback amplifiers'. Electronics Letters. 1995; **31**(4):246.

[63] Biolek D., Senani R., Biolkova V., Kolka Z. 'Active elements for analog signal processing: classification, review, and new proposals'. Radioengineering. 2008; **17**(4):15–31.

[64] Strukov D. B., Snider G. S., Stewart D. R., Williams R. S. 'The missing memristor found'. Nature. 2008; **453**:80–3.

[65] Kiranon W., Pawarangkoon P. 'Floating inductance simulation based on current conveyors'. Electronics Letters. 1997; **33**(21):1748–9.

[66] Pal K. 'Modified current conveyors and their applications'. Microelectronics Journal. 1989; **20**(4):37–40.

[67] Kacar F., Kuntman H. 'CFOA-based lossless and lossy inductance simulators'. Radioengineering. 2011; **20**(3):627–31.

[68] Smith K. C., Sedra A. 'Realization of the Chua family of new nonlinear network elements using the current conveyor'. IEEE Transactions on Circuit theory. 1970; **17**(1):137–9.

[69] Senani R. 'Novel Analog Circuits Using, Op-amps, OTAs and Current Conveyors'. LAP LAMBERT Academic Publishing, Deutschland, Germany; 2014. p. 123–5.

Further reading

[1] Anderson B. D. O. 'Impossibility of a class of gyrator realizations'. IEEE Transaction on Circuit Theory. 1967; **CT-14**(3):343.

[2] Antoniou A. '3-Terminal gyrator circuits using operational amplifiers'. Electronics Letters. 1968; **4**(26):591–2.

[3] Antoniou A. 'Novel RC-Active-network synthesis using generalised-immittance converters'. IEEE Transactions on Circuit Theory. 1970; **CT-17**(2): 212–17.

[4] Bendik J. 'Equivalent gyrator networks with nullors and norators'. IEEE Transactions on Circuit Theory. 1967; **CT-14**(1):90.

[5] Bhattacharyya B. B., Swamy M. N. S. 'Network transposition and its application in synthesis'. IEEE Transactions on Circuit Theory. 1971; **CT-18**(3):148–9.

[6] Biako M. 'Selective network with nonideal gyrator'. Electronics Letters. 1967; **2**(12):471–472.

[7] Biako M. 'On Q factor and Q sensitivity of an inductor simulated by a practical gyrator'. Electronics Letters. 1967; **3**(4):168–9.

[8] Biako M. 'Realization of inductive and capacitive gyrators'. IEEE Transaction on Circuit Theory. 1968; **CT-15**(2):158–60.

[9] Bickley D. T. 'Design of nonlinear networks using gyrators'. Electronics Letters. 1967; **3**(11):519–21.

[10] Bose N. K., Soliman A. M. 'Lossless multiport with termination problems'. Electronics Letters. 1970; **6**(4):100–2.

[11] Carlin H. J., Rosinski J. 'Equivalent circuit and loss invariant for helicon mode semiconductor device'. IEEE Transactions on Circuit Theory. 1969; **CT-16**(3):365–73.

[12] Cheung L. K., Ho C. F. 'Synthesis using active constant resistance 3-terminal networks'. Electronics Letters. 1969; **5**(15):328–30.

[13] Civalleri P. P., Ridella S. 'Impedance and admittance matrices of distributed three-layer N-ports'. IEEE Transactions on Circuit Theory. 1970; **CT-17**(3): 392–8.

[14] Deboo G. J. 'Application of a gyrator-type circuit to realize ungrounded inductor'. IEEE Transactions on Circuit Theory. 1967; **CT-14**(1):101–2.

[15] Delansky J. P. 'Some synthesis methods for adjustable networks using multivariable techniques'. IEEE Transactions on Circuit Theory. 1969; **CT-16**(4):435–43.

[16] Dewilde P., Silverman L. M., Newcomb R. W., A passive synthesis for time-invariant transfer functions'. IEEE Transactions on Circuit Theory. 1970; **CT-17**(3):333–4.

[17] Fischman M., Matarese J. 'A gyrator horizontal sync system for TV'. IEEE Transactions on Broadcast and TV Receivers. 1970; **BTR-16**(3):213–16.

[18] Garg J. M., Carlin H. J. 'Network theory of semiconductor Hall-plate circuits'. IEEE Transactions on Circuit Theory. 1965; **CT-12**(1):59–73.

[19] Geffe P. R. 'Gyrator bandwidth limitations'. IEEE Transactions on Circuit Theory. 1968; **CT-15**(4):501.

[20] Girling F. E. J., Good E. F. 'The Q of a capacitor loaded gyrator'. IEEE Journal of Solid State Circuits. 1968; **SC-3**(3):304–6.

[21] Golosman B. S., Newcomb R. W. 'Natural frequencies of negative resistance with parasitics'. IEEE Transactions on Communication and Electronics. 1963; **68**:527–31.

[22] Hozony D. 'Permutation network operations'. IEEE Transactions on Circuit Theory. 1970; **CT-17**(1):146–8.

[23] Ho C.-W., Balabanian N. 'Synthesis of active and passive compatible impedances'. IEEE Transactions on Circuit Theory. 1967; **CT-14**(2): 118–28.

[24] Horowitz I. M. 'Optimum design of single stage gyrator-RC-filters with prescribed sensitivity'. IEEE Transactions on Circuit Theory. 1961; **CT-8**(2): 88–94.

[25] Karni S. 'A note on N-port networks terminated with N gyrators'. IEEE Transactions on Circuit Theory. 1963; **CT-10**(4):525–6.

[26] Levan N., Lampard D. G., Anderson B. D. O., Newcomb R. W. 'Step-up N-port networks'. IEEE Transactions on Circuit Theory. 1969; **CT-16**(3): 359–65.

[27] Longley S. R. 'Multioctive tunable 3-port circulator using a Y.I.G. Sphere'. Electronics Letters. 1970; **6**(13):406–8.

[28] Marsocci V. A. 'Realization of magnetoresistive gyrator circuit'. IEEE Journal of Solid-State Circuits. 1967; **SC-2**(2):44–6.

[29] May J. T., Pierce J. F. 'Bias-Level latch-up problems in gyrators'. IEEE Journal of Solid-State Circuits. 1971; **SC-6**(30):130–1.

[30] McCalla W. J., Pederson D. O. 'Elements of computer-aided circuit analysis'. IEEE Transactions on Circuit Theory. 1971; **CT-18**(1):14–26.

[31] McGregor I. M. "Theoretical and Realizable sensitivities of a two-pole gyrator active filter'. IEEE Transactions on Circuit Theory. 1968; **CT-15**(2): 94–100.

[32] Morse A. S., Huelsman L. P. 'A gyrator realization using operational amplifiers'. IEEE Transactions on Circuit Theory. 1964; **CT-11**(2):277–8.

[33] Murdoch J. B. 'RC-gyrator cascade synthesis'. IEEE Transactions on Circuit Theory. 1964; **CT-11**(2):111–16.

[34] Newcomb R. W. 'A Bayard-type nonreciprocal n-port synthesis'. IEEE Transactions on Circuit Theory. 1963; **CT-10**(1):85–90.

[35] Newcomb R. W. 'Topological analysis with transformers'. IEEE Transactions on Circuit Theory. 1963; **CT-10**(3):457–8.

[36] Oono Y., Koga T. 'Synthesis of a variable parameters one-port'. IEEE Transactions on Circuit Theory. 1970; **CT-10**(2):213–27.

[37] Rao T. N. 'Stability of wideband gyrator circuits'. IEEE Journal of Solid State Circuits. 1970; **SC5(3)**:129–32.

[38] Rollett J. M. 'Circulators-capacitor networks'. Electronics Letters. 1968; **4**(26):599–601.

[39] Schwartz E. 'Two Theorems on lossy two-ports'. IEEE Transactions on Circuit Theory. 1970; **CT-17**(1):129–31.

[40] Silverman J. M. 'Synthesis of Impulse response matrices by internally stable and passive realizations'. IEEE Transactions on Circuit Theory. 1968; **CT-15**(3):238–45.

[41] Smith W. E. 'An extension of Vratsano's Theorem to nonreciprocal networks'. IEEE Transactions on Circuit Theory. 1971; **CT-18**(2):276–7.

[42] Smith W. E. 'Element sensitivity and energy storage of a passive impedance'. IEEE Transactions on Circuit Theory. 1971; **CT-18**(3):337–42.

[43] Su K. L. 'FET-circuit realization of the inductance'. Electronics Letters. 1966; **2**(12):469–70.

[44] Tow J. 'Realization of the matrix for portless passive networks'. IEEE Transactions on Circuit Theory. 1969; **CT-16**(1):132–4.

[45] Tow J. 'The explicit form of Bashkow's a matrix for a class of linear passive networks'. IEEE Transactions on Circuit Theory. 1970; **CT-17**(1):113–1s5.

[46] Van Leeij H. Th., Adams K. M. 'Phase-compensation in electronic gyrator circuits'. Electronics Letters. 1968; **4**(20):430–1.

[47] Van Leeij H. Th., Adams K. M. 'Wideband electronic gyrator circuit'. Electronics Letters. 1968; **4**(20):431–2.

[48] Varaiya P. P., Liu R. 'Normal form and stability of a class of coupled nonlinear networks'. IEEE Transactions on Circuit Theory. 1966; **CT-13**(4): 413–8.

[49] Vasudeva M. S. 'Gyrator realization'. Electronics Letters. 1966; **2**(6):201–2.

[50] Woodard J., Newcomb R. W. 'Sensitivity improvement of inductorless filters'. Electronics Letters. 1966; **2**(9):349–50.

[51] Yanagiaawa T. 'Realization of a lossless transistor gyrator'. Electronics Letters. 1967; **3**(4):167–8.

[52] Yanagiaawa T. 'Direct-coupled transistor lossless gyrator'. Electronics Letters. 1967; **3**(7):311–12.

[53] Elwan H. O., Mahmoud S. A., Soliman A. M. 'CMOS voltage controlled floating resistor'. International Journal of Electronics. 1996; **81**(5):571–6.

[54] Radwan A. G., Soliman A. M., El-Sedeek A. L. 'An inductorless CMOS realization of Chua's circuit'. Chaos, Solitons & Fractals. 2003; **18**(1): 149–58.

[55] Soliman A. M. 'Two new LC mutators and their realizations'. IEEE Transactions on Circuit Theory. 1972; **19**(4):371–2.

[56] Elwakil A. S., Soliman A. M. 'Two modified for chaos negative impedance converter op amp oscillators with symmetrical and anti-symmetrical non-linearities'. International Journal of Bifurcation and Chaos. 1996; **8**(06): 1335–46.

[57] Soliman A. M. 'Three port gyrator circuits using transconductance amplifiers or generalized conveyors'. International Journal of Electronics and Communications (AEU). 2012; **66**(4):286–93.

[58] Soliman A. M. 'Realizations of ideal FDNC and FDNR elements using new types of mutators'. International Journal of Electronics. 1978; **44**(3):317–23.

[59] Mahmoud S. A., Elwan H. O., Soliman A. M. 'Generation of CMOS voltage-controlled floating resistors'. Microelectronics Journal. 1997; **28**(6–7): 627–40.

[60] Mahmoud S. A., Elwan H. O., Soliman A. M. 'Grounded MOS resistor'. Electronic Engineering. 1997; **69**:22–4.

[61] Mahmoud S. A. Elwakil A. S., Soliman A. M. 'CMOS current feedback op amp-based chaos generators using novel active nonlinear voltage controlled resistors with odd symmetrical characteristics'. International Journal of Electronics. 1999; **86**(12):1441–51.

[62] Soliman A. M. 'New LR mutators and their models'. International Journal of Electronics. 1973; **34**(6):753–6.

[63] Said L. A., Madian A. H., Isamil M. H., Soliman A. M. 'CMOS digitally programmable lossless floating inductor'. 2010 IEEE International Conference of Electron Devices and Solid-State Circuits, Hong Kong, China, 15–17 December 2010.

[64] Soliman A. M. 'Active circulator circuits using OA, CCII, CFOA and DVCC'. Journal of Circuits, Systems, and Computers. 2009; **18**(04):629–45.

[65] Elwan H. O., Mahmoud S. A., Soliman A. M. 'Voltage controlled square law grounded MOS resistor'. Electronic Engineering. 1995; **4**:34–8.

[66] Said L. A., Madian A. H., Ismail M. H., Soliman A. M. 'Current feedback operational amplifier (CFOA) based programmable lossless floating inductor realization'. 2014 International Conference on Engineering and Technology (ICET), 2014; 1–4.

[67] Soliman A. M., Fawzy M. 'Active R simulation of an inductor and a series resonator'. Proceedings of the 20th Midwest Symposium on Circuits and Systems, 1977:348–352.

[68] Soliman A. M. 'Synthesis of generalized impedance converter and inverter circuits using NAM expansion'. Analog/RF and Mixed-Signal Circuit Systematic Design. 2013:223–42.

[69] Soliman A. M. 'A new realization of the FDNC using the DVCCS/DVCVS'. International Journal Electronics and Communication (AEU). 1979; **33**(10): 423–4.

[70] Khalil N. A., Said L. A., Radwan A. G., Soliman A. M. 'General fractional order mem-elements mutators'. Microelectronics Journal. 2019; **90**:211–21.

[71] Yuce E. 'New low component count floating inductor simulators consisting of a single DDCC'. Analog Integrated Circuits and Signal Processing. 2009; **58**(1):61–6.

[72] Yuce E., Tokat S., Yücel F. 'A new wideband electronically tunable grounded resistor employing only three MOS transistors'. Turkish Journal of Electrical Engineering & Computer Sciences. 2015; **24**(4):2442–53.

[73] Yuce E., Tokat S., Alpaslan H. 'Grounded capacitor-based new floating inductor simulators and a stability test'. Turkish Journal of Electrical Engineering & Computer Sciences. 2015; **23**:2138–49.

[74] Alpaslan H., Yuce E. 'New grounded inductor simulator using unity gain cells'. Indian Journal of Pure and Applied Physics. 2013; **51**:651–6.

[75] Alpaslan H., Yuce E. 'Bandwidth expansion methods of inductance simulator circuits and voltage-mode biquads'. Journal of Circuits, Systems, and Computers. 2011; **20**(3):557–72.

[76] Yuce E. 'Reply to comment on "Novel lossless and lossy grounded inductor simulators consisting of a canonical number of components"'. Analog Integrated Circuits and Signal Processing. 2012; **72**(2):505–7.

[77] Dogan M., Yuce E. 'Supplementary single active device based grounded immittances function simulators'. International Journal of Electronics and Communications (AEU). 2018; **94**:311–21.

[78] Alpaslan H, Yuce E. 'Modified current follower-based immittance function simulators'. International Journal of Electronics. 2017; **104**(12):1974–91.

[79] Dogan M., Yuce E. 'CFOA-based a new grounded inductor simulator and its applications'. Microelectronics Journal, 2019; **90**:297–305.

[80] Çam U., Kaça F., Cicekoglu O., Kuntman H., Kuntman A. 'Novel grounded parallel immittance simulator topologies employing single OTRA'. International Journal of Electronics and Communications (AEU). 2003; **57**(4):287–90.

[81] Toker A., Zeki A., Cicekoglu O. 'Generalized gyrator implementation techniques using unity-gain cells'. International Journal of Electronics. 2001; **88**(5):517–28.

[82] Abuelma'atti M. T., Khan M. H., Al-Zaher H. A. 'Simulation of active-only floating inductance'. Frequenz. 1998; **52**(7–8):161–4.

[83] Al-Absi M. A., Al-Suhaibani E. S., Abuelma'att M. T. 'A new compact CMOS C-multiplier'. Analog Integrated Circuits and Signal Processing. 2017; **90**(3):653–8.

[84] Al-Absi M. A., Abuelma'atti M. T. 'A novel tunable grounded positive and negative impedance multiplier'. IEEE Transactions on Circuits and Systems II: Express Briefs. 2018; **66**(6):924–7.

[85] Nandi R. 'Novel active-R ideal frequency-dependent negative-resistance simulation'. Proceeding of IEEE. 1979; **67**(7):1080.

[86] Senani R., Bhaskar D. R., Gupta S. S., Singh V. K. 'Chapter 3: Current feedback op-amps, their applications, bipolar and CMOS implementations and their variants'. In E. Tlelo-Cuautle (Ed.). Integrated Circuits for Analog Signal Processing. Springer, New York; 2012. p. 61–84.

[87] Senani R., Bhaskar D. R., Singh A. K., Singh V. K. 'Chapter 9: Synthesis of electronically-controllable signal processing and signal generation circuits using modern active building blocks'. In M. Fakhfakh, E. Tlelo-Cuautle (Eds.). RF/Analog and Mixed Signal Circuits Systematic Design. USA: Springer, Berlin Heidelberg; 2013, 195–221.

[88] Senani R., Bhaskar D. R., Gupta S. S. Kumar P., Sharma R. K. 'Chapter 5: On the realization of sinusoidal oscillators using nullor'. In E. Tlelo-Cuautle, M. Fakhfakh, L. G. de la Fraga (Eds.). Analog Circuits: Fundamentals, Synthesis and Performance. New York, USA: Nova Publisher; 2017.

[89] Abuelma'atti M. T. 'Comment on novel lossless and lossy grounded inductor simulator consisting of a canonical number of passive components'. Analog Integrated Circuits and Signal Processing. 2011; **68**:139–41.

[90] Minaei S., Yuce E. 'A tunable circuit for realizing arbitrary floating impedances'. Journal of Circuits, Systems and Computers. 2008; **17**(2):1–12.

[91] Singh V. 'Floating operational transconductance amplifier based grounded impedance'. IEE Proceedings – Circuits, Devices and Systems. 2003; **150** (1):27–30.

[92] Minaei S., Yuce E., Cicekoglu O. 'Lossless active floating inductance simulator'. Proceeding of the Third IEEE International workshop on Electronics Design, Test and Applications (DELTA' 06), 2005.

[93] Kumar N., Vista J., Ranjan A. 'A tuneable active inductor employing DXCCTA: grounded and floating operation'. Microelectronics Journal. 2019; **90**:1–11.

[94] Yesil A., Babacan Y., Kacar F. 'Electronics tunable memristor based on VDCC'. International Journal of Electronics and Communication (AEU). 2019; **107**:282–90.

[95] Al-Absi M. A. 'Realization of a large values floating and tunable active inductor'. IEEE Access. 2019; **7**:42609–13.

[96] Kapoulea S., Psychalinos C., Elwakil A. S., Radwan A. G. 'One-terminal electronically controlled fractional-order capacitor and inductor emulator'. International Journal of Electronics and Communication (AEU). 2019; **103**:32–45.

[97] Ledvina J., Horsky P. 'A fully integrated digitally controllable grounded inductor simulator with a large inductance range for damping of ultrasonic transducers'. Analog Integrated Circuits and Signal Processing. Doi:10.1007/s10470019-01396-z.

[98] Jantakun A., Pisutthipong N., Siripruchyanun M. 'A synthesis of temperature insensitive/electronically floating simulator based on DVCCTAs'. *Proceeding of the 6th International Conference on Electrical Engineering/ Electronics, Computer, Telecommunication and Information Technology (ECTI-CON'09)*. 2009, 560–3.

[99] Aghazadeh S. R., Martinez H., Saberkari A., Alarcon E. 'Tunable active inductor-based second order all pass filter as a time delay cell for multi-GHz operation'. Circuits, Systems, and Signal Processing. 2019; **38**(8):3644–60.

[100] Verma R., Pandey N., Pandey R., 'Novel CFOA based capacitance multiplier and its application'. International Journal of Electronics and Communication (AEU). 2008; **107**:192–8.

Index

active circuits 179, 295, 473
active gyrator 44–5, 56, 152–3, 212–13, 216, 281, 308, 369–72, 381, 383, 404, 480, 494, 510
active inductor 56–7, 390, 473–95
active transformer 192–3, 411–12, 492–5
active-R 28, 74–7, 84, 86, 106, 128, 279–81, 321
admittance matrix 5, 10, 28, 107, 139, 171, 192, 199, 201, 214, 306, 309, 394, 405, 477
analog circuits 74, 76, 99–100, 134, 148, 249, 251, 279, 296, 325, 337, 367–8, 375, 392, 414, 424, 428, 463, 473, 531
analog multiplier 359, 361–2, 526
analog signal 209
analog signal generation 414, 464, 527, 531
analog signal processing 76, 140, 207, 368, 374, 380, 414, 464, 527, 531, 532
average circuit 376

back to back 59–61, 63, 71–3, 82, 84, 102, 105, 134–6, 151, 255, 317, 324, 369, 480, 483, 488, 508
bandwidth 15, 49, 148, 182, 251, 312, 339, 361, 368, 433, 437, 455, 483
BiCMOS 181, 183, 194, 283
bipolar junction transistor (BJT) 5, 9, 27, 102, 297–9, 413, 423–39, 463, 473–80, 495, 506, 508, 512
biquad filter 198, 383

capacitive gyrator 406–8
cascode current mirror 304
chaotic oscillator 148
CMOS 18, 19, 100, 128, 183, 194, 195, 211, 217, 290, 322, 382, 473–95
CMOS active inductor 473–95
CMOS active transformer 492–5
CMOS inductor 480–1, 489–92
complimentary output current follower (COCF) 371, 372
complimentary outputs 194
current amplifier (CA) 12, 371, 373, 380, 425, 439
current backwards transconductance amplifier (CBTA) 368, 410–12, 513, 514
current controlled current conveyor (CCC) 181–93
current controlled current conveyor Transconductance amplifier (CCCCTA) 393, 395
current controlled current differencing buffer amplifier (CCCDBA) 384–5
current controlled current differencing transconductance amplifier (CCCDTA) 381–2
current controlled current feedback operational amplifier (CC-CFOA) 282–3, 290
current controlled current source (CCCS) 258, 298, 380
current controlled differential difference current conveyor (CCDDCC) 213–15

current controlled Voltage source (CCVS) 258, 298, 374, 475
current conveyor (CC) 85, 137, 147–223, 249, 286, 302, 314, 327, 351–3, 363, 424, 480, 506, 512, 532
current conveyor transconductance amplifier (CCTA) 368, 392–6, 506, 532
current differencing buffer amplifier (CDBA) 368, 382–5, 396, 506, 532
current differencing current conveyor (CDCC) 506
current differencing transconductance amplifier (CDTA) 368, 380–2, 384, 385, 388, 396, 506
current differencing trans-resistance amplifier (CDTRA) 413
current differencing unit (CDU) 382, 385, 396
current differential amplifier (CDA) 412
current feedback amplifier (CFA) 363, 509
current feedback operational amplifier (CFOA) 9, 62, 85, 170, 249–56, 258–65, 267, 270, 271, 273, 274, 279–82, 284–6, 290, 302, 327, 337, 351–2, 355–7, 359–64, 367, 368, 374, 382–3, 427, 432, 444, 506, 509, 511, 512, 522–4, 526
current follower (CF) 72, 257, 299, 368–74, 385, 413, 506
current follower current conveyor (CFCC) 368, 402–5
current follower transconductance amplifier (CFTA) 385–7, 393, 506
current gain 163, 182–3, 199, 211, 304, 413, 439, 483, 529
current mirror 9, 100, 182, 303, 304, 314, 315, 316, 320, 323, 426, 428, 430, 438, 448–51, 455, 478, 479, 481, 483
current mode (CM) 9, 10, 147, 167, 198, 212–13, 312, 380, 494–5
current operational amplifier (COA) 406
current repeater 450, 478–9

degenerate element 8, 295
differential current controlled voltage source (DCCVS) 374
differential difference amplifier (DDA) 42, 325, 368, 375–80, 532
differential difference current conveyor (DDCC) 193, 200, 515–16
differential difference current conveyor transconductance amplifier (DDCCTA) 414
differential difference operational floating amplifier (DDOFA) 325–7
differential input dual output operational transconductance amplifier (DOTA) 508–9, 526–7
differential–input-single-output (DISO) 508–9, 517
differential voltage current controlled current conveyor (DVCCC) 206–9
differential voltage current controlled current conveyor transconductance amplifier (DVCCCTA) 395
differential voltage current conveyor (DVCC) 193–5, 206–7, 209, 217–18, 394, 412, 518–19
differential voltage current conveyor transconductance amplifier (DVCCTA) 393–5, 514, 519
differentiator 41, 49, 126, 527
digitally controlled current follower (DCCF) 413
direct-coupled 476, 477
dual complementary output DVCC (DO-DVCC) 412

dual output current controlled current conveyor (DOCCCIIs) 190, 191, 202–3, 223
dual output current conveyor (DOCC) 172–3, 194, 200, 525
dual output current follower (DOCF) 371–2
dual output differential difference current controlled current conveyor (DO-DDCC) 515
dual output differential difference current conveyor (DO-DDCC) 515
dual-X current controlled current conveyor (DXCCII) 193, 197–8, 203–5, 211, 213, 223
dual-X current controlled differential input transconductance amplifier (DXCCDITA) 393
dual-X current conveyor (DXCC) 211

emulator 505, 521, 525–6

FGMOS 463
field Effect Transistor (FET) 5, 212, 335–6, 338–9, 345, 347–9, 352, 355, 363, 475
filter 5, 12, 52, 72, 130, 153, 190, 209, 380, 453, 492
first generation current conveyor 147
floatation 61, 67, 71, 85, 118, 119, 122, 124
floatator 269
floating capacitance 67, 125, 154, 213–15, 269, 288, 320, 386–7, 390–2, 399, 411, 513, 517, 520
floating capacitance multiplier 125, 213–14, 386–7, 390–1, 399, 517
floating current controlled positive resistor (FCCPR) 184–6, 430–3
floating GIC 267, 307, 312–17, 356
floating immittance 80, 116, 132, 166, 199–200, 223, 305–12
floating immittance simulator 62–7, 171, 410–12
floating Impedance (FI) 28, 59–63, 71, 74–6, 80–2, 84–5, 100, 119, 122, 133, 151, 156–80, 187–92, 194, 195, 201–3, 207–9, 211, 222–3, 263, 266, 268, 296, 302, 305–7, 309–10, 314–17, 320, 323, 327, 355, 358, 361, 367–414, 478, 512, 519–20, 528, 531
floating impedance 28, 59–63, 71, 74–6, 80–2, 84–5, 100, 119, 122, 133, 151, 156–80, 187–92, 194–5, 201–3, 207–9, 211, 222, 223, 263, 266, 268, 296, 302, 305–7, 309–10, 314, 315–20, 323, 327, 355, 358, 361, 367–414, 478, 512–20, 528, 531
floating inductor 12, 72, 105, 138, 156, 169, 194, 256, 257, 288, 305, 325–7, 370, 375, 378–80, 382, 387, 405, 412, 453, 507–8
floating NIC 74
four terminal floating nullor (FTFN) 9, 28, 80–4, 86, 151, 223, 295–327, 358, 367, 374, 506, 532
fractional order 505, 512, 527–31
frequency dependence negative capacitance (FDNC) 36, 48, 61, 104, 105, 110, 126–9, 132, 140, 148, 156, 169, 192, 209, 263, 265, 269, 307, 308, 391, 510
frequency dependence negative resistance (FDNR) 28, 35–7, 47, 48, 61, 72, 84–5, 104–5, 111–15, 126, 148, 151–6, 158, 161–3, 166, 170–3, 177, 179, 185–7, 192, 194–6, 199, 200, 205–11, 219–21, 256–8, 263–4, 269, 273, 287, 289–90, 307–8, 314–15, 318, 320, 375, 386, 391–4, 399, 403–4, 406–7, 507, 510–11, 513, 519–20, 522

frequency shift keying (FSK) 432
fully balanced voltage differencing buffer amplifier (FB-VDBA) 396, 399
fully differential current conveyor (FDCCII) 193, 209–11, 223
fully differential voltage current conveyor (FDVCC) 217–18

gain bandwidth product (GBP) 35–6, 45, 75, 84, 128, 130, 312, 320, 341, 349, 406, 507
generalised negative impedance converter (GNIC) 263–9, 347, 374
generalised negative impedance inverter (GNII) 263–9, 347
generalised positive impedance inverter (GPIC) 166–70, 263–9, 308, 348, 355–6
generalised positive impedance inverter (GPII) 166–70, 263–9, 308, 348, 355–6
generalized Impedance Convertor (GIC) 33–7, 42–4, 60–3, 71–2, 84–5, 116–17, 123, 169, 267, 300, 305–8, 312–17, 342–5, 355–7, 374, 413, 507
grounded capacitor (GC) 40, 53–5, 65–6, 104–5, 107, 125–6, 134, 140, 151, 154, 159–61, 167–9, 173–5, 188, 190, 192, 194, 197, 200, 202–3, 206, 212–17, 223, 269, 277, 285, 326, 359–71, 374, 377–8, 382, 384, 387, 405, 410, 414, 460, 480, 488, 507–8, 510–11, 513, 515, 519, 520
gyrator 1–19, 27–86, 99–140, 148, 151, 156, 168, 251, 281, 295, 298, 367, 369, 414, 473, 495, 506

immittance 47, 62–7, 72, 75, 84, 116, 140, 163, 164, 166, 171, 193–218, 223, 274, 279, 384, 393, 394, 396, 410–12, 531, 532
impedance converter 7, 85, 170
impedance inverter 8
indefinite matrix 134
inductor 1–19, 27–86
input impedance 4, 6–7, 29–30, 32–4, 36–7, 40, 43–6, 49, 50, 54, 57, 75, 99, 102–4, 114, 123, 126, 148, 151, 153, 169, 183–4, 198, 204, 210, 215–16, 250, 252, 255, 258, 264, 281, 283, 287, 296, 342, 344–5, 347, 374, 378, 381, 382, 384, 388, 396, 398, 406, 453, 480, 482, 485, 488–9, 514, 518
integrated circuit 2, 27, 158, 166, 302, 367, 423, 473, 474
integrated inductor 1–19
integrator 35, 41, 48, 52, 75, 84, 103, 104, 107, 126–8, 132, 189, 255, 256, 270, 320, 406, 423, 440, 527
inverting current follower 475

junction field effect transistor (JFET) 335–7, 342, 344, 346–7, 357

ladder network 72
lossless floating Inductor (LLFI) 105, 106–9, 116, 118–19, 127, 132, 137, 140, 256, 325, 378–80, 382, 405, 507, 508, 510–14, 516, 517, 519, 524, 526
lossless Inductor 5, 33, 45, 47, 48, 52, 56, 84, 103, 106, 126, 156, 173, 211, 219, 273, 370, 374–5, 377–8, 400, 413, 474, 506, 508, 519
lossy floating inductor (LFI) 507, 512, 516
lossy inductor 48, 49, 54–5, 72, 79, 112–15, 153, 155, 196, 197, 222, 269, 274, 277, 378, 383, 399, 410, 511, 523, 530
low-component-count 36, 55, 79, 155, 178, 271–3, 461, 530, 531

memcapacitor 505, 521–2, 524–5, 527
memcapacitor emulator 525
meminductor 505, 522–5
meminductor emulator 525
memristor 2, 3, 505, 512, 521–4, 526–7
metal Oxide semiconductor (MOS) 18, 38, 196, 206, 348–9, 383, 407, 414, 423, 424, 438, 445, 453, 459, 475, 492, 495
metal oxide semiconductor field effect transistor (MOSFET) 9, 42, 197, 203, 204, 298, 299, 326, 335–7, 410, 423, 440, 449, 459, 461, 473, 481, 484, 487–90, 492, 494, 518
micro-miniaturization 13–15
mixed translinear cell (MTC) 181, 251, 252, 424, 426, 437, 463
modified current feedback operational amplifier (MCFOA) 282, 285–90, 512–13
modified DO-DDCC (MDO-DDCC) 215, 223
modified inverting second generation CC (MICC) 193
modified voltage differencing voltage transconductance amplifier (MVDVTA) 518
modified Wilson current mirror 252, 304, 358
MOS resistive circuit 348–9
MOS transistor 206, 349
MOS-C 325, 326, 327
MOSFET-C 32, 38, 197, 198, 204, 205, 212, 339, 441
M-R mutator 522–3
MR-MC mutator 523–4
multi gyrator 10–12, 28, 138
multi port gyrator 12, 62, 138–40
multifunction filter 259
multiplier 28, 36–8, 40, 100, 123–5, 140, 156, 161, 173, 212–15, 269, 287–8, 290, 359–63, 386–7, 390–1, 393–4, 396–7, 399, 413–14, 508, 517, 526, 532
multi-port immittance converter/inverter 62
multiport Inverter/converter 12
multivibrator 353
mutator 62, 72, 156, 521–4, 526, 529
mutually coupled circuit 138, 179–80, 395–6

negative Impedance Converter (NIC) 7, 31, 48, 51–2, 55, 73–4, 137, 149–50, 171, 200, 263–9, 271, 299, 318, 347, 374–5, 435–6, 452, 475, 510
negative memristance 526–7
NMOS 203, 443–4, 446, 448, 450–1, 454, 459, 462–3, 486–7, 493
nonreciprocal 6
norator 8–10, 44, 177, 295–8, 304, 309, 477
NPN 209, 429
nullator 8–10, 44, 177, 295–9, 306, 308, 309, 477
nullor 9, 10, 43, 44, 63, 80, 81, 85, 170–2, 177–8, 295–300, 306, 311, 323, 477–8, 506, 510

operational Amplifier (Op-amp) 9, 18, 27, 72, 337
operational floating amplifier (OFA) 223, 295–327
operational mirrored amplifier (OMA) 302–5, 312–23, 327, 357, 358
operational transconductance amplifier (OTA) 66, 81, 85, 99–140, 174, 176, 181, 201–3, 298, 299, 337, 367, 380, 388, 393, 396, 399, 401, 406, 508, 517–19, 528, 529
operational transimpedance amplifier (OTIA) 511
operational transresistance amplifier (OTRA) 368, 374–5, 506, 523, 532

oscillator 1, 27, 43, 44, 57, 58, 72, 75–7, 99, 100, 113–14, 118, 126, 127, 140, 148, 149, 219, 279, 305, 341, 342, 363, 423, 428, 434, 439, 484, 486, 494, 527

parasitic 16, 113, 128, 149, 156, 163, 169, 170, 186, 188, 200, 202, 250, 252–3, 262, 270, 272–4, 282, 285, 290, 403, 463, 480, 487, 489
passive inductor 5, 15, 17, 18, 201, 219
pathological element 8–10, 296
PID controller 527
pinch hysteresis 3, 522
PMOS 443, 445, 446, 448–51, 460, 462, 481, 486, 487, 493
PNP 209, 429
positive Impedance converter (PIC) 7, 33, 71, 507
positive memristance 526–7
power gyrator 16–17
PSPICE 380

quality factor 14–15, 17–18, 50, 84, 112, 132, 287, 430, 433, 482–4, 486, 487, 492

reciprocal 5, 12, 361, 522
relaxation oscillator 100, 148
resistively variable capacitor (RVC) 36, 38–9
resonator 75–6, 280

saturation 335, 407, 414, 444, 451, 458, 460–4, 482, 488
second generation current conveyor 9, 147, 217, 337
semi indefinite matrix 134
semi-floating 133, 134, 136–7
simulated inductor 1–19, 27–86, 197, 203, 219, 273, 289, 305, 367, 369, 473–5, 495, 505, 531
simulator 28, 35–6, 38–40, 189, 253–5
single-input-dual-output (SIDO) 517
sinusoidal oscillator 44, 57, 58, 76, 113, 127, 305, 527
slew rate 148, 181, 251, 351, 511
square wave generator 353
switched-Capacitor 28, 76–80, 86, 100
synthetic inductor 2, 18, 76, 290, 378

third generation current conveyor 193
three port gyrator 42, 66, 67, 125, 139, 140, 166–70, 174, 176, 204, 207–8, 260, 386, 390–1, 509–11
three port mutator 529–30
three terminal nullor 9, 170–1, 297–8
three transistor gyrator 475, 477–8
threshold voltage 336, 455
trans-capacitance 407
transfer function 48, 54, 84, 103, 256, 257, 278, 341, 528
transformer 2–5, 15–16, 18, 62, 100, 138, 192, 295, 411–12, 473–95
translinear cell 182, 478, 495
translinear circuit 148, 478
translinear operational current amplifier (TOCA) 380
translinear principle 430
transmission matrix 3, 6, 9, 29, 60, 61, 65, 71, 105–6, 296, 370
tri-amplifier 413
triode region 203, 326, 335, 344, 347–8, 352, 440, 442, 447, 449, 452, 460–2, 464, 482

unity gain amplifier (UGA) 49, 53–4, 57, 58, 72, 84, 116, 271, 349, 414
unity gain differential amplifier (UGDA) 54–5, 126–7, 399, 401
unity gain summer (UGS) 42, 66–7, 107, 127, 379

virtual ground 255, 374, 440
voltage and current gained second generation current conveyor (VCG-CCIIs) 211, 223

voltage controlled current source (VCCS) 32–3, 102, 107, 139, 253, 254–8, 266, 298, 312, 318, 369, 393, 432, 473, 477, 492, 508
voltage controlled GIC/GII 355–7
voltage controlled gyrator 382, 452
voltage controlled impedance (VCZ) 264, 337–52, 355–6, 359–64
voltage controlled negative resistance 338
voltage controlled oscillator (VCO) 342–3, 346, 484, 492
voltage controlled positive resistance 354
voltage controlled resistor 203, 339, 440, 445
voltage controlled voltage source (VCVS) 40, 99, 250, 257–8, 298
voltage differencing buffered amplifier (VDBA) 368, 396–9, 402
voltage differencing current conveyor (VDCC) 368, 396, 408, 506, 517, 520
voltage differencing differential input buffered amplifier (VD-DIBA) 368, 396, 399–402
voltage differencing inverting buffered amplifier (VDIBA) 368, 396, 402, 408–10
voltage differencing transconductance amplifier (VDTA) 368, 388–92, 399, 506, 516–17
voltage doubler 376
voltage follower (VF) 12, 79, 101, 154, 250, 299, 312, 320, 336, 353, 368–9, 376, 413, 508, 524
voltage gain 182, 199, 296, 410
voltage inverter 376
voltage mirror 9, 10
voltage mode (VM) 131, 312

waveform generator 181, 363
Wilson current mirror 100, 183, 252, 304, 358

Z-copy current controlled current-inverting transconductance amplifier (ZC-CCCITA) 393
Z-copy current follower current controlled current conveyor (ZC-CFCCC) 404–5
Z-copy current follower current conveyor (ZC-CFCC) 402–3
Z-copy voltage differencing buffer amplifier (ZC-VDBA) 397–8
Z-copy voltage differencing transconductance amplifier (ZC-VDTA) 516

www.ingramcontent.com/pod-product-compliance
Lightning Source LLC
LaVergne TN
LVHW011857200926
848484LV00002B/467